# INTEGRATED CHEMISTRY

## A TWO-YEAR GENERAL AND ORGANIC CHEMISTRY SEQUENCE

## VOLUME TWO, SECOND YEAR

## PRELIMINARY EDITION

Timothy R. Rettich
David N. Bailey
Jeffrey A. Frick
Forrest J. Frank

Illinois Wesleyan University

Houghton Mifflin Company   Boston  New York

TIMOTHY R. RETTICH:
>
> To Barbara, and to all my students, especially Tony,
> for having taught me so much.

DAVID N. BAILEY:
>
> To Sally and my students who always seem to be there to pick me up when I
> am down—they make it all worthwhile.

JEFFREY A. FRICK:
>
> To Phyllis, Abbey, and my students for making each day worthwhile.

FORREST J. FRANK:
>
> To Dottie, Sharon, Nathan, and Brenden who are still teaching me about life.

Printed in the U.S.A.

Library of Congress Catalog Card Number: 99-71922

ISBN: 0-395-98093-3

123456789–HS–03 02 01 00 99

## ACKNOWLEDGEMENTS

The following students made significant contributions to the material in this text by providing the drawings and other illustrations, running spectra, and doing the layout of the book in *QuarkXPress*: Sarah Beth Anderson, Michael Davis, Sarah Grigsby, Laura Hornbeck, David Martel, Dustin Mergott, Kristel Monroe, Angie Nettleton, Scott Robowski, and Sarah Studnicki.

Thanks to Marc Featherly for taking the photographs used in the book.

The data for figures 15.43, 15.44, 15.45, and 15.61 are from the Protein Data Base. These data were obtained via the web site maintained by the Brookhaven National Laboratory.

We thank Dr. Anthony Otsuka at Illinois State University for allowing us to photograph the model of DNA which appears as the frontice piece on chapter 16.

The authors also express their appreciation to the many chemistry teachers who have offered comments before and during the writing. We are especially thankful for the services of our Advisory Committee: Doris Kolb, Ken Kolb, William Bordeaux, Pat Holt, Graham Ellis, Maury Ditzler, and John Moore.

We are also grateful to the faculty at the cooperating institutions who believed enough in this new concept that they agreed to adopt the new curriculum and provide us feedback as it was class tested. The institutions are: Bellarmine College, Huntington College, Illlinois College, and Illinois Wesleyan University. The authors also thank the students at these institutions for the enthusiasm with which they have accepted the first book in the series and the invaluable feedback they have given us that will result in an even better text in the revisions.

The authors also express their appreciation to the National Science Foundation for financial support of this project through Undergraduate Course and Curriculum Grant number DUE 9455718.

This project was supported, in part, by the National Science Foundation

Opinions expressed are those of the authors and not necessarily those of the Foundation.

This book is the second volume of *Integrated Chemistry: A Two–Year General and Organic Sequence*. It is designed to extend the material covered in volume one of that text to encompass the rest of the topics normally covered in General and Organic Chemistry courses. As in volume one, this book integrates material from all major branches: analytical, bio–, inorganic, organic, and physical chemistry into one book. For those who use these books, this produces a curriculum without the artificial divisions of the standard curriculum. Also carried on from the first volume is the emphasis on "real world" applications and the introduction of the experimental facts prior to the theoretical explanations for those facts.

We hope that students will enjoy learning chemistry from this text and will find it both interesting and informative. It is designed for a second year course and assumes mastery of the material in a first year college course taught from the first volume of *Integrated Chemistry*.

In addition to the material normally covered in traditional General and Organic Chemistry courses, this text includes some material not normally covered in those courses. The coverage of inorganic topics, especially, is at a level above that of the normal General Chemistry course. This is possible because the arrangement of the topics ensures that students already have an understanding of substantial parts of organic chemistry. The new topics can then build on this understanding to provide a greater depth than would have been otherwise possible.

Despite our efforts to be as complete as possible, no single textbook can ever contain all the material that is relevant to the study of the subject. Like other texts, this textbook provides an overview; it presents a limited number of specific examples, and develops theory from one (or occasionally two or three) perspectives. This is even more true because this is a preliminary version. Students may find it helpful, or even necessary, to consult other textbooks to find different approaches to (or levels of) theoretical concepts, or more problems to test one's understanding. Therefore they should make use of the college library, in addition to this text, as they study.

Each chapter contains some "concept check" problems, that are worked out for you. It is best if, when you read the book, you first attempt your own solution to the problem before reading the solution.

There are problems at the end of each chapter. The answers to some of these problems are given in Appendix D. Again, it is advisable to try to work through to an answer on your own before checking the index. Learning to solve problems backwards (i.e., starting with the answer) may not translate into good problem solving skills on exams where, in general, the answers are not presented on the last page. There is a range of difficulty in these end of chapter problems, from simply following the paradigm laid out in a prior concept check to some problems designed to challenge the most capable students.

Most chapters also have "overview problems". These are typically multiple concept problems that incorporate some aspect of the material presented in the current chapter with material or concepts treated earlier in the course. These are not necessarily the hardest problems in the chapter, but, because the study of chemistry must be cumulative, they do require that you maintain a command of the important concepts from each chapter.

Especially because this is a preliminary version, the authors welcome your comments and suggestions.

We wish all who use this text an interesting and profitable journey through the fascinating world of chemistry.

# INTEGRATED CHEMISTRY
## A Two-Year General and Organic Chemistry Sequence
## **Volume Two** Second Year

CHAPTER 8:    MECHANISMS II: SUBSTITUTION AND ELIMINATION

CHAPTER 9:    COMPLEX CHEMISTRY: MORE THAN FOUR NEIGHBORS

CHAPTER 10:    CHEMICAL BONDING: MORE THAN JUST s & p

APPENDICES

# CHAPTER one

# REACTION
# Energetics

## HOT ENOUGH FOR YOU?

# REACTION
## Energetics

I n the first book in this two–part series, we emphasized the structure, properties, and chemical reactivity of matter. We learned how chemical reactions change matter by breaking and making bonds, turning reactants into products. Yet chemical reactions involve more than a simple change in matter. Energy is required to break the bonds in the reactants, energy is released in making the bonds in the products, and a net quantity of energy is associated with nearly every chemical change.

In our earlier studies, we explained that reactions occurred in the fashion described because of certain "driving forces", "lower energy states" or simply "stabilization". The goal of the early chapters in this book is to clarify and expand the meaning of such terms through the study of thermodynamics. Thermodynamics is a combination of two Greek words, *therme*, meaning **heat**, and *dynamikos*, meaning **work**. These are both types of energy and both can be involved in the way that energy is produced or consumed by chemical reactions.

Energy can be categorized in many different fashions. One way of categorizing energy that you probably learned in physics is as either potential energy (the energy associated with position or condition, *e.g.*, a stone teetering at top of cliff, or a charged battery) or as kinetic energy (the energy associated with motion, *e.g.*, the stone actually rolling down the hill with an energy of $1/2 \, mv^2$.) A bouncing ball illustrates the interconversion of potential and kinetic energy, as the potential energy (when the ball is at the top of its bounce) is converted into energy of motion as it descends, then into potential energy as the ball is compressed at the bottom of its fall, and finally back into kinetic energy as it bounces back. In this chapter we will learn about another kind of energy conversion: heat and work, especially as related to chemical reactions.

When people speak of "chemical energy", they are likely referring to a type of potential energy. As we have already noted, a certain amount of energy is associated with every chemical reaction. Understanding the production and conversion of chemical energy is fundamental to understanding why and how things work: from bio–energetics in our own bodies to the global production and consumption of energy resources. We begin with some basic definitions.

## 1.2  HEAT AND TEMPERATURE

How often have you or someone you know said "It's hot!" or, "It's cold!"? Probably often, but what do such expressions really mean? The "it" referred to is probably understood to be the air in the room, the water in the pool, *etc.* But what does "hot" signify? Certainly, different things to different people, as anyone who has tried to set a thermostat in shared living quarters can testify. In fact, even the same person at the same time can have differing views of hot and cold, as in the following often–cited example. A person puts one hand into ice cold water (0°C) and the other hand into very hot water (50°C). Then the person puts both hands into lukewarm water (20°C). What they "feel" is confusion, since the water feels cool to one hand, but warm to the other!

In a very real sense, heat is subjective. That is, heat only has meaning in terms of an interaction between two bodies. Whether an object you touch feels hot or cold depends upon whether it loses heat to you, or you lose heat to the object. A single body does not "have" heat. What it "has" is *energy*, or on a more experimentally practical basis, **temperature**.

Common usage frequently interchanges the terms "heat" and "temperature". But they are significantly different. Consider one drop of water and one gallon of water, both boiling at 100°C. Being spattered with a single drop of boiling hot water is a much different experience than, say, dipping your hand into a pot full of boiling water. Both are at the same temperature, yet the amount of heat we experience when contacting them differs quantitatively. Similarly, a wooden match and a bonfire burn at approximately the same temperature, but clearly the heat evolved differs. Here we see a clear distinction between heat and temperature. Temperature is an **intensive property**, independent of the amount of substance. Water boils at 100°C, regardless of how much water one has. Heat, on the other hand, is an **extensive property**, and depends on the amount of material.

For our purposes in this chapter, we can define temperature as simply what a thermometer measures. We note, however, that temperature is also a measure of the internal energy of an object (related to the kinetic energy of the molecules of the material as we learned in chapter 15 of last year's book.) What then is heat, and how is it measured? We know that it is related to temperature; specifically it is related to a difference in temperature $\Delta T$ (*e.g.*, between an object and ourselves.) The greater the temperature difference, the greater the amount of heat transferred, symbolized by "q". We write this proportionality in the form:

$$q \propto \Delta T$$

EQ. 1.1

We also note that as the mass, m, of the object increases, the heat that flows to or from that object increases. Thus:

$$q \propto m \, \Delta T$$

EQ. 1.2

But two different objects with the same mass and with the same temperature transmit different amounts of heat to a third common body. Consider, for example, finding a metal ice cube tray and a plastic ice cube tray in the freezer. Experience tells us that the metal ice cube tray "feels" colder to our hand than does the plastic tray. Both are at the same temperature, and may have the same mass, but the heat transferred from our hand is significantly different between the two. Metal and plastic differ in many properties, one of which is the rate at which heat is transferred between two bodies, called the thermal conductance. While this can be (and is) important in determining how hot or cold a material feels to us, another property, called the **specific heat** is even more important and is the property that we wish to discuss here. Specific heat, symbolized by (lower case) c, is the amount of heat required to raise one gram of material one degree (Celsius or Kelvin). The specific heat of water, for example is about **1.00 calorie** per gram degree. Calories (abbreviated as cal) are units of heat energy. If **joules** (abbreviation J) were used, the specific heat of water would be about 4.18 joules per gram degree. Specific heat is the proportionality constant that changes the expression in equation 1.2 into the following:

$$q = c \, m \, \Delta T$$

EQ. 1.3

Since the heat capacities of metals are typically larger than the heat capacity of plastics, a given mass of metal will transfer more heat than the same mass of plastic at the same initial temperature.

As a final illustration, consider a cool fall day. A stroll in the air at 60°F may be invigorating; a plunge into water at that same temperature would likely be considered unpleasant. Again, the specific heat of water is greater than that of air, and the heat the body loses to water at 60°F is much greater than the heat lost to air of the same temperature.

## 1.3  CALORIMETERY

Heat and temperature are now clearly seen to be different. Heat is extensive, and has units of calories or joules. Temperature is intensive, and has units of Kelvin (K) or degrees Celsius (°C). Experimentally, however, heat is commonly measured via a temperature change, utilizing equation 1.3. This experimental process is known as calorimetry. Two common experimental devices for determining heat are shown in figures 1.1 and 1.2.

In fig. 1.1, we see a vessel open to the atmosphere. Whatever is the process by which heat is produced or consumed, it is occurring at a constant ambient pressure. Heat generated in such a fashion is symbolized $q_p$, with the subscript p indicating constant pressure conditions.

In fig. 1.2, we see a rigid vessel containing the chemical reaction (usually a combustion reaction occurring in excess oxygen.) Since the metal container holds reactive materials at high pressures, it is commonly referred to as a "bomb", and the process as bomb calorimetry. Since the vessel is closed and inflexible, the heat is produced under constant volume conditions. Such heat is symbolized as $q_v$.

FIGURE 1.1: Coffee Cup Calorimeter (Constant pressure)

In order to use a calorimeter to determine a quantity of heat, the calorimeter must first be calibrated. This can be done in a variety of ways. An electric resistance heater can be used to put a known quantity of heat into a measured amount of water, as shown in fig. 1.3. A particular advantage of this method is that the quantity of heat added is easily calculated from measurement of time and electrical values made during the calibration.

By knowing the heat input (q), and measuring the change in temperature ($\Delta T$), the heat capacity, C, of the **system** can be determined.

$$C = \frac{q}{\Delta T}$$

EQ. 1.4

Note that the heat capacity, C, has units of energy per degree (e.g., J/Kelvin). This is similar to the specific heat, c, which typically has units of J/g Kelvin. If the mass of the calorimetric apparatus shown in fig. 1.3 was measured, then the specific heat could be determined by rearranging eq. 1.3 to solve for c:

$$c = \frac{q}{m\Delta T}$$

EQ. 1.5

Specific heat is commonly used only for pure substances. That is, we would commonly refer to the specific heat of water, but a sample composed of both water and plastic would commonly be treated in terms of heat capacity. Even the plastic container by itself (*e.g.*, a foamed and molded mixture of polymers and gases) would be described in terms of heat capacity, as opposed to specific heat. The following sample problems illustrate these issues.

FIGURE 1.2: Bomb Calorimeter (Constant volume)

**PROBLEM 1:** Consider the calorimeter shown in fig. 1.3. A 100.0 watt heating coil was operated for 30.0 seconds, and resulted in a temperature rise of the water (and the cup holding the water) of 2.50°C. Calculate the heat capacity of the calorimeter (i.e., the water and the container combined).

**SOLUTION:** First, we need to remember the definition of a watt as a joule per second. Consequently, the heat put into the system can be determined:

$$q = (100.0 \text{ watt})\left(\frac{1 \text{ Joule/sec}}{1 \text{ watt}}\right)(30.0 \text{ sec}) = 3.00 \times 10^3 \text{ J}$$

The heat capacity may now be calculated:

$$C = \frac{q}{\Delta T} = \frac{3.00 \times 10^3 \text{ J}}{2.50 \text{ deg}} = 1.20 \times 10^3 \text{ J deg}^{-1}$$

**PROBLEM 2:** Assume that 200.0 g of water was used in the preceding problem, and that the water had a constant specific heat of 4.18 J/g How much heat was absorbed by the water, and how much absorbed the container?

FIGURE 1.3: Coffee Cup Calorimeter With Electric Heating Coil

Solution: The heat absorbed by the water can be calculated using equation 1.3, since the specific heat, mass and temperature change are all known.

$$q = c \, m \, \Delta T = \left(4.184 \, \frac{J}{g \, deg}\right)(200.0 \, g)(2.50 \, deg) = 2.09 \times 10^3 \, J$$

Since the total amount of heat absorbed was 3.00 kJ, the heat absorbed by the cup can be determined:

$$3000 \, J - 2090 \, j = 910 \, J = 9.1 \times 10^2 \, J$$

Note that the answer to the second part of this problem is known to only 2 significant figures because of the subtraction.

If the cup shown in figure 1.3 were to be used in another test when the liquid held was not water, or a different amount of water was to be used, it would be necessary to know the thermal contribution of the cup itself.

PROBLEM 3: Calculate the heat capacity of the cup used in the previous problems.

SOLUTION: We calculate the heat capacity of the cup from the amount of heat the cup absorbs (Problem #2) and the temperature change undergone by the cup.

$$C = \frac{q}{\Delta T} = \frac{910 \, J}{2.50 \, deg} = 3.6 \times 10^2 \, J \, deg^{-1}$$

This means that for every 360 (2 sig. fig.) J of heat absorbed by this specific cup, its temperature will increase 1.0 degrees. Adding 720 (2 sig. fig.) J of heat to the cup will cause a 2.0 degree temperature rise. Note that the heat capacity was determined, not the specific heat. If one wished to calculate the specific heat of the cup, it still could not be determined without knowing the mass of the cup. Even if the specific heat of the cup was determined, the result could not be extrapolated to other, similar cups, even to cups of the same mass. Water is a pure substance, and one gram of it will always increase one degree in temperature when it absorbs one calorie of heat. The same is not true of a foamed plastic cup, which, depending on the specifics of its construction, may have differing densities, varying sizes of gas-filled cavities, *etc*. This is likely to be enough to affect the specific heat of the material.

In doing calorimetry problems, it is often useful to write the heat terms in a sum that equals zero. This is simply an application of the Law of Conservation of Energy that states that, in events not involving nuclear transformations, energy is neither created nor destroyed. Application of the Law of Conservation of Energy to calorimetry is considered more fully in section 1.5.

It is common in the laboratory to provide the heat for warming the calorimeter by adding hot water to the calorimeter containing cool water, instead of providing it electrically as in the previous example. If the Law of Conservation of Energy applies here, then the absolute quantities of heat gained and lost must be equal. However, we notice that application of equation 1.3 will result in opposite signs for the heat gained by the calorimeter and that given up by the hot water because in one case the value of $\Delta T$ is positive and in the other case it is negative. Since the absolute values of these two heats must be equal, the sum of the signed quantities will be zero exactly as demanded by the Law of Conservation of Energy. Let us consider a problem utilizing this approach.

**PROBLEM 4:** Consider a different calorimeter that contains 50.0 g of water at 18.6°C. When 45.0 g of water, initially at 52.0°C, are added to the cold water container, the final temperature of the calorimeter and water is 29.5°C. What is the heat capacity of the calorimeter?

**SOLUTION:** We assume that water has a constant specific heat of 4.18 J/g deg. We calculate the $\Delta T$ for the hot water, the cold water and the calorimeter itself in a consistent manner. That is, $\Delta T = T_{final} - T_{initial}$ for every substance. We then set the total energy change, which equals the sum of all the heat terms, equal to zero:

$$0 = q_{(hot\ water)} + q_{(cold\ water)} + q_{(calorimeter)}$$

Then each of the heat terms is calculated:

$$q_{(hot\ water)} = \left(45.0\ g\right)\left(4.18\ \frac{J}{g\ deg}\right)\left(29.5 - 52.0\ deg\right) = -42\overline{3}2\ J$$

$$q_{(cold\ water)} = \left(50.0\ g\right)\left(4.18\ \frac{J}{g\ deg}\right)\left(29.5 - 18.6\ deg\right) = +22\overline{7}8\ J$$

$$q_{(calorimeter)} = 0 - [q_{(hot\ water)} + q_{(cold\ water)}]$$

$$q_{(calorimeter)} = 0 - \left[\left(-42\overline{3}2\right) + \left(+22\overline{7}8\right)\right] = +1.95 \times 10^3\ J$$

Knowing the heat absorbed by the calorimeter, the heat capacity of the calorimeter is determined using the temperature change of the calorimeter. Again, $\Delta T$ must be the final minus the initial temperature.

$$C_{(calorimeter)} = \frac{q_{(calorimeter)}}{\Delta T_{(calorimeter)}} = \frac{19\overline{5}4\ J}{29.5 - 18.6\ deg} = 179\ J\ deg^{-1}$$

Note that a positive value of heat capacity is obtained. A negative heat capacity would imply that an object that absorbed heat decreased in temperature (an illogical supposition but one that sometimes occurs in laboratory situations because of the magnitude of the errors involved in making all of the measurements.)

**PROBLEM 5:** Exactly 100.0 g of water was added to the empty calorimeter that was calibrated in problem 4. The initial temperature of the water and calorimeter was 20.0°C. Then a 155 g piece of metal, heated to 100.0°C was added. The final temperature of the metal, water and calorimeter was 34.9°C. Calculate the specific heat of the metal.

**SOLUTION:** A constant specific heat of water of 4.18 J/g deg is assumed. Heat terms are calculated for the water, the calorimeter, and the metal.

$$q_{(cold\ water)} = \left(100.0\ g\right)\left(4.18\ \frac{J}{g\ deg}\right)\left(34.9 - 20.0\ deg\right) = +62\overline{2}8\ J$$

$$q_{(calorimeter)} = \left(179\ \frac{J}{deg}\right)\left(34.9 - 20.0\ deg\right) = +26\overline{6}7\ J$$

$$q_{(metal)} = \left(155\ g\right)\left(c\right)\left(34.9 - 100.0\ deg\right) = -10\overline{0}90\ (c)\ J$$

Setting the sum of all the heat terms to zero yields:

$$0 = 62\overline{2}8\ J + 26\overline{6}7\ J + (-10\overline{0}90\ g\ deg)(c)$$

from which the specific heat is calculated.

$$c = \frac{-\left(62\overline{2}8\ J + 26\overline{6}7\ J\right)}{-10\overline{0}90\ g\ deg} = 0.882\ J\ g^{-1}\ deg^{-1}$$

## 1.4   HEAT AND WORK

Before we can apply the principles of thermodynamics to chemistry, we will first need to better understand heat and work. Heat is a phenomenon observed when what we are studying (called the system) undergoes a change that results in a change in the temperature of its **surroundings**. When the change in the system results in an increased temperature of the surroundings, heat is leaving the system and entering the surroundings. Since the system is losing heat, we assign a negative value to this quantity. This is known as a system-centered sign convention. For example, in problem 5 if we called the metal the system, and the water and calorimeter the surroundings, the surroundings gained heat, and the system (the metal) lost heat. This is reflected in positive values of q for the cold water and the calorimeter, and a negative value of q for the metal. Again, the sum of the heat terms is zero. Note also that heat spontaneously flows from regions of higher temperature (the hot metal) to regions of lower temperature (the cold water).

Work is a phenomenon observed when the system undergoes a change that results in the raising or lowering of a mass in the surroundings (or its mechanical equivalent). Like heat, work is observed only when the system is undergoing a change. Like heat, it is observed only as an interaction between system and surroundings. And like heat, work is determinable *via* a change in the surroundings.

From physics, you may have learned that work is force times a distance, and that force is a mass times an acceleration. Thus:

EQ. 1.6

$$\text{work} = (\text{mass})\,(\text{acceleration})\,(\text{distance}).$$

For example, the work to move a mass (m) in a gravity field (with an acceleration constant g) a distance (h) is given by

EQ. 1.7

$$\text{work} = mgh.$$

Consider the cartoon person in figure 1.4 to be the system. The mass m in the surroundings is being lifted via a mechanical connection between the system and the surroundings. The system is said to be doing work on the surroundings.

Now let's consider the system to be a metal spring attached to a platform, as shown in the first part of figure 1.5. The spring at rest assumes some equilibrium shape, specifically length along the y axis. Then let a weight be placed upon the platform.

The mass of the weight, m, compresses the spring a certain distance h. From eq. 1.7 the work associated with this is mgh. In this case, the surroundings are said to have done work on the system.

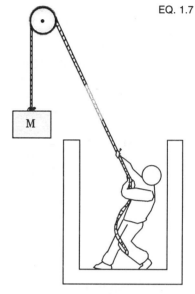

FIGURE 1.4: Man in a Box Lifting Mass Outside of Box *via* Rope and Pulley

The work associated with fig. 1.4 and the work associated with fig. 1.5 are both calculated via the product mgh. Yet there is a difference. The system (*i.e.*, the person) in fig. 1.4 has expended energy to lift the mass in the surroundings. The system in fig. 1.5 (the compressed spring) has had its energy increased as a result of the lowering of a mass in the surroundings. This increased potential energy of the system has been stored in the spring as potential energy. In the first case (man in a box), the system does work on the surroundings, and amount of energy is lost, and work is given by - mgh. In the second case (*i.e.*, the spring), the surroundings do work on the system, an amount energy is gained, and work is given by + mgh.

Both work and heat are manifestations of energy changes. When the system gains energy as a result of heat flowing into the system, the sign of q is positive. When the system loses energy as a result of heat flowing out of the system, the sign of q is negative. When the system gains energy as a result of work being done by the surroundings on the system, the sign of the work is positive. When the system loses energy as a result of work being done by the surroundings, the sign of the work is negative. We have a consistent sign convention, based upon the point of view of the energy of the system.

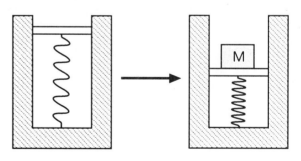

FIGURE 1.5: Work Related to a Mechanical Spring

## 1.5 CONSERVATION OF ENERGY

Consider a system of an ideal gas inside a container with a variable volume, such as the cylindric piston shown in fig. 1.6.

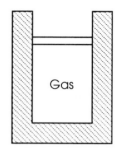

FIGURE 1.6: Gas Piston

The top of the piston is a floating disk, which is assumed to have no mass and to be able to move in a frictionless fashion (neat trick, huh?)  The disk will go up or down depending on whether the pressure inside the cylinder is greater or less than the external pressure.  The disk will come to rest in an equilibrium position when the pressures inside and outside are equal because the forces on the two sides of the disk will then offset each other.

Let us assume that the piston pictured in fig. 1.6 has reached equilibrium, and that subsequently a mass, m, is placed on top of the disk.  This adds an additional force pushing down on the disk — the force of gravity acting on the added mass.  Consequently, the disk will move downward a distance h, as shown in fig. 1.7.  Since a mass has been lowered in the surroundings, work has been done on the system.  The amount of work is mgh, and the sign of the work is positive.

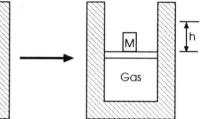

FIGURE 1.7: Gas Piston with added mass m

As shown in fig. 1.7, this squeezing of our system of an ideal gas has caused the volume to change an amount $\Delta V$. The cylinder cross section area is $\pi r^2$, where r is the radius of the cylinder.  The change in volume is thus $\pi r^2 h$.  Pressure, as we saw last year, is a force per unit area.  The force exerted by the mass is the product mg.  Putting this all together, we see that:

$$w = \left(\text{force}\right)\left(\text{distance}\right) = \left(\frac{\text{force}}{\text{area}}\right)\left(\text{area}\right)\left(\text{distance}\right) = \left(P\right)\left(\pi r^2\right)\left(h\right) = -P\Delta V \qquad \text{EQ. 1.8}$$

The negative sign before the $P\Delta V$ term is included to keep a consistent sign convention.  Note that the change in volume of the system ($V_{final} - V_{initial}$) is a negative number, since the final volume of gas is smaller as a result of the change.  Since we know work to be positive (a mass was lowered in the surroundings), we obtain a positive work by changing the sign of $P\Delta V$.  Note the close similarity between the work involved when a mass was added to a spring system in fig. 1.5, and the work involved when a mass was added to a gaseous system in fig. 1.6.  The two works are equivalent:

$$w = mgh = -P\Delta V. \qquad \text{EQ. 1.9}$$

Note that these expressions hold regardless of the direction of the change. Consider what happens when the system of an ideal gas expands, lifting a mass m, as shown in fig. 1.8.

The final volume of the system is now larger than the initial volume, so $\Delta V$ is a positive value.  Work, equal to $-P\Delta V$, must then be a negative value.  In this case, the system is doing work on the surroundings, a weight is raised in the surroundings, the system is losing energy, and work should, in fact, be negative. Under conditions when only work is involved, the energy of the system increases or decreases an amount equal to the work. **Internal energy** is symbolized by U, so the change in energy is given by:

$$\Delta U = w \text{ (when the only change is work)}$$ 

EQ. 1.10

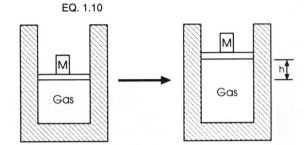

FIGURE 1.8: Gas Piston with mass being lifted

Now consider what happens when our system of an ideal gas is constrained in such a manner that the volume does not change and the system is heated.

Since the volume is constant, there is no $\Delta V$ possible, and the work associated with this change is zero. The flow of heat into the system of the ideal gas increases the temperature of the system. Temperature, as noted earlier, is a measure of energy. Last semester, when we studied the behavior of gases and the kinetic molecular theory, we showed the relationship between the average kinetic energy of the gas molecules and the temperature. The flow of heat from the surroundings into the system is a positive q, and corresponds to an increase in internal energy, a positive $\Delta U$.

$$\Delta U = q \text{ (when the only change is heat)}$$ 

EQ. 1.11

Note that the change in the system shown in figure 1.9 has resulted in the system having more <u>energy</u>. The system does <u>not</u> have more <u>heat</u>. For example, consider what would happen if a small weight, m, were added to the piston and the restraining brackets holding the piston in place were removed.

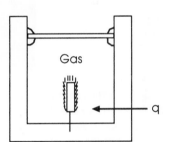

FIGURE 1.9: Pinned Gas Piston with heating element

The volume of the gas expands with increased temperature, according to Charles Law. This expansion lifts a weight in the surroundings, so the system is doing work. Thus the increase in internal energy can be seen as an increased capacity for work. The system pictured in figure 1.9 does not have more work; it does not have more heat; it has an increased energy.

If the system is insulated from heat flows and is mechanically isolated from doing work, then the internal energy of the system is constant. If the system does interact with the surroundings, then the amount of energy gained or lost by the system during a change in the system depends upon both the heat and the work associated with that change. Both of the preceding sentences are statements of conservation of energy, which is also a statement of the **First Law of Thermodynamics**. This is expressed in the equation:

$$\Delta U = q + w$$ 

EQ. 1.12

This says that the change in internal energy of the system is the sum of the heat flow into the system and the work done on the system. Note that a sign convention that regarded work done on the surroundings as negative was in common use until a few years ago. This resulted in a first law equation that was $\Delta U = q - w$. Even today some books still use

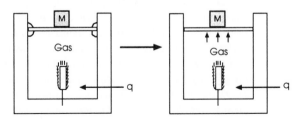

FIGURE 1.10: Brackets Off, Gas Cylinder Expanding, Lifting Weight

the older convention. You must, therefore, be certain what sign convention the book is using when you look at any book about thermodynamics. In this book, we shall use only the current convention.

### SIDEBAR: HEAT, WORK AND HEAT ENGINES;

*With the exceptions of muscular work and the motion of tides, heat engines are our primary sources of power. The ability to move objects and perform work mechanically is dependent on the conversion of heat into work. The earliest example of this was Hero's Steam Engine.*

FIGURE 1.11:
Hero's Engine

*Although little more than an amusing novelty in its day (about 2,000 years ago) the idea of turning heat into motion was important in the development of more advanced steam engines, which in turn facilitated the industrial revolution. Schematically, we picture the conversion as follows:*

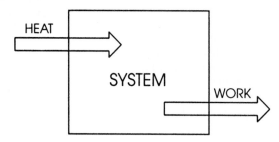

FIGURE 1.12: Heat Into
System, Work Out

*The development from coal fired steam engines to current automobile engines and modern power stations, wherein electricity is produced from nuclear reactions or the combustion of fossil fuels, is more an improvement in efficiency (maximizing the work output compared to fuel input) than a redesign of the fundamental relationships pictured in the above figures and explained in terms of the First Law of Thermodynamics. The figure below shows more of the real–life complexity involved in getting the most work out for a given energy input (your money.)*

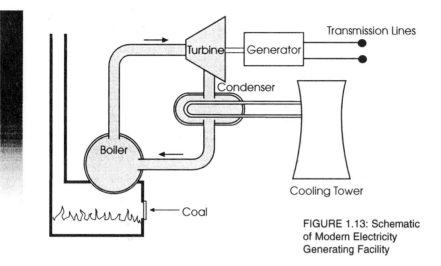

FIGURE 1.13: Schematic
of Modern Electricity
Generating Facility

## 1.6 STATE FUNCTIONS AND PATH FUNCTIONS

The state of the system can be characterized by several properties. For a system that is an ideal gas, the state is characterized by values such as P(ressure), V(olume), T(emperature) and U (Internal Energy). Since these are functions of the state of the system, they are called **state functions**, and are written in capital letters. When the system undergoes a change in state, there is a specific change in the value of the state function, generically symbolized below by "X."

$$\Delta X = X_{final} - X_{initial}$$

EQ. 1.13

That is, a change in the state of the system results in a $\Delta P$, $\Delta V$, $\Delta T$, and/or $\Delta U$. Their values depend only on the initial and final state of the system. Sometimes the initial and final states of the system are the same with respect to a certain state function. For example, the change may occur at constant pressure. In this case the value of $\Delta P$ is zero.

Work (w) and heat (q) are not state functions. The symbolism of lower case letters here is not accidental; it indicates that heat and work are **path functions**. The value of a path function does not depend upon the initial and final state of the system, but does depend upon the path taken between the initial and final state.

For example, consider elevation above sea level as a property. Consider your initial state to be where you now are, and the final state to be the very top of Mount Everest. If you were to travel from where you now are to the top of Mount Everest, there is a definite change in elevation above sea level:

$\Delta$(elevation) = height of Mt. Everest - current elevation.

Note that unless you are reading this while flying in an airplane above 30,000 ft., the value of $\Delta$(elevation) is positive, and is given by the final elevation minus the initial elevation. Elevation above sea level is a state function. But now consider the same initial and final states, but compute the work done in going from here to there. The amount of work done differs dramatically on the route taken and the means of travel (*e.g.*, flying to Nepal and helicoptering to the summit versus walking cross continent, rowing across the ocean and climbing to the top.) Since the value of work depends on the path taken (as opposed to simply the initial and final states) work is a path function.

Some special paths of chemical change are of particular interest. These include paths with pressure held constant, and with volume held constant. As indicated earlier, heat flowing under conditions of constant volume is symbolized $q_v$; heat flowing under constant pressure conditions is symbolized $q_p$. From our definition of work equaling -P$\Delta$V, conditions of constant volume imply a zero value of work since $\Delta$V is zero. From a statement of the first law, this means that, under constant V conditions:

$$\Delta U = q + w = q - P\Delta V = q_v \qquad \text{EQ. 1.14}$$

Remember that the kind of heat measured in fig. 1.2, a bomb calorimeter, was a constant volume heat. Thus a chemical reaction taking place in a bomb calorimeter has an associated heat that equals the change in internal energy as a result of the reaction.

In practice, however, most of the chemical reactions of practical interest occur under conditions of constant pressure (*i.e.*, exposed to the atmosphere.) The heat associated with this constant pressure process can be determined from the definition of work and the first law:

$$\Delta U = q + w; \ \ q_p = \Delta U - w = \Delta U - (-P\Delta V) = \Delta U + P\Delta V \qquad \text{EQ. 1.15}$$

The flow of heat under constant pressure conditions is thus simply:

$$q_p = \Delta U + P\Delta V \qquad \text{EQ. 1.16}$$

Note that we have a path function (q) related to an expression involving only state functions **U**, **P** and **V**. That is, once the initial and final states of the system are specified, the right hand side of equation 1.16 is specified. The value of the path function q is thus specified by the initial and final states — in this case because the path taken (constant pressure) is also specified. Since the terms on the right hand side of equation 1.16 are all state functions, they can be combined into another defined state function. This is called "enthalpy" (pronounced either en' thal pe or en thal' pe). Enthalpy is symbolized by **H** and it is defined as follows:

$$H \equiv U + PV \qquad \text{EQ. 1.17}$$

The change in enthalpy is

$$\Delta H = \Delta U + P\Delta V + V\Delta P \qquad \text{EQ. 1.18}$$

Note that under conditions of constant pressure, $\Delta P$ is zero and eq. 1.18 simplifies to

$$\Delta H_p = \Delta U + P\Delta V = \Delta U - w = q_p \qquad \text{EQ. 1.19}$$

Equation 1.19 is a round about way of saying that the amount of heat flow under constant pressure conditions is equal to the change in enthalpy for the system. Note that the value $\Delta H$ exists for any change of state of the system, regardless of the conditions. But when the specific condition is constant pressure, the value of $\Delta H$ equals the heat.

## 1.7 ENTHALPY AND CHEMICAL REACTIONS

So far in this chapter, what has been presented may not look much like chemistry. But chemical reactions are always accompanied by a change in energy as well as a change in the substances involved in the reaction. Some of this change is reflected in heat. Reactions that produce heat are termed "**exothermic**". Those reactions that consume heat are termed "**endothermic**". For reactions occurring at constant pressure, the heat equals $\Delta H$. Heat that is produced during the change in state of the system leaves the system and enters the surroundings. For example, when we touch the flask in which a reaction is taking place, and it feels hot, heat is leaving the system. This is a negative value of q (a loss of energy from the system) and implies a negative value for $\Delta H$.

Exothermic → negative value of $\Delta H$

When the products of the reaction are at a lower energy than the reactants, energy is given off. It could be considered to be an additional product of the reaction. Assuming the reaction occurs at constant pressure, only the initial and final states of the system (the reactant energy and product energy) determine the value of $\Delta H$. The shape of the path connecting the initial and final states is irrelevant to the value of $\Delta H$ (or to the change in any other state function.)

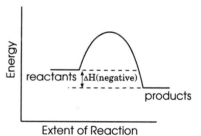

FIGURE 1.14: Energy *vs.* Extent of Reaction for Exothermic Reaction

Similarly, for reactions that consume energy in the form of heat, heat enters the system from the surroundings. When touched, the reaction flask feels cold. An input of heat indicates a gain in internal energy, a positive sign for heat, and a positive enthalpy change.

Endothermic → positive value of $\Delta H$

When the products of the reaction are at a higher energy level than the reactants, heat could be considered to be consumed as an additional reactant. The value of $\Delta H$ is positive, and again, since enthalpy is a state function, the value of $\Delta H$ depends only on the identity of reactants and products, not the path taken during the course of the reaction.

Thermodynamics, as applied to chemistry, has great predictive power because changes in state functions depend only upon the initial and final states, that is, the reactants and products. A detailed knowledge of how reactants become products is not necessary to predict whether a specific reaction consumes or

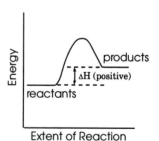

FIGURE 1.15: Energy *vs.* Extent of Reaction for an Endothermic Reaction

produces heat, or whether those reactants will under certain conditions form products.

Exact knowledge of the state of the reactants and products of a certain reaction would allow such calculations. For example, a table of absolute enthalpy values for various chemical species would allow the direct calculation of the $\Delta H$ for any reaction involving the tabulated species.

$$\Delta H_{reaction} = H_{products} - H_{reactants} \qquad \text{EQ. 1.20}$$

Such a table listing absolute values of H would be extremely valuable; unfortunately it does not (and will not) exist. We can never know the absolute enthalpy (or energy) of any specific substance. But as we will see in the next section, we will nonetheless be able to calculate the values we want.

## 1.8   STANDARD STATES AND ENTHALPY OF FORMATION

Every reaction has associated with it a specific value of $\Delta H$, regardless of the experimental conditions. Under the special condition of constant pressure, however, the enthalpy change is equal to the heat produced or consumed by the reaction. We know that heat is an extensive quantity; that is, the amount of heat produced by a chemical reaction depends upon the amount of reaction that occurs. For a specific balanced equation, we can refer to one mole of reaction. For example, consider the reaction of solid carbon and oxygen gas.

$$C_{(s)} + O_{2(g)} \rightarrow CO_{2(g)} \qquad \text{EQ. 1.21}$$

The amount of heat produced when one mole of carbon is oxidized is half the amount of heat produced when two moles of carbon is oxidized. To clear up this point, we refer to the molar enthalpy change. This is the enthalpy change for one mole of reaction as written. The molar enthalpy change of reaction is symbolized by $\Delta \overline{H}_{reaction}$ or simply $\Delta \overline{H}$. The bar notation over thermodynamic quantities will symbolize molar values (per mole of reaction as written.)

The measured value for the heat produced by oxidizing one mole of carbon according to equation 1.21 is -393.5 kJ/mol of reaction. The negative sign indicates that heat is produced by the reaction; *i.e.*, the reaction is exothermic. If the reaction were written in another form, such as

$$2\,C_{(s)} + 2\,O_{2(g)} \rightarrow 2CO_{2(g)} \qquad \text{EQ. 1.22}$$

The value of $\Delta \overline{H}_{reaction}$ would be -787.0 kJ/mol of reaction. Since one mole of the reaction shown in eq. 1.22 oxidizes two moles of carbon, the molar enthalpy change of reaction will be twice the value of the molar enthalpy change for eq. 1.21, in which one mole of carbon was oxidized.

It does seem strange to speak of "one mole of reaction", especially when the reaction is not written in its lowest terms as is the case for equation 1.22, but it is necessary to keep the energy quantities correct. The key to understanding this is to pay particular attention to the "as written" in the phrase "per mole of reaction as written." The units of all thermodynamic state function changes will involve "per mole".  Although usually "of reaction as written" is not included when the unit is written, it must be understood to be there when working with thermodynamic quantities.  Only when "of reaction as written" is understood to be present in the unit will the units work out properly in a factor–label analysis of the problem.

There are many different types of chemical reactions.  A small subset of these reactions form a substance from its constituent elements.  The following reactions all produce carbon dioxide gas, using carbon and oxygen.

$$C_{(g)} + 2\ O_{(g)} \rightarrow CO_{2(g)} \hspace{4cm} \text{EQ. 1.23}$$
$$C_{(s)} + 2\ O_{(g)} \rightarrow CO_{2(g)} \hspace{4cm} \text{EQ. 1.24}$$
$$C_{(s)} + 2/3\ O_{3(g)} \rightarrow CO_{2(g)} \hspace{4cm} \text{EQ. 1.25}$$
$$C_{(s)} + O_{2(g)} \rightarrow CO_{2(g)} \hspace{4cm} \text{EQ. 1.26}$$
$$2\ C_{(s)} + 2\ O_{2(g)} \rightarrow 2CO_{2(g)} \hspace{4cm} \text{EQ. 1.27}$$

Although reactions 1.23, 1.24, 1.25 and 1.26 all have identical products, the reactions produce differing amounts of heat.  The first four reactions differ in the physical state of one of the elements, or the specific allotrope of one of the elements; the fifth is a multiple of the fourth.  Which, if any, of these is what is meant by a **formation reaction**— the reaction to which many of our thermodynamic tables refer?

We define a formation reaction for a substance as one that makes one mole of the substance from its constituent elements.  If this is a formation reaction to which a change in some standard thermodynamic function is to apply, all materials in the reaction must be present in their standard states.  The thermodynamic standard state is not the same thing as standard temperature and pressure (STP) of a gas, which you may have learned earlier.  Besides denoting a certain temperature and pressure, (usually 298.15K and 1 bar pressure) the thermodynamic standard state convention is such that the phase and the specific allotrope of the reactant elements must be their most energetically stable form at that temperature and pressure.  Thus carbon must be considered to be a solid (not the gas form indicated in eq. 1.23.)  Oxygen in its most stable form is the diatomic molecule, not atomic oxygen (in eq. 1.23 and 1.24) and not ozone (eq. 1.25.)  Equation 1.26 is the only reaction that forms one mole of carbon dioxide from its constituent elements in their standard state.  The symbol used for standard molar enthalpy changes of formation is $\Delta \overline{H}_f^o$ —H is the symbol for enthalpy;  the delta before the H and the bar notation over the H indicate an enthalpy change for one mole of a reaction.  The subscript f indicates that the reaction in question is a formation reaction;  the superscript o indicates that reaction conditions are standard state.

From the definition of state functions, and particularly from the definition of the standard molar enthalpy of formation, it should be clear that the $\Delta \overline{H}_f^o$ of an element in its standard state is zero.  For example, making the element oxygen gas from the element oxygen gas, both at standard state conditions, indicate the following:

$$O_{2(g)} \rightarrow O_{2(g)}$$

EQ. 1.28

For this reaction, the value of $\Delta \overline{H}_f^o$ is zero, since the final and initial states are identical. The standard molar enthalpy change of formation for any element in its standard state is, by definition, zero.

The definition of a standard state, and particularly the chemical convention of standard state molar enthalpy changes of formation, $\Delta \overline{H}_f^o$, although logical is arbitrary, and is a matter of convenience. In addition to that, however, this definition establishes a reference point from which the relative enthalpies of all chemical species may be determined. Let us return to the analogy of geographical elevation as a state function. Sea level is set as an arbitrary reference point, and elevation above (or below) that mark allows calculation of changes in elevation between two points. If Death Valley is 282 feet below sea level, and if Mount Everest is 29,028 feet above sea level, a climb from the former to the latter will result in a change in elevation of 29,028 - (-282) = +29,310 feet. Note that an absolute knowledge of the elevation of the sea was not necessary; we were able to accurately calculate changes in elevation relative to this arbitrary standard.

In a fashion similar to listing the elevation of various points on a map using this arbitrary reference, tables of enthalpy changes of formation and other thermodynamic values (such as those found in appendix A) are similarly constructed. These tables of thermodynamic values prove extremely useful in determining energy relationships between reactants and products.

CONCEPT CHECK

A student ran the following reaction in a calorimeter:

$$H_2SO_{4(aq)} + 2\,NaOH_{(aq)} \rightarrow Na_2SO_{4\,(aq)} + 2\,H_2O_{(l)}$$

EQ. 1.29

A student placed 50.00 ml of 2.00 **M** NaOH and 50.00 ml of 1.01 **M** $H_2SO_4$ into a calorimeter. The initial temperature of both solutions was 22.5°C. After reaction was finished, it was determined that the solution had reached a temperature of 39.5°C immediately after mixing. The solution was then allowed to cool to 25.2°C, a 50.0 Watt electrical heating coil was immersed into the solution and an electric current was run through the heating coil for exactly 80.0 seconds. The final temperature of the solution was 37.1°C. Calculate the standard enthalpy change for the reaction in terms of kJ per mole of water formed.

In order to do this reaction we must know how many moles of reaction occurred and how many J of heat were released by the reaction of sodium hydroxide and sulfuric acid. We can determine the former from the information given about the volumes and concentrations of the solutions that were mixed. We can determine the latter from the temperature rise experienced by the solution once we know the heat capacity of the solution and calorimeter. The latter can determined from the temperature rise accompanying the addition of electrical heat. Let us determine these quantities in the reverse order.

First the heat capacity of the solution:

The unit Watt can be replaced by its equivalent: $J\ sec^{-1}$. Therefore, a 50.0 Watt heater operating for 80.0 seconds has delivered 4000 J (3 SF) of electrical energy to the solution as heat.

$$\left(50.0\frac{J}{sec}\right)(80.0\ sec) = 4.00 \times 10^3 J \qquad \text{EQ. 1.30}$$

The addition of this 4000 J of heat caused a rise in temperature of the solution of: $37.1°C - 25.2°C = 11.9°C$. The heat capacity of the solution can then be calculated:

$$\frac{4.00 \times 10^3 J}{11.9\ deg} = 366\frac{J}{deg} \qquad \text{EQ. 1.31}$$

Using this information and the rise in temperature that accompanied the chemical reaction, $39.5°C - 22.5°C = 17.0°C$, we can calculate the amount of heat released by the reaction:

$$\left(17.0\ deg\right)\left(336\frac{J}{deg}\right) = 5.71 \times 10^3 J \qquad \text{EQ. 1.32}$$

This amount of heat was released by the reaction of 50.00 ml of 2.00 **M** NaOH and 50.00 ml of 1.01 **M** $H_2SO_4$. We must determine the limiting reagent to determine the number of moles of water formed in this reaction. Since the reaction requires 2 moles of NaOH per mole of $H_2SO_4$, we can see that the limiting reagent in this case is the NaOH. This means that

$$\left(50.00\ ml\right)\left(\frac{1\ l}{1000\ ml}\right)\left(2.00\ \frac{mol\ NaOH}{l}\right)\left(\frac{1\ mol\ reaction}{2\ mol\ NaOH}\right) = 0.0500\ mol\ reaction \qquad \text{EQ. 1.33}$$

0.100 mol of water was formed by the reaction. The enthalpy change for the reaction was:

$$\Delta\bar{H} = \frac{5.71 \times 10^3\ J}{0.0500\ mol\ reaction}\ \frac{1\ mol\ reaction}{2\ mol\ H_2O} = 5.71 \times 10^4\ J/mol\ H_2O = 57.1\ kJ/\ mol\ H_2O \qquad \text{EQ. 1.34}$$

Since the solution was warmed by the reaction, heat was liberated by the reaction and the enthalpy change is, therefore, negative. The answer to the problem is $\Delta\bar{H} = -57.1\ kJ\ mol^{-1}$.

## 1.9   HESS' LAW AND ENTHALPY CHANGE OF CHEMICAL REACTION

As noted earlier, formation reactions comprise only a small fraction of all possible chemical reactions. For example, when two compounds react to form one or more new compounds, the reaction is not a formation reaction. Yet tables listing enthalpy changes of formation of the compounds involved provide the information needed to calculate the enthalpy change for this reaction.

Consider the following combination reaction:

$$CaO_{(s)} + H_2O_{(l)} \rightarrow Ca(OH)_{2(s)}$$

EQ. 1.35

Although this reaction "forms" calcium hydroxide, it is not the formation reaction of calcium hydroxide because this reaction makes calcium hydroxide from calcium oxide and water instead of from calcium, hydrogen, and oxygen. The formation reaction whose enthalpy change is tabulated as the standard enthalpy of formation for calcium hydroxide is

$$Ca_{(s)} + H_{2(g)} + O_{2(g)} \rightarrow Ca(OH)_{2(s)}$$

EQ. 1.36

The standard state enthalpy change of formation of solid calcium hydroxide is given in Table 1.1, along with corresponding values for the formation reactions of solid calcium oxide and liquid water.

**STANDARD STATE ENTHALPIES OF FORMATION**

TABLE 1.1

| Compound | $\Delta \overline{H}_f^{\circ}$ (kJ/mol) |
|---|---|
| $CaO_{(s)}$ | -635.09 |
| $H_2O_{(l)}$ | -285.83 |
| $Ca(OH)_{2(s)}$ | -986.09 |

From this information, the standard state enthalpy change for the reaction given in eq. 1.35 can be calculated, taking advantage of the definition of state functions: *i.e.*, the change can be calculated from the initial and final states, and is independent of the path. Specifically, although what we want to know is the enthalpy change for the direct reaction:

$$CaO_{(s)} + H_2O_{(l)} \rightarrow Ca(OH)_{2(s)}$$

EQ. 1.35

we can determine the enthalpy change going from the initial state of calcium oxide and liquid water to a final state of calcium hydroxide via a multi-step process:

$$CaO_{(s)} \rightarrow Ca_{(s)} + 1/2\, O_{2(g)}$$

EQ. 1.37

$$H_2O_{(l)} \rightarrow H_{2(g)} + 1/2\, O_{2(g)}$$

EQ. 1.38

$$Ca_{(s)} + H_{2(g)} + O_{2(g)} \rightarrow Ca(OH)_{2(s)}$$

EQ. 1.39

Adding equations 1.37, 1.38 and 1.39 together and canceling terms common to both sides yields the desired reaction, equation 1.35. We can picture what this means in terms of enthalpy changes in the following diagram.

FIGURE 1.16: Enthalpy Levels of Reactants, Elements, and Products

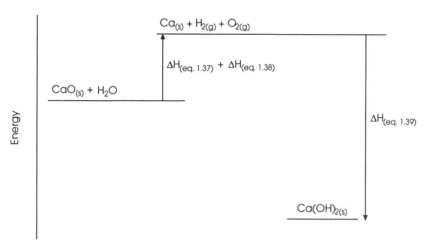

Note that eq. 1.37 is the reverse of the formation reaction of calcium oxide. By reversing a reaction, the final state becomes the initial state and vice versa. This results in a change of sign of the enthalpy change. Specifically, the enthalpy change for eq. 1.37 is - $\Delta\bar{H}_f^o$ for $CaO_{(s)}$, or - (-635.09) = +635.09 kJ/mol. That is, calcium metal and oxygen gas are at a higher enthalpy level than calcium oxide. This is as indicated in fig. 1.16. Similarly, eq. 1.38 is the reverse of the formation reaction of water. The enthalpy change for eq. 1.38 is - $\Delta\bar{H}_f^o$ for $H_2O_{(l)}$, or - (-285.83) = +285.83 kJ/mol. Again, this is pictured in fig. 1.16 as hydrogen gas and oxygen gas being at a higher enthalpy level than liquid water. Equation 1.39 is the formation reaction of calcium hydroxide. As shown in Table 1.1, this reaction is exothermic: the product, calcium hydroxide, is 986.09 kJ lower in enthalpy than its constituent elements in their standard states.

The net process shown in fig. 1.16 is a combination of steps to form the overall reaction of eq. 1.35. The total enthalpy change going from reactants to products via the intermediate stage of standard state elements is the sum of the individual enthalpy changes: (+635.09) + (+285.83) + (-986.09) = -65.17 kJ/mol. Solid calcium oxide and liquid water react to form solid calcium hydroxide and 65 kJ of heat in an exothermic process. The process by which the overall enthalpy change is calculated from a sum of real or imaginary steps that add to the overall reaction is the result of what is known as **Hess' Law**.

Consider another example, given by the decomposition of nitroglycerine:

$$4\ C_3H_5(NO_3)_{3(l)} \rightarrow 10\ H_2O_{(l)} + 12\ CO_{2(g)} + 6\ N_{2(g)} + O_{2(g)} \qquad \text{EQ. 1.40}$$

The enthalpy change of this reaction can be calculated from the enthalpy changes of formation of the reactants and the products, again using the elements in their standard states as an imaginary intermediate state.

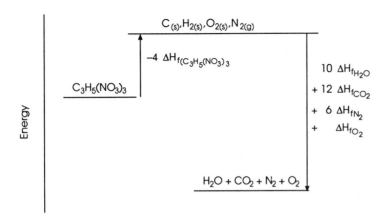

FIGURE 1.17:
Enthalpy Diagram
for Decomposition
of Nitroglycerine

As seen in the previous example, we could rationalize each enthalpy step along the way from reactants to products. Hess' law can also be stated in the following equation:

$$\Delta \overline{H}_{reaction} = \sum \left( \Delta \overline{H}^{o}_{f, \, products} \right) - \sum \left( \Delta \overline{H}^{o}_{f, \, reactants} \right)$$

EQ. 1.41

The sum of the enthalpies of formation of the reactants is given a negative sign in eq. 1.41 because these reactions need to go from the compound to the elements (as shown in fig. 1.17.) Another simple way of remembering the signs involved in eq. 1.41 is to note that products minus reactants is the same convention we used earlier in calculating the change in any state function: *i.e.*, final - initial.

Applying Hess' law and using the data in appendix A yields

$$\Delta \overline{H}_{reaction} = 10\Delta \overline{H}^{o}_{f, \, H_2O(l)} + 12\Delta \overline{H}^{o}_{f, \, CO_2(g)} - 4\Delta \overline{H}^{o}_{f, \, C_3H_5(NO_3)_3(l)}$$

$$\Delta \overline{H}_{reaction} = 10(-285.83) + 12(-393.51) - 4(-1509.7) = -1541.4 \text{ kJ/mol}$$

Note that in this example, the molar enthalpies of formation of water and carbon dioxide are multiplied by their stoichiometric coefficients in the balanced chemical equation. Nitrogen gas and oxygen gas were not included in the calculation because elements in their standard states are defined to have a zero value for the enthalpy of formation. Also note that the final answer, -1541.1 kJ/mol, is not per mole of nitroglycerine reacted; it is per mole of reaction ( which means four moles of nitroglycerin) as written in eq. 1.40.

Nitroglycerine is highly explosive: one mole (about 143 ml) of liquid nitroglycerine produces 4.75 moles (about 116 liters at STP) of gaseous products. In addition, over 1,541 kJ of heat are produced in this exothermic reaction. Remember: energy is consumed in breaking the bonds in nitroglycerine, and energy is released in the formation of bonds in the products. It is because more energy is produced in making the bonds in the products than is consumed in breaking the bonds in the reactants that makes this or any reaction exothermic. It is the magnitude of the enthalpy change for this reaction, accompanied by the much larger volume of the products as compared to the reactants, and the rapidity of the reaction that makes nitroglycerin (or any other material) an explosive.

## CONCEPT CHECK

The heat released from the combustion of 3.70 g of decane $C_{10}H_{22}$ under constant pressure conditions, was captured in 1.00 kg of water held in a calorimeter bucket. The heat capacity of the calorimeter bucket was 50.0 J deg-1. The initial temperature of the water and calorimeter bucket was 20.0°C, the final temperature was found to be 62.0°C. Given that the standard enthalpies of formation for $CO_{2(g)}$ and $H_2O_{(l)}$ are –393.51 kJ mol$^{-1}$ and –285.83 kJ mol$^{-1}$ respectively, calculate the standard enthalpy of formation for decane.

In order to do this problem we must first find the enthaply change for the combustion reaction:

$$C_{10}H_{22(l)} + 15.5\ O_{2(g)} = 10\ CO_{2(g)} + 11\ H_2O_{(l)}$$

EQ. 1.42

This is done by first determining the enthalpy change for the reaction in eq. 1.42 using the calorimetric information provided. The total heat capacity for the water and the calorimeter bucket is

$$50.0\ J\ deg^{-1} + \left(1000.g\right)\left(4.184\ Jg^{-1}deg^{-1}\right) = 4234\ J\ deg^{-1}$$

EQ. 1.43

From the observed change in temperature of the calorimeter, the enthalpy change for the calorimeter must have been

$$\Delta H = C\Delta T = \left(4234\ J\ deg^{-1}\right)\left(62.0 - 20.0\ deg\right) = 1.7\overline{7}8 \times 10^5 J = 17\overline{7}.8\ kJ$$

EQ. 1.44

The enthalpy change for the reaction must then have been – 177.8 kJ. Since the reaction consumed only 3.70 g (0.2604 mol) of decane, the molar enthalpy change for the reaction is

$$\Delta \overline{H}_{reaction} = \frac{17\overline{7}.8\ kJ}{0.02604\ mol} = 68\overline{2}9\ kJ\ mol^{-1}$$

EQ. 1.45

The Hess' law expression for the enthalpy change for the reaction is

$$\Delta \overline{H}_{reaction} = \sum \Delta \overline{H}_{products} - \sum \Delta \overline{H}_{reactants}$$

$$= \left(10\ \Delta \overline{H}^o_{f_{CO_2}} + 11\ \Delta \overline{H}^o_{f_{H_2O}}\right) - \left(\Delta \overline{H}^o_{f_{decane}} + 15.5\Delta \overline{H}^o_{f_{O_2}}\right)$$

$$= \left(10\left(-393.51\right) + 11\left(-285.83\right)\right) - \left(\left(\Delta \overline{H}^o_{f_{decane}}\right) - 15.5\left(0\right)\right)$$

$$-6829 = \left(\left(-3935.1\right) - \left(-3144.1\right)\right) - \left(\Delta \overline{H}^o_{f_{decane}}\right)$$

$$-6829 = \left(-7079.2\right) - \left(\Delta \overline{H}^o_{f_{decane}}\right)$$

$$\Delta \overline{H}^o_{f_{decane}} = -250\ kJ\ mol^{-1}$$

EQ. 1.46

**ENTHALPY OF PHASE CHANGE:**

An enthalpy change is associated with any change in the state of a material, either physical or chemical. A physical change, such as a phase transition, usually has a definite value for $\Delta H$. But instead of the heat input causing a change in the temperature of the material, it causes a physical change. For example, at its normal boiling point of 373.15 K water has an enthalpy of **vaporization** of +40.656 kJ/mol:

$$\text{heat} + H_2O_{(l)} \rightarrow H_2O_{(g)}$$

EQ. 1.47

The "reaction" in which liquid water is converted into steam is an endothermic process: *i.e.*, it requires heat. But at one atmosphere pressure, the temperature is constant. Instead of raising the temperature of liquid water, the added heat changes the liquid water at 100°C to steam at 100°C.

The reverse of this reaction has an associated enthalpy change known as the enthalpy of **condensation**, with a value of - 40.656 kJ/mol. Again, reversing the initial and final states changes the sign of $\Delta H$. When steam condenses, it releases a tremendous amount of heat. Steam burns are especially dangerous because of this extra heat, sometimes known as "latent heat". In comparison to a closely related compound, $H_2S$, water clearly has a significantly different enthalpy of vaporization and boiling point.

**BOILING POINT AND ENTHALPY OF VAPORIZATION**

TABLE 1.2

| Compound | Normal Boiling Point (K) | $\Delta H_{vap}$ (kJ/mol) |
|----------|--------------------------|---------------------------|
| $H_2O$   | 373.15                   | 40.656                    |
| $H_2S$   | 212.8                    | 18.67                     |

In general, boiling point and enthalpy of vaporization both reflect the strength of the intermolecular forces in the liquid. Oxygen is much more electronegative than sulfur, and thus $H_2O$ forms strong hydrogen bonds and $H_2S$ does not. The exceptional hydrogen bonding ability of water makes separating liquid phase water molecules from one another far enough to produce gaseous water molecules more difficult (*i.e.*, more endothermic) than separating liquid hydrogen sulfide molecules. Conversely, when steam condenses, much more energy is released because of the formation of the strong hydrogen bonds in water than is released when gasseous hydrogen sulfide condenses. Other types of phase changes have specific enthalpy values as well. The **fusion** of water is represented by:

$$H_2O_{(s)} \rightarrow H_2O_{(l)} \qquad \Delta H_{fusion} = 6.008 \text{ kJ/mol}$$

Changing solid ice into liquid water requires heat, and thus is an endothermic process. Note that the magnitude of energy required to melt a mole of ice is significantly less than the energy required to vaporize a mole of liquid water. In addition to a difference in temperature, these phase changes differ in the amount of molecular rearrangement that must be done. Since melting the solid requires less extensive breaking down of intermolecular forces than does vaporizing the liquid, $\Delta H_{fusion}$ values are generally significantly lower than $\Delta H_{vaporization}$ values.

**Sublimation** also requires a specific input of energy, as the intermolecular forces in the solid must be overcome in order to move the molecules to the greater distance of separation occurring in the gas phase. The most common example, the sublimation of $CO_{2(s)}$, dry ice, occurs at 194.6 K:

$$CO_{2(s)} \rightarrow CO_{2(g)} \qquad \Delta H_{sublimation} = 25.23 \text{ kJ/mol}$$

### ENTHALPY OF SOLUTION:

Often when separate components are mixed to form a solution, a quantity of heat is involved. Many of the solution processes are exothermic. For example, diluting concentrated sulfuric acid to form 1 liter of six molar aqueous sulfuric acid liberates several hundred kilojoules of heat. The directions for preparing dilute aqueous sulfuric acid call for concentrated sulfuric acid to be added slowly to a quantity of ice cold water with good mixing. Ice water is used because otherwise the heat evolved could cause significant local boiling and spattering and might even thermally stress the container enough to break it. Instead much of the heat is used to raise the temperature of the water and less is available for localized boiling. The chance of boiling is further reduced by adding the sulfuric acid slowly and mixing thoroughly to distribute the heat evolved evenly throughout the solution.

Less commonly, components form a solution in an endothermic process. This is the essence of the instant cold packs found in some first aid kits. Commonly a solid salt such as ammonium nitrate is packed along with liquid water in a container that, when ruptured, causes a significantly lower temperature.

$$\text{Heat} + NH_4NO_{3(s)} + H_2O_{(l)} \rightarrow NH_4^+{}_{(aq)} + NO_3^-{}_{(aq)}$$

**KEY WORDS**
**&Concepts**

**Calorie:** a non-metric unit of energy corresponding to the heat required to raise the temperature of one gram of water from 14.5°C to 15.5°C.

**Condensation:** the conversion of a gas to a liquid, or a gas to a solid. (The same name is applied to both processes by some authors.)

**Deposition:** the conversion of a gas to a solid.

**Endothermic:** a process that requires an input of heat, and thus has a positive value for the enthalpy change.

**Enthalpy:** a thermodynamic potential defined as the sum of the internal energy and the pressure volume product. The change in enthalpy equals the heat under constant pressure conditions.

**Exothermic:** a process that liberates heat, and thus has a negative value for the enthalpy change.

**Extensive property:** a property, such as energy or moles of matter, that depends upon the amount of material present.

**First law of thermodynamics:** a formal statement of the principle of the conservation of energy: the change in energy of the system is the sum of the heat and the work.

**Formation reaction:** the reaction that forms one mole of the specified substance from its constituent elements in their standard states.

**Fusion:** the conversion of a solid to a liquid. ("Melting")

**Heat:** a flow of energy during a change in state of a system. The flow of energy is due to a temperature difference between the system and the surroundings.

**Heat capacity:** the amount of heat required to raise the temperature of an object one degree.

**Hess' law:** the statement that the overall change in a state function (like enthalpy) for a multi-step process is the sum of the changes for each step.

**Intensive property:** a property, such as temperature or density, that does not depend upon the amount of material present.

**Internal energy:** a thermodynamic potential that expresses the amount of energy of the system. The change in internal energy equals the heat under constant volume conditions.

**Joule:** a metric unit of energy corresponding to the work associated with a force of one Newton acting over a distance of one meter.

**Path function:** a function, such as heat or work, whose value depends on the means by which the initial state of the system becomes the final state.

**Specific heat:** the amount of heat required to raise the temperature of one gram of a substance one degree.

**Standard state:** conditions including a pressure of one bar ($10^5$ Pascals) and a specified temperature, usually 298.15K. Elements specified as being in their standard state exist in the form of the most stable allotrope and most stable state of matter (solid, liquid, gas) at the standard temperature and pressure.

**State function:** a property that characterizes the state of the system (*e.g.*, T, V, P.) A change in state results in a specific value of $\Delta$T, *etc.*, regardless of the path taken.

**Sublimation:** the direct conversion of a solid to a gas.

**Surroundings:** everything in the universe except that part currently under study. See system.

**System:** that portion of the universe under study; everything else is called the surroundings. See surroundings.

**Temperature:** a measure of the internal energy of the system.

**Vaporization:** the conversion of a liquid to a gas

**Work:** a flow of energy during a chemical change in state of a system that can be mechanically related to the raising or lowering of a mass in the surroundings.

# HOMEWORK
## Problems

1. The specific heats of substances are often listed with units of $J\ g^{-1}\ K^{-1}$. That is, the temperature is specified in Kelvin instead of Celsius. Clearly explain why a measured specific heat has the same numerical value, whether the units are $J\ g^{-1}\ °C^{-1}$, or $J\ g^{-1}\ K^{-1}$.

2. For each of the following, which would be a more useful expression of their individual thermal characteristics: specific heat or heat capacity?

        a. a large pepperoni pizza
        b. water in the form of ice
        c. a textbook
        d. a specific type of stainless steel

3. Given that the enthalpy of fusion of water is 6.008 kJ/mol, calculate the total heat required to change 100.0 g of ice at 0.0°C to liquid water at 0.0°C.

4. Assume that the specific heat of liquid water is constant at $4.18\ J\ g^{-1}\ K^{-1}$, and that the enthalpy of vaporization of water is 40.656 kJ/mol. Calculate the total heat required to change 100.0 g of liquid water at 0.0°C to water vapor at 100.0°C.

5. Water can evaporate at temperatures less than the boiling point. From the data in appendix A, calculate the molar enthalpy of vaporization of water at 25.0°C. How does this compare to the value at 100°C?

6. Write the balanced formation reaction for each of the substances named below. Note: one mole of substance must be formed from the constituent elements in their standard states.

        a. chlorous acid
        b. chloroacetic acid
        c. ammonium bicarbonate
        d. 1-propanol

7. Name the compounds and write the balanced formation reaction for each of the following. Note: one mole of substance must be formed from the constituent elements in their standard states.

        a. HONO
        b. $CH_3CH_2NO_2$
        c. $H_2SO_3$
        d. $CH_3CH_2CHO$

8. Write the balanced formation reaction for each of the substances named below. Note: one mole of substance must be formed from the constituent elements in their standard states.

  a. sodium acetate
  b. potassium thiocyanate
  c. magnesium phosphate
  d. 2-propanol

9. Name the compounds and write the balanced formation reaction for each of the following. Note: one mole of substance must be formed from the constituent elements in their standard states.

  a. $H_3AsO_3$
  b. $C_6H_5CHO$
  c. $C_6H_5OH$
  d. $MgC_2O_4$

10. The reaction of a strong base like aqueous sodium hydroxide with either aqueous hydrofluoric acid or aqueous hydrochloric acid is exothermic. Which acid reacts in a more exothermic fashion, and why? (Hint: write a balanced net ionic reaction for both neutralization reactions, then consider bond breaking and bond making.)

11. The reaction of either aqueous sulfuric acid or aqueous sulfurous acid with excess aqueous sodium hydroxide is exothermic. Which acid reacts in a more exothermic fashion, and why? (Hint: write a balanced net ionic reaction for both neutralization reactions, then consider bond breaking and bond making.)

12. The standard state molar enthalpy change of formation of $P_4O_{10(s)}$ is -2984.0 kJ/mol. How much heat is evolved when 10.00 grams of elemental phosphorus reacts with excess oxygen to form $P_4O_{10(s)}$ ?

13. The standard state molar enthalpy change of formation of $Lu_2O_{3(s)}$ is -1878.2 kJ/mol. How much heat is evolved when 10.00 grams of elemental lutetium reacts with excess oxygen to form $Lu_2O_{3(s)}$?

14. Gaseous acetylene (ethyne) and liquid benzene have enthalpies of formation of + 226.73 and + 49.0 kJ/mol respectively. What is the $\Delta H$reaction for converting acetylene into one mole of benzene? Is this exothermic or endothermic? (Hint: write a balanced equation.)

15. Ethanol and acetic acid have enthalpies of formation of -277.69 and -484.5 kJ/mol respectively. When these compounds react to form one mole of ethyl acetate, the $\Delta H_{reaction}$ is -2.64 kJ/mol. What is the standard state molar enthalpy of formation of ethyl acetate? (Hint: remember water!)

16. Methanol and formic acid have enthalpies of formation of -238.66 and -424.72 kJ/mol respectively. When these compounds react to form one mole of methyl formate, the $\Delta H_{reaction}$ is -1.52 kJ/mol. What is the standard state molar enthalpy of formation of methyl formate? (Hint: remember water!)

17.  Isooctane (2,2,4-trimethylpentane) has a molar enthalpy of formation of -255.1 kJ/mol.  Write the balanced combustion reaction for isooctane reacting with excess oxygen to form liquid water and gaseous carbon dioxide.  Using the data in appendix A, calculate the standard enthalpy change for the combustion of one mole of isooctane.

18.  Acetone has a molar enthalpy of formation of -248.1 kJ/mol.  Write the balanced combustion reaction for acetone reacting with excess oxygen to form liquid water and gaseous carbon dioxide.  Using the data in appendix A, calculate the standard enthalpy change for the combustion of one mole of acetone.

19.  The molar enthalpy change for the reaction in which one mole of ethanol is combusted with excess oxygen to form liquid water and gaseous carbon dioxide is -1368 kJ/mol.  Calculate the standard state molar enthalpy of formation of ethanol.

20.  The molar enthalpy change for the reaction in which one mole of sucrose ($C_{12}H_{22}O_{11}$) is combusted with excess oxygen to form liquid water and gaseous carbon dioxide is -5645 kJ/mol.  Calculate the standard state molar enthalpy of formation of sucrose.

## CHAPTER ONE

OVERVIEW
# Problems

21.  What volume of methane, measured at 20.0°C and 800.0 Torr, would have to be combusted in excess oxygen to heat 160. L (about 40 gal.) of water from 15°C to 58°C (approximately the rise in temperature in a hot water heater)?  The heat capacity of the hot water heater tank is $3.2 \times 10^3$ J deg$^{-1}$.  You may assume that all of the heat from the combustion of the methane is captured in the water and the hot water heater tank.

22.  Consider the reaction of solid calcium hydroxide with liquid sulfuric acid to produce solid calcium sulfate and water.  How much heat would be released if 50.0 g of calcium hydroxide were reacted with 100.0 g of sulfuric acid?

23.  Monochromatic radiation at 450. nm. was incident upon a beaker of water for 1.00 hour.  The beaker was coated with a black, non–reflective coating in order that all of the energy in the beam of light was absorbed and converted into heat.  The beaker contained 55.3 g of water and its heat capacity was 5.42 J deg$^{-1}$.  The initial temperature of the beaker and the water was 22.8° C and its final temperature was 32.8° C.  Assuming no loss of heat to the environment and that the light intensity was constant over the time interval, what was the intensity of the light in photons per second?

24.  When potassium chlorate is formed from its elements, the standard enthalpy change of formation is –397.7 kJ/mole.

   a.  Write the balanced formation reaction making potassium chlorate.
   b.  What mass of oxygen and mass of chlorine are present in a 12.5

liter gas sample at 25.0°C, with a partial pressure of 300.0 torr of chlorine and 200.0 torr of oxygen?

c. What mass of potassium chlorate can be made from 5.00 g of potassium and the sample of gas described in part (b)?

d. What quantity of heat is associated with the reaction in part (c) assuming that it goes to completion under standard conditions?

25. When sodium nitrate is formed from its elements, the standard enthalpy change of formation is –467.8 kJ/mole.

a. Write the balanced formation reaction making sodium nitrate.

b. What mass of oxygen and mass of nitrogen are present in a 15.0 liter gas sample at 25.0°C, with a partial pressure of 200.0 torr of nitrogen and 400.0 torr of oxygen?

c. What mass of sodium nitrate can be made from 5.00 g of sodium and the sample of gas described in part (b)?

d. What quantity of heat is associated with the reaction in part (c) assuming that it goes to completion under standard conditions?

26. Liquid cyclohexene (MW = 82.15, density = 0.8102 g/ml) may react with gaseous hydrogen under the right conditions to form cyclohexane (MW = 84.16, density = 0.7785.) The standard state enthalpy of formation of cyclohexene and cyclohexane are –38.5 and –156.4 kJ/mole respectively.

a. Write the balanced formation reaction of cyclohexane.

b. Write the balanced reaction that makes cyclohexane from cyclohexene.

c. Starting with 100.0 ml of cyclohexene, what is the maximum volume of cyclohexane that could be formed?

d. What quantity of heat is associated with the reaction in part (c) assuming that it goes to completion under standard conditions?

27. Liquid cyclohexene (MW = 82.15, density = 0.8102 g/ml) may react with liquid water under the right conditions to form cyclohexanol (MW = 100.16, density = 0.9624 g/ml.) The standard state enthalpy of formation of cyclohexene and cyclohexanol are –38.5 and –348.2 kJ/mole respectively.

a. Write the balanced formation reaction of cyclohexanol.

b. Write the balanced reaction that makes cyclohexanol from cyclohexene.

c. Starting with 100.0 ml of cyclohexene, what is the maximum volume of cyclohexane that could be formed?

d. What quantity of heat is associated with the reaction in part (c) assuming that it goes to completion under standard conditions?

28. Solid calcium hydride ($CaH_2$, MW = 42.0960) reacts with liquid water to form solid calcium hydroxide ($Ca(OH)_2$, MW = 74.0948) and hydrogen gas. The standard enthalpy of formation of calcium hydride and calcium hydroxide are –186.2 and –986.09 kJ/mole respectively.

     a. Write the balanced reaction between solid calcium hydride and liquid water forming solid calcium hydroxide and gaseous hydrogen.
     b. Determine the standard state molar enthalpy change for the reaction in part (a).
     c. If 100.0 g of calcium hydride react with excess water, what volume of dry hydrogen gas, measured at 745.0 torr and 20.0°C could be formed?
     d. What quantity of heat is associated with the reaction in part (c) assuming that it goes to completion under standard conditions?

29. Solid calcium nitride ($Ca_3N_2$, MW = 148.25) reacts with liquid water to form solid calcium hydroxide ($Ca(OH)_2$, MW = 74.0948) and ammonia gas. The standard enthalpy of formation of calcium nitride and calcium hydroxide are –431.0 and –986.09 kJ/mole respectively.

     a. Write the balanced reaction between solid calcium nitride and liquid water forming solid calcium hydroxide and gaseous ammonia.
     b. Determine the standard state molar enthalpy change for the reaction in part (a).
     c. If 100.0 g of calcium nitride react with excess water, what volume of dry ammonia gas, measured at 730.0 torr and 18.0°C could be formed?
     d. What quantity of heat is associated with the reaction in part (c) assuming that it goes to completion under standard conditions?

30. Solid benzoic acid (MW = 122.13, DHf = -385.2 kJ/mole) under certain conditions may react with liquid methanol to form methyl benzoate (MW = 136.14, density = 1.094 g/ml, DHf = -343.5 kJ/mole)

     a. Write the balanced reaction between solid benzoic acid and liquid methanol forming methyl benzoate.
     b. Determine the standard state molar enthalpy change for the reaction in part (a).
     c. If 50.0 g of benzoic acid react with excess methanol, and eventually 5.0 ml of methyl benzoate is isolated, what is the percent yield?
     d. What difference(s) would one expect between IR spectra for benzoic acid and methyl benzoate?

# CHAPTER 2 + TWO

# REACTION Spontaneity

## WHICH WAY DOES IT GO?

# REACTION
## Spontaneity

ne of the central questions of chemistry is how to determine whether or not a specific reaction will occur. A closely related question is what factors determine if a
reaction "goes as written". Chemists pose such questions in terms of spontaneity.
In common usage, something that occurs spontaneously is thought to occur rapidly.
Newspapers will sometimes refer to certain industrial accidents in terms of a "spontaneous combustion," meaning an unexpected and very rapid redox reaction. In chemical
thermodynamics, however, spontaneity deals only with direction, not time. For example, consider the standard state reaction of solid carbon and gaseous oxygen to form carbon dioxide gas. Chemists refer to the following reaction:

$$C(s) + O_2(g) \rightarrow CO_2(g)$$

EQ. 2.1

as **spontaneous** under standard conditions. This means that carbon and oxygen react to
form carbon dioxide at 25°C and $10^5$ Pascals; the reverse of this reaction (carbon dioxide decomposing to form carbon and oxygen) does not happen under standard conditions. A reaction is said to be spontaneous if it proceeds as written (from reactants to
products). This does not mean such a change will occur quickly, or even noticeably.
Pure carbon in the presence of an atmosphere of oxygen will not instantly combust. In
spite of the book by Ian Fleming, thermodynamics can prove that diamonds are not for-
ever, but thermodynamics cannot predict when they will cease to be. As we shall see
later, the question "when?" is related to the specifics of the path taken between reactants
and products, and will be examined in terms of kinetics. The question of "which direction?" can be determined by state functions, and thus is most readily answered by thermodynamics.

In the example of carbon and oxygen, we lugged around the cumbersome phrase "under
standard conditions" when speaking of spontaneity. This is because the spontaneity of a
reaction will depend upon specific conditions such as temperature, pressure and concentration. We can usually adjust experimental conditions to make a reaction proceed in
either direction. Much of the art of synthetic chemistry lies in adjusting conditions to
produce a product that would not form with a reasonable yield if reactants were left to
their own devices. Consider the simple example of the phase change between ice and
water. Ice above 0°C will spontaneously melt to form liquid water:

$$H_2O(s) \rightarrow H_2O(l)$$

EQ. 2.2

But at a temperature below 0°C, the reaction is non-spontaneous. The reverse reaction
is then spontaneous. Many reactions will be seen to be spontaneous at one temperature
and non-spontaneous at another temperature. In this chapter we will examine the forces
that determine spontaneity, and we will learn to predict how some experimental variables will influence spontaneity.

## 2.2 THE CRITERION OF HEAT

The determination of spontaneity lies outside the province of the first law of thermodynamics. The principle of conservation of energy alone cannot predict whether A goes to B or B goes to A spontaneously. From the first law, energy must be conserved in either direction. Ice at 0°C mixed with water at 100°C results in a temperature between those extremes because energy is conserved. Lukewarm water forming ice cubes while simultaneously increasing the temperature of the remaining liquid would similarly conserve energy. Yet such a process never is observed to occur naturally. Hot and cold objects move toward a common intermediate temperature spontaneously; the reverse is non-spontaneous.

As a further example, consider a rubber ball dropped from a height above the floor.

The potential energy of the suspended ball is converted into kinetic energy as the ball accelerates downward. After several bounces, each interconverting kinetic and potential energy, the ball eventually comes to rest on the floor. Energy is not lost in the process; it has been converted through friction to heat the ball and the floor. Some energy was also "lost" in the sound waves associated with the sound of the ball hitting the floor. Overall, however, we recognize that energy was conserved. Yet we would be greatly surprised to see a ball initially at rest on the floor start to bounce up and down when slightly heated. Everyday observations lead us to correctly predict the direction of change (spontaneity) without regard to energy conservation. Some other principle is at work.

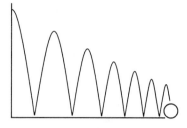

FIGURE 2.1: Bouncing Ball

One of the earliest suggested explanations was that a specific reaction would be spontaneous if it evolved heat (*i.e.*, had a negative enthalpy change.) It is tempting to draw an analogy between the change of reactants to products in figure 2.2 to the ball in figure 2.1.

The conversion of the initial state to the final state in both of these figures puts the system at a lower final energy level. Of course since energy is conserved, the energy "lost" is accounted for by a transfer of heat to the surroundings. The analogy, though far from exact, helps us picture how exothermicity and spontaneity may be related through the system seeking a lower energy level. In fact the majority of spontaneous reactions <u>are</u> exothermic. But clearly there are many endothermic reactions that also occur spontaneously under standard conditions. In the previous example of solid ice and liquid water, we know that ice at 25°C does melt, even though the process requires an input of heat. The solution reaction used in the instant cold packs described in the last chapter is both spontaneous and endothermic. Evolution of heat by the system, although positively correlated with spontaneity, cannot be the sole criterion for spontaneity.

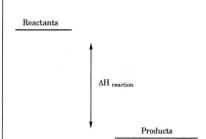

FIGURE 2.2: Energy Levels of Reactants and Products for an Exothermic Reaction

## 2.3 THE CRITERION OF DISORDER

Changes that are both endothermic and spontaneous have an important feature in common: the system becomes more disordered. Figure 2.3 shows what happens when a solid melts, or a liquid boils (both endothermic processes.) When continuously heated, the orderly array of the solid gives way to the jumble of the liquid, which in turn becomes the chaos of the gas phase. A plot of heat versus temperature is shown in figure 2.4.

The constant temperature plateaus correspond to the phase transitions. The width of the plateaus is related to the magnitude of the enthalpy of the phase change. On these plateaus, heat is being added to the system, but the temperature of the system remains constant. The heat is "turned into" something other than increased temperature. The Greek word for "turning into" is **"entropy"**. What the heat is transformed into is disorder or randomness. Entropy is a state function, symbolized by S. A positive $\Delta S$ corresponds to an increase in disorder or randomness. A negative $\Delta S$ corresponds to a decreased disorder (or increased order.) Therefore heat input that either causes an increase in the temperature of a material or that causes a material to undergo a phase change will result in a positive $\Delta S$ for the material. Recall from our study of the kinetic–molecular theory that raising the temperature of a material also causes an increase in the average kinetic energy of the molecules. One consequence of this increased motion is an increase in the randomness of the system. Therefore, even the heat that simply causes an increase in temperature also causes an increase in the entropy of a material.

When we picture the addition of solid ammonium nitrate to liquid water (the endothermic process used in the instant cold packs), it is easy to picture the process as leading to increased disorder, as shown in figure 2.5.

One has to be careful, however, about making blanket predictions based on such simple pictures. Specifically, the process involves more than simply breaking up an orderly crystalline solid and dispersing the ions randomly throughout an aqueous solution. The solvent molecules interact with the ions as shown in figure 2.6.

The water molecules that were in random motion in the pure liquid are more orderly in the immediate neighborhood of a solute species. This acts to decrease the overall entropy and may, in some cases, offset the increased entropy due to breaking the orderly lattice structure of the solid.

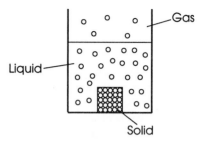

FIGURE 2.3: Phase Changes

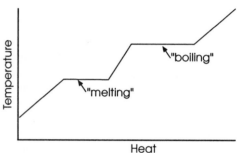

FIGURE 2.4: Heating Curve for a Material

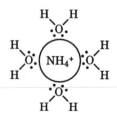

FIGURE 2.5: Solid Ionic Compound Dissolving in Liquid Water

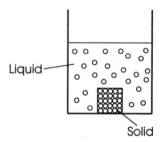

FIGURE 2.6: Ordering of Water Molecules Around Ions

## 2.4 ENTROPY, PROBABILITY AND THE SPREAD OF MATTER AND ENERGY ——

**MATTER SPREAD**

Nature favors processes that increase disorder. One of the two general categories of such processes may be termed "matter spread". For example buildings, given enough time, spontaneously turn into a pile of rubble. The reverse process, a pile of stones assembling itself into a building, is a very large entropy decrease and does not occur spontaneously. This is due to statistical probability. Consider the example of a variable number of rocks randomly positioned in a two-compartment box. Figure 2.7 shows the different ways that one can place a different number of particles into a two compartment box. If there is only one particle, the first line in the drawing shows that there are two possible situations. On the second line, the four possible arrangements of two particles are shown. Three particles give rise to the eight different arrangements shown on the last two lines.

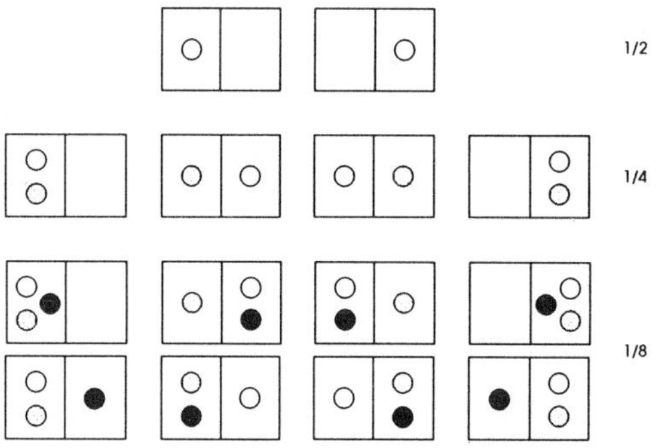

FIGURE 2.7: Particles in Two-Compartment Box

The probability of finding all "n" rocks on one particular side of the box (which is a minimum criterion for, say, making a wall out of the rocks) would be

$$\text{Probability} = \left(\frac{1}{2}\right)^n$$

EQ. 2.3

In a complicated system, where the number of rocks (n) becomes very large, the probability of finding all the rocks on one side becomes vanishingly small. The vast majority of all possibilities are those with roughly equal distributions. Statistically, there are many ways for a haphazard pile of rocks to exist, but only a very small fraction of those arrangements correspond to something resembling a wall, let alone a building!

As a final example of the role of "matter spread" in determining the direction of a reaction, consider a tornado. When this violent energy is unleashed on a trailer park (as it is so often inclined to do) the tornado forms a junkyard. The same amount of energy in a violent storm could, simply from the First Law standpoint, equally well turn a junkyard into a trailer park. The latter never happens because of the statistical improbability of organizing dispersed matter into a more ordered form.

**ENERGY SPREAD**

The other general category of disorganization is termed energy spread. The example of the bouncing ball is a good illustration. Initially we raised a ball in a gravity field so that it had potential energy as a result of the work done on it. When released, the kinetic energy moved all the atoms in the ball in the same direction. As we saw last chapter, such organized motion of a mass is termed work. But after several collisions between the ball and the floor, the associated friction caused an end to the bouncing and an increased thermal energy in the ball and the floor. Heat, as we saw earlier, is a random motion. Thus the organized motion of work is degraded into the random motion of heat. This is an example of energy spread. Going from work to heat is a conservative process. One hundred percent of the energy of the work associated with lifting the ball can be converted into heat. Nature favors such a process, since it randomizes motion. But the reverse of the bouncing ball process, *i.e.*, heating a ball and the floor on which it rests, never results in the ball starting to bounce. That would correspond to turning random motion into organized motion, a trend in opposition to the probability arguments advanced earlier. One important practical consequence of this drive to spread energy is that heat can never be 100% converted into work. The statement of the impossibility of completely converting heat into work is one formulation of the Second Law of Thermodynamics.

## 2.5  RULES OF THUMB FOR ENTROPY AND ENTROPY CHANGES

Unlike other state functions, absolute values of standard state molar entropies are available: $\bar{S}^o$ not $\Delta \bar{S}_f^o$. Table 2.1 shows standard state absolute entropies for a variety of substances.

TABLE 2.1  STANDARD STATE MOLAR ENTROPY VALUES  $\bar{S}^o$  (J/MOL K)

### COMMON INORGANIC SUBSTANCES

| | | | | | |
|---|---|---|---|---|---|
| $F_2(g)$ | 202.78 | NaF(s) | 51.46 | HF(g) | 173.78 |
| $Cl_2(g)$ | 223.07 | NaCl(s) | 72.13 | HCl(g) | 186.91 |
| $Br_2(g)$ | 245.46 | NaBr(s) | 86.82 | HBr(g) | 198.70 |
| $I_2(g)$ | 260.69 | NaI(s) | 98.53 | HI(g) | 206.59 |

### COMMON ORGANIC SUBSTANCES

| | | | |
|---|---|---|---|
| Methane(g) | 186.2 | Methanol(l) | 126.8 |
| Ethane(g) | 229.5 | Ethanol(l) | 160.7 |
| Propane(g) | 269.9 | Acetic acid(l) | 159.8 |
| *n*–Butane(g) | 310.0 | $CF_4(g)$ | 262.3 |
| *i*–Butane(g) | 293.6 | $CBr_4(g)$ | 358.2 |
| Ethylene | 219.8 | $CH_3Cl(g)$ | 234.2 |

### EFFECT OF PHASE

| | Liquid | Gas |
|---|---|---|
| $CCl_4$ | 216.40 | 309.85 |
| $SiCl_4$ | 239.7 | 330.73 |
| $GeCl_4$ | 245.6 | 347.72 |
| $SnCl_4$ | 258.6 | 365.8 |

Several generalizations may be drawn from these data. The standard state absolute entropies of all elements and compounds are positive. (Contrast this with the definition of $\Delta \bar{H}_f^o \equiv 0$  This is because disorder is associated with motion. At a temperature of 25°C, even a pure crystalline substance has some disorder caused by thermal motion and the elements of the crystal vibrate around their equilibrium positions in the crystal lattice. [Note: Only a perfect crystal at a temperature of absolute zero would be perfectly ordered and thus have a zero value of entropy. This is a statement of the Third Law of Thermodynamics.] We see that solids do generally have the lowest absolute entropy values, followed by liquids, then gases. This trend compares well with our images of the degree of disorder in the states of matter pictured in figure 2.3. Another trend is that within comparable substances (*e.g.*, halogens $X_2$; salts NaX; gases HX), as the size of an atom increases, so does the disorder. Finally we notice that the entropy increases as the number of atoms in the substance increases.

The values of absolute entropies for both products and reactants allow calculation of $\Delta S$ for a reaction. The entropy of vaporization of carbon tetrachloride at 25°C is given by

$$\Delta \bar{S}^o_{vap} = \bar{S}^o_{CCl_4(g)} - \bar{S}^o_{CCl_4(l)} = 309.85 - 216.40 = 93.45 \; \text{J}/_{mol \cdot K}$$
EQ. 2.4

Note that the positive entropy change corresponds to the increased disorder found in a gas compared to a liquid. Also note that entropy has units of energy per degree (or molar entropy has units of energy per degree per mole.) This ratio of energy per degree is similar to heat capacity that related how heat caused a change in temperature of a specific substance. But as shown in figure 2.4, a phase transition occurs at a constant temperature. For such a phase transition, the entropy change can be calculated as

$$\Delta \bar{S}^o_{vap} = \frac{\Delta \bar{H}^o_{vap}}{T}$$
EQ. 2.5

Note that $\Delta \bar{H}^o_{vap}$ (and thus $\Delta \bar{S}^o_{vap}$) can be computed from literature values for the $\Delta \bar{H}^o_f$ of the liquid and gaseous substance for the standard temperature of 25°C.

Chemical reactions can result in an increase in entropy by spreading energy, matter, or both. Reactions that produce more moles of products than moles of reactants will increase disorder, and thus generally be favored by entropy. Similarly, if more moles of gaseous products are formed than there are moles of gaseous reactants, matter will be spread and the reaction will be favored by entropy. Consider the following reaction:

$$N_2O_4(g) \longrightarrow 2 \, NO_2(g)$$
EQ. 2.6

$$\Delta \bar{S}^o_{reaction} = 2 \, \bar{S}^o_{NO_2(g)} - \bar{S}^o_{N_2O_4(g)}$$
EQ. 2.7
$$= 2 \, (240.0) - (304.2) = 175.8 \; \text{J}/_{mol \cdot K}$$

A process by which one molecule is broken into two molecules increases the spread of matter, and, therefore, increases disorder. This is shown in the positive value calculated for the entropy change for the reaction in equation 2.6.

The process shown in equation 2.8 actually decreases the number of particles as reactants become products, but the number of moles of gas increases.

$$SiO_2(s) + 4\,HF(aq) \longrightarrow SiF_4(g) + 2\,H_2O(l)$$

EQ. 2.8

The calculated standard state molar entropy change for the reaction in equation 2.8 is 430.6 J/K mol, showing the great increase in disorder made possible when the number of moles of gas increases during a reaction even if the overall number of moles decreases.

## 2.6 THE ENTROPY OF THE UNIVERSE AND FREE ENERGY

We have seen two trends positively correlated with spontaneity in chemical reactions. First is the general trend to change potential energy into thermal energy: exothermic reactions tend to be spontaneous. Second is the general trend towards greater disorder: reactions with a positive $\Delta S$ tend to be spontaneous. In cases where both principles favor a reaction, the reaction is always spontaneous. In cases where both principles oppose the reaction (*i.e.*, favor the reverse reaction) the reaction is always non-spontaneous. In cases where the two trends are in opposition, spontaneity is determined by the temperature. These experimental results are summarized in table 2.2.

TABLE 2.2    THE DEPENDENCE OF SPONTANEITY ON $\Delta H$, $\Delta S$, AND T

| If $\Delta H$ is | and $\Delta S$ is | the reaction is |
|---|---|---|
| − | + | spontaneous at all T |
| + | − | non-spontaneous at all T |
| − | − | spontaneous at low T |
| + | + | spontaneous at high T |

Remember that for exothermic reactions, the system loses heat and the surroundings gain heat. Thus an exothermic reaction is expected to increase the thermal motion, and thus the entropy, of the surroundings.

A single criterion for spontaneity is that the reaction causes the entropy of the universe to increase. The entropy change of the universe is given by the sum of the changes in the system and surroundings:

$$\Delta S_{system} + \Delta S_{surroundings} = \Delta S_{universe}$$

EQ. 2.9

A positive value of $\Delta S_{universe}$ means spontaneity. By means that are beyond the scope of this course, it can be shown that this means, for spontaneous reactions at constant temperature and pressure,

$$\Delta H - T\Delta S < 0$$

EQ. 2.10

The thermodynamic value of H - TS is given the special symbol G, and is called **Gibbs Free Energy**. This is a state function, so the value of $\Delta G$ is determinable from Hess' Law and standard tables of $\Delta \overline{G}_f^o$. As long as $\Delta G$ is negative for the reaction occurring at constant T and P, the reaction is spontaneous. Substitution of positive and negative values for $\Delta H$ and $\Delta S$ confirms the various outcomes noted in table 2.2. The equation below is an expression of the Second Law of Thermodynamics in terms of a formula:

$$\Delta H - T\Delta S = \Delta G \qquad \text{EQ. 2.11}$$

**EXAMPLE 1**

Consider the following reaction:

$$2\ NO(g)\ +\ O_2(g) \longrightarrow 2\ NO_2(g) \qquad \text{EQ. 2.12}$$

In this reaction we see 3 moles of gaseous reactants converted to two moles of gaseous products. We expect to see a decrease in entropy. Substitution of literature values confirms this:

$$\Delta \overline{S}^o_{reaction} = 2\ \overline{S}^o_{NO_2(g)} - \left[ 2\ \overline{S}^o_{NO(g)} + \overline{S}^o_{O_2(g)} \right]$$

$$= 2\,(240.0) - \left[ 2(210.7) + 205 \right] = -146.4\ \text{J}/_{mol \cdot K} \qquad \text{EQ. 2.13}$$

The reaction is disfavored by entropy. The enthalpy change for the reaction is similarly calculated:

$$\Delta \overline{H}^o_{reaction} = 2\ \Delta \overline{H}^o_{NO_2(g)} - \left[ 2\ \Delta \overline{H}^o_{NO(g)} + \Delta \overline{H}^o_{O_2(g)} \right]$$

$$= 2\,(33.2) - \left[ 2(90.25) + 0 \right] = -114.1\ \text{kJ}/_{mol} \qquad \text{EQ. 2.14}$$

Thus the formation of the product (nitrogen dioxide) is energetically favored but entropically disfavored. Under standard conditions, the temperature is 298.15 K, and the decisive factor is seen to be enthalpy.

$$\Delta \overline{G}^o_{reaction} = \Delta \overline{H}^o_{reaction} - T\Delta \overline{S}^o_{reaction}$$

$$= \left( -114.1\ \text{kJ}/_{mol} \right) - \left( 298.15\ \text{K} \right)\left( -146.4\ \text{J}/_{mol \cdot K} \right)\left( 1\ \text{kJ}/_{1000\ J} \right) = -70.4\ \text{kJ}/_{mol} \qquad \text{EQ. 2.15}$$

The standard state reaction is seen to have a negative free energy change, and thus will be spontaneous. Under standard conditions, nitric oxide will spontaneously react with oxygen to form nitrogen dioxide (as opposed to the reverse.) Note that the unfavorable entropy term is enhanced by temperature. The reverse reaction (decomposing $NO_2$ into NO and $O_2$) can be made spontaneous by increasing T. We could calculate the exact T at which the spontaneous direction of the reaction reverses if we assume the values of ΔH and ΔS do not change much with T.

$$\Delta\bar{G} \geq 0 \;\rightarrow\; \Delta\bar{H} - T\Delta\bar{S} \geq 0; \text{ if } \Delta\bar{G} = 0,$$

EQ. 2.16

$$T = \frac{\Delta\bar{H}}{\Delta\bar{S}} = \frac{-114.1 \text{ }^{kJ}/_{mol}}{-0.1464 \text{ }^{kJ}/_{mol \cdot K}} = 779.4 \text{ K}$$

Thus we expect that significantly increasing the temperature will change equation 2.12 from a spontaneous reaction to a non-spontaneous reaction.

**EXAMPLE 2**

Ethene reacts with chlorine, forming 1,2-dichloroethane with a large negative standard state free energy change (–147.67 kJ/mol).

$$C_2H_4(g) \;+\; Cl_2(g) \longrightarrow ClCH_2CH_2Cl(l)$$

EQ. 2.17

This reaction converts two gas phase molecules into one liquid phase molecule. This is most definitely a decrease in entropy ($\Delta S = -234.10$ J/K mol), a condition faced by all such addition reactions. The spontaneity of such reactions must be due to enthalpy concerns. Breaking the relatively weak π bond to form two stronger σ bonds (one to each carbon atom) will normally be favored by enthalpy. In fact, for equation 2.17, $\Delta H_{reaction} = -217.49$ kJ/mol. Addition reactions will typically have a favorable enthalpy term, unless the π bond in the reactant molecule is part of an unusually stable system (*e.g.*, an aromatic compound like benzene). In such a case, halogen molecules will not add across a double bond. Instead, as we have previously seen, halogenation proceeds via electrophilic aromatic substitution, not direct addition.

## 2.7 FREE ENERGY CHANGES UNDER NON-STANDARD CONDITIONS

So far we have considered only those reactions that occur under standard conditions. In addition to experimental conditions of 25°C and $10^5$ Pascals (one bar), such conditions assume pure, unmixed reactants and products each at **unit activity**. For our purposes, we can assume this means about one molar concentration for all species in solution. In fact, chemists never run real reactions under these conditions. What then becomes of our ability to predict spontaneity for real world conditions? A decrease in free energy still proves to be the decisive factor in determining whether a specific reaction mixture is headed toward the reactant side or the product side, *i.e.*, whether it is spontaneous or non-spontaneous as written. But as we will see, there is a difference between $\Delta\bar{G}^\circ$ (the standard state molar free energy change) and $\Delta\bar{G}$ (the molar free energy change for the actual experimental conditions.)

Consider the gas phase reaction in which 'a' moles of A is converted into 'b' moles of B at 298.15K:

$$a A(g) \longrightarrow b B(g)$$

EQ. 2.18

$\Delta\overline{G}°$ corresponds to the difference in free energy between pure B and pure A. This can be calculated using Hess' law and literature values for the standard state free energies of formation of A and B:

$$\Delta\overline{G}°_{reaction} = b\left(\Delta\overline{G}°_{f(B)}\right) - a\left(\Delta\overline{G}°_{f(A)}\right)$$

EQ. 2.19

If the standard state free energy of the reaction is negative, the reaction taking pure A to pure B is spontaneous. This can be pictured graphically in figure 2.8.

Since pure B is at a lower free energy than pure A, $\Delta\overline{G}°$ is negative, and represents the force driving the reaction from pure A towards pure B. But the path by which pure A becomes pure B takes the system through conditions where there is neither pure A nor pure B, but a mixture of the two. At some non-standard pressure, $P_A$, the free energy of A(g) can be shown to be:

$$\overline{G}_A = \overline{G}°_A + RT\ln(P_A)$$

EQ. 2.20

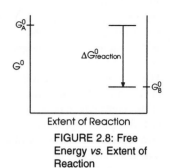

FIGURE 2.8: Free Energy *vs.* Extent of Reaction

Note that the pressure of pure A started at $10^5$ Pascals, or one bar (essentially one atmosphere). When $P_A$ numerically equals one, equation 2.20 shows that the actual free energy equals the standard state free energy. Note also that during the course of the reaction, as the pressure of A becomes less than one, the value of $\Delta\overline{G}_A$ decreases. This is because the logarithm of a number less than one is negative.

A similar expression for B can be written as equation 2.21:

$$\overline{G}_B = \overline{G}°_B + RT\ln(P_B)$$

EQ. 2.21

Combining Equations 2.19, 2.20 and 2.21 yields

$$\Delta\overline{G}_{reaction} = \left[b\overline{G}°_B - a\overline{G}°_A\right] + RT\left(b\ln P_B - a\ln P_A\right)$$

EQ. 2.22

which simplifies to

$$\Delta\overline{G}_{reaction} = \left[b\overline{G}°_B - a\overline{G}°_A\right] + RT\ln\left(\frac{(P_B)^b}{(P_A)^a}\right)$$

EQ. 2.23

The ratio of the pressure (or concentration) of products over the pressure (or concentration) of reactants, each taken to the power of their stoichiometric coefficient, is known as the **proper quotient of reaction**, symbolized Q.

$$\Delta\overline{G}_{reaction} = \Delta\overline{G}°_{reaction} + RT\ln Q$$

EQ. 2.24

Figure 2.9 shows the variation of $\Delta G$ with extent of reaction. At the start of the reaction, when there is pure A and no B, the value of Q is zero. The logarithm of zero is negative infinity, and thus $\Delta\overline{G}_{reaction}$ is guaranteed to be negative. Thus, when only pure A is present, the reaction forming B is spontaneous. But at the opposite extreme, when only pure B is present, the value of Q is positive infinity, and the value of $\Delta\overline{G}_{reaction}$ is guaranteed to be positive. Thus, when only pure B is present, the reverse reaction forming A is spontaneous. Pure A reduces its free energy by partially becoming B, and pure B reduces its free energy by partially becoming A. At some point, intermediate between pure A and pure B, equilibrium is reached. Exactly where along the reaction coordinate equilibrium is reached depends upon which material, A or B, has the higher free energy and the magnitude of the difference in their free energies.

The driving force for chemical reactions is a decrease in free energy. As seen in figure 2.9, this decrease in free energy will occur until the system reaches some intermediate point (i.e., neither pure A nor pure B). The mixture of A and B at this point represents the minimum free energy the system can attain. It also represents what we will see later to be the equilibrium condition of the reaction. The real driving force for a reaction is thus $\Delta\overline{G}$, the difference in free energy between the current composition and the composition corresponding to equilibrium conditions. The $\Delta\overline{G}$ of reaction thus varies according to the extent of reaction, as seen in several examples in figure 2.10:

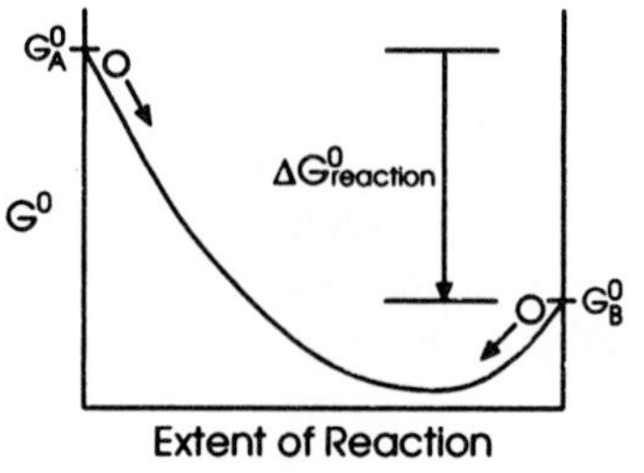

For the general reaction:

$$a\,A + b\,B \rightarrow c\,C + d\,D$$

EQ. 2.25

the free energy change at any point along the path converting reactants to products is given by

$$\Delta\overline{G}_{reaction} = \Delta\overline{G}^{o}_{reaction} + RT \ln\left(\frac{[C]^{c}[D]^{d}}{[A]^{a}[B]^{b}}\right)$$

EQ. 2.26

When the system reaches minimum free energy, no further net change takes place, and $\Delta\overline{G}_{reaction} = 0$. Thus, for a system in which the reactants and products are in solution, equation 2.26 becomes

$$\Delta\overline{G}^{o} = -RT \ln\left(\frac{[C]^{c}_{eq}[D]^{d}_{eq}}{[A]^{a}_{eq}[B]^{b}_{eq}}\right)$$

EQ. 2.27

The square brackets subscripted by "eq" mean equilibrium concentration. The proper quotient of reaction Q, when determined at equilibrium, has a specific value called the **equilibrium constant**. It is given the symbol upper case K. (Do not use a lower case k; as we will see in chapter 5, this is reserved for something else.) There is an almost identical expression for the equilibrium constant for reactions involving gases. We shall not consider such systems at this time.

$$\Delta\overline{G}^{o} = -RT \ln K$$

EQ. 2.28

Thus the actual free energy change for any reaction can be written as

$$\Delta\overline{G}_{reaction} = -RT \ln K + RT \ln Q = RT \ln\left(\frac{Q}{K}\right)$$

EQ. 2.29

FIGURE 2.9: Free Energy vs. Extent of Reaction

and spontaneity can be determined by the ratio of Q/K.

## TO SUMMARIZE:

Free energy is an enormously useful thermodynamic function. The value of the standard state free energy change determines the value of the equilibrium constant. A large negative value of $\Delta G^0$ indicates a large value of K (more products than reactants); a large positive value of $\Delta G^0$ indicates a small value of K (more reactants than products). The sign of the (non-standard) free energy change determines whether a reaction is spontaneous ($\Delta G$ negative), non-spontaneous ($\Delta G$ positive), or at equilibrium ($\Delta G = 0$). The magnitude of $\Delta G$ equals the amount of useful work that can be obtained from the reaction under conditions of constant temperature and pressure.

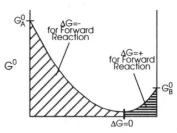

Extent of Reaction

FIGURE 2.10: Free Energy *vs.* Extent of Reaction for Different Compositions

### CONCEPT CHECK

Consider the following reaction:

$$H_2(g) \;+\; I_2(g) \longrightarrow 2\,HI(g) \qquad\qquad \text{EQ. 2.30}$$

When this reaction was allowed to come to equilibrium, the concentrations of hydrogen and iodine gases were found to be 0.0783 M and the concentration of HI gas was found to be 1.92 M. Given that the standard free energy of formation of HI(g) is 1.7 kJ mol$^{-1}$, what is the standard free energy of formation of iodine gas?

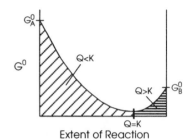

Extent of Reaction

FIGURE 2.11: Free Energy *vs.* Extent of Reaction Showing Q/K Regions

Our approach here will be to first find the value for the equilibrium constant for the reaction, then use that value to determine the standard free energy for the chemical reaction, and finally use that value along with the free energy of formation of HI to get the final answer.

The value of the equilibrium constant can be found by setting up the equilibrium constant expression and substituting into it the equilibrium concentrations:

$$K_{eq} = \frac{[HI]^2}{[H_2][I_2]}$$

$$K_{eq} = \frac{(1.92)^2}{(0.0783)(0.0783)} = 60\overline{1}.3 \qquad\qquad \text{EQ. 2.31}$$

This value can then be placed into equation 2.28 to yield

$$\Delta \overline{G}^0_f = -RT \ln K_{eq}$$

$$\Delta \overline{G}^0_f = -(8.314)(298.15)\ln(60\overline{1}.3)$$

$$\Delta \overline{G}^0_f = -(8.314)(298.15)(6.39\overline{9}) \qquad\qquad \text{EQ. 2.32}$$

$$\Delta \overline{G}^0_f = -1586\overline{2}\ \text{J mol}^{-1} = -15.86\ \text{kJ mol}^{-1}$$

Note that the value of the standard free energy change for the reaction is good to four significant figures because the rules for the number of significant figures in a logarithm state that only those figures behind the decimal point are significant. The third significant figure of the logarithm is the third digit to the right of the decimal point. This means that the number actually has four significant figures when used in subsequent calculations.

Using Hess' Law we can then obtain the final value:

$$\Delta \bar{G}^o_{reaction} = \sum \Delta \bar{G}^o_{f, products} - \sum \Delta \bar{G}^o_{f, reactants}$$
$$\Delta \bar{G}^o_{reaction} = 2 \Delta \bar{G}^o_{f, HI(g)} - \Delta \bar{G}^o_{f, H_2(g)} - \Delta \bar{G}^o_{f, I_2(g)}$$
$$-15.86 = 2(1.7) - (0) - \Delta \bar{G}^o_{f, I_2(g)}$$
$$-19.\overline{2}6 = - \Delta \bar{G}^o_{f, I_2(g)} \; ; \Delta \bar{G}^o_{f, I_2(g)} = 19.3 \text{ kJ mol}^{-1}$$

EQ. 2.33

Notice that the free energy of formation is not zero because the iodine reactant is not in its standard state (solid) in this reaction.

## 2.8   SOME SPECIAL APPLICATIONS OF FREE ENERGY

Chemical reactions are often used to provide energy. The use of chemical reactions to produce energy occurs nearly everywhere, from fuel cells in outer space to organisms growing near thermal vents in the ocean floor. Let us specifically look at two types of reactions: those that occur in internal combustion engines, and those that occur inside our bodies.

### THE INTERNAL COMBUSTION ENGINE

The internal combustion engine burns gasoline, a complex mixture composed mostly of hydrocarbons, in the presence of air. Air provides the source of oxygen, which is used to oxidize the carbon atoms in the fuel to form carbon dioxide. Consider the example of the combustion of octane:

$$C_8H_{18}(g) + \tfrac{25}{2}O_2(g) \longrightarrow 8\,CO_2(g) + 9\,H_2O(l)$$

EQ. 2.34

From literature values of $\Delta \bar{H}^o_f$ for reactants and products, the enthalpy change of reaction can be determined:

$$\Delta \bar{H}^o_{reaction} = \left[ 8(-393.51) + 9(-285.83) \right] - \left[ (-208.45) + 12.5(0) \right]$$
$$= -5512.10 \text{ kJ/mol}$$

EQ. 2.35

The $\Delta S^o$ of reaction can be similarly determined from literature values of absolute standard state entropies:

$$\Delta \bar{S}^o_{reaction} = \left[ 8(213.74) + 9(69.91) \right] - \left[ (466.84) + 12.5(205.138) \right]$$
$$= -691.96 \text{ J/mol}$$

EQ. 2.36

The fact that the reaction is highly exothermic favors the reaction, but decreasing the number of moles of gas formed causes a decrease in disorder. Using the definition of the second law (equation 2.11), the standard state free energy change can be determined:

$$\Delta \overline{G}^o_{reaction} = \left(-5512.10 \, {}^{kJ}\!/_{mol}\right) - (298.15 \, K)\left(-691.96 \, {}^{J}\!/_{mol \cdot K}\right)\left(\frac{1 \, kJ}{1000 \, J}\right)$$
$$= -5305.79 \, {}^{kJ}\!/_{mol}$$

EQ. 2.37

The negative value for $\Delta G^o$ indicates a spontaneous reaction under standard conditions. When the values of $\Delta \overline{G}^o_f$ are available for all reactants and products, $\Delta G^o$ can also be calculated from literature values using Hess' law. The magnitude of $\Delta G^o$ can be used to calculate the value of the equilibrium constant at 298.15 K by rearranging equation 2.28:

$$\ln K = \frac{-\Delta G^o}{RT} = \frac{-\left(-5305.79 \, \frac{kJ}{mol}\right)}{\left(8.3145 \, \frac{J}{mol \, K}\right)\left(\frac{1 \, kJ}{1000 \, J}\right)(298.15 \, K)} = 2140.3$$

EQ. 2.38

which ultimately yields the enormous value of $K = 3 \times 10^{929}$. Besides being such a large number that a calculator cannot handle it directly, such a large equilibrium constant means that the reaction goes essentially the entire way to completion. Thus the real free energy change approximates the large negative standard state value of $\Delta G^o$. This means that a large amount of free energy is released in the combustion of octane. It is this energy that is tapped to make the engine run and the car go.

In practice, the combustion reaction does not take place at 25°C and 1 bar pressure, but the answer we get is approximately right for the real conditions.

It is also worth noting that in an auto engine, the combustion reaction is initiated by an electrically generated spark. This input of electrical energy is negligible in terms of the overall energy balance. Without the spark, however, neither the reaction nor the car should be expected to go. The spark provides sufficient energy to initiate the reaction of a few molecules. The reaction of the first few molecules then causes the reaction of a number of additional molecules which, in turn, causes reaction of yet more molecules. This process continues at a very rapid rate until all of the materials in the automobile engine cylinder have reacted (reached equilibrium). The initial and final states are the same with or without the spark; the reaction is simply initiated by the spark. The spark allows the reaction to occur more rapidly. Since the spark does not change the initial or final states, it can not change the value of $\Delta G^o$, and thus can not influence the value of K.

## APPLICATION OF THERMODYNAMICS TO BIOCHEMICAL PROCESSES

Next, let us consider the application of thermodynamics to biochemical processes. The first step in the metabolism of glucose is its conversion to glucose-6-phosphate.

$$C_6H_{12}O_6 + HPO_4{}^{2-} \rightarrow C_6H_{11}PO_9{}^{2-} + H_2O$$

EQ. 2.39

A **biochemical standard state** (one that includes specific concentrations of various ions, especially $H^+$ and $Mg^{2+}$) is denoted by a prime marker. That is, the biochemical standard state free energy changes are noted as $\Delta G^{0'}$. The direct phosphorylation of glucose is energetically unfavorable, with $\Delta G^{0'}$ for equation 2.39 being +16.7 kJ/mol. In order to force the reaction to occur, it is coupled with the hydrolysis of adenosine triphosphate, which has a large negative biochemical standard state free energy change:

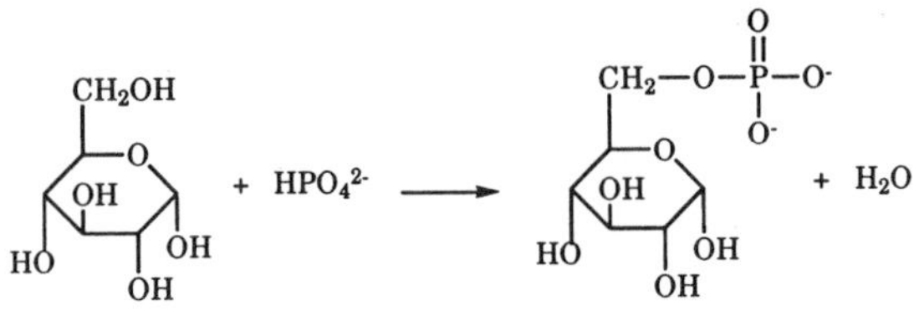

FIGURE 2.12:
Biochemical Conversion of Glucose to Glucose-6-phosphate

$$ATP + H_2O \rightarrow ADP + P_i \qquad \Delta G^{0'} = -31.0 \text{ kJ/mol} \qquad \text{EQ. 2.40}$$

(Here $P_i$ stands for inorganic phosphate; ATP and ADP stand for adenosine triphosphate and adenosine diphosphate respectively.) The coupling of the energetically unfavorable reaction in equation 2.39 ($\Delta G^{0'} = +16.7$ kJ/mol) with the energetically highly favorable reaction in equation 2.40 ($\Delta G^{0'} = -31.0$ kJ/mol) is an example of Hess' law, which permits addition of chemical reactions to yield the sum of the individual changes in the state functions. Specifically, the reaction

$$\text{glucose} + ATP \rightarrow \text{glucose-6-phosphate} + ADP \qquad \text{EQ. 2.41}$$

has a total $\Delta G^{0'} = -14.3$ kJ/mol. Thus although phosphorylation with inorganic phosphate (equation 2.39) is non-spontaneous, phosphorylation via ATP (equation 2.41) is spontaneous under biochemical standard state conditions. Such biochemical coupling, using the negative free energy change of one reaction to drive another non-spontaneous reaction is often shown in the form of tangent curved arrows, seen in figure 2.13.

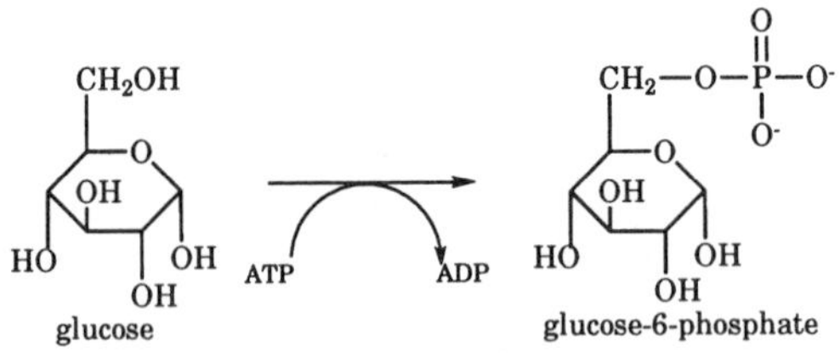

FIGURE 2.13:
Biochemical Curved Arrow Representation for the Reaction in Equation 2.41

Finally, let us consider the role that concentration plays in biochemical reactions. One of the steps in the photosynthetic production of glucose is the following:

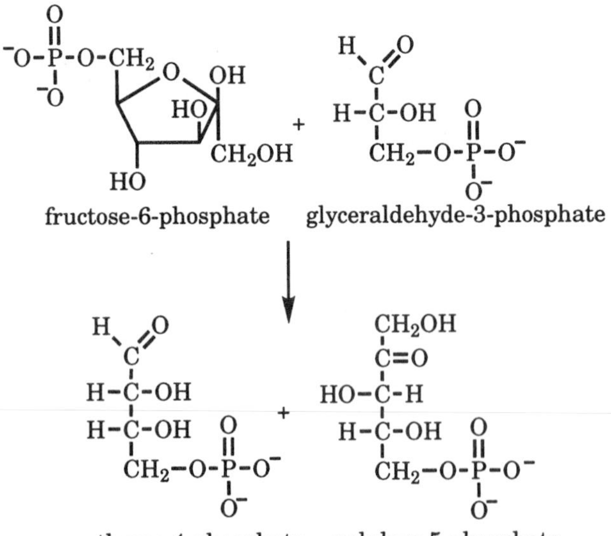

fructose-6-phosphate glyceraldehyde-3-phosphate

EQ. 2.42

erythrose-4-phosphate xylulose-5-phosphate

Under standard state conditions (when all four of these species have a concentration of 1.0 M), the value of $\Delta G^{o'} = 6.3$ kJ/mol. That is, the reaction is non-spontaneous under standard state conditions. In the chloroplast, where the reaction occurs, the approximate concentrations are the following: fructose-6-phosphate, $5 \times 10^{-4}$ M; glyceraldehyde-3-phosphate, $3 \times 10^{-5}$ M; erythrose-4-phosphate, $2 \times 10^{-5}$ M; and xylulose-5-phosphate, $2 \times 10^{-5}$ M. The value of Q, the proper quotient of reaction, is thus calculated to be 0.026 . Substituting the values of $\Delta G^o$ and Q into equation 2.26 yields:

$$\Delta G = 6.3 + (.0083145)(298.15) \ln(0.026) = -2.7 \text{ kJ/mol} \qquad \text{EQ. 2.43}$$

The negative value for $\Delta G$ indicates that under the actual conditions in the chloroplast, the reaction is spontaneous.

In most biochemical processes, we will find that reactions are catalyzed by enzymes. These biological catalysts are neither reactants nor products, but provide a lower energy pathway between the initial and final state. Since the value of the equilibrium is determined by the initial and final states of the system, enzymes do not influence how far a reaction goes. Nevertheless, we will see that they play a crucial role in the rates of reactions.

**Biochemical standard state:** thermodynamic state with conditions including a pressure of one bar ($10^5$ Pascals), a specified temperature (usually 310.6 K), a pH of 7, and a solute activity equal to the total concentration of all species of the specified molecule at a pH of 7.

**Entropy:** a thermodynamic state function, symbolized by S, that is the measure of the amount of disorder in the system.

**Equilibrium constant:** symbolized by K, representing the ratio of the equilibrium concentration (or pressure) of products over reactants, with each term raised to the power of its stoichiometric coefficient. K is a subset of Q, determined when the reaction has reached equilibrium.

**Free energy:** a thermodynamic state function, symbolized by G, and defined to be H - TS. The sign of $\Delta G$ determines spontaneity of a reaction, and the magnitude of $\Delta G$ determines the amount of usable work associated with the process. Under standard state conditions, the value of $\Delta G^O$ determines the equilibrium constant.

**Proper quotient of reaction:** symbolized by Q, representing the ratio of the concentration (or pressure) of products over reactants, with each term raised to the power of its stoichiometric coefficient.

**Spontaneous:** a reaction is said to be spontaneous if, under the specified conditions, there is net conversion of reactants into products. This occurs when $\Delta G$ is negative.

**Unit activity:** a thermodynamic variable used to specify that the "activity" of a material present at the time of reaction has a particular value. For mixtures of gases this corresponds, roughly, to a partial pressure of 1 bar of the gas; for materials in solution this corresponds, roughly, to a concentration of 1 Molar. For pure liquids and solids, it corresponds to the presence of the pure liquid or solid.

1. What would be the sign of $\Delta H$ for the following phase changes?
(Hint: no calculations are needed!)

(a) $CH_3OH(l) \rightarrow CH_3OH(g)$
(b) $C_6H_6(l) \rightarrow C_6H_6(s)$
(c) $Hg(s) \rightarrow Hg(l)$
(d) $NH_3(g) \rightarrow NH_3(l)$

2. What would be the sign of $\Delta S$ for the phase changes in question #1?
(Hint: no calculations are needed!)

3. As a rule, the standard state entropies of compounds are positive, and the entropy changes of formation of compounds are negative. Explain both.

4. A fertilized chicken egg develops within its shell to form a chicken embryo. What is the sign of $\Delta S_{system}$ for this process? (The system is the egg and its contents.)

5. Since the process described in question 4 is spontaneous, what can be said about the sign of $\Delta H$ for the process described? Is this related to the need for heat in incubating the egg?

6. Equation 2.3 gives the probability of n particles all finding their way into one of two equally likely compartments in a box. Calculate the probability of Avogadro's number of particles all ending up in one specified compartment.

7. The liquid and gaseous phases of vinyl chloride, $CH_2CHCl$, have standard molar enthalpies of formation of 14.6 and 35.6 kJ/mol respectively. Calculate the standard molar entropy change of vaporization of vinyl chloride, and explain the significance of the sign of your answer.

8. From the values of absolute standard state molar entropies given in table 2.1, calculate the standard molar enthalpies of vaporization for $SiCl_4$, $GeCl_4$ and $SnCl_4$. Find literature values for these three compounds, and rank these compounds in order of increasing enthalpies of vaporization and increasing boiling points. What does this imply about the relative order of the strength of intermolecular forces between these molecules?

9. Predict whether the following reactions will have a positive or negative $\Delta S$. (Hint: calculations are not required!)

(a) $2\ O_3(g) \rightarrow 3\ O_2(g)$
(b) $CaCO_3(s) \rightarrow CaO(s) + CO_2(g)$
(c) $2\ N_2(g) + O_2(g) \rightarrow 2\ N_2O(g)$
(d) $Ca_3P_2(s) + 6\ H_2O(l) \rightarrow 3\ Ca(OH)_2(s) + 2\ PH_3(g)$

10. Predict whether the following reactions will have a positive or negative $\Delta S$. (Hint: calculations are not required!)

$\qquad$ (a) $2 H_2O(g) \rightarrow 2 H_2(g) + O_2(g)$
$\qquad$ (b) $C_2H_2(g) + H_2(g) \rightarrow C_2H_4(g)$
$\qquad$ (c) $4 Fe(s) + 3 O_2(g) \rightarrow 2 Fe_2O_3(s)$
$\qquad$ (d) $PbSO_4(s) \rightarrow PbO_2(s) + SO_2(g)$

11. Write the expression for Q in terms of concentration for each of the following reactions:

$\qquad$ (a) $C_2H_2(g) + H_2(g) \rightarrow C_2H_4(g)$
$\qquad$ (b) $2 O_3(g) \rightarrow 3 O_2(g)$
$\qquad$ (c) $2 N_2(g) + O_2(g) \rightarrow 2 N_2O(g)$

12. For the following reactions, use the values of standard molar entropies and standard molar enthalpies of formation in the appendix to calculate the molar enthalpy and entropy changes of reaction.

$\qquad$ (a) $C_2H_2(g) + H_2(g) \rightarrow C_2H_4(g)$
$\qquad$ (b) $CaCO_3(s) \rightarrow CaO(s) + CO_2(g)$
$\qquad$ (c) $PbSO_4(s) \rightarrow PbO_2(s) + SO_2(g)$

13. From the results from problem #12, calculate the standard state free energy change for each reaction at 25°C.

14. Use the values of standard molar entropies and standard molar enthalpies of formation in the appendix to calculate the molar enthalpy and entropy changes of reaction for:

$\qquad$ (a) $CaH_2(s) + 2 H_2O(l) \rightarrow Ca(OH)_2(s) + 2 H_2(g)$
$\qquad$ (b) $3 MgO(s) + 2 Fe(s) \rightarrow Fe_2O_3(s) + 3 Mg(s)$
$\qquad$ (c) $CH_3CH_2OH(l) + 3 O_2(g) \rightarrow 2 CO_2(g) + 3 H_2O(l)$

15. From the results from problem #14, calculate the standard state free energy change for each reaction at 25°C.

16. $NO_2(g)$ and $N_2O_4(g)$ have standard state free energies of formation of 51.31 and 97.89 kJ/mol respectively. For the reaction:

$$N_2O_4(g) \rightarrow 2 NO_2(g)$$

calculate the following:

$\qquad$ (a) the standard state molar free energy change;
$\qquad$ (b) the equilibrium constant at 25°C;
$\qquad$ (c) the value of Q, if at a certain time the partial pressures of $N_2O_4$ and $NO_2$ are .800 and .300 bars respectively;
$\qquad$ (d) the value of $\Delta G$ for the conditions described in (c);
$\qquad$ (e) if the reaction is spontaneous for the conditions described in (c).

17. $N_2O_5(g)$ and $N_2O_4(g)$ have standard state free energies of formation of 115.1 and 97.89 kJ/mol respectively. For the reaction

$$2 N_2O_5(g) \rightarrow 2 N_2O_4(g) + O_2(g)$$

calculate the following:

(a) the standard state molar free energy change;
(b) the equilibrium constant at 25°C;
(c) the value of Q, if at a certain time the partial pressures of $N_2O_4$, $N_2O_5$ and $O_2$ are .800, .400 and 0.200 bars respectively;
(d) the value of $\Delta G$ for the conditions described in (c);
(e) if the reaction is spontaneous for the conditions described in (c).

**CHAPTER TWO**

OVERVIEW
# Problems

18. Metathesis reactions were characterized in book one as having three possible driving forces.

(a) Name each of these "driving forces" for metathesis reactions.
(b) The driving force for any chemical reaction at a specified temperature and pressure is actually free energy, which we have seen is composed of enthalpy and entropy terms. Explain which of these two terms is more significant for each of the "driving forces" listed in (a), and explain the likely sign of that term. (You may wish to compare your qualitative answer to this question to the quantitative answer to the following three problems.)

19. Aqueous silver ion ($\Delta H_f$ = 105.579 kJ/mol, $S°$ = 72.68 J/K mol, $\Delta G_f$ = 77.107 kJ/mol) reacts with aqueous chloride ion ($\Delta H_f$ = -167.159 kJ/mol, $S°$ = 56.5 J/K mol, $\Delta G_f$ = -131.228 kJ/mol).

(a) Predict the product(s), including the phase(s), and write the balanced reaction.
(b) Is the reaction spontaneous under standard state conditions?
(c) Calculate the values of the change in enthalpy and the change in entropy for the reaction.
(d) In a few sentences, explain the chemical reason for the sign of the enthalpy change and the sign of the entropy change determined in (c).

20. Liquid nitric acid reacts with solid sodium fluoride ($\Delta H_f$ = -573.647 kJ/mol, $S°$ = 51.46 J/K mol, $\Delta G_f$ = -543.494 kJ/mol) to form gaseous hydrogen fluoride and solid sodium nitrate ($\Delta H_f$ = -467.85 kJ/mol, $S°$ = 116.52 J/K mol, $\Delta G_f$ = -367.00 kJ/mol)

(a) Write the balanced chemical reaction, including phases.
(b) Is the reaction spontaneous under standard state conditions?
(c) Calculate the values of the change in enthalpy and the change in entropy for the reaction.
(d) In a few sentences, explain the chemical reason for the sign of the enthalpy change and the sign of the entropy change determined in (c).

21. Aqueous sodium acetate ($\Delta H_f$ = -726.13 kJ/mol, $S°$ = 145.6 J/K mol, $\Delta G_f$ = -631.20 kJ/mol) reacts with aqueous hydrochloric acid ($\Delta H_f$ = -167.159 kJ/mol, $S°$ = 56.5 J/K mol, $\Delta G_f$ = -131.228 kJ/mol) to form aqueous sodium chloride ($\Delta H_f$ = -407.27 kJ/mol, $S°$ = 115.5 J/K mol, $\Delta G_f$ = -393.133 kJ/mol) and aqueous acetic acid ($\Delta H_f$ = -485.76 kJ/mol, $S°$ = 178.7 J/K mol, $\Delta G_f$ = -396.46 kJ/mol).
   (a) Write the balanced chemical reaction, including phases.
   (b) Is the reaction spontaneous under standard state conditions?
   (c) Calculate the values of the change in enthalpy and the change in entropy for the reaction.
   (d) In a few sentences, explain the chemical reason for the sign of the enthalpy change and the sign of the entropy change determined in (c).

22. Acid chlorides are especially reactive towards water. When liquid acetyl chloride ($CH_3COCl$, MW = 78.499, $\Delta G_f$ = -208.1 kJ/mole) reacts with liquid water, liquid acetic acid and gaseous hydrogen chloride form.
   (a) Write the formation reaction of acetyl chloride.
   (b) Write the balanced reaction for the hydrolysis of acetyl chloride.
   (c) Determine the molar free energy change for the reaction in (b).
   (d) If 100.0 g of acetyl chloride reacts with 25.00 g of water, what is the maximum volume of dry HCl gas that could be collected at 745.0 torr and 20.0°C?

23. Benzyl alcohol ($C_7H_8O$, $\Delta G_f$ = -27.49 kJ/mole) reacts in acidic aqueous solution with permanganate ion ($MnO_4^-$, $\Delta G_f$ = -447.2 kJ/mole) to form liquid water, $Mn^{2+}$(aq) ($\Delta G_f$ = -228.1 kJ/mole) and benzoic acid ($C_7H_6O_2$, $\Delta G_f$ = -245.3 kJ/mole). (Note that $H^+$(aq) has $\Delta G_f$ = 0 kJ/mole.)
   (a) Write the balanced redox reaction.
   (b) Determine the molar free energy change for the reaction in (a).
   (c) How will the IR spectra of benzyl alcohol and benzoic acid differ?
   (d) What reagent(s) could be used to convert the benzoic acid back into benzyl alcohol?

24. Liquid benzene can undergo an electrophilic aromatic substitution reaction with liquid nitric acid to form liquid water and nitrobenzene ($C_6H_5NO_2$, $\Delta Gf$ = 146.23 kJ/mole). The reaction is usually run with sulfuric acid catalysis, but since a catalyst is neither a reactant nor a product, it can not influence the initial or the final state. Thus the change in free energy for the reaction does not depend on the catalyst.
   a. Write the mechanism for the nitration of benzene under sulfuric acid catalysis conditions.
   b. Write the net balanced reaction.
   c. Determine the molar free energy change for the reaction in (b).

# CHAPTER
# 3three

# CONFORMATIONAL
# Equilibrium
### THAT'S GAUCHE

# CONFORMATIONAL Equilibrium

## 3.1 CONFORMATIONAL ANALYSIS OF ACYCLIC ALKANES

Since we know that molecules are almost never completely motionless and that rotation about carbon-carbon single bonds is facile, it is important to have a rather detailed look at this process. We will begin our study by examining the rotation about the carbon-carbon bond of ethane, the simplest hydrocarbon that contains a carbon-carbon single bond, and then extend our study of this phenomenon to more complex alkanes and cycloalkanes.

As you may remember, rotation about carbon-carbon bonds leads to an infinite number of possible **conformers** (structures that differ by rotation about a carbon-carbon bond), but we are usually most interested in the extreme cases. If we sight down the carbon-carbon bond of ethane, we note that the two extreme cases are the **staggered conformation** and the **eclipsed conformation**. Several different types of drawings of these two configurations are presented as figures 3.1 through 3.3.

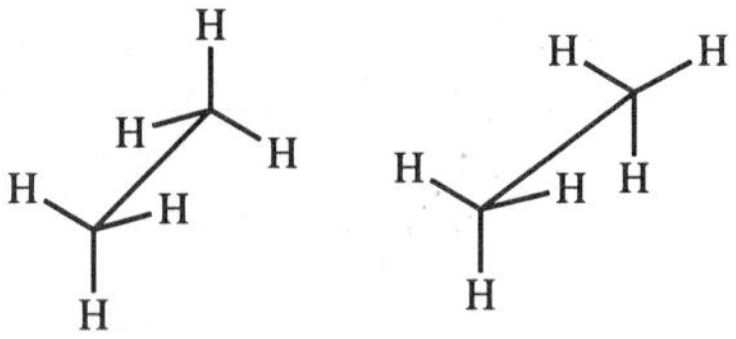

You would be correct in assuming that, since the two conformations are different, they have different energies. Let us look at the energy difference by looking at the potential energy as a function of **dihedral** (torsional) **angle**. As the angle between the hydrogen atoms on adjacent carbon atoms is varied, the potential energy changes in the following manner:

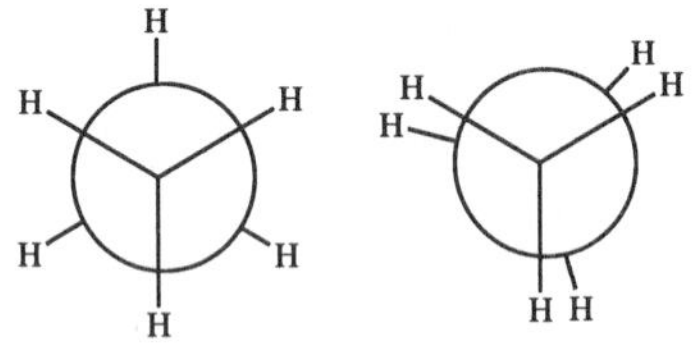

a. staggered        b. eclipsed

FIGURE 3.1: Ball and Stick Drawings Showing the Staggered and Eclipsed Conformations of Ethane

a. staggered        b. eclipsed

FIGURE 3.2: Sawhorse Drawings Showing the Staggered and Eclipsed Conformations of Ethane

a. staggered        b. eclipsed

FIGURE 3.3: Newman Projections Showing the Staggered and Eclipsed Conformations of Ethane

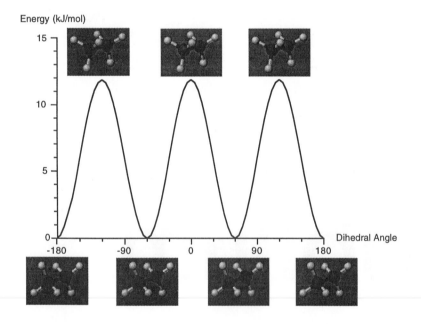

FIGURE 3.4: Potential Energy of Ethane as a Function of Dihedral Angle

As you can see from the curve, the staggered conformation is about 12 kJ/mol more stable than the eclipsed conformation. Such a small energy difference is insignificant at room temperature because of the energy associated with normal molecular motion. (It is approximately the same as the energy needed to excite the molecule from one vibrational level to the next higher level, *i.e.*, the energy of an IR photon.) Therefore the energy barrier separating the two conformers is low enough that essentially free rotation occurs around the carbon–carbon bond. As a result, interconversion of the conformers occurs rapidly at all but very low temperatures where the molecule does not have enough kinetic energy to overcome the barrier.

In the case of ethane, all of the staggered conformations are the same and all of the eclipsed conformations are the same. Let us now consider a situation where this is not the case. Consider what you would see if you were to sight down the C2-C3 bond of butane. We will use the methyl groups on the front and back carbon atoms as our point of reference. If we begin with the methyl-methyl eclipsed conformation (a) and rotate the back carbon six times by 60°, we end up with the six conformations shown in figure 3.5.

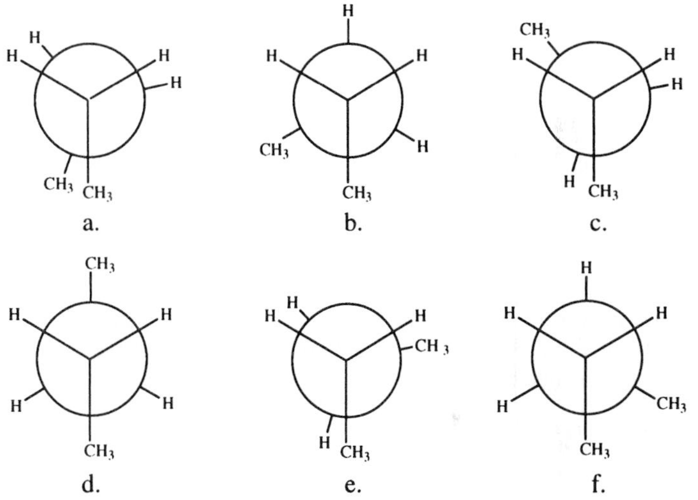

If you look carefully at the drawings, you will note that of the conformations we have drawn there are four unique possibilities; the methyl-methyl eclipsed conformation (a), the **gauche conformation** (b and f), the methyl-hydrogen eclipsed conformation (c and e), and the **anti conformation** (d). As you can probably guess, each unique conformation has a different potential energy. Another way of stating this is that the conformers differ with respect to stability. The least stable conformer is, of course, the methyl-methyl eclipsed (a), and the most stable conformer is the anti conformer (d). If you look at space filling representations (figure 3.6) of these conformers, you can see that the difference in stability is primarily a result of the distance between the two methyl groups. In the methyl-methyl eclipsed conformation, the methyl groups are very close to each other and their electron clouds are close enough that a relatively large amount of repulsion occurs between the two methyl groups. This type of interaction is referred to as steric hindrance. (In this conformation, replusion between the eclipsed hydrogen atoms on the carbon atoms will also be maximized. While normally these interactions are quite small, to the extent that they occur at all, they are maximized in this conformation.) In the other case, the anti conformation, the methyl groups are as far apart from each other as is possible. In this case the electron clouds of the two methyl groups are now as far away from each other as possible and repulsive interactions between them are now as small as they can be. (In addition, this configuration also minimizes hydrogen–hydrogen interactions.) In other conformations the methyl–methyl, methyl–hydrogen, and hydrogen–hydrogen repulsive interactions between the electron clouds are

FIGURE 3.5: Six Conformations of *n*-Butane

a. anti        b. eclipsed

FIGURE 3.6: Space Filling Representations of the Methyl-Methyl Anti Confomation and the Eclipsed Conformation of *n*-Butane

intermediate between these two extremes. Figure 3.7 shows a plot of potential energy vs. dihedral angle (the angle between the methyl groups as viewed down the $C_2$–$C_3$ bond) for the conformers of butane. This plot depicts the relative stabilities of these different conformations.

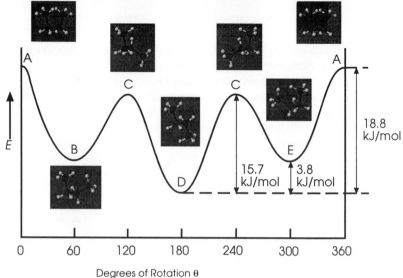

FIGURE 3.7: Potential Energy of Butane vs. Dihedral Angle About the $C_2$-$C_3$ Bond

As you saw in Chapter 2, the free energy change associated with a process can be used to calculate the equilibrium constant for that process. The equation for this calculation is written as

$$\Delta \overline{G}^{\,o}_{process} = - RT \ln K \qquad \text{EQ. 3.1}$$

Thus, we should be able to use this equation to determine what fraction of butane exists in given conformations. For the sake of simplicity, we will assume that the butane exists in only two conformations, the anti conformation and the gauche conformation. The equilibrium we need to consider then is

$$\text{anti} \rightleftharpoons \text{gauche} \qquad \text{EQ. 3.2}$$

In order to calculate the equilibrium ratio for this process we need to first calculate the change in free energy for the process using the following equation:

$$\Delta \overline{G} = \Delta \overline{H} - T \Delta \overline{S} \qquad \text{EQ. 3.3}$$

We know that the standard state change in enthalpy ($\Delta H$) for the process is approximately 3.8 kJ/mol and that the standard state change in entropy is approximately 5.76 J/mol K. Using equation 3.3 we arrive at the following for the standard state change in free energy of the process.

$$\Delta \overline{G} = 3800 \;^{J}/_{mol} - \left(298.15 \text{ K}\right)\left(5.76 \;^{J}/_{mol \cdot K}\right) = 2\bar{0}83 \;^{J}/_{mol} \qquad \text{EQ. 3.4}$$

Using equation 3.1 and solving for K allows us to calculate an equilibrium ratio.

$$K = e^{\frac{-\Delta \overline{G}}{RT}} = e^{\frac{\left(-2\bar{0}83 \;^{J}/_{mol}\right)}{\left(8.314 \;^{J}/_{mol \cdot K}\right)\left(298.15 \text{ K}\right)}} = 0.43 \qquad \text{EQ. 3.5}$$

This value corresponds to an equilibrium mixture of approximately 70% of the anti conformer and 30% of the gauche conformer. This corresponds to slightly more than a 2-fold preference for the anti conformation over the gauche conformation.

## 3.2  GEOMETRIC ISOMERISM IN CYCLOALKANES

Recall that **geometric isomerism** exists when rotation about a bond or a number of bonds is restricted such that there exists a defined arrangement of the groups attached to that bond(s). We have already studied one example of geometric isomerism; that is the case of 2-butene which exists both as *cis*-2-butene and as *trans*-2-butene.

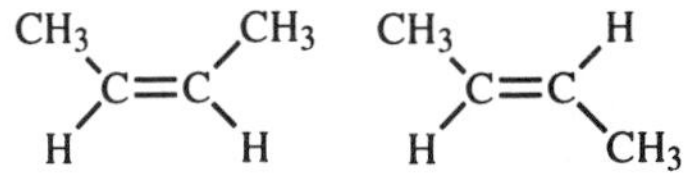

Since free rotation about double bonds cannot happen without breaking and reforming the π bond — a process that would have an energy barrier that is approximately 250 kJ/mol, the methyl groups are locked into place with respect to one another.

This same phenomenon is observed in cycloalkanes where the ring structure restricts rotation about the carbon-carbon bonds of the ring. Consider, for example, 1,2-dibromocyclopentane. Since rotation is restricted about the C1-C2 bond, the bromine atoms can be positioned either on the same side of the ring (the *cis* form) or on opposite sides of the ring (the *trans* form). Representations of these molecules are presented in figure 3.9.

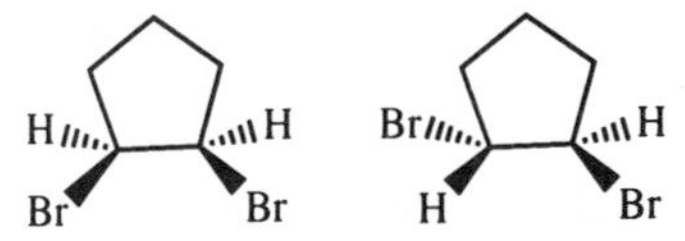

We also need to examine whether or not this restricted rotation gives rise to chirality. Recall that for a molecule to be **chiral**, it must be non-superimposable on its mirror image. Also, if a molecule contains an internal plane of symmetry it is considered to be achiral. Let us consider *cis*-1,2-dibromocyclopentane first. As you can see in figure 3.10, this molecule does indeed contain an internal plane of symmetry and is, therefore, achiral. (Recall that compounds like this are termed *meso*.) On the other hand, the *trans*-1,2-dibromocyclopentane does not contain an internal plane of symmetry. Furthermore, if we look at a molecule of the *trans* isomer and its mirror image (figure 3.11) we see that the two structures are non-superimposable. Thus, *trans*-1,2-dibromocyclopentane is chiral.

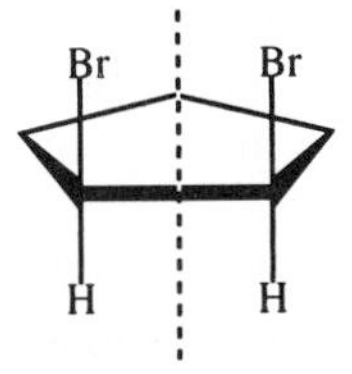

## 3.3  THE CHAIR CONFORMATIONS OF CYCLOHEXANE

Up to this point our drawings of cyclohexane may have led you to believe that it is a flat molecule; this is not really correct. If cyclohexane were flat, the bond angle between the carbon atoms would have to be the hexagonal angle, 120˚, instead of the tetrahedral angle, 109.5˚. By adopting a puckered conformation such as the one known as the **chair conformation**, the correct tetrahedral bond angles can be achieved. At the same time the additional instability caused by the eclipsing hydrogen atoms that would occur in the flat molecule is reduced. In fact, unlike many of the other small ring systems, cyclohexane is able to achieve a conformation in which it has no **ring strain**. The two chair forms of cyclohexane are shown in figure 3.12. If you use your imagination, you may be able to see that this conformation of cyclohexane actually looks somewhat like a lounge chair; complete with a headrest and a footrest.

The two forms are in equilibrium with each other and this interconversion between forms is known as a **ring-flip**. In the case of unsubstituted cyclohexanes the equilibrium between the two forms is rapid and each form is present

FIGURE 3.8: *cis*-and *trans*-2-butene

FIGURE 3.9: *cis*- and *trans*-1,2-dibromocyclopentane

FIGURE 3.10: The Internal Plane of Symmetry in *cis*-1,2-dibromocyclopentane

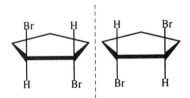

FIGURE 3.11: The Mirror Images of *trans*-1,2-dibromocyclopentane

equally. However, as we will see,
introducing a substituent onto the ring
can cause one form to predominate
over the other form.

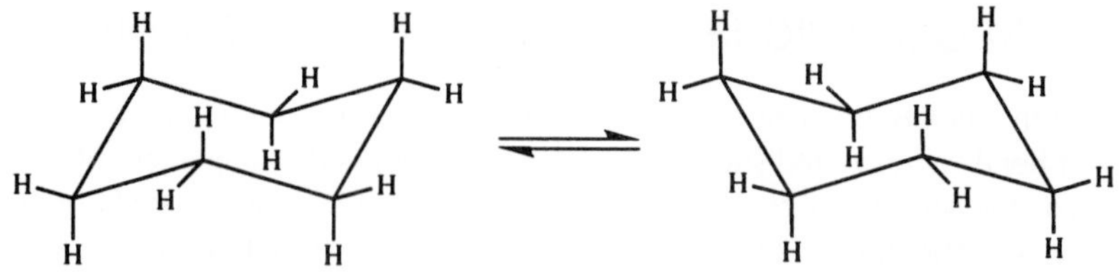

FIGURE 3.12: The
Chair Forms of
Cyclohexane

You may have also noticed that there
are two different positions for the
hydrogen atoms in the chair form of
cyclohexane. In each form in Figure 3.12, three of the hydrogen atoms are
pointing directly up and three are pointing directly down. The positions occupied by these hydrogen atoms are called **axial positions** and the hydrogen
atoms are termed axial hydrogens. The other six hydrogen atoms point away
from the ring and are in **equatorial positions**. These two
positions are emphasized in figure 3.13.

You should note that the direction each hydrogen is pointing changes as you proceed around the ring. If we look at
the axial hydrogen atoms first, we see that the hydrogen
atoms attached to carbon 1, 3, and 5 are all pointing
straight up while those attached to carbons 2, 4, and 6 are
all pointing straight down. On the other hand, the equatorial hydrogen atoms on carbons 1, 3, and 5 are all angled out and down away
from the ring and those attached to carbons 2, 4, and 6 are all angled away
from the ring and upward.

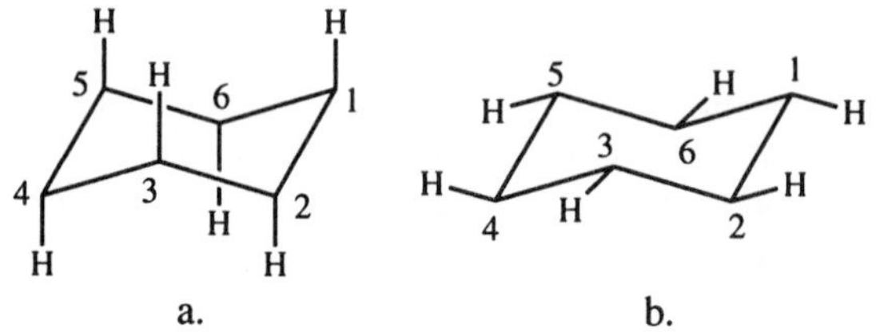

FIGURE 3.13: The Chair Form
of Cyclohexane Showing a) the
Axial Hydrogens and (b) the
Equatorial Hydrogens

We mentioned that the chair form of cyclohexane contains no strain introduced
by eclipsing hydrogen atoms. This is best seen by looking at the Newman projection of the chair form of cyclohexane in which it is clear that the bonds are
all staggered with respect to one another.

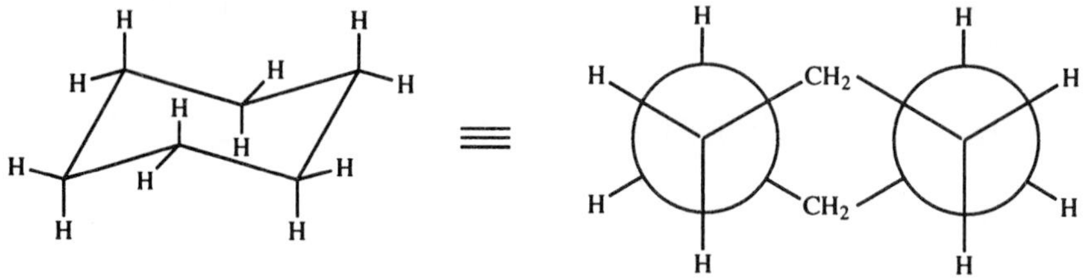

FIGURE 3.14: The Newman
Projection of the Chair Form of
Cyclohexane

## 3.4 CONFORMATIONAL ANALYSIS OF MONOSUBSTITUTED CYCLOHEXANES—

We begin our study of substituted cyclohexanes by looking at methylcyclohexane. The two forms of this compound are shown at right.

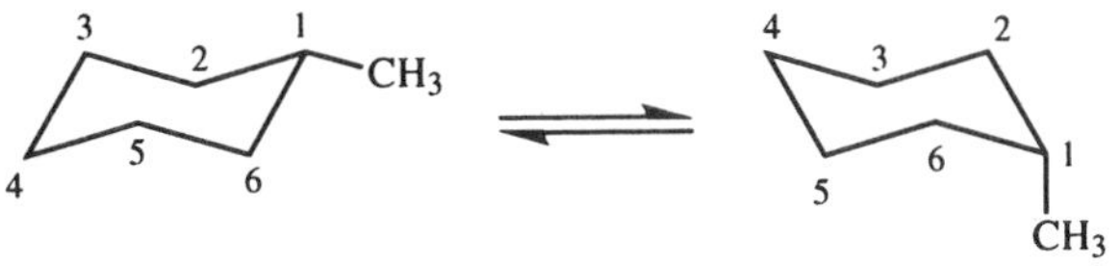

FIGURE 3.15: The Two Chair Forms of Methylcyclohexane

In the first case, the methyl group is in the equatorial position, but in the ring-flipped form the methyl group is in the axial position. As you can probably guess, one form is more stable than the other. This is probably easier to see if we look at the Newman projections of both forms, shown in figure 3.16.

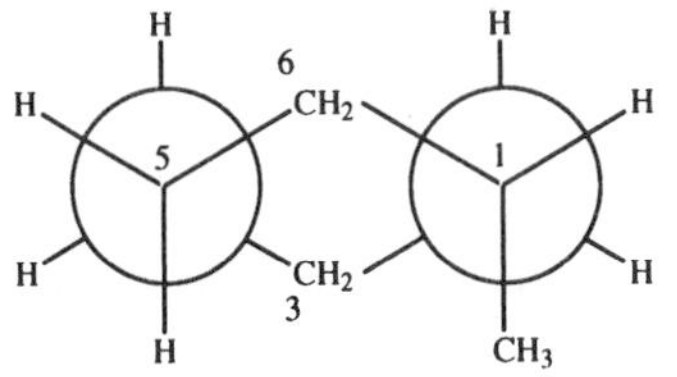

(b)

The methyl group in (a) is anti to the C-5 methylene group in the conformer in which it is in the equatorial position. On the other hand, in the conformer in which the methyl group occupies the axial position, there are two gauche interactions: one with the C-3 methylene group (b) and one with the C-5 methylene group (c) (you should probably convince yourself of this with molecular models). Recall that in the case of butane each gauche interaction added another 3.7 kJ/mol of energy. Thus we would expect the axial conformation of methylcyclohexane to be approximately 7.4 kJ/mol less stable than the equatorial conformation. If we wish to calculate the equilibrium composition of methylcyclohexane at 25° C we can use equation 3.1 to solve for K. Since the equatorial form is the more stable, the free energy change going from the axial form as a reactant to the equatorial form as product is -7.4 kJ/mol. Solving for the value of the equilibrium constant for this reaction, we obtain

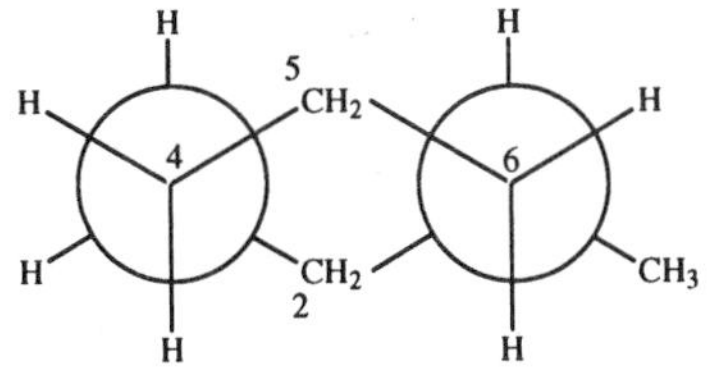

(a)

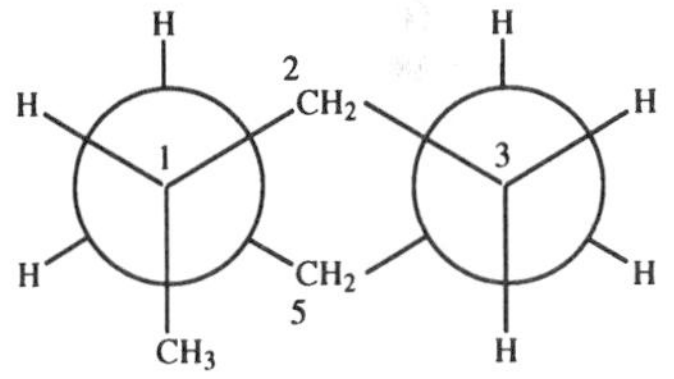

(c)

FIGURE 3.16: Newman Projections of Methylcyclohexane

$$K = e^{\frac{-\left(-7400\ ^{J}/_{mol}\right)}{\left(8.314\ ^{J}/_{mol\cdot K}\right)\left(298.15\ K\right)}} = 20 \qquad \text{EQ. 3.6}$$

So, in the case of methylcyclohexane, the equilibrium composition is approximately 95% equatorial and 5% axial. This is a much greater difference than we saw earlier for the two forms of *n*–butane. The energy differences for a variety of substituents are supplied in Table 3.1.

TABLE 3.1

**TABLE 3.1  ENERGY DIFFERENCES FOR ALTERNATE FORMS OF MONOSUBSTITUTED CYCLOHEXANES**

| Substituent | Energy Difference (kJ/mol) |
|---|---|
| –F | 0.8 |
| –Cl | 2.1 |
| –Br | 2.5 |
| –OH | 4.2 |
| –COH | 5.8 |
| –$CH_3$ | 7.1 |
| –$CH_2CH_3$ | 7.5 |
| –$CH(CH_3)_2$ | 8.8 |
| –$C_6H_5$ | 13 |
| –$C(CH_3)_3$ | 22 |

The gauche interactions in the axial conformation of methylcyclohexane also lead to steric interactions with the hydrogens on carbons 3 and 5. These steric interactions are referred to as **1,3-diaxial interactions**, because they refer to the interaction of the axial substituents on a particular carbon atom (C1) with axial substituents that are on the third carbon atom (C3). As you can imagine, the more sterically demanding a substituent is, the more likely it is to occupy an equatorial position. Thus, for example, *t*-butylcyclohexane exists almost exclusively in the equatorial conformation.

## 3.5  CONFORMATIONAL ANALYSIS OF DISUBSTITUTED CYCLOHEXANES

Let us now look at an example of a disubstituted cyclohexane. Consider 1,2-dimethylcyclohexane. As you know there are two geometric isomers of 1,2-dimethylcyclohexane, the *cis* isomer and the *trans* isomer. Shown at right are the two chair conformers of *cis*-1,2-dimethylcyclohexane.

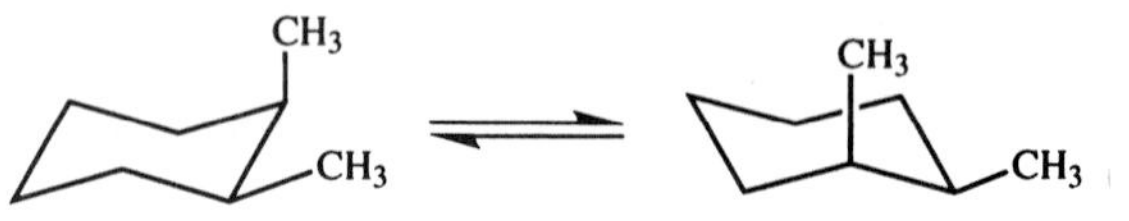

FIGURE 3.17: Chair Conformations of *cis*-1,2-dimethylcyclohexane

In each conformation, one of the methyl groups occupies the axial position and the other occupies the equatorial position. Thus, there is no difference in energy between these two conformations and we would expect to find an equal amount of both forms in an equilibrium mixture of *cis*-1,2-dimethylcyclohexane. On the other hand, if we look at the two conformations of *trans*-1,2-dimethylcyclohexane we note that in one conformation both methyl groups occupy axial positions, and in the other conformation they both occupy equatorial positions. Obviously the more stable conformation is the one in which both methyl groups are in the equatorial position and an equilibrium mixture of *trans*-1,2-dimethylcyclohexane would be expected to contain more of the diequatorial conformation. You would probably benefit by going through a similar analysis of the 1,3- and 1,4-dimethylcyclohexane conformations on your own.

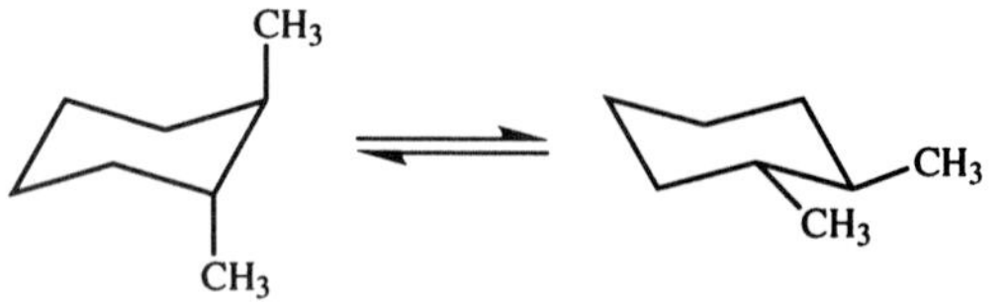

FIGURE 3.18: Chair Conformations of *trans*-1,2-dimethylcyclohexane

It is sometimes difficult for students to convert between planar representations of cyclohexane molecules and the chair conformations of these same molecules. Let us look at one example to help you understand this conversion. Consider the following representation of *cis*-1-bromo-2-chlorocyclohexane.

Both carbon–halogen bonds are directed upward in the planar representation, therefore, both bonds must also be directed upward in the chair conformation. Note that this places one substituent in an equatorial position and the other in an axial position.

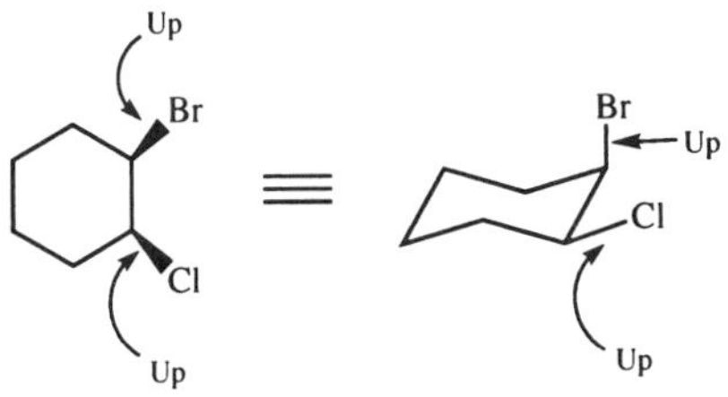

FIGURE 3.19: Alternate Representations of *cis*-1-bromo-2-chlorocyclohexane

**CONCEPT CHECK**

Draw one confomer of *cis*-1-*t*-butyl-4-methylcyclohexane. Now draw the ring flip of this conformation and use arrows to indicate the direction of equilibrium between the two conformers.

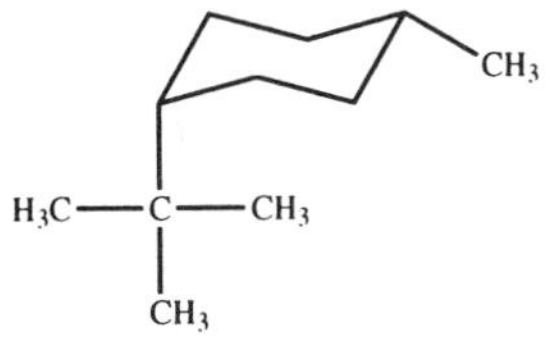

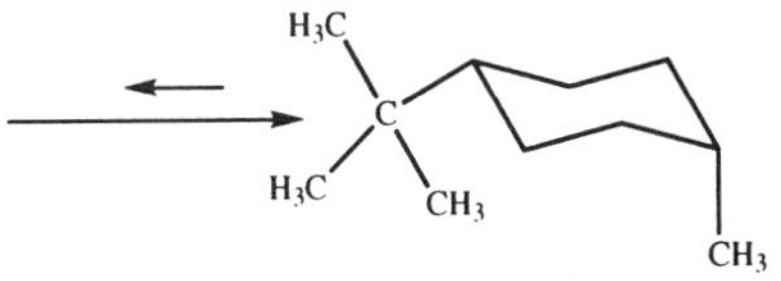

If we start with a position in which the axial position is straight down, call that carbon atom number 1 and place the *t*-butyl group at that point, the methyl group on carbon atom 4 must be in the equatorial position for the two groups to be cis to one another. After ring flip, the *t*-butyl group will be in an equatorial position and the methyl group in an axial position. These conformations are shown in figure 3.20. In both cases one of the groups will be in an axial position and the other will be in an equatorial position. Because there is more room in equatorial positions, and because the *t*-butyl group is a much larger group than is the methyl group, the preferred conformation will be the one in which the *t*-butyl group is in the equatorial position.

FIGURE 3.20: Conformations of *cis*-1-*t*-butyl-4-methylcyclohexane

## 3.6  CONFORMATIONAL ANALYSIS AND STEREOCHEMISTRY: APPLICATIONS TO REACTIONS

Now that we have a reasonably good understanding of conformational analysis and stereochemistry let us use this information to examine the stereochemical outcome of some reactions. In particular, we would like to explain the results of these reactions using our knowledge of stereochemistry.

**CATALYTIC HYDROGENATION OF ALKENES**

Recall that hydrogen adds readily to alkenes in the presence of a catalyst, typically palladium on carbon, to produce the saturated alkane. The process is known as catalytic hydrogenation. For example, ethane is formed by the addition of hydrogen to ethylene.

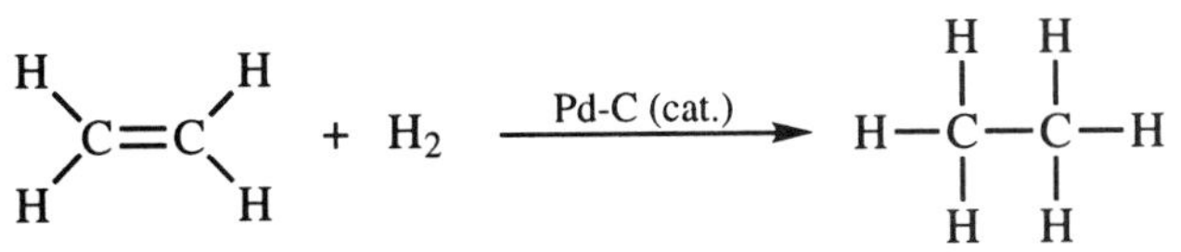

EQ. 3.7

As you might expect, deuterium is also able to add across the carbon-carbon double bond of alkenes. Consider the addition of $D_2$ to cyclohexene (equation 3.8). When cyclohexene is treated with deuterium gas in the presence of an appropriate catalyst, we note that the product is the cis isomer. That is, the deuterium atoms have ended up on the same side of the cyclohexane ring.

$$\text{cyclohexene} + D_2 \xrightarrow{\text{Pd-C (cat.)}} \text{cis-1,2-dideuteriocyclohexane} \qquad \text{EQ. 3.8}$$

We can explain the stereochemical outcome of this reaction by considering the role that the catalyst plays. While it is not appropriate to enter into a detailed discussion of catalysis at this point, it is sufficient to mention that one function of a catalyst is to bring reacting species into proximity with one another so that they may react more easily. In the case of catalytic hydrogenation, the catalyst also functions to direct the reaction in a specific orientation. In this particular reaction, both hydrogen atoms add to the same face of the alkene; a process known as **syn addition** (figure 3.21).

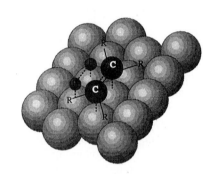 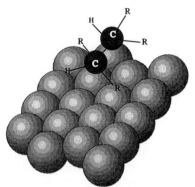

FIGURE 3.21: Catalyst directs the syn- Addition of Hydrogen to an Alkene

### ELECTROPHILIC ADDITION OF HALOGENS TO ALKENES

As we saw in book 1, chapter 12, alkenes act as Lewis bases. There we considered the addition of hydrogen halides to alkenes. As another example, consider another reaction that we have encountered before — the addition of halogens to alkenes. Recall that bromine adds to ethene to produce 1,2-dibromoethane.

$$\begin{array}{c} H \\ H \end{array} C=C \begin{array}{c} H \\ H \end{array} + Br_2 \longrightarrow Br-\underset{\underset{H}{|}}{\overset{\overset{H}{|}}{C}}-\underset{\underset{H}{|}}{\overset{\overset{H}{|}}{C}}-Br \qquad \text{EQ. 3.9}$$

Let us now examine another example of this reaction — the addition of bromine to 2-butene.

2 chiral centers

$$CH_3-\overset{\overset{H}{|}}{C}=\overset{\overset{H}{|}}{C}-CH_3 + Br_2 \longrightarrow CH_3-\underset{\underset{Br}{|}}{\overset{\overset{H}{|}}{C}}-\underset{\underset{Br}{|}}{\overset{\overset{H}{|}}{C}}-CH_3 \qquad \text{EQ. 3.10}$$

Note that this reaction results in the formation of two stereocenters in the product. Furthermore, we must remember that there are two geometric isomers of 2-butene; *cis*-2-butene and *trans*-2-butene. It has been observed that when bromine adds to *cis*-2-butene the product is composed of the enantiomeric pair of 2,3-dibromobutanes.

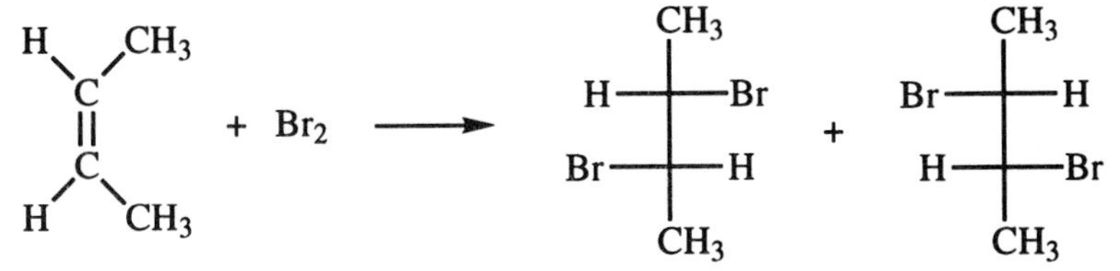

EQ. 3.11

On the other hand, when bromine adds to *trans*-2-butene the product consists of the *meso*–2,3-dibromobutane.

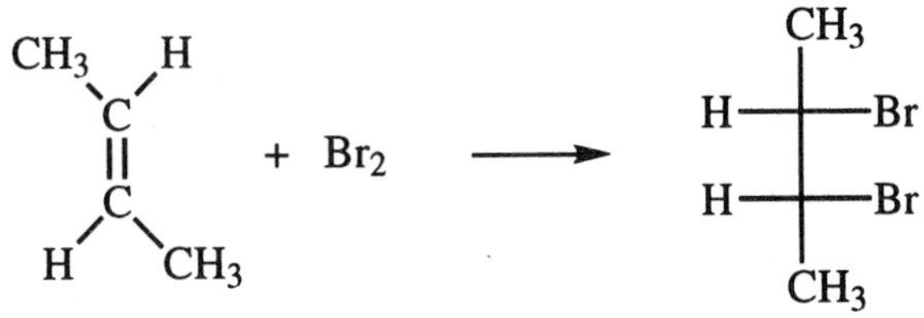

EQ. 3.12

The question is now, can we explain these results by looking at possible reaction pathways? Since we were able to rationalize the stereochemistry of hydrogen addition to alkenes as syn addition, it makes sense to consider a similar pathway here. Figure 3.22 illustrates the expected outcome of the syn addition of bromine to *cis*-2-butene.

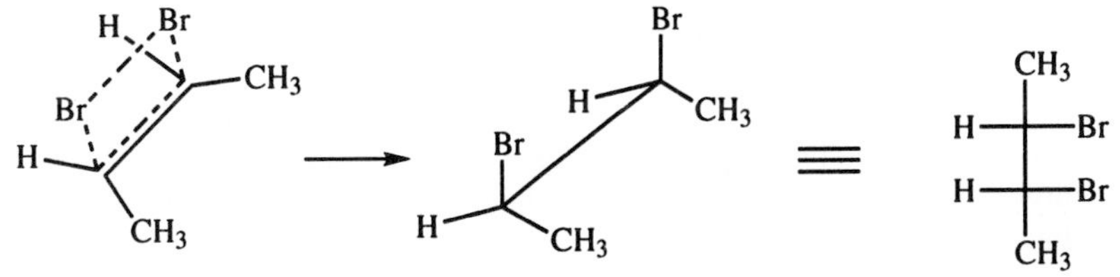

FIGURE 3.22: Possible syn- Addition of Bromine to *cis*-2-butene

Since this process would produce the *meso*-2,3-dibromobutane from the *cis*-2-butene, we know that this is not the correct pathway for this reaction.

Another possibility is that a bromide ion adds to the alkene to form a carbocation that then reacts with another bromide ion to form the product.

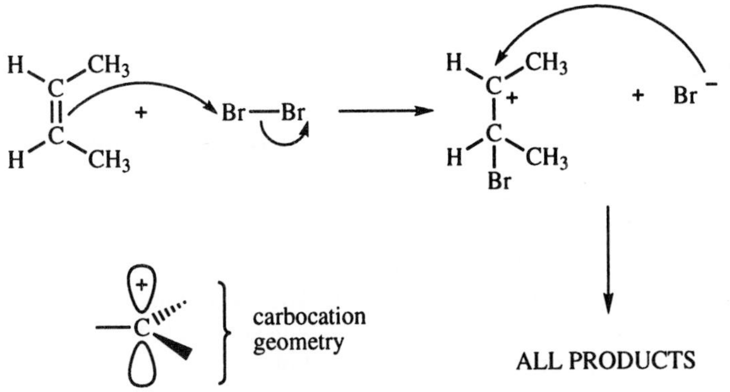

Again, this reaction pathway does not explain the formation of the observed products. Recall that the local geometry around a double bond is planar. In the first step of the reaction the bromine could add from either the top or the bottom. The carbocation formed in the first step of the reaction also has $sp^2$ hybridized carbon atoms. As a

carbocation geometry

ALL PRODUCTS

FIGURE 3.23: Possible Carbocation Addition of Bromine to *cis*-2-butene

result it is also planar. The bromide ion could then add either to the top or bottom face of the carbocation. This pathway would lead to a mixture of all products whether either the cis or the trans alkene was used. This is not what is observed and is, therefore, not a possible pathway either.

A final possibility is depicted in figure 3.24. In this case the bromine binds to both carbon atoms of the alkene distributing the positive charge throughout the resulting three-membered ring. This positively charged compound is known as a **bromonium ion**. Subsequent addition of a bromide ion to the back carbon leads to the product as shown. Addition to the front carbon would lead to the enantiomer. Repeating this exercise with *trans*-2-butene (which you should do) shows that this pathway predicts formation of the meso compound. Thus, it is consistent with the observed stereochemistry of the reaction. It also highlights the fact that the addition of halogens to alkenes proceeds through **anti-addition**.

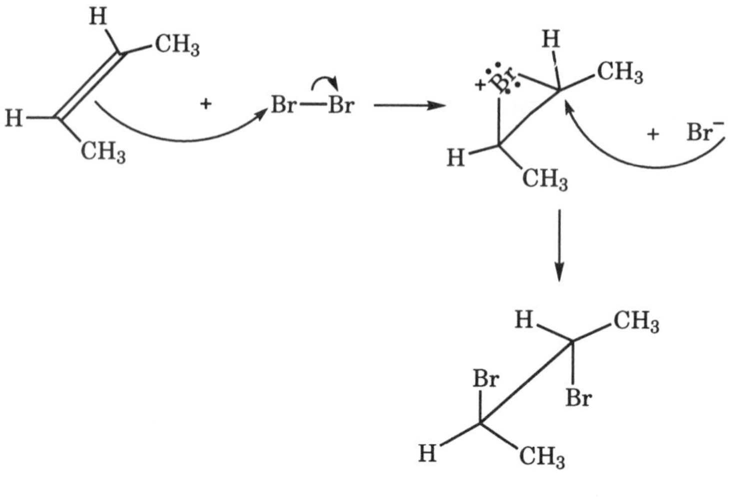

FIGURE 3.24: The Bromonium Intermediate Pathway for the Addition of Bromine to *cis*-2-butene

## OXYMERCURATION/DEMERCURATION

You have already learned that water can add across a double bond in the presence of an acid catalyst to produce an alcohol. For example, water adds to 1-methylcyclohexene in the presence of a few drops of sulfuric acid to form 1-methylcyclohexanol (equation 3.13). Note that the -OH ends up on the carbon that was more substituted in the alkene, so the product is the so-called Markovnikov addition product.

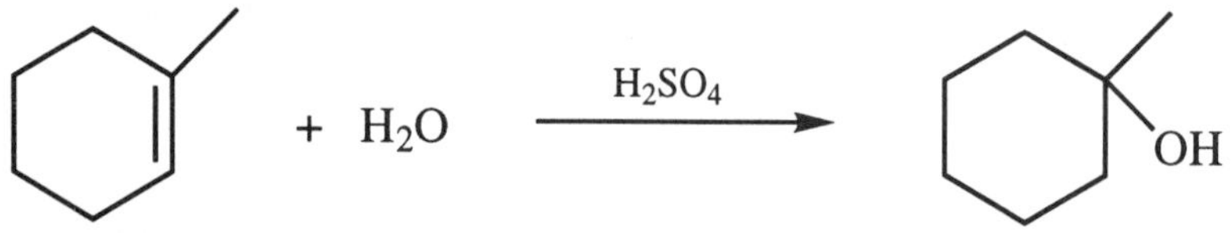

EQ. 3.13

Recall that the acid catalyzed hydration of alkenes proceeds through a carbocation intermediate that forms when the double bond is protonated by the acid. As we will see in Chapter 8, carbocation intermediates are prone to rearrangements that can lead to unwanted products. Another process that hydrates alkenes, and does so without the potential for rearranged products, is the process of oxymercuration-demercuration. The process involves treating an alkene first with mercuric acetate and then with sodium borohydride. An example of the type of reaction is shown below.

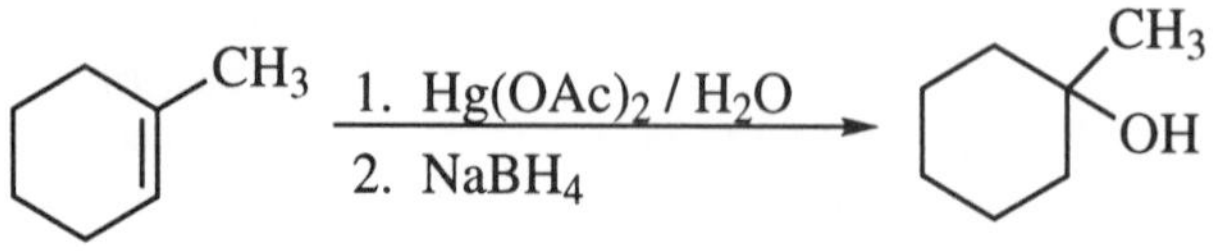

EQ. 3.14

Note that the Markovnikov addition product (1–methylcyclohexanol, not 2–methylcyclohexanol) is the observed product of this reaction. This is not too surprising if we examine the course of the reaction and the intermediates that form during the reaction.

Based on the name of this reaction, oxymercuration-demercuration, it is probably clear to you that the process involves two steps. The first step involves mercuration of the double bond by an electrophile generated from the mercuric acetate, $Hg(OAc)_2$. It is postulated that the electrophile forms by the dissociation of mercuric acetate into an acetate ion and a positively–charged mercury–containing species as shown below.

$$Hg(OAc)_2 \longrightarrow OAc^{\ominus} + {}^{\oplus}HgOAc \qquad \text{EQ. 3.15}$$

As we have seen, the double bond in an alkene can behave as a Lewis base by reacting with the electrophilic mercury containing species. This forms a mercurium ion. The mercurium ion is quite similar to the bromonium ion encountered in the previous section in that it is composed of a three membered ring bearing a positive charge.

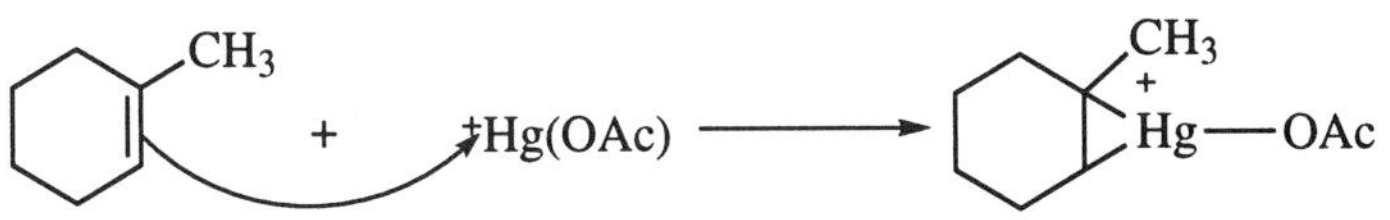

EQ. 3.16

Since the starting alkene is not symmetrical, we might also expect that the mercurium ion would not be symmetrical either. It is this prediction that allows us to understand why the Markovnikov product is the observed product from this series of transformation. Given the unsymmetrical nature of the mercurium ion, we might expect that the mercurium ion might look more like the structure in figure 3.25, where a partial positive charge exists on the tertiary carbon and a partial negative charge exists on the mercury.

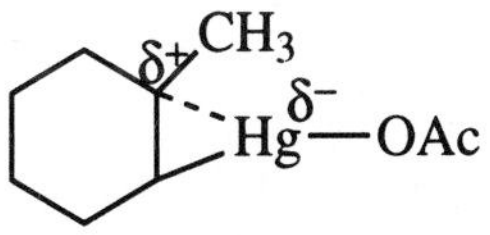

FIGURE 3.25:
Unsymmetrical
Mercuration
Intermediate

This is the most stable intermediate. The other possible intermediate would contain a partial positive charge on a secondary carbon. Once the mercurium ion is formed, it is subject to nucleophilic opening by water (which is present as one of the reaction solvents). Just as in the opening of bromonium ions, the water attacks from the side opposite the mercury in an anti–addition process. The ring opening reaction is followed by rapid deprotonation and results in the formation of an organomercurial alcohol. This process is shown in equation 3.17.

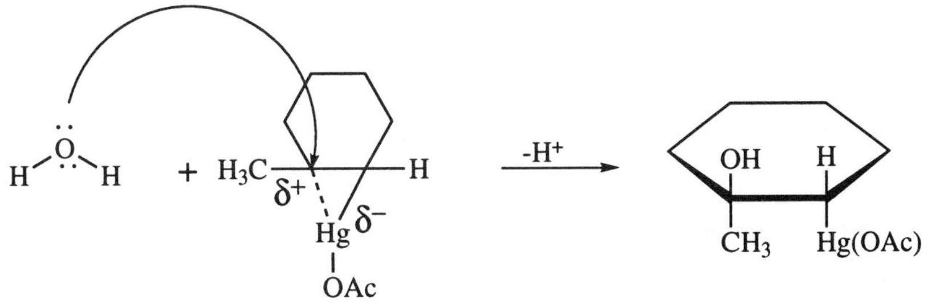

EQ. 3.17

Finally, the organomercurial alcohol is reduced with sodium borohydride to give the alcohol.

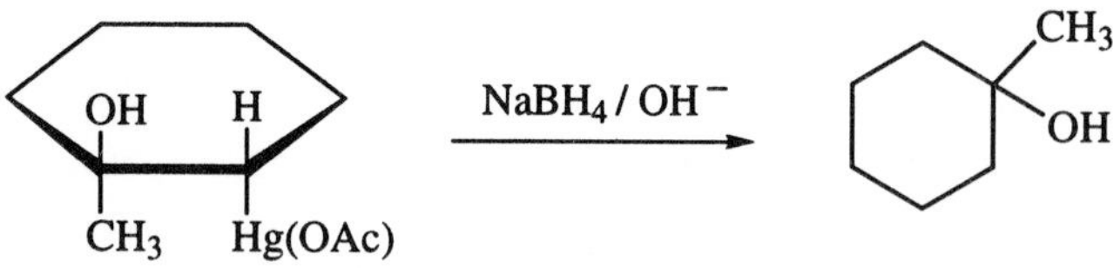

EQ. 3.18

## HYDROBORATION

We now turn our attention to a hydration process that produces the **anti–Markovnikov product**, that is, 2–methylcyclohexanol. We will first examine the reaction by looking at a case where the regiochemistry is unimportant and then consider the case of 1–methylcyclohexene.

In 1959 a chemist from Purdue University, H.C. Brown, and his group reported that the compound borane ($BH_3$) added to alkenes to produce compounds known as organoboranes. Recall that borane is a Lewis acid; it is reasonable that the alkene, acting as a Lewis base, would react readily with the borane.

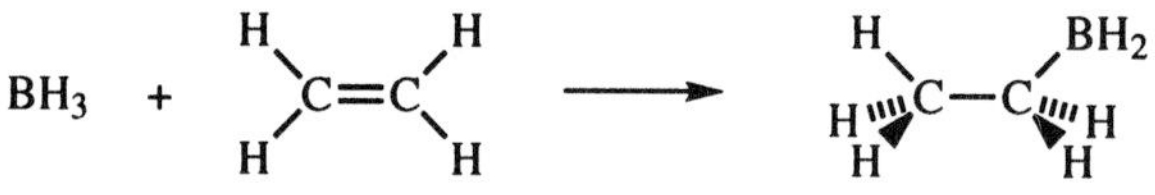

Organoborane

EQ. 3.19

Notice that the boron atom in the organoborane still has two remaining hydrogen atoms, thus the stoichiometry of the reaction involves the addition of three molecules of alkene for every one borane molecule to form a trialkylborane.

Triethylborane

EQ. 3.20

Subsequent oxidation of the trialkylborane with hydrogen peroxide in basic solution results in cleavage of the carbon-boron bond and formation of an alcohol. Therefore, the result of this two-step (hydroboration/oxidation) process is that the double bond is hydrated.

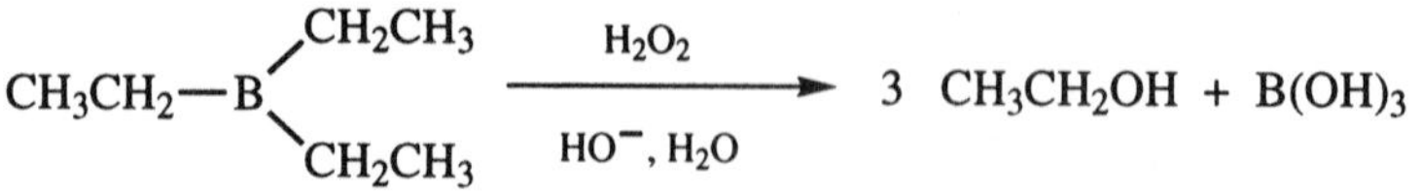

$$CH_3CH_2-B\begin{matrix}CH_2CH_3\\ \\CH_2CH_3\end{matrix} \xrightarrow[HO^-,\ H_2O]{H_2O_2} 3\ CH_3CH_2OH + B(OH)_3$$

EQ. 3.21

Let us look at another example. Consider the addition of borane to 1-methylcyclohexene followed by oxidation. The observed product of this reaction sequence is *trans*-2-methylcyclohexanol.

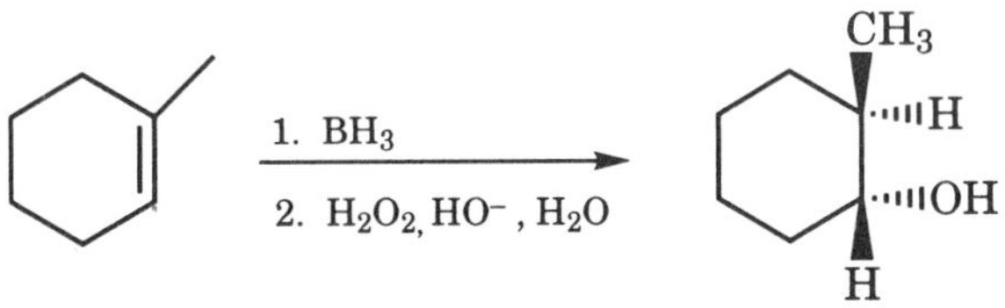

EQ. 3.22

We know that the first step involves the addition of borane to the double bond. It is possible that the boron and the hydrogen both add to the alkene at the same time, that is, in a concerted process. If this is the case, we would expect the boron and the hydrogen to end up on the same side of the double bond. We must also consider that the boron could add to either of the olefinic carbons. Thus, we can draw two possible pathways for the formation of the organoborane.

In path A we note that a positive charge develops on a secondary carbon while in path B a positive charge develops on a tertiary carbon. Since the tertiary carbon is better able to accommodate the positive charge, path B is favored. This explains the observed regiochemistry of this reaction. Oxidation of the organoborane results in replacement of the boron with oxygen on the same side, thus the observed stereochemistry is consistent with the proposed pathway.

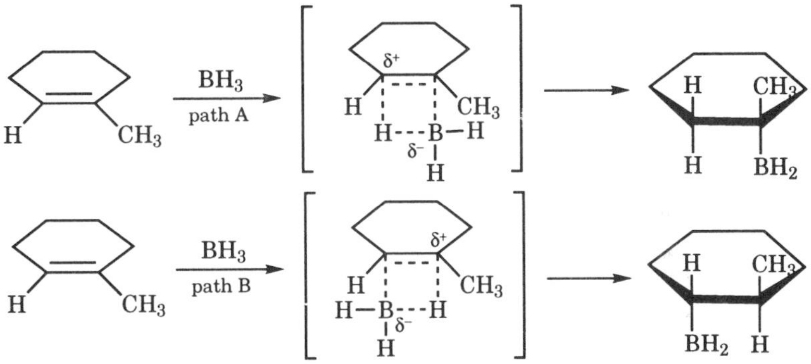

FIGURE 3.26: The Formation of the Alkylborane from 1-Methylcyclohexene

**CONCEPT CHECK**

What organic product would you expect to isolate from the following reaction?

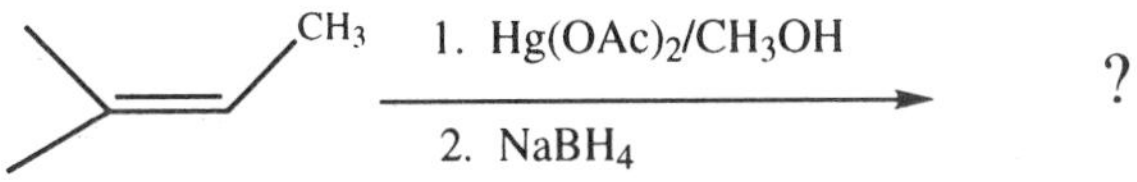

EQ. 3.23

Recall that this series of reactions is one that hydrates the double bond of an alkene and gives the Markovnikov product. Thus we would expect the product to be 2–methyl–2–methoxybutane.

Explain the formation of the expected product with a step–by–step mechanism.

This product can be rationalized by looking at the reaction mechanism shown in figure 3.27.

In this reaction, the first step is an electrophilic attack on the double bond by a cation comprised of a mercury (II) ion with one acetate ion bound to it. This forms a cyclic intermediate in which there is a positive charge on one of the carbon atoms that were formerly involved in

the double bond.
Since the positive
charge is better carried
by the more substitut-
ed carbon atom, this is
the site at which
nucleophilic attack by
the oxygen atom of a
methanol molecule
then occurs. After
formation of a bond
between that carbon
atom and the oxygen
atom, a proton is lost
from the oxygen atom,
leaving an organomer-
curial ether. Finally,
the mercury can be
expelled by treatment
with sodium borohydride.

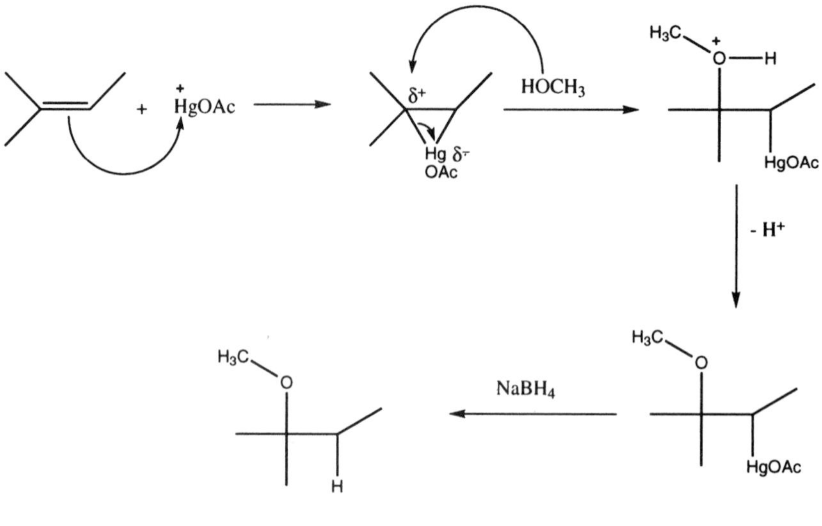

FIGURE 3.27:
Reaction Mechanism
for Oxymercuration-
Demercuration in
Alcohol

## HYDROXYLATION OF ALKENES

Finally, let us consider two reactions in which two hydroxyl groups are added
to the double bond of alkenes. Alkenes react with $OsO_4$ or $KMnO_4$ to form
cyclic osmate or manganate esters, respectively. These esters are then cleaved
under appropriate conditions (hydrogen peroxide in the case of osmate esters
and base/water in the case of permanganate esters) to produce 1,2-diols. Since
the esterification reaction is also a concerted process the hydroxyl groups end
up syn to each other. Examples of these reactions are provided below.

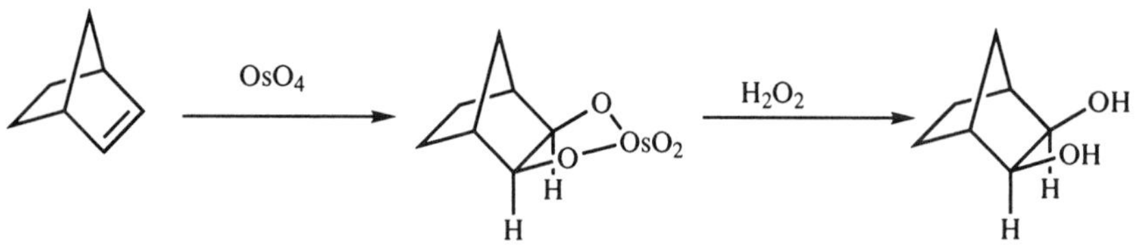

EQ. 3.24

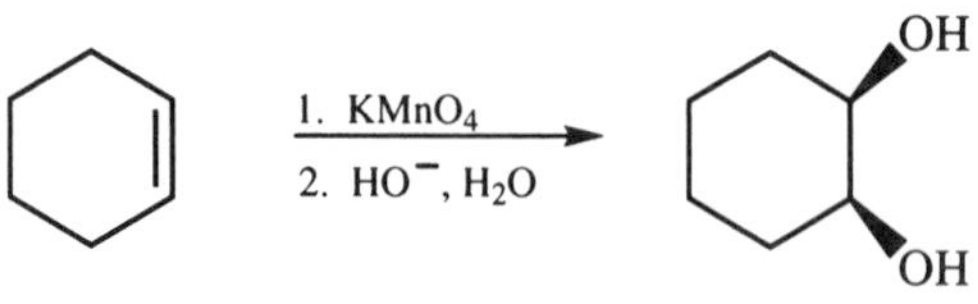

EQ. 3.25

**Anti Addition:** addition from opposite sides of a planar functional group such as an alkene.

**Anti Conformation:** the conformation in which the substituents on two adjacent atoms are oriented 180° with respect to each other.

**Anti-Markovnikov Product:** the alkene addition product in which a hydrogen atom ends up on the more substituted carbon of the alkene.

**Axial Position:** the positions on the chair conformation of cyclohexane that are directly up or down relative to the ring.

**Bromonium ion:** a three membered, positively charged ring that contains bromine and two carbon atoms.

**Chair Conformation:** the most stable conformation of cyclohexane because it contains no strain due to eclipsing hydrogen atoms.

**Chiral:** a chiral object is an object that is not superimposable on its mirror image.

**Conformers:** representations of molecules that differ by rotation about a carbon-carbon single bond.

**1,3-Diaxial Interactions:** steric interactions between a substituent on C1 and axial substituents on the C3 carbon atom.

**Dihedral Angle:** the angle between two substituents on adjacent atoms. The dihedral angle between the methyl groups in the anti conformation of butane is 180°.

**Eclipsed Conformation:** the conformation in which the substituents on adjacent atoms are directly aligned with each other.

**Equatorial Position:** a position in the chair conformation of cyclohexane that always points away from the ring.

**Gauche Conformation:** the conformation in which substituents on two adjacent atoms have a 60° dihedral angle.

**Geometric Isomerism:** isomerism that results from restricted bond rotation.

**Ring-Flip:** the interconversion of the two chair forms of cyclohexane.

**Ring Strain:** the strain that is present in many small rings because the proper bond angles about carbon cannot be achieved.

**Staggered Conformation:** the conformation in which all substituents on two adjacent atoms are as far apart as possible.

**Syn Addition:** addition to the same side of a planar functional group such as an alkene.

HOMEWORK
**Problems**

1. Draw the potential energy *vs.* dihedral angle diagrams along the C2-C3 bonds for:

    a. 2-methylbutane
    b. 2,3-dimethylbutane
    c. 2,2-dimethylbutane

2. Compare the stability of each of the conformations of the compounds in problem 1 with those of butane (fig. 3.7).

3. Cyanocyclohexane at room temperature exists in an axial:equatorial ratio of 1:35. What is the equilibrium constant and the free energy difference between the two conformations?

4. N,N-dimethylaminocyclohexane at room temperature exists in an axial:equatorial ratio of 1:35. What is the equilibrium constant and the free energy difference between the two conformations?

5. Assign R/S to all the chiral centers in the following compounds:

a.

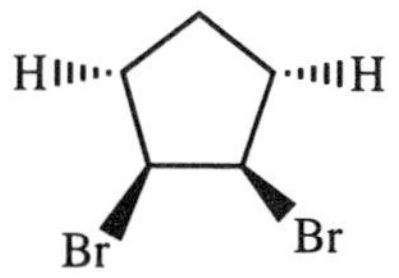

b.

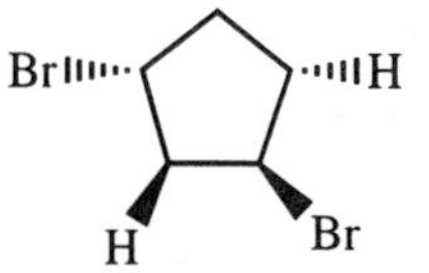

c.

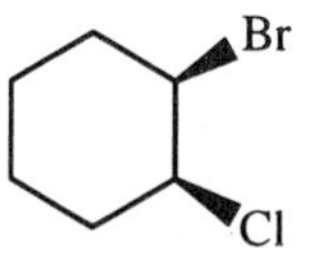

d.

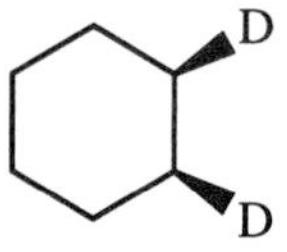

e.

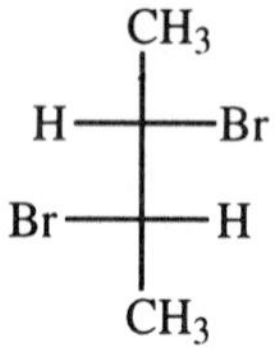

f.

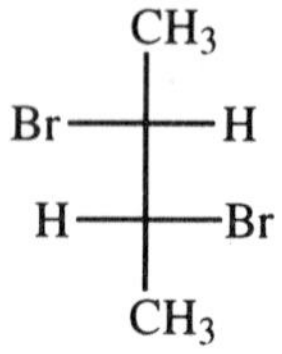

g.

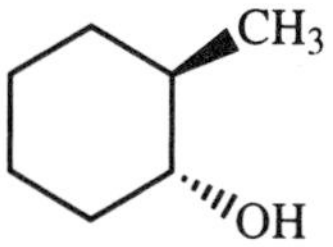

6. Draw the Newman projections of the diaxial and diequatorial 1,2-dimethyl cyclohexanes (hint: see figure 3.16). Be sure to sight down the C1-C2 bond.

7. Draw one conformation of each of the following compounds. Then, draw the ring flip of this conformation. For each pair, indicate which conformation, if either, is more stable.

    a. *cis*-1,2-diethylcyclohexane      d. *trans*-1,2-diethylcyclohexane
    b. *cis*-1,3-diethylcyclohexane      e. *trans*-1,3-diethylcyclohexane
    c. *cis*-1,4-diethylcyclohexane      f. *trans*-1,4-diethylcyclohexane

8. Draw the answers to the previous problem in their planar representations (hint: see figure 3.19).

9. Predict the products of the following reactions of (Z)-3,4-dimethyl-3-hexene. When appropiate, indicate stereochemistry.

    a. with $H_2$ and catalyst
    b. with $D_2$ and catalyst
    c. with bromine
    d. with borane followed by hydrogen peroxide, base, and water

10. Predict the products of the following reactions of (E)-3,4-dimethyl-3-hexene. When appropiate, indicate stereochemistry.

    a. with $H_2$ and catalyst
    b. with $D_2$ and catalyst
    c. with chlorine
    d. with borane followed by hydrogen peroxide, base and water

11. Write the mechanism for the addition of chlorine to (Z)-2-pentene.

12. Write the mechanism for the addition of chlorine to (E)-2-pentene.

13. Write a mechanism for the formation of *n*-butylborane from 1-butene.

14. Show reactions for the preparation of the following from any alkene you choose.

    a. (R,R)-1,2-dichlorocyclopentane  d. *meso*-1,2-cyclopentanediol
    b. 2-ethyl-1-cyclohexanol         e. 1-hexanol
    c. 1-ethyl-1-cyclohexanol        f. 2-hexanol

15. Predict the products of the following reactions.

    a. 3-methyl-1-butene + water with sulfuric acid catalyst to give A
    b. 2-methyl-2-butene + borane to give B
    c. B + hydrogen peroxide, sodium hydroxide and water to give C
    d. 3-methyl-1-butene + bromine to give D
    e. 2-methyl-2-butene + osmium tetroxide and hydrogen peroxide to give E

16. A compound with the formula $C_5H_{10}$ was reacted with borane followed by hydrogen peroxide, sodium hydroxide and water to give a product with the following HNMR: 0.9 ppm (d, 6H), 1.5 ppm (q, 2H), 1.7 ppm (multiplet of 9 peaks, 1H), 2.8 ppm (s, 1H), 3.7 ppm (t, 2H).

     a. Give the structure of this product. Predict the CNMR spectrum of this product.
     b. A small amount of another product was isolated from the above reaction. Its CNMR consisted of peaks at 9, 29, 36, and 71 ppm. Give the structure of this product.

17. When 2-butanol ($\Delta G_f = -167.3$ kJ/mol) reacts with heat and an acid catalyst, two possible organic dehydration products can occur in addition to liquid water: 1-butene ($\Delta G_f = 71.29$ kJ/mol) and 2-butene ($\Delta G_f = 62.96$ kJ/mol).

     a. Which of the two possible products is theoretically favored based upon Saytzeff's rule?
     b. Calculate the standard state free energy change and the equilibrium constant for the reaction forming 1-butene.
     c. Calculate the standard state free energy change and the equilibrium constant for the reaction forming 2-butene.
     d. Explain your results for (b) and (c) in terms of the predictions in (a).

18. When hydrogen bromide gas adds to propene ($\Delta G_f = 62.72$ kJ/mol) either 1-bromopropane ($\Delta G_f = -22.47$ kJ/mol) or 2-bromopropane ($\Delta G_f = -27.23$ kJ/mol) is the major product.

     a. Which of the two possible products is theoretically favored based upon Markovnikov's rule?
     b. Calculate the standard state free energy change and the equilibrium constant for the reaction forming 1-bromopropane.
     c. Calculate the standard state free energy change and the equilibrium constant for the reaction forming 2-bromopropane.
     d. Explain your results for (b) and (c) in terms of the predictions in (a).

19. In Book One, chapter 5, we noted that alkyl bromides could react with water to yield the alcohol, and that alcohols could react with hydrogen halides to yield the alkyl halide. It was noted that tertiary alkyl bromides had better yields of the corresponding alcohol than primary alkyl bromides. Confirm this by considering the hydrolysis (with liquid water) of 1-bromobutane ($\Delta G_f = -12.89$ kJ/mol) or 2-bromo-2-methylpropane ($\Delta G_f = -28.16$ kJ/mol) to form aqueous HBr ($\Delta G_f = -103.96$ kJ/mol) and 1-butanol ($\Delta G_f = -150.66$ kJ/mol) or tertiary butyl alcohol ($\Delta G_f = -191.04$ kJ/mol).

     a. Write the balanced reaction when 1-bromobutane is hydrolyzed to form 1-butanol and aqueous hydrogen bromide.
     b. Calculate the standard state free energy change and the equilibrium constant for the reaction in (a).
     c. Write the balanced reaction when 2-bromo-2-methylpropane is hydrolyzed to form tertiary butyl alcohol and aqueous hydrogen bromide.
     d. Calculate the standard state free energy change and the equilibrium constant for the reaction in (c).

# SOLUTION
# Equlilibria

# SOLUTION
## Equilibria

## 4.1   THE EQUILIBRIUM CONSTANT

One of the inescapable consequences of the laws of thermodynamics is that no chemical reaction ever goes 100% of the way to completion and all chemical reactions go at least a small way toward completion.  Hence, if we consider the hypothetical reaction:

$$A + B \rightarrow C + D$$

EQ. 4.1

where A and B represent reactants and C and D represent products, there must be at least a small amount of all species present when the reaction has finished. It may well be that the quantities of one or more of the species will be so small that it is impossible, as a practical matter, to measure the concentration, but thermodynamics tells us that at least some of all species must be present.  In chapter two you learned that the non–standard and the standard free energy changes, ($\Delta G$ and $\Delta G°$, respectively) were related by an equation that had an expression in it that was called the reaction quotient.  You also learned that the non–standard free energy change ($\Delta G$) was a measure of the driving force for the reaction.

For equation 4.1, the expression relating $\Delta G$ and $\Delta G°$ is given by:

$$\Delta G = \Delta G° + RT \ln\frac{[C][D]}{[A][B]}$$

EQ. 4.2

If the reaction has reached **equilibrium**, there is no further driving force for reaction in either direction.  This means that the free energy of the system cannot be further lowered by reaction in either direction; therefore $\Delta G = 0$. Placing that value into equation 4.2 and rearranging the equation, gives us the following expression in which the concentrations are those at equilibrium.

$$\Delta G° = - RT \ln\left(\frac{[C][D]}{[A][B]}\right)_{eq}$$

EQ. 4.3

Since, at any given temperature, this can only occur if $\left(\frac{[C][D]}{[A][B]}\right)_{eq}$ has a fixed value,  that value is termed the **equilibrium constant** and is given the symbol $K_{eq}$ (upper case K is used by convention in order to distinguish the equilibrium constant for a reaction from its rate constant).

## 4.2   THE SOLUBILITY PRODUCT

One of the simplest uses of the equilibrium constant is to aid in the determination of how much of a sparingly soluble, ionic substance will dissolve in water. To a first approximation (sometimes not a very good one because of the formation of complex ions that will be discussed in Chapter 9), we expect that the dissolution of an "insoluble salt" such as silver chloride can be described by the chemical equation:

EQ. 4.4

$$AgCl_{(s)} \rightleftharpoons Ag^+ + Cl^-$$

(Note: Since we will be dealing with aqueous solutions almost exclusively in this chapter, when there is no indication of state for an ionic species in this chapter, it will be considered to be in aqueous solution, *i.e.*, $Ag^+$ actually means $Ag^+(aq)$.) If a quantity of solid silver chloride is placed into water and the solution is stirred until no more silver chloride dissolves, then the solution is at equilibrium with the solid, and the equilibrium constant expression

EQ. 4.5

$$K_{sp} = \frac{[Ag^+][Cl^-]}{[AgCl]}$$

applies to the situation. In this equation, as in all equilibrium constant expressions, terms within square brackets are the molar concentrations of the species *at equilibrium*. By the conventions that govern the way equilibrium constants are written, the effective concentration of any solid or pure liquid is defined as unity. Thus, the term in the denominator is equal to 1 provided only that there is some solid silver chloride in contact with the solution. The expression therefore simplifies to

EQ. 4.6

$$K_{sp} = [Ag^+][Cl^-]$$

This equilibrium constant is referred to as the **solubility product** for silver chloride, and the K is usually subscripted with the letters 'sp' to indicate this fact. Solubility products for a variety of salts are tabulated in appendix B.

We can use the value of this equilibrium constant to predict the solubility of silver chloride in water. In order to solve this, and other equilibrium problems, we shall use a variation of the method the we used to solve limiting reagent problems. We will set up the same table of reactants and products that we did for those problems. The only difference will be that the entries in the table will be the concentrations of the various species instead of the number of moles of the species.

If we apply the method to this problem, we generate the following table:

**TABLE 4.1:** EQUILIBRIUM SOLUTION TABLE

TABLE 4.1

|  | AgCl(s) | Ag⁺ | Cl⁻ |
|---|---|---|---|
| initial conc. | — | 0 | 0 |
| reacts | $x$ | — | — |
| forms | — | $x$ | $x$ |
| at equilibrium | — | $0 + x$ | $0 + x$ |

From this we determine that the equilibrium concentration of both silver ion and chloride ion at equilibrium must be $x$. Because the solubility product value is known ($1.7 \times 10^{-10}$, appendix B), it is only a short step to substitute $x$ for the concentration of those ions in the solubility product constant expression and solve for the value of $x$:

$$K_{sp} = 1.7 \times 10^{-10} = (x)(x)$$

EQ. 4.7

$$x = \sqrt{1.7 \times 10^{-10}} = 1.3 \times 10^{-5}\,M$$

EQ. 4.8

If we now ask: "What weight of silver chloride will dissolve in 100. mL of water at 25°C ?" we can quickly answer that question because we know, from the table, that $x$ moles of AgCl must dissolve in a liter of water. Therefore:

$$\left(\frac{1.3 \times 10^{-5}\,\text{mol AgCl}}{\text{liter solution}}\right)\left(0.100\ \text{liter solution}\right)\left(\frac{143.34\ \text{g AgCl}}{\text{mol AgCl}}\right) = 1.\overline{8}7 \times 10^{-4}\,g$$

EQ. 4.9

Our estimate is that $1.9 \times 10^{-4}$ g (0.19 mg) of silver chloride will dissolve in 100. mL of water. The actual solubility of silver chloride will be somewhat greater than this because silver ion forms complexes with a number of species in the solution. The formation of these complexes, unaccounted for by this simple equilibrium expression, means that the actual solubility of silver chloride will be a bit larger. (In later courses you may well learn how to handle this complication.)

Solving the silver chloride problem is quite easy because the salt is a 1:1 salt, and, as a consequence, the concentrations of the silver and chloride ions must be equal. The situation is only slightly more complicated, but easily handled, if this is not the case. Consider the solubility of bismuth iodide:

$$BiI_3(s) \rightleftharpoons Bi^{3+} + 3\,I^-$$

EQ. 4.10

**TABLE 4.2**  EQUILIBRIUM SOLUTION TABLE

TABLE 4.2

|  | $BiI_3(s)$ | $Bi^{3+}$ | $3\,I^-$ |
|---|---|---|---|
| initial conc. | — | 0 | 0 |
| reacts | $x$ | — | — |
| forms | — | $x$ | $3x$ |
| at equilibrium | — | $0 + x$ | $0 + 3x$ |

$$K_{sp} = 8.1 \times 10^{-19} = [Bi^{3+}][I^-]^3 = (x)(3x)^3 = 27x^4$$

EQ. 4.11

$$x = \sqrt[4]{\frac{8.1 \times 10^{-19}}{27}} = 1.\overline{3}2 \times 10^{-5}M$$

EQ. 4.12

In this case the chemistry of the situation means that the value calculated is likely to be less valid because many ions present in the solution, including iodide and hydroxide ion, often form strong complexes with cations, including bismuth ion. Nevertheless, the value calculated in this manner is an approximation that is often quite useful even if not totally accurate.

A minor complication arises if the salt is placed not in pure water but in a solution containing one of the ions of the salt (this is a so–called "common ion" problem.) Although these problems appear to be more difficult at the start, in practice they are usually even easier to solve than the ones we have just attempted. Consider the situation in which a quantity of solid lead iodate is added to a solution containing 0.10 **M** sodium iodate.

$$Pb(IO_3)_2(s) \rightleftharpoons Pb^{2+} + 2\,IO_3^-$$

EQ. 4.13

**TABLE 4.3** Equilibrium Solution Table

TABLE 4.3

|                | $Pb(IO_3)_2$ | $Pb^{2+}$ | $IO_3^-$ |
|----------------|:---:|:---:|:---:|
| Init conc.     | —   | 0   | 0.10 |
| Reacts         | $x$ | –   | —    |
| Forms          | —   | $x$ | $2x$ |
| At equilibrium | —   | $0 + x$ | $0.10 + 2x$ |

EQ. 4.14

$$K_{sp} = 2.6 \times 10^{-13} = \left[Pb^{2+}\right]\left[IO_3^-\right]^2 = (x)(0.10 + 2x)^2$$

This equation looks messy to solve until we look at the value of the solubility product and realize that lead iodate is a sparingly soluble salt. This means that the value of $x$ is likely to be quite small. Placing it in a solution containing iodate ion will repress the solubility even further by LeChatelier's Principle. We are likely to be on safe ground if we make the assumption that $(0.10 + 2x) \approx 0.10$. This is to say that so little lead iodate dissolves that the additional iodate ion from its dissolution does not change the overall iodate concentration appreciably. If this is true, equation 14 can be greatly simplified:

$$2.6 \times 10^{-13} = (x)(0.10)^2$$

EQ. 4.15

or

$$x = \frac{2.6 \times 10^{-13}}{0.010} = 2.6 \times 10^{-11} \text{ M}$$

EQ. 4.16

To finish this problem we have only to check that our assumption is correct. Is $(0.10 + 5.2 \times 10^{-11}) \approx 0.10$? Since this is obviously true, we have solved the problem and the calculated solubility is $2.6 \times 10^{-11}$ M. (You may be concerned about how to handle a problem if such an assumption is not valid. Although those problems do not arise very often, it is possible to solve them. Often they are solved by iterative methods that are more properly taken up in later chemistry courses.)

## 4.3    ACIDITY AND BASICITY IN WATER

### THE AUTOPROTOLYSIS OF WATER

When we discussed acid/base reactions last year, we looked at the reaction:

$$H_2O \ + \ H_2O \ \longrightarrow \ H_3O^+ \ + \ OH^-$$

EQ. 4.17

This reaction is known as the **autoprotolysis** reaction for water and is, perhaps, the most important equilibrium reaction in the universe for life as we know it. Equation 4.17 governs much of what happens in many aqueous reactions, especially biochemical reactions, since the hydronium or hydroxide ion concentration plays a large part in determining the rate or the product(s) in many aqueous chemical reactions.  Specifically how this happens will be taken up in later chapters in this book.

The equilibrium constant expression for reaction in equation 4.17 is

$$K = \frac{[H_3O^+][OH^-]}{[H_2O]^2}$$

EQ. 4.18

Since this reaction occurs in what is essentially pure water, the term in the denominator takes on a value of 1 because of thermodynamic standard state conventions.  This results in the following expression:

$$K_w = [H_3O^+][OH^-]$$

EQ. 4.19

At 25°C the value of this equilibrium constant is $1.0 \times 10^{-14}$.  Because of the value of this equilibrium constant, the concentrations of both $H_3O^+$ and $OH^-$ ions can vary over a wide range.  Since practical concentrations of either ion can reach approximately 10 **M**, we can easily calculate using equation 4.19, that if $[H_3O^+] = 10$ **M**, then $[OH^-]$ must be $1 \times 10^{-15}$ M and *vice versa*.

### THE pH SCALE

The wide range of possible concentrations to which we have just referred has led chemists to invent a different way of representing them in numbers that are more easily understood by people unaccustomed to working with numbers in exponential notation.  This was done by defining a quantity called pH (pronounced "P", break, "H", not "fffffff") as the negative of the logarithm (to the base 10) of the hydronium concentration:

$$pH = -\log_{10}[H_3O^+]$$

EQ. 4.20

Therefore, if $[H_3O^+] = 5.6 \times 10^{-7}$ **M**, the pH may be easily calculated as:

$$pH = -\log_{10}(5.6 \times 10^{-7}) = 6.25$$

EQ. 4.21

The lower case 'p' in front of the 'H' indicates that the negative log of 'H' is to be obtained. This notation carries over to many other quantities and we will often see references to pOH (negative log of OH) and pK (negative log of K) in many places.

At this point, a word concerning significant figures as they are applied in the representation of logarithmic values is in order. Since the number before the decimal point in a log value signifies the power to which 10 must be raised when the number is converted into its "normal", *i.e.*, non–logarithmic, value and expressed in scientific notation, the number before the decimal point in a log value is never significant. It is the numbers behind the decimal point that actually carry the significance. Therefore *all* numbers behind the decimal point in a log are significant (even leading zeros after the decimal point and before the first non–zero digit following the decimal point). In the result from equation 4.21, since the number $5.6 \times 10^{-7}$ had 2 significant figures, we must write its log representation (6.25) with two significant figures. Therefore the digits 2 and 5 following the decimal point in the pH are the digits in this number that are significant. The 6 before the decimal point is not significant. [Note: Had the value from equation 4.21 been 6.05 instead of 6.25, the '0' would have been significant.]

### THE pH OF PURE WATER

It is now time to ask: "What is the pH of pure water?". Once again we should first write the equilibrium table.

TABLE 4.4

**TABLE 4.4** EQUILIBRIUM SOLUTION TABLE

|          | $H_2O$ | $H_2O$ | $H_3O^+$ | $OH^-$ |
|----------|--------|--------|----------|--------|
| init con | —      | —      | 0        | 0      |
| reacts   | $x$    | $x$    | —        | —      |
| forms    | —      | —      | $x$      | $x$    |
| at eq.   | —      | —      | $0 + x$  | $0 + x$ |

$$K_w = 1.0 \times 10^{-14} = (x)(x) = x^2$$

EQ. 4.22

The solution is again quite easily obtained as:

$$x = \sqrt{1.0 \times 10^{-14}} = 1.0 \times 10^{-7} \text{ M}$$

EQ. 4.23

This leads to a pH of 7.00 for pure water at 25°C. The presence of acids in the water decreases the pH, conversely the presence of bases in the water raises the pH. pH values below 7 are acidic; those above 7 are basic. The pH values of solutions of a number of common substances are given below.

**TABLE 4.5**  pH of Several Common Substances

TABLE 4.5

| Substance | Approximate pH Value |
|---|---|
| Blood | 6.9–7.2 |
| Saliva | 6.5–7.5 |
| Stomach Contents | 1.0–3.0 |
| Milk | 6.6–7.6 |
| Apples | 2.9–3.3 |
| Beer | 4.0–5.0 |
| Cider | 2.9–3.3 |
| Grapefruit | 3.0–3.3 |
| Lemon | 2.2–2.4 |
| Lime | 1.8–2.0 |
| Dill Pickles | 3.2–3.6 |
| Soft Drinks | 2.0–4.0 |
| Tomatoes | 4.0–4.4 |
| Vinegar | 2.4–3.4 |
| Drinking Water | 6.5–8.0 |
| Wine | 2.8–3.8 |

## 4.4   SOLUTIONS OF A SINGLE SUBSTANCE

### pH Values for Solutions of Strong Acids and Bases

It is important to understand that the concentrations of $H_3O^+$ and $OH^-$ may vary when other materials are placed in the solution, but that the product of these concentrations will always be such that the equilibrium constant value is obeyed.  This means that placing something in the solution that results in an increase in the $OH^-$ concentration will simultaneously cause a reduction in the $H_3O^+$ concentration and *vice versa*.

Let us first consider what happens when we place a strong acid or a strong base in water.  From our discussion of acids and bases in chapter 12 last year, you will recall that the definition of a strong acid or a strong base includes that the acid or base be completely dissociated in water.  There are only a limited number of strong acids and bases.  The entire list of common strong acids and bases is contained in table 4.6.  You should make certain that you know this list. (While essentially all metal hydroxides would be classified as strong bases, only those whose solubilities in water are high enough to make strongly basic solutions are included in this table.)

TABLE 4.6

TABLE 4.6    LIST OF COMMON STRONG ACIDS AND BASES

| Acids | Bases |
|-------|-------|
| HCl | all alkali metal hydroxides |
| HBr | $Ca(OH)_2$ |
| HI | $Sr(OH)_2$ |
| $HNO_3$ | $Ba(OH)_2$ |
| $HClO_4$ | TlOH |
| $H_2SO_4$ (1st proton only) | |

Let us consider, then, what happens when we dissolve a **strong acid** in water:

Problem 1: What are the concentrations of $H_3O^+$ and $OH^-$ as well as the pH and the pOH of a solution that is 0.050 **M** in $HNO_3$?

We approach this problem as we do any other equilibrium problem of a similar type. We should first write the equilibrium table:

TABLE 4.7    EQUILIBRIUM SOLUTION TABLE

TABLE 4.7

| | $H_2O$ | $H_2O$ | $H_3O^+$ | $OH^-$ |
|-----------|------|------|---------|------|
| init conc | — | — | 0.050 | 0 |
| reacts | $x$ | $x$ | — | — |
| forms | — | — | $x$ | $x$ |
| at eq. | — | — | $0.050 + x$ | $0 + x$ |

We can now place these values into the equilibrium constant expression:

$$K_w = 1.0 \times 10^{-14} = [H_3O^+][OH^-] = (0.050 + x)(x)$$

EQ. 4.24

As we did in the lead iodate problem, we will attempt to make a mathematical assumption that will help us solve the problem without sacrificing any chemical significance to the answer. In this case we could reason that placing a strong acid that contributes $H_3O^+$ to the solution will decrease the amount of water that dissociates by the LeChatelier Principle (the common ion effect). This will make the value of $x$ quite small, almost certainly so small that the quantity $(0.050 + x) \approx 0.050$. If this is true, equation 24 can be rewritten.

$$K_w = 1.0 \times 10^{-14} = [H_3O^+][OH^-] = (0.050)(x)$$

EQ. 4.25

This is very easily solved to give

$$x = \frac{1.0 \times 10^{-14}}{0.050} = 2.0 \times 10^{-13}$$

EQ. 4.26

Before proceeding further we should check that our assumption is correct by asking if $0.050 + (2.0 \times 10^{-13}) \approx 0.050$. Of course it is. This means that the hardest part of the problem is now past. From here we simply need to go to the equilibrium table to determine the actual concentrations of $H_3O^+$ and $OH^-$. These are 0.050 **M** and $2.0 \times 10^{-13}$ **M** respectively. To calculate the pH and pOH we need to obtain the negative logs of each of these numbers. These are pH = 1.30 and pOH = 12.70.

## FURTHER MUSINGS ON pH

Perhaps you noticed that when we did the problem to find the pH of the nitric acid solution, the sum of the pH and pOH values was 14.00. This was not a coincidence; rather it is a consequence of the water constant expression:

$$K_w = [H_3O^+][OH^-]$$  EQ. 4.27

Since the laws of mathematics dictate that the log of a product is the sum of the logs of the factors of the product, we can rewrite this equation as

$$\log K_w = \log[H_3O^+] + \log[OH^-]$$  EQ. 4.28

This, if multiplied through by the factor -1, the expression becomes

$$-\log K_w = -\log[H_3O^+] + (-\log[OH^-])$$  EQ. 4.29

We recognize the two terms on the right as pH and pOH respectively. The value of $-\log K_w$ (or $pK_w$) is 14.00 at 25°C. Therefore:

$$14.00 = pH + pOH \quad (at\ 25°C)$$  EQ. 4.30

This is a very handy way to compute pOH or pH if the other is already known. It is not necessary to convert the pH or pOH to a "normal" number, divide that number into $1 \times 10^{-14}$, and then take the negative log of the result. Simply subtract the pH or pOH from 14.00 to obtain the answer.

## WEAK ACIDS AND BASES

Except for the limited number of strong acids and bases mentioned in table 4.6, all other acids and bases are considered **weak acids** or **weak bases**. The term "weak" simply means that they do not ionize completely when dissolved in water. There is still a very wide variation in acid or base strength for these materials.

While there is great variety in the types of organic compounds that are weak acids, the most typical of this category are the carboxylic acids. These contain the functional group, —COOH. In addition to this class of compounds, however, there are many other organic functional groups that are capable of losing a proton. This produces the very wide variation in acid strength available. Finally, in addition to the weak organic acids, there are also a number of common inorganic materials with acidic properties. These include the binary acids of the Group V elements from P to Bi and the lower valent oxo–acids of the Group IV, V and VI elements. These include, for instance, nitrous acid ($HNO_2$), phosphoric acid ($H_3PO_4$), sulfurous acid ($H_2SO_3$), and carbonic acid ($H_2CO_3$). All of these are also weak acids.

## WEAK ACID DISSOCIATION CONSTANTS

If we focus only on the ability of the acid to act as a Brønsted-Lowry acid, we can write a generic reaction that describes the ionization of any of these weak acids:

$$HA(aq) + H_2O(l) \rightleftharpoons H_3O^+ + A^-$$

EQ. 4.31

The equilibrium constant expression for this reaction is

$$K_a = \frac{[H_3O^+][A^-]}{[HA]}$$

EQ. 4.32

Since the species HA is dissolved in water and is not the solvent, it remains in the expression. The term for the unionized water does not appear because it is the solvent. To symbolize that the **dissociation constant** being described is an acid dissociation constant, the K is usually subscripted with a lower case 'a'. This text book contains the dissociation constants for a wide variety of different acids and bases in appendix B. A glance at these dissociation constants will show that acid strengths can vary widely. Some acids are extensively ionized ($pK_a \approx 1$) while other acids ionize only to a very small extent ($pK_a \approx 14$). Unless some strong electron–withdrawing group or element is close to the carboxyl group, the $pK_a$ of a carboxylic acid, the most common type of weak acid, is approximately 5.

## WEAK BASE DISSOCIATION CONSTANTS

Common weak bases are organic compounds to which a proton may attach *via* an unshared pair of electrons. The most common atom to which the proton becomes attached is a nitrogen atom, either contained in an amino group or as a heterocyclic nitrogen atom in a ring compound. Except for the ammonia molecule ($NH_3$) and its close relatives hydrazine ($NH_2NH_2$) and hydroxylamine ($NH_2OH$), there are no common, soluble, neutral inorganic weak bases. Weak bases produce hydroxide ions in solution by undergoing a Brønsted-Lowry reaction with the solvent. Using ammonia as an example and water as the solvent, this reaction is

$$NH_3(aq) + H_2O(l) \rightleftharpoons NH_4^+ + OH^-$$

EQ. 4.33

Writing this in more general form gives

$$B(aq) + H_2O(l) \rightleftharpoons BH^+ + OH^-$$

EQ. 4.34

Here the dissociation constant expression becomes

$$K_b = \frac{[BH^+][OH^-]}{[B]}$$

EQ. 4.35

Once again the term for the solvent does not appear in the dissociation constant expression. By convention, this K is usually subscripted with a lower case 'b' to indicate that it is a base dissociation constant. Dissociation constants for a variety of weak bases are also contained in appendix B. In this appendix you will note that there is the same wide variety in the dissociation constants that is exhibited by the acids. As a class, organic aliphatic amines have dissociation constants for which the $pK_b$ is approximately 5.

## DISSOCIATION CONSTANTS FOR CONJUGATE ACID/BASE PAIRS

Brønsted-Lowry theory tells us that when a weak acid dissociates, one of the products is the conjugate base of the acid, *viz.*:

$$HA(aq) + H_2O(l) \rightleftharpoons H_3O^+ + A^-$$

EQ. 4.36

We have already written the dissociation constant expression for this reaction:

$$K_a = \frac{[H_3O^+][A^-]}{[HA]}$$

EQ. 4.37

As we learned last year, the species $A^-$ is termed the conjugate base. It is, indeed, a base in its own right. The equation for its reaction with water in a Brønsted-Lowry fashion would be

$$A^- + H_2O(l) \rightleftharpoons HA(aq) + OH^-$$

EQ. 4.38

If we now write the base dissociation expression for the reaction in equation 4.38, we find

$$K_b = \frac{[HA][OH^-]}{[A^-]}$$

EQ. 4.39

If we now multiply equation 4.37 by equation 4.39 we obtain

$$K_a K_b = \left( \frac{[H_3O^+][A^-]}{[HA]} \right) \left( \frac{[HA][OH^-]}{[A^-]} \right) = [H_3O^+][OH^-] = K_w$$

EQ. 4.40

The product of the acid dissociation constant and the base dissociation constant for a **conjugate acid/base pair** is the water constant. This means that if one of the values is known, the other can be easily obtained. *It is important to note, however, that the dissociation constants that are multiplied together must be the acid dissociation constant for the acid and the base dissociation constant for the base of species that form a Brønsted-Lowry conjugate acid/base pair.* (That the product of the two is the water constant holds true only when the two are dissolved in water. If they are in a different solvent, the product of the two dissociation constants is the solvolysis constant of the solvent.)

## DETERMINATION OF DISSOCIATION CONSTANT FROM pH

In an earlier section it was mentioned that pH is an extremely important variable in many life processes, affecting many of them by causing conformational changes in biological molecules, chiefly proteins. Exactly how does this occur? How does the body regulate these changes to achieve, in most cases, the conditions necessary for life itself? It is time to examine the topic of weak acid/base equilibria in more detail so that these questions can, ultimately, be answered.

Probably the most important characteristic of a weak acid or base is its dissociation constant. This equilibrium constant governs the extent to which the molecule will dissociate in pure water and, therefore, the pH of the resulting solution. It also governs the extent to which it will dissociate in solutions whose pH is determined by other acids or bases. One means of determining the dissociation constant of an acid or base is to determine the pH of a solution containing a known concentration of that acid or base.

Let us consider a 0.10 **M** solution of lactic acid in water. Measurement of the pH of this solution yields a value of 2.43. Let us use this information to determine the dissociation constant of the acid. To do this we will first write the equilibrium table that we have used in the past. Although it is only necessary to know that lactic acid is a monoprotic acid in order to do the problem, figure 4.1 shows a structural representation of the dissociation of the lactic acid molecule into lactate ion and hydrogen ion (a shorter means of writing what is really hydronium ion and one that we will use in most places from now on).

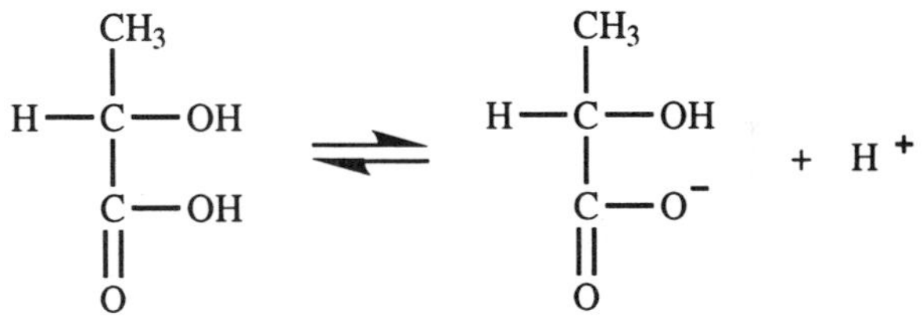

FIGURE 4.1: Dissociation of Lactic Acid

A shorter way of writing this equation, representing lactic acid by the abbreviation HLac, is used in the equation 4.41 and the equilibrium table:

$$HLac(aq) \rightleftharpoons H^+ + Lac^-$$

EQ. 4.41

**TABLE 4.8**  EQUILIBRIUM SOLUTION TABLE

TABLE 4.8

|  | HLac | $H^+$ | $Lac^-$ |
|---|---|---|---|
| init concentration | 0.10 | 0 | 0 |
| reacts | $x$ | — | — |
| forms | — | $x$ | $x$ |
| at equilibrium | $0.10 - x$ | $0 + x$ | $0 + x$ |

We know that the pH of this solution is 2.43. This tells us what the concentration of the hydrogen ion must be:

$$[H^+] = 10^{-2.43} = 3.72 \times 10^{-3}$$

EQ. 4.42

Looking in the table we note that the concentration of hydrogen ion in the final solution is $x$. This allows us to determine the concentrations of all of the species in the solution:

$$\left[H^+\right] = x = 3.\overline{7}2 \times 10^{-3} \text{ M}$$

EQ. 4.42A

$$\left[Lac^-\right] = x = 3.\overline{7}2 \times 10^{-3} \text{ M}$$

EQ. 4.42B

$$\left[HLac\right] = 0.10 - x = 9.\overline{6}3 \times 10^{-2} \text{ M}$$

EQ. 4.42C

We now have all of the numbers that we need to solve for the dissociation constant:

$$K_a = \frac{\left[H^+\right]\left[Lac^-\right]}{\left[HLac\right]} = \frac{\left(3.\overline{7}2 \times 10^{-3}\right)\left(3.\overline{7}2 \times 10^{-3}\right)}{9.\overline{6}3 \times 10^{-2}} = 1.43 \times 10^{-4}$$

EQ. 4.43

Note that the rules of significant figures that we have been following allow us to write this value to only one significant figure. Therefore, we should report that value as $1 \times 10^{-4}$. Fortunately there are better, more precise means of determining these values. The value in the appendix is $1.4 \times 10^{-4}$.

The procedure is only slightly different for the determination of the dissociation constant of a weak base. To illustrate, let us consider the case of a 0.050 M solution of the weak base pyridine that was found to have a pH of 8.96. Once again, it is not necessary to know the chemical formula for pyridine in order to do this problem.

The chemical reaction for the base dissociation of pyridine is

EQ. 4.44

If pyridine is represented by the abbreviation Py, the equilibrium table would be

**TABLE 4.9**    EQUILIBRIUM SOLUTION TABLE

TABLE 4.9

|  | Py | $H_2O$ | $PyH^+$ | $OH^-$ |
|---|---|---|---|---|
| init. conc. | 0.050 | — | — | — |
| reacts | $x$ | $x$ | — | — |
| forms | — | — | $x$ | $x$ |
| at eq. | 0.050 - $x$ | — | $0 + x$ | $0 + x$ |

In this case the pH indirectly provides us the value of $x$. We must first find the value of pOH, then convert that value to [OH⁻]:

$$pOH = 14.00 - pH = 14.00 - 8.96 = 5.04$$

EQ. 4.45

$$[OH^-] = 10^{-5.04} = 9.\overline{2}1 \times 10^{-6} \text{ M}$$

EQ. 4.46

This leads to:  $$[OH^-] = x = 10^{-5.04} = 9.\overline{2}1 \times 10^{-6} \text{ M}$$

EQ. 4.46A

$$[PyH^+] = x = 10^{-5.04} = 9.\overline{2}1 \times 10^{-6} \text{ M}$$

EQ. 4.46B

$$[Py] = 0.050 - x = 0.050 - 9.\overline{2}1 \times 10^{-6} = 0.04\overline{9}99 \text{ M}$$

EQ. 4.46C

We now have all the values needed to calculate the dissociation constant:

$$K_b = \frac{[PyH^+][OH^-]}{[Py]} = \frac{\left(9.\overline{2}1 \times 10^{-6}\right)\left(9.\overline{2}1 \times 10^{-6}\right)}{0.050} = 1.\overline{6}9 \times 10^{-9}$$

EQ. 4.47

We should report this value to the 2 significant figures justified by the pH value. The result is $1.7 \times 10^{-9}$.

## WEAK MONOPROTIC ACIDS AND BASES

An important use of the dissociation constant in simple acid/base problems is to determine the pH of a solution of a weak acid or base. This is really a fairly straightforward equilibrium problem that presents few real challenges beyond those we have already mastered. Let us examine a few typical problems of this sort.

First let us consider a 0.50 M solution of acetic acid. The chemical equation for this dissociation is:

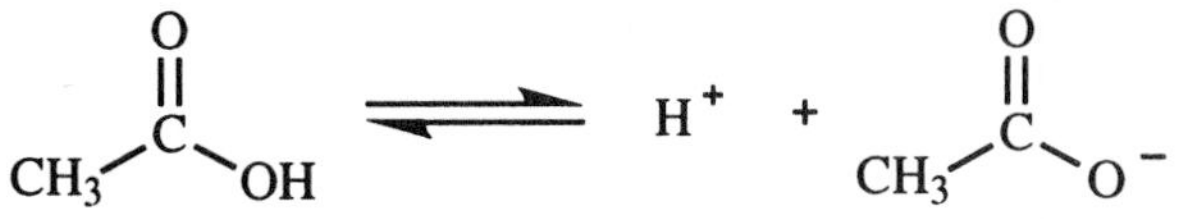

EQ. 4.48

This allows us to set up the equilibrium table (abbreviating acetic acid as HAc):

TABLE 4.10   EQUILIBRIUM SOLUTION TABLE

TABLE 4.10

|           | HAc        | $H^+$ | $Ac^-$ |
|-----------|------------|-------|--------|
| init conc | 0.50       | 0     | 0      |
| react     | $x$        | —     | —      |
| form      | —          | $x$   | $x$    |
| at eq.    | 0.50 - $x$ | $x$   | $x$    |

From this we can directly substitute into the acid dissociation constant:

$$K_a = 1.8 \times 10^{-5} = \frac{[H^+][Ac^-]}{[HAc]} = \frac{(x)(x)}{(0.50 - x)}$$

EQ. 4.49

While this would produce a quadratic equation that would be fairly easily solved, we can make the arithmetic easier by thinking about the chemistry of the situation and noting that the size of the dissociation constant means that the acid will be only slightly ionized in solution. This means that the value of $x$ will be small and, as a consequence, the value of $0.50 - x$ is likely to be so close to 0.50 that we can easily make the approximation that $(0.50 - x) \approx 0.50$. Having made this approximation, equation 4.49 then becomes:

$$1.8 \times 10^{-5} = \frac{(x)(x)}{0.50}$$

EQ. 4.50

which can be readily solved:

$$x = \sqrt{\left(1.8 \times 10^{-5}\right)\left(0.50\right)} = 3.0 \times 10^{-3}$$

EQ. 4.51

Before accepting this as the answer, we must check the assumption:

$$0.50 - 0.003 = 0.497$$

EQ. 4.52

which, to the 2 significant figures allowed by the data rounds to 0.50. The assumption is good and, therefore the pH = 2.52.

As a second example, let us use a base with a base dissociation constant larger than the acid dissociation constant of acetic acid. This would mean that it would be more likely to ionize enough so that the assumption of a small $x$ value would not be valid. Consider a 0.050 M solution of piperidine, a mono-protic weak base with $K_b = 1.6 \times 10^{-3}$.

EQ. 4.53

Abbreviating piperidine as Pip, the equilibrium table is

**TABLE 4.11**   EQUILIBRIUM SOLUTION TABLE

TABLE 4.11

|          | Pip       | $H_2O$ | PipH$^+$ | OH$^-$ |
|----------|-----------|--------|----------|--------|
| init conc | 0.050     | —      | 0        | 0      |
| react    | $x$       | $x$    | —        | —      |
| form     | —         | —      | $x$      | $x$    |
| at eq.   | $0.050 - x$ | —    | $0 + x$  | $0 + x$ |

Substituting these into the base dissociation constant expression gives

$$K_b = 1.6 \times 10^{-3} = \frac{[PipH^+][OH^-]}{[Pip]} = \frac{(x)(x)}{0.050 - x}$$

EQ. 4.54

If we made the same type of assumption that we made before $((0.050 - x) \approx 0.050)$ and then solved the equation, the value of $x$ would be $8.9 \times 10^{-3}$. The value of $0.050 - (8.9 \times 10^{-3})$ does not round to the 0.050 that we used in the solution. Therefore the assumption is incorrect. We must solve the problem the long way, which can done using the quadratic equation. Multiplying out the equation and putting it into standard quadratic form, we obtain

$$x^2 + (1.6 \times 10^{-3})x - (8.0 \times 10^{-5}) = 0$$

EQ. 4.55

Since the standard form of the quadratic equation is $ax^2 + bx + c = 0$, we recognize the value of $a = 1$, $b = 1.6 \times 10^{-3}$ and $c = -8.0 \times 10^{-5}$. This can be put into the quadratic equation to obtain the following expression:

$$x = \frac{-b \pm \sqrt{b^2 - 4ac}}{2a} = \frac{-1.6 \times 10^{-3} \pm \sqrt{\left(1.6 \times 10^{-3}\right)^2 + 3.2 \times 10^{-4}}}{2}$$

EQ. 4.56

Doing the arithmetic provides us with the solutions: $x = 8.2 \times 10^{-3}$ and $x = -9.8 \times 10^{-3}$. Since it is impossible for a real concentration to be negative, the first value must be the correct one. Because we made no assumption to solve this problem, there is no assumption to check and this is the value of $x$ that is the solution to the problem. We must convert only this value of $x$ into the pOH. The pOH of this solution is 2.09, which results in an answer of pH = 11.91.

### SALTS OF WEAK MONOPROTIC ACIDS AND BASES

When the salt of a weak acid or base is placed in solution, it affects the pH of the solution. This is not surprising, since the salt is a base or acid in its own right because it must be the conjugate base or acid of the original acid or base. This presents only one additional step in solving a pH problem — the determination of the value to use for the dissociation constant. For example, consider the following problem:

Calculate the pH of a 0.10 **M** solution of pyridinium chloride.

From the name of the species we should recognize that this is the hydrogen chloride salt of the weak base pyridine. The pyridinium ion is, then, the conjugate acid of pyridine. The structures of the two species are presented in figure 4.2.

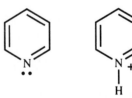

Pyridine        Pyridinium Ion

FIGURE 4.2:
Structure of Pyridine and
Pyridinium Ion

If we abbreviate pyridine as Py, we note that there are two chemical reactions that occur when pyridinium chloride is dissolved in water, *viz.*:

$$PyHCl \longrightarrow PyH^+ + Cl^-$$

EQ. 4.57

$$PyH^+ \rightleftharpoons Py(aq) + H^+$$

EQ. 4.58

The first of these reactions (equation 4.57) simply indicates that the salt dissolves completely in water, and, since it is a strong electrolyte, it undergoes complete dissociation in water. There is no equilibrium constant to be considered for this reaction. The second reaction (equation 4.58) is an equilibrium reaction. It represents the weak acid dissociation of the pyridinium ion. The only difference between the pyridinium ion and other, more conventional, acids such as acetic acid, is that this weak acid has a positive charge and a different dissociation constant. In terms of solving the problem, however, there are no differences as long as we use the value of the acid dissociation constant for the pyridinium ion and not the base dissociation constant for pyridine in solving the problem. With this in mind, let us set up the equilibrium table for this problem:

TABLE 4.12

**TABLE 4.12** EQUILIBRIUM SOLUTION TABLE

|  | $PyH^+$ | Py | $H^+$ |
|---|---|---|---|
| init conc | 0.10 | 0 | 0 |
| reacts | $x$ | — | — |
| forms | — | $x$ | $x$ |
| at eq | $0.10 - x$ | $0 + x$ | $0 + x$ |

We can obtain the value that we need for the acid dissociation constant of the pyridinium ion from the base dissociation constant of its conjugate base, pyridine, by using equation 4.40. If we now place these values into the expression for the acid dissociation constant, we obtain

$$K_a = \frac{K_w}{K_b} = \frac{1.0 \times 10^{-14}}{1.7 \times 10^{-9}} = 5.9 \times 10^{-6} = \frac{[Py][H^+]}{[PyH^+]} = \frac{(x)(x)}{0.10 - x}$$

EQ. 4.59

Looking at the value of the dissociation constant, we expect that we should be able to make the simplifying assumption that $(0.10 - x) \approx 0.10$. Making this assumption provides the solution, $x = 7.7 \times 10^{-4}$. Testing the assumption we find that $0.10 - (7.7 \times 10^{-4}) = 0.09923$ which rounds to 0.10. Obviously the value of $x$ is small enough that the assumption is valid. Finishing the problem, the pH of the solution is 3.12.

This type of problem is often called an "hydrolysis of a salt" problem. Giving it a special name calls attention to the reaction of the active species with water. This is, of course, exactly what our understanding of Brønsted-Lowry theory of acid/base behavior would lead us to expect. However, our understanding of Brønsted-Lowry theory also tells us that this active species is an acid or a base in its own right and, therefore, we already know how to solve these problems. We must not fall into the trap of believing that this type of problem is somehow fundamentally different from an ordinary weak acid or weak base problem.

## 4.5 SOLUTIONS CONTAINING A CONJUGATE ACID/BASE PAIR————

### BUFFER SOLUTIONS

Let us now look at the effect of a common ion on the equilibrium of a weak acid or base. To do this, let us consider a solution that contains 0.10 **M** acetic acid and 0.050 **M** sodium acetate. Since sodium acetate is a strong electrolyte, it is completely ionized in solution. Before taking into consideration the equilibrium reaction of the weak acid, the solution contains 0.10 **M** acetic acid molecules and 0.050 **M** acetate ions (the conjugate base of the weak acid). Once again, using HAc as the abbreviation for the acetic acid molecule, the chemical equation for the equilibrium reaction is

$$HAc(aq) \rightleftharpoons H^+ + Ac^-$$

EQ. 4.60

We now proceed to write the equilibrium table:

**TABLE 4.13**  EQUILIBRIUM SOLUTION TABLE

TABLE 4.13

|           | HAc        | $H^+$   | $Ac^-$      |
|-----------|------------|---------|-------------|
| init conc | 0.10       | 0       | 0.050       |
| reacts    | $x$        | —       | —           |
| forms     | —          | $x$     | $x$         |
| at eq     | $0.10 - x$ | $0 + x$ | $0.050 + x$ |

Note that this equilibrium table differs only from those we had written before in having a non–zero initial concentration for the acetate ion. The effect this has on the final line in the table is to have a value added to the unknown ($x$) term for the acetate concentration at equilibrium. If we now place these values into the equilibrium constant expression for acetic acid, we obtain

$$K_a = 1.8 \times 10^{-5} = \frac{[H^+][Ac^-]}{[HAc]} = \frac{x(0.050 + x)}{0.10 - x}$$

EQ. 4.61

While this equation is only a quadratic equation and would be possible to solve without making assumptions, the application of a bit of chemical reasoning will make the mathematics much easier (and, after all, why should we make the mathematics any harder than it needs to be?). If we note the size of the equilibrium constant, we expect that the value of $x$ will be very small. In other problems we have already seen that the value of $x$ is usually small enough that we can say that $(0.10 - x) \approx 0.10$. In this case we expect the value to be even smaller because of the presence of the common ion (acetate ion). Therefore we shall make the approximations $(0.10 - x) \approx 0.10$ and $(0.050 + x) \approx 0.050$. With these changes, equation 4.61 reduces to

$$1.8 \times 10^{-5} = \frac{x(0.050)}{0.10}$$

EQ. 4.62

which can be easily solved to give

$$x = \left(1.8 \times 10^{-5}\right)\left(\frac{0.10}{0.050}\right) = 3.6 \times 10^{-5} \, \mathrm{M}$$

EQ. 4.63

Looking back at our assumptions, we note that both were, in fact, justified. This produces a pH value of 4.44.

## The Henderson–Hasselbalch Equation and Buffers

If we now take a careful look at equations 4.61 and 4.62, we see that we have just assumed that the effect of the ionization of the acid on the concentrations of acetic acid and acetate ion is negligible. This means that we are able to use the initial concentrations of both in the final equilibrium expression. In mathematical symbolism, if $C_x$ stands for the initial (before equilibrium) concentration of species $x$ and $[x]$ stands for the equilibrium concentration of that species, then we have made a substitution of $C_x$ for $[x]$ in the equilibrium constant expression. In the specific case of the acetic acid/sodium acetate solution that was done as the example, it has been shown that this substitution was chemically (and mathematically) justified. While there are a few cases in which such a substitution may not be justified, certainly the majority of real cases are those in which the substitution is justified. The effect of this on the equilibrium constant expression for the typical weak acid HA is easily seen (equation 4.64).

$$K_a = \frac{[H^+][A^-]}{[HA]} = \frac{[H^+]C_{A^-}}{C_{HA}}$$

EQ. 4.64

If we rearrange this equation we obtain

$$[H^+] = K_a \left(\frac{C_{HA}}{C_{A^-}}\right)$$

EQ. 4.65

which, if we take the base 10 logarithm of both sides, gives

$$\log[H^+] = \log K_a + \log\left(\frac{C_{HA}}{C_{A^-}}\right)$$

EQ. 4.66

After multiplying through by -1, we have

$$-\log[H^+] = -\log K_a - \log\left(\frac{C_{HA}}{C_{A^-}}\right)$$

EQ. 4.67

which, after recognizing the definitions of the terms and inverting the log term becomes:

$$pH = pK_a + \log\left(\frac{C_{A^-}}{C_{HA}}\right)$$

EQ. 4.68

This equation is known as the **Henderson–Hasselbalch equation** (or, sometimes, the Buffer equation) and is the equation most often used to calculate the pH of a **buffer** solution. While the pH values calculated with this equation are not always totally accurate because some of the simplifying assumptions are not valid, it will usually provide a sufficiently accurate estimate of the pH for practical purposes.

## 4.6   SOLUTIONS OF TWO UNRELATED SUBSTANCES

### ADDITION OF STRONG ACID TO A WEAK BASE

When a solution is made by adding two different, unrelated acid/base species together, the pH of the final solution is, in almost all practical cases, determined by only one (or two, in the case of a buffer) species. The species that will control the pH is the one that is the strongest acid or base present. The only situation in which this might not be true is the case in which a *very* small concentration of a strong acid/base is present along with a large concentration of a weak acid/base. These situations arise so rarely that they are not worth the time to consider at this point.

Before attempting to solve a problem in which two unrelated acid or base species are present in a single solution, it is necessary to think about any chemical reactions they might undergo, what the products of those reactions might be, and what the concentrations of those products will be. Only after this step is complete is it possible to attempt to solve the problem.

Let us illustrate by considering the situation in which 100.0 mL of 0.100 **M** ammonia is added to 50.0 mL of 0.15 **M** hydrochloric acid. Since ammonia is a weak base and hydrochloric acid is a strong acid, we must first consider the chemical reaction that will occur when the two are mixed. Ammonia and hydrochloric acid react to form ammonium chloride *via* the following, net ionic, reaction:

$$H^+ + NH_3 \longrightarrow NH_4^+$$

EQ. 4.69

Next, we must write the limiting reagent table.

**TABLE 4.14**   LIMITING REAGENT TABLE

TABLE 4.14

|            | $NH_3$       | $H^+$        | $NH_4^+$ |
| ---------- | ------------ | ------------ | -------- |
| init moles | 0.0100       | .0075        | 0        |
| reacts     | $x$          | $x$          | —        |
| forms      | —            | —            | $x$      |
| remaining  | .0100 - $x$  | .0075 - $x$  | $x$      |
| poss $x$   | .0100        | .0075        | —        |
| final quant| .0025        | 0            | .0075    |

From this table we determine that the final solution will contain 0.0075 moles of ammonium and chloride ions and 0.0025 moles of ammonia in 150.0 mL of total solution. Since this solution has both a base and its conjugate acid present, this is a buffer solution and we can utilize the Henderson–Hasselbalch equation to determine the pH.

$$pH = pK_a + \log\left(\frac{C_{NH_3}}{C_{NH_4^+}}\right) = 9.26 + \log\left(\frac{\frac{.0025}{150}}{\frac{.0075}{150}}\right) = 8.78$$

EQ. 4.70

There are two things that we should note before we leave this problem: (1) although a conventional way to think of this problem is as a weak base and its salt, we still applied the Henderson–Hasselbalch equation written in its usual sense by using the $pK_a$ of the conjugate acid of the pair ($NH_4^+$), and (2) since both the $NH_3$ and the $NH_4^+$ are in the same solution and the Henderson–Hasselbalch equation utilizes only the ratio of the two concentrations, it was not really necessary to convert the results of the limiting reagent table to concentrations before solving the problem because the mole ratio is the same as the concentration ratio. (Notice that the volume (150 mL) can be canceled in the log term.)

The problem just solved became a buffer problem because there was less (fewer moles) strong acid present than there was weak base. This meant that the weak base was not totally consumed in the chemical reaction and the final solution contained both the weak base ($NH_3$) and its conjugate acid ($NH_4^+$). Had the results been otherwise, the solution would have contained a strong acid (HCl) and a weak acid ($NH_4^+$). In such a case, the pH of the solution would be controlled by the concentration of the strong acid.

### ADDITION OF SMALL QUANTITIES OF ACIDS TO A BUFFER

Buffers play a very important role in maintaining the pH of their systems after the addition of small quantities of acids or bases. This permits pH-sensitive reactions to proceed without interference. Nowhere is this more important than in biological systems in which many of the chemical reactions produce or consume hydrogen ions. pH control is especially important because these reactions are catalyzed by enzymes whose function is very pH-dependent. Let us look at a problem that will illustrate just how well a buffer system does resist a change in pH caused by the addition of an acid.

Consider the addition of 1.00 mL of 0.10 M HCl to 100.0 mL of a buffer that is 0.050 M in both acetic acid and sodium acetate. What would be the change in pH caused by this addition?

We must first calculate the pH of the buffer before the addition of the HCl. Since this is a buffer, we can apply the Henderson–Hasselbalch equation directly:

$$pH = pK_a + \log\left(\frac{C_{Ac^-}}{C_{HAc}}\right) = 4.76 + \log\left(\frac{.050}{.050}\right) = 4.76$$

EQ. 4.71

We must then do a limiting reagent calculation to determine what the new quantities of the buffer components will be after addition of the HCl. We approach this problem by first considering what the concentrations of the HAc and the Ac⁻ would be after reaction of the weak base (Ac⁻) with the strong acid (H⁺). Only after we determine these concentrations will we consider the effect of equilibrium. This is entirely justified because reaction 4.72 goes essentially to completion (its equilibrium constant has a value of $\approx 10^5$).

$$H^+ \ + \ Ac^- \ \rightleftharpoons \ HAc(aq)$$

EQ. 4.72

**TABLE 4.15**   LIMITING REAGENT TABLE

TABLE 4.15

|            | $H^+$      | $Ac^-$     | $HAc$       |
|------------|------------|------------|-------------|
| init quant | 0.00010    | .0050      | .0050       |
| reacts     | $x$        | $x$        | —           |
| forms      | —          | —          | $x$         |
| remaining  | $.00010 - x$ | $.0050 - x$ | $.0050 + x$ |
| poss x     | .00010     | .0050      |             |
| final quant| 0          | .0049      | .0051       |

This solution is still a buffer solution since it still contains rather large quantities of both an acid and its conjugate base. Therefore we can still apply the Henderson–Hasselbalch equation. Here we shall make use of the fact that the mole ratio is the same as the concentration ratio and not convert back to concentrations.

$$pH = pK_a + \log\left(\frac{C_{Ac^-}}{C_{HAc}}\right) = 4.76 + \log\left(\frac{.0049}{.0051}\right) = 4.74$$

EQ. 4.73

Therefore the solution underwent a change of only 0.02 pH units after the addition of this rather large quantity of acid to the buffer. In the same way the small additions of acids or bases that accompany biochemical reactions cause very small changes in the pH of the physiological solutions in which they occur. It is only in advanced disease states or under the influence of outside agents when things go seriously wrong that the body's buffer systems fail to maintain the proper pH balance. Then, the failure to maintain the proper pH becomes extremely serious and may actually be the reason for many of the serious effects of the condition. In extreme cases it may even be a cause of death.

# 4.7 POLYPROTIC ACIDS AND BASES

### DISSOCIATION REACTIONS AND EQUILIBRIUM CONSTANTS

To this point, all of the weak acids or bases that we have considered have had only one ionization site. Such acids or bases are referred to as monoprotic. While there are many such monoprotic acids and bases, it is equally true that there are also a large number of acids and bases that have more than one site at which a proton may or may not be present depending upon the pH. This class of **polyprotic acids and bases** includes all of the **amino acids** — arguably the most important class of acid/base species involved in life processes.

First let us consider the chemical ionization reactions of this class of compounds. Taking phosphoric acid as a typical polyprotic acid, we can write three acid dissociation reactions:

$$H_3PO_4(aq) \rightleftharpoons H^+ + H_2PO_4^-$$

EQ. 4.74

$$H_2PO_4^- \rightleftharpoons H^+ + HPO_4^{2-}$$

EQ. 4.75

$$HPO_4^{2-} \rightleftharpoons H^+ + PO_4^{3-}$$

EQ. 4.76

From these we can also write three equilibrium constants:

$$K_{a1} = \frac{[H^+][H_2PO_4^-]}{[H_3PO_4]}$$

EQ. 4.77

$$K_{a2} = \frac{[H^+][HPO_4^{2-}]}{[H_2PO_4^-]}$$

EQ. 4.78

$$K_{a3} = \frac{[H^+][PO_4^{3-}]}{[HPO_4^{2-}]}$$

EQ. 4.79

The number of dissociation constants will equal the number of ionizable protons. In order to distinguish them, it is customary to include the dissociation constant number in the subscript. In all known cases the values of the dissociation constants decrease with the next dissociation constant in the series always being smaller than the previous one. In the case of phosphoric acid: $K_{a1} = 7.4 \times 10^{-3}$, $K_{a2} = 6.2 \times 10^{-8}$, and $K_{a3} = 4.8 \times 10^{-13}$.

The dissociation of the amino acids differs from that of phosphoric acid only in that the fully protonated species is always a cation. Alanine is a typical amino acid. The structures of all three acid/base forms of alanine are shown in figure 4.3. It is easy to see that alanine has only two ionization sites and, therefore, only two dissociation constants. The majority of amino acids have two ionization sites; a few have three.

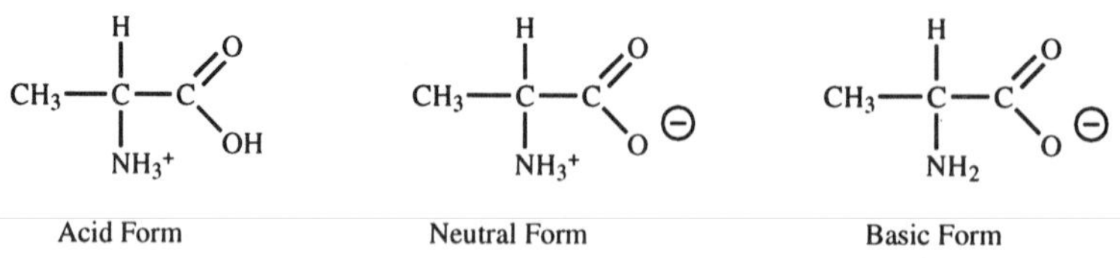

Acid Form    Neutral Form    Basic Form

FIGURE 4.3:
The Three Acid/Base
Forms of Alanine

Calculation of the pH of a solution of a polyprotic acid or base has the additional complication of the 2nd, 3rd, etc., ionizations of the acid or base. Fortunately these complications are easily handled and the calculation is, in practice, no different from that for a monoprotic acid or base in almost all real cases. This is because subsequent ionization steps are always weaker than the first and, therefore, do not contribute significantly to the total hydrogen ion concentration of the solution.

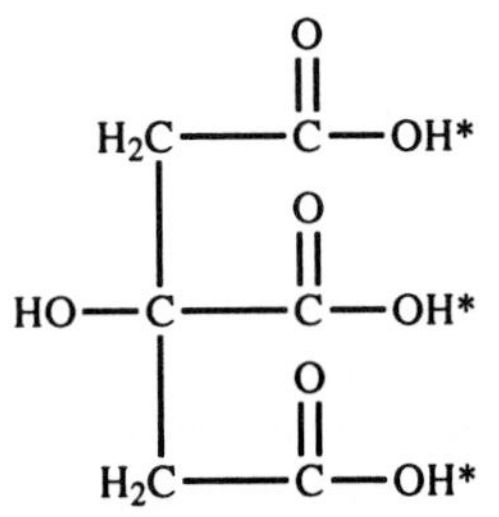

Let us illustrate by calculating the pH of a 0.10 **M** solution of citric acid, a typical weak triprotic acid. The structure of citric acid is given in figure 4.4.

Abbreviating citric acid as $H_3Cit$, we can write the three ionization steps of citric acid as:

$$H_3Cit(aq) \rightleftharpoons H^+ + H_2Cit^-$$ EQ. 4.80

$$H_2Cit^- \rightleftharpoons H^+ + HCit^{2-}$$ EQ. 4.81

$$HCit^{2-} \rightleftharpoons H^+ + Cit^{3-}$$ EQ. 4.82

Let us temporarily consider citric acid as if it were a monoprotic weak acid. Then we can set up the equilibrium table.

**TABLE 4.16** EQUILIBRIUM SOLUTION TABLE

|  | $H_3Cit$ | $H^+$ | $H_2Cit^-$ |
|---|---|---|---|
| init conc | 0.10 | 0 | 0 |
| reacts | $x$ | — | — |
| forms | — | $x$ | $x$ |
| at eq | 0.10 - $x$ | $x$ | $x$ |

Using the results of this table, we then write the equilibrium constant expression and solve it *via* the quadratic equation:

$$K_{a1} = 7.4 \times 10^{-4} = \frac{[H^+][H_2Cit^-]}{[H_3Cit]} = \frac{(x)(x)}{0.10 - x}$$ EQ. 4.83

$$x^2 + (7.4 \times 10^{-4})x - (7.4 \times 10^{-5}) = 0$$ EQ. 4.84

This provides a value of $x = 8.2 \times 10^{-3}$ as the solution to equation 4.84. According to the equilibrium table that we set up above, this value is the concentration of both the hydrogen ion and the dihydrogen citrate ion. If those values are to be correct, then there must be no significant hydrogen ion concentration contributed by the second ionization. If that is true, then the concentration of the monohydrogen citrate ion will have to be negligible compared to the

*Ionizable Proton

FIGURE 4.4:
Structure of Citric Acid

TABLE 4.16

hydrogen ion concentration because the chemical equation for the second ion-ization step shows that a monohydrogen citrate ion is produced every time the second ionization produces a hydrogen ion. We can calculate the concentration of the monohydrogen citrate ion using the equilibrium constant for the second dissociation step:

$$K_{a2} = 1.7 \times 10^{-5} = \frac{[H^+][HCit^{2-}]}{[H_2Cit^-]} = \frac{(8.2 \times 10^{-3})[HCit^{2-}]}{8.2 \times 10^{-3}}$$

EQ. 4.85

Equation 4.85 is easily solved to provide a value of $1.7 \times 10^{-5}$ **M** for the mono-hydrogen citrate concentration, and, of course, the additional hydrogen ion con-centration contributed by the second dissociation. Since $1.7 \times 10^{-5}$ **M** is negli-gible compared to the $8.2 \times 10^{-3}$ **M** contributed by the first step, we were justi-fied in making the assumption that we could treat citric acid as if it were a monoprotic acid for the purpose of calculating the pH of the solution. Finishing the problem by converting the value of $8.2 \times 10^{-3}$ **M** to pH gives our answer of pH = 2.09 for this citric acid solution.

While this assumption will not hold for all polyprotic acids or bases, the actual number of cases in which it does not rigorously hold is so small that we will not be concerned with those in this course. Obviously the assumption becomes better as the difference of the values between the first and second ionization constants becomes larger. In most polyprotic acids or bases, there is an even greater difference between the first and second ionization constant values than is the case for citric acid. Even in those cases where the difference in the val-ues is smaller and the assumption cannot be rigorously made, corrections due to the second and subsequent ionizations are small and can be ignored for all but the most detailed calculations.

## pH OF THE AMPHIPROTIC SPECIES

All dissociation reactions of acids and bases produce the conjugate base or acid of the original species. When dealing with a monoprotic species, the conjugate base of the acid is a base — period, and the conjugate acid of the base is an acid — period; end of discussion. The situation is different for a polyprotic species. Consider the polyprotic acid, phosphoric acid ($H_3PO_4$). There is no question that the conjugate base of phosphoric acid is the dihydrogen phos-phate ion ($H_2PO_4^-$). That ion can act as a base by reacting with a proton to reform phosphoric acid. But it can also donate a proton to a solvent molecule to form the monohydrogen phosphate ion ($HPO_4^{2-}$). In this reaction it is behaving as a Brønsted-Lowry acid. Similarly the monohydrogen phosphate ion can act as a base by accepting a proton to become the dihydrogen phos-phate ion or as an acid by donating a proton and becoming the phosphate ion ($PO_4^{3-}$). The phosphate ion can act only as a base; for all practical purposes the phosphoric acid molecule acts only as an acid. The two intermediate species, the dihydrogen phosphate ion ($H_2PO_4^-$) and the monohydrogen phos-phate ion ($HPO_4^{2-}$), can act both as acids and as bases. A species which can act both as an acid and as a base is called an **amphiprotic** species.

We already know how to calculate the pH of a solution containing either an acid or a base. We simply write an acid or base dissociation reaction, whichever is the appropriate reaction for the species, and then set up the problem from an equilibrium table set up around that reaction. How, then, do we deal with a species that can react in either manner?

To answer this question, let us consider the case of the hydrogen carbonate ion ($HCO_3^-$) and calculate the pH of a solution containing 0.10 **M** $HCO_3^-$. If we write the two chemical reactions this species can undergo, we have

$$HCO_3^- \rightleftharpoons H^+ + CO_3^{2-}$$

EQ. 4.86

$$HCO_3^- + H^+ \rightleftharpoons H_2CO_3$$

EQ. 4.87

Since the bicarbonate ion is a stronger base than water, when one bicarbonate ion donates a proton, that proton is much more likely to be donated to another bicarbonate ion than it is to a water molecule. This means that the number of carbonate ions and the number of unionized carbonic acid molecules must be equal in the solution. This, in turn, means that the concentrations of those two species must be equal. If we now write the two dissociation constant expressions for carbonic acid:

$$K_{a1} = \frac{[H^+][HCO_3^-]}{[H_2CO_3]}$$

EQ. 4.88

$$K_{a2} = \frac{[H^+][CO_3^{2-}]}{[HCO_3^-]}$$

EQ. 4.89

Solving equation 4.88 for the concentration of carbonic acid and equation 4.89 for the concentration of carbonate ion and equating the two expressions gives

$$[H_2CO_3] = \frac{[H^+][HCO_3^-]}{K_{a1}} = K_{a2}\left(\frac{[HCO_3^-]}{[H^+]}\right) = [CO_3^{2-}]$$

EQ. 4.90

Equation 4.90 can be solved for the hydrogen ion concentration:

EQ. 4.91

$$[H^+] = \sqrt{K_{a1}K_{a2}}$$

Notice that, surprisingly, the pH of the solution does not even depend upon the concentration of the species (provided, of course, that the concentration is not so small as to be negligible). An alternate form of equation 4.91 that is quite useful is

EQ. 4.92

$$pH = \frac{pK_{a1} + pK_{a2}}{2}$$

Employing the values of the dissociation constants for carbonic acid (appendix B) provides the answer pH = 8.34 for the bicarbonate problem stated above.

In the case of a tri, tetra, *etc.*, protic acid, the pH of the solution is midway between the two $pK_a$s that involve the species whose solution is being considered. Therefore the pH of a solution of dihydrogen phosphate is given by

$$pH = \frac{pK_{a1} + pK_{a2}}{2}$$

EQ. 4.93

because both $K_{a1}$ and $K_{a2}$ involve the concentration of the dihydrogen phosphate ion, and that of a solution of monohydrogen phosphate is given by

$$pH = \frac{pK_{a2} + pK_{a3}}{2}$$

EQ. 4.94

## 4.8  pH EFFECTS ON DISTRIBUTION OF ACID/BASE SPECIES

It is now time to begin a consideration of the effects of pH on the distribution of species in an acid/base system. In this chapter until now we have assumed that the pH of the solution was determined by the concentration(s) of the acid and/or base under consideration. In this section we shall consider what is probably even more important to our understanding of the effect of pH on reactions, namely the effect that the pH, determined by one acid/base system at high concentration, has on the distribution of the acid and base forms of another, less concentrated, acid/base system. If we begin with the familiar Henderson–Hasselbalch equation,

$$pH = pK_a + \log\left(\frac{[B]}{[A]}\right)$$

EQ. 4.95

in which [B] and [A] refer to the concentrations of the conjugate base and conjugate acid forms of a conjugate acid/base pair respectively, we can rearrange it into a form that is more directly applicable to the present situation.

$$\frac{[B]}{[A]} = 10^{\left(pH - pK_a\right)}$$

EQ. 4.96

This equation clearly shows that the ratio of the base and acid forms of a conjugate acid/base pair depends upon the difference between the pH of the solution and the $pK_a$ of the acid. If we now look at the distribution of forms for the acetic acid/acetate ion system, we find the distribution depicted in figure 4.5. In this figure we note that the unionized acetic acid molecule predominates at $pH < pK_a$ and the acetate ion predominates at $pH > pK_a$. It is only when the pH is close to the $pK_a$ that significant quantities of both species exist in the solution at the same time. When the pH and $pK_a$ are equal (pH = 4.76), the concentrations of the two forms are equal.

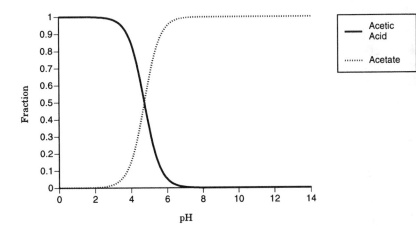

FIGURE 4.5:
Distribution of Species in
the Acetic Acid System

In a polyprotic system, such as any amino acid, the situation is only slightly more complicated. The amino acid alanine is quite typical of such systems. The structure of all three acid/base forms of the alanine system was previously shown in Figure 4.3.

The distribution of the three forms (cation, neutral, and anion) is shown in Figure 4.6. In this situation, since alanine has two $pK_a$s, we see that each species predominates over a range of pH values. The cation predominates at $pH < pK_{a1}$, the anion at $pH > pK_{a2}$, and the neutral species at pH values between the two $pK_a$ values. This figure clearly shows that a reaction requiring the neutral form of alanine, perhaps because it is the only form that properly binds to an enzyme, will proceed with a reasonable rate only between pH 3 and 8. When the pH is outside these values, there is so little of the neutral alanine species present that the reaction rate will be very slow. (This, of course, assumes that the structure of the enzyme catalyzing the reaction is pH–independent. In fact the structure of the enzyme is also pH–dependent. This would further affect the reaction rate.)

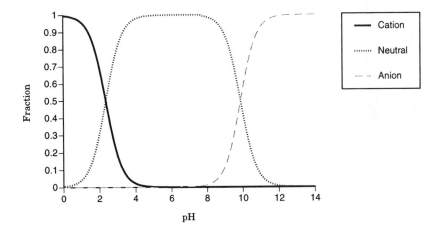

FIGURE 4.6:
Distribution of Species in
the Alanine System

## 4.9   TITRATION CURVES

Once we have a clear understanding of the variation of the forms of an acid/base system with pH, we can begin to understand what happens during an acid/base titration. **Titration** is one of the most important means of analysis from the standpoint of the number of times the procedure is carried out. No doubt you have done many titrations during your time in the chemistry laboratory. In a conventional titration, a known volume of a reagent, the **titrant**, is added to a solution containing the material, the **analate**, whose quantity is to be determined. Addition of the titrant continues until some there is some indication, usually provided by the change in color of a material called an **indicator**, that exactly enough titrant has been added to react with all of the analate. When this happens the titration is said to have come to the **endpoint**. In this section, we wish to use our new knowledge of equilibria to determine exactly how an acid/base titration works.

Let us first consider the titration of a strong acid, such as HCl, with a strong base, NaOH.

The chemical equation for the reaction between the analate and the titrant, written in net ionic form, is

$$H^+ + OH^- \rightleftharpoons H_2O$$

EQ. 4.97

The equivalence point in this titration occurs when exactly the same number of moles of sodium hydroxide (in solution and represented in equation 4.97 as hydroxide ion) have been added as there were initially moles of hydrochloric acid (hydrogen ion in equation 4.97). At this point in the titration, the solution is essentially one of sodium chloride dissolved in water. Since neither sodium ion nor chloride ion has any significant Brønsted acid/base properties, the pH of the solution at the equivalence point is 7. If the pH of the solution is calculated after addition of various volumes of the titrant both before and after the equivalence point, a procedure that is tedious but not difficult, a plot of pH vs. the volume of titrant added to the solution looks like this:

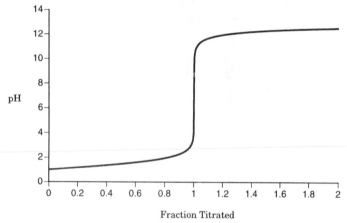

FIGURE 4.7:
Titration Curve for an
HCl *vs.* NaOH Titration

The distinctive S–shape is the result of plotting a logarithmic value (pH) against a linear value (volume of titrant). If the dilution that occurs because of the addition of the titrant is ignored, prior to the endpoint the hydrogen ion concentration decreases linearly with the addition of the titrant. After the end-point, the hydroxide ion concentration changes linearly with the volume of additional titrant. In both cases this causes a logarithmic change in the pH. The result is the very rapid change in the pH, centered about the equivalence point. It is this rapid change in pH that occurs over a very small volume change that makes possible the use of a visual indicator to determine the end-point of the titration.

A visual indicator is simply a weak acid or base in which the acid and base forms are different colors. The most commonly used indicator for the HCl/NaOH titration is phenolphthalein (which, incidentally, is the active ingredient in many, recently banned, laxatives). This particular weak acid has a $pK_a$ value of approximately 10. A species diagram for phenolphthalein, along with the structures of the acid and base forms of phenolphthalein and their colors, is shown in figure 4.8.

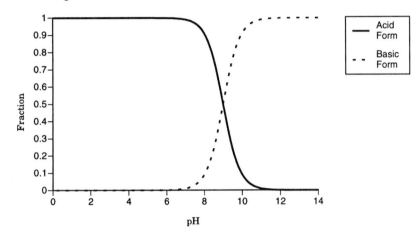

acidic form - colorless

basic form - pink

FIGURE 4.8:
Phenolphthalein Forms in
Solution as a Function of pH

In acidic solution, for instance at pH 6, there is so little of the basic form of the indicator present that the solution appears colorless. As the pH of the solution increases, especially as it rises above pH 8, the fraction of the indicator in the basic form increases until the solution begins to appear red to the eye. Note that, the solution does not appear red to the chemist until approximately pH 8. This means that the endpoint will occur slightly later than the true equivalence point. This is of no real consequence, however, because, in almost all real situations, the volume of NaOH that is required to change the pH from the equivalence point (pH 7) to the endpoint (approximately pH 8) is so small that it is not possible to read the buret accurately enough to tell the difference.

Phenolphthalein is a typical indicator in all respects except one — one of its forms is colorless. This means that the normal eye can detect much smaller concentrations of the one (the colored) form than would be possible if both forms were colored. For the average indicator the color change is noticeable for approximately 1 pH unit on either side of the $pK_a$ of the indicator. There are a wide variety of indicators with color change regions that span the entire normal pH range for 0 to 14. In addition many different color changes are possible. This is fortunate, not only for helping persons with various types of color–blindness, but also because not all acid/base titrations have an equivalence point pH of 7.

The titration curve of a weak acid with a strong base differs from the HCl/NaOH titration curve in several important ways. All of these differences occur because one of the products formed in the titration reaction, the conjugate base of the weak acid being titrated, does affect the pH of the solution. You will recall that the conjugate base of the strong acid (chloride ion in the HCl/NaOH example above) did not affect the pH in the strong acid/strong base titration. Therefore we did not need to account for its presence in solution at any time during the titration. However, in a weak acid/ strong base titration we must take the presence of the conjugate base into consideration at titrant volumes prior to and at the equivalence point. After some titrant has been added but still prior to the equivalence point, the solution contains significant quantities of the conjugate base along with the unreacted weak acid. This means that the solution is a buffer solution and its pH may be calculated using the Henderson–Hasselbalch equation. At the equivalence point, the pH is affected by the presence of the conjugate base — a weak base in its own right. Thus the pH of the solution is not 7 at the endpoint. Rather it is somewhat above 7.

Once again the calculations required to determine the pH at various points along the titration curve are tedious but not difficult. They are left for the student who wishes to try them out for himself or herself. The result of these calculations is a titration curve shown as figure 4.9. This titration curve is typical of that for a weak acid and was developed for an acid with a $pK_a$ of 6. Note that it differs from the HCl/NaOH curve at three points: (1) the beginning pH is higher, reflecting the fact that the acid is not completely ionized; (2) over the middle part of the titration curve the pH at first rises rather sharply, then levels off, then finally begins another rather sharp rise. This reflects the formation of, operation of, and destruction of the buffer. And, (3) the pH at the equivalence point is greater than 7, reflecting the fact that the species in the solution at this point is a weak base.

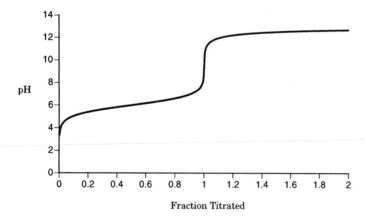

FIGURE 4.9:
Titration Curve for a
Weak Acid

Although the "break" at the endpoint is not as pronounced in this titration as it was in the HCl/NaOH titration, there is still a region of sharply rising pH at the endpoint that would be sufficient to use for endpoint determination. In fact, because the pH at the equivalence point is higher than 7, the phenolphthalein color change corresponds even more closely to the equivalence point in this titration than it did in the HCl/NaOH titration. There is one other point of note on this titration curve: the pH at the halfway titrated point (fraction titrated = 0.5) is numerically equal to the $pK_a$ of the acid.

## 4.10   PRACTICAL PROBLEM-SOLVING STRATEGIES

Perhaps the biggest problem that students have in solving acid/base equilibrium problems is in classifying the problem into the proper category in order to apply the problem-solving techniques to it. In some senses this is a matter of recognizing the species in the statement of the problem as weak or strong acids or bases. In this regard there is little that can be said except to recommend that the student become thoroughly familiar with naming conventions in use in chemistry so that a salt such as sodium acetate is recognized as the salt of a strong base and a weak acid and, therefore, that the pH determining species is the acetate ion: a weak base. Likewise ethylamine hydrochloride must be recognized as the salt of a strong acid and a weak base and the pH determining species is the ethylammonium ion which is a weak acid.

Of somewhat more difficulty, however, is the situation when two or more species are present in solution at the beginning of the problem. This presents no problem if they are related to one another as a conjugate acid/base pair, for the problem should then be immediately recognized as being a buffer problem. If the acid and base are not related as a conjugate acid/base pair, then consideration must be given to reaction of the acid with the base until what remains in the solution is only one or two acids, one or two bases, or a conjugate acid/base pair. *The only time that one may leave both an acid and a base in solution without further reacting them is if they are a conjugate acid/base pair.*

### CONCEPT CHECK

**Problem:** Calculate the pH of a solution made by mixing 30.00 ml of 0.100 **M** NaOH with 50.00 ml of 0.0500 **M** acetic acid.

In this problem we find an acid and a base described. These are not a conjugate acid/base pair, therefore it must be considered that they will react before the problem is attempted as an equilibrium problem. The net ionic equation for the chemical reaction will be

$$HAc + OH^- \longrightarrow Ac^- + H_2O$$

EQ. 4.98

To finish this problem we need only determine the amount of sodium hydroxide and acetic acid present, then determine what would be present when the reaction goes to completion. There are three possibilities:

**1.** There is more acetic acid present than sodium hydroxide. In this case there will be excess acetic acid in a solution which contains acetate ion. This solution contains both an acid and its conjugate base and the problem is, then, a buffer problem.

**2.** There is exactly the same quantity of acetic acid as there is sodium hydroxide. In this case there is only acetate ion in the final solution and the problem is one of determining the pH of a solution of a weak base.

**3.** There is more sodium hydroxide present than there is acetic acid. In this case the final solution contains acetate ion and excess hydroxide ion. These are both bases; one is a strong base and the other a weak base. Although we have not looked at such a situation in the past, in most cases the presence of the weak base can be ignored and the problem can be solved by assuming that only the strong base is present.

In the current situation we find that there are $(30.00)(0.100) = 3.00$ mmoles of sodium hydroxide and $(50.00)(0.0500) = 2.50$ mmoles of acetic acid present before reaction. After reaction this leaves us with 2.50 mmoles of acetate ion and 0.50 mmoles of excess hydroxide ion. The concentrations of acetate and hydroxide will be 2.50/75.00 $= 0.0333$ **M** and 0.50/75.00 $= 0.00667$ **M**, respectively. Since the hydroxide ion is in excess, we can simply conclude the problem by determining the pH of a solution with this concentration of hydroxide ion:

$$pOH = -\log [OH^-] = -\log (0.00667)$$
$$pOH = 2.18; \ pH = 14.00 - pOH = 11.82$$

EQ. 4.99

Here is a second problem that we could use to illustrate this situation:

**Problem:** Calculate the pH of a solution made by mixing 25.00 ml of 0.200 **M** phosphoric acid and 50.00 ml of 0.250 **M** sodium hydroxide.

In this problem we find that we have $(25.00)(0.200) = 5.00$ mmoles of $H_3PO_4$ and $(50.00)(0.250) = 12.5$ mmoles of NaOH.

The equation for the first reaction that will occur between this acid and base is

$$OH^- + H_3PO_4 \longrightarrow H_2O + H_2PO_4^-$$

EQ. 4.100

After this reaction proceeds the solution will contain 7.5 mmoles of $OH^-$ and 5.00 mmoles of $H_2PO_4^-$. These are a base and an acid, respectively. They will react as represented by equation 4.101.

$$OH^- + H_2PO_4^- \longrightarrow H_2O + HPO_4^{2-} \qquad \text{EQ. 4.101}$$

This reaction will leave the solution containing 2.5 mmoles of $OH^-$ and 5.00 mmoles of $HPO_4^{2-}$. Again, we have an acid and a base which will further react *via* the following equation:

$$OH^- + HPO_4^{2-} \longrightarrow H_2O + PO_4^{3-} \qquad \text{EQ. 4.102}$$

This reaction will leave a solution containing 2.5 mmoles each of $HPO_4^{2-}$ and $PO_4^{3-}$. Since these make up a conjugate acid/base pair the problem is now that of finding the pH of a buffer solution containing 2.5 mmoles each of $HPO_4^{2-}$ and $PO_4^{3-}$ in a total volume of 75.00 ml. Utilizing the Henderson-Hässelbalch equation gives us the result:

$$pH = pK_{a3} + \log\left(\frac{\left[PO_4^{3-}\right]}{\left[HPO_4^{2-}\right]}\right) = 12.32 + \log\left(\frac{2.50}{2.50}\right) = 12.32 \qquad \text{EQ. 4.103}$$

We shall accept this answer as valid even though we realize that the assumptions on which this equation is based are not entirely accurate. Notice that, because the log term involved a ratio, as long as both the numerator and denominator were in the same units, the units dropped out and we could use the number of moles of each without having to convert the values into actual concentrations.

## 4.11   pH IN NON-AQUEOUS SOLUTIONS

All of the discussions thus far in this chapter have concerned happenings in aqueous solution. While quite important for many processes, indeed probably for most chemical processes, a discussion of acidity and basicity would not be complete without some mention of what happens in non–aqueous solutions. Is there some analog to 'pH' in non–aqueous solution? How do we discuss the ability of the solvent to contribute protons to or extract protons from some species dissolved in the solution? These are the questions that we shall answer in this section.

First we must recall that this topic has been previously discussed, although very briefly. When we first asked the questions, "What is an acid?" and "What is a base?" in chapter 12 last year, we discussed the solvent theory of acids and bases and saw some examples of strange–looking acids and bases. Of more immediate interest and importance, we also discussed Lewis Acid/Base Theory — a theory which is not solvent-based. While these are interesting and useful in some ways, neither the Solvent Theory nor the Lewis Theory are well–suited to discuss the ability of a solvent to supply or absorb protons in a chemical reaction. Brønsted-Lowry Theory, with its concept of acid/base reactions being proton–transfer reactions, is much more useful in this situation. However, although we shall use some of the aspects of Brønsted-Lowry Theory in our discussion, we shall concentrate principally upon the development of a working definition of pH in situations for which our previous definition of pH is inadequate.

The definition of pH expressed in equation 4.20 works well in aqueous solutions provided that the concentration of hydrogen ion is 0.01 **M** or less. A major reason for this seems to be that, at higher hydronium ion concentrations, a significant number of water molecules are involved in solvating the protons and the "concentration" of water can no longer be assumed to be that of pure water. (There are other reasons for this problem as well. They may be taken up in later chemistry courses.)

A number of semi–empirical equations that work well can correct for these problems until the concentration of hydronium ions reaches approximately 1 **M**. At higher concentrations even the best of these equations begin to break down and can not predict accurately the pH of the solution, *i.e.*, ability of the solvent to protonate materials dissolved in it.

At very high concentrations of strong acids a working definition of pH can be obtained only empirically. The **Hammett Acidity Function** does this quite adequately for the transfer of a proton to an uncharged weak base. While this was used initially in solvents that were mixtures of concentrated sulfuric acid and water, the concept has been extended to completely non-aqueous, high-dielectric-constant, organic solvents as well.

To understand how the Hammett acidity function is determined, let us consider the reaction:

$$BH^+ \rightleftharpoons B + H^+ \qquad\qquad \text{EQ. 4.104}$$

where B is some uncharged weak base such as *p*–nitroaniline. Hammett defined the acidity function, $H_o$, as:

$$H_o = pK_a + \log\left(\frac{[B]}{[BH^+]}\right) \qquad\qquad \text{EQ. 4.105}$$

Note the similarity between this equation and the Henderson-Hasselbalch equation (eq. 4.68). Indeed, the two are very similar and the Hammett equation becomes the Henderson-Hasselbalch equation in low-concentration, aqueous solutions. However, equation 4.105 permits the measurement of $H_o$ under conditions that produce a large error in pH measurement. For reasons that are beyond the scope of this course, the equation works only with uncharged bases. Operationally, a base whose $pK_a$ in water is well-established is used to measure the value of $H_o$ in a relatively dilute sulfuric acid/water solution (one in which the Henderson-Hasselbalch equation still works well.) This is done by measuring the concentrations of both B and BH$^+$, usually with a spectrophotometer. A weaker base is then placed into the same sulfuric acid/water solution and the concentrations of its basic and acidic forms are also measured. This allows the calculation of the p$K_a$ of the weaker base. This second base can then be placed into a more concentrated sulfuric acid/water solution (one in which the Henderson-Hasselbalch equation does not work well) to determine the value of $H_o$ for that solution. This technique can be extended to even weaker bases and more concentrated sulfuric acid/water solutions, eventually ending with pure sulfuric acid. If necessary the Hammett acidity function can be measured in other, non-aqueous solvents by the same technique. This approach to the empirical measurement of a pH analog in non-aqueous solution works well provided that uncharged bases are used and the solvent has a relatively high dielectric constant.

Once the Hammett acidity function value for a particular solvent is known, this solvent can be used to determine the dissociation constants of very weak, uncharged bases. These measurements would be extremely difficult, if not impossible to carry out, in aqueous solution. The values obtained are important in helping to establish relative basicities for series of compounds. These can then help to predict relative reactivities of the compounds in reactions that depend upon the basicity to influence the outcome of the reaction.

**Amino Acid:** a species having both a carboxyl and an amino group in the same molecule. Naturally occurring amino acids have both groups attached to the same carbon atom. They are the building blocks of proteins.

**Amphiprotic:** describing a species that can either donate or accept a proton.

**Analate:** in an analytical procedure, the material whose quantity or identity is being determined.

**Autoprotolysis:** a chemical reaction in which one molecule transfers a proton to a chemically identical molecule.

**Brønsted-Lowry Theory/Reaction:** an acid/base theory that considers an acid to be a proton donor and a base to be a proton acceptor. Acid/base reactions consist of the transfer of a proton from an acid to a base to form a new acid/base pair.

**Buffer:** a term applied to a solution that contains relatively large quantities of an acid and its conjugate base. Such solutions resist changes in pH upon the addition of small quantities of other acids or bases.

**Conjugate Acid/Base Pair:** a term applied to two chemical species that differ by only one proton.

**Dissociation Constant:** the equilibrium constant that describes the extent to which a species dissociates when dissolved in a solvent.

**Endpoint:** that point in a titration at which the chemist believes is the equivalence point. This is normally indicated by a change in the color of an indicator.

**Equilibrium:** a chemical reaction state reached when concentrations of reactants and products no longer change even if an infinite time is allowed for them to do so.

**Equilibrium Constant:** a mathematical expression relating the concentrations of all species in a chemical reaction at the time at which there is no longer any change in those concentrations.

**Equivalence point:** that point in a titration when the number of moles of titrant added is exactly equal to the number of moles of analate in solution.

**Hammett acidity function:** an empirical function able to predict the ability of a solvent to transfer a proton to an uncharged weak base in high–acidity, high–dielectric–constant solvents. This provides an operational equivalent of a pH scale in the solvent.

**Henderson–Hasselbalch Equation:** an equation that permits the calculation of the pH of a buffer.

**Indicator:** a compound, added in very small quantities to a titration flask, that signals the end of the titration by a color change.

**p—:** the negative log (base 10) of the term following the 'p'.

**pH:** the negative log (base 10) of the hydrogen ion concentration. Used to describe the acidity of a solution.

**Polyprotic Acid/Base:** an acid or base that has more than one site at which a proton may bind.

**Solubility Product:** the equilibrium constant describing the dissolution of an ionic species.

**Strong Acid/Base/Electrolyte:** an acid, base, or electrolyte that dissociates completely into ions when dissolved in a solvent.

**Titrant:** in a titration, the solution that is in the buret. Usually the concentration of this solution is precisely known.

**Titration:** an analytical procedure in which a solution of known concentration is added to a solution containing a substance whose concentration is to be determined until some indication is given that the reaction is complete.

**Weak Acid/Base/Electrolyte:** an acid, base, or electrolyte that only partially dissociates into ions when dissolved in a solvent.

1. Calculate the solubility of CuI in pure water.

2. Calculate the solubility of $Ba(IO_3)_2$ in pure water.

3. A very concentrated solution of LiCl is added slowly to a solution initially containing 0.10 **M** $Hg_2^{2+}$ and 0.20 **M** $Ag^+$. (You may ignore dilution effects in this problem.)

       a. What salt precipitates first?
       b. What is the concentration of chloride when that first salt just begins to precipitate?
       c. What is the concentration of chloride when the second salt begins to precipitate?
       d. What fraction of the first metal has precipitated when the second begins to precipitate?

4. Using radio-isotope techniques, it was found that a saturated solution of $BiPO_4$ in distilled water contains 1.096 ng ($1.096 \times 10^{-9}$ g) of bismuth phosphate per liter of solution. Calculate the $K_{sp}$ for bismuth phosphate.

5. Calculate the pH of the following solutions:

    a.  0.10 **M** $HNO_3$       b.  0.010 **M** KOH
    c.  0.20 **M** $Ba(OH)_2$     d.  $1.0 \times 10^{-4}$ **M** HCl

6. Calculate the pH of the following solutions:

    a. 0.25 **M** Acetic Acid       b. 0.010 **M** Boric Acid
    c. 0.0010 **M** Chloroacetic Acid    d. 0.050 **M** Phenol
    e. $1.0 \times 10^{-4}$ **M** Sulfamic Acid   f. 0.25 **M** Hydrocyanic Acid

7. Calculate the pH of the following solutions:

    a. 0.25 **M** Ammonia       b. 0.010 **M** Piperidine
    c. 0.0010 **M** Pyridine      d. 0.050 **M** Triethylamine
    e. $1.0 \times 10^{-4}$ **M** Hydrazine    f. 0.25 **M** Ethanolamine

8. Calculate the pH of the following solutions:

    a. 0.25 **M** Malonic Acid       b. 0.010 **M** Arsenic Acid
    c. 0.0010 **M** Citric Acid      d. 0.050 **M** Tartaric Acid
    e. $1.0 \times 10^{-4}$ **M** Hydrosulfuric Acid   f. 0.25 **M** Oxalic Acid

9. Calculate the pH of the following solutions:

    a. 0.25 **M** Ethylenediamine    b. 0.010 **M** Sodium Acetate
    c. 0.0010 **M** Ammonium chloride  d. 0.050 **M** Trisodium phosphate
    e. $1.0 \times 10^{-4}$ **M** Sodium sulfite   f. 0.25 **M** Sodium picrate

10. Calculate the pH of the following solutions:

   a. 0.25 **M** Isoleucine      b. 0.010 **M** Sodium Bicarbonate
   c. 0.0010 **M** Proline       d. 0.050 **M** Disodium phosphate

11. Calculate the pH of a solution saturated with $Ca(OH)_2$.

12. Calculate the solubility of $Al(OH)_3$ in a solution buffered at pH 6.50.

13. Calculate the pH of a solution containing 0.10 **M** acetic acid and 0.050 **M** sodium acetate.

14. Calculate the pH of a solution containing 0.050 **M** ethanolamine and 0.025 **M** ethanolammonium chloride.

15. What volume of 0.50 **M** NaOH must be added to 100.0 mL of 0.10 **M** propionic acid to make a solution whose pH is 4.95?

16. What volumes of 0.50 **M** lactic acid and 0.25 **M** NaOH must be mixed to produce 250.0 mL of a solution with a pH of 4.00?

17. Calculate the pH of a solution made by mixing equal volumes of 0.15 **M** phosphoric acid and 0.25 **M** NaOH.

18. Calculate the pH of a solution made by mixing equal volumes of 0.10 **M** aspartic acid and 0.20 **M** KOH.

19. Calculate the pH of a solution made by mixing 50.0 mL of 0.15 **M** oxalic acid and 25.0 mL of 0.50 **M** NaOH.

20. A 0.200 **M** solution of the amino acid arginine is buffered to pH = 8.00. Calculate the concentrations of all species in solution (*i.e.* $[H_3Arg^{2+}]$, $[H_2Arg^+]$, $[HArg]$, $[Arg^-]$, $[H^+]$, and $[OH^-]$.

*21. Calculate the solubility of $Ag_3PO_4$ in a solution buffered at pH 11.50.

*22. It is desired to separate copper and nickel ions by selective precipitation using sulfide ion. Assuming that both $Cu^{2+}$ and $Ni^{2+}$ are present in solution at a concentration of 0.1 **M**:
   a. at what pH would CuS just begin to precipitate in a solution saturated with $H_2S$? (You may assume that $[H_2S] = 0.12$ in a solution saturated with $H_2S$.)
   b. at what pH would NiS just begin to precipitate?
   c. what would the concentration of the copper and nickel ions be if the solution were buffered at a pH midway between your answers for parts (a) and (b)?

23. Consider the inductive effect on the relative strengths of α-chloropropionic acid and β-chloropropionic acid.

   a. Draw the structures of the two acids, and explain which is expected to be the stronger acid based upon the inductive effect.

   b. The $pK_a$ of the alpha and beta chloropropionic acids are 2.83 and 3.98 respectively. Calculate the corresponding $K_a$ values.

   c. What would be the pH of a 0.100 $M$ solution of α-chloropropionic acid?

   d. What would be the pH of a 0.100 $M$ solution of β-chloropropionic acid?

24. Consider the inductive effect on the relative strengths of bromoacetic acid and iodoacetic acid.

   a. Draw the structures of the two acids, and explain which is expected to be the stronger acid based upon the inductive effect.

   b. The $pK_a$ of bromo and iodo acetic acids are 2.69 and 3.12 respectively. Calculate the corresponding $K_a$ values.

   c. What would be the pH of a 0.100 $M$ solution of bromoacetic acid?

   d. What would be the pH of a 0.100 $M$ solution of iodoacetic acid?

25. Consider the relative acidity of cyclohexanol versus phenol.

   a. Draw the structure of each compound, and the structure of the conjugate base formed by acid dissociation.

   b. Show resonance delocalization of charge in the conjugate base by drawing the possible resonance forms.

   c. Phenol has a $pK_a$ of 9.99, and cyclohexanol has a $pK_a$ of 16. Calculate the $K_a$ of both compounds, and explain this relative to your results in (a) and (b).

   d. An aqueous solution of 0.100 $M$ cyclohexanol has an experimental pH of essentially 7.0. Calculate the value of the pH of this solution from the $K_a$ of cyclohexanol determined in (c), and explain the difference between that value and the experimentally measured value.

26. The standard state molar free energy of formation of $I_2$ gas is 19.327 kJ/mol. Consider the reaction where one mole of hydrogen gas and one mole of iodine gas react to form two moles of hydrogen iodide gas.

   a. Write the balanced reaction.

   b. Determine the standard state free energy change and the equilibrium constant for the reaction in (a) at 25°C.

   c. Initially, 2.00 moles of hydrogen and 2.00 moles of iodine gas are reacted at 25°C in a 10.0 liter vessel. Some time later, the concentration of iodine is measured to be 0.198 molar. At that time, what are the concentrations of hydrogen and of hydrogen iodide?

   d. What is the value of Q at the time specified in (c)?

   e. What is the value of ΔG for the reaction at the time specified in (c)?

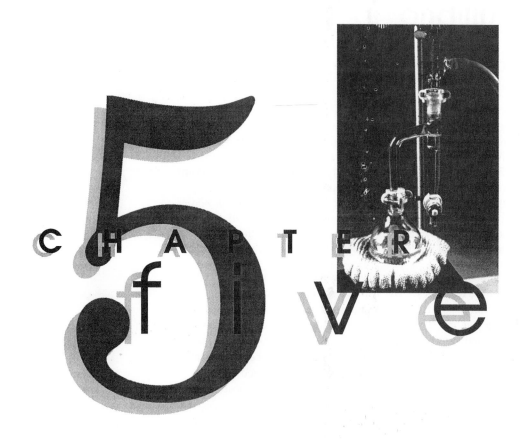

# SYNTHETIC Equilibria

# Synthetic Equilibria

## 5.1    BACKGROUND

**I**n chapter four you were introduced to the equilibrium constant and its relationship to the standard state free energy change undergone when reactants are converted to products. The magnitude of the standard state free energy change, and thus the value of the equilibrium constant, determines how far toward completion a given reaction will go. All chemical reactions, including those responsible for the regulation of the many biochemical processes in our bodies, must obey the laws of chemical equilibrium. In this chapter, we will apply equilibrium principles to the synthesis of compounds to choose the conditions so that we obtain the desired product in the best possible **yield**.

Two reactions will be used to illustrate how we can often manipulate reaction conditions to favor the reaction we want. The first reaction, presented in equation 5.1, is the reaction in which an **ester** and water are formed from a carboxylic acid and an alcohol. The second reaction, illustrated by equations 5.2 and 5.3, illustrate reactions between an alcohol and either an aldehyde or a ketone. The products of these two reactions are termed **hemiacetals** or **acetals** depending upon whether one or two moles of alcohol have reacted with the carbonyl compound. Earlier terminology distinguished between those compounds formed from a ketone and alcohol from those formed from an aldehyde and alcohol by calling the former compounds **hemiketals** or **ketals**. Current IUPAC nomenclature uses the term "acetal" or "hemiacetal" for both types of compounds. In both cases we shall consider both the forward and reverse reactions, *i.e.*, both the formation and degradation of the products. Emphasis will be placed upon using equilibrium considerations to improve the yield of the desired compound.

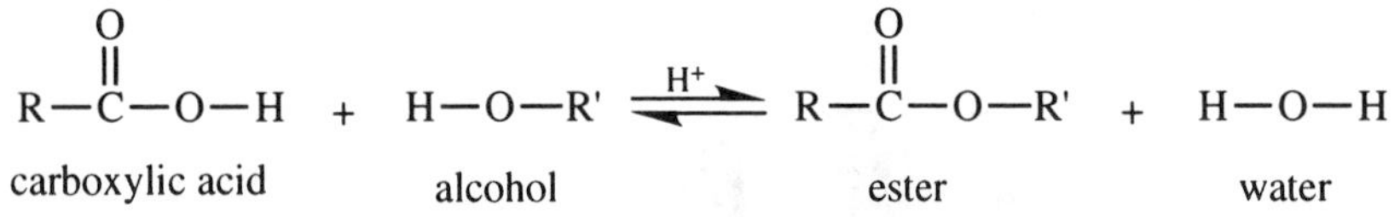

$$
\underset{\text{carboxylic acid}}{R-\overset{\overset{\displaystyle O}{\|}}{C}-O-H} \;+\; \underset{\text{alcohol}}{H-O-R'} \;\underset{}{\overset{H^+}{\rightleftarrows}}\; \underset{\text{ester}}{R-\overset{\overset{\displaystyle O}{\|}}{C}-O-R'} \;+\; \underset{\text{water}}{H-O-H} \qquad \text{EQ. 5.1}
$$

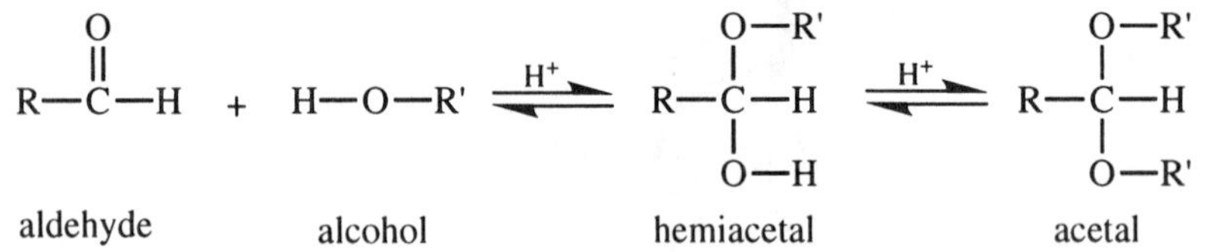

$$
\underset{\text{aldehyde}}{R-\overset{\overset{\displaystyle O}{\|}}{C}-H} \;+\; \underset{\text{alcohol}}{H-O-R'} \;\overset{H^+}{\rightleftarrows}\; \underset{\text{hemiacetal}}{R-\overset{O-R'}{\underset{O-H}{C}}-H} \;\overset{H^+}{\rightleftarrows}\; \underset{\text{acetal}}{R-\overset{O-R'}{\underset{O-R'}{C}}-H} \qquad \text{EQ. 5.2}
$$

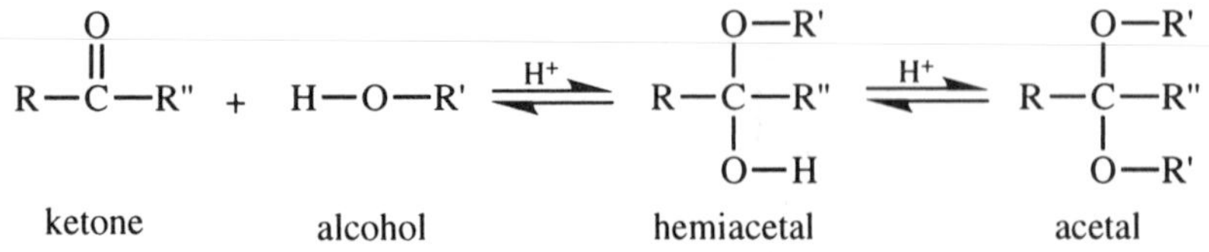

$$
\underset{\text{ketone}}{R-\overset{\overset{\displaystyle O}{\|}}{C}-R''} \;+\; \underset{\text{alcohol}}{H-O-R'} \;\overset{H^+}{\rightleftarrows}\; \underset{\text{hemiacetal}}{R-\overset{O-R'}{\underset{O-H}{C}}-R''} \;\overset{H^+}{\rightleftarrows}\; \underset{\text{acetal}}{R-\overset{O-R'}{\underset{O-R'}{C}}-R''} \qquad \text{EQ. 5.3}
$$

One of the common methods for preparing esters is to react a carboxylic acid and alcohol in the presence of strong acid catalyst. This reaction is normally done at elevated temperature, often at **reflux**, in order to increase the rate at which the reaction occurs. This reaction is known as the **Fischer esterification** reaction in honor of Emil Fischer (1852–1919) who first studied the reaction in detail in the 19th century. The reaction is carried out in the presence of a small amount of a strong acid (usually sulfuric acid) as a catalyst.

The equilibrium constant for the reaction can be written as

$$K = \frac{[\text{ester}][\text{water}]}{[\text{carboxylic acid}][\text{alcohol}]}$$

EQ. 5.4

The values of the equilibrium constants for the formation of the acetate esters of alcohols containing 1-4 carbon atoms are given in table 5.1. Let us see what the magnitude of the equilibrium constants tells us about the relationship between the structure and the equilibrium constants for the synthesis of the esters formed by these alcohols.

**TABLE 5.1**  EQUILIBRIUM CONSTANTS FOR SYNTHESIS OF ACETATE ESTERS

TABLE 5.1

| Alcohol | K |
|---------|------|
| methyl | 5.24 |
| ethyl | 3.96 |
| *n*–propyl | 4.07 |
| isopropyl | 2.35 |
| *n*–butyl | 4.24 |
| *t*–butyl | 0.0049 |

Note especially that the equilibrium constant for the formation of *t*-butyl acetate is so small that it would be impractical to attempt to synthesize it in this manner. This is common when working with tertiary alcohols and these alcohols require other methods for the synthesis of their esters. This table also indicates that esterification reactions for primary alcohols have the larger equilibrium constants than those for secondary alcohols. However even the values of the equilibrium constants for the primary alcohols are small enough that poor yields are obtained unless we apply the lessons learned in chapter four to increase the yield. For example, consider the reaction between acetic acid and methyl alcohol. This reaction has the largest equilibrium constant of those listed in table 5.1. It is, therefore, the reaction that is most product–favored by the equilibrium constant, and should be the one from which we should get the best yield.

$$CH_3-\overset{\overset{\displaystyle O}{\|}}{C}-O-H \ + \ CH_3-O-H \ \underset{\longleftarrow}{\overset{H^+}{\longrightarrow}} \ CH_3-\overset{\overset{\displaystyle O}{\|}}{C}-O-CH_3 \ + \ H_2O \qquad\qquad \text{EQ. 5.5}$$

acetic acid          methyl alcohol          methyl acetate          water

We would like reaction conditions that result in an **actual yield** for the production of the ester that is at least 90% of the **theoretical yield**. Thus if we start with 1 mole each of acetic acid and methyl alcohol, we wish to obtain at least 0.9 mole each of the ester and water. For a number of reasons, not all of them related to equilibrium considerations, organic reactions rarely achieve the theoretical yield. Let us first consider the effect of equilibrium on the potential yield for this reaction before we go on to determine how we could improve the yield by altering the conditions.

First, let us determine what yield we would expect if we react equimolar quantities of acid and alcohol. To make calculations easier, let us react 1.00 mole each of acetic acid and methyl alcohol in a total volume of solution, V. Setting up the equilibrium table as in chapter four, we obtain for the reaction:

TABLE 5.2   EQUILIBRIUM TABLE

TABLE 5.2

| | acid | alcohol | ester | water |
|---|---|---|---|---|
| init. conc | 1.00/V | 1.00/V | 0 | 0 |
| reacts | $x$/V | $x$/V | — | — |
| forms | — | — | $x$/V | $x$/V |
| at equil. | (1.00–$x$)/V | (1.00–$x$)/V | $x$/V | $x$/V |

Placing these values into the equilibrium constant expression provides equation 5.6:

$$K = 5.24 = \frac{\left(\dfrac{x}{V}\right)\left(\dfrac{x}{V}\right)}{\left(\dfrac{1.00-x}{V}\right)\left(\dfrac{1.00-x}{V}\right)} \qquad\qquad \text{EQ. 5.6}$$

Equation 5.6 can be easily reduced to equation 5.7 by recognizing that the reactants and products are all contained in a common volume (V) that can be factored out of the expression to give

$$K = 5.24 = \frac{(x)(x)}{(1.00-x)(1.00-x)} \qquad\qquad \text{EQ. 5.7}$$

Solution of this equation yields an answer of $x = 0.70$ mol. Thus we note that 0.70 mol of ester and water can be formed with 0.30 mol acid and alcohol left unreacted. If the reaction had gone to completion we would have expected that 1.00 mole of each of the products would have been formed; this means that the potential yield is only 70% of the theoretical yield.

## 5.3 ACID-CATALYZED ESTER FORMATION: MAXIMIZING YIELD

By LeChatelier's principle we would predict that amounts of ester would increase if the equilibrium were shifted by adding excess of acetic acid or methyl alcohol. Usually the less expensive reagent, in this case methyl alcohol, is used in excess. Let us increase the methyl alcohol quantity from 1.00 to 5.00 mol and recalculate the potential yield.

**TABLE 5.3**   EQUILIBRIUM TABLE

TABLE 5.3

|              | acid           | alcohol        | ester   | water   |
|--------------|----------------|----------------|---------|---------|
| init. conc.  | 1.00/V         | 5.00/V         | 0       | 0       |
| reacts       | $x$            | $x$            | —       | —       |
| forms        | —              | —              | $x$     | $x$     |
| at equil.    | (1.00–$x$)/V   | (5.00–$x$)/V   | $x$/V   | $x$/V   |

Placing these values into the equilibrium constant expression and canceling out the common V term produces equation 5.8:

$$K = 5.24 = \frac{(x)(x)}{(1.00 - x)(5.00 - x)}$$

EQ. 5.8

Thus 0.96 mol of the ester and water could be produced, leaving 0.04 mol of acid and 4.04 mol of the alcohol unreacted. This is 96% of the theoretical yield of 1.00 mol.

This is a satisfactory method for synthesis of many esters. Before we go on to consider some of the other methods that may also be used, let us examine the apparatus used to perform this synthesis of methyl acetate. Because the rate at which the reactants reach equilibrium would be very slow at room temperature, the reactants are heated to boiling. If this were done in an open flask, eventually all of the solution would boil away — reactants and products. In order to prevent the major loss of liquid and still be able to keep the solution at the boiling point, we use a flask to which a condenser is attached so that the vapors are condensed and returned to the reaction vessel. The term for the process we are using is reflux. The apparatus in which it is carried out is pictured in figure 5.1.

FIGURE 5.1:
Reflux Apparatus

LeChatelier's principle also suggests that yields may be improved by removing the products as they are formed. In the current situation we might try displacing the equilibrium by removal of the ester and/or water formed in the reaction. If the boiling point of the ester and/or water is enough lower than the boiling point of the acid and alcohol, one or the other (or both) may be removed using the distillation apparatus shown in figure 5.2.

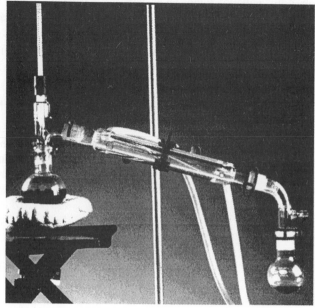

FIGURE 5.2:
Distillation Apparatus

This method of shifting the equilibrium in favor of the products works best when a high molecular weight acid and a high molecular weight alcohol are reacted. Because of the hydrogen bonding interactions present in acids and alcohols, the product ester usually boils at a temperature far enough below that of the reactants that it can be removed by distillation without also removing significant quantities of the reactants. Fortunately for those situations involving low molecular weight acids and/or alcohols, water can also be removed another way. We might think that one way to remove water would be to add one of the common drying agents such as $CaCl_2$ or $Na_2SO_4$ to the refluxing solution. These drying agents, as well as most others, work because of the conversion of the solid anhydrous salt to a solid hydrate. A chemical equation for the conversion of anhydrous calcium chloride to its hydrate is

$$CaCl_2 \;+\; x\,H_2O \;\longrightarrow\; CaCl_2 \cdot x\,H_2O$$

solid       liquid       solid (removed by decanting)

EQ. 5.9

The value of $x$ varies with the amount of water present. Each mole of anhydrous calcium chloride is capable of taking up to six moles of water from the system. However, while this may work well for removing unwanted water from product mixtures at room temperature, this does not work at reflux temperatures with many drying agents because the equilibrium for the formation of metal salt hydrates is shifted back to the left at higher temperatures. Another type of drying agent, known as molecular sieves, does work. Molecular sieves are complex metal silicates with structures containing many small holes throughout the extended structure of the lattice (almost a "macromolecule"). They work by trapping the small water molecules in these cavities in their structures. Molecular sieves will not release the water until they are heated above about 250°C. Therefore, unless the boiling point of the ester is above that temperature, this is one way to remove water and, thereby, shift the equilibrium in favor of products.

Another way to remove water easily is to add a compound, such as benzene, that forms a low–boiling **azeotrope** with water. Fortunately, even with the lowest molecular weight acids and alcohols, this azeotrope is the lowest boiling liquid and vaporizes first. A slightly modified reflux apparatus, now containing what is termed a **Dean Stark trap,** is then used to remove the water. This type of trap is shown in figure 5.3.

The benzene–water azeotrope boils at 69°C and consists of 91% benzene and 9% water. Therefore, at 69°C the azeotrope boils off, turns to liquid in the condenser, and falls into the left-hand arm of the apparatus. In the trap at lower temperatures the azeotropic mixture separates into two layers with the more dense water on the bottom. As the trap fills up, the less dense benzene is the first material to overflow and return to the reaction flask. Provided that the volume of the trap is large enough, all of the water, except that small amount soluble in benzene, can be removed in this fashion. As expected, the removal of the water shifts the equilibrium of the reaction in favor of the products and the ester is synthesized in higher yield than would be the case without the Dean Stark trap.

FIGURE 5.3:
Azeotropic Removal
of Water

## 5.4   ACID-CATALYZED ESTER FORMATION: THE MECHANISM

Let us begin our look at the mechanism of ester formation by generalizing the reaction as illustrated in equation 5.10.

$$\text{RCOOH} + \text{R'OH} \xrightarrow{\text{H}^+} \text{RCOOR'} + \text{HOH}$$

EQ. 5.10

The reaction requires an acid catalyst because the alcohol (R′OH) is only a weak nucleophile. As a result, the strength of the Lewis acid to be attacked by the nucleophile must be increased. This is done when the proton adds to the oxygen atom of the carbonyl group in the first step of the reactions illustrated in equation 5.11. This step of the reaction is a reversible acid/base reaction. The species formed when the carbonyl oxygen is protonated is in resonance with the second form in the reaction sequence. This second form has the positive charge on the C atom rather than on the more electronegative O atom and would be expected to be the more stable resonance form.

EQ. 5.11

The weak nucleophile, R′OH, can now attack the positively charged carbon atom to form a bond between the oxygen atom of the alcohol and the carbon atom. The intermediate formed in this step has a positive charge on the oxygen atom that formerly belonged to the alcohol.

At this stage there are a number of possibilities for further reaction. All involve loss of one of the protons and one of the oxygen atoms. To determine which oxygen atom is lost we need to look at what the products of the reaction are. In the ester formation reaction, there are two OH groups and one OR′ group attached to the carbon atom. Losing one proton, one of the oxygen atoms, and the entity attached to the oxygen atom would result either in loss of $H_2O$ or in loss of HOR′. If the alcohol is lost, the reactants are regenerated. Loss of $H_2O$, however, leads to a protonated ester. This protonated ester quickly loses the proton attached to the oxygen atom and the ester has been formed. The steps of the mechanism with each proton transfer detailed are illustrated in equation 5.12:

$$\text{EQ. 5.12}$$

What is the evidence that this mechanism is correct? Although we can never prove beyond a doubt that a mechanism is correct, we can gather enough evidence consistent with a particular set of steps that no other mechanism appears likely. In this case, the evidence comes from running the reaction in the presence of an alcohol with the oxygen atom labeled with the non–radioactive isotope, $^{18}O$. Looking at the reaction mechanism outlined above, decide where the $^{18}O$ would end up — either in the water molecule eliminated during the reaction or in the product. The experimental evidence shows that the ester is enriched in $^{18}O$. This, then, provides the evidence for the mechanism outlined. Now test yourself by determining whether you would expect a racemic mixture if the alcohol used was (R)-2-butanol.

Esterification reactions performed by the procedures studied earlier are not always successful because of the small value of the equilibrium constants for the esterification. This is especially true if the alcohol involved is a tertiary alcohol. Equation 5.13 shows the proposed reaction of *t*–butyl alcohol with acetic acid to form *t*–butyl acetate. Esterification carried out by the techniques studied earlier is unsuccessful with these reactants because, as we have seen in table 5.1, the equilibrium constant for the reaction is only 0.0049. Thus, no matter what method we use to force this equilibrium toward completion, almost no product is formed.

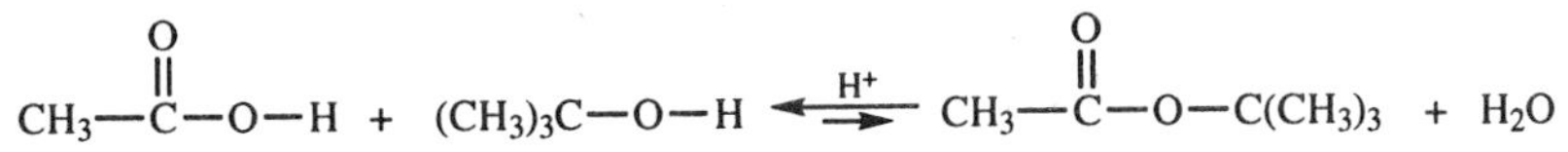

$$CH_3-\overset{\overset{\displaystyle O}{\|}}{C}-O-H \ + \ (CH_3)_3C-O-H \ \underset{\longrightarrow}{\overset{H^+}{\longleftarrow}} \ CH_3-\overset{\overset{\displaystyle O}{\|}}{C}-O-C(CH_3)_3 \ + \ H_2O \qquad \text{EQ. 5.13}$$

Especially when using tertiary alcohols, the key to producing good yields of esters is to use starting materials other than the simple acid. This procedure produces such good yields that it is commonly used even when esters involving primary and secondary alcohols are to be made. Most commonly used are the corresponding **acid chloride** or **acid anhydride**. This changes the esterification reaction to one for which the equilibrium constant is much larger. The structures of acid chlorides and anhydrides are presented in figure 5.4.

Use of either of these in place of acetic acid produces excellent yields of the desired ester by the reactions shown in equations 5.14 and 5.15.

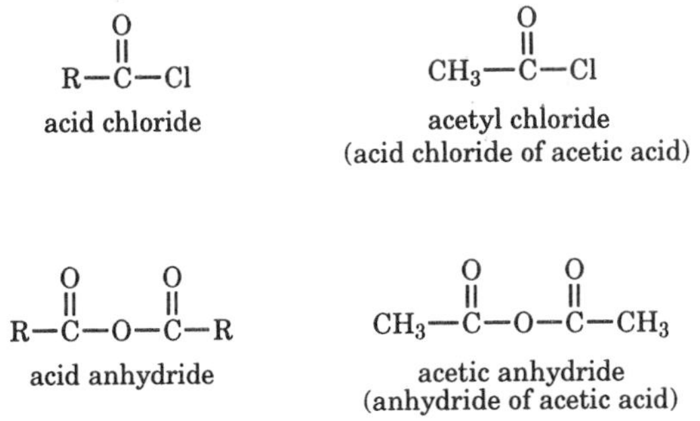

$$R-\overset{\overset{\displaystyle O}{\|}}{C}-Cl$$
acid chloride

$$CH_3-\overset{\overset{\displaystyle O}{\|}}{C}-Cl$$
acetyl chloride
(acid chloride of acetic acid)

$$R-\overset{\overset{\displaystyle O}{\|}}{C}-O-\overset{\overset{\displaystyle O}{\|}}{C}-R$$
acid anhydride

$$CH_3-\overset{\overset{\displaystyle O}{\|}}{C}-O-\overset{\overset{\displaystyle O}{\|}}{C}-CH_3$$
acetic anhydride
(anhydride of acetic acid)

FIGURE 5.4:
Acid Chlorides and
Acid Anhydrides

$$CH_3-\overset{\overset{\displaystyle O}{\|}}{C}-Cl \ + \ (CH_3)_3C-O-H \ \longrightarrow \ CH_3-\overset{\overset{\displaystyle O}{\|}}{C}-O-C(CH_3)_3 \ + \ HCl \qquad \text{EQ. 5.14}$$

$$CH_3-\overset{\overset{\displaystyle O}{\|}}{C}-O-\overset{\overset{\displaystyle O}{\|}}{C}-CH_3 \ + \ (CH_3)_3C-O-H \ \longrightarrow \ CH_3-\overset{\overset{\displaystyle O}{\|}}{C}-O-C(CH_3)_3 \ + \ HO-\overset{\overset{\displaystyle O}{\|}}{C}-CH_3 \qquad \text{EQ. 5.15}$$

Understanding what products are formed in these reactions is most easily accomplished by looking at what chemists call "lasso reactions". In these representations of reactions, portions of two molecules are "lassoed," removed, and the final product is formed by the combination of what is left. Note that in both cases in equation 5.16, the molecule being removed (the one formed by the combination of the "lassoed" parts) is a very stable, small molecule. It is the formation of this stable molecule that drives the desired reaction forward.

It is important, though, that it be understood that this does not necessarily represent a mechanism for the reaction. It is simply a means of determining products for the reaction.

$$CH_3-\overset{O}{\overset{\|}{C}}-O-\overset{O}{\overset{\|}{C}}-CH_3 \quad -HO-\overset{O}{\overset{\|}{C}}-CH_3 \quad CH_3-\overset{O}{\overset{\|}{C}} \quad \xleftarrow{-HCl} \quad CH_3-\overset{O}{\overset{\|}{C}}-Cl$$

$$(CH_3)_3C-O-H \qquad\qquad\qquad O-C(CH_3)_3 \quad (CH_3)_3C-O-H$$

from the acid anhydride                          from the acid chloride

EQ. 5.16

In both cases the mechanism of the reaction involves nucleophilic attack of the alcohol on the carbonyl carbon of the acid derivative followed by the departure of the small molecule. The general reaction of an acid chloride with an alcohol is illustrated in equation 5.17. Note the nucleophilic attack on the carbonyl carbon atom by the oxygen atom of the alcohol. The carbon atom is made more available for nucleophilic attack by attachment of the chlorine atom. After formation of a tetrahedral intermediate, loss of a chloride ion followed by abstraction of the hydrogen atom (initially part of the alcohol) forms the final product. The mechanism of the anhydride reaction is similar. The student is left to work out the details as an exercise.

$$R-\overset{O}{\overset{\|}{C}}-Cl + R-\overset{..}{\underset{..}{O}}-H \longrightarrow R-\overset{:\overset{\ominus}{O}:}{\underset{\underset{R'\quad H}{\overset{\oplus}{O}}}{\overset{|}{\underset{|}{C}}}}-Cl \longrightarrow R-\overset{O}{\overset{\|}{C}}-\overset{\oplus}{\underset{H}{O}}\overset{R^-}{\quad} + :\overset{\ominus}{Cl}: \longrightarrow R-\overset{O}{\overset{\|}{C}}-O-R' + HCl$$

EQ. 5.17

These reactions work as well as they do because the molecule furnishing the acid portion of the ester is much less stable (has an enthalpy of formation that is more positive) than the corresponding enthalpy of formation for the parent acid molecule. The starting materials are made from the parent acids by reaction with another, even less stable reagent. Acid chlorides are prepared from the carboxylic acids using $PCl_3$ or $SOCl_2$ as reagents. This reaction is shown in equation 5.18. Because two of the products of the reaction are gases, the effective equilibrium constant for this reaction is such that the point of equilibrium lies quite far on the side of the products. Alternatively, other halogenating agents such as one of the phosphorous chlorides or phosphorous oxychloride ($PCl_3$, $PCl_5$ and $POCl_3$) can be used in place of thionyl chloride. These also form the acid chloride as one product and a phosphorous oxoacid as the other product. Which of the phosphorous oxoacids is formed depends upon which of the phosphorous chlorides is used.

$$\overset{O}{\overset{\|}{R C}}-OH \quad + \quad \overset{O}{\overset{\|}{\underset{Cl\quad Cl}{S}}} \quad \longrightarrow \quad \overset{O}{\overset{\|}{R C}}-Cl \quad + \quad SO_2(g) \quad + \quad HCl(g)$$

EQ. 5.18

Acid chlorides are very reactive and must be protected from moisture to prevent hydrolysis with formation of gaseous HCl and the parent acid. Acid anhydrides are not quite as reactive as acid chlorides. Some are even stable enough to be found in nature. They are usually prepared in the laboratory by the reaction of an acid chloride with the salt of a carboxylic acid as follows. The reaction is shown in equation 5.19.

$$CH_3-\overset{\overset{O}{\|}}{C}-Cl \xrightarrow{NaO-\overset{\overset{O}{\|}}{C}-CH_3} CH_3\overset{\overset{O}{\|}}{C}-O-\overset{\overset{O}{\|}}{C}-CH_3$$

EQ. 5.19

Acetic anhydride has widespread industrial uses and is thus both plentiful and inexpensive. One example of its use is in the synthesis of aspirin. That reaction is shown as equation 5.20.

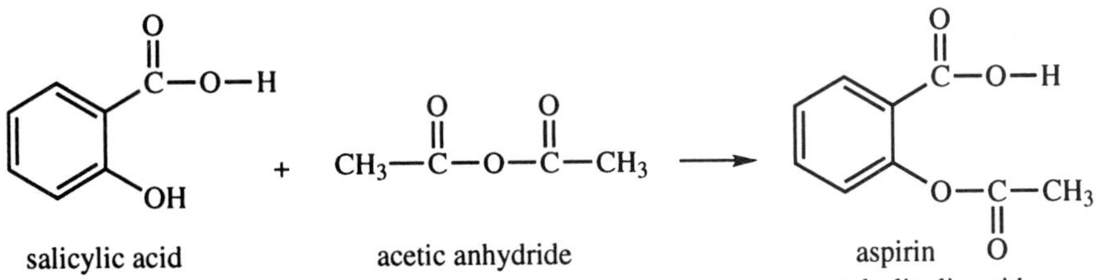

salicylic acid    acetic anhydride    aspirin
acetylsalicylic acid

EQ. 5.20

Another important industrial use of acetic anhydride is in the formation of cellulose acetate, one of the first "plastic" films available, and in the production of a closely related compound, rayon. Both of these materials are produced by esterification of the hydroxyl groups of the polysaccharide cellulose with acetic anhydride to form acetate esters. The processes in the formation of these materials are complex, but the composition of the rayon formed indicates that only two of the hydroxyl groups of each glucose monomer in the cellulose are esterified in the final product. A drawing of the polymer is shown in figure 5.5.

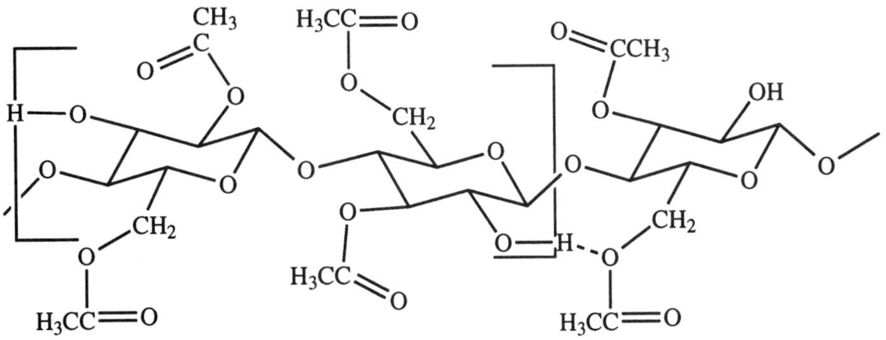

FIGURE 5.5:
Structure of the
Cellulose Acetate
Polymer

## 5.6   ESTER HYDROLYSIS

The reverse reaction of esterification is hydrolysis — the conversion of an ester to its corresponding acid and alcohol. It is illustrated in the following chemical equation with the mechanism of the acid–catalyzed reaction shown in figure 5.6 and the mechanism of the more common base–catalyzed reaction shown in figure 5.7. Notice the presence of a tetrahedral intermediate in both mechanisms.

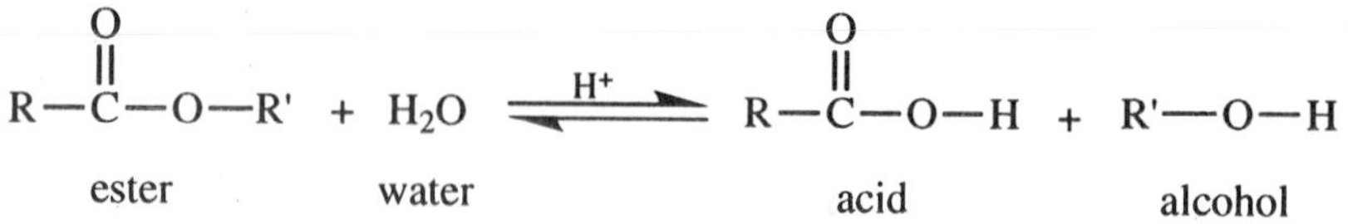

$$
\underset{\text{ester}}{R-\overset{\overset{\textstyle O}{\|}}{C}-O-R'} + \underset{\text{water}}{H_2O} \; \underset{}{\overset{H^+}{\rightleftharpoons}} \; \underset{\text{acid}}{R-\overset{\overset{\textstyle O}{\|}}{C}-O-H} + \underset{\text{alcohol}}{R'-O-H}
\qquad \text{EQ. 5.21}
$$

Specific instances of this reaction are important to the economy of the country and to life. Probably the most important use of this reaction in our economy is the production of soaps by the hydrolysis of animal fats and vegetable oils that are naturally occurring esters, usually of long–chain $(C_{10}-C_{20})$ acid esters of the triol, glycerol. Of more direct importance to life, however, this hydrolysis of an ester is the reaction that breaks down one of the products of nerve transmission, acetylcholine. Without hydrolysis of this product there would be continual stimulation of the nerve. Many extremely potent insecticides, as well as a number of the nerve gases developed for gas warfare, work by blocking the action of the enzyme, acetylcholinesterase, that catalyzes this hydrolysis in the body. The eventual result of the inactivation of acetylcholinesterase is complete paralysis and death. Ester hydrolysis is also important in the metabolic pathways for deactivating and excreting certain drugs, e.g., cocaine, that, if not eliminated from the body, would eventually reach toxic levels. In most cases the hydrolysis products (acid and alcohol), because they are much more capable of hydrogen bonding interactions with water, are much more water-soluble than the original ester.

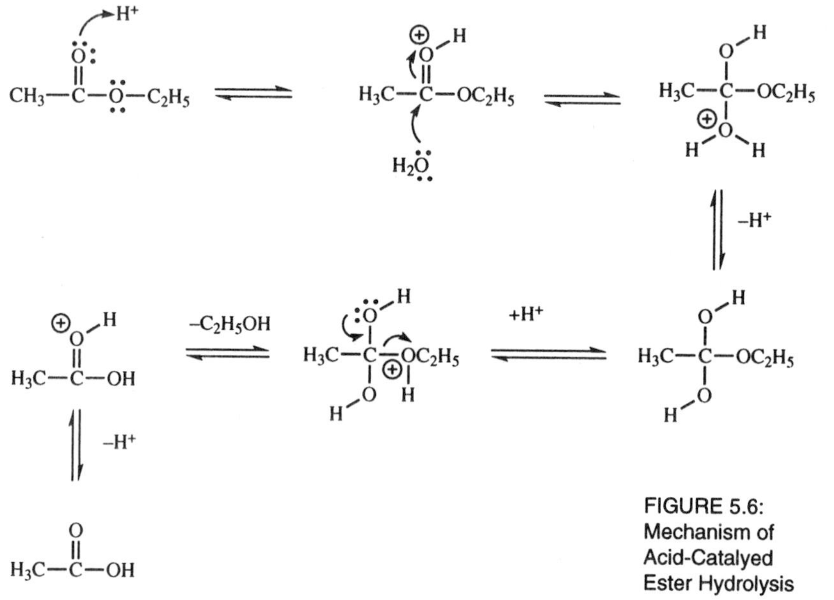

FIGURE 5.6:
Mechanism of
Acid-Catalyed
Ester Hydrolysis

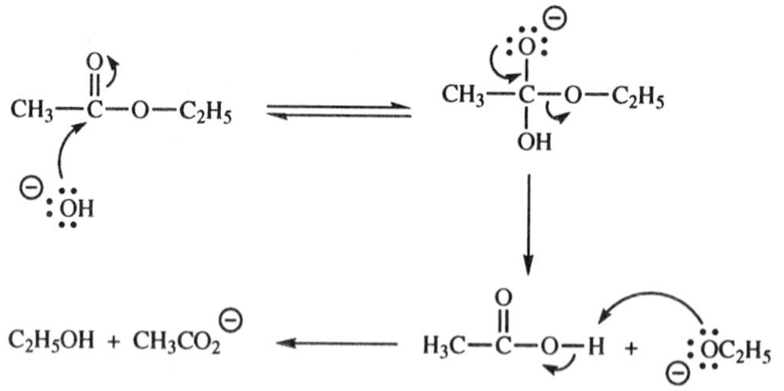

FIGURE 5.7:
Mechanism of
Base-Catalyzed
Ester Hydrolysis

LeChatelier's principle again predicts that an excess of reactants favors hydrolysis. In this case, since water is a reactant and is so inexpensive, large excesses are used, provided that the solubility of the reactants and products will permit such. Although the hydrolysis can be, and sometimes is, performed in acidic solution, it is normally done in basic solution. Why should this be? What effect would base have on the hydrolysis equilibrium? As shown in figure 5.7, the ethoxide ion reacts with the acid to form the carboxylate salt. This displaces the hydrolysis equilibrium to the right, favoring the products.

If the ester is an animal fat (a tri–ester of long chain, $C_{10}$ - $C_{20}$ acids and the triol glycerol), the mixture of the salts of the acids and glycerin resulting from the hydrolysis is called 'soap'. For many hundreds of years before soap was produced commercially in factories, each household would collect animal fats to be boiled with lye. The reaction, run under such strongly basic conditions, is called a 'saponification' reaction. A typical ester found in animal fat, the glycerol triester of stearic acid, is shown with reactions for complete saponification in equation 5.22. One of the products, the sodium salt of stearic acid, is easily obtained in a relatively pure state because it is water–insoluble and forms a waxy film on the surface of the water. Excess sodium hydroxide and glycerol remain in the water.

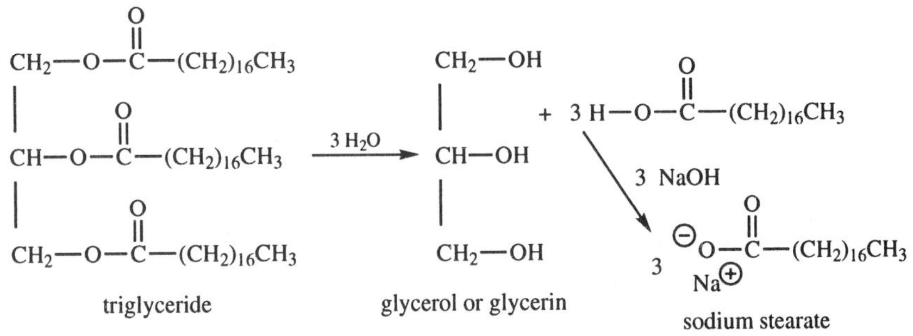

EQ. 5.22

In a related reaction, let us now consider the hydrolysis of nitriles. Nitriles can be hydrolyzed, eventually to carboxylic acids, in a two–step reaction that is very similar to the hydrolysis of esters. Equation 5.23 shows the overall reaction that occurs in this, synthetically very important, reaction. The ultimate products shown here are ammonia and the salt of the carboxylic acid — a compound that can easily be converted to the acid by treatment with a strong acid such as HCl.

EQ. 5.23

This reaction can be best thought of as occurring in two steps: (1) formation of an amide initiated by nucleophilic attack by the oxygen atom of an hydroxide ion on the nitrile carbon atom, followed by (2) nucleophilic displacement of the amino group of the amide by a second hydroxide ion. The mechanism of the first step is illustrated in equation 5.24; that of the second step is shown in equation 5.25.

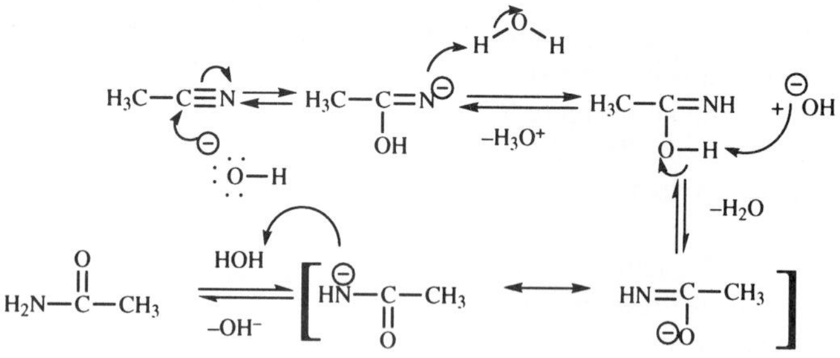

EQ. 5.24

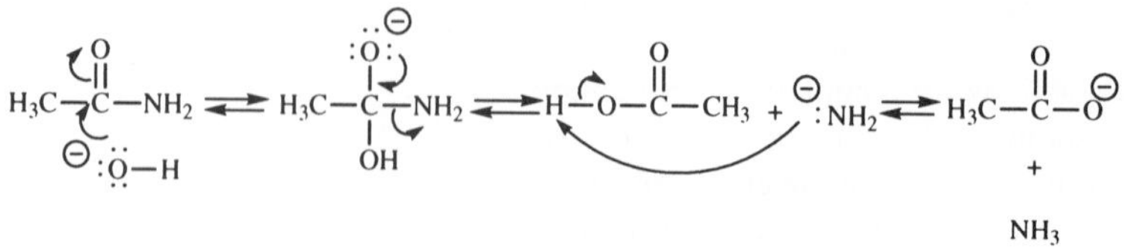

EQ. 5.25

As we shall see in a later chapter, the hydrolysis of nitriles is of great synthetic utility because of the ability to easily add one carbon atom to a molecule by introducing a cyano group for a halogen atom and then converting the nitrile into any number of functional groups accessible through a carboxylic acid.

## CONCEPT CHECK

Draw a mechanism that explains the following observations:

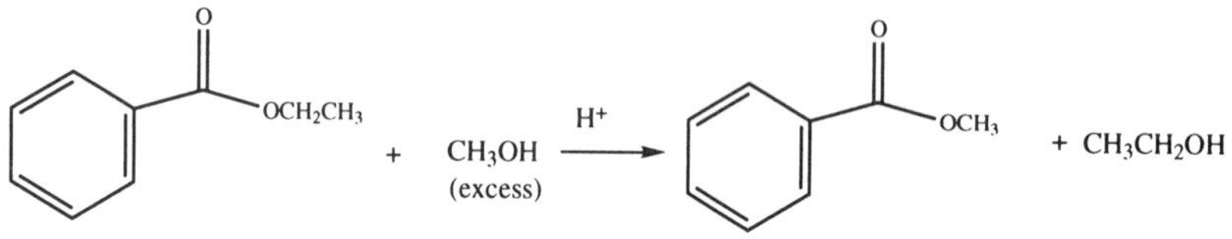

EQ. 5.26

Since the only difference between this reaction and an acid–catalyzed esterification reaction is that one of the starting materials is an ester rather than a carboxylic acid, we might expect the mechanism to be similar to that as well. Therefore we posit that the first several steps are similar to those previously seen in equation 5.11 and are illustrated for this particular example in figure 5.8.

At this point another hydrogen ion can add to the species at any of the three oxygen atoms. Because the alkyl groups are somewhat more electron–releasing than a simple hydrogen atom, it is more likely that the hydrogen atom will add to one of the two alkoxy groups. This will produce a good leaving group — a neutral alcohol molecule. Figure 5.9 pictures the hydrogen adding to the ethoxy group because it is the removal of this group

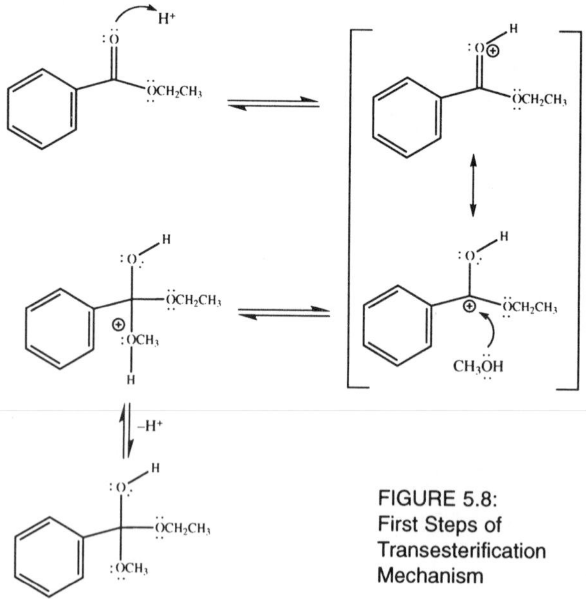

FIGURE 5.8:
First Steps of
Transesterification
Mechanism

that leads to the observed product. After the
ethanol molecule leaves, loss of the hydrogen atom
on the hydroxyl group as a hydrogen ion produces
the observed product. The excess methanol, probably present as the solvent, ensures that if the
methyl benzoate product of figure 5.9 is protonated
and undergoes the first steps of the mechanism in
figure 5.8, it will be a methanol molecule that will
add, producing a dimethoxy intermediate that will
produce methyl benzoate when an alcohol molecule is eliminated in the steps shown in figure 5.9.

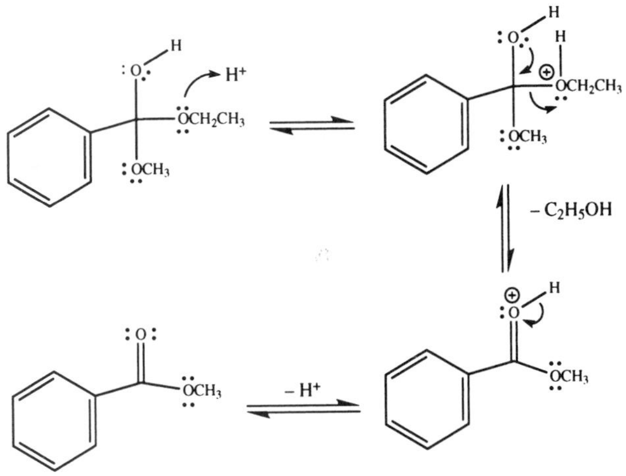

**FIGURE 5.9:**
Final Steps of
Transesterification
Mechanism

## 5.7 SYNTHESIS OF LACTONES: INTRAMOLECULAR ESTER FORMATION

Thus far the acid and alcohol involved in the formation of the ester have been
different molecules. What if the acid and alcohol groups are both attached to
the same backbone, *e.g.*, 4-hydroxybutanoic acid? There are now two different
possibilities for ester formation: (1)
intermolecular ester formation possibly
leading to a polymer, and (2) intramolecular ester bond formation. Are both
even possible? Which will be favored?
What ratio of products should we
expect if both are possible? We shall
examine each of those questions in turn.

Both are possible, although, as you
probably have guessed, both are not
equally favored. Compounds formed by
intramolecular ester formation comprise
a special class of esters known as **lactones**. An example of lactone formation
by 4-hydroxybutanoic acid is presented
in figure 5.10.

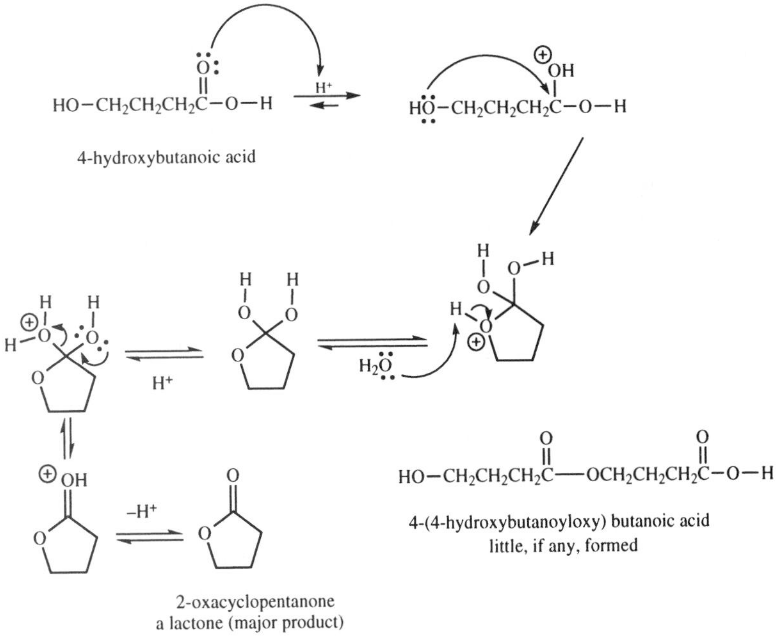

Product ratios show that the lactone is
the major product formed from 4-
hydroxybutanoic acid. None, or at best
only very little, of the 4-(4–hydroxybu-
tanoyloxy)butanoic acid (the ester

4-hydroxybutanoic acid

2-oxacyclopentanone
a lactone (major product)

4-(4-hydroxybutanoyloxy) butanoic acid
little, if any, formed

**FIGURE 5.10:**
Lactone Formation

formed by intermolecular reaction) is found among the products of the reaction. On the other hand, reactions of some other, similar, difunctional com-

pounds produce primarily or even exclusively the intermolecular ester. Why should there be such a dramatic difference with such closely related compounds?

The reason lies in the geometry of the final compounds formed and in the effect of the geometry on the thermodynamics of ester formation. It is very easy to explain why compounds such as 2–hydroxyacetic acid do not form lactones: to do so would result in formation of a 3–membered ring. This compound, if formed, would have a very large amount of ring strain — even more than in ethylene oxide because one of the carbon atoms must be $sp^2$ hybridized. Even though it is possible to form the lactone of 3–hydroxyacetic acid, its formation results in a 4–membered ring. Once again, the ring strain makes formation difficult and yields poor. The resultant bond angles in 3– or 4– membered ring compounds would be approximately 60° or 90°. This is too far from the 109.5° expected for the bond angles with $sp^3$ hybridization and the 120° expected for $sp^2$ hybridization for the formation of strong, stable bonds. This is illustrated in the following reaction:

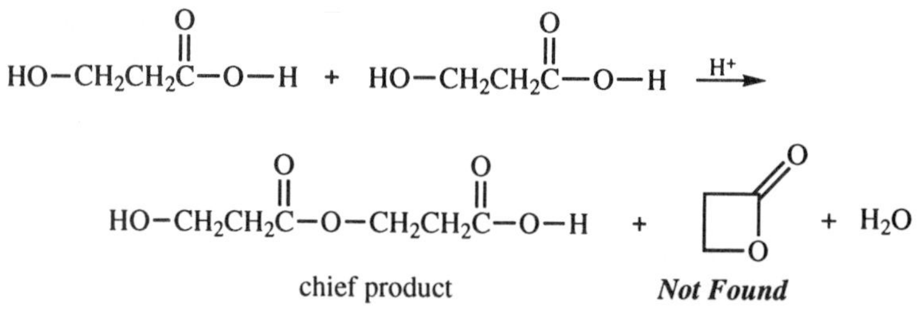

$$HO-CH_2CH_2\overset{\overset{\displaystyle O}{\|}}{C}-O-H \quad + \quad HO-CH_2CH_2\overset{\overset{\displaystyle O}{\|}}{C}-O-H \quad \xrightarrow{\ H^+\ }$$

$$HO-CH_2CH_2\overset{\overset{\displaystyle O}{\|}}{C}-O-CH_2CH_2\overset{\overset{\displaystyle O}{\|}}{C}-O-H \quad + \qquad\qquad + \quad H_2O$$

chief product            *Not Found*

EQ. 5.27

With 4–hydroxybutanoic acid and 5–hydroxypentanoic acid, the compounds formed would contain 5– or 6–membered rings respectively. If the compounds were planar, the bond angles expected would be 108° and 120° respectively. Both are relatively close to the required angles but can actually come even closer because the molecules assume conformations similar to those of cyclohexane, *i.e.*, non–planar conformation. This means that formation of the lactone is not disfavored from a geometric standpoint. But, this fact alone does not explain why the lactone is the major product. There must be some additional reason that favors lactone formation over intermolecular ester formation. What is this?

The equilibrium constant for any chemical reaction is directly related to the standard free energy change for the reaction. The standard free energy change has two components — the enthalpy and entropy changes. The enthalpy component is largely determined by the difference in bond energies between the products and the reactants. Provided that formation of an intramolecular compound does not involve a great deal of strain in the bonds due to small ring formation, there should not be a great deal of difference in the enthalpy component of the free energy change for intra– or inter–molecular ester formation. Any differences, therefore, must come from the entropy term.

The reason for the difference in the entropy term is that formation of an intermolecular ester bond requires that two molecules react to form two molecules - one larger molecule, the ester, and one small molecule, water. On the other hand, formation of an intramolecular ester, a lactone, results in the formation of two molecules where there had been one molecule before. While the entropy of the lactone itself is slightly smaller than that of the parent hydroxy acid because of the geometric constraints imposed by the ring in the lactone, this decrease is overcome by the formation of the second molecule. Therefore, ring formation should always be entropically favored over intermolecular ester formation — even the formation of 3– or 4–membered rings. It is only because the much more important enthalpy term overshadows the entropy term that prevents the formation of these lactones.

There is also another factor that comes into play in most instances — a kinetic factor. For any type of reaction to occur, the two portions of the molecule(s) that are to react must come into close proximity. The fact that the hydroxyl group and the carboxyl groups are attached by a short carbon chain means that they are constrained to be close to one another. The normal flexing and bending of a longer carbon backbone that occur at room temperature means that there is a much greater chance that the carboxyl group will encounter the hydroxyl group on the same molecule at least as frequently as it will the hydroxyl group on another molecule moving randomly throughout the solution. Thus, both kinetics and thermodynamics team up to make the formation of 5– and 6– membered lactones preferred over intermolecular esters formed by the same molecules.

## 5.8   ACETALS: EQUILIBRIA AND MECHANISMS

Another use of equilibrium reactions for synthesis involves the hemiacetal and acetal groups. The hemiacetal group is formed when an aldehyde or a ketone reacts with an alcohol. This initial reaction is then followed by reaction with a second molecule of the alcohol to form the acetal group. These reactions are illustrated in the following chemical equations:

$$
\underset{\text{aldehyde}}{R-\overset{\overset{\textstyle O}{\|}}{C}-H} \;+\; 2\,\underset{\text{alcohol}}{H-O-R'} \;\underset{\longleftarrow}{\overset{H^+}{\longrightarrow}}\; \underset{\text{hemiacetal}}{R-\overset{\overset{\textstyle O-H}{|}}{\underset{\underset{\textstyle O-R'}{|}}{C}}-H} \;\underset{\longleftarrow}{\overset{H^+}{\longrightarrow}}\; \underset{\text{acetal}}{R-\overset{\overset{\textstyle O-R'}{|}}{\underset{\underset{\textstyle O-R'}{|}}{C}}-H} \;+\; H_2O
\qquad \text{EQ. 5.28}
$$

$$
\underset{\text{ketone}}{R-\overset{\overset{\textstyle O}{\|}}{C}-R} \;+\; 2\,\underset{\text{alcohol}}{H-O-R'} \;\underset{\longleftarrow}{\overset{H^+}{\longrightarrow}}\; \underset{\text{hemiacetal}}{R-\overset{\overset{\textstyle O-H}{|}}{\underset{\underset{\textstyle O-R'}{|}}{C}}-R} \;\underset{\longleftarrow}{\overset{H^+}{\longrightarrow}}\; \underset{\text{acetal}}{R-\overset{\overset{\textstyle O-R'}{|}}{\underset{\underset{\textstyle O-R'}{|}}{C}}-R} \;+\; H_2O
\qquad \text{EQ. 5.29}
$$

Both esterification and acetal formation are equilibrium–controlled reactions. Formation of an acetal is normally done in a large excess of alcohol to drive the reaction to completion. Another similarity to an esterification reaction comes in that acetal formation reactions are also catalyzed by acids. The overall reaction for the formation of an acetal is shown in equation 5.30.

$$\text{RCHO} + \text{R'OH} \xrightarrow{\text{H}^+} \text{RCH(OH)OR'} \longrightarrow \text{RCH(OR')}_2 \qquad \text{EQ. 5.30}$$

This reaction also requires the addition of an acid catalyst to increase the strength of the Lewis acid that will be attacked by the nucleophile. Addition of the proton is again a reversible acid/base reaction. The first several steps in the mechanism are very similar to those of the acid esterification mechanism. Protonation of the carbonyl oxygen atom forms a positively charged carbon atom that is much more likely to be subject to nucleophilic attack.

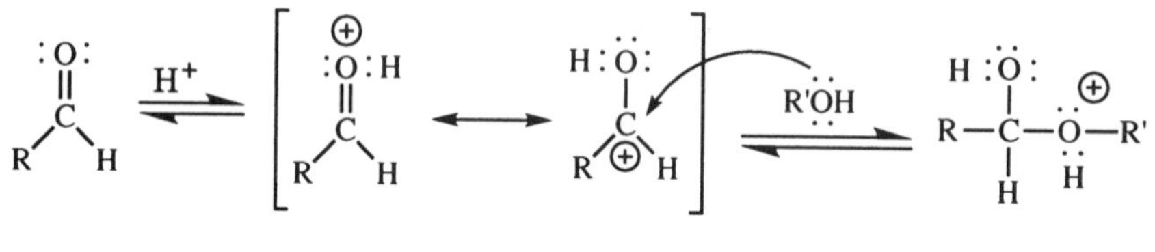

EQ. 5.31

The weak nucleophile, R′OH, can now attack the positively charged carbon atom to form a bond between the oxygen atom of the alcohol and the carbon atom. The intermediate formed in this step has a positive charge on the oxygen atom that formerly belonged to the alcohol.

At this stage the intermediate loses a molecule of water in the first three steps of the reaction outlined in equation 5.32. This produces another intermediate that is attacked by another molecule of alcohol in step four. The product of this addition then simply loses a proton to form the acetal.

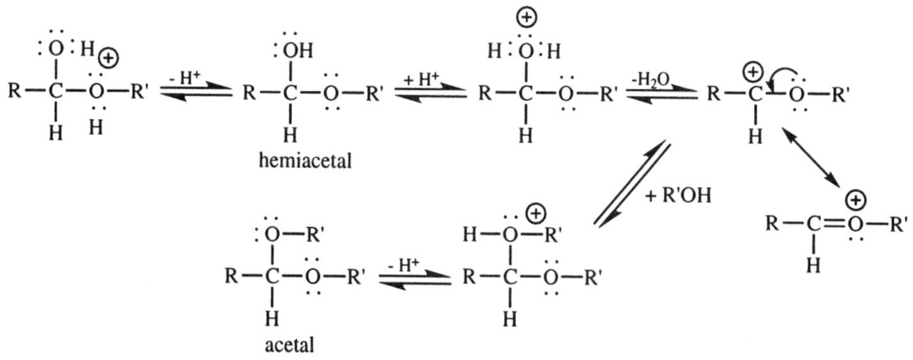

EQ. 5.32

As in the esterification reaction, notice that every step is written as an equilibrium reaction. Moving toward products requires that the conditions are appropriate to shift the equilibrium in that direction. The difference between the two reactions is that, in the acetal reaction, the hemiacetal formed after the first series of proton transfers goes on to react with a second alcohol molecule to form the acetal while the other reaction terminates with the formation of the ester. While the mechanisms of acetal formation and ester formation are similar, one reaction gives a substitution product (—OR for —OH), the ester, while the other gives an addition product, the acetal.

Both hemiacetals and acetals are common in carbohydrate chemistry and many reactions in the metabolic pathways of carbohydrates involve formation of these groups. Let us now look at the acetal group — a group that is both easy to form and to remove. This makes it very useful in synthesis as a protecting group.

## 5.9   ACETALS AS PROTECTING GROUPS

In many synthetic reactions there are often cases in which the reagents necessary for putting on one functional group will react with and destroy another functional group already on the target molecule. For example, reduction of the ester group in methyl 4-oxopentanoate to the corresponding alcohol would be done easily using $LiAlH_4$. The major problem with the use of $LiAlH_4$ with this molecule is the ketone group also present in the molecule. Use of this reagent would lead chiefly to the first product shown below rather than the desired keto–alcohol.

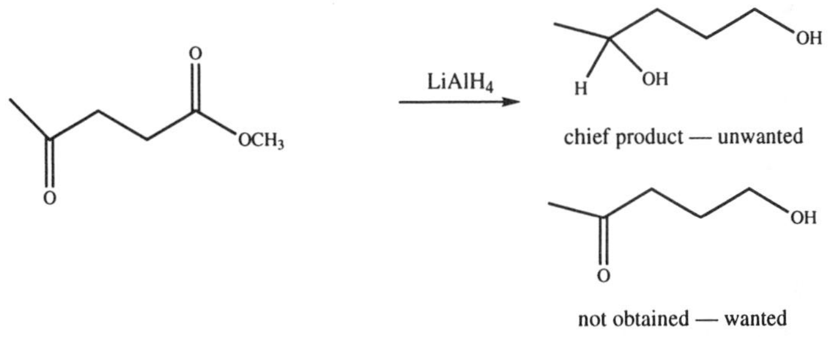

chief product — unwanted

not obtained — wanted

EQ. 5.33

To prevent the simultaneous reduction of the ketone group, we can make good use of an acetal as a protecting group. After reduction of the acetal–ester the protecting group can be removed to form the desired product. The general requirements for a good protecting group are: (1) the protecting group must be put on easily in excellent yields, (2) the protecting group must be inert to the proposed reaction, and (3) the protecting group must be removed easily in excellent yields. It might appear as if the first and third requirements are mutually exclusive. They are not if the principles of equilibrium are used to our advantage. There are a number of equilibrium reactions in which these work especially well. Acetal reactions fall into this latter category because the reaction conditions can be easily manipulated so that the reaction goes either way. Both formation and removal of the protecting group proceed under mild conditions in excellent yields. Let us now look at the specific reaction in equation 5.34.

The first step is to prepare the acetal. This means that we must decide which acetal of the ketone to prepare. Since the formation of the acetal involves reaction of two moles of alcohol with one mole of the ketone and since we want to use the mildest conditions possible, we frequently choose a difunctional alcohol, 1,2–ethanediol. Use of this difunctional alcohol forms a five–membered ring similar to the five–membered ring of the lactones discussed earlier in the chapter.

The formation of the cyclic acetal is shown as a two step reaction in which the hemiacetal formed by the first reaction is explicitly shown as an intermediate. In fact, the intermediate hemiacetal is never isolated and the reaction goes on, uninterrupted, to the acetal.

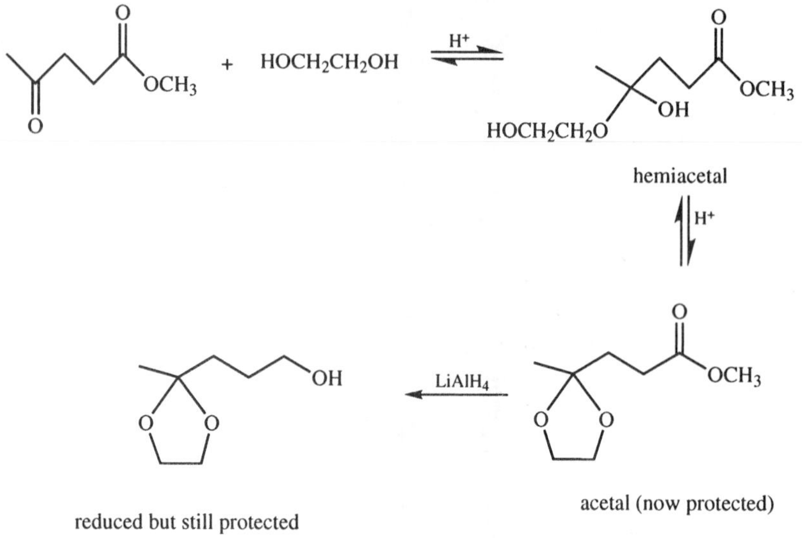

hemiacetal

acetal (now protected)

reduced but still protected

EQ. 5.34

In the equilibrium step between the hemiacetal initially formed and the acetal, the acetal is favored because the unreacted hydroxy group of the hemiacetal is held close to the acetal carbon. This means that a greater amount of the acetal will be formed than if a mono–functional alcohol such as methanol were used instead of the ethanediol. (You may recognize 1,2-ethanediol if you knew the common name is ethylene glycol. It is used as a permanent antifreeze for automobile radiators.)

The second step in the reaction is the reduction of the ester group. Now that the ketone group is protected as an acetal, the reduction can be done with $LiAlH_4$. The acetal is stable under the basic conditions in which this reduction reaction is run. Acidic conditions are not used for two reasons: one is to avoid removing the acetal protecting group before the reduction is finished, and the other is that the $LiAlH_4$ would react violently with any $H^+$ in solution.

After reduction is finished, acid is used to remove the acetal to reform the ketone. The use of water as solvent or cosolvent is also used to push the equilibrium toward the ketone. This reaction is shown as equation 5.35.

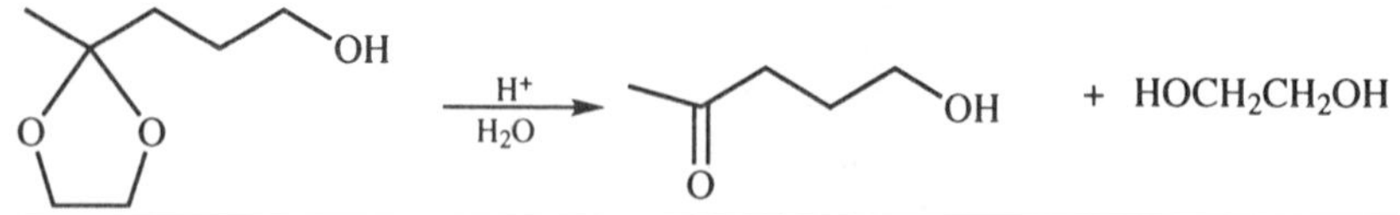

EQ. 5.35

There are many additional cases in which ketone or aldehyde functional groups must be protected during another reaction. One such example is the oxidation of the alcohol group on the difunctional cyclohexane in equation 5.36. Unless the aldehyde is protected, the $KMnO_4$ will oxidize the aldehyde group to a carboxyl group in addition to the hydroxyl group to a carbonyl (ketone) group. The details of the conversion are left to the student as an exercise.

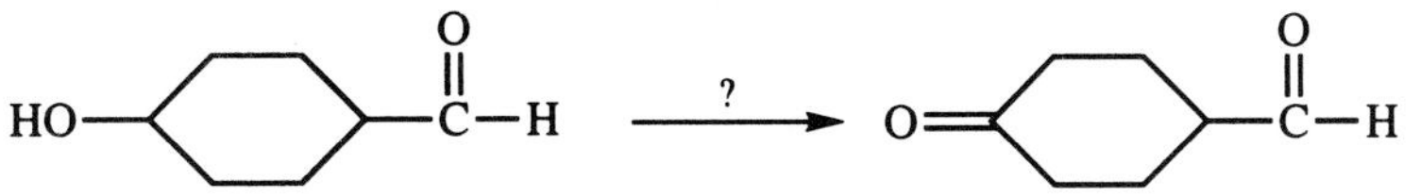

EQ. 5.36

**CONCEPT CHECK**

Draw the structures of the acetals that you would expect to form between the following carbonyl compounds and alcohols:

a. 4–methylcyclohexanone and methanol

b. p–nitrobenzaldehyde and ethylene glycol

c. 2,5–dimethyloctanal and propanol

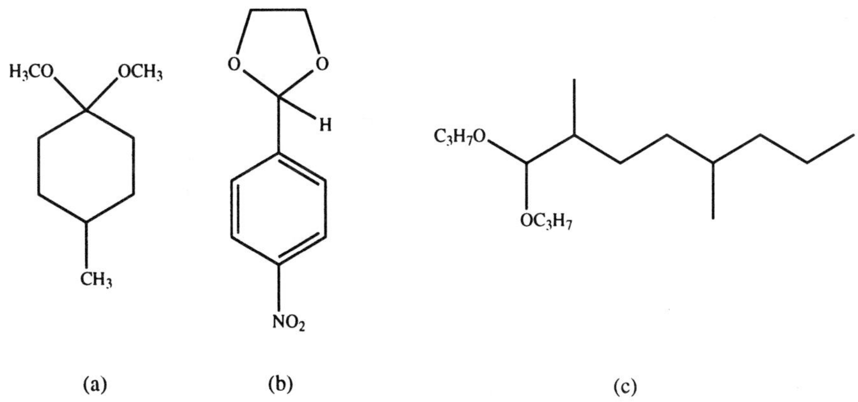

(a)      (b)            (c)

FIGURE 5.11:
Acetal Structures

## 5.10 CARBOHYDRATES

Now that we have looked at a few reactions involving acetals, let us briefly look at carbohydrate chemistry and the important part that acetals and hemiacetals play in it. Indeed as we shall see, the biochemistry of carbohydrates is mainly the reactions of carbonyls, hemiacetals and acetals.

Carbohydrates are polyhydroxy aldehydes or polyhydroxy ketones. The commonly encountered carbohydrates have from three to six carbon atoms. Even though the word "carbohydrate" literally means "hydrate of carbon" and the formulas of all carbohydrates can be written $C_n(H_2O)_m$, writing the formula in this fashion would be misleading. Let us consider the most important of the carbohydrates, glucose.

D-glucose has the Fischer structural formula shown in figure 5.12a. Note that it contains five hydroxy groups and one aldehyde group. This puts it in a category with a number of other carbohydrates known as the aldohexoses (6–carbon, aldehyde–containing carbohydrates). However, under most conditions, the Fischer structure is not an accurate representation of the structure of the glucose molecule. In most situations, the structure of the glucose molecule is that of a hemiacetal as shown in figure 5.12b. A better way of showing the ring structure actually formed in this hemiacetal is shown as figure 5.12c (the Haworth projection). To show the stereochemistry even better, figure 5.12d shows a structure similar to that of the chair form of cyclohexane with hydroxyl groups in the correct axial or equatorial positions around the ring. By convention, the oxygen atom in the ring is shown at the back, right position of the ring. Notice that, in glucose, all hydroxy groups are equatorial in its most stable beta (ß) form. The less stable form, called the alpha (α) form, has all the —OH groups equatorial except the hemiacetal group which is in the axial position. At equilibrium, the actual mixture is about 64:36::ß:α (64% ß and 36% α.)

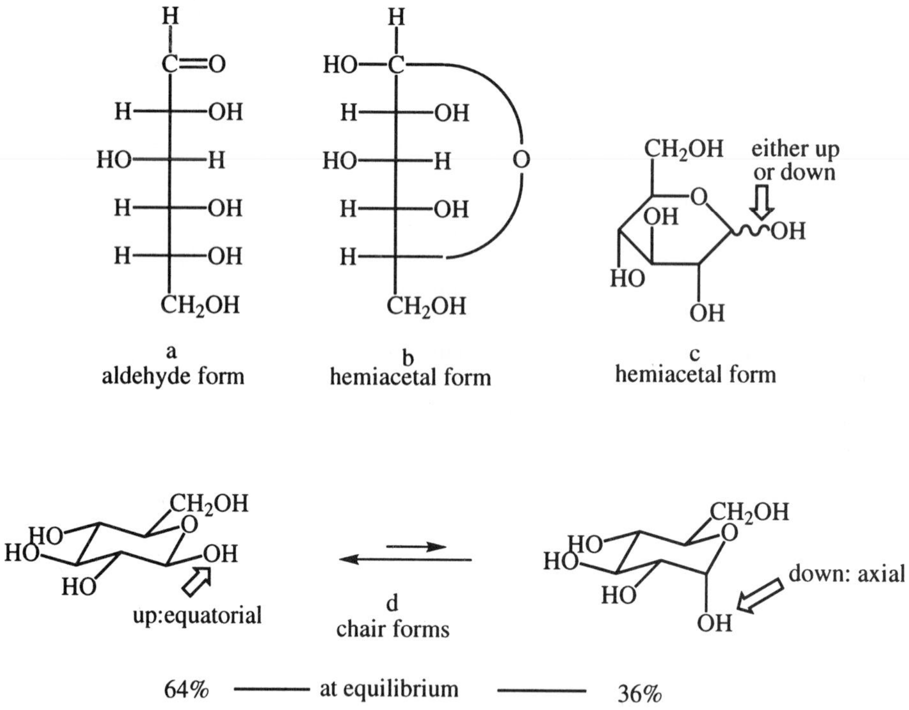

a
aldehyde form

b
hemiacetal form

c
hemiacetal form

up:equatorial

d
chair forms

down: axial

64% ———— at equilibrium ———— 36%

FIGURE 5.12:
Different Forms
of Glucose

Additional evidence of the dynamic equilibrium among these forms is presented by a number of the reactions of glucose in which the aldehyde group must be used to explain the reaction. Although the IR spectrum of glucose shows no trace of a peak in the area in which an aldehydic carbonyl would be expected to absorb, glucose does react with a number of mild reducing agents, e.g., $Ag(NH_3)_2^+$ in the Tollen's test, in a manner analogous to reactions of *bona fide* aldehydes. Other, naturally occurring, monosaccharides (single–ring sugars) include mannose and galactose. The latter two differ from glucose in the orientation of the hydroxyl groups at carbons number 2 and 4. In mannose the hydroxyl group is on the opposite side of carbon atom 2 as it is in glucose. The stereochemistry about all other carbon atoms remains the same. In galactose it is the stereochemistry around carbon atom 4 that is different from glucose. Galactose is one component of the disaccharide lactose, commonly known as "milk sugar". The components of lactose are glucose and galactose. The structures of galactose and mannose are given in figure 5.13.

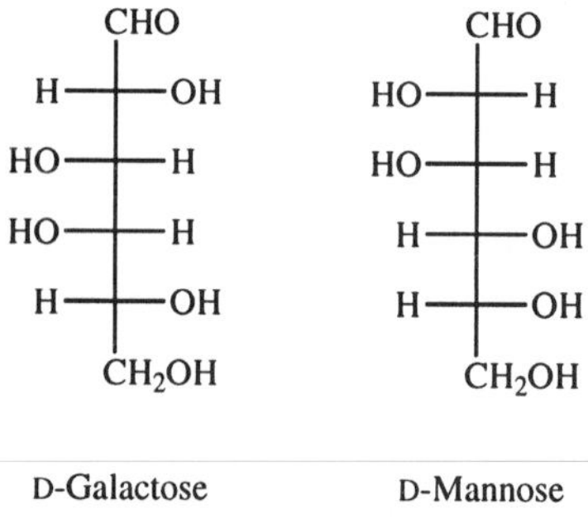

D-Galactose

D-Mannose

FIGURE 5.13:
Fischer Structures of
Galactose and Mannose

In addition to the aldohexoses, natural sugars include a number of aldoses with fewer than six carbon atoms. The most important of these is ribose and its cousin (non–carbohydrate) deoxyribose. Both of these contain five carbon atoms. Their structures are given in figure 5.14. Note that deoxyribose is not a carbohydrate because it is missing a hydroxyl group on carbon atom number 2. It is, however, still classified as a sugar.

In addition to the aldoses, one additional monosaccharide is of some importance: fructose — a component of sucrose, the common disaccharide we know otherwise as table sugar. Fructose is a ketohexose whose various structural representations are shown as figure 5.15.

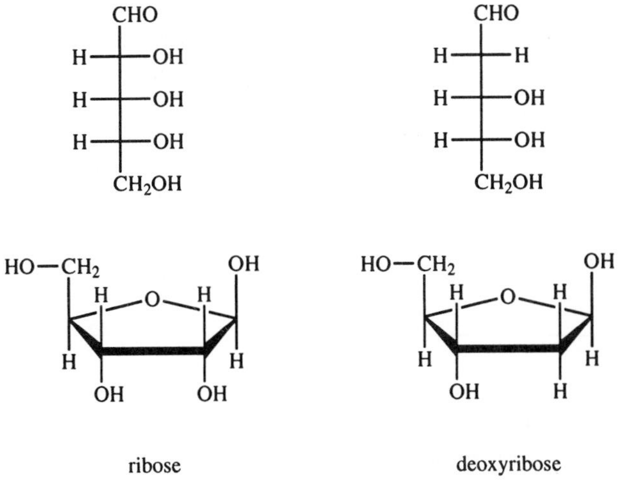

The common feature of all these monosaccharides is the presence of the cyclic structure necessary for formation of the polymeric saccharides including sucrose, glycogen, starch, and cellulose. These will be discussed further in Chapter 15. It needs only to be emphasized here that the formation of the cyclic structure is an important example of hemiacetal formation and that, because of the equilibrium between cyclic and open–chain forms, the monosaccharides exhibit properties common to both forms.

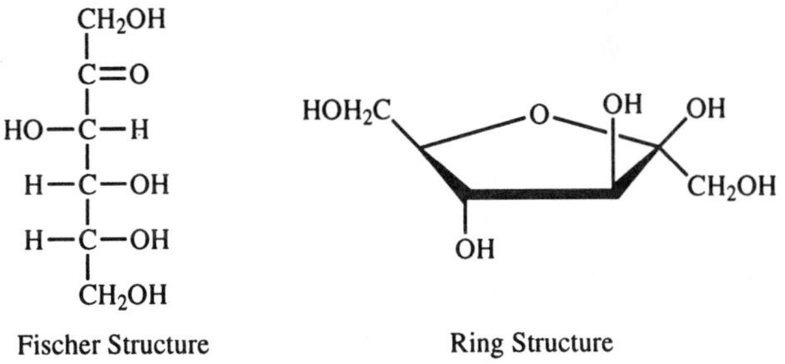

ribose

deoxyribose

FIGURE 5.14:
Fischer and Ring Structures of Ribose and Deoxyribose

Fischer Structure

Ring Structure

FIGURE 5.15:
Fischer and Ring Structures of Fructose

## 5.11  INTRAMOLECULAR RING FORMATION

The formation of cyclic acetals and hemiacetals is not limited to the saccharides but occurs in other compounds that have ketone or aldehyde functional groups located 3 or 4 carbon atoms away from hydroxyl groups. Such difunctional compounds have the two groups located in just the right position so that formation of an acetal produces a 5 or 6 membered ring. As we have discussed before, these rings are especially stable and are quite likely to form given the correct conditions. A specific example of intramolecular hemiacetal formation, to form one member of a class of compounds known as lactols, is presented in equation 5.37. Notice that the equilibrium mixture contains 89% of the cyclic form (2-hydroxytetrahydrofuran) and only 11% of the straight–chain form (4-hydroxybutanal). This corresponds to an equilibrium constant with a value of approximately 8. While this is not as large a value as many of the constants with which we have previously dealt, it is large enough that we can expect ring formation to affect the chemistry of the compound by making reactions of the cyclic compound much more likely and reactions of the open chain compound much less likely than would be expected if we ignored the cyclization.

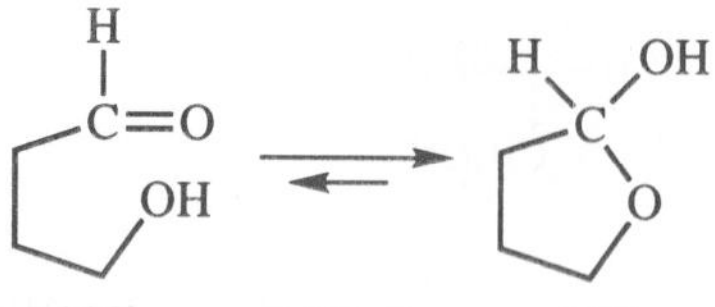

11% at equilibrium   89%

4-hydroxybutanal      2-hydroxytetrahydrofuran

EQ. 5.37

## 5.12   COMMERCIALLY IMPORTANT ESTERS: FIBERS

One of the most important synthetic fibers, Dacron®, is a polymer of *p*–phthalic acid (a di–carboxylic acid) and ethylene glycol (a diol). A section of the polymer is shown as figure 5.16. The reaction occurs when phthalic anhydride is reacted with the glycol and the reaction mixture is forced through spinnerets to form the fiber. The resulting fibers are strong and resilient and make an excellent synthetic fabric. Dacron® is only one of a number of polymeric esters (polyesters) now being used in many applications. The world–wide value of the polyester business is now approximately $30 billion per year with the capacity of just one company making polyesters recently having been increased to 1 million pounds per year. The amount of polyester made each year makes phthalic anhydride one of the most important commercial chemicals. Polymers will be studied in much greater detail in chapter 15.

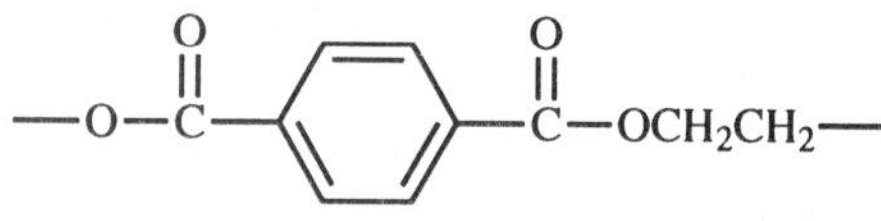

FIGURE 5.16:
Repeating Unit
of Dacron®

## 5.13   COMMERCIALLY IMPORTANT ESTERS: EXPLOSIVES

In addition to esters of carboxylic acids, important in nature as many of the flavoring and odor agents of living materials, there are important esters of inorganic acids. A number of these, esters of phosphoric acid, are of biochemical importance. Others, esters of sulfuric, nitric, and perchloric acids, have a number of important commercial uses. Let us look first at some esters of nitric acid.

A commercially important use of a nitric acid ester is seen for the tri–ester of glycerol (glycerin) with nitric acid. The chemical equation for this reaction is given as equation 5.38.

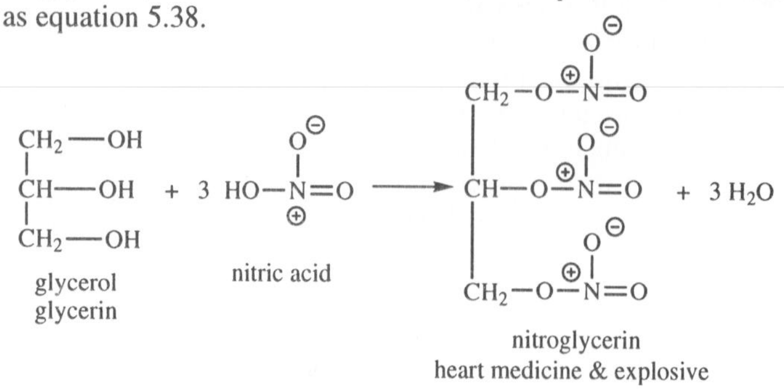

EQ. 5.38

glycerol
glycerin     nitric acid

nitroglycerin
heart medicine & explosive

Nitroglycerin is a dangerously unstable compound that undergoes a very rapid, highly exothermic decomposition to produce water, carbon dioxide, and nitrogen gas. The decomposition reaction is initiated by only a small shock. The fact that the reaction produces large quantities of hot gases (1 mole of nitroglycerin produces ~5.5 moles of gaseous products) makes it a very useful explosive. Its extreme instability, however, makes it a very dangerous material to handle. The Swedish chemist and industrialist Alfred Nobel (1833–1896) found that the stability of the material was markedly improved when it was mixed with an inert material such as kieselguhr (diatomaceous earth.) This mixture is known as dynamite. Nobel became both famous and rich from his invention. He eventually left a large portion of his fortune to establish the Nobel prizes awarded annually.

Nitroglycerin is such an unstable compound because it has, incorporated into the same molecule, both an oxidizing agent (the nitrate portion) and a reducing agent (the hydrocarbon part). Thus one molecule is capable of a self–sustaining internal oxidation–reduction reaction once enough energy is provided to begin the reaction. Another strong oxidizing agent, perchloric acid, also forms dangerously unstable esters with many organic compounds. These, too, are explosive. Unlike the nitrate esters, however, these perchlorate esters have not found commercial application.

## 5.14 BIOCHEMICALLY IMPORTANT ESTERS

With the exception of the fats discussed earlier, biochemically important esters are almost exclusively those of phosphoric acid. Phosphate esters of ribose and deoxyribose form the backbone of the molecules that contain and also those which transcribe the genetic information for all living things. Figure 5.17 shows the backbone structures of DNA and RNA. Note the phosphate group between the two ribose units. These biopolymers will be discussed in much more detail in chapter 15.

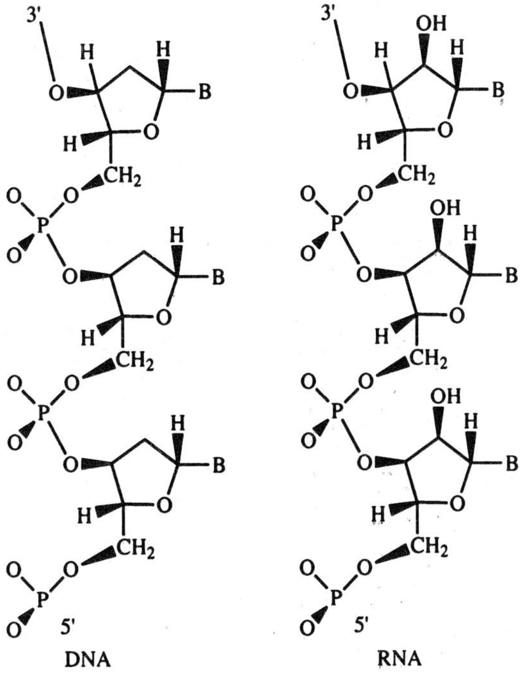

Besides their use in transmission of genetic information, phosphate esters are also involved in the metabolic pathways involved with the production of energy from food. Phosphate esters are ideal for this purpose because of the ease with which the reactions can be made to go in either direction, the relatively large energies involved in the reaction, and the ease with which the compounds can be transported from the sites where they are synthesized to the sites where they are needed for the production of energy.

The chief energy–transport molecule in the body is adenosine triphosphate (ATP) — a molecule composed of the nucleotide base adenine, a ribose sugar, and a polymeric form of phosphoric acid termed triphosphate. The structure of ATP is shown in figure 5.18. Hydrolysis of the terminal phosphate group produces adenosine diphosphate (ADP), inorganic phosphate ion and energy. This reac-

**FIGURE 5.17:** Backbone Structures of DNA and RNA

tion occurs primarily in cells where energy is required either to produce mechanical force (*i.e.*, the muscle cells) or to drive chemical reactions involved in other biosynthetic pathways (*i.e.*, virtually any cell). The chemical equation for the conversion of ATP to ADP is shown in equation 5.39.

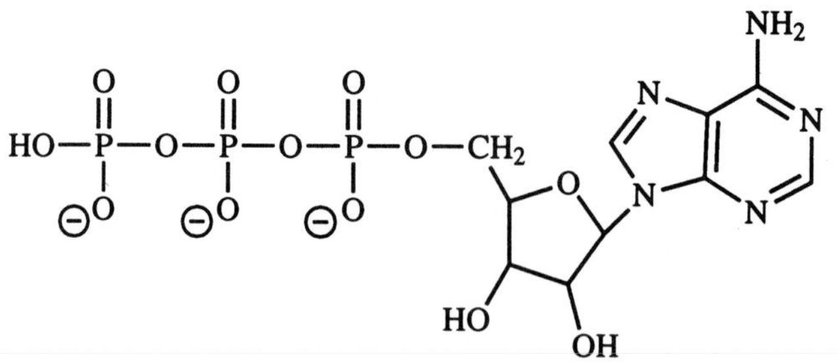

The standard free energy change for equation 5.39 is −32.9 kJ mol$^{-1}$, measured under conditions that approximate those of the living system. This is the amount of energy that is provided by hydrolysis of one mole of ATP at the site where the energy must be delivered.

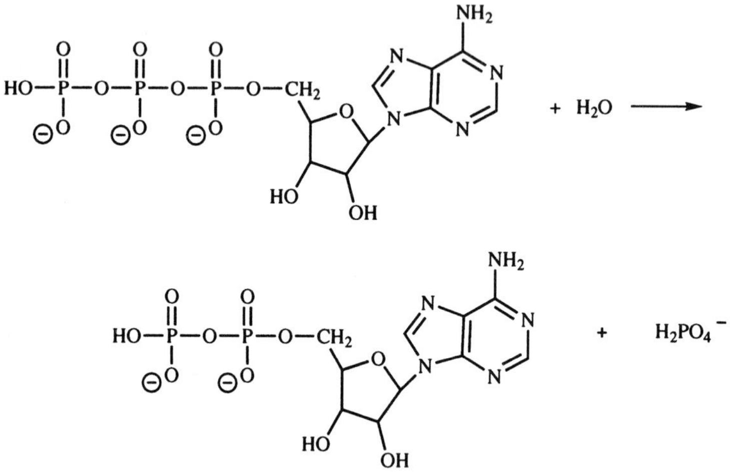

**FIGURE 5.18:**
**Structure of ATP**

EQ. 5.39

ATP is actually synthesized in the cell by the reverse of the reaction in equation 5.39. The energy to perform the synthesis comes from metabolic pathways that involve a number of other phosphate esters of biochemical importance. The structures of a number of other biochemically important phosphate esters are given in figure 5.19. The full details of reactions involving these compounds will be left to biochemistry courses.

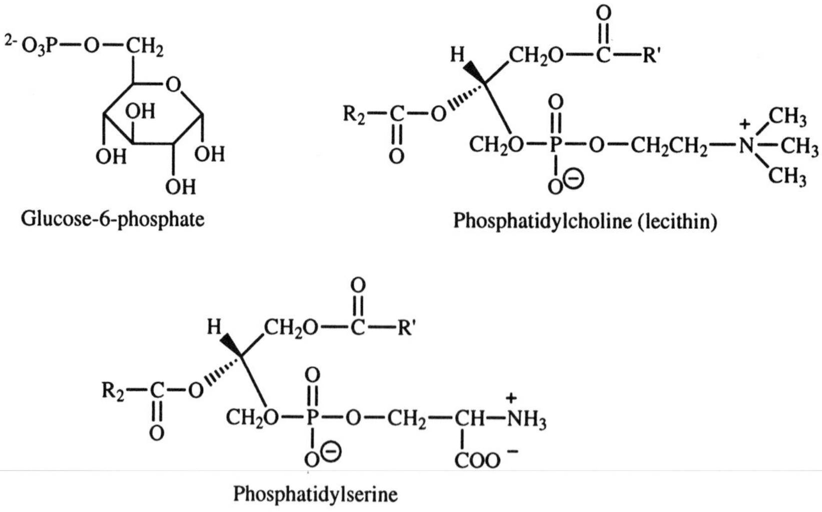

Glucose-6-phosphate

Phosphatidylcholine (lecithin)

Phosphatidylserine

**FIGURE 5.19:**
**Other Biochemically Important Phosphate Esters**

## 5.15   PESTICIDES: BIOCHEMICALLY AND ECONOMICALLY IMPORTANT ESTERS

Esters in the metabolic pathways have become inviting targets for molecules designed to interfere with those pathways in undesirable organisms. A number of pesticides work by mimicking certain molecules in the metabolic chain, thereby binding to enzymes that are involved in their production or degradation and blocking the chain at that point. The organism then dies either because some key intermediate cannot be synthesized or some toxic by-product cannot be converted into a harmless form. As a specific example, let us look at the pesticides malathion and parathion — materials that are quite toxic to certain insect species but which have a lower toxicity to mammalian species.

These pesticides work by interfering with the action of the enzyme acetyl-cholinesterase. Acetylcholine is a molecule released at the synapse of many nerves during nerve transmission. Until the acetylcholine is destroyed, the nerve is continuously stimulated. The body, therefore, has an extremely effective enzyme present near the nerve synapse, acetylcholinesterase, whose function is to catalyze the immediate hydrolysis of acetylcholine into choline and acetic acid as shown in equation 5.40. At one step during the hydrolysis mechanism the acyl group becomes attached to the enzyme. In a subsequent step the acyl group is released from the enzyme by the attack of a water molecule and the enzyme is regenerated.

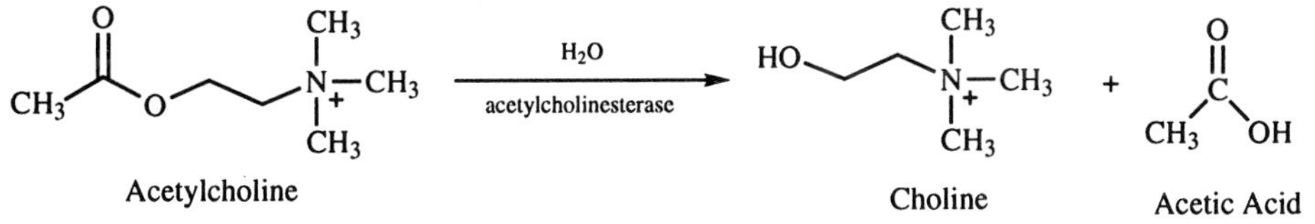

Acetylcholine    Choline    Acetic Acid    EQ. 5.40

As can be seen from the structures of malathion and parathion in figure 5.20, these compounds also have an ester linkage, albeit a phosphorous ester rather than a carbon ester. For reasons we won't explore here, the compounds themselves are relatively non–toxic to humans but become toxic when the sulfur atom is replaced by an oxygen atom. This can happen when the compound is exposed to elevated temperatures or light of the appropriate wavelength. During the step in the mechanism at which the acyl group becomes attached to the enzyme, the phosphate analog of the acyl group undergoes attachment to the enzyme. This group is not nearly as easily hydrolyzed as is the acyl group, and regeneration of the enzyme is prevented. Eventually taking enough enzyme molecules out of circulation results in the organism's inability to destroy acetylcholine.

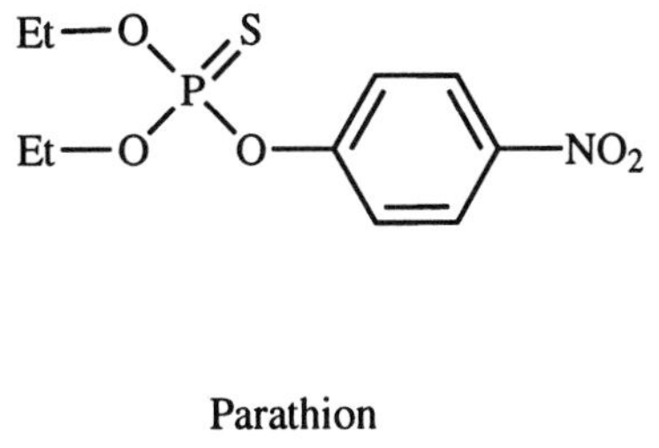

Parathion

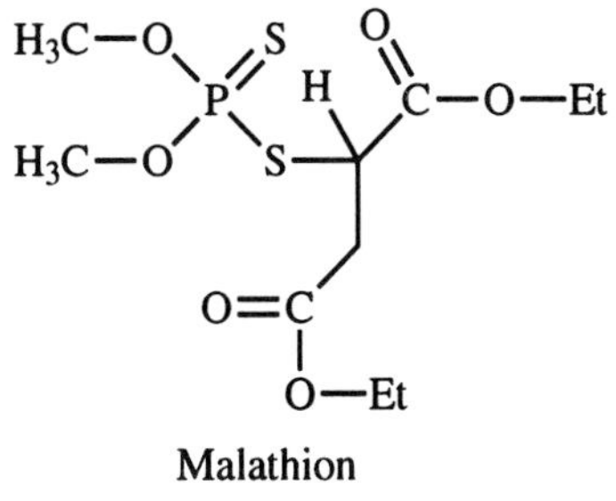

Malathion

FIGURE 5.20:
Structures of
Malathion and Parathion

In addition to the pesticides, a number of other materials are known which are acetylcholinesterase inhibitors. Unfortunately for the human species many of these are equally effective on mammals. Some of these are the very effective nerve gases Sarin and diisopropylfluorophosphate (DFP). The structures of these compounds are shown in figure 5.21. Although banned by International Law since after World War I, nations of all sizes have produced and maintained large stocks of these weapons of war in order, so they say, to be prepared in case an enemy should use them first. Only recently have the superpowers agreed to destroy their stocks of these weapons. As this is being written, the world is attempting to see that the stocks of these (and other) weapons have been eliminated in the country of Iraq.

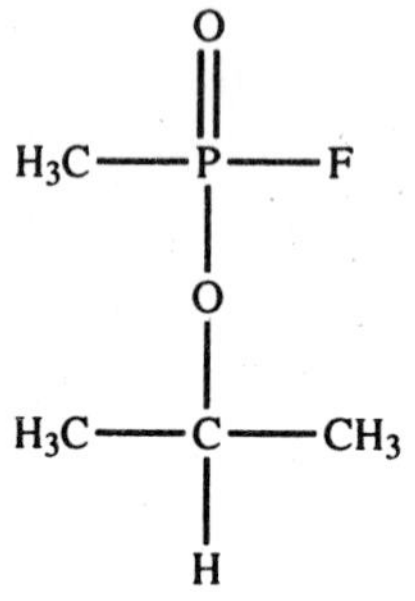

Sarin

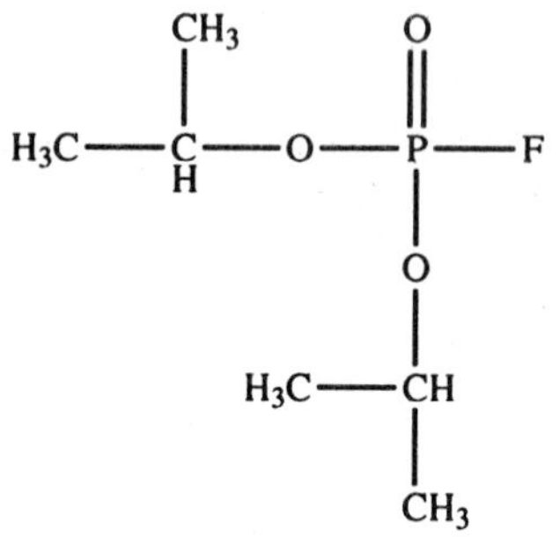

DFP

FIGURE 5.21:
Structures of
Sarin and DFP

**Acetal:** a compound, formed by reaction of an aldehyde or a ketone with excess alcohol, having 2 alkoxide groups bonded to the same carbon atom. See hemiacetal.

**Acid chloride:** a compound with a —COCl functional group. When hydrolyzed, this functional group produces a carboxylic acid by replacement of the chlorine atom with an hydroxyl group.

**Acid anhydride:** a compound having a —COOCO— functional group. It can be thought of as being formed by removal of a water molecule from the carboxyl groups on two carboxylic acid molecules. Hydrolysis produces two molecules of the parent acid.

**Azeotrope:** a mixture of two (or more) liquids which boils at a temperature either higher (termed a high–boiling azeotrope) than or lower (termed a low–boiling azeotrope) than any of the components of the mixture.

**Dean Stark trap:** a piece of glassware designed to remove water from a reaction mixture of water and a less dense, immiscible liquid. The solution is boiled, condensed, and the condensate allowed to run into this trap. The more dense water settles to the bottom and the less dense organic liquid overflows the trap and flows back into the pot.

**Ester:** usually the term refers to a compound with the functional group —COOR formed by replacing the —OH group of the carboxyl group with an alkoxide group. There are also esters of inorganic oxoacids formed by replacing one of the —OH groups of the oxoacid with an alkoxide group.

**Fischer esterification:** a reaction that produces an ester by reacting an alcohol and a carboxylic acid in the presence of a sulfuric acid catalyst.

**Hemiacetal:** a compound, formed by reaction of an aldehyde or a ketone with an alcohol, which has both one alkoxide and one hydroxide functional group bonded to the same carbon atom. See acetal.

**Hemiketal:** an older term for a hemiacetal formed from an alcohol and a ketone. The current practice is to call these compounds hemiacetals.

**Ketal:** an older term for an acetal formed from an alcohol and an aldehyde. The current practice is to call these compounds acetals.

**Lactone:** a cyclic ester formed from a difunctional hydroxy–acid.

**Reflux:** a term meaning to boil a reaction mixture, condense the vapors, and return the condensate to the reaction pot.

**Yield—actual/theoretical:** the actual/theoretical amount of product obtained from a reaction. Actual yield is often expressed as a percent of that which could be theoretically obtained by complete reaction.

1. Fill in the blanks with the structure of the organic compound.

    a. acetic acid + 1-propanol ⟶ **?** + water

    b. **?** + 2-butanol ⟶ 2-butylpropanoate + water

    c. benzoic acid + **?** ⟶ isopropylbenzoate + water

    d. **?** + methanol ⟶ [structure: CH with OH and OCH₃] + MeOH ⟶ **?**

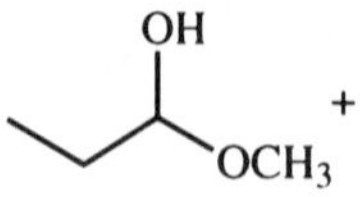

    e. propanal + ethanol ⟶ **?** + EtOH ⟶ **?**

    f. acetone + methanol ⟶ **?** + MeOH ⟶ **?**

2. Give the reactions for three different methods of preparing ethyl 3-methylbutanoate using ethanol.

3. Draw the structures of the esters which could be prepared from methanol and the following acids:

    a. 1,3-propanedioc acid
    b. nitric acid
    c. phosphoric acid

4. Give the hydrolysis products of the following compounds after hydrolysis in acid and in base.

    a. ethyl butanoate
    b. ethanenitrile
    c. benzoyl chloride
    d. acetic anhydride
    e. butyrolactone

5. a. Circle and lable all the hemiacetal and acetal linkages in figures 5.13, 5.14, and the following.

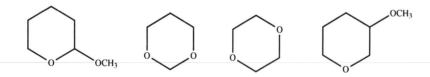

b. For each of the hemiacetal and acetal linkages in problem 5a, write the structures of the corresponding aldehydes or ketones and alcohols from which these compounds could be made.

6. Indicate the ester and anhydride bonds in the structures in figures 5.17, 5.18, and 5.19.

7. Draw structures of the hemiacetal and acetals which can result from reaction of 1-propanol and

     a. benzaldehyde
     b. diphenylketone
     c. 3-oxobutanal

8. Explain at least six ways in which you could maximize the yield of isobutyl 3-methylbenzoate prepared from the carboxylic acid and alcohol.

9. Draw the structure of the pentaacetate of glucose.

10. What is the structural difference between ribose and deoxyribose?

11. Calculate the theoretical yield for the ester formed by Fischer esterification using 0.12 mole of methanol and 0.26 mole of acetic acid. If 4.50 g of ester is actually isolated, what is the percent yield?

12. When the following compounds undergo esterification, sometimes ester polymers are formed, sometimes lactones are formed, and sometimes both are formed. For each of the following compounds tell whether the compound would be likely to form a polymer, a lactone, or both. Explain your answers.

     $HOCH_2COOH$
     $HO(CH_2)_2COOH$
     $HO(CH_2)_3COOH$

13. A test for the aldehyde group known as the Tollen's test uses silver ion, a very weak oxidizing agent, to oxidize the aldehyde group to a carboxyl group. Reduction of the silver ion during this test results in formation of a mirror on the inside of the test tube thus giving the test the name "the silver mirror test." Explain why alpha-methyl glycoside does not give a silver mirror while glucose does.

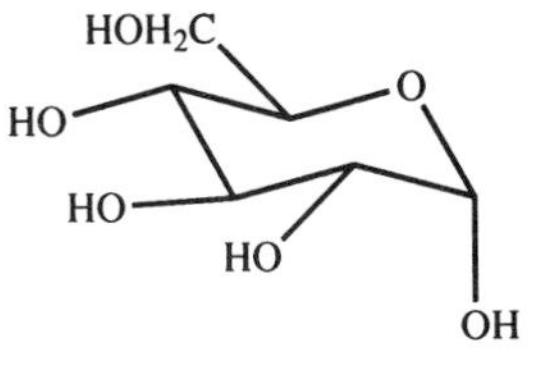

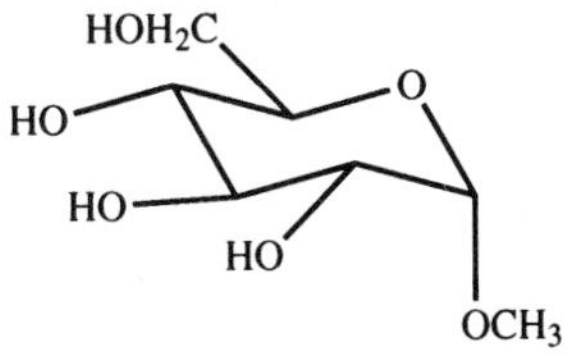

14. When pure alpha-glucose is dissolved in water containg a drop of HCl, isomerization occurs to give a mixture of alpha-glucose and beta-glucose. Explain how this occurs.

OVERVIEW
# Problems

15. Ethene (b.p. −103.7°C, $\Delta G_f$ = 68.12 kJ/mol) reacts with 1,3-butadiene (b.p. −4.4 °C, $\Delta G_f$ = 150.7 kJ/mol) to form gaseous cyclohexene (b.p. 83.0 °C, $\Delta G_f$ = 106.8 kJ/mol). The measured energy of activation for this synthesis of cyclohexene was 115 kJ/mole.

  a. Write the balanced reaction, showing the structure of reactants and products.
  b. Calculate the change in the standard state free energy for this reaction.
  c. Calculate the equilibrium constant for this reaction.
  d. Estimate the energy of activation for the reverse reaction.
  e. What sign do you expect for the entropy change for this reaction and why?
  f. What does the answer to part (e) imply about the selection of temperature for the synthesis of cyclohexene?
  g. The free energy of formation of liquid cyclohexene is 101.6 kJ/mole. What would this imply about the equilibrium constant for the reaction forming liquid product compared to that for the gaseous product?

16. Methyl formate ($\Delta G_f$ = -297.2 kJ/mol, $S°$ = 301.2 J/K mol) and gaseous water are formed when gaseous methanol and gaseous formic acid ($\Delta G_f$ = -351.0 kJ/mol, $S°$ = 248.7 J/K mol) react.

  a. Write the balanced reaction, showing the structure of reactants and products.
  b. Calculate the standard state change in free energy and entropy for the reaction.
  c. Calculate the equilibrium constant for the reaction.
  d. Calculate the enthalpy change for the reaction.
  e. If methyl formate was produced as a liquid, with the other species remaining in the gas phase, should the reaction be run at elevated or reduced pressure to optimize the yield of ester?

17. Gas phase ethylene glycol (1,2-dihydroxyethane, $\Delta G_f$ = -304.5 kJ/mol, $\Delta H_f$ = -389.3 kJ/mol) can be formed by reacting gaseous water with gaseous ethylene oxide ($\Delta G_f$ = -13.10 kJ/mol, $\Delta H_f$ = -52.63 kJ/mol).

  a. Write the balanced reaction, showing the structure of reactants and products.
  b. Calculate the standard state change in free energy and enthalpy for the reaction.
  c. Calculate the equilibrium constant for the reaction.
  d. Calculate the entropy change for the reaction.
  e. Discuss how the sign of the enthalpy and entropy changes should influence the choice of temperature to optimize the production of ethylene glycol.

# CHAPTER SIX

# 6

# REACTION
# Kinetics

HOW FAST DOES IT GO?

# REACTION

## Kinetics

## 6.1    INTRODUCTION TO CHEMICAL KINETICS

**E**arlier we determined from thermodynamics whether a specific reaction would be spontaneous or non-spontaneous.  The thermodynamic values also determined how far a particular reaction goes by establishing a value for the equilibrium constant.

However, knowing only thermodynamic state functions gives no insight concerning the path by which reactants become products.  Such chemical paths are known as mechanisms.  Understanding the mechanism by which reactants become products enables chemists to organize into a few general categories the myriad of known chemical reactions.  Understanding mechanisms also allows the chemist to model how currently unavailable products might eventually be made.

Our previous answers to "which way" and "how far" does a reaction go were supplied by thermodynamics.  Questions about "how" and "how rapidly" reactants become products will be answered through our examination of chemical kinetics in the next several chapters.

## 6.2    THE RATE OF CHEMICAL REACTIONS

The rate at which something happens is a ratio of a change in some parameter divided by a change in time.  The rate at which you drive a car is the change in position divided by the change in time (*e.g.*, miles per hour).  The rate at which you read is given by the number of words per minute.  In general, the **average rate** is given by the following equation:

$$\text{Average rate} = \frac{\Delta x}{\Delta t}$$

EQ. 6.1

This is an average rate measured over the relatively the time interval $\Delta t$.  The value of $\Delta t$ is usually relatively large.  Using the example of driving a car, the average rate is what is measured using the mileage reading on the odometer and a the time reading on a wrist watch;  *i.e.*, driving 25 miles in 30 minutes means an average speed of 50 miles per hour.  While you averaged 50 mph during the half hour trip, this does not mean that your speedometer was constantly at 50 mph for the entire trip.  The speedometer measures what is known as the instantaneous speed (or rate).  This measurement of the ratio of a change in position to the change in time is determined over a vanishingly small time increment, symbolized by *dt*.  Thus the **instantaneous rate** (that would be measured on the speedometer) is given by equation 6.2.

$$\text{Instantaneous rate} = \frac{\mathrm{d}x}{\mathrm{d}t}$$

EQ. 6.2

During the course of a chemical reaction, what is changing with time is the concentration of reactants and products. Using the square bracket symbolism for molarity, the average rate of the chemical reaction A → B is given by

$$\text{Average rate} = \frac{\Delta[B]}{\Delta t} = -\frac{\Delta[A]}{\Delta t}$$

EQ. 6.3

where $\Delta[B]$ is $[B]_{final} - [B]_{initial}$. Since B is being formed as a product, its concentration is increasing and the average rate is positive. Similarly the symbolism $\Delta[A]$ means $[A]_{final} - [A]_{initial}$. Since A is being destroyed as a reactant, its concentration is decreasing. Thus the negative sign is placed prior to the $\Delta[A]$ term in equation 6.3 in order to make the rate positive (we do not like to think of negative rates of reaction and it does not make sense that a reaction should have two different rates). If the reaction is simply A → B, then an average rate of one mole of B appearing per liter per minute means that there is one mole of A disappearing per liter per minute. This is reflected in Table 6.1 and Figure 6.1.

**TABLE 6.1**   CONCENTRATION AND TIME

| time (minutes) | [A] | [B] |
|---|---|---|
| 0.00 | 7.00 | 0.00 |
| 1.00 | 6.00 | 1.00 |

In figure 6.1, we see that the average rate of one mole per liter per minute equals the slope of the line for the concentration of B vs. time. The average rate is also given by the negative of the slope of the line for the concentration of A vs. time.

The situation is made more complicated when there are multiple reactants and products, and/or when the coefficients differ. Consider the general reaction:

A + 2 B ⟶ 3 C + 4 D          EQ. 6.4

for which the data table and graph are given below.

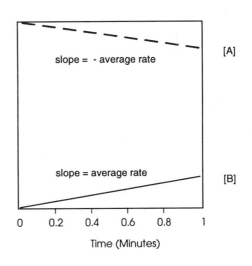

FIGURE 6.1:
Plot of Concentration vs.
Time for Data in Table 6.1

**TABLE 6.2**   CONCENTRATION AND TIME

| time(minutes) | [A] | [B] | [C] | [D] |
|---|---|---|---|---|
| 0.00 | 7.00 | 7.00 | 0.00 | 0.00 |
| 1.00 | 6.00 | 5.00 | 3.00 | 4.00 |

Since there is only one reaction going on, it really does not make sense to say that the average rate is one mole per liter per minute for A, two moles per liter per minute for B, three moles per liter per minute for C and four moles per liter per minute for D. Instead, chemical rates are in terms of moles of reaction as written per volume per time (just as the changes in thermodynamic state functions were given per mole of reaction as written.) This means that for the generic reaction:

$$a\,A \; + \; b\,B \quad \longrightarrow \quad c\,C \; + \; d\,D \qquad \text{EQ. 6.5}$$

the average rate is given by

$$\text{Average rate} = -\frac{1}{a}\frac{\Delta[A]}{\Delta t} = -\frac{1}{b}\frac{\Delta[B]}{\Delta t} = \frac{1}{c}\frac{\Delta[C]}{\Delta t} = \frac{1}{d}\frac{\Delta[D]}{\Delta t} \qquad \text{EQ. 6.6}$$

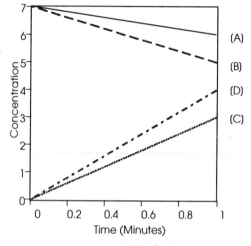

FIGURE 6.2:
Plot of Concentration *vs.*
Time for Data in Table 6.2

## 6.3   INSTANTANEOUS AND INITIAL RATES

In figure 6.2, the reaction between A and B to form C and D was imagined to cause a linear change in concentration with respect to time. It is the average rate over the entire time interval that is related to the slope of that imagined line, and given by equation 6.6. Since the data tables 6.1 and 6.2 show only what exists at time 0.00 and time 1.00, we have no knowledge of what happened in between those measurements. Currently it is possible to follow many reactions continuously by instrumental means. Following the absorbance of light (and thus the concentration of an absorbing reactant or product) as a function of time is a common example. Complications can occur, such as more than one species absorbing light at any available wavelength. But many times only a single reactant or product absorbs a specific wavelength of light, and we can

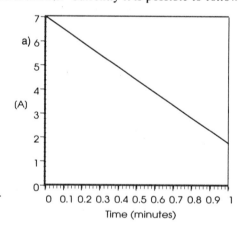

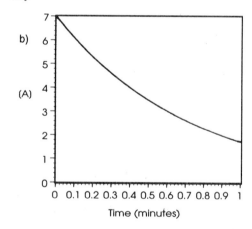

FIGURE 6.3:
Plots of Concentration *vs.* Time
for A → B
a) only 2 points measured
b) data continuously measured

"fill in the holes" in the data tables by measuring the concentration of a particular reactant or products continuously. In the large majority of such cases, what is observed between the two points defining the average rate is not a straight line, but a curved line.

The average rate is related to the slope of the line connecting the two points in figure 6.3(a). But the slope of the actual curve shown in figure 6.3(b) varies with time. The slope of the curve is steep at first, then becomes less so at later times. The slope of the line tangent to the curve at various times is related to the instantaneous rate of the reaction at those particular times. For the large majority of reactions, the steepness of the slope (and thus the instantaneous rate) is greatest at the start of the reaction. As the concentration of reactants decreases with time, the rate of reaction (steepness of the slope) decreases.

When instantaneous rate is measured, the defining relationship is still that outlined in equation 6.6, but with infinitesimal changes (d) as opposed to finite changes (Δ). This change is shown in equation 6.7:

$$\text{Instantaneous rate} = -\frac{1}{a}\frac{d[A]}{dt} = -\frac{1}{b}\frac{d[B]}{dt} = \frac{1}{c}\frac{d[C]}{dt} = \frac{1}{d}\frac{d[D]}{dt} \qquad \text{EQ. 6.7}$$

The instantaneous rate typically varies with time. The instantaneous rate determined at time zero is given the special name of the **initial rate**. Experimentally, this corresponds to steepness of the slope of the line tangent to the concentration-time plot at time zero, as seen in figure 6.4.

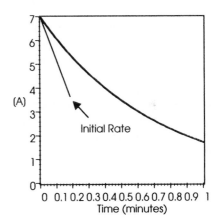

FIGURE 6.4:
Plots of Concentration *vs.* Time for A → B with Initial Rate

## 6.4   FACTORS THAT INFLUENCE CHEMICAL REACTION RATES

How fast a specific reaction goes depends upon several factors. These factors include (a) the presence of a **catalyst**, (b) the nature of the reaction, (c) the concentration (or physical state) of reactants/products, and (d) temperature.

Some changes in matter happen very quickly. They can occur so fast that, after a point, it is unclear if the process should be termed chemical or physical. Changes occurring in less than $10^{-12}$ seconds (such as absorption of light energy) are probably best classified as physical as opposed to chemical changes. At the other extreme are geochemical and cosmochemical processes that have apparently been creeping along for a few billion years (since the big bang). Some of these are very slow processes occurring within the nucleus of the atom, and again they are routinely relegated to the physicists. Between these extremes are reactions whose rates can be readily measured and for which variables can be controlled to determine their influence on the rate.

## 6.5  FACTOR ONE: THE PRESENCE OF CATALYSTS

Reactions that are thermodynamically spontaneous (have a negative $\Delta G$) may still take so much time that the reactions appear non-spontaneous. In Chapter 2 we examined the thermodynamics of a mixture of octane and oxygen. These reactants are "thermodynamically unstable" in the sense that they will ultimately form carbon dioxide and water in a spontaneous process, and release considerable energy in doing so. Yet the mixture of octane and oxygen is "stable" in the sense that a mixture containing only octane and oxygen will remain unreacted indefinitely. This is a "kinetic stability", meaning that although thermodynamically the reactants spontaneously form products, there is a barrier in the path connecting reactants to products. Again, thermodynamic state functions, like free energy, can not inform us about the path taken between reactants and products. But on the basis of kinetics, we picture the process to appear similar to figure 6.5.

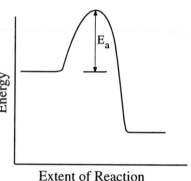

FIGURE 6.5:
Activation Energy
Diagram

The normal path connecting reactants to products includes a (usually) significant energy barrier, which is termed the **activation energy** ($E_a$). Before reactants can become products, bonds in the reactants must be broken. In the example of octane and oxygen, this bond breaking requires more energy than is typically provided by the thermal energy available at standard conditions. That is, the temperature determines the average kinetic energy of the gas molecules. If a collision between reactant molecules at 25°C lacks sufficient total kinetic energy to break the necessary bonds in the reactants, the reactants cannot pass over the energy barrier, and no reaction will occur.

A catalyst is a substance whose presence in a reaction mixture increases the rate of a reaction, but is itself neither a reactant nor a product. A catalyst is not consumed in the course of a reaction, and it is not a component of the initial or final energy states. A catalyst participates in the path connecting reactants to products. When an alternate pathway with a lower activation energy is provided by a catalyst, reactants become products more easily and faster. Although not a catalyst in the traditional sense of the word, in the case of the reaction between octane and oxygen, we can think of the spark as serving the role of the catalyst for that reaction. When "catalyzed" by a spark, the reaction occurs at such a rapid rate that the mixture is said to explode. In this particular case, the spark has opened an alternate pathway for the reaction by adding enough energy to one or two molecules to break bonds and form species known as radicals. These readicals are then involved in an exceedingly complex (and rapid) self–sustaining chain reaction. Radicals, species with unpaired electron spin, are formed in an initiation step in the spark. Chain propagation steps occur which convert reactants to products without changing the net number of radicals. The reaction terminates when one of the reactants is effectively exhausted, or when the chain is discontinued by having radicals combine to form spin-paired species. Leaded gasoline typically contains organolead compounds like tetraethyl lead, which readily decompose to provide a constant supply of radicals for maintaining a smooth combustion process as shown in equation 6.8.

$$Pb(C_2H_5)_4 \longrightarrow 4\ C_2H_5^{\bullet} + Pb \qquad \text{EQ. 6.8}$$

We have seen many examples of catalysts in the past two semesters. Recall that the addition of water to alkenes is normally done in the presence of a small quantity of a strong acid. The acid participates in the reaction by undergoing electrophilic attack on the pi cloud to form a carbocation intermediate. This intermediate then adds water to form a second intermediate which then loses a proton to reform the catalyst. The entire reaction sequence is outlined in figure 6.6.

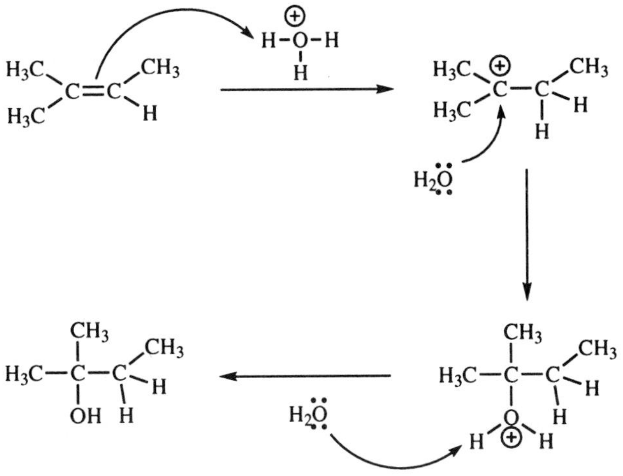

When catalyzed reactions involve large covalent species, as is often the case in biochemistry, subtle changes in structure can dramatically change reaction rates, often with serious consequences. Consider the examples of two polysaccharides, cellulose and amylose (a component of starch).

FIGURE 6.6:
Addition of Water
to an Alkene

Both cellulose and starch consist of polymeric chains of glucose. Cellulose, the basic structural component of plants, is a linear polymer of glucose units with β–1,4–linkages. Amylose is a linear polymer of glucose units with α–1,4–linkages. Meat–eating animals, such as humans, have enzymes that can hydrolyze the α–1,4–linkages but not the β–1,4–linkages. Thus humans can absorb the glucose formed from amylose as food. But dietary cellulose (*e.g.*, the roughage, or fiber in vegetables and whole grains) can not be hydrolyzed due to the small structural difference seen in figures 6.6 and 6.7. Thus cellulose passes through the human's digestive tract unaltered, despite the great similarity in structure to the readily hydrolyzed amylose. Both hydrolysis reactions are energetically favored. But cellulose is kinetically stable in the absence of effective catalysts. The β–1,4–linkages of cellulose can be hydrolyzed in a timely fashion only by certain bacteria and molds. The digestive tracts of ruminant animals (and termites) contain bacteria capable of this hydrolysis.

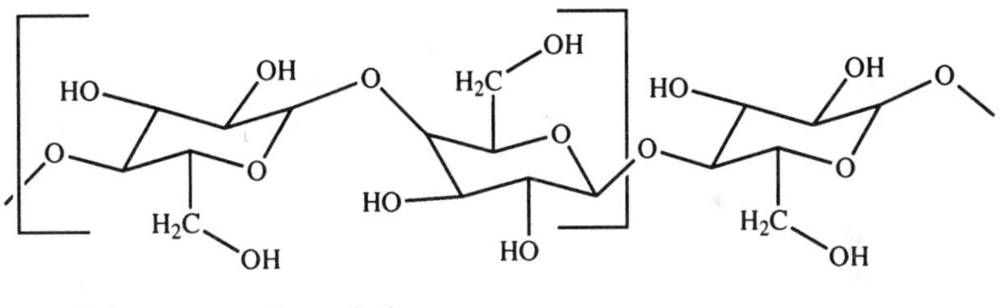

FIGURE 6.7:
Structure of Cellulose

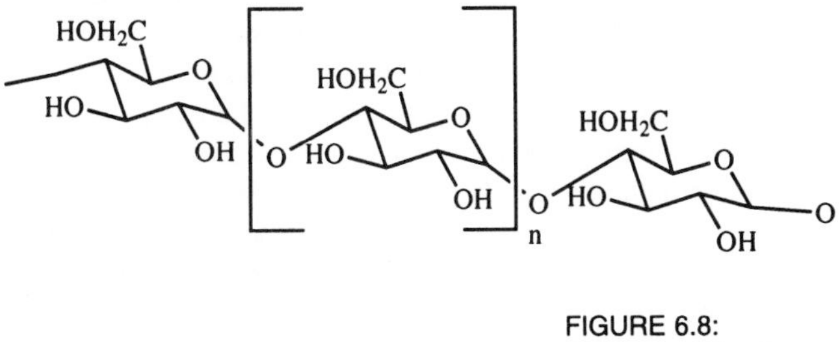

FIGURE 6.8:
Structure of Amylose

## 6.6 FACTOR TWO: THE NATURE OF THE REACTION

The rate of chemical change varies enormously from one reaction to the next. One key characteristic that separates the incredibly fast from the unbearably slow is the general type of reaction. Changes that involve a minimum of rearrangement of matter can (but will not necessarily) happen very rapidly. Such reactions may involve only the exchange of an electron as in the following redox example:

$$Ce^{4+} + Fe^{2+} \longrightarrow Ce^{3+} + Fe^{3+}$$

EQ. 6.9

or the exchange of a proton in the generic acid-base reaction:

$$HA + B \longrightarrow A^- + HB^+$$

EQ. 6.10

Thermodynamically spontaneous reactions between oppositely charged ions in solution also tend to occur rapidly, as in the net ionic precipitation reaction:

$$Ba^{2+}(aq) + SO_4^{2-}(aq) \longrightarrow BaSO_4(s)$$

EQ. 6.11

The same reaction in the solid phase, however, has a negligibly small rate:

$$Ba(NO_3)_2(s) + K_2SO_4(s) \longrightarrow BaSO_4(s) + 2\,KNO_3(s)$$

EQ. 6.12

In order to react, the reactants must physically contact one another; thus some motion is necessary to first bring reactants together, and then to move products out of the way of further reaction. The motion of ions in liquid solution provides this reaction opportunity. The relative lack of motion in the mixed solids greatly inhibits reaction.

Now let us consider the gas phase reaction between hydrogen and oxygen:

$$2\,H_2(g) + O_2(g) \longrightarrow 2\,H_2O(l)$$

EQ. 6.13

The reaction is definitely thermodynamically spontaneous, but very slow under standard conditions of temperature and pressure. The motion in the gas phase assures repeated collisions between hydrogen and oxygen. Yet, as was the case with octane and oxygen, collisions between the reactants at 25°C lack sufficient kinetic energy to achieve the activation energy, break bonds in the reactants, and initiate the reaction. A balloon filled with hydrogen and oxygen at room conditions is unreactive (in the absence of a spark or other catalyst, like finely divided platinum). Under different temperature and pressure conditions, however, the reaction proceeds explosively. Figure 6.8 shows the region of temperatures and pressures over which a hydrogen–oxygen mixture is explosive.

Such complex behavior is indicative of a reaction that is not simply a single collision between reactants directly forming products. Instead it indicates a complex mechanism, a stepwise series of smaller events leading to the overall reaction. More typically, the temperature is smoothly related to chemical rate, as shown in section 6.8.

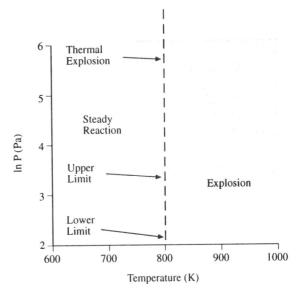

FIGURE 6.9:
Reactive Regions for a
Hydrogen-Oxygen Mixture

## 6.7   FACTOR THREE: PHYSICAL STATE AND CONCENTRATION

**PHYSICAL STATE**

Heterogeneous reactions (that involve more than one phase) have an additional factor influencing the rate: the physical state of the species involved. For example, the reaction between a solid substance and either a liquid or a gas is greatly dependent on the amount of exposed surface area of the solid. We know that in dissolving a powdered solid in a liquid, stirring facilitates the rate of solution (*e.g.*, dissolving powdered gelatin mix in water). This stirring increases the contact area between the solid and liquid phase species.

Similarly, a reaction between a gaseous reactant and a solid depends upon the surface of the solid. Finely ground solid reagents tend to react more rapidly than larger lumps of solid. Consider for example a cube 1.0 cm on a side, shown in figure 6.11. The exposed surface area of one side of the cube is 1.0 $cm^2$, so for six sides, a total of 6.0 $cm^2$ is available for reaction. If the same cube is chopped up into smaller cubes, each $10^{-2}$ cm per side, each of the $10^6$ small cubes has an area of 6.0 x $10^{-4}$ $cm^2$, for a total area of 6 x $10^2$ $cm^2$.

FIGURE 6.10:
Stirred Mixture of a
Solid in a Liquid

Further dividing the small cubes continues to increase the area, until a fine powder results. The mass initially in 1 cubic centimeter can then have many square kilometers of reactive surface area. This explains the phenomena of explosions in grain elevators and some coal mine explosions. We normally do not think of flour or coal as explosive. Yet when finely divided, the dust from these materials provides an enormous surface area for a combustion reaction in air. Such extremely rapid exothermic reactions cause explosions. Thus ventilation and other means of decreasing the concentration of combustible dust in the atmosphere are used in grain elevators and coal mines.

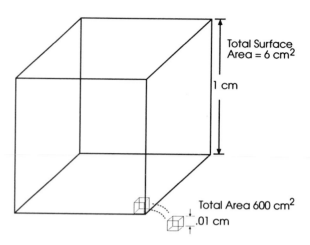

FIGURE 6.11:
Comparative Surface Areas of Same Volume of Material

## CONCENTRATION

In the case of the simple homogeneous solution reaction A $\rightarrow$ B, the rate could be followed in terms of the disappearance of A. The instantaneous rate R equals negative the slope of the tangent (at some time t) to the [A] versus time plot. Thus, as we saw earlier in Fig 6.3(b), the rate slows down as the reaction progresses. If we assume that the rate is dependent upon the concentration of A as shown below, then a rate that changes as reactant A is used up makes sense.

$$R \propto \left[A\right]^x$$

EQ. 6.14

For a general reaction a A + b B $\rightarrow$ c C + d D, the initial rate is assumed to have the form

$$R = k\left[A\right]^x\left[B\right]^y$$

EQ. 6.15

The equation above is termed a **rate law**, and expresses how the instantaneous rate R varies with concentration. The power "*x*" is called the **order** with respect to A; the power "*y*" is called the order with respect to B. The values of *x* and *y* must be experimentally determined, and should not be confused with the stoichiometric coefficients a and b. The sum *x* + *y* is referred to as the overall order of the reaction. Although a rate law of the form given in equation 6.15 is most common, it is also possible for the concentration of products and the concentration of catalysts to appear in the rate law. Although the orders of species appearing in the rate law are usually integral and positive, they may also be non-integral and/or negative. A zero order power for a particular concentration term indicates that its concentration does not influence the rate of the reaction. Unless there is some good reason to the contrary, species that do not affect the rate of the reaction (those species whose order is zero) are not written as part of the rate law. Of course, a negative order for a chemical species indicates that as its concentration increases, the reaction rate decreases. This is an uncommon situation, but one that does occur in some very complicated reaction mechanisms.

The proportionality constant k in equation 6.15 is called the **rate constant**. Its value also must be experimentally determined. For a given reaction mechanism, the orders *x* and *y* are constant, but the value of k will vary depending upon temperature.

## 6.8   FACTOR FOUR: TEMPERATURE

Reactions tend to happen more rapidly at higher temperatures. This makes sense because we have assumed that reactants need to collide with each other for a reaction to occur. Kinetic molecular theory indicates that the higher the temperature, the more rapidly the reactants are moving, and thus the more frequently collisions occur. A careful analysis shows that increasing the temperature of gas phase reactants by 10 °C may increase the number of collisions by a few percentage points (with the frequency of collision proportional to the square root of temperature). But for many typical reactions around room temperature, an increase of 10 °C causes a doubling of the reaction rate. The change in temperature influences the rate law by dramatically changing the rate constant k, as shown in figure 6.12.

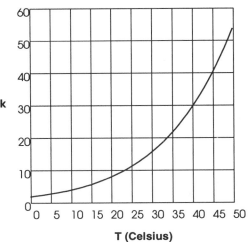

FIGURE 6.12:
Rate Constant *vs.*
Temperature

This large increase in the rate constant with increase in temperature means that the collison rate cannot be the only thing that is responsible for increasing the value of the rate constant as the temperature increases. The more important influence of T on k is not increasing the number of collisions, but increasing the fraction of those collisions with energy in excess of a specific value $E_a$ (the activation energy). Theory shows

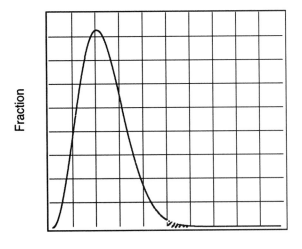

FIGURE 6.13:
Fraction of Collisions with
Specified Energy *vs.* Energy
at a Given Temperature

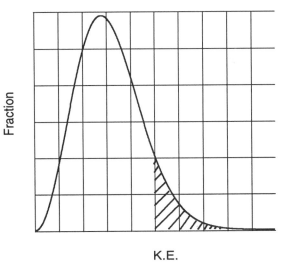

FIGURE 6.14:
Fraction of Collisions with
Specified Energy *vs.* Energy
at a Higher Temperature

that the distribution of kinetic energy among molecules has the majority with an energy near the average kinetic energy, and only a few molecules having a significantly higher or lower energy than the average energy. As temperature increases, the high energy "tail" skews to the right as shown in figures 6.13 and 6.14.

The fraction of collisions (f) with energy in excess of the energy $E_a$ is given by

$$f \propto e^{\left(-E_a/RT\right)}$$

EQ. 6.16

The equation that relates the rate constant k to the energy of activation is known as the **Arrhenius equation**:

$$k = Ae^{\left(-E_a/RT\right)}$$

EQ. 6.17

The proportionality constant A includes terms related to the number of collisions and any specific required orientation of the collisions in order for the reaction to proceed. Reactions inhibited by a large amount of steric hindrance have a correspondingly low value of A. For most reactions under most conditions it is safe to assume that A is essentially constant for a specific reaction (*i.e.*, independent of temperature), and that a plot of ln k versus 1/T will be linear.

$$\ln k = \ln A - \frac{E_a}{RT}$$

EQ. 6.18

You may notice that equation 6.18 is in the form y = b + m*x*, with y = ln k, *x* = $T^{-1}$, the y-intercept being ln A, and the slope being -$E_a$/R. If at temperature $T_1$ the rate constant is determined to be $k_1$, and at $T_2$ determined to be $k_2$, the change in ln k can be calculated as follows:

$$\ln k_2 - \ln k_1 = \left(\ln A - \frac{E_a}{RT_2}\right) - \left(\ln A - \frac{E_a}{RT_1}\right) = -\frac{E_a}{R}\left(\frac{1}{T_2} - \frac{1}{T_1}\right)$$

EQ. 6.19

and dividing by the change in reciprocal temperature yields the slope:

$$\text{slope} = \frac{\Delta y}{\Delta x} = \frac{\ln k_2 - \ln k_1}{\frac{1}{T_2} - \frac{1}{T_1}} = -\frac{E_a}{R}$$

EQ. 6.20

If we know any four of the five variables ($T_1$, $T_2$, $k_1$, $k_2$, and $E_a$), the fifth can be determined.

## CONCEPT CHECK

The rate constant for the decomposition of dinitrogen pentoxide into dinitrogen tetraoxide and oxygen has been determined at a number of temperatures. The data are shown in table 6.3.

TABLE 6.3    RATE CONSTANT DATA                                  TABLE 6.3

| k  (sec$^{-1}$) | T($^\circ$C) |
|---|---|
| 7.87 x 10$^3$ | 0 |
| 3.46 x 10$^5$ | 25 |
| 4.98 x 10$^6$ | 45 |
| 4.87 x 10$^7$ | 65 |

We can choose any two of these points and utilize equation 6.20 by placing into it values for $T_1$, $T_2$, $k_1$, and $k_2$. However, since we have more than two data points we will obtain a more statistically valid answer by plotting the data in the form of a straight line and determining the slope of the line. Whichever method we use, we must not forget that the temperature must be first converted to Kelvin. Although these calculations can easily be handled with a calculator, using a spreadsheet program, especially one that can plot, is simple and convenient. Output from such a program is shown as table 6.4.

TABLE 6.4    ARRHENIUS DATA FOR PLOTTING                         TABLE 6.4

| k (sec$^{-1}$) | T($^\circ$C) | T(K) | ln  k | 1/T |
|---|---|---|---|---|
| 7.87 x 10$^3$ | 0 | 273 | 8.971 | 0.003663 |
| 3.46 x 10$^5$ | 25 | 298 | 12.754 | 0.003356 |
| 4.98 x 10$^6$ | 45 | 318 | 15.421 | 0.003145 |
| 4.87 x 10$^7$ | 65 | 338 | 17.701 | 0.002959 |

When these data are graphed, the result is figure 6.15.

Choosing two points that are on the line, (10.00, 0.00355) and (17.5, 0.00300), we can obtain the slope:

$$\text{slope} = \frac{\Delta y}{\Delta x} = \frac{10.0 - 17.5}{0.00355 - 0.00300} = -1.\overline{3}64 \times 10^4 = -\frac{E_a}{R}$$

$$E_a = -\left(-1.\overline{3}64 \times 10^4\right)(8.314) = 1.\overline{1}34 \times 10^5 \, ^{J}/_{mol} = 1\overline{1}0 \, ^{kJ}/_{mol}$$

EQ 6.21

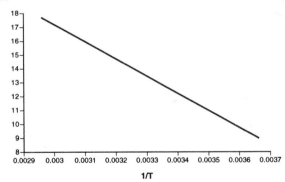

FIGURE 6.15:
Arrhenius Plot

If we did not have easy access to a spreadsheet to do the calculations and plot the graph, direct use of equation 6.20 using two points from the initial data would yield the work in equation 6.22.

$$E_a = -R\left(\frac{\ln k_2 - \ln k_1}{\frac{1}{T_2} - \frac{1}{T_1}}\right)$$

$$= -8.314\left(\frac{\ln\left(3.46 \times 10^5\right) - \ln\left(7.87 \times 10^3\right)}{\frac{1}{25 + 273} - \frac{1}{0 + 273}}\right)$$

EQ 6.22

$$= -8.314\left(\frac{12.75 - 8.97}{.003356 - .003663}\right) = 1.0\bar{2} \times 10^5 \text{ J}/_{\text{mol}}$$

$$= 1\bar{0}2 \text{ kJ}/_{\text{mol}}$$

## 6.9   EXPERIMENTAL DETERMINATION OF KINETIC ORDER

As we will see in chapter 7, a crucial clue in trying to determine the mechanism of a reaction is the rate law, and specifically the experimental orders of species involved in the reaction. There are numerous ways of determining the order with respect to each of the reactants, and we shall examine several.

### INTEGRATED RATE LAWS

The rate law for the reaction aA $\rightarrow$ bB may be of the form

$$R = k\left[A\right]^x$$

EQ. 6.23

Since the rate can be measured in terms of the disappearance of reactant A, this becomes

$$R = -\frac{1}{a}\left(\frac{d[A]}{dt}\right) = k\left[A\right]^x$$

EQ. 6.24

The concentration terms are put onto one side, and time on the other.

$$\frac{d[A]}{[A]^x} = -ak\,dt$$

EQ. 6.25

Using $[A]_o$ to represent the concentration of A at t(ime) = 0, and $[A]_t$ to represent the concentration of A at t(ime) = t, we can set up the following definite integral that we can then solve for various values of $x$:

$$\int_{[A]_o}^{[A]_t} \frac{d[A]}{[A]^x} = -ak\int_0^t dt$$

EQ. 6.26

If $x = 0$ (zero order with respect to A), the final result of the integration and evaluation at the limits yields an equation for which the rate does not depend upon the concentration of A:

$$[A] = [A]_0 - kt$$

EQ. 6.27

If $x = 1$ (first order with respect to A), integration, evaluation at the limits, and rearrangement yield:

$$\ln[A]_t = \ln[A]_o - akt$$

EQ. 6.28

This is known as the integrated first order equation. While it can be written in several forms, the equation above is in the form of $y = b + mx$. That is, a plot of ln [A] vs. time will be linear (with a slope of –ak) provided the reaction aA → bB is first order with respect to A.

If $x \neq 1$ then integration of equation 6.26 yields:

$$\frac{1}{[A]_t^{x-1}} = \frac{1}{[A]_o^{x-1}} + (x-1)\, akt$$

EQ. 6.29

Thus if the reaction aA → bB is second order in A, a plot of $[A]^{-1}$ versus time will be linear; if third order in A, a plot of $[A]^{-2}$ versus time will be linear; if zero order with respect to A, a plot of $[A]$ versus time will be linear. Determination of [A] as a function of time, when plotted in various fashions against time, will indicate the order with respect to A. These are summarized in the graphs below.

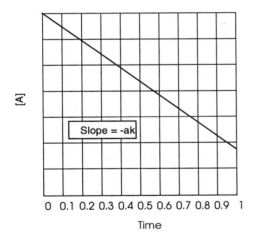

FIGURE 6.16:
Plot of [A] *vs.* t
for a Zero Order Reaction

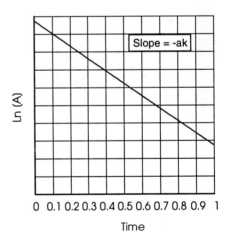

FIGURE 6.17:
Plot of ln [A] *vs.* t
for a First Order Reaction

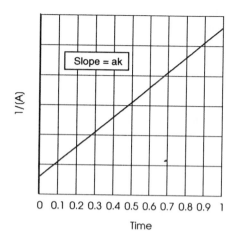

FIGURE 6.18:
Plot of 1/[A] *vs.* t
for a Second Order Reaction

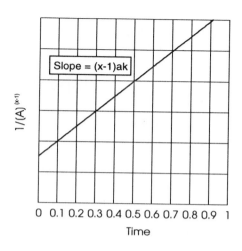

FIGURE 6.19:
Plot of 1/[A]$^{x-1}$ *vs.* t
for an **x**-Order Reaction

## CONCEPT CHECK

We wish to determine the rate law for the cyclization of butadiene at a certain temperature. The reaction was studied by determining the concentration of butadiene as a function of time. The following data were collected:

TABLE 6.5   TIME - CONCENTRATION DATA

TABLE 6.5

| Time (sec) | [Butadiene] (M) |
|---|---|
| 166 | 0.0156 |
| 513 | 0.0141 |
| 1059 | 0.0124 |
| 1853 | 0.0106 |
| 3519 | 0.0081 |
| 6915 | 0.0055 |

Since there is only one reactant for which data are available, we must assume a rate law of the following form:

$$\text{Rate} = \Big[\text{butadiene}\Big]^{x}$$

EQ. 6.30

In order to determine the order for the reaction, we must plot the data according to the integrated rate laws in equations 6.28 and 6.29. This is most easily done if a spreadsheet program with graphing capabilities is available. If these data are placed into a spreadsheet program and the appropriate calculations performed, the following table is generated:

TABLE 6.6

**TABLE 6.6** SPREADSHEET VALUES FOR PLOTTING

| Time (sec) | [Butadiene] (M) | ln [Butadiene] | 1/[Butadiene] |
|---|---|---|---|
| 166 | 0.0156 | −4.160 | 64.103 |
| 513 | 0.0141 | −4.262 | 70.922 |
| 1059 | 0.0124 | −4.390 | 80.645 |
| 1853 | 0.0106 | −4.547 | 94.340 |
| 3519 | 0.0081 | −4.816 | 123.457 |
| 6915 | 0.0055 | −5.203 | 181.818 |

If you do not have access to a spreadsheet program, the same calculations can be done using your calculator.

Plotting the data in three different graphs produces the following:

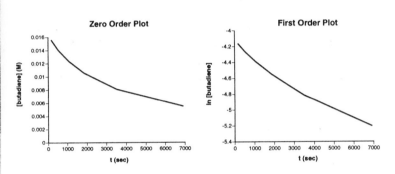

FIGURE 6.20:
Rate Law Plots

Examing these graphs shows that the best fit occurs with the second order plot. The zero order plot is definitely curved; the first order plot shows some curvature upon closer examination. All that remains is then to determine the rate constant for the reaction. From equation 6.29 it can be seen that the slope of the line is ak. We need only to find the value of the slope and the value of 'a'. The chemical equation for the cyclization is

$$H_2C = \overset{\overset{\textstyle H}{|}}{C} - \overset{\overset{\textstyle}{|}}{\underset{\underset{\textstyle H}{|}}{C}} = CH_2 \longrightarrow \square$$

EQ. 6.31

This shows that the value of 'a' is 1. To find the value of the slope, we merely choose two points that are <u>on the best straight line</u> through the points. Let us choose the points (6000, 160) and (2000, 99). According to equation 6.32, the slope of the line is

$$\text{slope} = \frac{\Delta y}{\Delta x} = \frac{6000 - 2000}{160 - 99} = 6\overline{5}.57$$

EQ. 6.32

The complete rate law can now be written:

$$\text{Rate} = \left(66 \; \text{L/mol·sec}\right) [\text{butadiene}]^2$$

EQ. 6.33

## FRACTIONAL LIFE METHOD

The integrated rate laws presented earlier in this section can be rearranged to reflect the ratio of reactant remaining to that initially present. Thus, the integrated first order rate law, equation 6.28, becomes

$$\ln\left(\frac{[A]_o}{[A]_t}\right) = akt$$

EQ. 6.34

The fractional life $t_y$ (where $y$ is a fraction less than one) is the time it takes for sufficient reaction to occur to remove fraction y of the initial amount of A. If $y$ equals one-half, this time is called the half life. For a half life symbolized $t_{0.5}$ and a reaction aA $\rightarrow$ bB that is first order in A, equation 6.34 becomes

$$t_{1/2} = \frac{\ln\left(\dfrac{[A]_o}{[A]_t}\right)}{ak} = \frac{\ln\left(\dfrac{[A]_o}{0.5\,[A]_o}\right)}{ak} = \frac{\ln 2}{ak}$$

EQ. 6.35

There is nothing unique about a fraction of one half. A one-fifth life, (the time required for one-fifth of the initial amount of A to decompose, symbolized by $t_{0.2}$) would work equally well, and yield (note that when 20% of the initial amount of A has decomposed, the current concentration of A is $0.8[A]_o$):

$$t_{1/5} = \frac{\ln\left(\dfrac{[A]_o}{[A]_t}\right)}{ak} = \frac{\ln\left(\dfrac{[A]_o}{0.8\,[A]_o}\right)}{ak} = \frac{\ln 1.25}{ak}$$

EQ. 6.36

The truly unique feature of equations 6.35, 6.36, and their other fractional life equivalents, is that the fractional life is independent of the initial amount present. However, this is true only for a reaction that is first order overall. The most common use of half-life is in the description of radioactivity. All known radioactive decay processes are first order. For example, the half-life of $^{230}U$ is 20.8 days. Since radioactive decay follows first order kinetics, we do not have to specify how much $^{230}U$ is initially present. After 20.8 days, 50% will remain; after 41.6 days, 25% will remain; and after 62.4 days, (three half-lives) 12.5% will remain.

The methods used in equations 6.34 and 6.35 can be applied to reactions other than first order reactions to obtain equations that relate the fractional–life of the reaction to the rate constant. However, for all orders other than first order, the fractional–life of the reaction depends upon the initial concentration(s) of the reactant(s). This gives such equations limited utility. Therefore we shall not pursue this subject further.

## METHOD OF INITIAL RATES

For reactions that have more than one species appearing in the rate law, the integrated rate law and fractional life approaches, while possible to use, prove very cumbersome. That is, for a reaction such as: $a A + b B \rightarrow c C + d D$, an alternate method of determining the order of reaction is advised. The most common method used for this situation utilizes the initial rate of reaction (the instantaneous rate measured at time zero). The initial rate method is also commonly used for determining the rates of enzyme–catalyzed reactions.

While experimentally convenient, the experimental determination of the initial rate is subject to significant errors. Not the least of these is involved in determining the slope of the tangent line to a curve at a point where only one side of the curve is available (see figure 6.21). This is problematic at best.

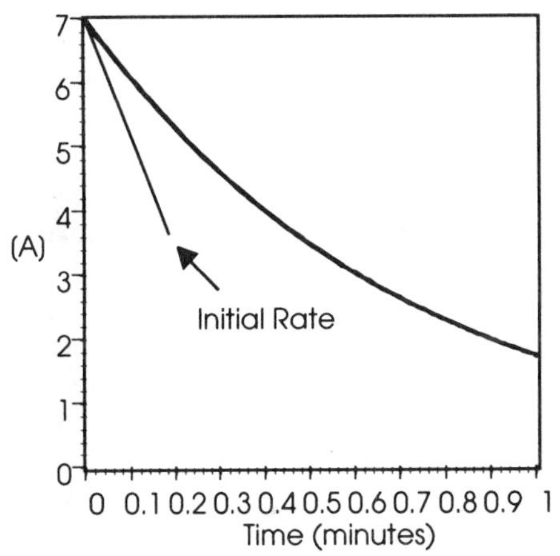

Added to this difficulty is the experimental problem of deciding what is time zero: Is it measured at the start of mixing? at the end of mixing? at the time of the first measurement? The first points measured are likely to be most error prone, since the system may still be mixing, thermal gradients will be at their largest, *etc.* Finally, using the method of initial rates means that multiple experiments at differing initial concentrations of some of the components must be done. At a minimum, the number of experiments must be one more than the number of species whose orders are being determined. Even with all of these problems, the use of the initial rate method proves quite beneficial in determining the orders of each of the multiple reactants.

Consider the data in table 6.7 below, for the reaction in equation 6.37.

$$2 H_2(g) + 2 NO(g) \longrightarrow 2 H_2O(g) + N_2(g)$$

FIGURE 6.21:
Determination of
Initial Rate

EQ. 6.37

TABLE 6.7

**TABLE 6.7**    INITIAL RATE DATA FOR NO + H$_2$ REACTION

| Trial # | [NO]$_o$ (M) | [H$_2$]$_o$ (M) | Initial Rate (mol / L sec) |
|---------|--------------|-----------------|----------------------------|
| 1 | 0.0060 | 0.0010 | 0.025 |
| 2 | 0.0060 | 0.0020 | 0.050 |
| 3 | 0.0060 | 0.0030 | 0.075 |
| 4 | 0.0010 | 0.0090 | 0.0063 |
| 5 | 0.0020 | 0.0090 | 0.025 |
| 6 | 0.0030 | 0.0045 | 0.028 |

The symbol [NO]$_o$ means the initial concentration of nitric oxide, after mixing but before reaction starts. The symbol R$_o$ is used for a a generic initial rate; R$_1$, [NO]$_1$ and [H$_2$]$_1$ symbolize the initial rate in trial one, the initial concentration of nitric oxide in trial one, and the initial concentration of hydrogen in trial one. We assume that the rate law for equation 6.37 has the following form:

$$R = k \left[NO\right]^x \left[H_2\right]^y$$

EQ 6.38

After substituting the initial rate and initial concentrations of nitric oxide and hydrogen into equation 6.37, it still looks hopeless. We have one equation, but three unknowns: k, $x$ and $y$. We can, however, use the data above to solve for all three unknowns, by taking the ratio of rate expressions. Carefully choosing which trials to compare will make this exercise possible. If we take the ratio of trial two to trial one, we see the following:

$$\frac{R_2}{R_1} = \frac{k\left(\left[NO_2\right]_2\right)^x \left(\left[H_2\right]_2\right)^y}{k\left(\left[NO_2\right]_1\right)^x \left(\left[H_2\right]_1\right)^y} = \frac{k\left(0.0060\right)^x\left(0.0020\right)^y}{k\left(0.0060\right)^x\left(0.0010\right)^y} = \left(\frac{0.0020}{0.0010}\right)^y$$

EQ 6.39

The rate constants, k, in equation 6.39 cancel; because, since temperature is constant, the rate constant is the same for both trials. The concentration terms for nitric oxide are numerically the same on top and bottom; this cancels the unknown value of $x$. Substituting the measured initial rates for trial one and trial two into equation 6.39 yields

$$\frac{R_2}{R_1} = \frac{0.050}{0.025} = 2.0 = \left(2.0\right)^y$$

EQ 6.40

It is probably intuitively obvious that $y = 1$ in equation 6.40. You will, however, need to be able to solve such expressions explicitly for situations in which it will not be intuitively obvious. This is done by taking the natural logarithm of both sides of the equation and rearranging:

$$\ln(2.0) = \ln(2.0)^y = y\ln(2.0)$$

$$y = \frac{\ln(2.0)}{\ln(2.0)} = 1.0$$

EQ 6.41

The reaction is first order with respect to hydrogen: doubling the concentration of hydrogen (without changing anything else) doubles the rate. Next is the determination of $x$, the order with respect to nitric oxide. In trying to determine the influence of nitric oxide, it is simplest to find a pair of experiments where that is the only variable changing. Only in trials 4 and 5 is the concentration of hydrogen held constant, so we would most likely choose those. It does not matter which order the trials are expressed in the ratio, as long as we are consistent on both sides of the equation:

$$\frac{R_5}{R_4} = \frac{k\left([NO_2]_5\right)^x\left([H_2]_5\right)^y}{k\left([NO_2]_4\right)^x\left([H_2]_4\right)^y} = \frac{k(0.0020)^x(0.0090)^y}{k(0.0010)^x(0.0090)^y} = \left(\frac{0.0020}{0.0010}\right)^x$$

EQ 6.42

Again, substituting measured values for the initial rates yields

$$\frac{R_5}{R_4} = \frac{0.025}{0.0063} = 3.\bar{9}68 = (2.0)^x$$

$$x = \frac{\ln(3.\bar{9}68)}{\ln(2.0)} = \frac{1.3\bar{7}83}{0.6\bar{9}3} = 1.\bar{9}8 = 2.0$$

EQ 6.43

Thus the reaction is determined to be first order with respect to hydrogen, second order with respect to nitric oxide, and third order overall. Also note that we did not have to choose trials four and five in determining the order for nitric oxide once the order for hydrogen was already known; any two trials in which the concentration of nitric oxide changed would work, even if hydrogen concentration changed. For instance, comparing trial six and four yields the same results:

$$\frac{R_6}{R_4} = \frac{k\left([NO_2]_6\right)^x\left([H_2]_6\right)^y}{k\left([NO_2]_4\right)^x\left([H_2]_4\right)^y} = \frac{k(0.0030)^x(0.0045)^y}{k(0.0010)^x(0.0090)^y} = \left(\frac{0.0030}{0.0010}\right)^x(0.50)$$

EQ 6.44

Substituting measured initial rates into this yields a second order for nitric oxide, the same as previously determined:

$$\frac{R_6}{R_4} = \frac{0.028}{0.0063} = 4.\bar{4}4 = (3.0)^x(0.50)$$

$$x = \frac{\ln(8.\bar{8}8)}{\ln(3.0)} = 1.\bar{9}8 = 2.0$$

EQ 6.45

Finally, the rate constant can be experimentally determined from these data once all the orders are known. Equation 6.38 is rearranged and values for rate and concentrations are substituted to give

$$k = \frac{R_1}{\left([NO]_1\right)^2\left([H_2]_1\right)^1} = \frac{0.025 \ ^{mol}/_{L \cdot sec}}{\left(0.0060 \ ^{mol}/_{L \cdot sec}\right)^2\left(0.0010 \ ^{mol}/_{L \cdot sec}\right)^1} = 6.9 \times 10^5 \ ^{L^2}/_{mol^2 \cdot sec}$$

EQ 6.46

Note that the rate constant has units that depend upon the overall order of reaction. If rate is expressed in moles per liter per second, then the rate constant will have units of $\frac{L^{(n-1)}}{mol^{(n-1)} \cdot sec}$ where n represents the overall order of reaction.

Finally, also note that, although the balanced reaction of equation 6.37 has coefficients of two for both nitric oxide and hydrogen, the orders for these are two and one respectively. In many mechanisms, orders and stoichiometric coefficients will not match.

## 6.10 KINETICS AND EQUILIBRIUM

As we saw in chapter 2, no reaction goes to completion. There is always a driving force in the reverse direction. Let us again consider the generic example of $aA + bB \rightarrow cC + dD$. Earlier we assumed a rate law in the form of $R = k[A]^x[B]^y$. That is, we assumed that C and D played no part in the reaction kinetics. What was being determined, in fact, was the kinetic expression for the forward reaction. The rate of the forward reaction, $R_f = k_f[A]^x[B]^y$, will decrease as A and B are consumed (assuming $x$ and $y$ are not negative). At the same time, the reverse reaction, perhaps with a rate law of the form $R_r = k_r[C]^u[D]^v$ has the concentration of products C and D increasing. Thus, as the reaction progresses, A and B are consumed (reducing the forward rate) and C and D are formed (increasing the reverse rate.) This means that the forward reaction rate is decreasing and the reverse reaction rate is increasing. Ultimately, at equilibrium, these two rates will become equal. If the rate of the reverse reaction is significant, it is possible that there will be significant *re*formation of reactants and the measurement of the forward rate will be in error. (One of the reasons for the popularity of the method of initial rates is to avoid this complication. Initially no product is present, and one determines the influence of only the reactants on the rate law.)

As we noted above, eventually, the reaction between A and B forming C and D reaches equilibrium. This does not mean that all reactions have stopped. Although the macroscopic concentrations of reactants and products do not change, both the forward and reverse reactions continue at equal and offsetting rates:

$$aA + bB \underset{R_r}{\overset{R_f}{\rightleftharpoons}} cC + dD$$

EQ 6.47

If the forward rate law is of the form

$$R_f = k_f [A]^a [B]^b$$

EQ 6.48

and the reverse rate law is of the form

$$R_r = k_r [C]^c [D]^d$$

EQ 6.49

then under equilibrium conditions, when the forward and reverse rates are equal,

$$k_f [A]_{eq}^a [B]_{eq}^b = k_r [C]_{eq}^c [D]_{eq}^d$$

EQ 6.50

The ratio of rate constants is the ratio of the concentrations of the products over the concentrations of the reactants at equilibrium, or the equilibrium constant:

$$\frac{k_f}{k_r} = K_{eq} = \frac{[C]_{eq}^c [D]_{eq}^d}{[A]_{eq}^a [B]_{eq}^b}$$

EQ 6.51

This equation makes intuitive sense in that, for a rapid forward reaction (large $k_f$) and a slow reverse reaction (small $k_r$), the equilibrium constant will lie far to the right. But as shown in examples at the start of this chapter, there is no guarantee that thermodynamic spontaneity (a large K) necessarily implies a fast forward reaction. A significant activation energy for both the forward and reverse reactions may make either direction appear non-spontaneous. See figure 6.22.

Finally, you may have noticed that the orders in the forward and reverse rate laws ($x$, $y$, $u$, $v$) magically transformed themselves into the stoichiometric coefficients $a$, $b$, $c$, $d$, despite our demonstration at the end of section 6.9 that the orders are not necessarily the same as the coefficients. Proof that this is allowed in this case lies beyond the scope of this course; suffice it to say that this result is due to the fact that the path taken going from reactants to products is necessarily the same path taken during the reverse trip, from products to reactants.

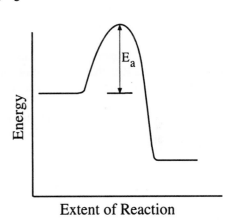

FIGURE 6.22:
Energy vs.
Extent of Reaction

**Activation energy:** the energy barrier (symbolized by $E_a$) separating reactants and products. The height of this barrier helps determine the rate constant of the reaction according to the Arrhenius equation.

**Arrhenius equation:** $k = A e^{(-E_a/RT)}$ by which the rate constant k is related to the activation energy $E_a$ and the temperature.

**Average rate:** the ratio of the macroscopic change in concentration to the macroscopic change in time. This is the chemical rate measured over a relatively long time frame.

**Catalyst:** a chemical species that influences the rate of a reaction, but is neither created nor destroyed by the reaction.

**Initial rate:** the instantaneous rate measured at time zero.

**Instantaneous rate:** the ratio of the infinitesimal change in concentration to the infinitesimal change in time. This is related to the slope of the tangent to the curve of concentration versus time.

**Order:** the order of a specific species is the power to which the concentration term of that species appears in the rate law. The order of each reactant in an overall equation must be experimentally determined. The overall order of a reaction is the sum of all orders of all species appearing in the rate law.

**Rate constant:** a proportionality constant (symbolized by k), that relates the instantaneous rate R to the concentration terms appearing in the rate law. The rate constant is a constant for a specific reaction at a specific temperature.

**Rate law:** the mathematical relation that equates the instantaneous rate, R, to the product of the rate constant and the concentration terms taken to the appropriate powers.

1. In one or two sentences, clearly distinguish among average rate, initial rate and instantaneous rate.

2. In one or two sentences, explain what is meant by (a) a rate law; (b) a rate constant.

3. The reaction $3 A \rightarrow 2 B$ changes the concentration of A from its initial value of 2.50 molar to a value of 1.85 molar after the reaction has progressed for 1.00 hours. Express the average rate of the reaction over this time interval in units of moles per liter per second.

4. If the reaction in problem 3 was determined to be first order with respect to A, write the integrated rate law for the reaction (*i.e.*, in the form which shows what should be plotted versus time in order to obtain a straight line).

5. Calculate the value of the rate constant appearing in problem #4. Calculate the value of the half-life for this reaction.

6. For the reaction dealt with in problems 3, 4 and 5, calculate the concentration of A remaining 4.00 hours after the start of the reaction. Hint: use the initial concentration and the integrated rate law and rate constant determined earlier.

7. The reaction $2 A \rightarrow B$ is second order in A, with a rate constant of $6.5 \times 10^{-5}$ liter $mol^{-1}$ $sec^{-1}$. If initially the concentration of A is 3.00 **M**, calculate the amount of A present after (a) 1.00 minutes; (b) 1.00 hours; (c) 1.00 day. (Hint: use the integrated rate law.)

8. The rate constant in problem #7 was measured at 25.0°C. Increasing the temperature to 60.0°C increased the rate constant to $1.0 \times 10^{-3}$ liter $mol^{-1}$ $sec^{-1}$. Calculate the activation energy for the reaction in problem #7 in units of kJ/mol.

9. From the results of problem #8, calculate the rate constant expected at 100.0°C.

10. The rate constant for the myosin catalyzed hydrolysis of ATP varies with temperature as indicated below. Calculate the $E_a$ and A terms in the Arrhenius equation for this reaction:

| T (°C) | k ($s^{-1}$) |
|---|---|
| 39.9 | $4.67 \times 10^{-6}$ |
| 43.8 | $7.22 \times 10^{-6}$ |
| 47.1 | $10.0 \times 10^{-6}$ |

11. The combination of myoglobin (Mb) with oxygen in solution to form oxymyoglobin ($MbO_2$) follows the rate law: $R = k [O_2] [Mb]$. In a 0.1 **M** phosphate buffer, the reaction has a rate constant of $5 \times 10^6$ lit/mol sec at 1.0°C, and $7 \times 10^6$ at 11.0°C. Calculate the $E_a$ and A terms in the Arrhenius equation for this reaction.

12. If the initial concentration of Mb and $O_2$ are both $3.0 \times 10^{-6}$ **M**, what is the initial rate of the reaction described in problem 11 at 1.0°C?

13. A "three minute egg" refers to the relative texture/hardness of an egg boiled for three minutes under standard atmospheric conditions (meaning at 100.0°C). At 8,000 feet elevation, the boiling point of water becomes 92.0°C. If the activation energy for the denaturation of egg protein (responsible for the texture of the boiled egg) is 40 kJ/mol, how long will one have to cook a "three minute egg" at this temperature?

14. The reaction $SO_2Cl_2 \rightarrow SO_2 + Cl_2$ is first order in $SO_2Cl_2$, with $k = 2.2 \times 10^{-5}$ s$^{-1}$ at 320°C. What % of $SO_2Cl_2$ remains after heating for 1.00 hour?

15. Butadiene dimerizes in the gas phase at elevated temperatures according to the reaction $2 C_4H_6(g) \rightarrow C_8H_{12}(g)$. The total pressure of the system is monitored, from the initial reading (when only butadiene is present) to subsequent readings, when the pressure reflects the combined pressure of both monomer and dimer. If the total pressure was initially $P_0$, and the total pressure at some later time t was $P_t$, derive an expression for the partial pressure of $C_4H_6$ at time t (in terms of $P_0$ and $P_t$).

16. From the data below, and using the relationship derived in problem #15, calculate the partial pressure of butadiene at each of the listed times.

| t (min) | $P_{total}$ (atm) | $P_{butadiene}$ |
|---------|-------------------|-----------------|
| 0.00    | .8315             | .8315           |
| 17.30   | .7464             | ?               |
| 42.50   | .6761             | ?               |
| 90.05   | .5864             | ?               |
| 259.5   | .5013             | ?               |

17. Using the integrated rate law, show that the data in #16 support a second order dependence (*i.e.*, draw the appropriate graph) and calculate the value of the rate constant in atm$^{-1}$ min$^{-1}$.

18. Use the method of initial rates to determine the orders and rate constant for the reaction $H_2SeO_3 + 6 I^- + 4 H^+ \rightarrow Se + 2 I_3^- + 3 H_2O$.

| $[H_2SeO_3]_o$ | $[I^-]_o$ | $[H^+]_o$ | $R_o$ (mole/lit sec) |
|---|---|---|---|
| $2.00 \times 10^{-4}$ | .0300 | .0200 | $1.2 \times 10^{-6}$ |
| $4.00 \times 10^{-4}$ | .0300 | .0200 | $2.4 \times 10^{-6}$ |
| $2.00 \times 10^{-4}$ | .0900 | .0200 | $3.3 \times 10^{-5}$ |
| $2.00 \times 10^{-4}$ | .0300 | .0500 | $7.6 \times 10^{-6}$ |

19. Use the method of initial rates to determine the orders and rate constant for the acid catalyzed reaction of iodine with acetone:

$$I_2 + CH_3COCH_3 \rightarrow CH_3COCH_2I + H^+ + I^-$$

| $[CH_3COCH_3]_o$ | $[I_2]_o$ | $[H^+]_o$ | $R_o$ (mole/lit sec) |
|---|---|---|---|
| $1.5 \times 10^{-3}$ | .0300 | .0200 | $2.1 \times 10^{-9}$ |
| $3.0 \times 10^{-3}$ | .0300 | .0200 | $4.3 \times 10^{-9}$ |
| $3.0 \times 10^{-3}$ | .0900 | .0200 | $4.4 \times 10^{-9}$ |
| $3.0 \times 10^{-3}$ | .0300 | .0500 | $1.1 \times 10^{-8}$ |

20. In the presence of base, the reaction of $OCl^- + I^- \rightarrow OI^- + Cl^-$ has the following initial rate data. Use the method of initial rates to determine the orders and rate constant for the reaction.

| $[OCl^-]_o$ | $[I^-]_o$ | $[OH^-]_o$ | $R_o$ (mole/lit sec) |
|---|---|---|---|
| $1.00 \times 10^{-3}$ | $1.00 \times 10^{-3}$ | 1.00 | $6.1 \times 10^{-5}$ |
| $4.00 \times 10^{-3}$ | $1.00 \times 10^{-3}$ | 1.00 | $2.4 \times 10^{-4}$ |
| $4.00 \times 10^{-3}$ | $5.00 \times 10^{-3}$ | 1.00 | $1.2 \times 10^{-3}$ |
| $2.00 \times 10^{-3}$ | $5.00 \times 10^{-3}$ | 0.500 | $1.2 \times 10^{-3}$ |

21. Radioactive $^{64}Cu$ in the form of copper acetate is used to study Wilson's disease. $^{64}Cu$ has a half-life of 12.7 hours. How much unreacted $^{64}Cu$ remains after exactly one day (24 hours)? How long does it take for 99% of the isotope to decompose?

### CHAPTER SIX

OVERVIEW
**Problems**

22. Isobutylene (2-methylpropene) undergoes acid catalyzed hydrolysis, with the concentration of isobutylene varying according to the table:

| t (min) | [isobutylene] |
|---|---|
| 0.00 | 0.0800 |
| 5.00 | 0.0747 |
| 10.0 | 0.0698 |
| 15.0 | 0.0651 |
| 20.0 | 0.0608 |

a. Write the balanced hydrolysis reaction, showing the structure of the expected major product.
b. Determine the order of the reaction with respect to isobutylene concentration.

23. Oxalic acid ($H_2C_2O_4$) decomposes when dissolved in concentrated sulfuric acid. The rate of the reaction can be determined by titrating a measured volume of the reaction mixture at varying times, using potassium permanganate. A 25.00 ml sample of the oxalic acid reaction mixture is titrated at the times indicated below, using the volumes of 0.2000 **M** $MnO_4^-$ listed.

a. Write the balanced redox reaction between oxalic acid and permanganate ion that occurs in acidic solution.

| t (hours) | volume $KMnO_4$ (ml) |
|-----------|----------------------|
| 0.00      | 12.50                |
| 2.00      | 9.83                 |
| 4.00      | 7.73                 |
| 6.00      | 6.08                 |
| 8.00      | 4.79                 |
| 10.0      | 3.77                 |

b. Calculate the concentration of oxalic acid present at each of the times listed in the table above.

c. Show graphically that the reaction is first order with respect to oxalic acid concentration.

24. Hydrogen peroxide decomposes to form water and oxygen gas. When a 100.0 ml sample of aqueous hydrogen peroxide (density 1.11 g/ml) is completely reacted, the resulting dry oxygen, collected at 30.0°C in a 12.50 liter vessel, generates a pressure of 740.2 torr. This shows that the initial concentration of hydrogen peroxide was 9.788 **M**.

a. Show that the data supports the claim that the initial concentration of hydrogen peroxide was 9.788 **M**. (Note: You must write a balanced chemical reaction!)

b. Given the data below, for a fresh 100.0 ml sample of aqueous hydrogen peroxide undergoing decomposition (and the oxygen measured as before),

| t (min) | P $O_2$ (torr) |
|---------|----------------|
| 0.00    | 0.00           |
| 4.00    | 75.8           |
| 8.00    | 143.8          |
| 12.00   | 204.8          |
| 16.00   | 259.7          |

calculate the moles of oxygen gas formed, the moles of hydrogen per-oxide reacted, the moles of hydrogen peroxide remaining, and the concentration of hydrogen peroxide remaining for each entry in the table.

c. Show graphically that the decomposition of hydrogen peroxide is first order, using the information tabulated in part (b).

25. For certain gas phase reactions, the total pressure can be a means of monitoring the rate of the reaction. Consider the thermal decomposition of ethylene oxide ($C_2H_4O$) to form methane and carbon monoxide. Initially, only ethylene oxide is present, but as time progresses, methane and carbon monoxide are formed. The moles of each product formed is directly related to the moles of ethylene oxide reacted. A series of measurements showed the following:

| t (min) | $P_{total}$ (torr) |
|---------|--------------------|
| 0.00 | 100.0 |
| 5.00 | 106.4 |
| 10.00 | 112.4 |
| 15.00 | 118.1 |
| 20.00 | 123.3 |

a. Show by Dalton's law how the total pressure is related to the partial pressures of ethylene oxide, methane, and carbon monoxide.

b. If the pressure of ethylene oxide decreases by $x$ torr over a certain time interval, the pressure of methane and the pressure of carbon monoxide both increase by $x$ torr. Using the equation determined in (a), calculate the value of $x$ for each entry in the table above.

c. Calculate the partial pressure of ethylene oxide at each time listed in the table above. (Note: the pressure of ethylene oxide at time 0 is 100.0 torr, and it is 100.0 torr − x at every listed time.)

d. Calculate ln P of ethylene oxide and plot that versus time. What does this linear plot indicate about the order of the reaction with respect to ethylene oxide?

e. Calculate the value of the rate constant from the plot in (d).

26. For the gas phase reaction $2\,HI \rightarrow H_2 + I_2$, the following rate constants for the forward rate constant ($k_f$) were determined.

| T (K) | k (lit mol$^{-1}$ sec$^{-1}$) |
|-------|-------------------------------|
| 500.0 | $3.26 \times 10^{-9}$ |
| 550.0 | $1.91 \times 10^{-7}$ |
| 600.0 | $5.65 \times 10^{-6}$ |
| 650.0 | $9.95 \times 10^{-5}$ |
| 700.0 | $1.16 \times 10^{-3}$ |

a. Calculate the standard state free energy change and equilibrium constant for the reaction, given that the standard state free energy of formation of gaseous iodine is 19.327 kJ/mole.

b. Calculate the value of A and $E_a$ in the Arrhenius equation from the tabulated data.

c. Calculate the forward rate constant at 25.0 °C.

d. Calculate the reverse rate constant ($k_r$) at 25.0 °C.

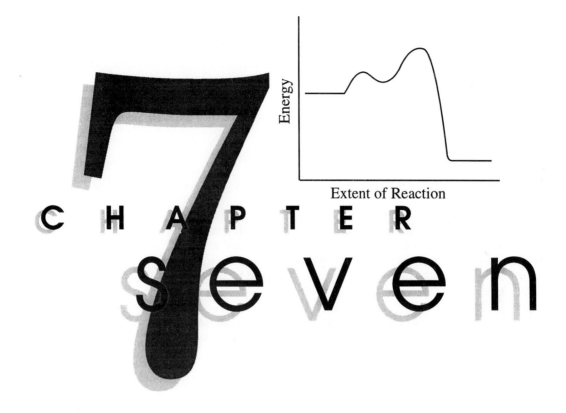

# REACTION
# Mechanisms
### WHAT REALLY HAPPENS?

# REACTION ─
## MECHANISMS: What Really Happens?

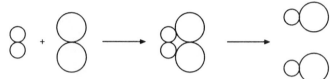

## 7.1   ELEMENTARY REACTIONS ─

Consider the general reaction:

$$A_2 + B_2 \rightarrow 2AB \qquad\qquad \text{EQ. 7.1}$$

The simplest picture of what could occur is that $A_2$ collides with $B_2$ and as a result of that collision two molecules of product AB form. A cartoon image of this process is shown in figure 7.1.

FIGURE 7.1:
Elementary reaction of two diatomic molecules.

A process wherein reactants collide and, in a single step, produce products is known as an **"elementary" reaction**. The entire process results from a single collision, described as one elementary step. The concentration versus time profiles for elementary reactions are always just as expected: the rate at which $A_2$ (or $B_2$) disappears is matched by the rate at which AB appears.

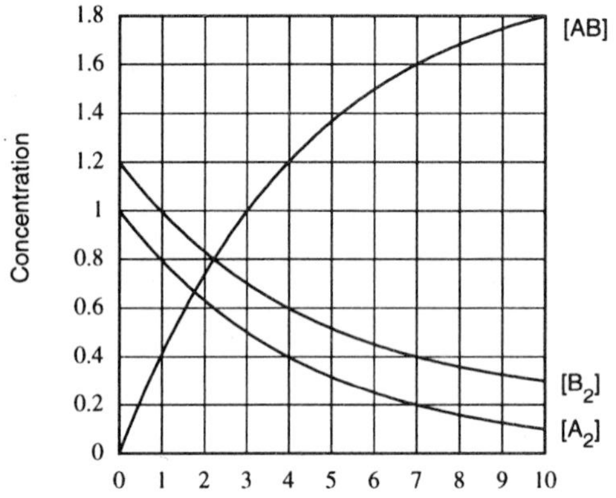

The rate of this elementary reaction is expected to be proportional to the rate at which $A_2$ and $B_2$ collide. Consider the figure below, which shows a motionless target molecule $B_2$, with a certain concentration of $A_2$ molecules moving around it (figure 7.3a). If we double the concentration of $A_2$ molecules in the vicinity of the $B_2$ target molecule and hold all other factors constant (figure 7.3b) we expect the number of collisions per unit time between the $B_2$ molecule and $A_2$ molecules to double. Since the temperature (and thus kinetic energy) was held constant in these two cases, the fraction of collisions leading to products remains constant. Thus doubling the concentration of $A_2$ doubles the rate of collisions, which doubles the chemical rate of reaction.

FIGURE 7.2:
Variation of [A], [B], and [AB] with time.

As we saw last chapter, if doubling the concentration of $A_2$ means doubling the rate, then the reaction is first order in $A_2$. Similarly, if we double the concentration of $B_2$ and keep everything else constant, there should be twice the number of collisions between $B_2$ and $A_2$ molecules per volume per time, and thus twice the rate of reaction. Again, this leads to a first order dependence of the reaction rate on $B_2$ in the rate law. Thus for the elementary reaction 7.1, we expect the rate law:

$$R = k[A_2]^1[B_2]^1 \qquad\qquad \text{EQ. 7.2}$$

That is, *for elementary reactions only*, the rate law can be predicted on the basis of reaction stoichiometry. As another example, consider the reaction of two monomer molecules, M, combining to form a dimer, D:

$$2M \rightarrow D \qquad\qquad \text{EQ. 7.3}$$

a)
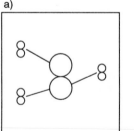

b)
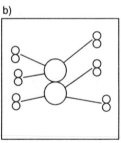

FIGURE 7.3:
The influence of concentration on collision frequency.

If this is an elementary reaction, it occurs in a single step exactly as written. That is, two molecules of M collide to form the molecule D. M plays a dual role in the collision: it is both the target molecule for the collision *and* the species hitting the target molecule. Doubling the concentration of M thus not only doubles the number of objects doing the hitting, it doubles the number of objects being hit. This results in a fourfold increase in reaction rate. As we saw with the method of initial rates, if doubling the concentration of M results in a four–fold increase in the reaction rate, then M appears in the rate law to the second power. For elementary reactions only, the stoichiometric coefficient of each reactant necessarily equals the order of that reactant in the rate law.

## 7.2 MECHANISMS

For elementary reactions, the decrease in concentration of reactants always exactly matches the concentration increase in products. That is, when the rate of destroying reactants is at its maximum, the rate of forming products is also at its maximum; when the rate of destroying reactants slows, so does the rate of product formation. For other types of reactions, however, the process by which reactants become products does not involve only a single collision, but rather these reactions occur by a series of elementary steps. This may be shown in terms of a concentration profile abnormality, an experimental rate law with orders not matching reaction coefficients, or by other means.

It is sometimes observed, for example in the reaction U → W, that the rate at which the reactant U disappears does not match the rate at which the product W appears. Figure 7.4 shows a concentration-time profile in which there is a delay in the production of product W. This time delay is called an induction period, and indicates the presence of at least one species (V) which is first produced and then destroyed during the **mechanism** (a series of elementary steps which sum to the overall reaction.)

Thus, although the observed stoichiometry of the balanced overall reaction is U → W, it is clear that the process is not elementary (*i.e.*, it does not occur in a single step). Instead we could propose a mechanism given below:

| U → V | elementary step one | EQ. 7.4 |
| V → W | elementary step two | EQ. 7.5 |
| | | |
| U → W | observed overall reaction | EQ. 7.6 |

The overall reaction (equation 7.6) is proposed to occur *via* a two step mechanism, given by equations 7.4 and 7.5.

Another type of evidence that a reaction is not elementary is an experimental rate law with powers that do not match the stoichiometric coefficients. The balanced reaction:

$$NO_2(g) + CO(g) \rightarrow NO(g) + CO_2(g) \qquad \text{EQ. 7.7}$$

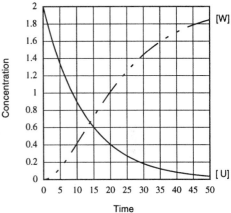

FIGURE 7.4: Concentration-time profile for U→W showing induction period.

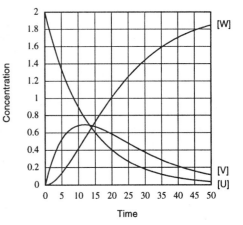

FIGURE 7.5: Concentration-time profile for U, V and W.

if it occurred as an elementary process would be predicted to have an experimental rate law of:

$$R = k[NO_2]^1[CO]^1$$

EQ. 7.8

The actual rate law for this reaction is experimentally determined to be:

$$R = k[NO_2]^2$$

EQ. 7.9

That is, the reaction is second order in nitrogen dioxide and zero order in carbon monoxide. Thus the reaction is clearly not simply the collision of a molecule of nitrogen dioxide with a molecule of carbon monoxide.

A possible mechanism for reaction 7.7 that satisfies the observed rate law (equation 7.9) is given by equations 7.10 and 7.11, which add up to form the reaction in equation 7.7.

$$NO_2 + NO_2 \rightarrow N_2O_4$$

EQ. 7.10

$$N_2O_4 + CO \rightarrow NO + NO_2 + CO_2$$

EQ. 7.11

The proposed two-step mechanism by which nitrogen dioxide and carbon monoxide become nitric oxide and carbon dioxide involves two separate collisions, each an elementary step. As we shall see later, if the first of the steps (7.10) is much slower than the second step (7.11), the rate of the overall mechanism will agree with the experimental rate law shown in equation 7.9.

## 7.3 ORDER AND MOLECULARITY: A SUMMARY

In discussing the kinetics of a particular mechanism, the terms "order" and "molecularity" are both used. Their meanings are sometimes erroneously interchanged, especially in older scientific literature. Both are described with numerical values, but they have very different meanings. The order with respect to a certain chemical species is the power of the concentration of that species appearing in the observed rate law, and must be experimentally determined. The order is a positive or negative integral or half-integral value: 1, 0, 1.5, -2, and so forth. Order is associated with each chemical species, and depends on the mechanism as a whole.

Molecularity, on the other hand, depends only on individual elementary steps, and does not depend on the individual species or the overall mechanism. Molecularity quantifies the number of particles on the left hand side of a specific elementary reaction, and commonly has values of two, one, or in rare circumstances, three. A unimolecular reaction is an elementary step with one reactant particle, *viz.*:

$$AB \rightarrow A + B$$

EQ. 7.12

A bimolecular reaction involves two particles colliding in an elementary process. The identity of the particles is not important; thus both elementary steps 7.13 and 7.14 are described as bimolecular.

$$A + B \rightarrow AB \qquad\qquad\qquad\qquad \text{EQ. 7.13}$$

$$A + A \rightarrow A_2 \qquad\qquad\qquad\qquad \text{EQ. 7.14}$$

Since molecularity refers only to elementary processes, the numerical value specifies the number of particles undergoing a single collision. It is possible, though statistically highly unlikely, that a simultaneous three body collision can occur. Equation 7.15, occurring as an elementary process, is referred to as termolecular.

$$A + B + C \rightarrow ABC \qquad\qquad\qquad\qquad \text{EQ. 7.15}$$

Note that the following sequence of steps, which comprise a possible mechanism for the overall reaction equation 7.15, consists of two bimolecular reactions, and is not equivalent to a single-step, termolecular reaction.

$$A + B \rightarrow AB \qquad\qquad\qquad\qquad \text{EQ. 7.16}$$

$$AB + C \rightarrow ABC \qquad\qquad\qquad\qquad \text{EQ. 7.17}$$

## 7.4 THE RATE DETERMINING STEP OF A MECHANISM

A mechanism is a series of elementary steps. In nearly all cases, one of the steps in the sequence is decidedly slower than all the others. Consider the following example of a "mechanism" for getting a college degree.

| | |
|---|---|
| get to campus | step 1 |
| find a place to live | step 2 |
| move in | step 3 |
| meet all degree requirements | step 4 |
| go to commencement | step 5 |

Steps 1, 2, 3 and 5 require a few minutes to a few hours. Step 4 may take four (or more) years, and is the slow step, or "rate determining step". The rates of all the other steps are negligible compared to the rate of step 4; if it takes five years to complete step 4, the total time spent is five years.

To demonstrate the principle using chemical reactions, consider the previous example mechanism:

**Mechanism A**

$$NO_2 + NO_2 \rightarrow N_2O_4 \qquad\qquad \text{step a; } k_1 \qquad\qquad \text{EQ. 7.18}$$

$$N_2O_4 + CO \rightarrow NO + NO_2 + CO_2 \qquad\qquad \text{step b; } k_2 \qquad\qquad \text{EQ. 7.19}$$

which sums to:

$$NO_2(g) + CO(g) \rightarrow NO(g) + CO_2(g) \qquad\qquad\qquad\qquad \text{EQ. 7.20}$$

If we assume that step a is slow, then the rate constant associated with step a ($k_1$) is much smaller than the rate constant ($k_2$) associated with step b. A slow step a means that it takes a long time for $N_2O_4$ to form. A fast step b means that as $N_2O_4$ is formed by step a, it is rapidly used up in step b.

The rate of step b is:

$$R_2 = k_2 [N_2O_4] [CO] \qquad\qquad \text{EQ. 7.21}$$

The rate of step a is:

$$R_1 = k_1 [NO_2]^2 \qquad\qquad \text{EQ. 7.22}$$

The exact rate of step b is not relevant to the overall rate, since the concentration of $N_2O_4$ at any time is vanishingly small. The overall rate R is determined by the rate of the rate determining step (often abbreviated as RDS) which is the rate for step a:

$$R_{(\text{proposed mechanism A})} = k_1 [NO_2]^2 \qquad\qquad \text{EQ. 7.23}$$

That is, the rate at which products are formed (and reactants disappear) in mechanism A is the rate at which $N_2O_4$ is formed, *i.e.*, the rate of step a.

An alternate mechanism could be made with the same two steps, but making the first step fast and the second step slow this time.

### Mechanism B

$$NO_2 + NO_2 \rightarrow N_2O_4 \qquad\qquad \text{step a; } k_1 \qquad\qquad \text{EQ. 7.24}$$

$$N_2O_4 + CO \rightarrow NO + NO_2 + CO_2 \qquad\qquad \text{step b; } k_2 \qquad\qquad \text{EQ. 7.25}$$

Now the overall rate equals the rate of step b, the RDS. The predicted rate law for mechanism B is:

$$R_{(\text{predicted mechanism B})} = k_2 [N_2O_4]^1[CO]^1 \qquad\qquad \text{EQ. 7.26}$$

Note that the concentration of $N_2O_4$ appears in the predicted rate law. Remember that the overall reaction is:

$$NO_2(g) + CO(g) \rightarrow NO(g) + CO_2(g) \qquad\qquad \text{EQ. 7.27}$$

and $N_2O_4$ is neither a reactant nor a product. $N_2O_4$ is called an "intermediate", a species formed and destroyed during a reaction mechanism. Before writing a rate law predicted by a mechanism, we must first be certain that the concentrations of any intermediates that appear in the rate law are replaced by equivalent expressions using measurable quantities (*e.g.*, the concentrations of reactants and/or products).

In mechanism B, step a is rapid compared to step b. $N_2O_4$ is formed rapidly in a and only slowly removed by step b. This means that the nitrogen tetroxide and nitrogen dioxide in step a have time to reach equilibrium. This means we could rewrite mechanism B such that:

## Mechanism B

$$NO_2 + NO_2 \rightleftarrows N_2O_4 \qquad \text{step a; } k_1, k_{-1} \qquad \text{EQ. 7.28}$$

$$N_2O_4 + CO \rightarrow NO + NO_2 + CO_2 \qquad \text{step b; } k_2 \qquad \text{EQ. 7.29}$$

The equilibrium in step a can be described by the equilibrium constant K:

$$K = k_1/k_{-1} = [N_2O_4]/[NO_2]^2 \qquad \text{EQ. 7.30}$$

Thus the concentration of the intermediate $N_2O_4$ can be written as:

$$[N_2O_4] = (k_1/k_{-1}) [NO_2]^2 \qquad \text{EQ. 7.31}$$

Thus the predicted rate law for Mechanism B then becomes:

$$R_{(predicted\ mechanism\ B)} = k_2(k_1/k_{-1}) [NO_2]^2[CO]^1 \qquad \text{EQ. 7.32}$$

To summarize, mechanism A predicts $R = k_1[NO_2]^2$. Mechanism B predicts $R = k_2(k_1/k_{-1}) [NO_2]^2[CO]^1 = k_{obs}[NO_2]^2[CO]^1$. The experimental rate law, as mentioned earlier, is second order in nitrogen dioxide and zero order in carbon monoxide. Thus mechanism B is definitely wrong. Note that we did not claim that mechanism A is definitely right; it is merely consistent with experimental evidence. Other mechanisms, however, can make a similar claim. Consider mechanism C:

## Mechanism C

$$NO_2 + NO_2 \rightarrow NO_3 + NO \qquad \text{step a; } k_1 \text{ (RDS)} \qquad \text{EQ. 7.33}$$

$$NO_3 + CO \rightarrow NO_2 + CO_2 \qquad \text{step b;} k_2 \qquad \text{EQ. 7.34}$$

Again, the predicted rate is the rate of the RDS, which is

$$R_{(predicted\ mechanism\ C)} = k_1 [NO_2]^2 \qquad \text{EQ. 7.35}$$

Both mechanisms A and C predict rate laws consistent with the experimentally observed rate law. Either (or even another mechanism not yet under discussion) could be what actually occurs. More experimental evidence (*e.g.*, finding evidence for the existence of $N_2O_4$ or $NO_3$ in the reaction mixture) would bolster the support for one or the other mechanism, but kinetics can never conclusively prove that a particular mechanism is the only possible explanation of what happens.

## 7.5 MECHANISMS AND RATE LAWS

Sometimes a rate law can be deceptively simple. Consider the reaction:

$$H_2(g) + I_2(g) \rightarrow 2 HI(g) \qquad \text{EQ. 7.36}$$

which is experimentally shown to follow the rate law:

$$R_{(experimental)} = k [H_2] [I_2] \qquad \text{EQ. 7.37}$$

Since the orders in the rate law correspond to the stoichiometric coefficients, it is possible that the reaction occurs as an elementary step, with collision between hydrogen and iodine leading directly to products. Careful examination of the reaction mixture indicates the presence of I and $H_2I$. These species could be intermediates in another mechanism, which is proposed below:

$$I_2 \rightleftharpoons 2\,I \qquad\qquad \text{EQ. 7.38}$$

$$I + H_2 \rightleftharpoons H_2I \qquad\qquad \text{EQ. 7.39}$$

$$H_2I + I \rightarrow 2\,HI \qquad\qquad \text{EQ. 7.40}$$

The rate of this mechanism is determined by the rate for the last step which is the RDS. The rate predicted by this mechanism is:

$$R = k_3\,[H_2I]^1[I]^1 \qquad\qquad \text{EQ. 7.41}$$

Neither species appearing in equation 7.41 is a reactant or a product. Both are intermediates, whose concentration must be expressed in terms of the preceding equilibria. Note that all steps in a mechanism that occur before the rate determining step are assumed to be at equilibrium. For the first equilibrium, we can write:

$$K_1 = \frac{k_1}{k_{-1}} = \frac{[I]^2}{[I_2]} \qquad\qquad \text{EQ. 7.42}$$

from which the concentration of I is obtained:

$$[I] = \sqrt{\frac{k_1}{k_{-1}}[I_2]} \qquad\qquad \text{EQ. 7.43}$$

The second equilibrium step can be solved for the concentration of $H_2I$:

$$K_2 = \frac{k_2}{k_{-2}} = \frac{[H_2I]}{[H_2][I]}; \quad [H_2I] = \frac{k_2}{k_{-2}}[H_2][I] \qquad\qquad \text{EQ. 7.44}$$

Substituting the result of equation 7.43 into equation 7.44 yields:

$$[H_2I] = \frac{k_2}{k_{-2}}[H_2]\sqrt{\frac{k_1}{k_{-1}}[I_2]} \qquad\qquad \text{EQ. 7.45}$$

Finally substituting 7.43 and 7.45 into 7.41 yields:

$$R_{predicted} = k_3\left(\frac{k_2}{k_{-2}}[H_2]\sqrt{\frac{k_1}{k_{-1}}[I_2]}\right)\left(\sqrt{\frac{k_1}{k_{-1}}[I_2]}\right) \qquad\qquad \text{EQ. 7.46}$$

which simplifies to:

$$R_{predicted} = \frac{k_1 k_2 k_3}{k_{-1} k_{-2}}[H_2][I_2] \qquad\qquad \text{EQ. 7.47}$$

Note that the rate law predicted for the mechanism (equation 7.47) matches the experimentally determined rate law of equation 7.37. The presence of the intermediates lends additional credence to the three-step mechanism over the one-step reaction.

As another example, consider the reaction between nitric oxide and hydrogen, which was seen in the last chapter to be second order in nitric oxide and first order in hydrogen.

$$2\ NO + 2\ H_2 \rightarrow 2\ H_2O + N_2$$ EQ. 7.48

$$R_{(experimental)} = k\ [NO]^2\ [H_2]^1$$ EQ. 7.49

Since the powers in the rate law do not match the coefficients, it can be conclusively stated that equation 7.48 does not happen by an elementary process. (Even if the orders agreed with the coefficients, a simultaneous four-body collision is such an unlikely event as to make the mechanism unrealistic.) Many additional potential mechanisms provide the necessary stoichiometry (*i.e.*, add up to the right overall reaction). Several of these mechanisms also predict a rate law in agreement with experiment. One such mechanism is:

$$NO + H_2 \rightleftharpoons NOH_2 \qquad \text{step a; } k_1, k_{-1}$$ EQ. 7.50

$$NOH_2 + NO \rightarrow N_2 + H_2O_2 \qquad \text{step b; } k_2 \text{ (RDS)}$$ EQ. 7.51

$$H_2O_2 + H_2 \rightarrow 2\ H_2O \qquad \text{step c; } k_3 \text{ (fast)}$$ EQ. 7.52

These three steps add up to the overall reaction, and the predicted rate is the rate of the slow step, step b. The rate of this step involves the concentration of an intermediate, $NOH_2$, but this is resolved using the equilibrium approach:

$$R_{predicted} = k_2[NO][NOH_2] = k_2[NO]\left(\frac{k_1}{k_{-1}}[NO][H_2]\right)$$ EQ. 7.53

which simplifies to a form that agrees with the experimental rate in equation 7.54

$$R_{predicted} = \frac{k_2 k_1}{k_{-1}}[NO]^2[H_2]$$ EQ. 7.54

## 7.6 AROMATIC NITRATION: AN EXAMPLE OF ADDITIONAL MECHANISTIC EVIDENCE

A mechanism that predicts a rate law that agrees with the experimentally established rate law is termed "possible" or "likely". Other types of evidence, both kinetic and non-kinetic, are used to examine and test the mechanism in more detail.

Consider the nitration of benzene, with the overall reaction shown in equation 7.55:

$$C_6H_6 + HNO_3 \rightarrow C_6H_5NO_2 + H_2O$$ EQ. 7.55

The mechanism proposed for this reaction is outlined below

$$2 \, HNO_3 \; \leftrightarrow \; NO_2^+ + NO_3^- + H_2O$$ EQ. 7.56

$$NO_2^+ + C_6H_6 \; \leftrightarrow \; C_6H_6NO_2^+$$ EQ. 7.57

$$C_6H_6NO_2^+ + NO_3^- \; \leftrightarrow \; C_6H_5NO_2 + HNO_3$$ EQ. 7.58

Equations 7.56—7.58 add up to the overall stoichiometry. The rate law for this reaction is fairly complicated, and it depends upon the specific reaction conditions. When run in nitric acid as the solvent, the rate law becomes simply:

$$R_{(experimental)} = k \, [C_6H_6]^1$$ EQ. 7.59

Under these conditions, we see first order dependence on benzene, and zero order dependence on everything else. Nitric acid is present in great excess as the solvent, and its concentration does not appear in the rate law. Since the concentration of nitric acid is essentially a constant value, it is subsumed into the rate constant, k. The process of simplifying a chemical rate law by having one or more reactants in large excess is known as the **isolation method**. Since nitric acid is the solvent, its concentration is large, constant, and $NO_2^+$ is rapidly formed in step 7.56; Under these conditions, step 7.57 is the slow step. When another solvent, for example glacial acetic acid, is used, nitric acid is not present in great excess, and step 7.56 becomes the rate determining step.

The three step mechanism for the nitration of benzene is, however, supported by more than the experimental rate laws. The evidence is multi-faceted, and includes evidence of the existence of the proposed intermediate, specific catalytic influences, and isotope effects. We will examine all three.

### EXISTENCE OF THE INTERMEDIATE

Note that the nitronium ion formed in the proposed mechanism is an intermediate, and as such it does not appear in the rate law. There is good evidence that the species does, in fact, exist under appropriate reaction conditions, and that it is involved in the mechanism. Like all polyatomic species, the nitronium ion has specific modes of vibration. One of these, pictured at right, is a symmetric stretch.

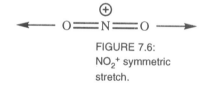

FIGURE 7.6:
$NO_2^+$ symmetric stretch.

While many such vibrations appear in the infrared spectrum (see chapter 3 of *Volume 1*), a few such stretches appear in the Raman spectrum (a different, complimentary technique we will not pursue at this time). Due to the peculiar symmetry of the nitronium ion stretch pictured in figure 7.6, it is predicted to be Raman-active (and IR inactive). In pure nitric acid, a weak Raman signal is observed at 1400 cm$^{-1}$, and this is attributed to the nitronium ion. Further evidence supporting the existence of the nitronium ion is its isolation as the cation in solid salts such as $(NO_2^+)(ClO_4^-)$.

### SPECIFIC CATALYTIC INFLUENCE

Sulfuric acid is known to catalyze aromatic nitration reactions. The mechanism thought to account for this effect was presented in *Volume 1*, and is repeated below:

$$HNO_3 + H_2SO_4 \rightleftharpoons \overset{\oplus}{N}O_2 + HSO_4^{\ominus} + H_2O \qquad \text{EQ. 7.60}$$

$$\overset{\oplus}{N}O_2 + C_6H_6 \rightleftharpoons C_6H_6NO_2^{\oplus} \qquad \text{EQ. 7.61}$$

$$C_6H_6NO_2^{\oplus} + HSO_4^{\ominus} \rightleftharpoons C_6H_5NO_2 + H_2SO_4 \qquad \text{EQ. 7.62}$$

These three elemental steps sum to the observed stoichiometry of equation 7.55. Note that as a catalyst, sulfuric acid is first used, then reformed in the mechanism, disappearing from the overall balanced reaction. Nevertheless its concentration influences the overall rate of the reaction. Evidence for the specific role of sulfuric acid in forming nitronium ion *via* equation 7.60 is again presented by Raman spectroscopy. The weak Raman signal at 1400 cm$^{-1}$ observed in pure nitric acid becomes a much stronger signal in mixtures of nitric acid and sulfuric acid. Sulfuric acid is thus thought to speed the reaction by increasing the rate of formation of the nitronium ion intermediate in cases where nitric acid is not the solvent.

## ISOTOPE EFFECTS

The three step nitration reaction mechanism involves a second step, the addition of the electrophile to the aromatic ring, and a third step, elimination of H$^+$ from the ring, restoring its aromaticity. An alternate mechanism might combine these two steps into one concerted process. For example, equations 7.61 and 7.62 could be replaced by:

$$NO_2^+ + C_6H_6 + HSO_4^- \rightleftharpoons C_6H_5NO_2 + H_2SO_4 \qquad \text{EQ. 7.63}$$

which is pictured in Figure 7.7:

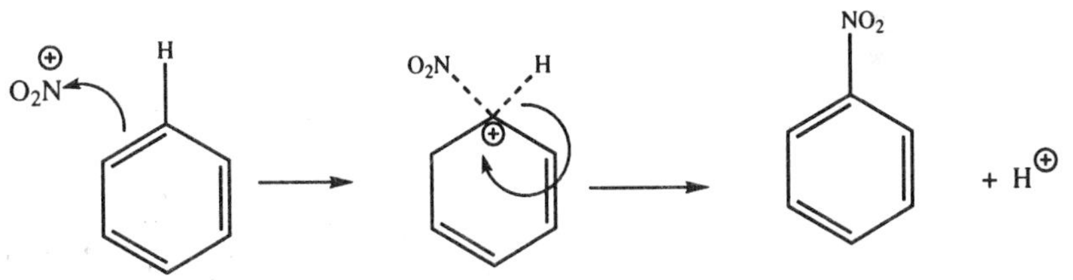

The concerted process (equation 7.63), shown in figure 7.7, proposes that the addition of the electrophile, NO$_2^+$, to the carbon coincides with the breaking of the C—H bond.

FIGURE 7.7: Potential concerted process.

When equation 7.63 is combined with equation 7.60, a two step mechanism results, which could yield the same rate law as the three step mechanism of equations 7.60 — 7.62. The concerted process is, however, ruled out by the measurement of **kinetic isotope effects**. In general the kinetic isotope effect is such that, when an isotope with a larger mass is used, the rate is slowed. This assumes, however, that the bond involving the substituted isotope is altered during the rate determining step. Frequently hydrogen can be substituted by deuterium or, more rarely, tritium. It has been observed that using deuterium for hydrogen can slow a reaction by a factor of between 3 and 12. Using tritium in place of hydrogen may slow a reaction by a factor between 5 and 30. In the case of the nitration of benzene, replacing the hydrogen atoms on benzene with deuterium or tritium atoms has no effect on the overall observed rate. Thus, a concerted process involving the breaking of the carbon – hydrogen bond can be eliminated.

**Isotopic labeling** can also be used to locate certain atoms within a reacting molecule. Following where the labeled atom in a reactant molecule appears in a product molecule often helps to establish the kinetic pathway. A specific example of this was examined earlier when we studied the mechanism of esterfication.

## 7.7  COMPLEX MECHANISMS

Many reaction mechanisms are quite complex, and beyond the scope of this course. As indicated in the previous chapter, even seemingly simple reactions, such as that between hydrogen and oxygen to form water, can have a surprising degree of complexity in their reaction. Over fifty different elementary steps are involved in the reaction mechanism that explains how hydrogen and oxygen react, and the exact mathematical solution of the rate law under varying conditions proves quite challenging. In many ways this is typical of radical reactions that occur in combustion (or general redox) reactions.

As an example, consider what happens when you light a Bunsen burner. A spark or a flame is used to "catalyze" a reaction between oxygen in the air and natural gas (a mixture of hydrocarbons, mostly methane). The first step in this type of radical chain mechanism is called "initiation", and it forms one or more radical species. The radical formed in the initiation step enters into a chain reaction. One or more steps in the chain involve "propagation": the process by which a radical reacts to form product plus another reactive radical. Directly or indirectly, the radical species formed at the start of the propagation process is regenerated, and the chain continues. In some reactions, a single radical reactant forms two or more radical products in an elementary step, usually *via* reaction with a molecule of oxygen. These are called "branching" reactions. The geometric growth in the number of radicals in the reaction mixture is often the cause of a rapidly escalating rate of reaction, which is called an explosion. Finally, the chain is terminated when two radical species combine to form a non-reactive species. The list below contains some of the elementary steps thought to be involved in the combustion of methane. R represents one of several possible organic radicals formed in the combustion process (*e.g.*, $CH_3$).

| | | |
|---|---|---|
| $CH_4 \rightarrow CH_3{\cdot} + H{\cdot}$ | initiation | EQ. 7.64 |
| $O_2 \rightarrow 2\,O{\cdot}$ | initiation | EQ. 7.65 |
| $CH_4 + H{\cdot} \rightarrow CH_3{\cdot} + H_2$ | propagation | EQ. 7.66 |
| $H{\cdot} + O_2 \rightarrow HO{\cdot} + O{\cdot}$ | branching | EQ. 7.67 |
| $CH_3{\cdot} + O{\cdot} \rightarrow CH_2O + H{\cdot}$ | propagation | EQ. 7.68 |
| $CH_2O + H{\cdot} \rightarrow CHO{\cdot} + H_2$ | propagation | EQ. 7.69 |
| $HO{\cdot} + CH_4 \rightarrow CH_3{\cdot} + H_2O$ | propagation | EQ. 7.70 |
| $CHO{\cdot} \rightarrow CO + H{\cdot}$ | propagation | EQ. 7.71 |
| $CO + HO{\cdot} \rightarrow CO_2 + H{\cdot}$ | propagation | EQ. 7.72 |
| $H{\cdot} + R{\cdot} \rightarrow RH$ | termination | EQ. 7.73 |

Another reaction that proceeds via a free radical mechanism is the substitution of a halogen atom for a hydrogen atom in saturated hydrocarbons. Such a reaction is called a halogenation reaction. For instance, when propane is treated with chlorine a series of halogenated propanes is formed as illustrated in equation 7.74.

$$C_3H_8 + x\,Cl_2 \rightarrow C_3H_{(8-x)}Cl_x + x\,HCl \qquad \text{EQ. 7.74}$$

If the amount of chlorine is controlled so that the mole ratio of chlorine to propane is 1:1, then the monochloro products are favored. Careful analysis of the monochloro products reveals that there is more 2–chloropropane formed than 1–chloropropane even though there are only 2 hydrogen atoms on the middle carbon atom *vs.* 6 on the terminal atoms. The reason for this is that secondary free radicals (free radicals in which the single electron resides on a secondary carbon atom) are more stable than primary free radicals because the effect of the missing bond can be spread out over two immediately adjacent carbon atoms. It would then be expected that tertiary free radicals are even more stable than secondary free radicals. They are. You will recall that this is the same stability order that is observed for carbocations.

It must be noted here, however, that free radicals are so reactive that it can be quite difficult to control the conditions that favor one product over another and in free radical reactions, more so than in most other types of reactions, a wide variety of different products are obtained.

## 7.8 REACTION MECHANISMS FOR OZONE IN THE ATMOSPHERE

The mechanisms for reactions involving ozone in the atmosphere have drawn much attention recently, notably the ozone hole over the Antarctic, and the implications of widespread depletion of the ozone layer. Ozone, $O_3$, is a reactive gas that, when formed in the troposphere (near ground level), is considered a pollutant. In the upper reaches of the stratosphere, however, ozone serves a vital purpose; it screens the surface of the earth from most of the sun's ultraviolet radiation.

Like all chemical species in a dynamic equilibrium, there is a balance between the factors influencing ozone concentration in the stratosphere. One such factor is the formation of ozone from oxygen and ultraviolet radiation:

$$O_2 + h\nu \rightarrow 2\,O \qquad \text{EQ. 7.75}$$

$$O + O_2 \rightarrow O_3 \qquad \text{EQ. 7.76}$$

Another factor is the destruction of ozone. This can happen photochemically:

$$O_3 + h\nu \rightarrow O + O_2 \qquad \text{EQ. 7.77}$$

or by the thermal decomposition of ozone to form oxygen:

$$2\,O_3 \rightarrow 3\,O_2 \qquad \text{EQ. 7.78}$$

The amount of solar radiation and the concentration of diatomic oxygen in the atmosphere are both nearly constant, so reactions 7.75, 7.76, and 7.77 create a stable concentration of ozone, about 10 ppm. The reaction of equation 7.78 is sufficiently slow that, without anthropogenic pollution, it does not influence the concentration of ozone in the stratosphere.

The uncatalyzed reaction in equation 7.78 has an experimental rate law of

$$R_{experimental} = k\frac{[O_3]^2}{[O_2]} = k[O_3]^2[O_2]^{-1}$$

EQ. 7.79

The negative one order of diatomic oxygen indicates that oxygen is acting to suppress the decomposition in the uncatalyzed ozone decomposition mechanism. The commonly accepted mechanism that explains the observed rate law is:

$$O_3 \rightleftharpoons O_2 + O \qquad k_1 \text{ forward, } k_{-1} \text{ reverse}$$

EQ. 7.80

$$O + O_3 \rightarrow 2\,O_2 \qquad k_2, \text{ RDS}$$

EQ. 7.81

The rate predicted by this two-step mechanism equals the rate of the RDS, which is:

$$R_{(predicted)} = k_2\,[O_3]\,[O]$$

EQ. 7.82

Since oxygen atom is an intermediate, its concentration is replaced by the equilibrium expression from the first step in the mechanism. This results in a predicted rate law that agrees with the experimental rate law, equation 7.79.

$$R_{predicted} = k_2[O_3]\frac{k_1[O_3]}{k_{-1}[O_2]} = k[O_3]^2[O_2]^{-1}$$

EQ. 7.83

The slow rate of reaction 7.78 becomes much more important when even traces of certain pollutants enter the stratosphere. Nitric oxide, NO, is produced by internal combustion engines, and can catalyze the destruction of ozone:

$$O_3 + NO \rightarrow O_2 + NO_2 \qquad \text{step 1}$$

EQ. 7.84

$$NO_2 + O_3 \rightarrow 2\,O_2 + NO \qquad \text{step 2}$$

EQ. 7.85

Note that steps 1 and 2 add up to equation 7.78. The $NO_2$ produced in step one is consumed in step two, and is thus an intermediate. The NO used in step one is reproduced in step two, and by dramatically increasing the rate of the overall reaction, serves as a catalyst.

Perhaps the most significant factor in explaining the rapid seasonal decline of ozone concentrations is related to chlorofluorocarbons. For example, $CCl_2F_2$ is a refrigerant, known as Freon 12. In 1928 the discovery of chlorofluorocarbons for use as refrigerants was hailed as a tremendous technological breakthrough. The chlorofluorocarbons were non-flammable and non-toxic, and provided a much safer alternative to the refrigerants in use at the time (flammable propane and poisonous ammonia). But the very unreactivity of chlorofluorocarbons proved to be a problem. Since they react with nothing in the troposphere, the chlorofluorocarbons survive for such a long time that they finally reach the stratosphere, where they can react with UV radiation to form chlorine atoms:

$$CCl_2F_2 + h\nu \rightarrow CClF_2 + Cl$$

EQ. 7.86

The chlorine atom can then react with atomic oxygen from equations 7.75 and 7.77:

$$Cl + O \rightarrow ClO$$

EQ. 7.87

Since the concentration of chlorine atoms is very small, reaction 7.87 by itself would not significantly influence ozone concentration. However the ClO, once formed, reacts with another oxygen atom to form molecular oxygen and regenerate the chlorine atom:

$$ClO + O \rightarrow Cl + O_2$$

EQ. 7.88

The combination of equations 7.87 and 7.88 is simply:

$$2\,O \rightarrow O_2$$

EQ. 7.89

which is catalyzed quite effectively by the chlorine atom. It is estimated that a single chlorine atom may cause reaction 7.89 to occur $10^5$ times before the chlorine atom is removed from the stratosphere by other means. The removal of oxygen atoms from the equilibrium with ozone in equation 7.70 causes a shift according to LeChatelier's principle. The actual process that occurs in the stratosphere is actually far more complicated than outlined here, and involves combinations of oxides of nitrogen and chlorine in heterogeneous reactions involving both the gas phase water and solid ice crystal surfaces. The net result of having chlorofluorocarbons in the stratosphere is a significant decrease in ozone concentration. A change of 1% in the concentration of ozone in the stratosphere translates into about a 2% increase in the sun's ultraviolet radiation that reaches the surface of the earth. While arguments continue about the exact magnitude of the ozone depletion that has occurred and the relation this has to skin cancer and worldwide climate, the news is almost certainly not good.

## 7.9 CHLOROFLUOROCARBONS AND OZONE: THE OUTLOOK

Worldwide production of chlorofluorocarbons was banned as of January 1, 1996, and new refrigerants are beginning to replace the older Freons. The good news is that one refrigerant replacement compound, $CF_3CFH_2$, has proven not to cause the problems seen with chlorofluorocarbons. The bad news is that the chlorofluorocarbons still in the environment slowly will circulate, end up in the stratosphere, and react with ozone. Also, the new refrigerants are not quite as energy efficient as the old Freons. This means you have to use more electricity to get the same amount of cooling in your refrigerator or air conditioner. More electricity means more burning of fossil fuels and higher concentrations of carbon dioxide, a greenhouse gas, in the atmosphere. Life is complicated!

**Elementary reaction (step)**: a single step process in which the collision of reactants leads directly to products.

**Freons**: any of a family of chlorofluorocarbons, commonly used as refrigerants prior to the discovery of their ozone-depleting characteristics.

**Induction period**: the interval between the start of the reaction (when the rate of disappearance of reactant is greatest) and the time when the rate of appearance of product is greatest. An induction period is evidence of the presence of one or more intermediates in a multi-step mechanism.

**Intermediate**: a chemical species that is produced, then consumed during the course of a chemical mechanism.

**Mechanism**: a series of elementary steps that sum to the overall stoichiometry. A complete mechanism must specify the relative rates of the different steps in the mechanism.

**Molecularity**: the number of particles colliding as reactant species in an elementary process.

**Rate determining step (RDS)**: the slowest step in a multi-step mechanism. The rate of this step determines the rate of the overall reaction.

HOMEWORK
## Problems

1. Determine the overall balanced reaction, given the following mechanism:

$$Cl_2 \rightleftharpoons 2\,Cl$$
$$CHCl_3 + Cl \rightarrow CCl_3 + HCl \quad RDS$$
$$CCl_3 + Cl \rightarrow CCl_4$$

2. Determine the overall balanced reaction, given the following mechanism:

$$2\,NO \rightleftharpoons N_2O_2$$
$$N_2O_2 + H_2 \rightarrow N_2O + H_2O \qquad RDS$$
$$N_2O + H_2 \rightarrow N_2 + H_2O$$

3. Determine the overall balanced reaction. Hint: some steps may have to be taken more than one time to arrive at a balanced reaction; nitric oxide and nitrogen dioxide do not appear in the final equation.

$$NO + N_2O \rightarrow N_2 + NO_2 \qquad RDS$$
$$2\,NO_2 \rightarrow 2\,NO + O_2$$

4. Classify the role(s) of nitric oxide and nitrogen dioxide in the mechanism in problem 3.

5. Determine the overall balanced reaction. Hint: some steps may have to be taken more than one time to arrive at a balanced reaction; water, oxygen, bromine and hydrogen bromide are the only species appearing in the final reaction.

$$HBr + O_2 \rightarrow HOOBr \qquad RDS$$
$$HOOBr + HBr \rightarrow 2\,HOBr$$
$$HOBr + HBr \rightarrow H_2O + Br_2$$

6. Determine the overall balanced reaction, given the following mechanism:

$$NO_2Cl \rightarrow NO_2 + Cl \qquad RDS$$
$$Cl + H_2O \rightarrow HCl + OH$$
$$OH + NO_2 + N_2 \rightarrow HNO_3 + N_2$$

7. Specify the molecularity of each of the three steps in the mechanism of problem #1.

8. Specify the molecularity of each of the three steps in the mechanism of problem #6.

9. Solve for the predicted rate law for the mechanism given in problem #1. Be certain that the concentrations of intermediate species do not appear in the rate law.

10. Solve for the predicted rate law for the mechanism given in problem #2. Be certain that the concentrations of intermediate species do not appear in the rate law.

11. Solve for the predicted rate law for the mechanism given in problem #3. Be certain that the concentrations of intermediate species do not appear in the rate law.

12. Solve for the predicted rate law for the mechanism given in problem #5. Be certain that the concentrations of intermediate species do not appear in the rate law.

13. Solve for the predicted rate law for the mechanism given in problem #6. Be certain that the concentrations of intermediate species do not appear in the rate law.

14. For the mechanism given below, (a) write the balanced net reaction, and (b) derive the predicted rate law.

$$H_2O_2 + I^- \rightarrow H_2O + OI^- \qquad \text{slow}$$
$$H^+ + OI^- \rightarrow HOI \qquad \text{fast}$$
$$HOI + H^+ + I^- \rightarrow I_2 + H_2O \qquad \text{fast}$$

15. From your rate law, predict if the rate of reaction in #14 will increase, decrease or remain unchanged in a solution with a higher pH.

16. For the mechanism given below, (a) write the balanced net reaction, and (b) derive the predicted rate law. [Note: $H^+$ and $OH^-$ may always appear in the final rate law for a reaction that is acid– or base–catalyzed.]

$$OCl^- + H_2O \rightleftharpoons HOCl + OH^- \qquad \text{fast}$$
$$I^- + HOCl \rightarrow HOI + Cl^- \qquad \text{slow}$$
$$HOI + OH^- \rightarrow H_2O + OI^- \qquad \text{fast}$$

17. From your rate law, predict if the rate of reaction in #16 will increase, decrease or remain unchanged in a solution with a higher pH.

18. For the overall reaction $S_2O_8^{2-} + 3\ I^- \rightarrow 2\ SO_4^{2-} + I_3^-$, the experimental rate law is $R_{(exp)} = k\ [S_2O_8^{2-}]^1[I^-]^1$. Predict the rate law if:

    (a) step a is the RDS
    (b) step b is the RDS
    (c) step c is the RDS
    (d) step d is the RDS
    (e) which of these is (are) consistent with the experimental rate law?

$$I^- + S_2O_8^{2-} \rightarrow IS_2O_8^{3-} \qquad \text{step a}$$
$$IS_2O_8^{3-} \rightarrow 2\ SO_4^{2-} + I^+ \qquad \text{step b}$$
$$I^+ + I^- \rightarrow I_2 \qquad \text{step c}$$
$$I_2 + I^- \rightarrow I_3^- \qquad \text{step d}$$

19. In the following mechanism, indicate which steps would be termed "initiation", which would be termed "termination" and which would be termed "propagation".

$$A_2 + h\nu \rightarrow 2\,A \qquad\qquad \text{step a}$$
$$A + B_2 \rightarrow AB + B \qquad\qquad \text{step b}$$
$$B + A_2 \rightarrow AB + A \qquad\qquad \text{step c}$$
$$A + A \rightarrow A_2 \qquad\qquad\qquad \text{step d}$$

20. Identify each species in #19 as a reactant, product, intermediate or catalyst.

21. Suppose 2–methylbutane and chlorine gas are mixed in a 1:1 mole ratio.
   a. What products would you expect from this reaction?
   b. Which of the products would you expect in the largest yield?
   c. Which of the products would you expect in the lowest yield?
   d. Justify your answers to (b) and (c) by reference to the intermediates in the reaction.
   e. Write a mechanim for the formation of the product formed in the greatest yield.

**CHAPTER SEVEN**

OVERVIEW
Problems

22. Consider the two step mechanism listed below:

$$1.\ CO + NO_2 \rightarrow CO_2 + NO$$
$$2.\ NO + 0.5\,O_2 \rightarrow NO_2$$

   a. Write the net balanced reaction
   b. For all 5 species present in the mechanism, specify if it is a reactant, product, intermediate or catalyst.
   c. Calculate the change in free energy and the change in entropy for the reaction, using the thermodynamic values in the appendix.
   d. Explain the significance of the signs in the two answers to part (c).
   e. Write the rate law, assuming the first step is slow.
   f. Write the rate law, assuming the second step is slow, and the first step is in equilibrium.

23. Stilbene (1,2-diphenylethene, $C_{14}H_{12}$) has two geometric isomers: *cis*-stilbene and *trans*-stilbene ($\Delta G_f = 317.6$ kJ/mole). The isomerization reaction from *cis* to *trans* has K = 4.8 at 25.0°C.
   a. Draw the structures of the two isomers.
   b. Calculate the value of the change in standard free energy of formation of the *cis* isomer.
   c. If the isomerization reaction of *cis* to *trans* was elementary (occurred exactly as written) what rate law is predicted for the reaction?

24. A key breakthrough in the development of organic chemistry was the discovery that ammonium ion and cyanate ion (from inorganic salts) could react to form urea (an organic compound.) The experimental rate law is $R = k [NH_4^+]^1[CNO^-]^1$. A possible mechanism is given below.

1. $CNO^- + H^+ \rightarrow HCNO$ $\qquad$ $K_{eq1}$
2. $NH_4^+ + OH^- \rightarrow NH_3 + H_2O$ $\qquad$ $K_{eq2}$
3. $H^+ + OH^- \rightarrow H_2O$ $\qquad$ $K_{eq3}$
4. $HCNO + NH_3 \rightarrow (NH_2)_2CO$ $\qquad$ slow

a. Determine the net overall reaction.
b. Show that the rate law for this mechanism agrees with the experimentally determined rate law.
c. Show how the equilibrium constants are subsumed within the apparent rate law constant for this mechanism
d. Calculate the numerical value of $K_{eq1}$, $K_{eq2}$, and $K_{eq3}$.

25. The reaction $2 N_2O \rightarrow 2 N_2 + O_2$ is thought to occur by the mechanism listed below.

1. $N_2O \rightarrow N_2 + O$
2. $O + N_2O \rightarrow N_2 + O_2$

a. Calculate the standard state free energy change and equilibrium constant for the overall reaction at 25.0°C.
b. If the first step of the mechanism is slow, determine the predicted rate law.
c. If the second step of the mechanism is slow, determine the predicted rate law.
d. How might one experimentally discriminate between the mechanism in (c) and a one step elementary mechanism?

26. *Trans*-1,2-diiodoethene reacts with iodide ion in methanol solvent as shown below. Doubling the initial concentration of either reactant resulted in doubling the rate of reaction. The Arrhenius equation has $A = 4 \times 10^{12}$, $E_a = 123$ kJ/mol.

$$ICH=CHI + I^- \rightarrow C_2H_2 + I_3^-$$

a. Draw the Lewis dot diagram of both product species.
b. Write the rate law for the reaction.
c. What is the value of the rate constant at 25.0°C?

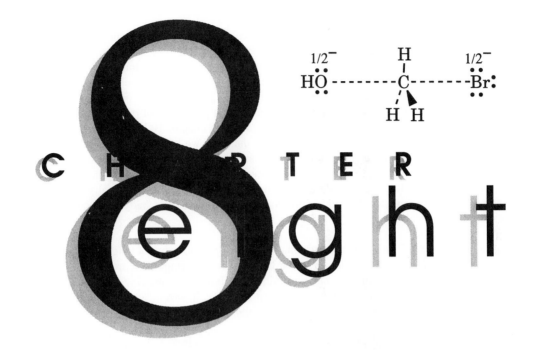

# CHAPTER 8 eight

# MECHANISMS (II)
## Substitution
### and Elimination

# MECHANISMS (II):
## Substitution and Elimination

O ut of all of the many millions of possible chemical reactions that different chemical substances can undergo, you may be wondering how it is possible for anybody
to learn enough to be able to decide the probable outcome of a reaction. The
answer is, of course, that chemists, like all scientists, spend much of their time
looking for similarities that can be generalized. In that manner, chemists only have to
learn a relatively few rules of chemical behavior that enable prediction of products and
reactivities for chemicals with a wide range of structures. In chapter 10 last year, we
saw this applied to the reactions of the elements. At that time we learned many of the
general principles governing the behavior of simple substances. Many of those rules,
especially those involving the properties of the elements are incorporated in the enormously useful periodic table. As compounds become more complex, and especially in
the realm of carbon chemistry, additional insights are often gained by a study of reaction
mechanisms.

Last year, in chapter 14, we began this systematic study of reactions by looking at the
mechanisms of nucleophilic and electrophilic **substitutions** occurring on an aromatic
ring. We learned that the products of many reactions involving aromatic rings can be
predicted after learning a few general principles. At that time, however, we were not yet
able to present much of the evidence used to deduce those mechanisms.

In chapter five this year, we continued our discussion of mechanisms by looking at reactions occurring at a carbonyl group. We found that nucleophilic attack on the carbonyl
group of ketones or aldehydes led to addition reactions while the same type of reaction
at the carbonyl group of an acid or ester led to a substitution reaction. The evidence
used in support of these generalizations was primarily that of tracking a specific isotope
incorporated into one of the reactants. In both previous cases, the reactions that were
studied were reactions occurring at an unsaturated carbon atom — one with $sp^2$ hybrid
orbitals. Now we begin our study of reactions at saturated carbon atoms. We shall use
a wide variety of different kinds of evidence to validate these mechanisms. In this chapter we will study both substitution and **elimination** reactions. Many of these are already
familiar to you from your study of chemical reactions in previous chapters both this year
and last year.

## 8.2 SUBSTITUTION REACTION MECHANISMS

The first two mechanisms we will study in this chapter are nucleophilic substitution mechanisms occurring at an sp$^3$ hybridized (saturated) carbon atom. In these reactions, an atom or group of atoms, termed a **leaving group**, is replaced by a **nucleophile** (Nu:). This is represented in schematic form in Figure 8.1. In this figure, the saturated carbon atom is considered to be part of the alkyl group, R.

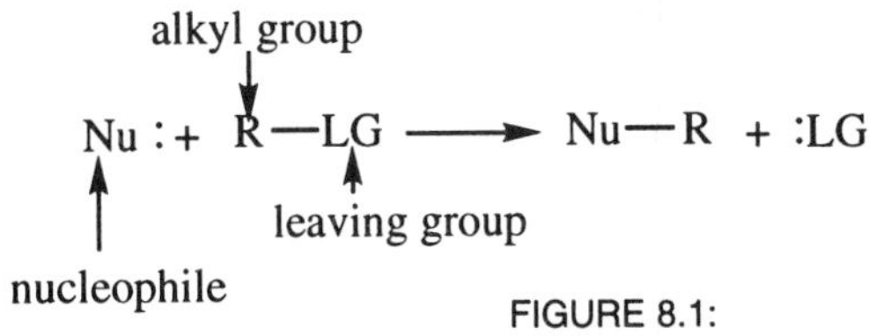

This is one of the most common reactions in all of carbon chemistry because of the many different combinations of reactants possible. The nucleophile can be almost any substance with a free pair of electrons. Common nucleophiles in this type of reaction range from hydroxide ions to the amine group of an amino acid to the water molecule. Although the leaving group (LG) is often a halide ion, it can be any group capable of stabilizing a pair of electrons. Besides halide ions, the anions of other strong acids often are good leaving groups. The alkyl group can, of course, be any alkyl group (*e.g.*, CH$_3$—, CH$_3$CH$_2$—, *etc.*).

alkyl group

Nu :+  R—LG  $\longrightarrow$  Nu—R  + :LG

leaving group

nucleophile

FIGURE 8.1:
Schematic
representation of
nucleophilic
substitution.

Figure 8.1 shows that the final outcome of this reaction is that the pair of electrons initially on the nucleophile becomes the pair of electrons involved in the bond between the nucleophile and the alkyl group. The leaving group carries away the old pair of electrons that initially formed the bond between it and the alkyl group. The result is substitution of the nucleophile for the leaving group. Such reactions are termed Nucleophilic Substitution reactions. The chemist's shorthand for such reactions is 'S$_N$'.

There are fundamentally two different S$_N$ mechanisms. The first one we shall study is termed an **S$_N$2** mechanism. In this case the '2' indicates that the reaction rate law is second order overall. The other nucleophilic substitution mechanism is termed **S$_N$1** because it follows first order kinetics. Besides the differences in the rate laws, the two reactions also differ in a number of other ways as well. Some of these we will develop in the course of studying the reaction mechanisms in more detail.

Many different useful products can be made using a variety of nucleophiles in a substitution reaction. A wide variety of leaving groups is available as well. Table 8.1 lists a number of nucleophiles and the products that can be obtained by the reaction of the nucleophile with the compound R–X. Table 8.2 lists a number of good leaving groups arranged in order of their ability to leave.

**TABLE 8.1** NUCLEOPHILES FOR SUBSTITUTION REACTIONS

TABLE 8.1

| NUCLEOPHILE | PRODUCT | NAME OF PRODUCT |
|---|---|---|
| I$^-$ | R—I | iodide |
| SH$^-$ | R—SH | thiol or mercaptan |
| OH$^-$ | ROH | alcohol |
| OR$^-$ | R—OR | ether |
| CN$^-$ | R—CN | nitrile |
| H$_2$O | R—OH | alcohol |

TABLE 8.2

**TABLE 8.2:** SOME COMMON LEAVING GROUPS

| LEAVING GROUP | NAME |
|---|---|
| 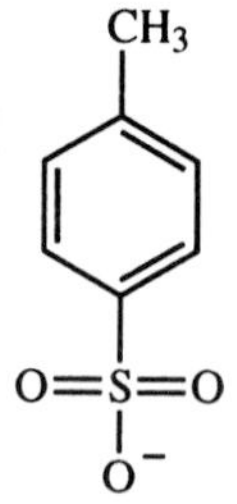 | p–toluenesulfonyl (tosyl) |

| | |
|---|---|
| I⁻ | iodide |
| Br⁻ | bromide |
| Cl⁻ | chloride |
| CN⁻ | cyano or cyanide |
| H₂O | water |
| OH⁻ | hydroxide |

## 8.3   S$_N$2 MECHANISM—A POSSIBLE MECHANISM

Let us consider the reaction of methyl bromide with hydroxide ion. Schematically, the reaction occurs as shown in figure 8.2. This figure shows the hydroxide ion approaching the carbon atom from the side opposite that of the chlorine atom. When this occurs, the result is that the bromine atom is "pushed out", leaving as a bromide ion. This is shown as a series of "stop action" steps in figure 8.3.

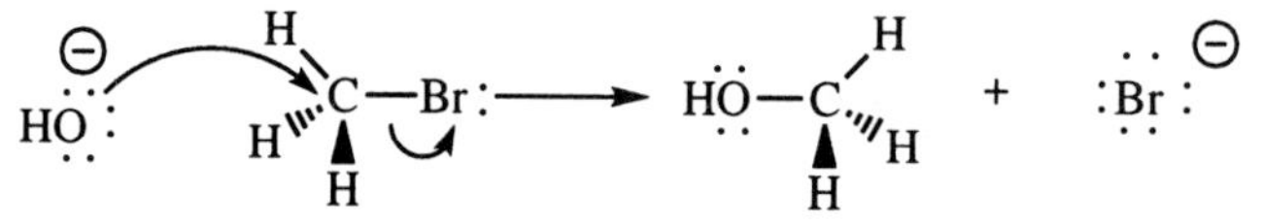

FIGURE 8.2:
Overall view of
S$_N$2 substitution.

The success of the reaction depends upon the nucleophile being a strong nucleophile and the leaving group being a "good" leaving group. Generally anionic species are stronger nucleophiles than are neutral compounds. Hence hydroxide ion and ethoxide ion work better in a reaction with methyl iodide to produce methyl alcohol and methyl–ethyl ether than do water and ethanol. Similarly, "good" leaving groups are those which can easily accommodate the extra electrons than once formed the bond to the carbon atom. Halide ions, carboxylate ions, the conjugate bases of strong acids, and the water molecule all make particularly good leaving groups.

HO⁻ approaches carbon
from backside, at 180⁰

HO⁻ begins to form bond with
carbon, bromide bond begins to
break

HO bond half formed and Br
bond half broken, "middle
point", TRANSITION STATE

HO bond almost formed, Br bond
almost broken

HO bond completely formed, Br
bond completely broken

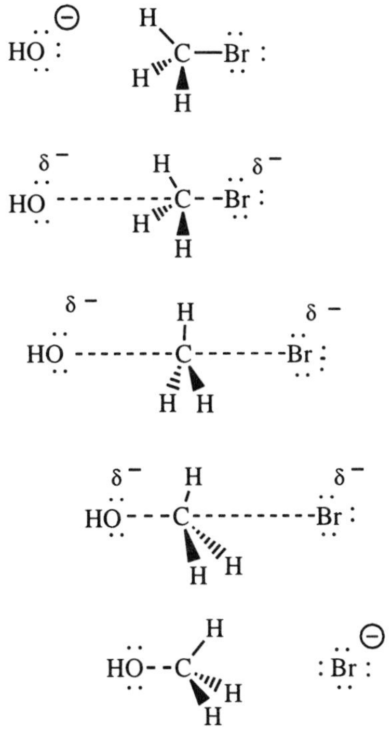

# 8.4  S$_N$2 MECHANISM—THE TRANSITION STATE

In chapter 6 we discussed the changes in energy that occur as a reaction proceeds from reactants to products.  We noticed that essentially all reactions have
an activation energy barrier that must be traversed during the reaction.  This
activation energy barrier exists because of the changes in electronic structure
and changes in molecular geometry which must accompany reaction.  In order
to completely understand how a reaction proceeds it is necessary to understand
the changes in electronic and molecular structure that accompany the reaction.

Figure 8.3 presented a depiction of those changes in the conversion of methyl
bromide into methyl alcohol.  The reaction coordinate diagram for this reaction
is presented in figure 8.4 with the structures of the various species  and their
energy relationships indicated at each point.  The structure that is key to our
understanding of the process is the one in the middle — the structure that is
half–way between reactants and products.  This structure is known as the transition state structure — commonly just called the transition state.  Transition
states are commonly enclosed in brackets and post superscripted with a double
dagger as shown in figure 8.4.  As would be expected for a five–coordinate
species, the transition state involved in S$_N$2 reactions has trigonal bipyramidal
geometry.  This can not be a stable species because carbon does not have
enough low–energy atomic orbitals to form the required hybrid orbitals for a
five–coordinate species.  In order for this transition state to form, therefore, the
electronic energy of the alkyl halide molecule must be raised quite far above its
initial energy.  It is this increase in energy that, in this case, is the cause of the
activation energy barrier.

In the $S_N2$ reaction, the transition state is located at the highest energy point along the reaction coordinate diagram depicted in figure 8.4. When the reaction has reached this point along the reaction coordinate, movement in either direction along the reaction coordinate reduces the energy of the system. If the reaction proceeds to the left along the reaction coordinate diagram, the reactants are reformed; when the reaction proceeds in the other direction, products are formed.

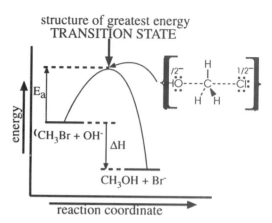

FIGURE 8.4:
Reaction
coordinate diagram.

## 8.5   $S_N2$ MECHANISM—STEREOCHEMICAL EVIDENCES

What evidence do we have that suggests that the mechanism outlined in figures 8.2 or 8.3 is actually what happens when methyl bromide reacts with hydroxide ion to produce methyl alcohol? What experiments have been devised to "prove" that the mechanism is reasonable? Of course, as we discussed in chapter 7, it is impossible to prove beyond the shadow of a doubt that a mechanism is correct. The best that we can ever do is to show that our proposed mechanism is consistent with observations made when experiments are done to test the mechanism. What are some of these observations?

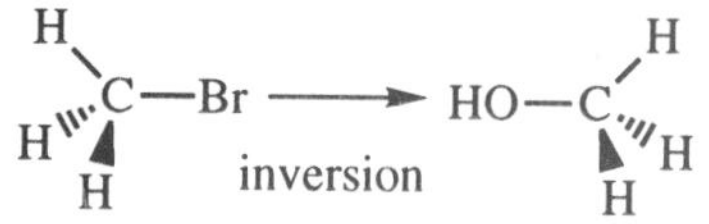

The first piece of evidence is the configuration of the groups about the carbon atom. If the above mechanism is correct, the configuration around the carbon must be inverted as shown in figure 8.5.

In the case of the reaction discussed thus far, it would not be possible to test this hypothesis because neither the reactant nor the product is chiral. However, if we use a molecule with a chiral carbon, react it under $S_N2$ conditions and determine the stereochemistry of the product, we should be able to answer this question. The reaction of (S)-2-bromobutane with hydroxide ion to obtain 2–butanol is shown as figure 8.6. If our mechanism is correct, this figure predicts that the product should be the $R$ isomer. This is what is experimentally found. Thus we have one piece of evidence to support the mechanism outlined above.

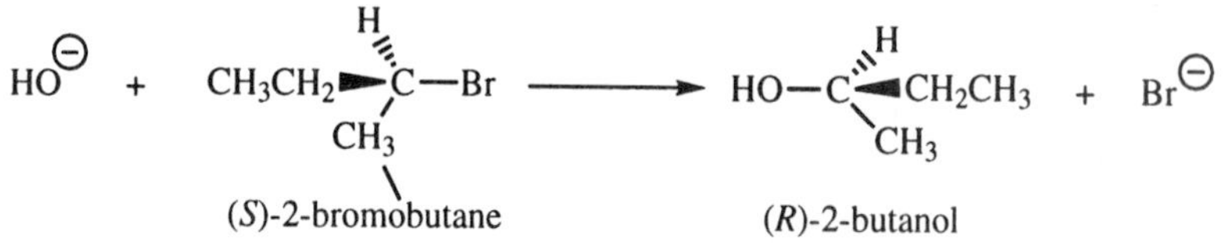

FIGURE 8.5:
Inversion of
configuraton for $S_N2$
reaction.

$$\underset{\text{(S)-2-bromobutane}}{HO^{\ominus} \quad + \quad CH_3CH_2\overset{H}{\underset{CH_3}{\diagup\!\!\!\!\diagdown}}C\!\!-\!\!Br} \quad \longrightarrow \quad \underset{\text{(R)-2-butanol}}{HO\!\!-\!\!\overset{H}{\underset{CH_3}{\diagup\!\!\!\!\diagdown}}C\!\!-\!\!CH_2CH_3} \quad + \quad Br^{\ominus}$$

FIGURE 8.6:
Inversion evidence
for $S_N2$ mechanism.

## 8.6 S$_N$2 MECHANISM—RATE LAW EVIDENCE

The reaction, if it occurs as suggested in figures 8.2 and 8.3, is a single step reaction involving collision of the alkyl bromide molecule and the hydroxide ion. That step is, therefore, bimolecular and the rate law for the step should be first order in the alkyl bromide and first order in hydroxide ion. The experimentally determined rate law for the reaction of 2–bromobutane with hydroxide ion is presented in equation 8.1

$$\text{Rate} = k[CH_3CHBrCH_2CH_3]^1[OH^-]^1$$

EQ. 8.1

The experimentally determined rate law is consistent with the proposed mechanism. A somewhat more detailed summary of the mechanism for this reaction is shown as figure 8.7. This figure shows the sequence of the happenings during the single step of the proposed mechanism — a sort of stop–action "movie" of this **concerted** process.

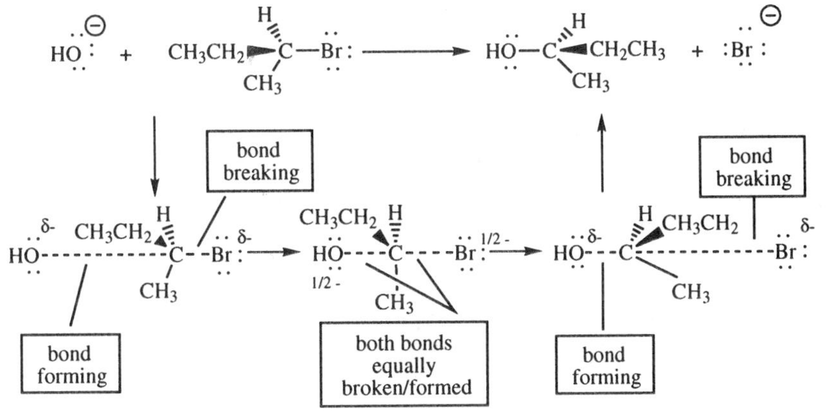

FIGURE 8.7:
Reaction mechanism summary.

## 8.7 S$_N$2 MECHANISM—STRUCTURE AND REACTIVITY EVIDENCE

In this mechanism, the nucleophile approaches the carbon atom on the side opposite the bromine atom. Not only does the inversion of conformation show this, but this must also be true since otherwise the nucleophile and the leaving group would interfere with each other's ability to move toward or away from the carbon atom. The maximum freedom of approach for the nucleophile and freedom of departure for the leaving group is accomplished by a backside approach of the nucleophile.

If this picture of the S$_N$2 mechanism is correct, the reaction should occur more easily if the approach of the nucleophile is unhindered. If we compare reactivities of different alkyl bromides toward hydroxide ion, the alkyl bromide with the smallest group(s) attached to the carbon atom carrying the bromine atom should react the fastest. Thus, *unless the mechanism changes*, a primary bromide should react faster than a secondary bromide that should, in turn, react faster than a tertiary bromide. In addition, within these groups faster reaction should occur when smaller alkyl groups are attached to the carbon atom carrying the bromine atom than when larger alkyl groups are attached. Thus we expect that the methyl bromide would react more rapidly than ethyl bromide even though both are primary alkyl bromides. This is what has been observed. Adding another methyl group as in *iso*–propyl bromide further slows the reaction. Yet another methyl group makes *t*–butyl bromide — a molecule which would react the most slowly. However, *t*–butyl bromide actually reacts primarily *via* another mechanism and it is difficult to really estimate the actual rate of reaction *via* this mechanism to make a good comparison. The value for *t*–butyl bromide should be taken as an estimate that has a fairly high error associated with it. Table 8.3 summarizes these findings and presents the relative rates of reaction for these compounds *via* the $S_N2$ mechanism.

TABLE 8.3

**TABLE 8.3:** SUMMARY OF STERIC EFFECTS

| STRUCTURE | SUMMARY | RELATIVE RATE |
|---|---|---|
| | The only groups that hinder approach of the hydroxide ion are three small hydrogen atoms. **Hindrance small** | 100 |
| | One methyl group and two hydrogen atoms hinder approach of the hydroxide ion. **Hindrance greater** | 1.31 |
| | One larger ethyl group and two hydrogen atoms hinder approach of the hydroxide ion. **Hindrance greater still** | 0.81 |
| | Two methyl groups and one hydrogen atom hinder appoach or the hydroxide ion. **Hindrance greater still.** | 0.015 |
| | Three methyl groups hinder approach of the hydroxide ion; most reaction occurs by another mechanism. **Hindrance greatest.** | 0.004 |

## CONCEPT CHECK

Draw the structure of the major organic product of the following reactions. Assume that these substitution reactions proceed *via* the $S_N2$ mechanism

a.

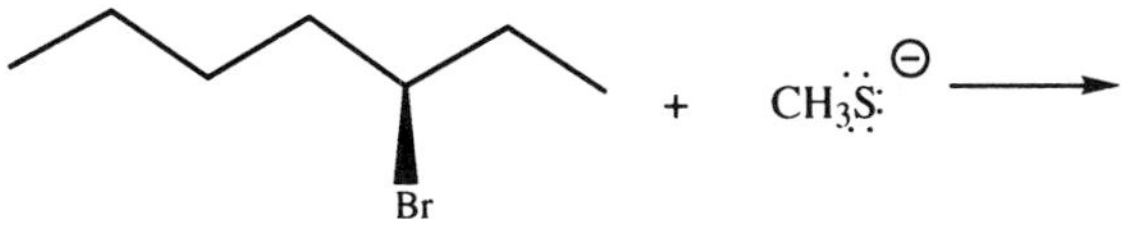

$+ \quad CH_3\overset{\ominus}{\underset{..}{\overset{..}{S}}}: \longrightarrow$

b.

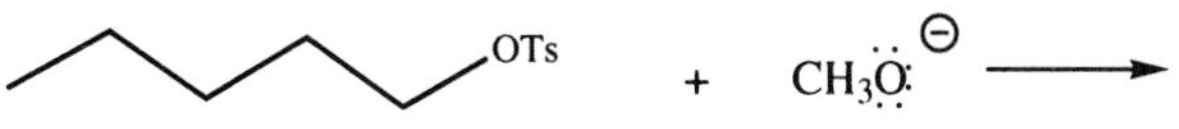

OTs $\quad + \quad CH_3\overset{\ominus}{\underset{..}{\overset{..}{O}}}: \longrightarrow$

c.

$+ \quad \overset{\ominus}{N_3} \longrightarrow$

All of these are simple substitution reactions in which the reactant nucleophile replaces the leaving group. However, since these reactions prodeed *via* the $S_N2$ mechanism, we need to keep in mind the inversion that occurs during the substitution. This will be noticeable in (a) since substitution is occurring at a secondary carbon atom. It will not be noticeable in (b) and (c) because substitution is occurring at a primary carbon atom.

a.

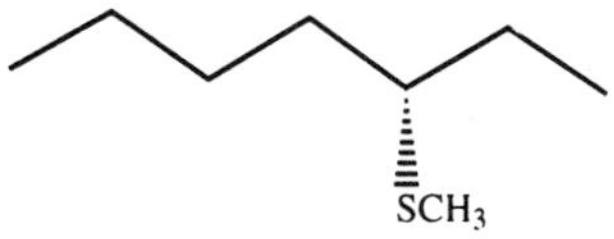

SCH₃

b.

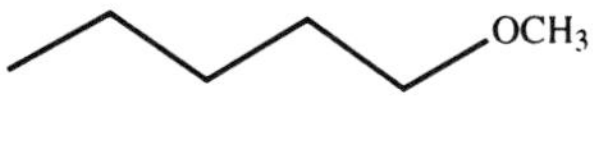

OCH₃

c.

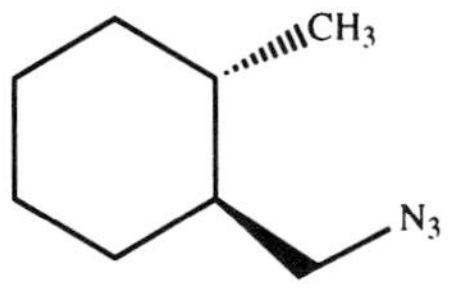

## 8.8  S$_N$1 MECHANISM—HYPOTHESIS

Even though tertiary halides are sufficiently hindered that the substitution reaction does not occur *via* an S$_N$2 mechanism, reaction of the halide in a similar substitution reaction does, indeed, occur. For reasons that will be taken up later in this chapter, hydroxide ion is too strong a base to use for making alcohols from tertiary alkyl halides. Conversion of tertiary halides to tertiary alcohols is normally accomplished by the use of water as the nucleophile. The mechanism of this reaction is quite different from that of the second order, one–step mechanism we have just considered. Instead, based upon evidence that will be considered in the next few sections, chemists have hypothesized a two–step mechanism in which the alkyl halide first ionizes into a carbocation and a halide ion. The nucleophile then combines with the carbocation to form the final product. Note that, if the first step is the rate determining step, the rate law for this mechanism would be first order in the alkyl halide and first order overall. Such a mechanism is called an S$_N$1 mechanism. The steps of this mechanism are given as equations 8.2— 8.4.

$$(CH_3)_3CBr \rightarrow (CH_3)_3C^{\oplus} + Br^{\ominus} \qquad \text{RDS} \qquad\qquad \text{EQ. 8.2}$$

$$(CH_3)_3C^{\oplus} + H_2O \rightarrow (CH_3)_3COH_2^{\oplus} \qquad \text{(fast)} \qquad\qquad \text{EQ. 8.3}$$

$$(CH_3)_3COH_2^{\oplus} \rightarrow (CH_3)_3COH + H^{\oplus} \qquad \text{(fast)} \qquad\qquad \text{EQ. 8.4}$$

The first (ionization) step is similar to the ionization of salts such as sodium chloride in water:

$$NaCl \rightarrow Na^{\oplus}_{(aq)} + Cl^{\ominus}_{(aq)} \qquad\qquad \text{EQ. 8.5}$$

However, since the carbocation is not as stable a species as the hydrated sodium ion, we should not expect the ionization of the alkyl halide to proceed nearly as far toward the right as does equation 8.5. This means that solutions of alkyl halides do not contain ions under normal conditions and, therefore, do not carry an electric current as do solutions of sodium chloride. What is more important, however, for the present situation, this also means that double replacement reactions involving alkyl halides occur more slowly than do reactions involving sodium chloride. (In fact, double replacement reactions involving inorganic, ionic materials such as sodium chloride are often diffusion–controlled reactions — which simply means that the reactions occur as rapidly as the reactants can come together.)

The second step of the mechanism is reaction of the carbocation with a nucleophile. In the present case there are two nucleophiles present: bromide ion and water. Reaction of the carbocation with the bromide ion would lead to formation of the starting material — hence no net reaction. Reaction of the carbocation with water would lead, after release of a hydrogen ion, to formation of an alcohol — the product.

## 8.9  S$_N$1 MECHANISM—KINETIC EVIDENCE

The S$_N$1 mechanism, as proposed, is a two step mechanism. This requires that a relatively stable (compared to transition states on either side of it) intermediate species be present. The reaction coordinate diagram that corresponds to such a mechanism is shown in figure 8.8. Note, especially, the presence of the two transition states, TS1 and TS2, and an intermediate, INT on the diagram. In this particular case, the hypothesis is that the intermediate is the short–lived t–butyl cation.

A series of structures corresponding to the state of the t–butyl bromide during the first step of the reaction is presented in figure 8.9. These structures correspond to what is expected at the various places during the reaction indicated by the letters on the diagram. Note that both the first and second transition states were placed closer on the reaction coordinate diagram to the intermediate than to the more stable species on the other side of the transition state. This is an example of the **Hammond Postulate** which states that:

> Related species that are similar in energy are similar in structure as well. Therefore, the structure of the transition state resembles the structure of the closest stable species.

An important consequence of this postulate is that the transition state lies close to products in endothermic reactions and close to reactants in exothermic reactions. In the current situation, the closest "stable" species is the intermediate and we expect the transition state on both sides of the intermediate to have more ionic character than covalent character, *i.e.*, the bond between the alkyl group and the group partially bonded to it is more than half broken. Since the carbocation can be considered to be a product of the first step and a reactant in the second step of this mechanism, the Hammond postulate can be applied to both steps to arrive at the placement of the transition states shown in figure 8.8.

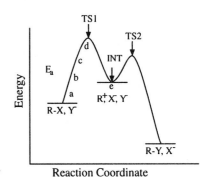

FIGURE 8.8:
S$_N$1 transition state energy diagram.

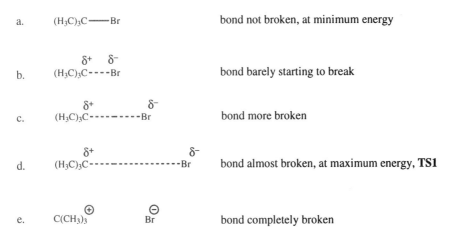

a.  $(H_3C)_3C$——Br       bond not broken, at minimum energy

b.  $\overset{\delta^+}{(H_3C)_3C}$----$\overset{\delta^-}{Br}$       bond barely starting to break

c.  $\overset{\delta^+}{(H_3C)_3C}$--------$\overset{\delta^-}{Br}$       bond more broken

d.  $\overset{\delta^+}{(H_3C)_3C}$----------------$\overset{\delta^-}{Br}$       bond almost broken, at maximum energy, **TS1**

e.  $C(CH_3)_3^{\oplus}$      $Br^{\ominus}$       bond completely broken

FIGURE 8.9:
Stages in the ionization of t-butyl bromide.

When the reaction arrives at the low energy point between the two transition states, it is in a local energy minimum. More stable intermediate states result in a lower local minimum in the energy at this point. This local minimum corresponds to a situation in which the carbocation and the bromide ion both exist in the solution. When the reaction reaches this point, two things can happen: the carbocation can react with the nearby bromide ion, going back through the first transition state and reforming the starting material, or the carbocation can react with a nearby water molecule proceeding through the second transition state and eventually forming products. In figure 8.8, this second transition state is indicated as having a lower energy than the first transition state. Applying our knowledge of kinetics and the relationship of the rate constant to the activation energy, we know that the rate constant for this second step will be larger than the rate constant for the reverse of the first step. This means that the rate of the second step should be faster than the rate of the first step. The result of reaction with water is a protonated alcohol molecule that quickly loses a proton to form the alcohol. The second and third steps in the mechanism are both very rapid with respect to the first step. This makes the first step rate–limiting with the rate law shown in equation 8.6. The reaction, therefore, has the overall first order kinetics expected of an $S_N1$ mechanism.

$$\text{Rate} = [(CH_3)_3CBr] \qquad \text{EQ. 8.6}$$

A further piece of kinetic evidence supporting this mechanism is that the rate of substitution reactions is independent of the concentration and identity of the nucleophile. This lends additional support to the hypothesis that the first, ionization, step is the rate–determining step.

## 8.10  $S_N1$ MECHANISM—STEREOCHEMICAL EVIDENCE

What does our picture of the $S_N1$ reaction mechanism require of the stereochemistry? We have seen that the stereochemistry around the carbon atom is inverted in an $S_N2$ reaction. If our mechanism for the $S_N1$ reaction is correct, a carbocation is formed as an intermediate in the reaction. This species would be planar with an unhybridized atomic p orbital perpendicular to the plane of the rest of the molecule. The carbon atom uses this p orbital to accept the electrons on the nucleophile in the initial stages of bond formation. Clearly there are lobes of this orbital on both sides of the plane. Figure 8.10 shows that the nucleophile can approach the carbocation from either side. If attack occurs with equal frequency from both sides, the product will be completely racemized.

FIGURE 8.10:
Stereochemistry of
the $S_N1$ mechanism.

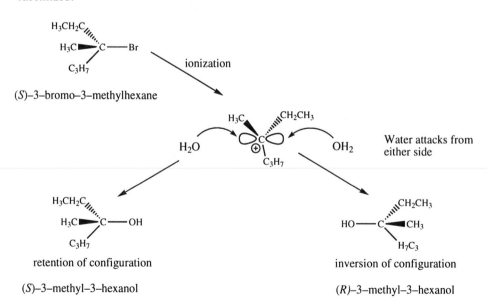

(S)–3–bromo–3–methylhexane

Water attacks from
either side

retention of configuration

(S)–3–methyl–3–hexanol

inversion of configuration

(R)–3–methyl–3–hexanol

Racemization does, indeed, occur in this reaction. However, in many cases, racemization is incomplete. This simply means that the carbocation is more easily approached by the nucleophile from one direction than the other. This is not too surprising if we consider that the bromide ion has not completely left the carbocation when the local minimum corresponding to the transition state is reached. If the carbocation is unstable and reacts very quickly, it is reasonable that attack from the side away from the departing bromide ion might be somewhat preferred over attack past the bromide ion. In these cases we would expect that incomplete racemization would occur since more of the inverted product than the product that retains the configuration of the original reactant is formed. This has, indeed, been observed in the hydrolysis of (S)–(–)–1–chloro–1–phenylethane. Table 8.4 shows the results of a series of experiments in which the compound was hydrolyzed in acetone—water mixtures. As can be seen, as the amount of water in the solvent decreases, racemization becomes more complete. In pure water it is likely that a water molecule attacks from the side opposite the chlorine before it has a chance to completely move away from the carbocation. As the amount of water decreases, the carbocation probably exists for a longer period of time. This allows the chloride ion to move away from the carbocaton which, in turn, allows water equal access to both faces of the carbocation. This results in almost complete racemization.

**TABLE 8.4** PRODUCT YIELDS FOR HYDROLYSIS OF (S)–(–)–1–CHLORO–1–PHENYLETHANE

TABLE 8.4

| Solvent Mixture | (S)–(–)–1–phenyl–1–ethanol | (R)–(+)–1–phenyl–1–ethanol |
|---|---|---|
| 100% H$_2$O | 41% | 59% |
| 40% H$_2$O 60% Acetone | 47% | 53% |
| 20% H$_2$O 80% Acetone | 49% | 51% |

## 8.11 S$_N$1 MECHANISM—REACTION ENVIRONMENT EVIDENCE

If the intermediates formed during the mechanism are ionic, we would expect that the rate of the reaction would be solvent dependent. Solvents capable of stabilizing ions to a greater extent should make the first, rate–determining, step in the mechanism faster. The rate of the reaction of t–butyl bromide with hydroxide is, indeed, dependent upon the solvent. Table 8.5 presents the rate constants for the hydrolysis of t–butyl chloride in a variety of mixtures of water with organic solvents. As expected, the reaction rate constant decreases with an increase in the concentration of the organic constituent.

**TABLE 8.5** RATE CONSTANTS AND SOLVENT COMPOSITION*  TABLE 8.5

| CO–SOLVENT | MOLE FRACTION OF CO–SOLVENT | RATE CONSTANT $(\times 10^4)/s^{-1}$ |
|---|---|---|
| None (pure water) | 0 | 41.45 |
| Ethanol | 0.075 | 14.37 |
|  | 0.11 | 7.113 |
|  | 0.15 | 6.579 |
|  | 0.25 | 2.153 |
| 1–Propanol | 0.02 | 25.45 |
|  | 0.05 | 23.56 |
|  | 0.075 | 13.01 |
|  | 0.10 | 8.353 |
|  | 0.20 | 4.649 |
| 1–Butanol | 0.02 | 34.85 |
|  | 0.05 | 18.72 |
|  | 0.10 | 12.64 |
|  | 0.20 | 5.100 |
| THF | 0.05 | 16.37 |
|  | 0.10 | 6.977 |
|  | 0.20 | 2.257 |
| Acetonitrile | 0.05 | 19.98 |
|  | 0.10 | 12.96 |
|  | 0.20 | 9.542 |

* M.J. Blandamer, J. Burgess and P.P. Duce, *J. Chem. Soc. Faraday Trans. I* 1981, **77**, 1999–2008

## 8.12 COMPETING MECHANISMS

Since the intermediate in the $S_N1$ mechanism is a carbocation, we would expect that anything that stabilizes the carbocation formed when the halide ionizes will favor that mechanism. Similarly, to the extent that the carbocation is not stable, the alternate, $S_N2$ mechanism should be favored. All other things being equal, as we learned last year, carbocations formed from tertiary halides are able to "spread out" the charge over a larger number of carbon atoms and, thus, are more stable than those formed from secondary halides. In turn, these secondary carbocations are more stable that those formed from primary halides. Thus we are not surprised to find that primary halides react almost exclusively by the $S_N2$ mechanism with complete inversion of configuration (although inversion can not be experimentally observed for primary halides) while tertiary halides show the much more complete racemization that is characteristic of reaction by an $S_N1$ mechanism. Indeed, with certain halides for which the rate constants for the first order and second order reactions are comparable, it is possible to manipulate solution conditions in order to favor one mechanism over the other. Solution conditions which favor the $S_N2$ mechanism will lead to a greater amount of inversion than those which favor the $S_N1$ mechanism. Since the overall rate of the $S_N2$ mechanism is dependent on both the alkyl halide and the hydroxide concentration, while the rate of the $S_N1$ mechanism depends only on the alkyl halide concentration, increasing the hydroxide concentration while holding the alkyl halide concentration constant leads to a greater fraction of total substitution reaction going *via* the $S_N2$ mechanism at higher hydroxide ion concentrations. (However, as we shall see, this is complicated by the fact that secondary and tertiary halides also undergo elimination reactions that compete with substitution, especially at higher hydroxide ion concentrations.)

## 8.13  CARBOCATION STRUCTURE AND STABILITY RELATIONSHIPS

The carbocation is a very strong Lewis acid, an electron deficient species that is very reactive toward nucleophiles. Let us now look more closely at the structure of the species that we have called the carbocation. A carbocation is a species made by breaking one of four sigma bonds to a carbon atom in such a way that both of the electrons forming the electron pair bond go with the departing atom or group of atoms. This leaves the carbon atom with three sigma bonds and no unshared electron pairs. Remembering the bonding theory that we learned last year in chapter 11, we predict that the carbon atom is in an $sp^2$ hybridization state. This requires that the carbon atom and the three other atoms to which it is bonded *via* the sigma bonds all lie in the same plane. We expect approximately 120° bond angles. The unhybridized p orbital left after hybridization is oriented perpendicular to this plane. This was, of course, one of the things that we discussed in section 8.10 to explain racemization in $S_N1$ reactions.

The stability of the carbocation is strongly affected by the groups to which it is bonded. All other things being equal, the more highly substituted the carbon atom, the more stable is the carbocation. Thus, the order of stability of the carbocations is as follows in figure 8.11.

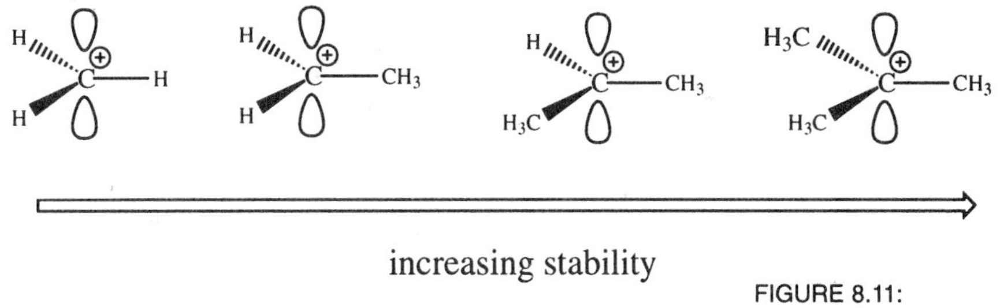

increasing stability

FIGURE 8.11:
Relative stability of alkyl carbocations.

There are two reasons for this increase in stability with increase in substitution. The first is the inductive effect of the alkyl group(s) attached to the carbon atom formally carrying the charge. The positive charge is said to induce a flow of electrons ("to pull the electrons") in the sigma bonds toward the charged atom to help partially neutralize the charge. The fact that an alkyl group is larger than a hydrogen atom, and therefore contains a larger number of electrons, makes the electrons on the alkyl group more easily drawn toward the charged carbon atom. This provides additional stabilization to the carbocation. It stands to reason that this stabilization will increase as the number of alkyl substituents increases. This is illustrated for the isopropyl carbocation in figure 8.12.

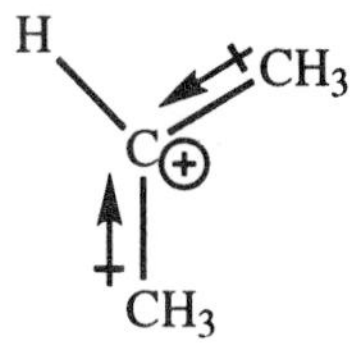

FIGURE 8.12:
Inductive effects of methyl groups in the isopropyl carbocation.

The inductive effect essentially distributes the charge over several carbon atoms by placing a partial positive charge on the carbon atoms adjacent to the one formally carrying the charge. This is illustrated for the primary, secondary, and tertiary butyl carbocations in figure 8.13.

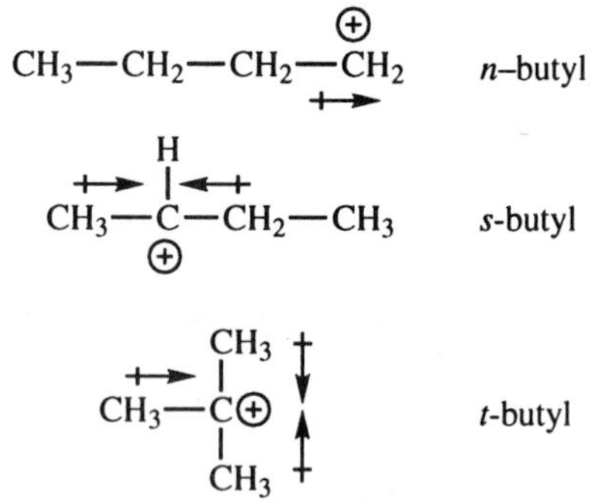

The second effect that increases the stability of substituted carbocations is weak back–bonding that occurs between the filled sigma orbitals and the unfilled p orbital. When the sigma bond between the charged carbon atom and one of the substituents is rotated such that the one of the bonds on the substituent is in the eclipsed position with respect to the vacant p orbital, there is a small degree of overlap between that sigma orbital and the vacant p orbital. This is illustrated in figure 8.14. This type of overlap is termed hyperconjugation. Because of the size of the atom and the resulting "tightness" with which the electrons are held in the bond, this type of back–bonding is not possible when the atom bonded directly to the charged carbon atom is a hydrogen atom. As the number of alkyl groups bonded directly to the charged carbon atom increases, the effects of hyperconjugation also increase.

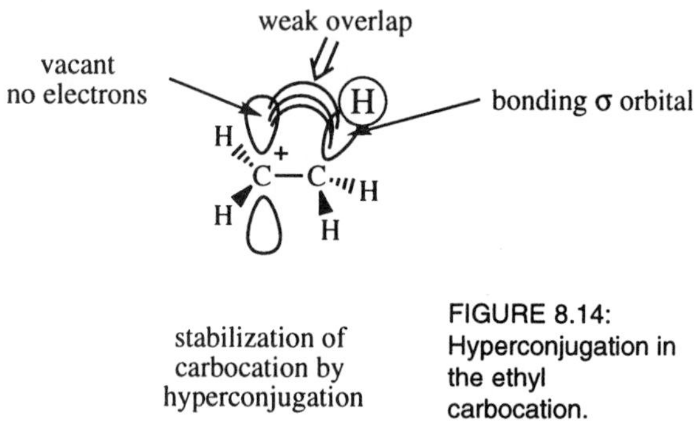

$CH_3$—$CH_2$—$CH_2$—$\overset{\oplus}{CH_2}$   *n*–butyl

$CH_3$—$\overset{H}{\underset{\oplus}{C}}$—$CH_2$—$CH_3$   *s*-butyl

$CH_3$—$\overset{CH_3}{\underset{CH_3}{C\oplus}}$   *t*-butyl

FIGURE 8.13:
Comparison of inductive effects in $C_4$ carbocations.

weak overlap

vacant
no electrons

bonding σ orbital

stabilization of
carbocation by
hyperconjugation

FIGURE 8.14:
Hyperconjugation in
the ethyl
carbocation.

## 8.14   EVIDENCE FOR CARBOCATIONS FROM REARRANGEMENTS

Carbocations are unstable and have only a fleeting existence. One possible fate of a carbocation is to rearrange to a more stable carbocation. The fact that **rearrangement** does occur is additional evidence for the existence of the carbocation. A specific example of such a rearrangement occurs in the reaction of 2-bromo-3-methylbutane with ethanol. Two products are formed in this reaction. The first is the expected product: 2-ethoxy-3-methylbutane. The other is a product that can only be formed by rearrangement of the secondary carbocation initially formed into the more stable tertiary carbocation. Addition of an ethanol molecule followed by loss of a proton to the rearranged carbocation forms 2-ethoxy-2-methylbutane.

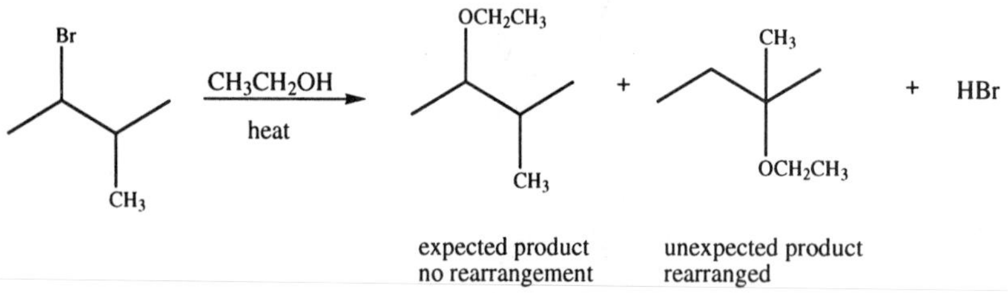

EQ. 8.7

expected product
no rearrangement

unexpected product
rearranged

The first step is formation of a secondary carbocation:

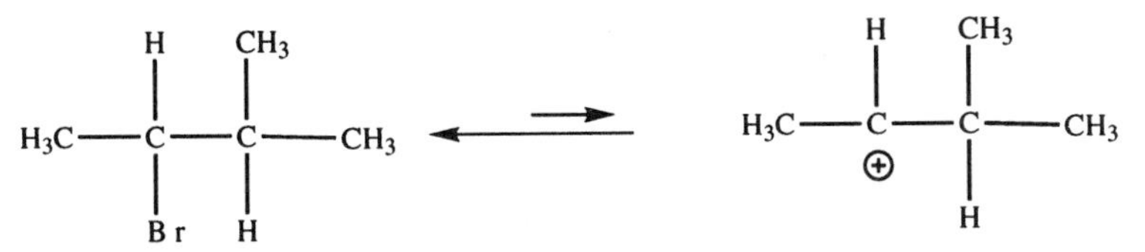

EQ. 8.8

A hydrogen atom and the two electrons involved in the sigma bond now shift from one carbon to the adjacent carbon. This is termed a hydride shift because the effect is the same as moving a hydride ion (a proton and <u>two</u> electrons) from one atom to the other. After the movement of the hydrogen atom and the electrons, the more stable tertiary carbocation now exists.

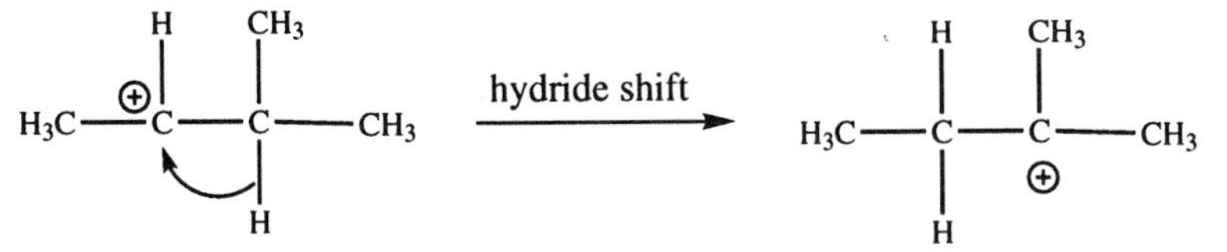

EQ. 8.9

The final step is the reaction of the tertiary carbocation with an ethanol molecule to form the ether linkage. This is followed immediately by loss of the proton to form the product ether. Attack, by ethanol, on the secondary carbocation leads to 2-ethoxy-3-methylbutane.

EQ. 8.10

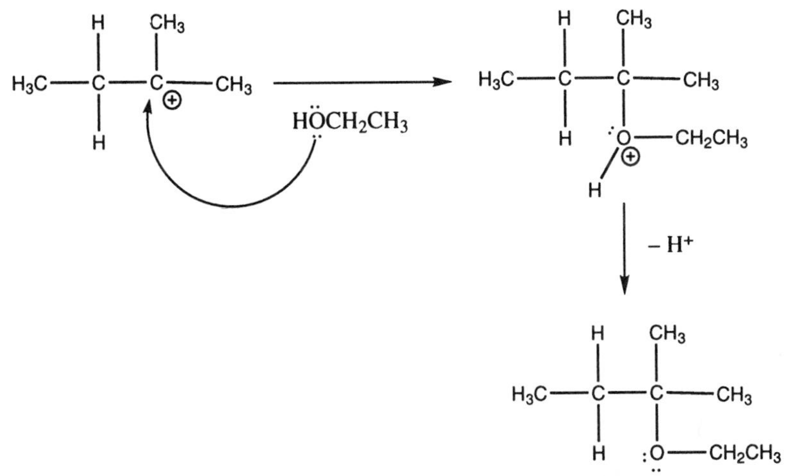

Although formation of primary carbocations is not favored because of their instability, it is possible to force the creation of some primary carbocations by reacting a primary alkyl halide with silver ion. The insolubility of the silver halide helps in the formation of the ion. These carbocations are, however, so reactive that they quickly react with anything that is close by to form stable neutral molecules. In some cases, the primary carbocation undergoes rapid rearrangement into a more stable carbocation before the product is formed. For instance, the reaction of 2,2 dimethyl–1–iodo–propane with aqueous silver ion has been found to yield 2–methyl–2–butanol as illustrated in equation 8.11. This rather exoergic reaction that occurs because of the formation of the very insoluble silver iodide.

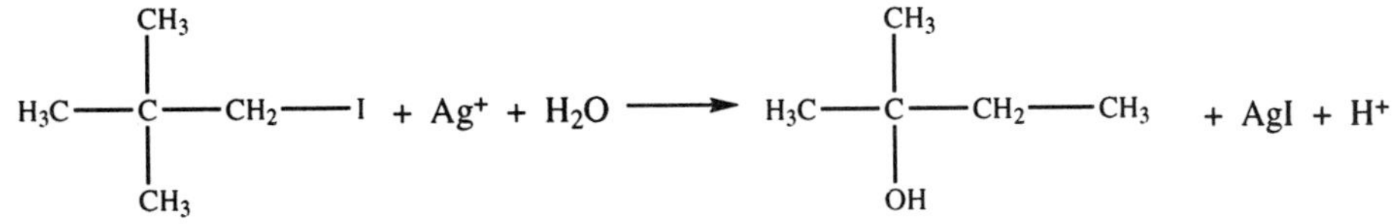

EQ. 8.11

As expected, the first step is formation of the primary carbocation:

$$H_3C-\overset{\overset{\displaystyle CH_3}{|}}{\underset{\underset{\displaystyle CH_3}{|}}{C}}-\underset{H_2}{C}-I \ + \ Ag^+ \ \longrightarrow \ H_3C-\overset{\overset{\displaystyle CH_3}{|}}{\underset{\underset{\displaystyle CH_3}{|}}{C}}-\overset{\oplus}{C}H_2 \ + \ AgI$$

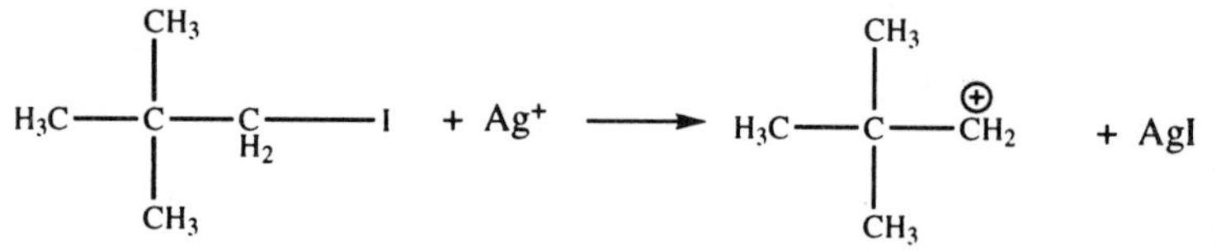

EQ. 8.12

This carbocation quickly rearranges into the more stable tertiary carbocation *via* the transfer of a methyl group and its bonding electrons from the central carbon atom to the charged carbon atom. This is a termed a methide shift because the ionic species formally transferred would be called the methide ion. After the shift, water reacts with the newly formed tertiary carbocation with the eventual formation of the tertiary alcohol. The entire course of the reaction following formation of the primary carbocation is shown as equation 8.13.

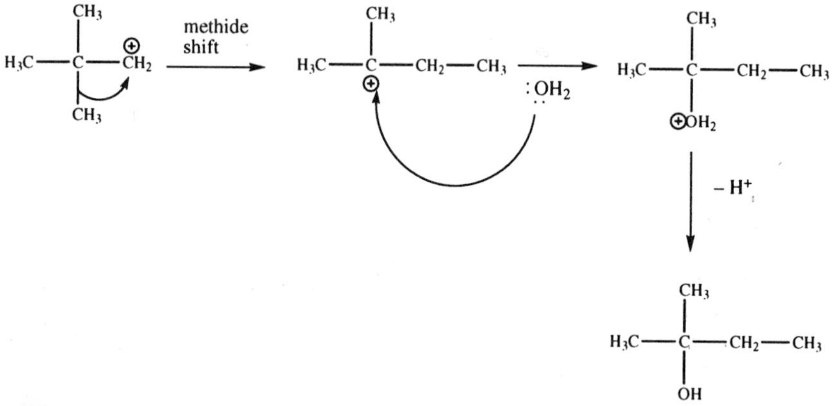

EQ. 8.13

**CONCEPT CHECK**

Draw a step–by–step mechanism that explains the following transformation.

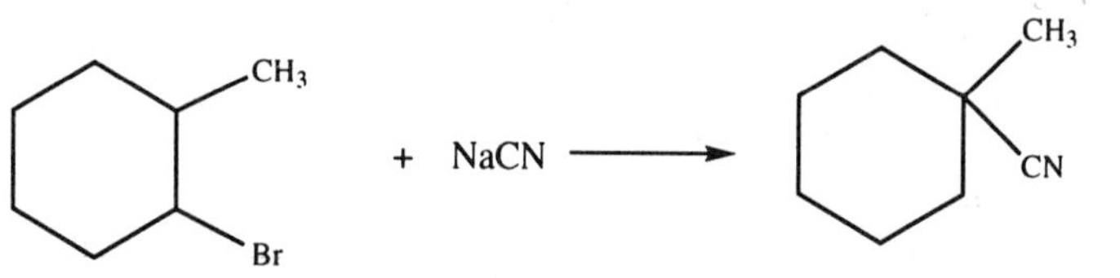

+ NaCN ⟶

EQ. 8.14

As we have just seen, rearrangement of products is most common when a carbocation intermediate is present in the mechanism. This argues for an $S_N1$ mechanism for this reaction. After the secondary carbocation is formed in the first step of the mechanism, it rearranges to the more stable tertiary carbocation *via* what we have called a hydride shift. When this newly formed tertiary carbocation adds a cyanide ion, the product shown in equation 8.14 results. The entire series of steps is shown in figure 8.15.

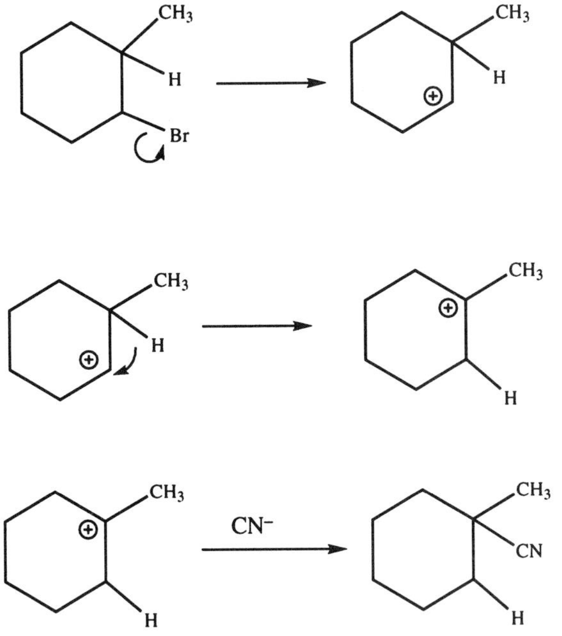

FIGURE 8.15:
Formation of 1-methylcyclo-
hexylnitrile from 1-methyl-2-
bromocyclohexane.

## 8.15   ELIMINATION REACTIONS

Many nucleophilic substitution reactions are accompanied by elimination reactions. This is especially true of the $S_N1$ reaction. Elimination, accompanying an $S_N1$ reaction, results when the intermediate carbocation, rather than reacting with the nucleophile to make the substitution product, instead loses a hydrogen ion from an adjacent carbon atom (termed the β carbon atom). The result of this elimination is the formation of an alkene. The overall scheme is shown in figure 8.16.

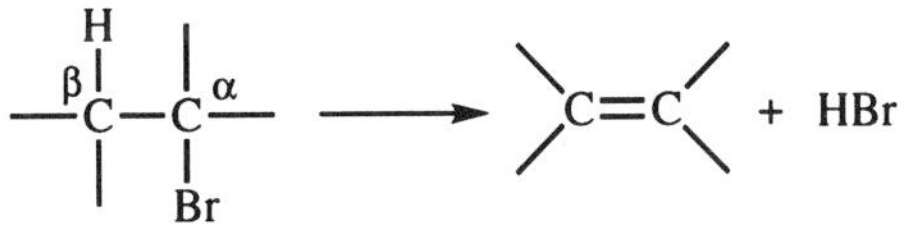

FIGURE 8.16:
Elimination of HBr
from an alkyl bromide.

In some cases, elimination of hydrogen halide from alkyl halides can result in the formation of more than one alkene. For example, elimination of hydrogen bromide from 2-bromobutane can give either 1-butene and 2-butene. In such situations, the more highly substituted alkene is usually the one predominately formed. This rule is known as the Saytzeff rule. Thus, in this example, 2-butene is formed with a higher yield. This is illustrated in figure 8.17.

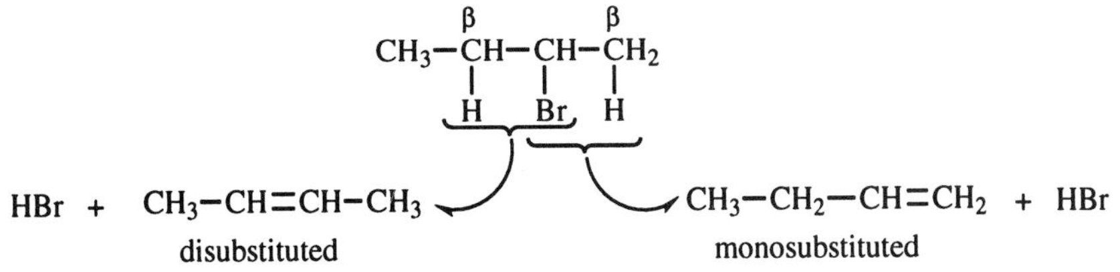

FIGURE 8.17:
Elimination products
from 2-bromobutane.

The order of alkene stability predicted by the Saytzeff rule is given in figure 8.18. While the Saytzeff rule works well in many cases, there is sometimes not a large predominance of the predicted product and, if the approach of the strong base to one of the hydrogen atoms is sterically hindered, it is possible that the rule may not be followed.

$$R_2C=CR_2 > R_2C=CHR > R_2C=CH_2 \ \& \ RHC=CHR > RHC=CH_2 > H_2C=CH_2$$
tetra     tri        di        di       mono      un

FIGURE 8.18:
Order of alkene
stability.

Although elimination is more common accompanying the $S_N1$ reaction, it can also accompany the $S_N2$ reaction. As with the substitution reactions, the mechanisms of these reactions can be either first or second order. They are termed **E1** and **E2** mechanisms respectively. The E1 mechanism often occurs under the same conditions as the substitution reaction since it shares with the $S_N1$ mechanism the formation of the carbocation. On the other hand, elimination *via* the E2 mechanism usually requires the presence a fairly strong base to make it competitive with substitution.

## 8.16  THE E1 ELIMINATION MECHANISM

The first step in the E1 mechanism is the same step as in the $S_N1$ mechanism — formation of the carbocation intermediate. In the second step, a strong base removes a proton on the β carbon atom to form an alkene. This is illustrated for the formation of 2–methyl–2–propene from *t*-butyl bromide in figure 8.19.

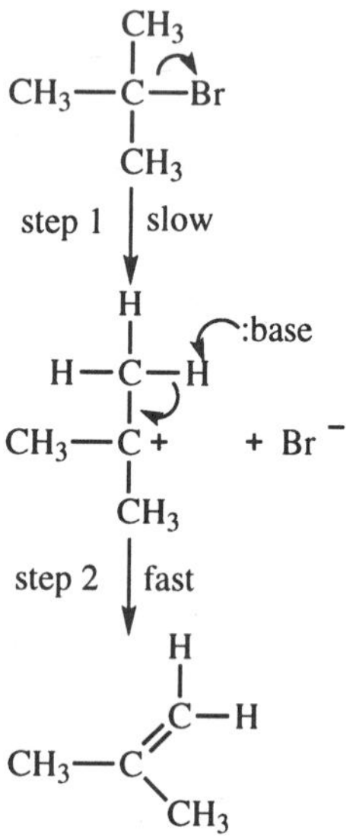

Note that the carbocation is the same one that was involved in the formation of *t*–butyl alcohol from *t*–butyl bromide. As in the substitution reaction, the first step is slow and, therefore, rate determining. This gives the reaction overall first order kinetics. The only role played by the base is to pull off a proton from the carbocation intermediate in the second step of the mechanism. Because acid/base reactions are normally quite fast, this second step does not affect the rate law. However, the relative amounts of substitution and elimination are determined by the relative rates of the second step in each mechanism. If conditions favor elimination, *i.e.*, if the rate of the second step in the E1 mechanism is faster than the second step in the $S_N1$ mechanism, then elimination will become the predominate route for the reaction and the yield of substitution product will suffer. Unfortunately for substitution, it is almost always the case that elimination occurs faster than substitution. This is especially true when strong bases are used. Thus tertiary halides, which would undergo substitution *via* an $S_N1$ mechanism, are almost never used as reactants in substitution reactions because of the poor yields caused by competition of substitution with elimination. Even elimination reactions, which predominate with tertiary halides, often do not result in the desired product because of rapid rearrangement of the carbocation intermediate. Nevertheless, dehydrohalogenation of alkyl halides is an often–used method for the formation of alkenes.

Another method for formation of alkenes is the acid–catalyzed dehydration of alcohols. The mechanism for this reaction is illustrated in figure 8.20. Since removal of the water molecule produces the same carbocation as is produced by loss of a halide ion from an alkyl halide, the same alkene is produced in this reaction. The Saytzeff rule predicts that the usual product is the most highly substituted alkene, however, in this case as in dehydrohalogenation, rearrangement of the carbocation is rather common and other products are often found.

FIGURE 8.19:
E1 reaction
mechanism.

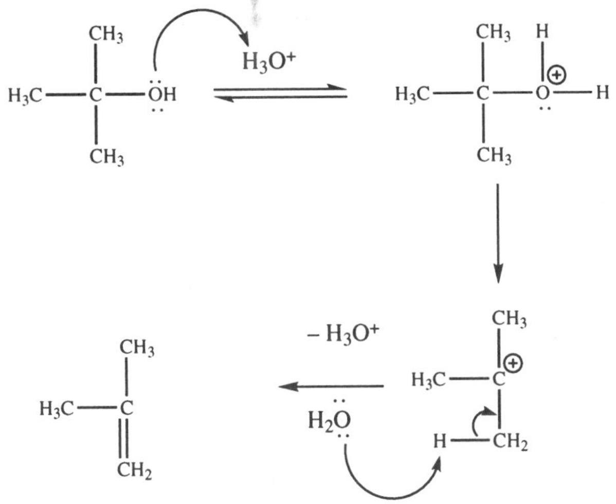

## 8.17  THE E2 ELIMINATION MECHANISM

We have seen that the E1 mechanism shares much in common with the $S_N1$
mechanism — so much that the two processes of elimination and substitution
are in competition with one another. This is not the case with the other major
elimination mechanism: the E2 mechanism. As one would expect from the
name of the mechanism, the rate law for the E2 mechanism is second order
overall. In that way it is similar to the $S_N2$ mechanism. The similarity ends
there, however.

As with the E1 mechanism, the end product is an alkene formed by removal of
a hydrogen atom and another atom, or group of atoms, from the adjacent carbon atom. The E2 mechanism, however, begins with the attack of a strong
base to remove a proton from the β carbon atom. At the same time, the leaving
group departs from the a carbon atom, taking with it both of the electrons that
initially formed the bond between it and the carbon atom. Figure 8.21 depicts
this concerted mechanism.

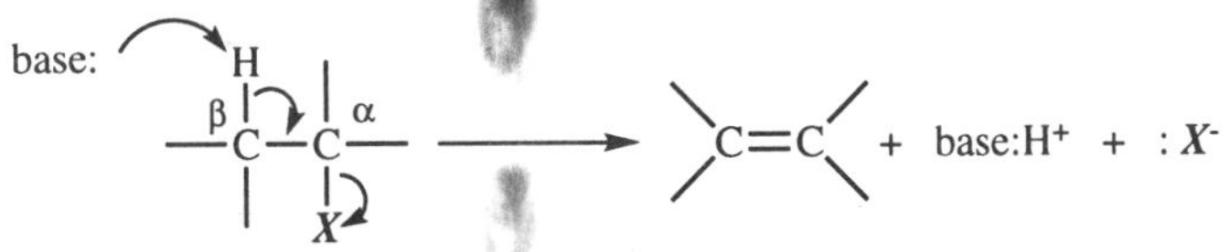

If the reaction truly occurs *via* a concerted mechanism, the rate of the reaction
must be dependent on both the concentration of the base and the alkyl bromide.
This leads to a rate expression that is first order in the alkyl halide and first
order in the base: Rate = k[RX][base]. Figure 8.22 depicts a transition state
that would be likely in such a situation. Note that the proton and the leaving
group are on opposite sides the carbon atom and are *in the same plane*. This
condition is termed anti-coplanar — anti because the base approaches from the
side opposite the leaving group; co–planar because hydrogen atom, both carbon
atoms, and the leaving group atom are all in the same plane. Co–planarity is
required because the orbitals bonding the hydrogen and the leaving group must
be situated in such a way that the old orbitals can smoothly form the new π
bond in the product alkene. This formation of a new orbitals from the old is
pictured in figure 8.23.

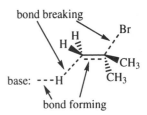

base: - - -H

bond breaking

bond forming

If this picture of the reaction is correct, we can infer a great deal about the stereochemistry of the product from the stereochemistry of the alkyl halide. In fact, the stereochemistry of the alkyl halide must determine the stereochemistry of the alkene. As an example, let us consider the dehydrobromination of the two stereoisomers of 1,2-dibromo-1,2-diphenylethane. Reaction of the meso isomer is shown in figure 8.24. To remove the bromine atom on the right–most carbon atom, the only configuration in which the bromine atom and the hydrogen atom are anti and coplanar is shown in the sawhorse diagram and the corresponding Newman diagrams on the left side of the figure. Removal of HBr when the molecule is in this configuration results in the E isomer of the alkene.

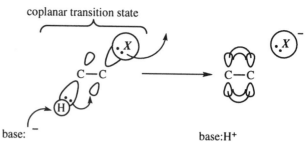

FIGURE 8.23:
Orbital reorganized during E2 mechanism.

On the other hand, when the 1S,2S form reacts with the base, the only configuration in which the bromide leaving group and proton can be anti and coplanar is shown in the sawhorse and Newman diagrams in figure 8.25. The resulting alkene has the Z configuration.

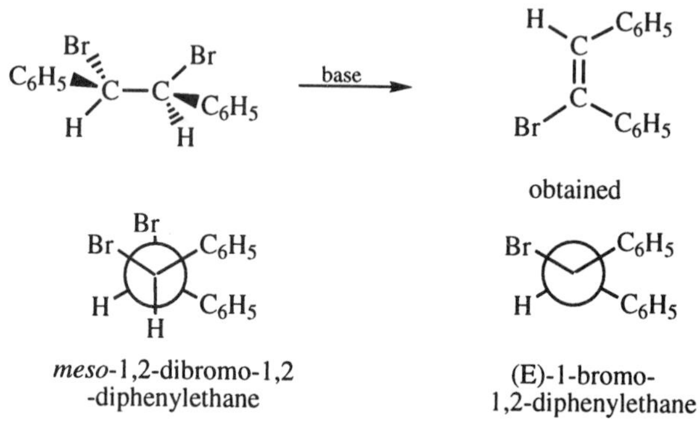

Since many strong bases are also nucleophiles, here too, there may be competition between substitution and elimination. Once again, whether substitution or elimination prevails depends upon the nature of the reactants and the reaction conditions. Since there is no second step in this mechanism as there is in the E1 mechanism, it can not be the relative rates of the "second steps" that determines the ratio of substitution to elimination products as it was in the E1 mechanism. We must look for another factor that governs the ratio. In this case the use of a stronger base as the nucleophile, or the use of a more hindered alkyl halide, and the use of more harsh reaction conditions all favor elimination over substitution. The reasons: (1) stronger bases are more likely to remove the proton rather than act as a nucleophile toward the carbon atom and, (2) hindered alkyl halides present steric problems for the approach of the nucleophile to the carbon atom undergoing substitution but do not hinder the movement of the proton toward the base.

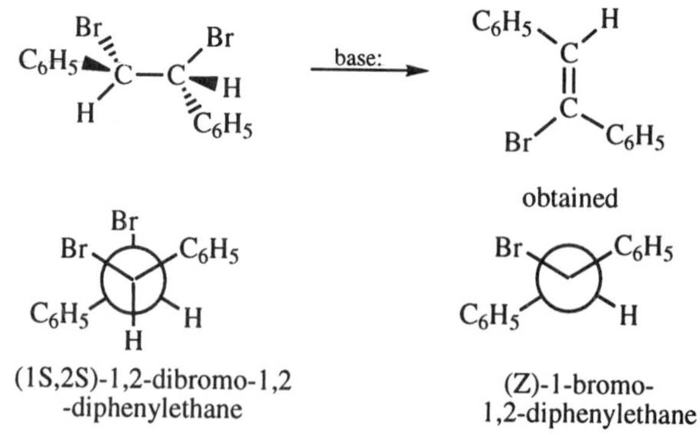

Acid–catalyzed dehydration of alcohols can also occur *via* an E2 mechanism. In this case the first step in the reaction, an equilibrium step, is the protonation of the hydroxyl group of the alcohol to form a better leaving group, the water molecule. The rate–determining step is the concerted removal of the proton from the β–carbon atom and the departure of the leaving group. Just as in the previously considered elimination of HX from an alkyl halide, the stereochemistry of the product an be determined from the configuration of the substrate. The mechanism of acid–catalyzed dehydration of (R)–2–butanol is shown in figure 8.26. The student is left to determine whether (E)–2–butene of (Z)–2–butene is the product.

*meso*-1,2-dibromo-1,2-diphenylethane

(E)-1-bromo-1,2-diphenylethane

FIGURE 8.24:
Dehydrobromination of *meso*-1,2-dibromo-1,2-diphenylethane.

(1S,2S)-1,2-dibromo-1,2-diphenylethane

(Z)-1-bromo-1,2-diphenylethane

FIGURE 8.25:
Dehydrobromination of (1S,2S)-dibromo-1,2-diphenylethane.

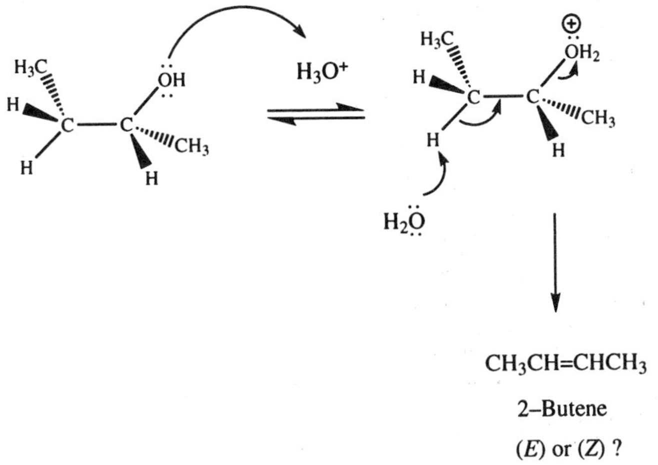

FIGURE 8.26:
Acid-catalyzed dehydration
of (R)-2-butanol.

CH₃CH=CHCH₃

2–Butene

(E) or (Z) ?

## 8.18  EPOXIDE FORMATION FROM α–HALOALCOHOLS

In a substitution reaction that is, in some ways, quite closely related to an elimination reaction, we now look at the formation of epoxides from α–haloalcohols.  The reaction occurs under basic conditions that make the stronger nucleophile, an aloxide ion, from the weaker nucleophile, the neutral hydroxyl group.  In this case, however, rather than reacting with and displacing the chloro group on another molecule, the presence of the chlorine atom on the α carbon atom presents a good leaving group close by and in the right position for an intramolecular reaction.  The result of this intramolcuelar displacement of the chlorine atom is a cyclic ether, termed an epoxide.  The course of the reaction is shown in figure 8.27.  The reaction shown is that for a primary alcohol.  However, unless there are steric problems, other, more highly substituted, species also form epoxides in good yield under these conditions.

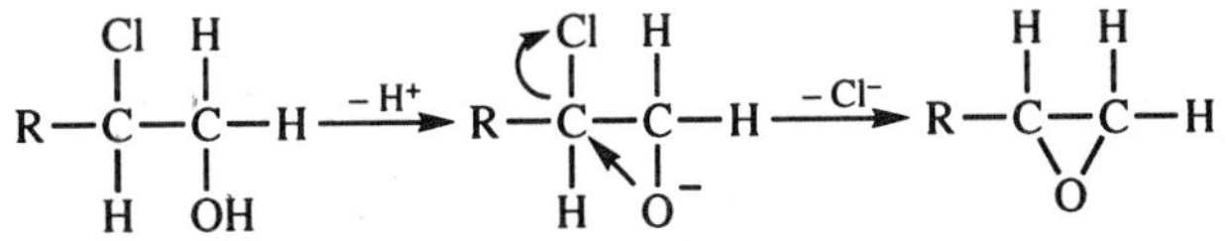

FIGURE 8.27:
Epoxide formation.

This same type of reaction is often used to make ethers and is known as the **Williamson Ether Synthesis**.   The reaction is typically restricted to primary alkyl halides to minimize competing elimination reactions.  For example, dibutyl ether can be prepared from 1–bromobutane and sodium butoxide as shown in equation 8.15.

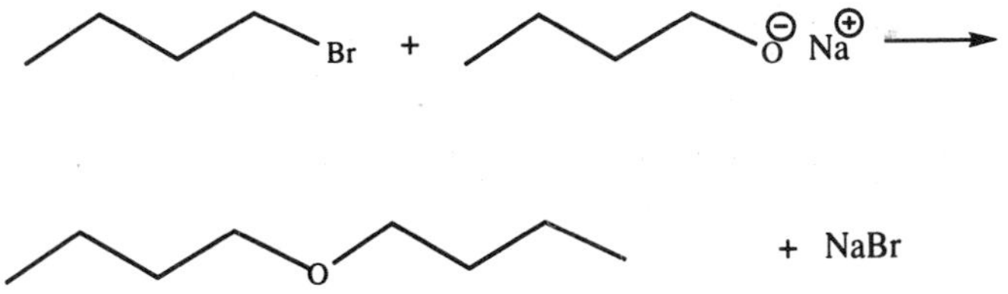

EQ. 8.15

+ NaBr

## 8.19  EPOXIDE OPENING REACTIONS

Epoxides, or oxiranes as they are formally known, are highly reactive and susceptible to ring–opening reactions.  In both cases the product is a substituted alcohol.  Let us examine the acid–catalyzed reaction first.

As expected, the first step in an acid–catalyzed ring opening reaction is the protonation of the epoxide oxygen atom by the acid catalyst. This forms an oxonium ion which is easily attacked at one of the carbon atoms attached to the oxygen atom by a nucleophile in an S$_N$2 fashion. This nucleophile is usually the conjugate base of the acid used to catalyze the reaction. If, for example, hydrobromic acid is used as the reactant, the result is an β–bromoalcohol. The overall reaction scheme is shown in figure 8.28.

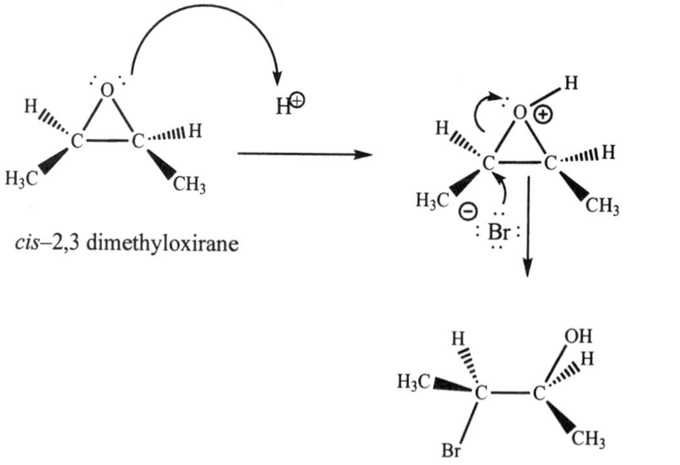

*cis*–2,3 dimethyloxirane

(2S,3S)–3–bromo–2–butanol

**FIGURE 8.28:** Acid-catalyzed epoxide ring-opening.

The stereochemistry of the product depends upon which carbon atom is attacked. Figure 8.28 depicts the carbon atom on the left undergoing the nucleophilic attack. If the carbon atom on the right were to undergo attack, a product with a different stereochemistry would be formed. The student should verify that the product formed in this instance is (2R,3R)–3–bromo–2–butanol. Since attack on the left or the right carbon atom would be equally likely in this case, a racemic mixture of the two isomers is formed in good yield.

Epoxides can also be directly attacked by nucleophiles. This normally occurs in basic solution as illustrated in figure 8.29. Consider the reaction of the same epoxide with sodium methoxide in methanol. In this reaction, the nucleophile, the methoxide ion, first attacks one of the carbon atoms bonded to the epoxide oxygen to open the ring and form an anion with the charge on the former epoxide oxygen atom. This atom quickly abstracts a proton from a methanol molecule to generate the product and regenerate the methoxide ion. Once again, since either carbon atom is susceptible to attack by the nucleophile, a variety of isomers can be formed. Only one such isomer is shown. The student is left to work out formation of the other isomers.

*cis*–2,3 dimethyloxirane

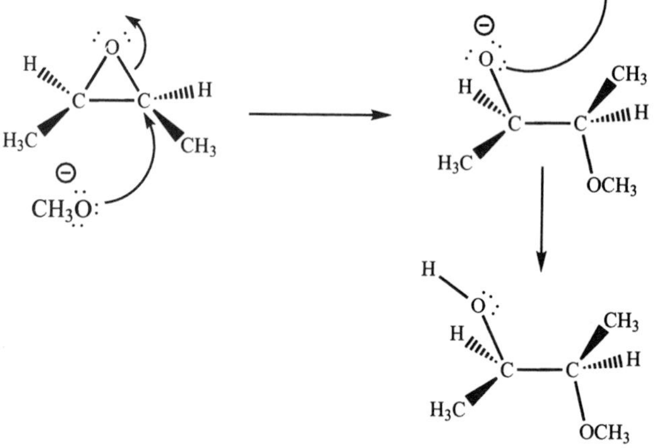

**FIGURE 8.29:** Nucleophilic opening of an epoxide ring.

We also need to consider the ring-opening reactions of unsymmetrical epoxides under both basic and acidic conditions. First let us look at the reaction between isobutylene oxide and sodium methoxide as shown in equation 8.16. The major product of this reaction is 1-methoxy-2-methyl-2-propanol.

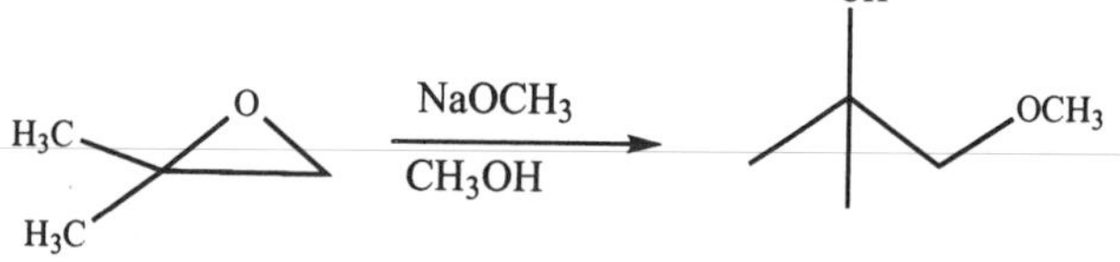

EQ. 8.16

This reaction is very much like a typical S$_N$2 reaction in that methoxide serves as a nucleophile and the epoxide oxygen serves as the leaving group. Although we don't usually think of an alkoxide as an especially good leaving group, the inherent ring strain of the epoxide makes this ring opening much more probable that we otherwise might predict. The mechanism for this reaction is shown in figure 8.30. Note that under basic conditions the nucleophile attacks the epoxide at the less-hindered carbon. This is primarily a function of sterics.

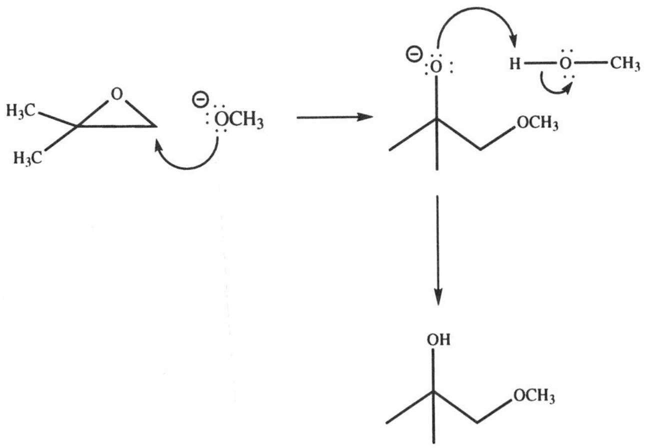

In acidic solution the situation is a bit more complex. If we now consider the reaction between isobutylene oxide and methanol in the presence of an acid catalyst, we observe that the major product is 2-methoxy-1-methyl-1-propanol.

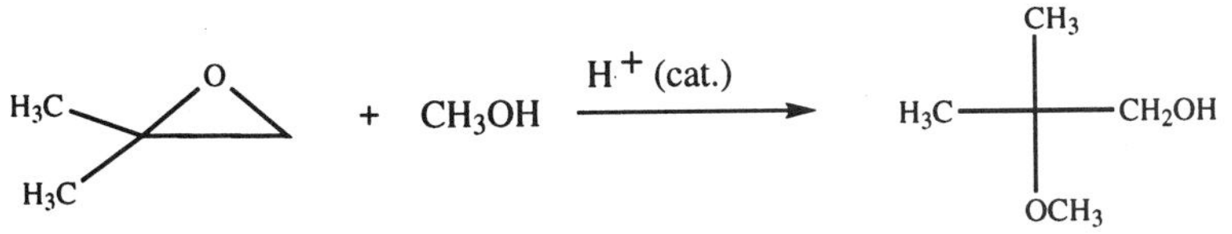

**FIGURE 8.30:**
Mechanism for epoxide ring opening under basic conditions.

EQ. 8.17

We are able to rationalize the formation of this other product by examining the mechanism of the acid catalyzed process. The first step in the acid catalyzed process involves protonation of the epoxide oxygen. Delocalization of the charge on the oxonium ion results in a partial positive charge developing on one or both of the carbon atoms in the epoxide. Since the tertiary carbon is better able to accommodate a developing charge than is the secondary carbon, it is more likely to do so. Thus the partial positive charge probably resides mostly on the tertiary carbon. In fact it has been postulated that the carbon oxygen bond may completely break. This would result in the formation of a full blown carbocation. Since the tertiary carbon likely contains at least partial positive character the incoming nucleophile (methanol in this case) is most attracted to this carbon. The final step in the process is loss of a proton. Thus the acid catalyzed opening of epoxides is more like an $S_N1$ reaction.

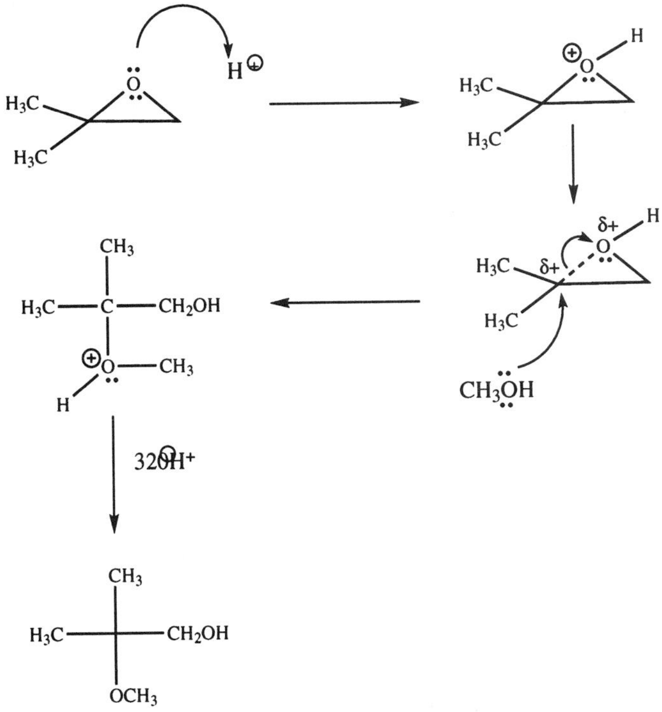

**FIGURE 8.31:**
Mechanism for epoxide ring opening under acidic conditions.

## 8.20 NUCLEOPHILICITY VS. BASICITY

It is likely that the student has been wondering about the use of the word "nucleophile" that has appeared so often in this chapter. Of course, the term nucleophile simply means "nucleus loving" — in deference to its unshared, and readily available, pair of electrons that participates in the reaction. This, of course, simply makes the species a Lewis Base (and often a Brønsted base as well). What, then, is the difference?

Simply put, the term "nucleophile" refers to a species whose electron pair is used in a substitution reaction and which, during the reaction, shares its electrons with a carbon atom. Ultimately there is formation of a bond between carbon and the nucleophile. "Nucleophile", then, is a term used when discussing the rates of substitution reactions. It has primarily Lewis base implications. The best nucleophile in any particular reaction is the species for which the substitution rate is the fastest. The term, therefore, has kinetic implications.

On the other hand, the term "basicity" normally is used to discuss the ability of a species to accept a proton, thus forming a bond between the base and the proton. The term, then, has primarily Lowry–Brønsted base implications. Better bases are those species for which the point of equilibrium lies further toward the conjugate acid. The term, therefore, has thermodynamic (equilibrium constant) implications.

Of course, species that are good nucleophiles are usually good bases. However, for steric reasons, the opposite is sometimes not true.

## 8.21 COMPETITION BETWEEN REACTION TYPES—THE EFFECT OF THE LEAVING GROUP

The common theme of all reactions in this chapter has been the departure of an atom, or group of atoms, termed the leaving group, from the reactant as the product is formed. The success of the reactions depends on the ability of this atom or atoms to leave cleanly and smoothly. It should come as no surprise to learn that some leaving groups are better that others. What, then, are the characteristics of a good leaving group?

In a nucleophilic substitution or elimination reaction, the leaving group must take with it both of the electrons that formerly were involved in its bond to the reactant. Thus, the more capable the leaving group is of accommodating those electron pairs, the better will be the leaving group. Halide ions are often found as leaving groups. All halide ions are stable species, quite capable of accommodating the eight electrons in the octet. Yet iodide is a much better leaving group than is fluoride. Hydroxide ion is also frequently encountered as a leaving group. It, too, is stable and able to accommodate the octet. However, halide ions are better leaving groups than hydroxide ion. Why is this?

There are two reasons: The first is the ability of the group to stabilize the pair of electrons (and, often, the charge the accompanies the pair of electrons). This means that, within a family, larger ions will be better leaving groups because of their ability to spread out the charge over a larger volume of space (*i.e.*, the charge density of the ion will be smaller). It also means that resonance-stabilized species will be better leaving groups that otherwise similar, but non-resonance-stabilized, species.

Of course both of these factors affect the basicity of the species. If a group is a strong base, such as a hydroxide ion, an unshared pair of electrons competes strongly for a nucleus. In an acid/base reaction, this nucleus is a proton. In a substitution or elimination reaction, this nucleus is the atom to which the group is attached. Hence, the same factors that make species stronger bases are the factors that make groups poorer leaving groups. Since iodide ion is the weakest base of the halide ions, this explains the ordering of the halide ions' ability to function as leaving groups. It also explains the fact that hydroxide is a poor leaving group compared to the halides.

One way to make a group a weaker base is to protonate it, thus forming the conjugate acid; reactions potentially involving hydroxide or alkoxide ions as the leaving group are often run in acid solution so that the water molecule or an alcohol molecule becomes the *de facto* leaving group. Of course, this approach can only be used when the nucleophile is a weaker base than the reactant. For instance, attempting to replace the hydroxyl group of an alcohol with a nitrile group can not be done in acid solution because the cyanide ion would pick up the proton to form hydrogen cyanide (with potentially deadly consequences for the chemist). Fortunately, because alcohols are such common and easily obtained compounds, it is possible to convert the hydroxyl group into an excellent leaving group, a sulfonic acid salt, by formation of a sulfonate ester. This is most commonly done by reacting the alcohol with the acid chloride of p-toluenesulfonic acid. This reaction is so common, that the p-toluenesulfonyl group has been given a non-IUPAC name: tosyl. The group is often abbreviated in reactions as —OTs. A typical reaction sequence is illustrated in figure 8.32.

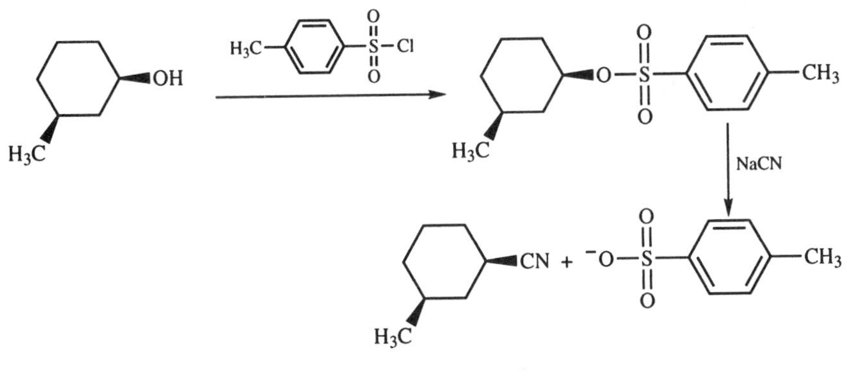

FIGURE 8.32:
Use of tosyl group as leaving group.

## 8.22  COMPETITION BETWEEN REACTION TYPES—SUBSTITUTION AND ELIMINATION

As we have already seen, substitution and elimination are always found together to some extent. For the E1 and $S_N1$ mechanisms, this is because they occur with the *same* intermediate — the carbocation. For the E2 and $S_N2$ mechanisms, although the intermediate is not exactly the same, the concerted mechanism requires that the base (nucleophile) approach the alkyl halide in a similar fashion regardless of whether elimination or substitution is to be the eventual outcome. If we wish to favor one process over another, we need to learn the conditions which favor one type of reaction over the others.

Since first order processes happen *via* a charged intermediate (a carbocation), conditions which stabilize the carbocation will favor first order processes. Of course, the most important of the conditions that determine the stability of the carbocation is the nature of the reactant itself: tertiary carbocations are much more stable than secondary which are more stable than primary. However, reaction conditions can also be altered to shift the balance toward one or the other. First, polar protic solvents with high dielectric constants, such as water and the small alcohols favor first order processes because the solvent is better able to stabilize the charged intermediate. Aprotic solvents favor second order processes, partially because they are usually less polar. Second, large concentrations of nucleophile tend to favor second order processes since the rate of a second order reaction is dependent upon the concentration of both of the rectants while the rate of the first order process is independent of the concentration of the base. Third, use of a strong base that is not a particularly good nucleophile (*e.g.*, a highly hindered, very large species) will favor elimination over substitution. And fourth, increases in temperature generally favor elimination over substitution. Therefore, treatment of *t*-butyl chloride with a 10% sodium ethoxide in ethanol at high temperatures will yield a much larger amount of the elimination product, 2-methylpropene, than of *t*-butyl alcohol, the substitution product. In contrast, treatment of *n*-propyl chloride with 50% aqueous ethanol containing 1% sodium hydroxide will result in much larger amounts of the substitution product, 1-propanol, the substitution product, than of 1-propene, the elimination product.

**Carbocation rearrangement** - a carbocation intermediate rearranges to a more stable carbocation by movement of a hydride or alkylide species to the adjacent carbon atom.

**Concerted step:** a step in a reaction mechanism in which one atom or group of atoms forms a bond simultaneously with the breaking of the bond to another atom or group of atoms.

**E1 reaction mechanism:** a unimolecular elimination mechanism having two steps and a carbocation intermediate.

**E2 reaction mechanism:** a bimolecular elimination mechanism involving attack of a base on a hydrogen of the beta-carbon.

**Elimination reaction:** a reaction resulting in loss of two atoms or groups on adjacent atoms.

**Hammond Postulate:** states that the structure of the transition state resembles the structure of the closest stable species

**Leaving group:** the group that leaves during a reaction.

**Nucleophile:** an atom or group of atoms containing a pair of electrons easily available for sharing during a reaction.

**$S_N1$ reaction mechanism:** a first order, nucleophilic substitution mechanism having two steps and a carbocation intermediate.

**$S_N2$ reaction mechanism:** a second order, concerted, nucleophilic substitution mechanism involving backside attack of the nucleophile.

**Substitution reaction:** reaction in which one atom or group of atoms replaces another atom or group of atoms.

**Williamson Ether Synthesis:** reaction in which an alkyl halide is reacted with an alkoxide to form an ether.

1. Arrange the compounds in each set in order of increasing reactivity to $S_N2$ reactions.

(a) 2-bromo-3-methylbutane, 1-bromo-3-methylbutane, 2-bromo-2-methylbutane

(b) bromocyclohexane, 1-bromo-1-cyclohexylmethane, 1-bromo-1-methylcyclohexane

(c) chlorocyclopentane, 3-chloro-1-cyclopentene, 1-chloro-1-cyclopentene

(d) 2,2-dimethyl-5-iodopentane, 2,2-dimethyl-4-iodopentane, 2,2-dimethyl-3-iodopentane

2. Arrange the compounds in each set in order of increasing reactivity to $S_N1$ reactions.

(a) 2-bromo-3-methylbutane, 1-bromo-3-methylbutane, 2-bromo-2-methylbutane

(b) bromocyclohexane, 1-bromo-1-cyclohexylmethane, 1-bromo-1-methylcyclohexane

(c) chlorocyclopentane, 3-chloro-1-cyclopentene, 1-chloro-1-cyclopentene

(d) 2,2-dimethyl-5-iodopentane, 2,2-dimethyl-4-iodopentane, 2,2-dimethyl-3-iodopentane

3. A solution of (R)-2-butanol retains its optical activity in base but when in dilute sulfuric acid, a racemic mixture is formed. Explain what happens.

4. When 3,3-dimethyl-2-butanol is heated with HCl, a mixture of 2,3-dimethyl-2-chlorobutane and 2,3-dimethyl-2-butene is formed. Explain these results with a mechanism.

5. Give reagents needed to convert 1-bromopropane into:

(a) methyl *n*-propyl ether
(b) *n*-propyl magnesium bromide
(c) *n*-propyl iodide
(d) *n*-propyl alcohol
(e) *n*-butanenitrile
(f) *n*-propylamine
(g) *n*-propanethiol (RSH)

6. Predict the elimination products of the following reactions.

(a) 2-bromobutane with potassium hydroxide
(b) 2-pentanol with 70% sulfuric acid
(c) 2-methyl-2-butanol with 90% sulfuric acid
(d) 1-bromo-1-methylcyclohexane with potassium *t*-butoxide

(e)

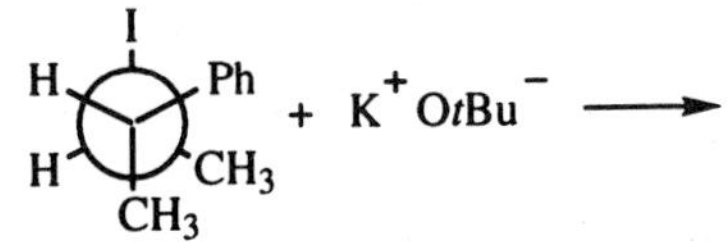

(f)

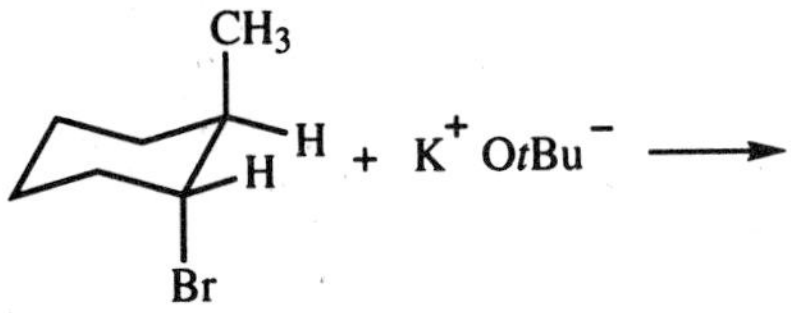

7. Many times experiments result in different results than expected. For example, when benzene and 1-chloropropane are reacted under Friedel Crafts conditions with $AlCl_3$, isopropylbenzene is found instead of *n*-propylbenzene. Explain this result with a mechanism.

8. One place that E1 reaction is seen is the synthesis of alkenes from alcohols. Write the mechanism for the reaction of *t*-butyl alcohol in 20% sulfuric acid and 80% water which produces 2-methylpropene.

9. In this chapter, you saw that chiral compounds such as (*R*)-2-bromobutane can be used to test whether $S_N2$ reactions undergo inversion of configuration. Substituted cyclopentanes are relatively easy to synthesize and have also been used to check inversion. They offer the advantage that the inverted product can be more easily separated from a noninverted product. Show with drawings how the following reaction supports the $S_N2$ mechanism.

   *cis*-1-Bromo-3-methylcyclopentane reacts with hydroxide to produce *trans*-3-methylcyclopentanol.

10. When 1-chloro-2-butene is placed in a warm solvent of 50% acetone and 50% water, two alcohols are formed, 3-buten-2-ol and 2-buten-1-ol. Write a mechanism to explain formation of the two alcohols.

11. If the rate of the $S_N2$ reaction of 0.1 **M** ethyl bromide with 0.2 **M** sodium hydroxide to form ethanol were 0.06 mol $L^{-1}$ $sec^{-1}$, what would be the rate of:

   (a) 0.2 **M** ethyl bromide and 0.2 **M** sodium hydroxide?
   (b) 0.1 **M** ethyl bromide and 0.1 **M** sodium hydroxide?
   (c) 0.2 **M** ethyl bromide and 0.4 **M** sodium hydroxide?

12. If the rate of the $S_N1$ reaction of 0.1 **M** *t*-butyl bromide with 0.2 **M** sodium hydroxide to form *t*–butyl alcohol were 0.06 mol $L^{-1}$ $sec^{-1}$, what would be the rate of:

   (a) 0.2 **M** *t*-butyl bromide and 0.2 **M** sodium hydroxide?
   (b) 0.1 **M** *t*-butyl bromide and 0.1 **M** sodium hydroxide?
   (c) 0.2 **M** *t*-butyl bromide and 0.4 **M** sodium hydroxide?

13. Provide the structures for each of the compounds at each stage in the reactions. In addition, provide reagents and conditions where they are not indicated.

(R)-2-Pentanol ———→ tosylate ester $\xrightarrow[\text{acetone}]{\text{Br}^-}$

a. Bromide ———→ Alcohol + olefin

(S)-1-Phenyl-1-butanol $\xrightarrow[\text{ether}]{\text{SOCl}_2}$ Alkyl chloride ————→

optically active acetate ———→ optically active alcohol ———→

b. tosylate ester ———→ (S)-1-Phenyl-1-Iodobutane

14. Explain why:

a. Benzyl magnesium bromide can be made in good yield only in dilute solution.

b. $(CH_3)_2CClCO_2C_2H_5$ is stable under conditions that lead to hydrolysis of $(CH_3)_3CCl$.

c.

reacts slowly with KI in acetone

d. the yields of substitution products for the reaction of 2-chlorobutane with $(CH_3)_2N^-$, $(CH_3)_3CO^-$, $CH_3O^-$, and $I^-$ increase in the order from left to right.

15. Provide the structures in the following synthetic sequences:

a.

$$H_3C-C\equiv CH \xrightarrow{\text{NaNH}_2} \underline{\hspace{2cm}} \xrightarrow{\overset{\text{H}_2\text{C}-\text{CH}_2}{\underset{\text{O}-\text{CH}_2}{|\quad|}}} \underline{\hspace{2cm}}$$

$$\xrightarrow{\text{SOCl}_2} \underline{\hspace{2cm}} \xrightarrow{\text{KCN}} \underline{\hspace{2cm}}$$

b. $CH_3CH_2CH_2Br \xrightarrow{\text{KN}_3} \underline{\hspace{2cm}} \xrightarrow{\text{H}_2, \text{Pt}} \underline{\hspace{2cm}}$

$$\xrightarrow[\text{excess}]{\text{CH}_3\text{I}} \underline{\hspace{2cm}}$$

c. $CH_2(CO_2C_3H_7) \xrightarrow{\text{Na}} \underline{\hspace{2cm}} \xrightarrow{\text{CH}_3\text{I}} \underline{\hspace{2cm}}$

$$\xrightarrow{\text{Na}} \underline{\hspace{2cm}} \xrightarrow{\text{CH}_3\text{CH}_2\text{CH}_2\text{Br}} \underline{\hspace{2cm}}$$

d.

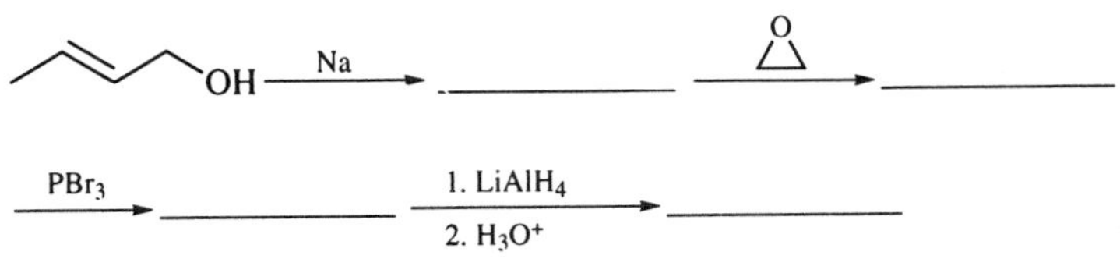

$$\xrightarrow{\text{PBr}_3} \underline{\hspace{2cm}} \xrightarrow[\text{2. } H_3O^+]{\text{1. LiAlH}_4} \underline{\hspace{2cm}}$$

16. Predict the major organic product for the following reactions. Pay attention to stereochemistry when necessary.

a.

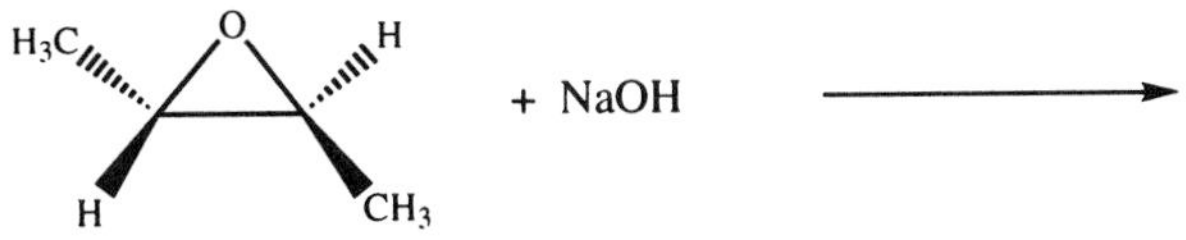

+ NaOH $\longrightarrow$

b.

+ $CH_3CH_2OH$ $\xrightarrow{H^+ \text{ (cat.)}}$

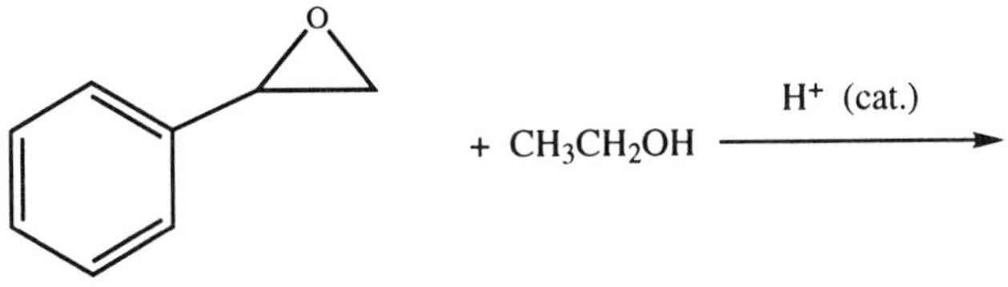

17. Show how the following compounds can be made from an alkene and any other necessary reagents.

a.

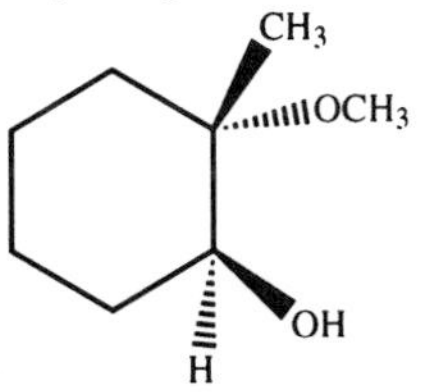

b.

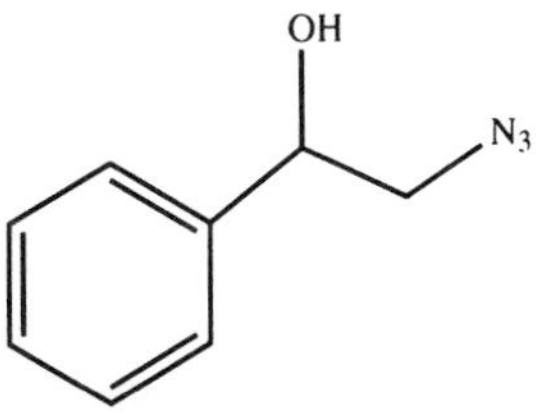

**CHAPTER EIGHT**

OVERVIEW
## Problems

18. When 2-bromo-2-methylpropane reacts with hydroxide ion to form tertiary butyl alcohol and bromide ion, the rate can be measured by the method of initial rates.

a. Doubling the concentration of the alkyl bromide is expected to have what effect on the initial rate?

b. Doubling the concentration of hydroxide ion is expected to have what effect on the initial rate?

c. The carbocation $(CH_3)_3C^+$ is an intermediate assumed to be present in the mechanism for this reaction. Draw the Lewis dot diagram of this species.

d. What hybrid orbitals does the central carbon in $(CH_3)_3C^+$ utilize?

e. What C-C-C bond angle is expected for this species?

19. When 1-bromobutane reacts with azide ion ($N_3^-$) to form $CH_3CH_2CH_2CH_2N_3$ and bromide ion, the rate can be measured by the method of initial rates.

    a. Doubling the concentration of the alkyl bromide is expected to have what effect on the initial rate?

    b. Doubling the concentration of azide ion is expected to have what effect on the initial rate?

    c. Draw a Lewis dot diagram of the azide ion.

    d. When the solvent for this reaction is changed from water to acetonitrile, the rate of reaction increases by a factor of 700. By how much does the activation energy of the reaction decrease going from water to acetonitrile?

20. In acetone solvent, bromomethane reacts with chloride ion to form chloromethane and bromide ion. The rate of reaction follows the Arrhenius equation, with $A = 2.0 \times 10^9$ lit/mol sec, and $E_a = 65.6$ kJ/mol.

    a. Calculate the rate constant, including units, at 25.0°C.

    b. The units of the rate constant indicate that the reaction is what overall order?

    c. Halving the initial concentration of chloride ion would have what effect on the initial rate?

    d. If the standard state free energy change of the reaction is + 67.0 kJ/mol, determine the equilibrium constant and the rate constant (including units) for the reverse reaction at 25.0°C.

21. When tertiary butyl bromide reacts with $OH^-$ in aqueous solution to form tertiary butyl alcohol, the rate of disappearance of the alkyl bromide is followed. When formate ion is introduced into the system at the start of the reaction, tertiary butyl formate is formed, but the rate of disappearance of the alkyl bromide remains the same. When chloroacetate ion is introduced into the system at the start of the reaction, tertiary butyl chloroacetate is formed, but the rate of disappearance of the alkyl bromide remains the same.

    a. Draw the structures of tertiary butyl alcohol, tertiary butyl formate, and tertiary butyl chloroacetate.

    b. Write a two-step mechanism for the formation of each of the three products. Label each step as either fast or slow.

    c. In one or two sentences, explain why the rate of the three different reactions are identical.

22. When comparing the reactivity of substitution by the ethoxide ion ($CH_3CH_2O^-$) with either methyl bromide or neopentyl bromide (1-bromo-2,2-dimethylpropane), it was noted that the activation energy for the reaction with methyl bromide was 26 kJ/mol lower than that for neopentyl bromide.

    a. Write the reaction for methyl bromide, showing the structure of the reactants and products.

    b. Write the reaction for neopentyl bromide, showing the structure for the reactants and products.

    c. Assuming that the pre-exponential terms (A) are about the same, which reaction is faster? By what factor? (i.e., what would be the ratio of the two rate constants.)

    d. What effect (if any) would tripling the concentration of ethoxide ion be expected to have on the measured rate of either reaction?

# 9

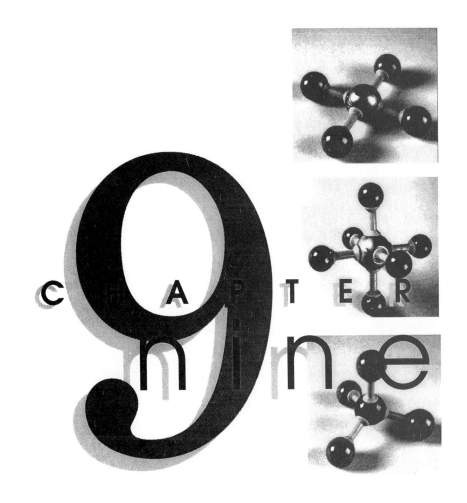

# COMPLEX
# Chemistry

## MORE THAN FOUR NEIGHBORS

# COMPLEX
## CHEMISTRY: More Than Four Neighbors

## 9.1   THE HISTORY OF COORDINATION CHEMISTRY

In the mid–1800s, after chemists had learned to determine the composition of materials accurately and were able to fix the atomic composition of compounds with little trouble, there was a class of compounds that puzzled chemists. These compounds seemed to be stoichiometric combinations of two or more other compounds.  One such group of compounds were the so–called "**double salts**".  For instance, $Al_2(SO_4)_3 \cdot K_2SO_4 \cdot 6H_2O$, one of a group of compounds known as alums, is a double salt.  In this class of compounds, aqueous solutions of the compounds produce ions that give reactions characteristic of the individual ions — in this particular case, $K^+$, $Al^{3+}$ and $SO_4^{2-}$.

A second group of stoichiometric compounds forms aqueous solutions that do not react as if all of the individual ions are present.  Examples of this second group are  $3KCN \cdot Fe(CN)_3$, $4KCN \cdot Fe(CN)_2$, $2NaF \cdot SiF_4$,  and $3NaF \cdot AlF_3$. Solutions of these compounds typically test positive for the alkali metal ions, but simple tests do not indicate the presence of the other ions ($Fe^{2+}$, $Fe^{3+}$, $Si^{4+}$, $Al^{3+}$, $F^-$, or $CN^-$).  Such chemical behavior clearly showed early investigators that these were not double salts but, rather, some entirely new class of compounds.

At the same time it was well known that an interesting series of reactions could be carried out by addition of certain neutral compounds to solutions of metal ions.  For instance, addition of aqueous ammonia to solutions of copper sulfate produced a deep blue color in the solution.  Addition of the same reagent to a solution of nickel (II) chloride, produced a deep purple color in the solution. Analysis of the solids formed after evaporation of the solutions gave the results $CuSO_4 \cdot 4NH_3$ and $NiCl_2 \cdot 6NH_3$ respectively.  It was also found that this phenomenon was not limited to the salts of only a few metals, but was rather typical behavior for virtually every metal ion whose charge was greater than +1.

Along with the efforts to synthesize such compounds, there was equally intense work on the determination of their properties.  This work led to some startling discoveries.  For instance, when a related series of platinum (IV)-ammonia–chlorides was studied, the compound $PtCl_4 \cdot 6NH_3$ was found to precipitate 4 moles of AgCl per mole of platinum compound when a solution of $AgNO_3$ was added to a solution of the platinum compound (equation 9.1).

$$PtCl_4 \cdot 6NH_{3\,(aq)} + 4\ Ag^+ \rightarrow 4\ AgCl_{(s)} + Pt^{4+} + 6\ NH_{3\,(aq)}$$

EQ. 9.1

This indicated to the investigators that all four of the chloride ions were "free". However, as the other compounds in the series were analyzed, it was found that as the number of moles of ammonia per mole of platinum was reduced, the number of moles of "free" chloride ions per mole of the platinum compound decreased until, with the compound $PtCl_4 \cdot 2NH_3$, there were no free chloride ions.  They went on to determine that the remaining two ammonia molecules could be replaced by potassium chloride without adding to the number of free chloride ions. Electrical conductivity measurements added to the evidence that there were different numbers of free chloride ions. These data are summarized in Tables 9.1 and 9.2

**TABLE 9.1** MOLES AgCl PRECIPITATED PER MOLE OF PLATINUM

TABLE 9.1

| PLATINUM COMPOUND | MOLES OF SILVER CHLORIDE PRECIPITATED PER MOLE OF PLATINUM |
|---|---|
| $PtCl_4 \cdot 6NH_3$ | 4 |
| $PtCl_4 \cdot 5NH_3$ | 3 |
| $PtCl_4 \cdot 4NH_3$ | 2 |
| $PtCl_4 \cdot 3NH_3$ | 1 |
| $PtCl_4 \cdot 2NH_3$ | 0 |
| $PtCl_4 \cdot NH_3 \cdot KCl$ | 0 |
| $PtCl_4 \cdot 2KCl$ | 0 |

**TABLE 9.2** MOLES OF IONS PRODUCED PER MOLE OF PLATINUM

TABLE 9.2

| PLATINUM COMPOUND | ELECTRICAL CONDUCTIVITY (S) | NUMBER OF IONS PRODUCED | CURRENT FORMULA OF COMPOUND |
|---|---|---|---|
| $PtCl_4 \cdot 6NH_3$ | 523 | 5 | $[Pt(NH_3)_4]^{4+}$; 4 $Cl^-$ |
| $PtCl_4 \cdot 5NH_3$ | 404 | 4 | $[Pt(NH_3)_3Cl]^{3+}$; 3 $Cl^-$ |
| $PtCl_4 \cdot 4NH_3$ | 229 | 3 | $[Pt(NH_3)_4Cl_2]^{2+}$; 2 $Cl^-$ |
| $PtCl_4 \cdot 3NH_3$ | 97 | 2 | $[Pt(NH_3)_4Cl_3]^+$; $Cl^-$ |
| $PtCl_4 \cdot 2NH_3$ | 0 | 0 | $[Pt(NH_3)_4Cl_4]$ |
| $PtCl_4 \cdot NH_3 \cdot KCl$ | 109 | 2 | $K^+$; $[Pt(NH_3)_4Cl_5]^-$ |
| $PtCl_4 \cdot 2KCl$ | 256 | 3 | 2 $K^+$; $[Pt(NH_3)_4Cl_6]^{2-}$ |

Continued study by many chemists, including the Danish chemist S. M. Jørgensen (1837–1914) and the Swiss chemist Alfred Werner (1866–1919), finally led Werner to propose his coordination theory in 1893. Werner won the Nobel prize in chemistry in 1913 for this theory. Of critical importance to Werner's theory was that some of these compounds are not simple stoichiometric mixtures of their individual components, but there are strong bonds formed between the components resulting in the formation of entirely new species. Werner called these compounds **coordination compounds** and postulated that the metal ion was situated in the center of a sphere on the surface of which ions or molecules were situated. This surface surrounding the central metal ion is the boundary of what is known as the **coordination sphere** or **coordination shell**. Groups on and within this sphere were not free to move about the solution when the compound dissolved but, rather, remained tightly bonded to the metal. For example, the structures of two of the platinum compounds discussed above are presented in figure 9.1. In this compound the ammonia molecules and the bound chloride ions are strongly bonded to the metal ion in the center of the species. The chloride ions not in the coordination sphere are free to move throughout the solution and can, therefore, be precipitated by addition of silver nitrate. Chloride ions in the coordination sphere can not be precipitated by silver nitrate because they are tightly bound to the platinum atom.

Werner postulated that each different metal ion could accommodate only a certain number of species in its coordination shell. That particular number of species was always present in the coordination shell. If necessary, water molecules (or another solvent molecule if the solvent is not water), occupy **coordination sites** that are not occupied by other groups. If additional non–solvent molecules are then added that can coordinate with the metal ion, these new species only substitute for one of the solvent molecules in the coordination sphere.

While a few of these ideas have changed since Werner originally proposed his theory, the modern idea of coordination compounds is remarkably close to that initially set forth by Werner. We currently recognize that there are groups, known as **ligand**s, that can occupy sites around the central metal ion. These ligands are bonded to the metal ion by covalent bonds in which both electrons in the bond were associated with the ligand before the bond was formed. This is, of course, a Lewis acid–base interaction in which the electron–deficient metal ion and the electron–rich ligand interact with the formation of a bond. A generalized equation for such a reaction is shown as equation 9.2. A specific example of equation 9.2 is the formation of a **complex** between silver ion and ammonia molecules that is often used as part of the separation of silver in the common qualitative analysis procedure for the determination of silver. This reaction is given in equation 9.3.

$$M \ + \ :L \ \rightarrow ML$$

EQ. 9.2

$$Ag^+ \ + \ 2\,NH_3 \ \rightarrow Ag(NH_3)_2{}^+$$

EQ. 9.3

We no longer believe that a particular metal ion has only one **coordination number** that it exhibits in all of its compounds. Rather the number exhibited by a particular metal ion can vary somewhat depending upon what ligands are present. With some ligands, capable of strong Lewis base behavior, the metal ion may exhibit a larger coordination number than with other ligands. Nevertheless most metal ions exhibit the same coordination number in most of their compounds.

## 9.2  THE 18-ELECTRON RULE

We have already seen how the Octet Rule can help in predicting the stoichiometry and structure of many compounds formed by the main group elements — especially those of periods two and three. Although exceptions to this rule are well–known, the general applicability of the rule to many covalently bonded compounds, and especially its universal applicability in carbon chemistry, make it one of the most useful rules learned by chemistry students everywhere. The word "octet" refers, of course, to the number of electrons in the outermost electron shell in that family of elements known as the Inert Gases. It is the filled s and p atomic orbitals that gives the inert gases their exceptional stability and reluctance to form compounds under any but the most extreme conditions.

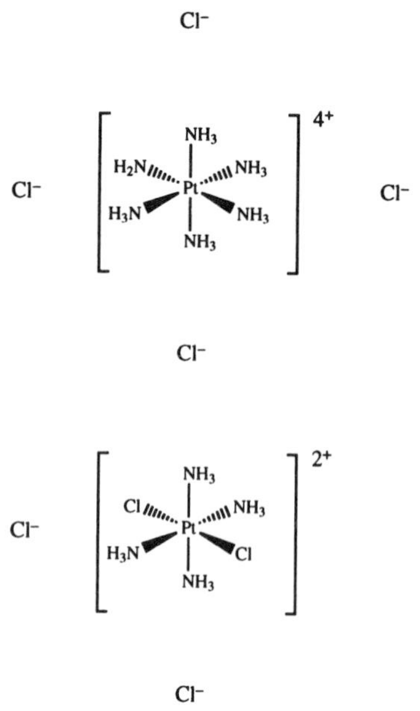

FIGURE 9.1:
Structures of
$Pt(NH_3)_6Cl_4$ and
$Pt(NH_3)_4Cl_4$.

When other main group elements react to form covalent compounds, in most cases they do so by sharing electons such that the end result is that each of the atoms is surrounded by 8 electrons.

One common exception to the octet rule occurs in compounds formed by the Boron–Aluminum group. In these compounds, the central atom does not have enough electrons to form enough covelant bonds to attain the octet. These compounds are, therefore, electron–deficient compounds and, therefore, good Lewis acids. When they do attain the octet, they do so by formation of coordinate covalent bonds, typically by strongly coordinating with a ligand in a Lewis acid–base reaction.

The other common exception occurs in covalent compounds formed by elements in the third or higher period. In these elements the central atom is large enough that atoms with more than 4 electrons in their outer shell often form more than 4 bonds. This results in more than 8 electrons surrounding the central atom. Size is not the issue, however. What is key is that these central atoms have low–energy unoccupied d orbitals available to involve in formation of hybrid orbitals. By involving the d orbitals in hybrid formation, enough hybrid orbitals can be formed, and, therefore, enough bonds can be formed, that the atom has more than 8 electrons surrounding it. Compounds of such elements are referred to as "expanded octet" compounds.

A similar rule exists that helps us to predict the stabilities and structures of compounds with metal–ligand bonds. That rule is known as the 18–Electron Rule. Like the Octet Rule, the 18–Electron Rule works because formation of the compound places around the central atom 18 electrons — exactly the number that is needed to achieve the next inert gas configuration for these elements. Using this rule it is possible to predict the number of ligands that will surround a metal ion in the most stable configuration.

Let us apply this rule to the ferrocyanide ion, $[Fe(CN)_6]^{4-}$, an ion formed by complexing a Fe(II) metal ion with six cyanide ions as ligands.

The elecron configuration of an iron atom is: $[Ar]\ 4s^2 3d^6$.

When the two electrons are removed to form the iron (II) ion, the electron configuration becomes $[Ar]\ 4s^0 3d^6$. Thus 6 electrons are contributed by the iron(II) toward the 18 electrons required by the rule. When the cyanide ions form coordinate covalent bonds to the iron ion, each one contributes both of the electrons in the bond. Six cyanide ions will, therefore, contribute 12 electrons. Thus the total of 18 electrons has been reached.

Unfortunately there are a large number of complexes which do not follow this rule, the previously cited $Ag(NH_3)_2^+$, for instance. Even the very stable ferricyanide ion, $[Fe(CN)_6]^{3-}$, does not follow the rule. One class of compounds which follows the rule much more faithfully is the class of compounds known as organometallic compounds — those compounds containing a metal–carbon bond — and especially those having carbon monoxide as a ligand. Many of these latter compounds are low-positive–valent, zero–valent, or even negative–valent compounds.

Let us consider two such compounds: $Fe(CO)_5$, and a compound with the empirical formula $Mn(CO)_5$. Both compounds, since they are compounds in which a neutral molecule, carbon monoxide, is the ligand, are compounds in which the metal is in the zero oxidation state. Let us now do the electron–counting:

TABLE 9.3

**TABLE 9.3** EIGHTEEN–ELECTRON RULE TABLE

$Fe(CO)_5$                          $Mn(CO)_5$

| SPECIES | # ELECTRONS CONTRIBUTED | SPECIES | # ELECTRONS CONTRIBUTED |
|---------|-------------------------|---------|-------------------------|
| Fe | 8 | Mn | 7 |
| CO | 5 x 2 = 10 | CO | 5 x 2 = 10 |
| Total electrons | 18 | | 17 |

This would tell us to expect that the compound $Fe(CO)_5$ would be a stable compound but that $Mn(CO)_5$ would not be stable. The experimental facts fit that prediction. However, there is a compound whose empirical formula is $Mn(CO)_5$. Such a compound would be possible if it were a dimer and had a manganese–manganese bond. Each manganese would contribute one electron to this bond. In the way of electron–counting that we are familiar with from Lewis Dot Structures, this means that both electrons (one from each manganese atom) are counted in the shells of both manganese atoms. A dimer with the molecular formula $Mn_2(CO)_{10}$ has been found and has been shown to contain a manganese–manganese bond. The eighteen electron rule also predicts that a monomer would exist if it were a mono–charged anion, $Mn(CO)_5^-$. Such a species has been found. Unless very unconventional oxidation numbers are given to the atoms in the carbon monoxide molecules, this requires that the manganese atom be given a –1 oxidation number — certainly not an oxidation number that is thought to be typical for a metal!

**CONCEPT CHECK**

What would you expect to be the formulas for neutral compounds formed between the metals Cr and V and the ligand CO? If either of these is most stable as a dimer, what would be the formulas for a similar complex with a single positive and a single negative charge?

Let us look at chromium first. The electronic structure of Cr is : [Ar] $4s^13d^5$ Therefore it has 6 electrons past the previous inert gas. This means it needs an additional 12 electrons to attain the 18 for the next inert gas. Since one carbon monoxide provides 2 electrons, this will require 6 carbon monoxide molecules. The formula would then be $Cr(CO)_6$. Such a molecule does exist. It is a volatile solid which decomposes at its melting point of 110°C.

The electronic structure of vanadium is: $[Ar]4s^23d^3$. With an odd number of electrons any neutral vanadium carbonyl must exist as a dimer with a V—V bond. Sharing an electron with another V atom will mean that each atom has effectively 6 electrons. An additional 12 electrons, or 6 carbon monoxide molecules will be required for each V atom. This means the formula would be $V_2(CO)_{12}$. Such a compound has not yet been synthesized. It would require that each V atom be surrounded by 6 CO ligands and have additional room to form a V—V bond. There is simply not enough room for this. Since a complex of V with 6 CO ligands would have 17 electrons and not have steric limitations, we expect that $V(CO)_6^-$ to be a possible stable species. Experiments have determined that a neutral $V(CO)_6$ does exist! Obviously this compound, with only 17 electrons surrounding the V atom, breaks the 18 electron rule. We would expect it to be much less stable than either its anion or the corresponding Cr complex. That is, indeed, true. It reacts about $10^{10}$ times more rapidly in substitution reactions than does chromium hexacarbonyl whereas the anionic species, $V(CO)_6^-$, is substitutionally quite stable.

## 9.3 GEOMETRY

### COORDINATION NUMBER TWO: LINEAR COMPLEXES

Although not common, this coordination number is exhibited in a few complexes involving Cu(I), Ag(I), Au(I), and Hg(II). Of these, the most common is probably the diammine complex of silver. Other two–coordinate complexes are the dichloro complexes of copper (I), gold (I), and Hg(II) (although all of these, especially Hg(II), will form 4–coordinate chloro complexes at much higher chloride concentrations). A case might be made for the existence of two–coordinate oxo complexes in some oxo–ions such as $UO_2^{2+}$, $UO_2^+$, $PuO_2^{2+}$, and a few others. However, these ions also have a number of other ligands coordinated at a somewhat longer distance and should probably not be considered true 2–coordinate metals. It is true, however, that the attraction of the metal for the oxide ion ligands in these complex ions is especially strong.

As would be expected from VSEPR as well as hybrid orbital theory for species that have two electron pairs around the central atom, the geometries of these species are linear. There are no geometric or optical isomers possible for these complexes unless the ligands themselves are optically active.

$$[\,H_3N : Ag : NH_3\,]^+$$

FIGURE 9.2:
Linear complex.

### COORDINATION NUMBER FOUR: TETRAHEDRAL COMPLEXES

There are a great number of four–coordinate complexes involving a wide range of metal ions. Those involving the main group elements as well as the pseudo–main group elements Zn(II), Cd(II), and Hg(II) are tetrahedral as expected from VSEPR considerations. Specific examples of these include $Zn(CN)_4^{2-}$ and $HgCl_4^{2-}$. In addition to these, the true transition metals sometimes show tetrahedral geometries in their complexes. For these elements, 4–coordinate complexes tend to be somewhat less stable than the 6–coordinate complexes that result from association with two additional ligands. Important examples of tetrahedral complexes in this group include tetrachloro complexes of Fe(III), Co(II), and the tetrabromo, iodo and thiocyanato complexes of Co(II). A few other tetrahedral complex ions exist in the crystalline state but seem to be quickly converted into complex ions with higher coordination numbers by association with water molecules when they dissolve. As we would expect from our earlier study of tetrahedral compounds, the only geometric or optical isomers of tetrahedral complexes are those involving 4 different ligands (unless the ligands themselves are optically active).

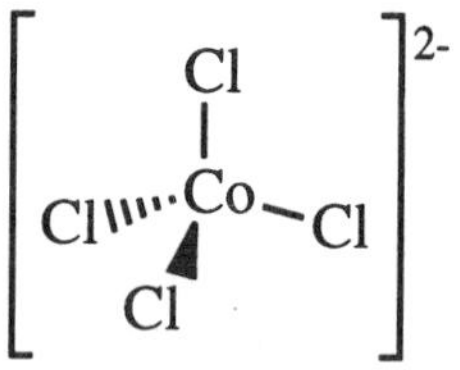

FIGURE 9.3:
Tetrahedral complex.

## COORDINATION NUMBER FOUR: SQUARE PLANAR COMPLEXES

There are a large number of square planar complexes of the transition metals. These complexes are particularly important in the chemistry of Pd(II), Pt(II), Au(III), Rh(I), and Ir(I), all of which, after ionization, have 8 d electrons in their valence shell. They are also important in the complex ion chemistry of Ni(II) (also $d^8$) and Cu(II) (a $d^9$ ion). The geometry is, of course, totally unexpected on the basis of VSEPR and can not be explained using that theory. An explanation for this geometry will be given in the next chapter when bonding in complexes will be discussed. As we shall see later in this chapter, a variety of geometric and optical isomers are possible for these complexes.

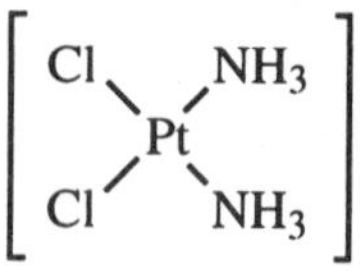

FIGURE 9.4:
Square planar complex.

## COORDINATION NUMBER FIVE: TRIGONAL BIPYRAMIDAL COMPLEXES

Trigonal bipyramidal geometry is common in a number of compounds formed by group V elements, *e.g.*, $PCl_5$. Figure 9.5 presents the trigonal bipyramidal structure. As you will recall from last year, the two positions in the structure are distinctly different positions and are termed equatorial and axial.

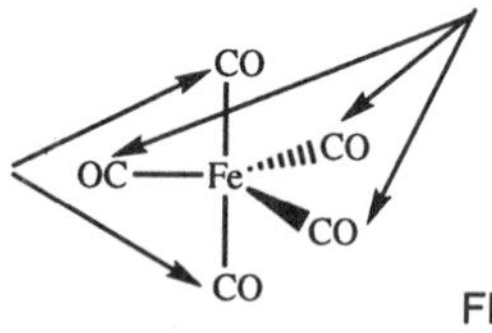

FIGURE 9.5:
Structure of $Fe(CO)_5$.

Since the positions in a trigonal bipyramid are not identical, the bond lengths and strengths are unequal, even when identical atoms occupy the two positions. Nevertheless, NMR spectra of $PF_5$ and $Fe(CO)_5$ have failed to show a distinction between the positions — presumably because of rapid interconversion between this trigonal bipyramidal structure and a square pyrimidal structure that, calculations show, is only energetically slightly less favorable. This rapid equilibrium is shown in figure 9.6. In this figure the As and B's are being used to distinguish between identical atoms initially at different positions in the structure.

Trigonal bipyramidal geometry, while not common for metal complexes, has been shown to exist for a number of species. The iron penta-carbonyl molecule previously discussed exhibits this structure as do a number of other species, particularly in the solid state. It is believed, however, that a number of these five–coordinate solid species associate with a solvent molecule to form a six–coordinate structure in solution.

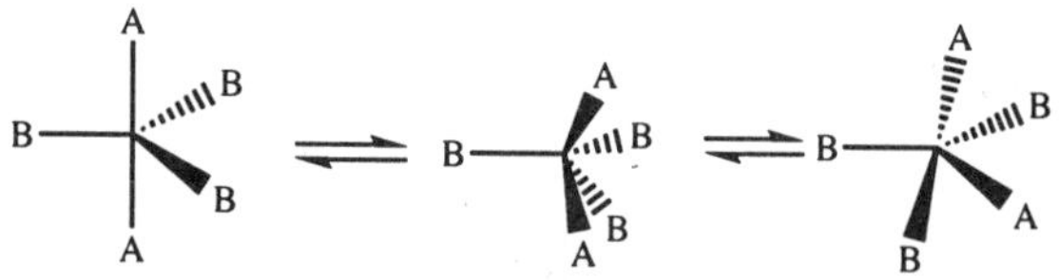

FIGURE 9.6:
Rapid interconversion of axial and equatorial positions in a trigonal bipyramidal structure.

## COORDINATION NUMBER SIX : OCTAHEDRAL COMPLEXES

The vast majority of complexes fall into this category. These include not only complexes of silicon and its relatives in the main group elements typified by $SiF_6^{2-}$ but also complexes of most of the transition metal ions. Several typical examples of transition metal complexes include $Cr(NH_3)_6^{3+}$, $Fe(CN)_6^{3-}$, and

$PtCl_6^{2-}$. Complexes with this coordination number are octahedral in geometry, as predicted by VSEPR theory. Octahedral complexes can have a wide variety of geometric and optical isomers depending upon the stoichiometry of the complex.

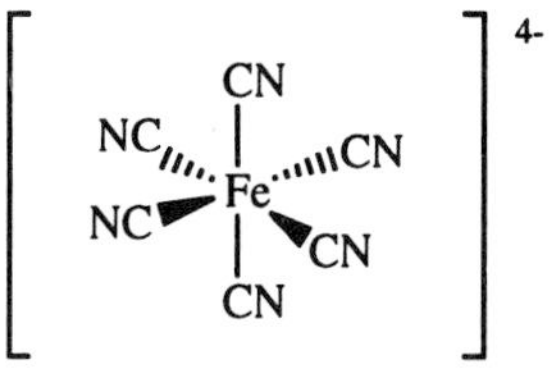

**OTHER GEOMETRIES**

Coordination numbers other than two, four, five, and six are relatively rare but do occur. There is some evidence for coordination numbers of three and five, mostly from stoichiometry. In most cases the evidence is sparse and these coordination numbers may, in fact, not exist.

On the other hand, there is good evidence for coordination numbers of seven, eight and nine, especially in **chelate** complexes involving the metals of the lanthanide and actinide series. This can be attributed to the large size of these ions and to the availability of f orbitals for bonding purposes. Specific examples include an 8–coordinate acetylacetonate (the anion of 2,4–pentanedione) complex of cerium: $Ce(acac)_4$ and the aquo complex of neodymium $(Nd(H_2O)_9^{3+})$.

FIGURE 9.7:
An octahedral complex.

## 9.4 POLYDENTATE LIGANDS AND CHELATES

To this point we have considered only the number of coordination sites available and their possible geometries. Our discussions have assumed that each site was occupied by a separate ligand. While a large number of complexes are, indeed, of this type, some ligands, *e.g.*, the acetylacetonate ligand mentioned in the previous paragraph, are capable of bonding to a metal at more than one site because they have two or more unshared electron pairs that they can donate as a Lewis base. These ligands are referred to as **polydentate** ligands from the Greek *poly* (many) and *dent* (tooth). One such ligand of real importance in chemistry is the molecule ethylenediamine (1,2–diaminoethane). Figure 9.8 shows that this molecule can be looked on as two ammonia molecules joined by a 2–carbon bridge of methylene ($CH_2$) groups. If this molecule coordinates to a metal ion at both unshared electron pair sites, the resulting complex ion would have the structure shown in figure 9.8. Coordination compounds with these ring structures formed by polydentate ligands are called *chelates*, again from the Greek *chele* meaning claw. Note that the length of the carbon chain joining the two amino groups is just long enough that a five–membered ring is formed when both amino groups occupy coordination sites on the metal. We have already seen that nature conveys extra stability to the formation of five and six membered rings because of the absence of ring strain in such compounds. Hence these compounds enjoy extra stability over their ammonia analogs. Extra stability is also conveyed because of the large, positive, entropy change that occurs when a single polydentate ligand replaces several monodentate ligands. Of course, if two coordination sites on a single ligand are good, more would be even better and polydentate ligands with six or more coordination sites are common.

$Cr(en)_3^{3+}$   $Cr(en)_3^{3+}$

$H_2NCH_2CH_2NH_2$

ethylenediammine (en)

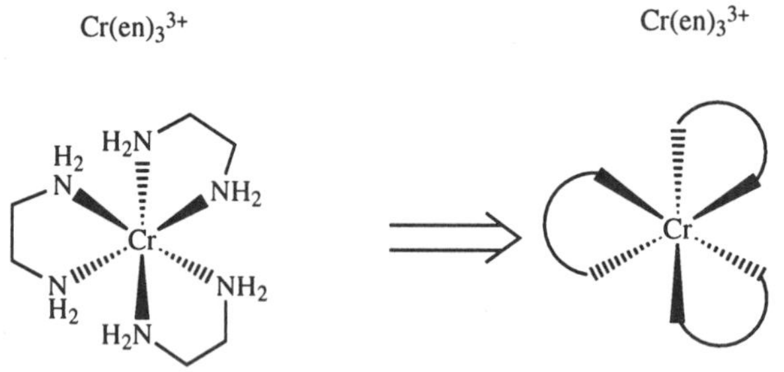

FIGURE 9.8:
Ethylenediamine and
two different ways of
representing the
Cr-En complex.

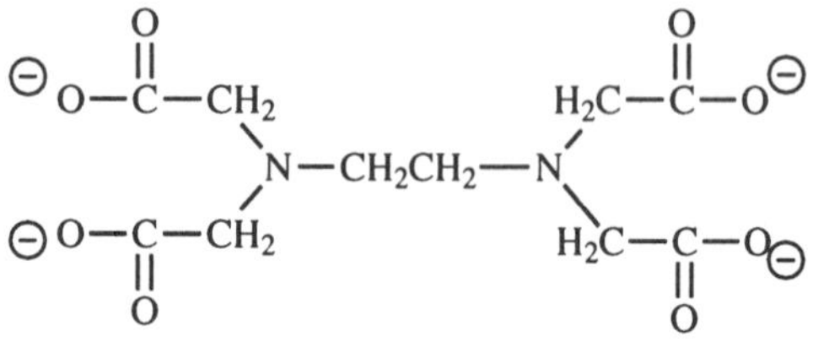

FIGURE 9.9:
Structure of EDTA ion.

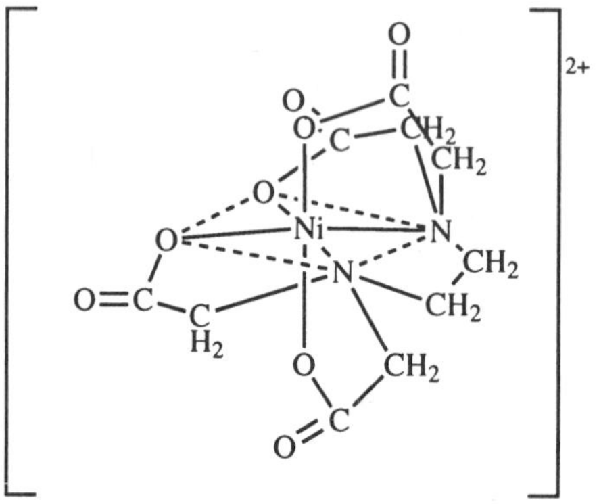

FIGURE 9.10:
EDTA—metal ion
complex structure.

Another important polydentate ligand is the hexadentate ligand shown in figure 9.9. The long common name of the neutral compound is ethylenediaminetetraacetic acid. Fortunately it is almost always referred to by its initials, EDTA. It is such an important ligand because its two amino groups and four carboxyl groups are all located exactly four atoms apart. It is able to occupy all of the coordination sites of a six–coordinate metal ion by forming five–membered rings. Metal ions in EDTA complexes are completely enveloped by the EDTA ligand as shown in figure 9.10. EDTA has a wide variety of uses in the fields of chemistry, medicine, and agriculture, in addition to uses in a wide variety of consumer products. In chemical analysis it is used in the quantitative determination of many metals by titration. Its medical uses involve administration of the calcium–EDTA complex as a treatment for lead poisoning and its use in certain types of blood–draw tubes as a clot–preventing reagent. Its use in lead poisoning treatment takes advantage of the formation equilibria of the calcium and lead complexes. Since the equilibrium lies far in the direction of the lead complex, the equilibrium point of the "chemical reaction" illustrated by equation 9.4:

$$Pb(body) + CaEDTA(solution) \rightleftharpoons Ca(body) + PbEDTA(solution)$$

EQ. 9.4

is far to the right. Lead that is solubilized is then excreted from the body in the form of the lead–EDTA complex. It agricultural uses include application of the Fe–EDTA complex to plants, especially trees, which are unable to pick up enough iron to satisfy their needs from other sources. As a consumer reagent, EDTA is often added to detergents as a **sequestering agent** to keep the ions that form "hard–water" in solution under the alkaline conditions necessary for proper cleaning. This prevents the build–up of the soap scum called "bathtub ring" by preventing the precipitation of the long–chain carboxylate salts of calcium and magnesium salts that form the scum (see figure 9.11). It also has uses as a food additive that acts to prevent spoilage.

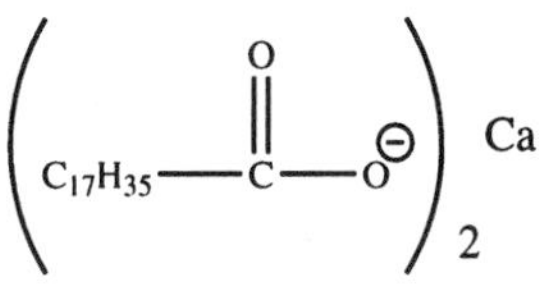

FIGURE 9.11:
Typical ingredient in soap scum.

## 9.5 ISOMERISM

As is the case with many other types of compounds, two metal complexes with the same chemical formula may have different properties because they have different structures. However, there are several types of isomerism possible in complexes that are not shown with other types of compounds. We shall consider only the three most common coordination numbers, two, four, and six, in our discussions. Of course, if a ligand is, itself, optically active, the complex will be optically active. For that reason we shall consider only isomerism caused by differences in the arrangements of the metal ion and ligands — not isomerism solely due to one (or more) of the ligands.

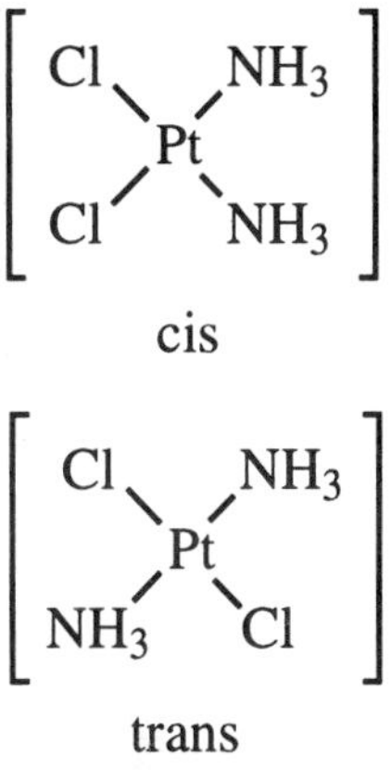

**GEOMETRIC ISOMERISM**

Geometric isomerism is only possible for complexes with four and six coordination sites. Since tetrahedral complexes show isomerism only if they are of the type MABCD and then the compounds are chiral, discussion of tetrahedral complexes will be deferred until later. Square planar complexes of the type $MA_2B_2$ or $MA_2BC$ may exist in two different geometric isomers known as *cis* and *trans* as depicted in figure 9.12. One very important example of this type of isomerism is the compound $Pt(NH_3)_2Cl_2$ which, in its *cis* form is the anti–cancer drug cisplatin.

When the number of coordination sites reaches six, the number of different geometric isomers increases dramatically. Of course there is only one isomer for compounds of the type $MA_5B$. There are two different compounds of the type $MA_4B_2$ (figure 9.13) once again these are *cis* and *trans* isomers. For compounds of the type $MA_3B_3$, there are also two different isomers — one in which the each of the two sets of ligands of the same type can be viewed as all lying along an imaginary line around the central metal ion (termed *mer* for meridional), and one in which there are two sets of identical ligands in which the individual ligands are at 90° to one another (termed *fac* for facial) (figure 9.14).

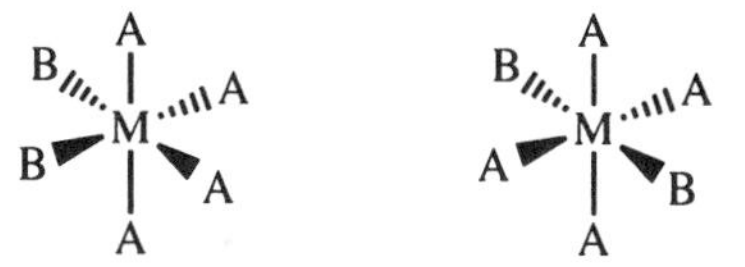

The number of possibilities increases as the number of different ligands increases. It is left as an exercise to the student to determine the number of possibilities for other possible combinations of ligands.

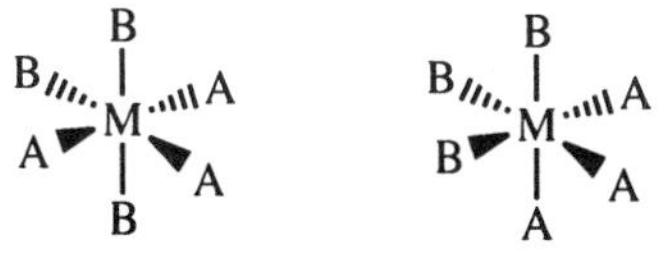

The presence of multi–dentate ligands complicates this situation somewhat. It is sufficient to understand that bidentate ligands almost always occupy adjacent coordination sites (sites that are 90° to one another) on the metal ion. Even with the smallest metal ions, the length of the chain between complexing sites on the ligand usually is not sufficient to span the distance between non–adjacent sites on the metal. Of course, when the number of coordination sites on

cis

trans

FIGURE 9.12:
Square planar *cis* and *trans* $Pt(NH_3)_2Cl_2$.

cis          trans

FIGURE 9.13:
*cis* and *trans* octahedral $MA_4B_2$.

mer          fac

FIGURE 9.14:
*mer* and *fac* octahedral $MA_3B_3$.

the ligand increases so do the possibilities. Even so, adjacent coordinating sites on the ligand are almost always bonded to adjacent sites on the metal. As one example of the complications that can arise with multi–dentate ligands, figure 9.15 shows two different $Cr(en)_3^{3+}$ structures in which the direction in which the ethylenediamine ligands are wrapped around the chromium atom differ. In one of the structures the ethylenediame ligands form a right–handed propeller (termed a $\Delta$ isomer); in the other a left–handed propeller(termed a $\Lambda$ isomer). Of course, additional complications are possible when the ligands are, themselves, assymetric.

## OPTICAL ISOMERISM

In the past we have used the non–superimposability of mirror images to be the criterion for chirality. Because the criteria for chirality are somewhat more complex than that, there are rare excepons to this rule. For the present, however, we shall continue to use this as the criterion for optical activity.

Except for complexes containing optically active or asymmetric ligands, optical activity is shown only in tetrahedral and octahedral complexes. One class of tetrahedral complexes that show optical activity are those, as expected, of type MABCD. This is completely analogous to the situation for the carbon compounds that we have already studied. Since it is difficult to prepare complexes with four different ligands, these tend to be of little practical importance. Optical activity is also possible in tetrahedral complexes of the form $ML_2$, in which the bidentate ligands are asymmetric.

The only square–planar complexes that are optically active are, likewise, those in which the ligands, themselves, are optically active or in which asymmetric multidentate ligands are incorporated. In both cases the asymmetry of the ligand means that the mirror images will not be superimposible. Optically active tetrahedral and square–planar complexes with asymmetric ligands are illustrated in figure 9.16.

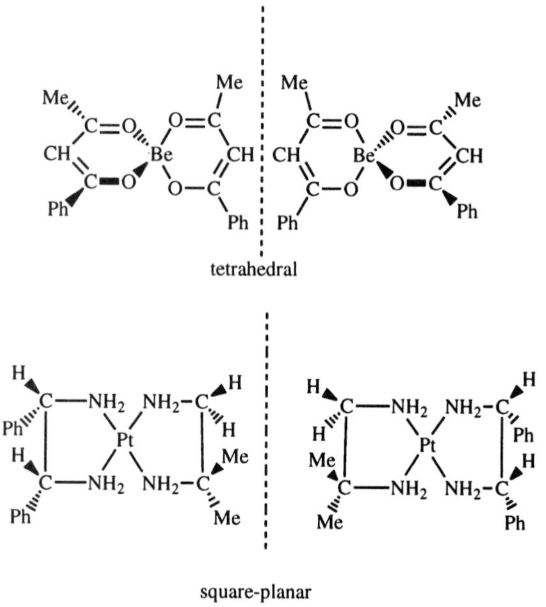

There is a wider variety of optically active octahedral complexes possible. Theoretically, complexes of the type MABCDEF should be optically active. However, since it is even more difficult to prepare complexes with six different ligands than it was to prepare tetrahedral complexes with four different ligands, these remain theoretically interesting but practically unimportant. The most important optically active octahedral complexes are those involving multidentate ligands. The *cis*, but not the *trans*, form of $M(en)_2A_2$ exists as optically active enantiomers as shown in figure 9.17 In addition, $M(en)_3$ exists in optically active forms as shown in figure 9.18. Optical activity should also be possible with complexes of ligands with even larger numbers of coordination sites. At least one such complex, ($CoEDTA^{4-}$), has been resolved into optically active enantiomers.

$Cr(en)_3^{3+}$

right–handed ( $\Delta$ )  left–handed ( $\Lambda$ )

**FIGURE 9.15:**
Two different helial structures of $Cr(en)_2^{3+}$.

tetrahedral

square-planar

**FIGURE 9.16:**
Tetrahedral $MA_2$ and Square-planar $MA_2$ with asymmetric ligands.

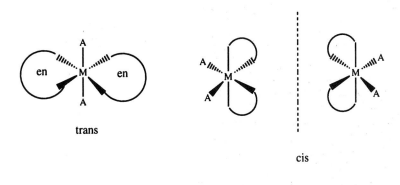

FIGURE 9.17:
*Trans* and two *cis*
forms of M(en)$_2$A$_2$.

FIGURE 9.18:
Two *cis* forms of
M(en)$_3$.

trans

cis

Lately there has been a great deal of interest in optically active metal chelate compounds as catalysts for stereospecific reactions. The Parkinson's disease drug *S*–(L)–DOPA and the artifical sweatener Aspartame® are made using an optically active rhodium catalyst to achieve an enantioselective hydrogenation. Optically active manganese chelates have been used in the stereoselective epoxidation of a number of alkenes. These are but a few examples of what is likely to become much a more important use of optical isomerism as an aid in the synthesis of stereochemically pure compounds.

## IONIZATION ISOMERISM

When the complex ion contains anions in the coordination sphere, ionization isomerism is possible. If the complex is still a cation, additional anions are necessary to render the entire compound electrically neutral. It is then possible for the counter ions and the coordinated ions to exchange places to form the different isomers. A specific example would be the compounds [Pt(NH$_3$)$_4$Cl$_2$]Br$_2$ and [Pt(NH$_3$)$_4$Br$_2$]Cl$_2$. In some cases, because exchange reactions are quite slow, the compounds can be isolated in relatively pure form. In other cases, exchange reactions are rapid enough that it is not possible to isolate the various isomers.

## LINKAGE ISOMERISM

This situation occurs with a select group of ligands that may coordinate through more than one site. The chief examples are the thiocyanate ion (SCN⁻) which may coordinate either through the sulfur atom (M:SCN) or through the nitrogen atom (M:NCS). Other examples include several nitrogen and oxygen–containing species such as the nitrite ion (NO$_2$⁻) that may coordinate either through the nitrogen or the oxygen atoms.

## HYDRATION ISOMERISM

This is a situation that can only occur in solid (crystals) and is similar to ionization isomerism. In this case a water molecule may either be within the coordination sphere or it may be outside the coordination sphere but present within the crystal as water of hydration. Specific examples include three forms of the compound whose empirical formula is $Cr(H_2O)_6Cl_3$. One of these has all of the waters within the coordination sphere of the chromium and is $[Cr(H_2O)_6]Cl_3$. The other two exchange one or two waters in the coordination sphere for chloride ions: $[Cr(H_2O)_5Cl]Cl_2 \cdot H_2O$ and $[Cr(H_2O)_4Cl_2]Cl \cdot 2H_2O$. The first is a violet–colored crystalline compound in which all three chloride ions are immediately precipitated by addition of $AgNO_3$. The other two are green and immediately yield only two–thirds and one–third of their chloride as AgCl upon treatment with $AgNO_3$. Because the ligand substitution reactions are equilibrium reactions, eventually all three chloride ions eventually will be precipitated after addition of $AgNO_3$. However, these substitution reactions are slow compared to precipitation reactions.

## COORDINATION ISOMERISM

This type of isomerism occurs when both the cation and the anion in a compound are complex ions. An illustrative example is: $[Cr(NH_3)_6][Co(CN)_6]$ and $[Co(NH_3)_6][Cr(CN)_6]$. Coordination isomerism can also occur when the metal ion is in two different oxidation states such as the situation in $[Pt(NH_3)_4Cl_2][PtCl_4]$ and $[Pt(NH_3)_4][PtCl_6]$. Finally, coordination isomerism is also possible in **binuclear complexes**. One possible example of such is shown in figure 9.19.

(a) $\left[ (NH_3)_4Co \underset{O\,H}{\overset{H\,O}{\diagdown\diagup}} Co(NH_3)_2Cl_2 \right]^{2+}$

(b) $\left[ Cl(NH_3)_3Co \underset{O\,H}{\overset{H\,O}{\diagdown\diagup}} Co(NH_3)_3Cl \right]^{2+}$

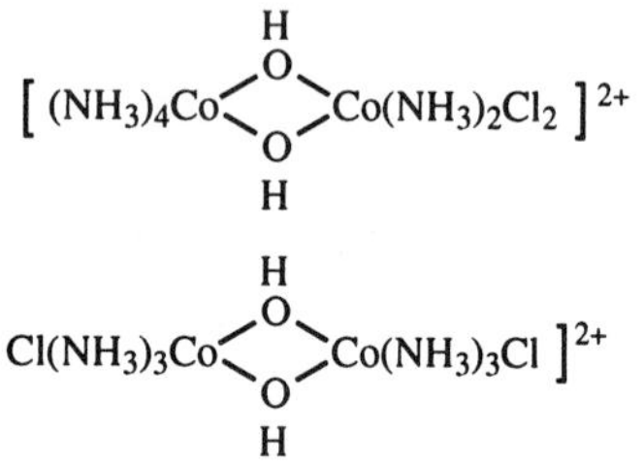

FIGURE 9.19: Coordination isomerism in a binuclear complex.

## 9.6   WRITING CHEMICAL FORMULAS

As in writing formulae for all charged species, the formula for the cation precedes the formula for the anion. The formula for a complex ion written by itself is always enclosed in square brackets with the overall charge of the ion written outside the right bracket in the normal fashion. Within the brackets, the symbol for the metal element is written first. If necessary to avoid ambiguity, the oxidation state is indicated in Roman numerals as a superscript. Ligands are then arranged in the order (1) anions, (2) neutral species, and (3) cations. Within the anions the ligands are arranged in alphbetical order in the order of the first symbol in the formula. Neutral species, and finally cationic species, if any, then follow — also in alphabetical order. In all cases the names of complicated organic ligands may be enclosed in parentheses in order to avoid ambiguity. The formula for a binuclear complex ion differs only in that metal ions are written in the center with the *bridging* ligands written between them. Ligands that are bonded to only one of the metal atoms are written to the outside. The entire binuclear complex is then enclosed within square brackets and the charge of the entire coordination compound is indicated in the usual manner. A number of examples are shown in figure 9.20:

FIGURE 9.20:
Structures and
formulas of
complexes.

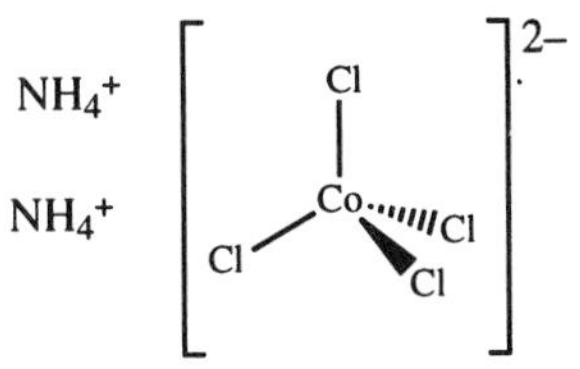

$= H_2 NCH_2CH_2NHCH_2CH_2NH_2$

$=$ dien

$(NH_4)_2[CoCl_4]$

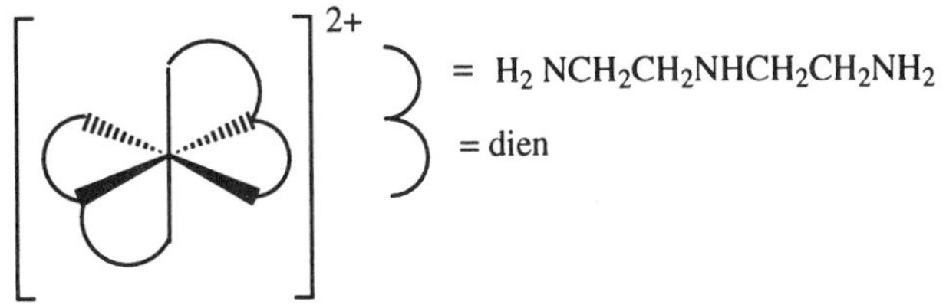

$[Co(dien)_2]^{2+}$

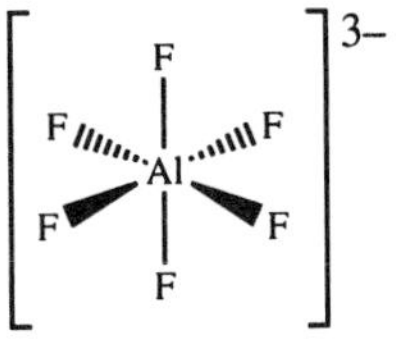

$[AlF_6]^{3-}$

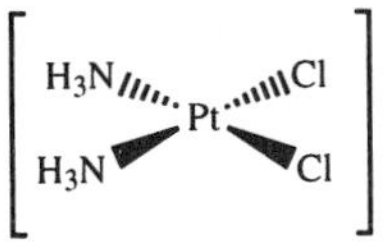

$[Pt^{II}Cl_2(NH_3)_2]$

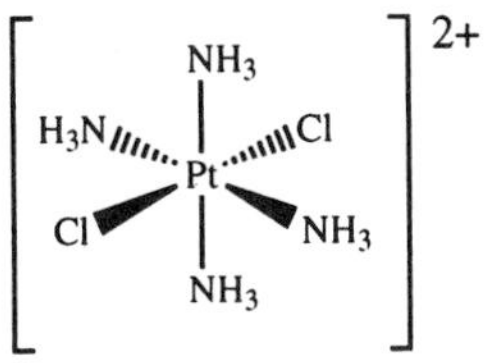

$[Pt^{IV}Cl_2(NH_3)_4]^{2+}$

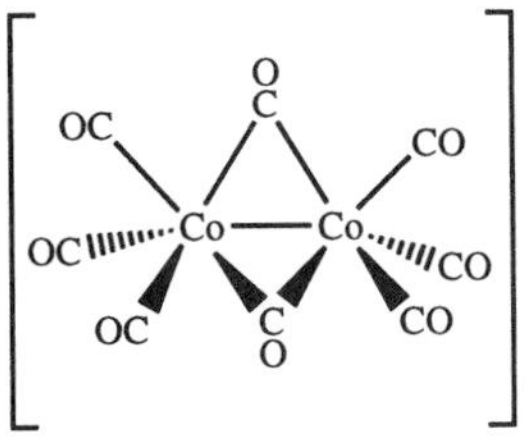

$[(CO)_3Co(CO)_2Co(CO)_3]$

## 9.7  NAMING COORDINATION COMPOUNDS

Complex ions are named according to a set of rules in a way similar to those used to name organic compounds.  These rules are:

1. The names of the ligands are written first; the nuclear atom is named last.

2. If the overall species is an anion, the suffix –ate is appended.  Cations and neutral species have no such suffix.

3. Ligands are cited in alphabetical order without regard to charge. Numerical prefixes necessary to designate the number of ligands are not considered in determining the order.  The names of anions are usually formed by using the stem of the anion followed by –o (*e.g.*, chloro, nitro, cyano, *etc.*).  Certain anions have different names depending upon the linkage (*e.g.*, when $NO_2^-$ is bonded through the nitrogen atom, the name is nitro; when bonded through one of the oxygen atoms it is nitrito; $SCN^-$ is thiocyanato when bonded through the sulfur and isothiocyanato when bonded through the nitrogen; $CN^-$ is cyano when bonded through the C and isocyano when bonded through the N).  The names of neutral and cationic ligands are the same as the species except for water (aquo), ammonia (ammine — note the spelling; it is not the same as the amine used to indicate an organic compound) nitric oxide (nitrosyl), carbon monoxide (carbonyl).

4. The prefixes mono, di, tri, *etc.*, are used to indicate the number of ligands of the particular type in the complex. If the ligand name itself begins with a similar prefix, or if the use of di, tri, etc. would create confusion (*e.g.*, bis(methylamine) would be used in place of dimethylamine), then bis, tris, tetrakis, *etc.*, are used and the ligand name is placed in parentheses.

5. If the coordination compound is a binuclear species, the name of the bridging group is separated from the rest of the ligands by the Greek letter mu ($\mu$).

6. Although formally discouraged by IUPAC conventions, prefixes such as *cis, trans, fac, mer, etc.* are used, where appropriate, to indicate specific geometric arrangements of ligands within structures. Similarly, indicators of the particular optical isomer [(+), (–)] are also used as prefixes.

Examples of formulae and the corresponding names are included in table 9.4.

TABLE 9.4

**TABLE 9.4:** NAMES AND FORMULAS OF SOME COMPLEX IONS

| FORMULA | NAME |
|---|---|
| $[AgCl_2]^-$ | dichlorosilverate(I) |
| $[Ag(NH_3)_2]^+$ | diamminesilver(I) |
| $[Fe(CN)_6]^{3-}$ | hexacyanoferrate(III) |
| $[PtCl_2(NH_3)_4]^{2+}$ | dichlorotetraammineplatinum(IV) |
| $[Co(H_2NCH_2CH_2NH_2)_3]^{3+}$ | tris(1,2diaminoethane)cobalt(II) |
| $[(CO)_4Os{-}CO{-}Os(CO)_4]$ | tetracarbonylosmium(0)–$\mu$–carbonyl–tetracarbonylosmium(0) |

## 9.8    ACID–BASE BEHAVIOR OF AQUO COMPLEXES

When metal salts dissolve in water in the absence of sufficient concentrations of complexing ligands, the "default" complex formed is the aquo complex. Hence, dissolving anhydrous iron(III) chloride in water produces hexaaquo iron(III) ions and chloride ions as shown in equation 9.5.

$$FeCl_3(s) \rightarrow Fe(H_2O)_6^{3+} + 3\ Cl^-$$

EQ. 9.5

Unless the anion has signifiant basic properties, these solutions are acidic because the hexaaquo iron(III) ion is a reasonably strong Lowry–Brønsted acid and undergoes ionization:

$$Fe(H_2O)_6^{3+} \rightleftharpoons FeOH(H_2O)_5^{2+} + H^+$$

EQ. 9.6

Indeed, the equilibrium constant for this reaction, which could be looked upon as the acid dissociation constant for the hexaaquo iron(III) ion, has a value that makes this ion a stronger acid than acetic acid. Estimates of the acid dissociation constants for a number of typical metal ions are given in table 9.5. Unlike acetic acid, however, the hexaaquo iron(III) ion is a polyprotic acid and is capable of undergoing more than one dissociation step. Subsequent ionization steps produce the species: $Fe(OH)_2(H_2O)_4^+$ and $Fe(OH)_3(H_2O)_3$. This last species is uncharged and, as a consequence, insoluble. In order to keep solutions containing iron(III) ions from precipitating ferric hydroxide, it is necessary to keep the pH of the solution sufficiently low that the dissociation reactions of the complex ion do not proceed far enough for the solution to contain significant concentrations of the neutral species. This, of course, can be generalized to many different metal ions.

**TABLE 9.5:** ACID DISSOCIATION CONSTANTS FOR TYPICAL METAL IONS

TABLE 9.5

| ION | $pK_a$ |
|---|---|
| $Al(H_2O)_6^{3+}$ | 5.1 |
| $Fe(H_2O)_6^{3+}$ | 2.2 |
| $Fe(H_2O)_6^{2+}$ | 9.5 |
| $Ni(H_2O)_6^{2+}$ | 10.6 |
| $Cu(H_2O)_6^{2+}$ | 6.8 |

The neutral hydroxo–aquo complexes of some metals are still sufficiently strong acids that still further ionization steps can occur. These form negatively charged species. If the pH of the solution is high enough to cause this degree of ionization, these metal hydroxides will re–dissolve with the formation of what are commonly called hydroxo complexes. Common examples of this include the formation of aluminate ($Al(OH)_4^-$), zincate ($Zn(OH)_4^{2-}$), and stannate ($Sn(OH)_6^{2-}$) ions in strongly basic solution of those metals.

Although exceptions can be found if one compares metals far apart on the periodic table, the higher the positive charge on the central metal atom and the smaller the size of the atom, the stronger the acid. Although electronegativity differences play some part if metals in different groups are considered, smaller cations are usually stronger acids than larger cations of the same charge because the higher charge density makes it easier to lose a proton. Hence the acid strength of the aquo ions increases from the bottom to the top of the periodic table within a group. In a similar fashion, in metals with variable oxidation states, the acid strength of a metal aquo complex increases as the oxidation number of the central metal ion increases. (In these cases the ion with the higher oxidation state will be smaller because the same nuclear charge is pulling in a smaller number of electrons.) Extreme examples of this latter effect include the permanganate and chromate ions that can be imagined as having lost all of the protons from a hypothetical aquo complex containing a Mn(VII) or Cr(VI) ion as its nucleus. These ions are such strong acids that there is no evidence for the parent aquo complex at all.

## 9.9  SUBSTITUTION REACTIONS IN COORDINATION COMPOUNDS

Many of the reactions of coordination compounds involve the substitution of one ligand for another. Whether or not such a reaction takes place and the extent to which such a reaction goes toward completion is determined by the thermodynamics of the reaction. There are a number of generalizations that can be made about the extent to which a substitution reaction will go: (1) Multidentate ligands always form a stronger complex with a metal than similar monodentate ligands. (2) Small metal ions with tightly held (non–polarizable) electron clouds form the strongest complexes with those ligands that are also relatively small and non–polarizable, while larger, more polarizable metal ions form the strongest complexes with those ligands that are also larger and more polarizable.

The class of small, non–polarizable metal ions includes the alkali and alkaline earth metal ions and those lighter members of the first transition series in their higher oxidation states (*e.g.*, $Ti^{4+}$, $Cr^{3+}$, $Fe^{3+}$, $Co^{3+}$). These have been termed **"hard acids"**. The corresponding group of **"hard bases"** includes those coordinating to the metal ion through an atom near the top of the periodic table (N, O, F, *etc.*). The group of **"soft acids"** includes those metals that are larger, lower–valent and more polarizable (*e.g.*, $Ag^+$, $Cu^+$, $Hg^{2+}$, $Pt^{2+}$, *etc.*). The **"soft bases"** are those which coordinate through atoms farther down on the periodic table (*e.g.*, $I^-$, $Br^-$, *etc.*). Hard acids prefer hard bases and *vice versa*.

Using the hard–soft acid–base classification gives us the ability to make some predictions about the relative strengths of a series of complexes in which either the metal or the ligand is kept constant while the other varies. For instance, if we look at the formation constants (the equilibrium constant for the reaction in equation 9.7) for a series of aluminum complexes with the halide ions, we find that there is a steady increase in the value of the equilibrium constant from iodide to fluoride. Table 9.6 lists the values of the formation constants for a series of metal – halide ion complexes. Note that the equilibrium constants sharply decrease from fluoride to iodide for aluminum, sharply increase for mercury (II), and vary much less for copper (II) or lead (II). This suggests that aluminum is quite a hard acid, mercury (II) is quite a soft, and copper (II) and lead (II) fall intermediate between the two extremes. Since the trend for copper (II) is slightly downward from fluoride to iodide, we could say that copper (II) has slightly more hard acid character. The opposite trend for lead (II) suggests that it has slightly more soft acid character. Neither copper (II) nor lead (II), however, are strongly hard or soft acids.

$$M^{n+} + L \rightleftharpoons ML^{n+}$$

EQ. 9.7

**TABLE 9.6:** FORMATION CONSTANTS FOR METAL–HALIDE COMPLEXES

TABLE 9.6

| METAL | $F^-$ | $Cl^-$ | $Br^-$ | $I^-$ |
|---|---|---|---|---|
| $Al^{3+}$ | $1.4 \times 10^6$ | $\sim 3 \times 10^1$ | $\sim 3 \times 10^0$ | n/a |
| $Cu^{2+}$ | $5 \times 10^0$ | $1 \times 10^0$ | $\sim 3 \times 10^{-1}$ | n/a |
| $Pb^{2+}$ | $\sim 1 \times 10^0$ | $1.2 \times 10^1$ | $4.4 \times 10^1$ | $1.5 \times 10^0$ |
| $Hg^{2+}$ | $1 \times 10^1$ | $5.5 \times 10^6$ | $8.7 \times 10^8$ | $57.4 \times 10^{12}$ |

The same trends that hold for the formation of the 1:1 complex of M and L, also hold for the formation of higher complexes between M and L. This would lead us to predict that if, for instance, a solution contained the complex $HgCl_4^{2-}$, addition of iodide ion to the solution would cause a substitution of iodide for chloride in the complex. The driving force for this reaction is the formation of stronger bonds between the soft acid mercury (II) and the soft base iodide than were present between the mercury (II) ion and the harder base chloride ion. The overall reaction is shown as equation 9.8.

$$HgCl_4^{2-} + 4\,I^- \rightleftharpoons HgI_4^{2-} + 4\,Cl^-$$

EQ. 9.8

## 9.10  LIGAND SUBSTITUTION MECHANISMS

There are two principal mechnisms for the substitution of one ligand by another in a complex. One mechanism, is termed the Dissociative Mechanism. This mechanism for the first order substitution reaction for replacement of a ligand, L, by a ligand, X, in the complex $ML_n$, can be written:

$$ML_n \rightarrow ML_{(n-1)} + L \qquad \text{(slow)} \qquad\qquad\qquad \text{EQ. 9.9}$$

$$ML_{(n-1)} + X \rightarrow ML_{(n-1)}X \qquad \text{(fast)} \qquad\qquad\qquad \text{EQ. 9.10}$$

According to this mechanism, the rate of the reaction is governed by the rate at which the old complex "ejects" one of its ligands from the coordination sphere. This forms a complex in which the coordination number is unsatisfied. This species almost always rapidly reacts with any other ligand nearby. If the activation energy for the first reaction is quite high, then the reaction will follow first–order kinetics and the rate of the reaction will depend only on the concentration of $ML_n$. Note the similarity of this mechanism to the $S_N1$ mechanism seen in the reaction of, for instance, $t$–butyl chloride with hydroxide ion. Since these complex substitution reactions also involve the reaction of a nucleophile (the new ligand) with a substrate, they, too, are referred to as $S_N1$ reactions.

The rate of any step in a reaction is related to the activation energy for that step. Almost all reactions can proceed by several different pathways. If, as is usually the case, one pathway has a much lower activation energy than another, then the reaction will go by that pathway. An alternate mechanism, termed the Associative Mechanism, for the substitution reaction might be:

$$ML_n + X \rightarrow ML_nX \qquad \text{(slow)} \qquad\qquad\qquad \text{EQ. 9.11}$$

$$ML_nX \rightarrow ML_{(n-1)}X + L \qquad \text{(fast)} \qquad\qquad\qquad \text{EQ. 9.12}$$

This mechanism requires that the metal form an (n+1) coordinated species as an intermediate. If this is more likely (*i.e.*, has a lower activation energy) than the formation of an (n-1) coordinated species, then the substitution reaction will proceed by this mechanism. This will give rise to a rate law that is first order in $ML_n$ concentration and first order in $X$ concentration and, therefore, second order overall. Such a mechanism is termed an $S_N2$ mechanism.

It is often the case that neither mechanism predominates because the activation energies of the two rate–determining steps are such that both proceed at approximately the same rate. This leads to kinetic plots that are neither first order nor second order but somewhere in between. To distinguish these situations, often these reactions are called simply Interchange reactions.

Whatever mechanism is used by the complex for exchange of ligands, the reaction may result in either fast or slow exchange. Complexes for which substitution reactions are rapid are called **labile** complexes; those for which the reactions are slow are termed **inert**. Many octahedral complexes in which the central ion has either a $d^3$ or $d^6$ electronic structure are inert if the substitution reactions proceed via an associative mechanism. Indeed $Cr(NH_3)_6^{3+}$ (a complex with a $d^3$ ion) and the corresponding $Co(NH_3)_6^{3+}$ (a $d^6$ complex) can sit in a strongly acidic solution for several days without undergoing aquation even though the removal of the ammonia molecules (and subsequent protonation to form ammonium ions) is highly favored thermodynamically. The reasons for the inertness of these complexes will become more clear in the next chapter after we discuss the electronic structures of metal ions in complexes.

In addition to the electronic structure of the central metal ion, a number of other factors influence the rate of the reaction. These include such expected things as steric hindrance and the nucleophilicity of the substituting ligand. In addition, for isoelectronic metal ions, the rate of substitution usually decreases as the charge on the ion increases. As a result, substitution reactions for complexes of $Fe^{3+}$ (a $d^5$ ion) are slower than those for the analogous complex of $Mn^{2+}$ (also $d^5$). The geometry of the complex also has an effect. Tetrahedral complexes and square planar complexes are rarely inert because there is enough room to form the (n+1) coordinate intermediate necessary for the associative ($S_N2$) mechanism to proceed. Finally, the solvent often has an effect on the rate of the reaction with more strongly ionizing solvents favoring the dissociative ($S_N1$) mechanism. The solvent may also serve as a pseudo, easily–removed ligand occupying the positions above and below the plane of a square planar complex making it more easily undergo substitution by a relatively simple replacement of one of the solvent molecules in this farther out position followed by rearrangement of the five–coordinate square pyramidal intermediate back to square planar geometry with the loss of one of the original ligands.

**Binuclear complexes:** Coordination compounds in which there are two metal ion centers.

**Chelate:** A coordination compound in which the ligand or ligands are polydentate ligands. Such species have one or more rings containing the central metal.

**Complex:** An older, now discouraged, name for a coordination compound. Although formally discouraged, this term is still in common use. See coordination compound.

**Coordination compound:** Compounds or ions consisting of a central metal ion surrounded by a number of ligands bound to the central ion.

**Coordination number:** The number of coordination sites shown by a metal ion in a particular coordination compound.

**Coordination shell:** See coordination sphere.

**Coordination sites:** Sites around a metal ion at which a ligand may bond to form a coordination compound.

**Coordination sphere:** An imaginary spherical surface around a metal ion that encloses the ligands bound to the metal ion. Also called coordination shell.

**Double salts:** A chemical combination of two or more simple ionic compounds that forms a distinct chemical substance.

**Hard Acid/Base:** A Lewis acid/base incapable of easily polarizing its electron pairs. Such acids or bases are small species with valence electron pairs relatively close to the nucleus. See Soft acid/base.

**Inert:** A term that means that a complex ion undergoes slow substitution of one ligand for another.

**Labile:** A term that means that a complex ion undergoes relatively rapid substitution of one ligand for another.

**Ligand:** A Lewis base capable of forming an electron–pair bond with a metal ion. Both of the electrons in the bond formally come from the ligand.

**Polydentate ligand:** A ligand with more than one site at which it can complex with a metal ion.

**Sequestering agent:** A ligand added to a solution with the purpose of "tying up" a metal ion or ions to prevent their reaction. For example, sequestering agents are used in laundry detergents to prevent the calcium and magnesium ions in "hard water" from precipitating in alkaline solution.

**Soft Acid/Base:** A Lewis acid/base whose electrons are easily polarized Such acids or bases are large species with valence electron pairs relatively far from the nucleus. See Hard acid/base.

1. Draw the possible isomers for the square–planar complex $PtCl_2Br_2^{2-}$.

2. Draw the possible isomers for an octahedral complex $NiCl_2(en)_2$. Of these, how many would be optically active?

3. Draw the possible isomers of the octahedral complex $Ni(NH_2CH_2CH_2NHCH_2CH_2NH_2)_2^{2+}$. Are any of these optically active?

4. How many different isomers are possible for a complex $MA_3B_2C$, where A, B, and C are monodentate ligands. Draw the structures and indicate which are optically active.

5. Silver forms a very strong complex with cyanide ion with a formula of $Ag(CN)_2^-$. The silver salt of this complex is insoluble. Given the following equilibrium constant values, estimate the equilibrium concentration of free cyanide ion in a saturated solution of silver cyanide ($Ag[Ag(CN)_2]$).

$$Ag[Ag(CN)_2]_{(s)} \rightarrow Ag^+_{(aq)} + Ag(CN)_2^-_{(aq)} \qquad K_{sp} = 5.0 \times 10^{-12}$$

$$Ag^+_{(aq)} + 2CN^-_{(aq)} \rightarrow Aq(CN)_2^-_{(aq)} \qquad K_f = 7.1 \times 10^{19}$$

6. Draw all of the diferent isomers (of all types) for a coordination compound that contains 43.40% Pt, 12.46% N, 2.69% H, 23.66% Cl, and 17.78% Br. The compound is know to contain only 1 platinum atom. Indicate which of the isomers, if any, are optically active. This is an octahedral complex of Pt(IV).

7. Name the following:

    a. $[Co(NH_3)_4Br_2]Cl$
    b. $K_3[FeF_6]$
    c. $[(H_3N)_4Co$—$(OH)_2$—$Co(en)_2]Cl_4$   (en = ethylenediamine)
    d. $[Zn(NH_3)_4]^{2+}$
    e. $SiF_6^{2-}$
    f. $[CO(H_2O)_4Cl_2]$
    g. $[Cr(NH_3)_6][Co(Cl_6)]$
    h. $[Fe(CN)_6]^{4-}$

8. Draw one structure of the following:

    a. Tetramminecobalt(III) chloride
    b. Silver dicyanosilverate(I)
    c. Iron (III) hexacyanoferrate(II)
    d. Iron (II) hexachloroplatinumate(IV) hexahydrate
    e. *cis*-dichlorobisethylenediamineiron(II)
    f. Sodium tetrahydroxozincate(II)
    g. *trans*-Dichlorotetramminechromium(III)
    h. Tetrafluorosilicon(IV)-μ-bishydroxo-tetrafluorosiliconate(IV)

9. For each of the following, state which complex you would expect to be the more stable. State your reasons.

    a. $AlF_6^{3-}$ or $AlI_6^{3-}$
    b. $WF_6^{3-}$ or $WI_6^{3-}$
    c. $Al(CN)_6^{3-}$ or $W(CN)_6^{3-}$
    d. $Cu(NH_3)_6^{2+}$ or $Cu(en)_3^{2+}$
    e. $TiF_6^{3-}$ or $ZrF_6^{3-}$

10. Draw Lewis Dot Diagrams for the thiocyanate ion, the nitrite ion, and the cyanide ion. Then draw Lewis Dot diagrams for two different hexacoordinate complexes of each of these with ferrous ($Fe^{2+}$) ion that differ by linkage isomerism. Name the complexes.

11. Draw Lewis Dot diagrams for the first species in problem 9a through 9e. Which of these obey the 18–electron rule?

12. On the basis of the 18–electron rule, what would you predict as the most likely coordination number for:

    a. $Cu^+$
    b. $Rh^{3+}$
    c. $Ni^{2+}$

13. Demonstrate that the 18–electron rule holds for the following metal carbonyls:

    a. $Os(CO)_5$

    b. $Ni(CO)_4$

    c. $Cr(CO)_6$

14. What is the oxidation number for nickel in $[Cr(en)_3][Ni(CN)_5]$? You may consider that the chromium atom is in the +3 oxidation state. How many electrons surround each of the metal atoms in this compound?

15. Draw the *mer* and *fac* isomers of $[Co(NO_2)_3(NH_3)_3]$.

16. Consider the following data for the formation of two different copper complexes:

$$[Cu(H_2O)_6]^{2+} + 2\ NH_3 \longrightarrow [Cu(NH_3)_2(H_2O)_4]^{2+} + 2\ H_2O$$

    $\Delta \bar{G}° = -44.7$ kJ mol$^{-1}$
    $\Delta \bar{H}° = -46$ kJ mol$^{-1}$
    $\Delta S° = -4$ J mol$^{-1}$ K$^{-1}$

$$[Cu(H_2O)_6]^{2+} + en \longrightarrow [Cu(en)(H_2O)_4]^{2+} + 2\ H_2O$$

    $\Delta \bar{G}° = -60.1$ kJ mol$^{-1}$
    $\Delta \bar{H}° = -55$ kJ mol$^{-1}$
    $\Delta S° = 25$ J mol$^{-1}$ K$^{-1}$

a. Calculate the equilibrium constants for each of the reactions.

b. Explain the difference between the equilibrium constants for the two reactions in terms of the entropy effect.

c. Calculate the equilibrium constant for the reaction in which diamminetetraaquocupper (II) is converted to ethylenediaminetetraaquocopper (II).

17. Use the following data to decide whether the mechanism of substitution is associative or dissociative for the reaction:

$$[MA_4] + B \longrightarrow [MA_3B] + A$$

| $[MA_4]_o$ (M) | [B] (M) | Initial rate (mol L$^{-1}$ sec$^{-1}$) |
|---|---|---|
| 0.0010 | 0.0010 | 5.26 |
| 0.0010 | 0.0020 | 5.26 |
| 0.0020 | 0.0010 | 10.52 |

18. Write the formula and name the compound from the following observations:

a. One mole of a compound that contains 6 moles of ammonia and 3 moles of nitrate ions per mole of chromium, dissolves in water to produce 4 ions in solution.

b. One mole of a compound that contains 2 moles of ammonia molecules and 4 moles of chloride ions per mole of platinum dissolves to produce a non–conducting solution.

## CHAPTER NINE

OVERVIEW
# Problems

19. Two moles of dimethylglyoxime (DMG, $C_4H_8N_2O_2$, MW = 116.12) reacts with one mole of $Ni^{2+}$ to form one mole of a red insoluble compound bis(dimethylgloximate)nickel (II) and two moles of $H^+$. This reaction is used to quantify the amount of nickel. For example, a 2.000 g sample of nickel-containing steel is dissolved in HCl. This solution is treated with a sufficient amount of DMG in alcohol solution so that all the nickel precipitates. [The DMG in alcohol solution is 1.0% (by mass) and has a density of 0.80 g/ml.]

a. Write the balanced complexation reaction, showing the structure of the product.

b. If 0.2948 g of red precipitate are obtained, what is the % by mass of nickel in the steel?

c. What is the minimum volume of alcoholic DMG that must be added to assure complete precipitation? [Note: in practice, an excess of DMG would be used.]

20. Acetylacetone (2,4-pentanedione) is a weakly acidic compound that loses a proton from the number 3 carbon to form the acetylacetonate anion (acac⁻). The acetylacetonate ion can readily complex metal ions such as $Cr^{3+}$ to form a water insoluble complex. This complex can be isolated and analyzed for purity *via* freeing the chromium from the complex and converting it to chromate ion ($CrO_4^{2-}$) by reacting the free $Cr^{3+}$ ion with hydrogen peroxide under basic conditions. The chromate ion can then be quantified by visible spectroscopy.

    a. Write the reaction of acetylacetone forming acetylacetonate ion, and show the structure of the reactant and product.

    b. Show the stability of the acetylacetonate ion by drawing the possible resonance forms of the ion.

    c. Write the balanced reaction between $Cr^{3+}$ ion and acetylacetonate ion to form a neutral compound.

    d. Write a balanced redox reaction showing the conversion of $Cr^{3+}$ ion to chromate ion via hydrogen peroxide in basic conditions.

21. Aqueous $Co(NH_3)_2^{2+}$ ($\Delta G_f = -127.5$ kJ/mol) reacts with aqueous ammonia ($\Delta G_f = -126.50$ kJ/mol) to form one mole of aqueous $Co(NH_3)_4^{2+}$ ($\Delta G_f = -189.3$ kJ/mol)

    a. Write the balanced reaction.

    b. Calculate the standard state molar free energy change for the reaction.

    c. Calculate the equilibrium constant for the reaction at 25.0°C.

    d. If the reaction were to occur as a single step elementary reaction, write the predicted rate law.

22. $Ni^{2+}$(aq) ($\Delta G_f = -45.6$ kJ/mol, S = -128.9 J/K mol) reacts with cyanide ion ($\Delta G_f = 172.4$ kJ/mol, S = 94.1 J/K mol) to form one mole of $Ni(CN)_4^{2-}$(aq) ($\Delta G_f = 472.1$ kJ/mol, S = 218 J/K mol).

    a. Write the balanced reaction.

    b. Calculate the standard state molar free energy change for the reaction.

    c. What is the significance of the sign of the answer to (b)?

    d. Calculate the standard state entropy change for the reaction.

    e. What is the significance of the sign of the answer to (d)?

    f. Calculate the standard state enthalpy change of the reaction.

23. Consider the following three step mechanism for the base hydrolysis of a cobalt coordination compound where step 1 is a rapid equilibrium, step 2 is slow, and step 3 is fast:

    1. $[Co(NH_3)_5Cl]_2^+ + OH^- \rightleftharpoons [Co(NH_2)(NH_3)_4Cl]^+ + H_2O$

    2. $[Co(NH_2)(NH_3)_4Cl]^+ \rightarrow [Co(NH_2)(NH_3)_4]^{2+} + Cl^-$
    3. $[Co(NH_2)(NH_3)_4]^{2+} + H_2O \rightarrow [Co(NH_3)_5(OH)]^{2+}$

    a. Write the net balanced reaction for the mechanism.

    b. Write an acceptable rate law for the proposed mechanism.

    c. If the rate constant for this reaction is 0.20 lit/mol sec at 25.0°C, and the initial concentration of $[Co(NH_3)_5Cl]^{2+}$ and hydroxide were both 0.050 **M**, what would be the initial rate of the reaction? [Hint: this reaction is essentially $S_N2$.]

    d. Why would it be easier for $[Co(NH_2)(NH_3)_4Cl]^+$ to lose a chloride than for $[Co(NH_3)_5Cl]^{2+}$ ?

24. The chelation reaction between $[Ni(H_2O)_6]^{2+}$ and en to form one mole of $[Ni(en)_3]^{2+}$ has an equilibrium constant of $1.9 \times 10^{18}$ at 25.0°C, and a standard enthalpy change of −12.1 kJ/mol.

    a. Write the balanced overall reaction.

    b. Calculate the standard state free energy change for the reaction.

    c. Calculate the standard state entropy change for the reaction at 25.0°C.

    d. Is the magnitude of the equilibrium constant primarily a function of the enthalpy change or the entropy change? Explain in terms of the chemistry of the reaction.

# CHEMICAL
# Bonding
## MORE THAN JUST S AND P

# CHEMICAL

## BONDING: More Than Just s & p

## 10.1   OVERVIEW OF HYBRID ORBITALS

Up to this point, all of the hybrid orbitals that have been discussed have involved only s and p orbitals. Third and higher period elements also have d orbitals available. These orbitals are often used in bonding; indeed, they must be involved in the bonding scheme if the central atom is to accommodate more than four pairs of electrons. In addition, we have seen in the last chapter that certain geometries exist in complexes that can not be explained using VSEPR theory. The explanation of these geometries must involve some sort of hybrid orbitals.

To fully understand the shapes and orientations of the hybrid orbitals, it is necessary to recall the geometries of all of the orbitals involved. Of course, the simplest orbital, the s orbital (figure 10.1), is spherically symmetric and imparts no directionality to hybrid orbitals. Any directionality must arise from combinations with other orbitals. The p orbitals (figure 10.2) are directed along the **Cartesian coordinate axes** and will, therefore, require that hybrid orbitals involving only one p orbital (the sp hybrid) have lobes directed along whichever of the Cartesian coordinate axes the p orbital used. When two or more of the p orbitals are involved, the hybrid orbitals end up with orientations that are determined by vector sums of the orientations of the p orbitals involved in the formation of the hybrids.

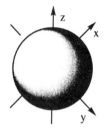

FIGURE 10.1:
The s orbital.

When d orbitals are involved, the number of orientations is increased dramatically. Recall that three of the d orbitals ($d_{xy}$, $d_{yz}$, and $d_{xz}$) have four lobes apiece lying in one of the Cartesian planes at 45° to the axes, one has its four lobes lying along two of the Cartesian coordinate axes (by convention these are the x and y axes and the orbital is designated $d_{x^2-y^2}$) and the fifth orbital has two lobes directed along the third Cartesian coordinate with a torus lying around its middle ($d_{z^2}$). Figure 10.3 presents a picture of all five of the d orbitals. Any hybrid orbital utilizing a d orbital will require that the directionality of that d orbital be mixed with the directionality of the other orbitals involved in the formation of the hybrid.

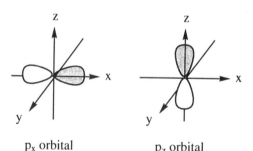

$p_x$ orbital          $p_z$ orbital          $p_y$ orbital

FIGURE 10.2:
The 3p orbital.

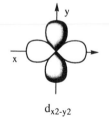      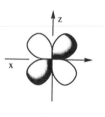   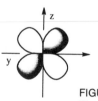   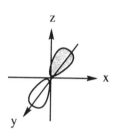

$d_{x2-y2}$          $d_{z2}$          $d_{xy}$          $d_{xz}$          $d_{yz}$

FIGURE 10.3:
The 5d orbitals.

## 10.2 HYBRID ORBITALS REVISITED: HYBRIDS INVOLVING s AND p ORBITALS —

In the first year course our primary motivation was a description of the geometries and bonding for first– and second–row elements with particular emphasis on carbon. Several approaches to the description of these forces were presented in the first year course. These included a discussion of the Valence Shell Electron Pair Repulsion theory (VSEPR), and then, later on, the formation of directed bonds using **hybrid orbitals**. Each of these methods has its usefulness; each has things that it cannot explain. It is now time for a further discussion of chemical bonding with a broader focus that includes the entire periodic table. Our focus in this chapter will be on the bonding that occurs in coordination compounds — bonding that often involves the d orbitals not present in the first and second period elements. However, before we do this, let us examine bonding involving only s and p orbitals in more detail.

The primary motivation for the development of hybrid orbital bonding descriptions was to explain the geometry of the molecules. This was done by picturing the chemical bond as the overlap of two atomic orbitals. The orbitals on the central atom must be oriented such that overlap of one of the hybrid orbitals with an orbital on each of the other atoms in the molecule would result in the observed geometry for the molecule. Consideration of the spatial arrangement of the s and p orbitals provides a set of **degenerate** hybrid orbitals bearing the spatial arrangements to one another set out in table 1 below. In each case the s orbital is hybridized with the designated p orbital or orbitals to form a set of the named hybrid orbitals. In the case of sp and $sp^2$ hybrids, the choice of the particular p orbital(s) to be used in forming the hybrids is arbitrary. If different p orbitals were used the orbitals would be the same in all aspects except that they would be oriented along different axes.

**TABLE 10.1** HYBRID ORBITALS INVOLVING s AND p ATOMIC ORBITALS          TABLE 10.1

| NUMBER OF DEGENERATE ORBITALS NEEDED | p ORBITAL(S) INVOLVED | HYBRID ORBITAL DESIGNATION | ANGLE BETWEEN ORBITALS | DESCRIPTION OF HYBRID ORBITALS |
|---|---|---|---|---|
| 2 | $p_x$ | sp | 180° | 2 orbitals located along the positive and negative x axis (linear electronic geometry) |
| 3 | $p_x$ & $p_y$ | $sp^2$ | 120° | 3 orbitals located in the xy plane (trigonal planar electronic geometry) |
| 4 | $p_x$, $p_y$, and $p_z$ | $sp^3$ | 109°28' | 4 orbitals directed toward the corners of a tetrahedron (tetrahedral electronic geometry) |

The procedure to determine which hybrid orbitals are used is: (1) determine the number of atoms to which the central atom is bonded and (2) add to this number the number of non–bonding electron pairs that must be accommodated around the central atom. The sum of these two numbers is, then, the number of equivalent orbitals that must be formed. This determines what hybrid orbitals must be used. For instance, a consideration of the bonding in the ammonia molecule leads us to expect that the nitrogen atom will bond using $sp^3$ hybrid orbitals because it must bond to three hydrogen atoms and accommodate one unshared electron pair. This means that it must have a total of 4 equivalent orbitals to use — one for bonding to each of the hydrogen atoms and one additional orbital to hold the unshared electron pair. Bonding in the water molecule would also be expected to use $sp^3$ orbitals on the oxygen atom because of the two unshared electron pairs to be accommodated. In both cases, then we would expect that the HNH or HOH bond angle to be the tetrahedral angle of 109°28'. In both cases the experimentally determined bond angle is approximately that — being 106°47' in ammonia and 104°27' in water. Certainly those numbers are close enough to the tetrahedral angle that we do not become concerned about the small difference.

If, however, we continue to look at the HXH bond angles of the Group V and VI hydrides, shown in table 10.2, some additional questions arise.

TABLE 10.2 BOND ANGLES OF THE GROUP V AND VI HYDRIDES

TABLE 10.2

| GROUP V COMPOUND | GROUP V ANGLE | GROUP VI COMPOUND | GROUP VI ANGLE |
|---|---|---|---|
| $NH_3$ | 106°47' | $H_2O$ | 104°27' |
| $PH_3$ | 93°30' | $H_2S$ | 92°16' |
| $AsH_3$ | 92°0' | $H_2Se$ | 91°0' |
| $SbH_3$ | 91°30' | $H_2Te$ | 89°30' |

VSEPR theory explains the small differences between the expected tetrahedral angles of the second period hydrides and the observed bond angles by suggesting that non–bonding electron pairs are more repulsive in nature and, therefore, require more room to minimize these repulsions. It would be expected, therefore, that this should become less a consideration as the central atom becomes larger. If this were true for the higher hydrides of the Group VI elements, the bond angles in molecules involving the larger central atoms should be closer to the ideal tetrahedral angle. A close look at the rest of the data does not, however, bear out this expectation unless non–bonding electron pairs become much more repulsive as their principal quantum number increases. The bond angles of the rest of the hydrides are such that it appears that a better description of the bonding would be to say that the central atom is using unhybridized atomic p orbitals for bond formation. Of course, if that were the case we would expect 90° bond angles.

This remains a problem as long as we insist that the hybrid orbitals be degenerate. This requirement means that each orbital must be composed of the same amount of s and p character. In the case of "pure" $sp^3$ hybrids, each orbital must have exactly 25% s and 75% p character. Hybrid orbitals formed in this manner are identical with exactly 109°28' angles between them. The real problem with imposing this restriction, is that it costs energy to hybridize the orbitals (figure 10.4). Unless this energy can be returned in the form of stronger bonds formed using the hybrid orbitals, nature will not use hybrid orbitals in the bonding scheme.

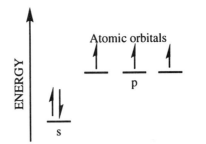

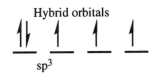

FIGURE 10.4: Energy relationships in sp³ hybrid orbital formation.

Let us now look at what happens to the **bond lengths** and **bond energies** in the group VA and VIA compounds as the central atom changes. These data are presented in table 10.3. They show, as expected, that the bond lengths get longer as the size of the central atom increases. The increase in bond length results in a decrease in bond energy. The net result is that there is less and less energy available to "pay back" the initial input of energy needed to hybridize the orbitals. Thus the benefit of hybridization is rapidly lost as the central atom becomes larger until, eventually, there is no benefit in hybridization.

**TABLE 10.3** BOND LENGTH AND STRENGTH FOR X–H BONDS

TABLE 10.3

| GRP V ATOM | X—H BOND LENGTH (pm) | X—H BOND ENERGY (kJ/mol) | GRP VI ATOM | X—H BOND LENGTH (pm) | X—H BOND ENERGY (kJ/mol) |
|---|---|---|---|---|---|
| N | 101 | 386 | O | 96 | 459 |
| P | 144 | 322 | S | 134 | 363 |
| As | 152 | 247 | Se | 146 | 276 |
| Sb | 169* | n/a | Te | 170 | 238 |

*estimated from sum of covalent radii

Of course, the real bonding description is neither pure one nor pure the other. Real molecules bond using a smooth continuum of hybrid and non–hybrid character in their orbitals in the way that produces the most stable final configuration. This can be accomplished by removing the restriction that all hybrid orbitals have the same amount of s and p character and permitting some orbitals to have more s character and some to have correspondingly more p character. When this happens the bond angles no longer have to be equal (and the orbital energies will no longer be degenerate). The closer the individual hybrid orbitals are to unhybridized orbitals, the less energy is required to hybridize them but the weaker the individual bonds will be. Using non–degenerate orbitals for bonding allows the molecule to seek the lowest overall energy for the electron pairs — both the bonding and non–bonding pairs — surrounding the central atom. For second row elements the stronger bonds formed from nearly equivalent orbitals win out over the extra hybridization energy cost. The opposite is the case for the elements in rows three and higher.

For s—p hybrids a simple relationship exists between the fractions of s orbital and p orbital contribution to an orbital and the angle between *equivalent* orbitals:

$$\cos \theta = \frac{s}{s-1} = \frac{1-p}{p}$$

EQ. 10.1

If we apply this to "pure" $sp^3$ hybrids (s= 0.25, p = 0.75), we find:

$$\cos \theta = \frac{.25}{.25-1} = -0.333$$

EQ. 10.2

From which we obtain $\Theta = 109°28'$. Applying this formula to the case of the water molecule for which $\Theta = 104.5°$, we find:

$$\cos 104.5 = -0.250 = \frac{s}{s-1}; \quad s = 0.20$$

EQ. 10.3

This suggests that the hybrids used to form the O—H bonds in the water molecule are composed of 20% s and 80% p character. In the $H_2Se$ molecule, the corresponding hybrids would be:

$$\cos 91.0 = -0.0175 = \frac{s}{s-1}; \quad s = 0.0172$$

EQ. 10.4

This says that these hybrids have 1.7% s and 98.3% p character — in essence the molecule bonds using unhybridized p orbitals. In this case it is clearly not worth the cost in energy to hybridize orbitals for bonding.

σ Bond Formation

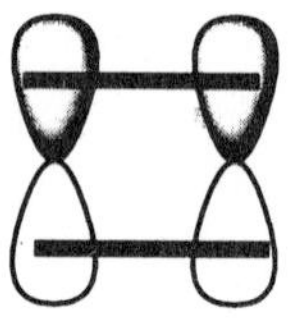

Before leaving this subject to consider formation of hybrid orbitals involving atomic d orbitals, we would be remiss if we did not mention that thus far we have not said anything about the formation of multiple bonds between atoms — a situation that occurs especially frequently in carbon compounds. Recall from last year's study that the determination of what hybrid orbitals are used, what atomic orbitals are involved in forming the hybrids, and the basic electronic and molecular geometry of the compound is no different than that already discussed at the beginning of this section. End–to–end overlap, as pictured in figure 10.5, of the singly–occupied hybrid orbitals then forms the first bond (the σ bond) between the two atoms and determines the geometry of the molecule. The only difference is that formation of hybrid orbitals will leave the unhybridized p atomic orbitals singly occupied. Sidewise overlap of these orbitals (also pictured in figure 10.5) then forms the second bond (the π bond) between the two atoms.

π Bond Formation

FIGURE 10.5:
σ and π bond
formation.

## 10.3  HYBRIDS AND SQUARE–PLANAR SPECIES

In the last chapter certain complex ions were discussed that had **square planar geometries**. These species obviously violate VSEPR rules, since **tetrahedral** electronic geometry is predicted for species surrounded by four electron pairs. This geometry can be explained by hybridization involving d orbitals. If we consider that the ligands surrounding the central atom are located along the positive and negative x and y axes, then we can see that there are three atomic orbitals with symmetry that could be used to make hybrids pointing in those directions — the $p_x$, $p_y$, and $d_{x^2-y^2}$ orbitals. Mixing those orbitals with the non–directional s orbital makes a degenerate set of four hybrid orbitals, termed $dsp^2$ (or $sp^2d$) orbitals, whose bonding lobes are located along the x and y axes as shown in figure 10.6 This is exactly the number of orbitals needed to form the four bonds to the ligands in the square planar complex.

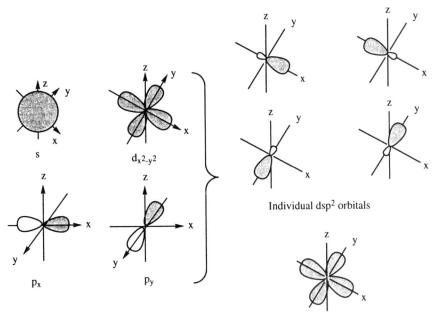

Individual dsp² orbitals

The set of all 4 dsp² orbitals

Note that each of the individual orbitals has a large lobe and a small lobe of opposite sign. Also note that, although the entire set of four orbitals (shown without the smaller lobes for clarity) superficially looks like the $d_{x^2-y^2}$ orbital, the drawing at the bottom right represents a set of four orbitals. This means that this set of orbitals will accommodate four pairs of electrons and not just the one pair that can be placed into the $d_{x^2-y^2}$ orbital.

Because often there are two different sets of d orbitals whose energy is close to the energy of the s and p orbitals that will be involved in hybrid formation (figure 10.7), there are two different possibilities for formation of these orbitals. One involves the use of the d orbitals in the shell under those of the s and p orbitals (*e.g.*, the 3d orbitals mixed with the 4s and 4 p orbitals). The other involves the use of d orbitals in the same shell as the s and p orbitals (*e.g.*, using the 4s, 4p, and 4d orbitals to form these hybrids). Hybrid orbitals formed in these two situations are designated dsp² and sp²d respectively. The dsp² hybrids are termed "inner orbital" hybrids because the d orbital involved is "inside" the s and p orbitals while the sp²d hybrids are called "outer orbital" hybrids.

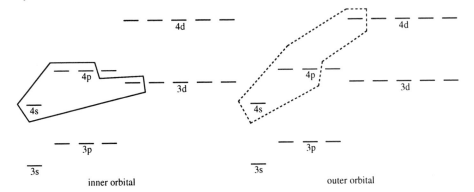

The central atom must have one completely empty d orbital to donate to hybrid formation because each resulting hybrid orbital must accept a pair of electrons from the ligand. Therefore, in order to form inner orbital hybrids, the central atom must have one completely empty d orbital in the inner shell. This means that the central ion must have four or fewer d electrons or something must have caused the pairing of enough electrons to free the required d orbital. The origin of this "pairing force" will be discussed later in this chapter. Most of the square planar complexes are those of $d^8$ ions (*ions* whose electron configuration ends in $d^8$) in which all electrons have been paired in the remaining d orbitals. Square planar complexes are most common with the ions $Ni^{2+}$, $Pd^{2+}$, and $Pt^{2+}$.

## 10.4 HYBRIDS AND TETRAHEDRAL SPECIES

The 3d orbitals are actually lower energy orbitals than the 4p orbitals in Mn and Cr. This makes the 3d orbitals more likely to be used in formation of tetrahedrally oriented hybrids in species such as $MnO_4^-$ or $CrO_4^{2-}$ than the 4p orbitals. In addition, since the energy of the np and nd electrons are very close in silicon and the heavier members of Group IVA, tetrahedral hybrids used in the bonding of such compounds as $GeH_4$ and $GeCl_4$ quite likely also involve some d orbital participation. In this case, the set of d orbitals involved would be the $d_{xy}$, $d_{xz}$, and $d_{yz}$ orbitals whose orientations would result in the hybrid orbitals having tetrahedral orientation. Hybrid orbital formation of $sd^3$ hybrids is shown in figure 10.8.

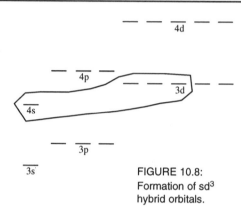

FIGURE 10.8: Formation of $sd^3$ hybrid orbitals.

## 10.5 HYBRIDS AND TRIGONAL–BIPYRAMIDAL SPECIES

Nature has no regular, three–dimensional object with five vertices. As a result, five–coordinate species cannot have all of the bonds equivalent as can two–, three–, four–, and six–coordinate species. Of the two possibilities, the more common arrangement of five groups around a central atom is the **trigonal bipyramid** (figure 10.9). This arrangement consists of three hybrid orbitals situated in a plane at 120° to one another with two additional orbitals above and below the plane and perpendicular to it. The three orbitals arranged in the plane are all equivalent in energy and are termed the **equatorial orbitals**. Groups bonded using these orbitals are referred to as equatorial groups. The other two orbitals are termed **axial orbitals** and groups bonded using these orbitals are termed the axial groups. Such a structure arises when an s, three p and the $d_z^2$ orbitals are used in the hybridization scheme. The hybrids are termed $dsp^3$ or $sp^3d$ orbitals depending upon whether an inner or outer d orbital is used in hybridization. Figure 10.10 shows the orbital formulation for the inner orbital $dsp^3$ hybrids.

Compounds with five electron pairs surrounding the central atom, principally those of the main group VA elements but also including a number of other species with central atoms in groups VIA, VIIA, or VIIIA, are normally bonded using these hybrids. For compounds

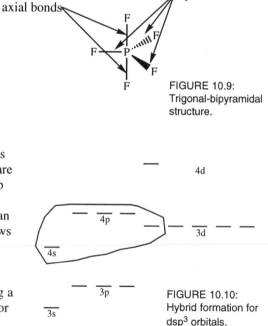

FIGURE 10.9: Trigonal-bipyramidal structure.

FIGURE 10.10: Hybrid formation for $dsp^3$ orbitals.

such as $PCl_5$ there is no ambiguity about what structure will result — the structure will be that of a trigonal bipyramid. Species such as $SF_4$, $XeF_2$, and $ICl_2^-$, however, could have different molecular geometries depending upon which positions, equatorial or axial, are occupied by the non–bonding electron pairs. The student is encouraged to draw the Lewis Dot diagrams for each of these compounds to verify that each is surrounded by 10 electrons.

VSEPR theory allows us to distinguish between these by giving us the rule that electron pairs are always arranged so as to minimize the repulsions. Repulsions between two non–bonding pairs are the strongest repulsions. Non–bonding — bonding pair repulsions are second strongest; bonding pair — bonding pair repulsions are the weakest. In order to determine which molecular structure is expected for $SF_4$, we need only determine whether the bonding — non-bonding repulsions will be stronger with the non–bonding pair in the equatorial or the axial position. If the non–bonding electron pair were to go in an axial position, the molecular geometry would be trigonal pyramidal; if it were to go in an equatorial position, the molecule would have a shape similar to a child's teeter–totter.

Placing the non–bonding pair in an axial position means that there are three 90° interactions with bonding pairs. Placing it in an equatorial position provides two 90° interactions and two 120° interactions. The latter situation is more energetically favorable and, thus, molecules or ions with an $MEA_4$ structure are always teeter–totter shaped (figure 10.11). The equatorial positions are still favored by the non–bonding electrons when more than one non–bonding pair is present in the molecule. Thus structures for $ME_2A_3$ molecules are T–shaped and those for $ME_3A_2$ are linear (figure 10.11).

## 10.6  HYBRIDS AND SQUARE PYRAMIDAL SPECIES

If the species hybridizes one s, three p, and one d orbital and uses the $d_{x^2-y^2}$ orbital rather than the $d_{z^2}$ orbital, the additional directional orientation in the xy plane and the loss of some directionality along the z axis results in a **square pyramidal** structure as shown in figure 10.12. This structure consists of four atoms (marked A in the figure) forming a square plane with a fifth atom (B) somewhat above that plane and a sixth atom (C) directly above the fifth. Here, as was also the case in the trigonal bipyramidal case, the five hybrid orbitals are not degenerate. This means that species in which all five ligands are identical will have one ligand (the one above the plane formed by the rest of the ligands) with a bond energy different from the other four.

A few five-coordinate species appear to favor the square pyramidal structure over the much more common trigonal bipyramidal structure. In those cases in which it does occur, the square pyramidal structure is always a slightly distorted structure with the metal atom situated slightly above the plane of the "base"

MEA$_4$ - Teeter-totter

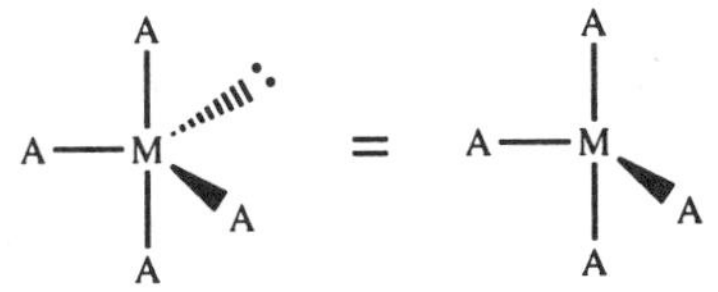

ME$_2$A$_3$ - T–shaped

ME$_3$A$_2$ - linear

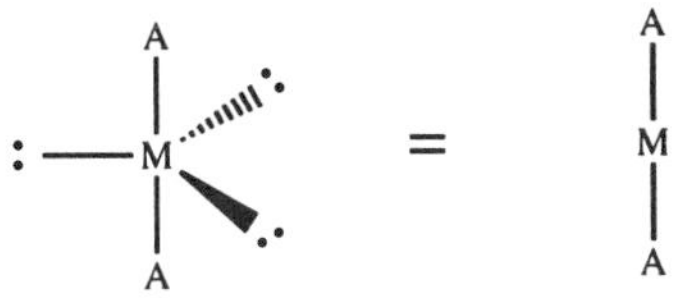

FIGURE 10.11:
Structures for MEA$_4$, ME$_2$A$_3$, and ME$_3$A$_2$.

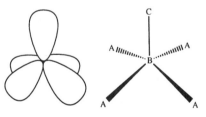

FIGURE 10.12:
Square pyramidal hybrid orbitals and structure.

four ligands. Such is the case with the pentacyanonickelate (II) ion [Ni(CN)$_5$]$^{3-}$ shown in figure 10.13. Note the difference in the bond length of the axial cyano group as compared to the planar cyano groups. In other cases it is unclear whether square pyramidal geometry is truly a different geometry or whether the complexes are really a distorted octahedral complexes with one very weakly coordinated ligand, perhaps a solvent molecule, below the plane.

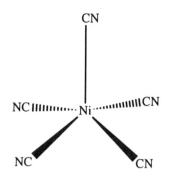

FIGURE 10.13:
Structures of the pentacyanonickelate (II) ion.

## 10.7  HYBRIDS AND OCTAHEDRAL SPECIES

Unlike the number five, nature does have a regular, six–coordinate figure. It is obtained by placing ligands at the vertices of a regular octahedron as shown in figure 10.14  This can be easily visualized by placing the metal atom at the origin and then considering six identical ligands placed the same distance from the metal atom in both directions along all three Cartesian axes. Planes connecting adjacent sets of three ligands will define a three–dimensional solid with eight sides — a figure known as an octahedron. In the structure shown as figure 10.14, the metal atom would be in the center with one ligand at each of the vertices.

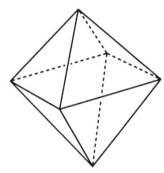

FIGURE 10.14:
Octahedron.

**Octahedral geometry** is found in a large number of transition metal complex ions as well as a smaller number of compounds of the main group elements. The necessary orbitals for bonding are produced by hybridizing one s orbital, three p orbitals, and the d$_{x^2-y^2}$ and d$_{z^2}$ orbitals. Hybridizing these orbitals produces a set of six degenerate orbitals known as d$^2$sp$^3$ (or sp$^3$d$^2$, if the outer d orbitals are used) orbitals. The directionality conveyed by the orbital set used in hybridization results in equivalent orbitals oriented along the positive and negative directions of all three Cartesian axes. Pictures of the octahedral orbitals and the octahedral structure are shown in figure 10.15.

FIGURE 10.15:
Octahedral hybrid orbitals and structure.

Since all of the hybrid orbitals are equivalent, it does not matter which site is occupied by a lone pair of electrons in MEA$_5$ structures. This results in a structure that is a square pyramid with a planar base. Since nature wishes to minimize lone pair–lone pair interactions, the second lone pair in ME$_2$A$_4$ structures is placed across the octahedron from the first, resulting in a square planar structure. The student is left to work out the other possible structures.

## 10.8  HYBRID ORBITALS OF OTHER TYPES

There are a few additional types of hybridization that occur occasionally and which should be mentioned here. We shall not take much time discussing any of these because they occur only rarely. One of these involves another hybrid for a two–coordinate species. In addition to the normally used sp hybrid, if there is an empty d orbital available, the central atom may hybridize that orbital with an s orbital to form an sd hybrid. Using the d$_{z^2}$ orbital produces two hybrid orbitals with bonding lobes oriented along the z axis. Since the d orbitals are usually quite close in energy to the p orbitals, it is likely that hybrids used to bond two–coordinate linear species have both some p and some d character to them if the d orbital is available for hybridization.

The majority of other hybrids will use f orbitals to some extent. These are commonly used for bonding ligands in lanthanide or actinide complexes. The size of these central ions and the availability of f orbitals for the bonding permit coordination numbers of more than six. Although bonding in such complexes no doubt involves hybrids with significant f orbital contribution, to date there has been relatively little work done to establish the exact identity of the hybrids. What theoretical work has been done has shown that a wide variety of hybrid geometries are possible.

## 10.9 COVALENT BONDING USING HYBRID ORBITALS

As we saw in Volume 1 of this text, the formation of a covalent bond in **valence bond theory** involves the overlap of two atomic orbitals to produce an area of space in which both electrons reside. We also saw that the atomic orbitals involved in the overlap to form the bond may either be "pure" atomic s, p, d, *etc.*, orbitals or may be hybrid atomic orbitals. The two orbitals involved in the bond can be directed with their principal axes in line with one another (*e.g.*, the overlapping of two $p_x$ orbitals between two atoms located on the x axis). Bonds formed by this type of overlap are referred to as sigma ($\sigma$) bonds. Alternatively, the orbitals may overlap with their principal axes parallel but displaced a short distance in space from one another (*e.g.*, overlap of the $p_y$ orbital on one atom with the $p_y$ orbital on another atom if the x axis is the internuclear axis). Such overlap results in a bond termed a pi ($\pi$) bond.

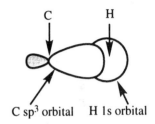

C sp³ orbital      H 1s orbital

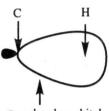

$\sigma$ molecular orbital

FIGURE 10.16:
Bonding in methane.

In the "normal" bonding case, indeed in every case that was discussed until we began talking about coordination compounds in the last chapter, each atom formally contributed one electron to the bond. This meant that each of the atomic orbitals that overlapped to form the bond was singly occupied. A simple example of this is shown in figure 10.16. This example depicts the formation of a $\sigma$ bond in methane by the overlap of an sp³ hybrid on the central carbon atom with a 1s atomic orbital on the four hydrogen atom. Four of these $\sigma$ bonds hold the methane molecule together. We could also depict this bonding situation using a box diagram to indicate the electrons before and after bond formation (figure 10.17).

The bonding situation in coordination compounds is different from these "normal" situations only in that both electrons forming the bond formally are contributed by one of the species forming the bond. The box diagram for the complex ion $[Cr(NH_3)_6]^{3+}$ then would be as seen in figure 10.18:

C sp³ Hybrid Orbitals

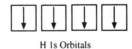

H 1s Orbitals

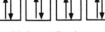

Methane $\sigma$ Bonds

FIGURE 10.17:
Box diagram for
bonding in methane.

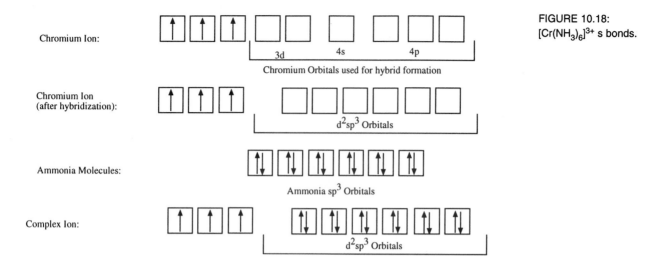

Chromium Ion:

3d 4s 4p

Chromium Orbitals used for hybrid formation

Chromium Ion (after hybridization):

$d^2sp^3$ Orbitals

Ammonia Molecules:

Ammonia $sp^3$ Orbitals

Complex Ion:

$d^2sp^3$ Orbitals

FIGURE 10.18: [Cr(NH$_3$)$_6$]$^{3+}$ s bonds.

Since the chromium ion had only 3 electrons in the d orbitals there was no doubt which d orbitals (the 3d or the 4d orbitals) would be used for hybrid formation. The lower energy 3d orbitals are the ones used. This results in $d^2sp^3$ hybrid orbitals and the complex is, therefore, an **inner orbital complex**. When the central ion contains four or more d electrons (and the complex ion has octahedral geometry), there are two possibilities. One of these is illustrated by the hexacyanocobaltate(III) ion [Co(CN)$_6$]$^{3-}$. This complex is an inner orbital complex and uses $d^2sp^3$ hybrids to bond. Since the Co(III) ion is a $d^6$ ion, the formation of the inner orbital hybrids requires that electron pairing occur. Such a complex, in which the total number of unpaired electrons is lower than in the uncomplexed metal ion because of spin pairing, is termed a **low spin complex**. Once spin pairing occurs, there are two completely free d orbitals for the formation of the hybrids. The box diagram for this species would appear as:

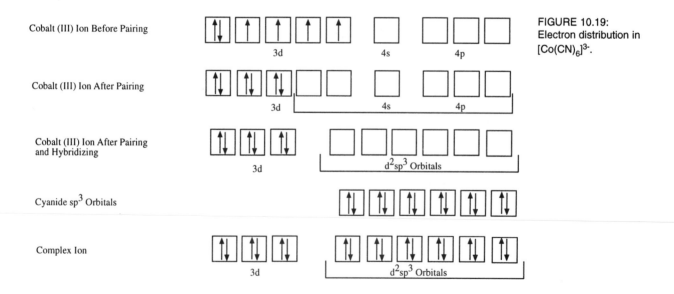

Cobalt (III) Ion Before Pairing

3d 4s 4p

Cobalt (III) Ion After Pairing

3d 4s 4p

Cobalt (III) Ion After Pairing and Hybridizing

3d $d^2sp^3$ Orbitals

Cyanide $sp^3$ Orbitals

Complex Ion

3d $d^2sp^3$ Orbitals

FIGURE 10.19: Electron distribution in [Co(CN)$_6$]$^{3-}$.

The other situation is illustrated by the hexafluorcobaltate(III) ion $[CoF_6]^{3-}$. In this situation spin pairing does not occur before hybrid formation and the **outer orbital, high spin complex** is formed as illustrated in figure 10.20. In the next section we shall discuss the reasons for the formation of inner orbital complexes in some cases and outer orbital complexes in other cases.

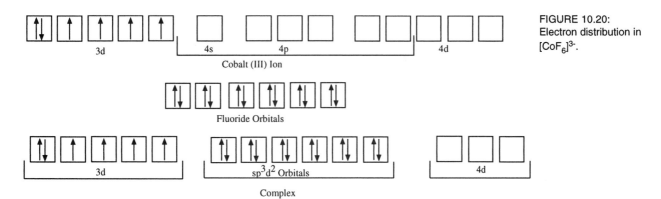

FIGURE 10.20: Electron distribution in $[CoF_6]^{3-}$.

## 10.10  CRYSTAL FIELD THEORY: ENERGY LEVELS OF d ORBITALS IN AN OCTAHEDRAL FIELD

One of the problems with the use of valence bond theory (VBT) and its involvement of hybrid orbitals in the bonding of ligands to a central atom is that it provides no real basis to explain the pairing of electrons that sometimes occurs allowing the formation of a low–spin, inner orbital complex and sometimes does not allow pairing forcing the formation of high–spin, outer orbital complexes. To explain these aspects of bonding, the earliest and still the simplest, theory is **Crystal Field Theory** (CFT) developed by Bethe and Van Vleck at about the same time that Valence Bond Theory was being set forth by Pauling. Although crystal field theory was used by physicists from the early 1930s, chemists did not begin using it until the 1950s. When chemists did turn to it to explain the formation of complexes, it quickly gained favor and, for a time, displaced Valence Bond Theory. Today it, in turn, has been replaced by **Molecular Orbital Theory** (MO) — a much more complete, and therefore much more difficult to apply, treatment for the bonding in all species. Indeed, Van Vleck pointed out in 1935 that CFT and VBT were both just special cases of MO Theory.

Crystal Field Theory looks at the interactions of electrons in the d orbitals with the electrons on the ligands as they approach the central atom. It treats the ligands as negative point charges and considers only the electrostatic interactions of the ions with the electrons. In order to understand this fully, it is necessary to recall the shapes and orientations of the 5 d orbitals. Conventionally the d orbitals consist of two sets of orbitals. One set, containing three orbitals ($d_{xy}$, $d_{yz}$, and $d_{xz}$) and which will be referred to as the $t_{2g}$ **orbitals** from this point forward, have their major lobes pointing at 45° to the Cartesian axes. The other group of two orbitals ($d_{x^2-y^2}$ and $d_{z^2}$), which will be referred to as the $e_g$ **orbitals** from now on, direct the major portion of the electron density along the Cartesian axes. In the absence of an asymmetric magnetic or electric field, these orbitals are degenerate.

An atom surrounded by charged ligands is, however, within an asymmetric electric field and, therefore, the energies of the orbitals are not equal. Let us consider how the orbital energies change as the ions approach the central atom.

The changes in energy levels will depend upon the geometry of the species. Let us first consider the octahedral case. In this case the ligands approach the central atom along both the positive and negative x, y, and z axes. The negatively charged ligands will interact with and affect the energy of the electrons in all the d orbitals. However, since the electrons occupying the $d_{x^2-y^2}$ and $d_{z^2}$ (the $e_g$) orbitals are pointing more directly at the incoming ligands, the energies of these orbitals will be more strongly affected than will the energies of the other group of orbitals (the $t_{2g}$ orbitals.) Thus the five previously degenerate d orbitals split into two groups of orbitals, one group of three orbitals, the $t_{2g}$ orbitals that drop in energy relative to the average orbital energy and the $e_g$ orbitals that increase in energy (relative to the average orbital energy) as shown in figure 10.21. This energy difference is called the **crystal field splitting energy**. The magnitude of the splitting is defined to be 10 $D_q$ (some authors use $\Delta$ for this value).

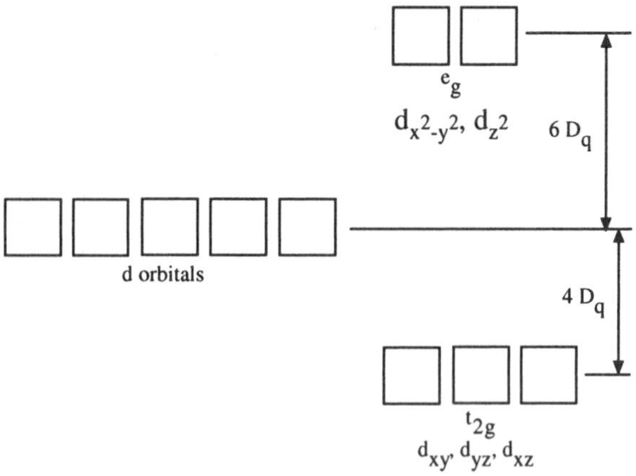

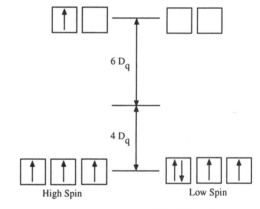

FIGURE 10.21:
Crystal field splitting diagram for an octahedral complex.

The amount of stabilization that an ion achieves by placing its electron(s) into the $t_{2g}$ orbitals rather than into the $e_g$ orbitals can now be determined in units of $D_q$. The total stabilization energy depends upon the electron configuration and is easily determined by drawing the orbital levels, placing the electrons in those levels and then computing the total energy. For instance, consider the case of a $d^4$ ion. Figure 10.22 shows the two different possibilities: one for the low spin case and one for the high spin case. In the case of the high spin configuration, there are three electrons in $t_{2g}$ orbitals, each of which experiences a -4 $D_q$ stabilization for a total of -12 $D_q$ stabilization. There is also one electron in an $e_g$ orbital. This electron is destabilized by +6 $D_q$. The total stabilization is, therefore -6 $D_q$. In the low spin case, all four electrons are in $t_{2g}$ orbitals for a stabilization of -16 $D_q$. In order to place all of these electrons into the $t_{2g}$ orbitals, one of them had to be spin paired. This costs an amount of energy that is symbolized by P. The total crystal field splitting energy is -16 $D_q$ + P. Crystal field stabilization energies for each electron configuration from $d^1$ to $d^{10}$ are given in table 10.4. The student is encouraged to verify these calculations, especially for the cases of $d^6$ through $d^8$.

FIGURE 10.22:
Orbital energy diagrams for $d_4$ ion in high spin and low spin case.

**TABLE 10.4** CRYSTAL FIELD STABILIZATION ENERGY

TABLE 10.4

| # D ELECTRONS | HIGH SPIN CONFIG. | HIGH SPIN STABILIZATION ENERGY | LOW SPIN CONFIG. | LOW SPIN STABILIZATION ENERGY |
|---|---|---|---|---|
| $d^1$ | $t_{2g}^{1}$ | $-4\,D_q$ | $t_{2g}^{1}$ | $-4\,D_q$ |
| $d^2$ | $t_{2g}^{2}$ | $-8\,D_q$ | $t_{2g}^{2}$ | $-8\,D_q$ |
| $d^3$ | $t_{2g}^{3}$ | $-12\,D_q$ | $t_{2g}^{3}$ | $-12\,D_q$ |
| $d^4$ | $t_{2g}^{3}e_g^{1}$ | $-6\,D_q$ | $t_{2g}^{4}$ | $-16\,D_q + P$ |
| $d^5$ | $t_{2g}^{3}e_g^{2}$ | $0$ | $t_{2g}^{5}$ | $-20\,D_q + 2P$ |
| $d^6$ | $t_{2g}^{4}e_g^{2}$ | $-4\,D_q$ | $t_{2g}^{6}$ | $-24\,D_q + 2P$ |
| $d^7$ | $t_{2g}^{5}e_g^{2}$ | $-8\,D_q$ | $t_{2g}^{6}e_g^{1}$ | $-18\,D_q + P$ |
| $d^8$ | $t_{2g}^{6}e_g^{2}$ | $-12\,D_q$ | $t_{2g}^{6}e_g^{2}$ | $-12\,D_q$ |
| $d^9$ | $t_{2g}^{6}e_g^{3}$ | $-6\,D_q$ | $t_{2g}^{6}e_g^{3}$ | $-6\,D_q$ |
| $d^{10}$ | $t_{2g}^{6}e_g^{4}$ | $0$ | $t_{2g}^{6}e_g^{4}$ | $0$ |

These stabilization energies are a first approximation to the energy of an atom or ion with this electron configuration. The exact energy of the species is determined by the exact orbitals into which the electrons are placed. For instance, let us consider the situation of an ion with octahedral geometry containing two d electrons. Of course, the ground state of this ion in an octahedral field is $(t_{2g})^2$, but let us consider the excited state configuration in which one electron is in one of the $t_{2g}$ orbitals and the other is in one of the $e_g$ orbitals. According to the orbital energies from the octahedral splitting diagram, the total energy of this configuration should be:

$$E = -4\,D_q + (+6\,D_q) = +2\,D_q.$$

EQ. 10.5

This situation does not take into account any of the electron–electron interactions that must be considered to determine the total energy of the configuration. While the specific methods used to calculate this energy are beyond the scope of this text, it is not difficult to understand the qualitative basis on which these calculations are based.

Let us consider an excited state in which the two electrons are in the $d_{xy}$ and the $d_{x^2-y^2}$ orbitals and compare this state with another in which the electrons are in the $d_{xy}$ and the $d_{z^2}$ orbitals. In the former situation the electrons "see" each other to a greater extent than they do in the latter case simply because of the geometric shapes of the orbitals they occupy. This means that there will be greater electron–electron repulsions in the former than in the latter case. The excited state in which the electron is in the $d_{z^2}$ orbital is, therefore, a lower energy situation than the one in which the promoted electron occupies the $d_{x^2-y^2}$ orbital. At least two different transitions will show up in the electronic spectrum of the ion. The exact situation is somewhat more complicated than this simple picture has suggested and the exact details will need to await formal study of quantum mechanics. For the rest of this discussion, we shall consider that these electron–electron repulsions need not be taken into account because the simple picture will be adequate to show what is necessary. The student must understand, however, that the results from such considerations are an approximation and slight deviations from this picture will become evident when real data are consulted.

## 10.11   EXPERIMENTAL DETERMINATION OF CRYSTAL FIELD SPLITTING ───────

At this point it is probably necessary to stop to consider how the value of 10 $D_q$ is determined.  To do this, let us consider the simplest case, that of a $d^1$ ion in an octahedral field.  A diagram of the electron energy levels is shown in figure 10.23.  In this diagram note that the lowest energy d→d electronic transition would be the promotion of the electron in the $t_{2g}$ orbital to one of the $e_g$ orbitals.  According to the diagram, the energy required for this transition is exactly 10 $D_q$.  Determination of the wavelength at which this transition occurs in the electronic spectrum of the species would, therefore, directly lead to the value of 10 $D_q$.  Electronic transitions involving the valence electrons occur in the UV–Vis portion of the electromagnetic spectrum — a portion of the spectrum that is quite easily accessed using instruments similar to the IR spectrophotometers used to determine the vibrational spectra we discussed last year.

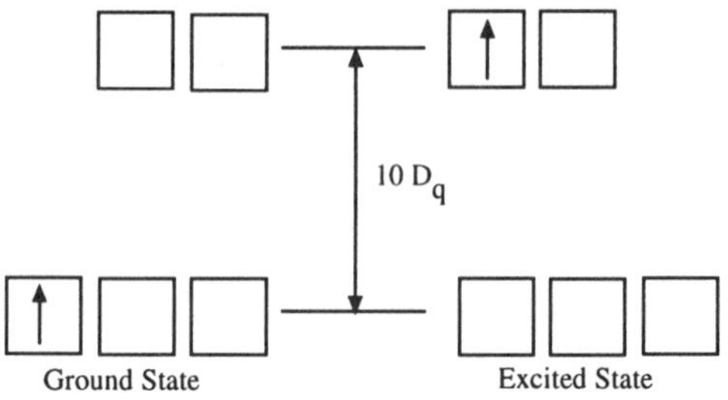

FIGURE 10.23:
Energy level diagram
for a $d^1$ ion.

This same approach is used for other ions containing more than one d electron.  However, in these cases the actual interpretation of the spectrum to obtain the value of 10 $D_q$ is somewhat more complicated because it is necessary to take into account electron–electron interactions in these multi–electron ions.  We shall not, therefore, attempt such an interpretation here.  Table 10.5 lists the 10 $D_q$ values obtained for the hexaquo complexes of a number of different metal ions.

**TABLE 10.5**  CRYSTAL FIELD SPLITTING ENERGIES OF A NUMBER OF METAL IONS

TABLE 10.5

| ELECTRON | ION | 10 $D_q$ VALUE (CM$^{-1}$) | SPLITTING CONFIGURATION (kJ mol-1) |
|---|---|---|---|
| $d^1$ | $Ti^{3+}$ | 20300 | 243 |
| $d^4$ | $Cr^{2+}$ | 13900 | 166 |
|  | $Mn^{3+}$ | 21000 | 251 |
| $d^5$ | $Mn^{2+}$ | 7800 | 93 |
|  | $Fe^{3+}$ | 13700 | 164 |
| $d^6$ | $Fe^{2+}$ | 10400 | 124 |
| $d^7$ | $Co^{2+}$ | 9300 | 111 |

These values of the crystal field splitting energy are typical of those found in many complexes (indeed some are considerably higher than the ones in this table).  Note that these splitting energies range from *ca.* 93 kJ mol$^{-1}$ to *ca.* 250 kJ mol$^{-1}$.  That is very comparable to bond energies in many molecules.  Of course some complexes are stabilized by more or less than 10 $D_q$, and promotion of an electron from a $t_{2g}$ orbital to an $e_g$ orbital may require an energy input of more than 10 $D_q$ because of the approximations that have been made to obtain this simple picture.  The important thing to note, however, is that the stabilization energies of the complexes are on the same order of magnitude as bond energies.

**CONCEPT CHECK**

The absorption maximum of the complex $Ti(H_2O)_6^{2+}$ occurs at 498 nm. What is the value of $10\,D_q$ for this complex?

We remember that the energy of a photon is related to its frequency by Planck's constant. We then remember that the frequency of a light wave is related to its wavelength by the speed of light. Putting these two together, we can calculate the energy in a photon of light at 448 nm. wavelength:

$$E = h\nu$$
$$c = \lambda\nu; \quad \nu = \frac{c}{\lambda}$$
$$E = \frac{hc}{\lambda}$$
$$E = \frac{(6.626 \times 10^{-34}\ \text{J sec})(2.998 \times 10^{8}\ \text{m sec}^{-1})}{498 \times 10^{-9}\ \text{m}}$$
$$E = 3.99 \times 10^{-19}\ \text{J}$$

However, this value is the value for the energy of *one* photon. To put this in terms with which we are somewhat more familiar, we should convert it to the energy of a mole of photons by multiplying it by Avogadro's number. This gives the final value of 240 kJ mol⁻¹. Notice that this value is close to many bond energies.

Remembering that the color that a compound appears is the complementary color to that which it absorbs, would you expect that value of $10\,D_q$ to be higher for a complex that was blue or one that was green?

A complex that is blue absorbs light in the yellow region of the spectrum. One that is green absorbs in the red region of the spectrum. Since red light occupies the portion of the electromagnetic spectrum at longer wavelengths than yellow light, red photons are less energetic than yellow photons and, therefore, the blue complex should have the larger value of $10\,D_q$.

## 10.12   THE SPECTROCHEMICAL SERIES

It will certainly not come as a great surprise that the value of the crystal field splitting depends upon the ligands as well as the central atom. This can be easily understood if one realizes that the color of a compound is directly related to its absorption spectrum and then recalls that different complexes of the same metal are often differently colored. If the values of the crystal field splitting energies are tabulated for a series of ligands bonded to the same metal ion (table 10.6), and if this is done for a variety of metal ions, the ligands may be ordered in terms of the size of the crystal field splitting that they cause. This ordered series of ligands is termed the **spectrochemical series**. A portion of that series is: $I^- < Br^- < S^{2-} < SCN^- < Cl^- < NO_3^- < F^- < OH^- < C_2O_4^{2-} < H_2O < NH_3 < en < CN^- < CO$. If water is assigned a value of 1.00 for the strength of the crystal field that it normally imparts, the values for the other ligands range from about 0.7 for iodide to about 1.7 for the cyanide ion. Again recall that, even with the iodide ion, the value of the energy difference between the orbitals caused by the crystal field is still approximately the same as a typical bond.

**TABLE 10.6** ABSORPTION WAVELENGTHS FOR A NUMBER OF COMPLEX IONS

TABLE 10.6

| COMPLEX | $\lambda$ (nm) |
|---|---|
| $[CrF_6]^{3-}$ | 664 |
| $[Cr(H_2O)_6]^{3+}$ | 575 |
| $[Cr(en)_3]^{3+}$ | 448 |
| $[Cr(CN)_6]^{3-}$ | 376 |
| | |
| $[Fe(H_2O)_6]^{3+}$ | 714 |
| $[Fe(ox)_3]^{3-}$ | 707 |
| $[Fe(CN)_6]^{3-}$ | 286 |
| | |
| $[CoF_6]^{3-}$ | 763 |
| $[Co(H_2O)_6]^{3+}$ | 482 |
| $[Co(NH_3)_6]^{3+}$ | 437 |
| $[Co(en)_3]^{3+}$ | 432 |

The presence of this spectrochemical series, however, presents a problem if we but take a moment to think about crystal field theory and recall that it is based upon the splitting of the d orbital degeneracies by the approach of a negative charge. Looking at the spectrochemical series, we note that most of the negative ions cause the smallest splittings and, apart from the cyanide ion which appears to be an exception to the rule, the species most effective in causing splitting are neutral ligands. Even among the neutral molecules there is a problem with ammonia. Ammonia has a smaller dipole moment than water but is above water in the spectrochemical series. One would have expected that a species with a larger dipole moment, indicating a less symmetric electron distribution, would be more ion–like, and therefore be more capable of causing splitting.

These things all raise doubts about the initial assumption that the interactions between the ligands and the central metal ion are purely electrostatic in nature. An extension of Crystal Field Theory that utilizes the same type of splitting argument to order the d orbitals but is based on covalent bonding instead of purely electrostatic interactions has been developed. This theory is called **Ligand Field Theory**. It helps to explain some of these differences. We shall leave more complete discussion of this theory to later courses in chemistry.

## 10.13   ENERGY LEVELS OF d ORBITALS IN TETRAHEDRAL FIELDS ———

The splitting of the d orbital degeneracies in tetrahedral fields are most easily determined if one notes that a tetrahedron may be considered as a cube that has had one–half of its corner elements removed. Figure 10.24 pictures just such a cube superimposed over drawings of the d orbitals.

The e$_g$ orbitals (d$_{z^2}$ and d$_{x^2-y^2}$), point directly along the
axes whereas the t$_{2g}$ orbitals sit in a plane with the lobes
of the orbitals between the axes. Therefore any ligand
approaching along the axis will interact most strongly
with the e$_g$ orbitals. This was the reason for the increase
of energy of the e$_g$ orbitals in an octahedral field. Notice
that, in a tetrahedral field, the orbitals do not approach the
central atom along the axes but, rather, approach from
between the axes. This means that the ligands will interact most strongly with the t$_{2g}$ orbitals. This exactly
inverts the splitting diagram that was found for the octahedral case, placing the t$_{2g}$ orbitals above the e$_g$ orbitals
as shown in figure 10.25. Experimentally the amount by
which the d orbitals are split (the value of 10 D$_q$) is about
one–half that of the comparable octahedral case because,
in the tetrahedral case an orbital has only 2 of its four
lobes interacting with ligands and because there is a
smaller total number of ligands involved. Therefore the splitting energy is not
large enough to cause the formation of low spin complexes. Hence only
outer–orbital, high–spin complexes are formed when the ions have more than
two d electrons and when those ions actually form tetrahedral complexes. In
the instances when a very high–field ligand is present, the energy of interaction
is large enough that complexes with higher coordination numbers and, therefore, other geometries are more stable.

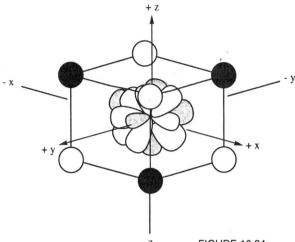

FIGURE 10.24:
Interaction of d
orbitals with ligands in
a tetrahedral field.

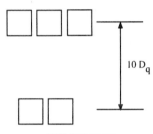

FIGURE 10.25:
Crystal field splitting
in a tetrahedral field.

## 10.14 ENERGY LEVELS OF d ORBITALS IN SQUARE PLANAR FIELDS

Square planar fields are best considered as octahedral fields in
which the two ligands along one of the axes (by convention, the z
axis) are removed. Removing the two ligands along the z axis
affects the electron–electron repulsions in d orbitals with a z–axis
component. There are three of these: the d$_{z^2}$, the d$_{xz}$, and the d$_{yz}$.
Of these, the d$_{z^2}$ is affected more than the other two. The other
two are affected by the same amount. Hence what initially were
two groups of orbitals (t$_{2g}$ and e$_g$) in an octahedral field are now
four groups of orbitals with an energy order as shown in figure
10.26.

Experimentally the only metal ions that are known to form square
planar complexes are those with d$^8$ electron configurations. Since
only one d orbital, the d$_{x^2-y^2}$, is needed to form the required dsp$^2$
hybrids, this complex may be either high–spin or low–spin in theory. However, the only square planar complexes of these ions are
those of strong field ligands because, in the absence of a strong
field ligand, there is no energy advantage to square planar geometry and such complexes are usually octahedral. The octahedral complexes benefit from the two additional bonds formed by the additional ligands.

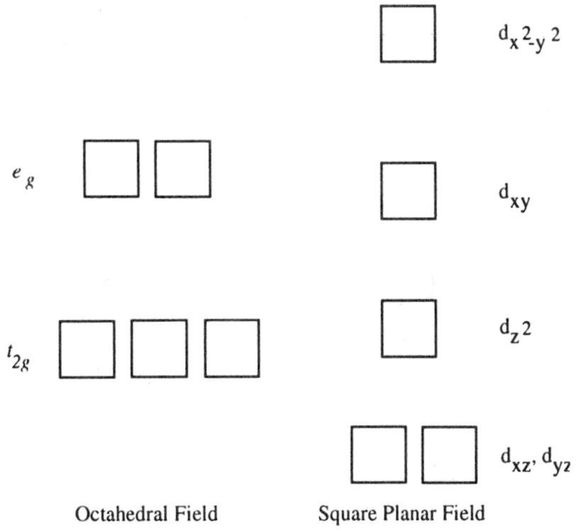

$e_g$ | | |
$t_{2g}$ | | | |

Octahedral Field

$d_{x^2-y^2}$
$d_{xy}$
$d_{z^2}$
$d_{xz}, d_{yz}$

Square Planar Field

FIGURE 10.26:
Orbital order diagram
for square planar
geometry.

**Axial orbitals:** Those orbitals located directly above and below the central atom in certain electron geometries. There are two such orbitals at 180° to one another. The term is usually used only with trigonal bipyramidal structures. See equatorial orbitals.

**Bond energy:** The amount of energy required to break a bond when the reactants and products are in the gas phase.

**Bond length:** The distance between the centers of the two atoms involved in the bond.

**Cartesian coordinate axes:** The commonly used x, y, and z axis system in which the three axes are at 90° to one another. Named for the French mathematician and philosopher Rene Descartes (1596–1650).

**Crystal field splitting energy:** The energy difference introduced into formerly degenerate orbitals (usually d orbitals) by the presence of charged ligands arranged around a central metal atom.

**Crystal field theory:** A theory that treats the interactions between the ligands and the central metal atom as purely an electrostatic problem.

**Degenerate:** A term used to mean that the energies of two or more different orbitals or electronic states are equal.

**$e_g$ orbitals:** The $d_{x^2-y^2}$ and $d_{z^2}$ orbitals. The name is derived from group theory. See $t_{2g}$ orbitals.

**Equatorial orbitals:** Those orbitals located in a plane around the central atom. The term is usually used only with trigonal bipyramidal structures. In such a structure the orbitals make angles of 120° with each other. See axial orbitals.

**High spin complex:** A complex that has the same total electron spin that the central metal ion had before complexation. See low spin complex.

**Hybrid orbitals:** Atomic orbitals made by combination of two or more "pure" atomic orbitals.

**Inner orbital complex:** A complex in which the hybrid orbitals involve d atomic orbitals from the (n–1) shell and s and p atomic orbitals from the n shell. See outer orbital complex.

**Ligand field theory:** An extension of crystal field theory that takes into account the covalent nature of the bonds formed between the central metal atom and the ligands.

**Low spin complex:** A complex that has a lower total electron spin than the central metal ion had before complexation. See high spin complex.

**Molecular orbital theory:** A bonding theory that describes a chemical bond in terms of molecular orbitals formed by combinations of atomic orbitals. These molecular orbitals exist only in the molecule.

**Octahedral geometry:** The regular geometric form exhibited by a central metal atom and six surrounding ligands. So–called because the solid figure has 8 equivalent triangular faces.

**Outer orbital complex:** A complex in which the hybrid orbitals involve d atomic orbitals from the n shell with s and p atomic orbitals from the same shell. See inner orbital complex.

**Spectrochemical series:** A series of ligands arranged in order of their ability to cause ligand field splitting.

**Square planar geometries:** A geometry in which the metal ion and 4 surrounding ligands are arranged in a plane with the 4 ligands at the corners of a square and the metal ion in the center of the square.

**Square pyramidal:** A geometry in which the metal and four ligands are arranged as in square planar geometry but a fifth ligand is located directly above the center of the plane. This forms a pyramid with a square base.

**$t_{2g}$ orbitals:** The $d_{xy}$, $d_{yz}$, and $d_{xz}$ orbitals. The name is derived from group theory. See $e_g$ orbitals.

**Tetrahedral:** Pertaining to a regular solid figure consisting of 4 triangular faces. It normally consists of an atom, in the center of the solid, and four surrounding atoms arranged at the vertices of the figure. The angles formed between lines drawn from any two of the atoms on the vertices to the central atom are all 109°28'.

**Trigonal bipyramid:** A non–regular solid figure formed by a central atom and five surrounding ligands. In this figure, 3 of the ligands are arranged in a plane at 120° to one another and the remaining two ligands are arranged above and below the center of the plane at 90° to the plane and at 180° to one another. The central atom is situated at the center of the plane.

**Valence bond theory:** A bonding theory which explains bonding between two atoms as occurring by the overlap of two atomic orbitals. These may be "pure" atomic orbitals or they may be hybrid orbitals.

HOMEWORK
**Problems**

1. For each of the following species (i) draw the Lewis Dot Diagram, (ii) describe the electronic geometry about the central atom, (iii) describe the atomic geometry about the central atom, (iv) describe the hybrid orbitals used by the central atom in forming bonds, and (v) describe the bond type between the central atom and each of the atoms to which it is bonded.

| | | | | |
|---|---|---|---|---|
| a. | $H_3BO_3$ | | f. | $COCl_2$ |
| b. | $SCl_4$ | | g. | $XeF_4$ |
| c. | $PF_5$ | | h. | $I_3^-$ |
| d. | $SeF_6$ | | i. | $H_2SO_4$ |
| e. | $IF_5$ | | | |

2. For each of the following, (1) state what type of hybrid orbitals are used by the central metal ion in bonding, (2) predict the number of unpaired electrons, and (3) compute the ligand field stabilization energy for both inner and outer hybrid formation (for octahedral complexes only).

| | | | | |
|---|---|---|---|---|
| a. | $Fe(C_2O_4)_3^{3-}$ | | e. | $Ni(CN)_4^{2-}$ (square planar) |
| b. | $Co(en)_2Cl_2^+$ | | f. | $Pt(NH_3)_2Cl_2$ (square planar) |
| c. | $Fe(CN)_6^{4-}$ | | g. | $MnCl_5^{2-}$ |
| d. | $Cr(NH_3)_6^{3+}$ | | h. | $MoCl_6^-$ |

3. How might you distinguish between the following pairs of solids:

   a.  $[Cr(NH_3)_3Cl_3]$ and $[Cr(NH_3)_6][CrCl_6]$
   b.  $[CoBr(NH_3)_5]SO_4$ and $[CoSO_4(NH_3)_5]Br$

4. Construct a table of ligand field stabilization energies similar to table 10.4 for all possible $d^n$ configurations in a tetrahedral field. Which of these configurations would lead to a diamagnetic species in a strong ligand field if the species were to form a tetrahedral complex?

5. Construct a table of ligand field stabilization energies similar to table 10.4 for all possible $d^n$ configurations in a square planar field. Which of these configurations would lead to diamagnetic species in a strong ligand field if the species were to form a square planar complex?

6. Consider the electron configuration ($1s^2$, $2s^2$, *etc.*) of the gas phase $Cr^{3+}$ and $Co^{3+}$ species, compared to the electron configuration of those species when in an octahedral complex, *e.g.*, $Cr(acac)_3$ and $Co(acac)_3$. Note that the d-orbitals would then be split into $t_{2g}$ and $e_g$ levels. Measurement of the magnetic moment of $Cr(acac)_3$ and $Co(acac)_3$ show that the chromium complex is paramagnetic, while the cobalt complex is diamagnetic.

     a. Give the electron configuration expected for gas phase $Cr^{3+}$. How many unpaired electrons are present in this species?
     b. Give the electron configuration expected for gas phase $Co^{3+}$. How many unpaired electrons are present in this species?
     c. Explain the magnetic moment observed in $Cr(acac)_3$. How many unpaired electrons are there in this complex?
     d. Explain the magnetic moment observed in $Co(acac)_3$.

7. $Al(acac)_3$ $[Al(CH_3COCHCOCH_3)_3]$ is a colorless compound. When analyzed by HNMR, it shows a singlet at 2.0 ppm and a singlet at 5.4 ppm. Integration shows that these two peaks are in a 6:1 ratio. The uncoupled CNMR spectrum shows a quartet at 27 ppm, a doublet at 98 ppm, and a singlet at 187 ppm. The mass spectrum of the compound shows many peaks besides the parent peak at 324, including peaks at 225 and 126.

     a. Draw the structure of $Al(acac)_3$. Transition metal acetoacetonates are usually intensely colored. Why might aluminum be an exception to this?
     b. Interpret the HNMR spectrum, assigning peaks to the protons shown in your answer for the structure in (a).
     c. Interpet the CNMR spectrum, assigning peaks to the carbons shown in (a).
     d. Interpret the peaks noted in the mass spectrum.

8. Cyclopentadiene ($C_5H_6$) reacts with strong bases to form the cyclopentadiene anion $C_5H_5^-$, abbreviated as cp. This species forms compounds with many metal ions, including ferrocene, $Fe(cp)_2$, and nickelocene, $Ni(cp)_2$. Both iron and nickel ions were in the +2 oxidation state when these sandwich compounds were formed; the $C_5H_5^-$ rings being the two slices of bread, the metal ion being the meat.

     a. Show the stability of $C_5H_5^-$, by drawing the various resonance forms.
     b. Show whether the species $C_5H_5^-$, is aromatic or non-aromatic.
     c. How many electrons occupy the d-orbitals of $Ni^{2+}$? Of $Fe^{2+}$?
     d. All the delocalized pi electrons in both cp rings can be assumed to be donated to the metal ion in forming the coordination compound. How many valence electrons would result in ferrocene? In nickelocene?
     e. Ferrocene is essentially air stable, while nickelocene rapidly oxidizes when exposed to air. Explain this based upon your answer to (e).

9. The molecule ferrocene, $FeC_{10}H_{10}$, is described in the previous problem. It was named due to some similarity to benzene. Like benzene, ferrocene can undergo electrophilic aromatic substitution. Acetyl chloride, in the presence of $AlCl_3$ catalyst, will react with ferrocene to form acetylferrocene. In fact, the rate of acylation under the same conditions is far greater for ferrocene than for benzene.

    a. Show the mechanism for the reaction making acetylbenzene.

    b. Show the mechanism for the reaciton making acetylferrocene.

    c. Explain why the rate of electrophilic aromatic substitution in (b) might be faster than the reaction in (a).

    d. When the stoichiometry is adjusted so that two acetyl groups replace two protons in ferrocene, the substitutions invariably occur such that one acetyl group is on both rings, never two on one ring. Explain.

# CARBON
## AS A Nucleophile

# CARBON ─────── AS A NUCLEOPHILE

Ⓘn previous chapters we have studied a wide variety of nucleophiles. Many of these were anions, such as the halide ions or the hydroxide ion, while others were neutral molecules such as ammonia or water. All of these had one feature in common: an unshared pair of electrons that makes the species a Lewis base. In only a few instances, *e.g.* the cyanide ion, has the atom carrying the unshared pair of electrons been a carbon atom. This is as expected because, except for the hydrogen atom, the carbon atom is less electronegative than the other atoms to which it is commonly attached. Therefore we would expect the electron pair to be unequally shared in a way that gives the carbon atom less than an equal share of the electron pair. In this chapter we look at several classes of reactions in which the nucleophilic pair of electrons is attached to a carbon atom. In many of these cases the nucleophile is a species with the negative charge formally carried on the carbon atom. Such a species is termed a **carbanion**. Reaction of this nucleophile with another atom forms a bond between that atom and the carbon atom of the nucleophile. In many cases the atom undergoing nucleophilic attack is another carbon atom. Thus a new carbon–carbon bond is formed.

Before prodeeding it would be good to recall that in most cases the stronger the base the stronger the nucleophile. Strong nucleophiles need readily available electron pairs for bond formation. Stronger bases have more readily available electron pairs. Last year we learned that there was an inverse relationship between the strength of an acid and the strength of its conjugate base for Brønsted-Lowry acid–base pairs. The strongest bases are, therefore, the ones from which it is most diffiult to remove a proton, *i.e.*, those which are least stable having an unshared pair of electrons. Since our attention at that point was on acid strength, we looked at the factors that affect acid strength, we found that one factor affecting acid (and therefore base) strength was the resonance stabilization of the conjugate base. Acids, such as benzoic acid, in which the electron pair can be delocalized throughout an aromatic ring are stronger acids than the aliphatic carboxylic acids.

A second factor affecting acid strength is the size of the atom to which the proton is bonded. As the size of the X atom increases, the X—H bond becomes weaker and the acid becomes stronger. A third factor is the number and identity of other groups attached to the X atom. As these groups become more and more electron withdrawing, the electrons in the X—H bond are drawn away from the H atom and toward the X atom. This was termed the inductive effect and is responsible for the differences in acid strength between, for example, acetic acid and its mono, di, and trichloro derivitives. The student is urged to review the material covered in chapter 12 of the first year book to obtain a complete understanding of acid/base strength and the factors affecting it.

## 11.2   CARBANION GENERATION

Formation of carbanions almost always results directly from one of two reactions: (1) removal of an acidic (often only extremely weakly acidic) proton by reaction with a strong base, or (2) rupture of a carbon–metal bond. Perhaps the carbanion with which most people are familiar is the cyanide ion. One of the reasons the cyanide ion is so familiar is that its parent acid, hydrogen cyanide, contains a relative weak carbon–hydrogen bond because of the attraction of the electrons surrounding the carbon atom to the more electronegative nitrogen atom. As a consequence of this weak carbon–hydrogen bond, the hydrogen atom is relatively easily ionized. In fact, hydrogen cyanide is a sufficiently strong acid ($pK_a$ ~ 10) that its aqueous solutions are slightly acidic. Very few other compounds have an easily ionizable hydrogen atom bonded directly to a carbon atom.

Less familiar carbanions include those formed by removal of an even less acidic proton bonded directly to a carbon atom. We shall consider a number of such situations later in this chapter. At this point let us consider just one such example: a terminal alkyne.

Terminal alkynes are much less acidic ($pK_a$ ~ 26) than hydrogen cyanide. This means that a base much stronger than the hydroxide ion must be used to cause the loss of the proton. Typically the strong base sodium amide ($NaNH_2$) is used. Such reactions can only occur in non–aqueous, usually **aprotic**, solvents. This is because the water molecule is a sufficiently strong acid that it will react with bases stronger than the hydroxide ion. Placing an acetylide ion in water would, then, quantititatively regenerate the acetylene and hydroxide ion. Equations 11.1 and 11.2 show reactions of hydrogen cyanide and propyne with a generic base (B) to form the carbanion. Note that this is a Brønsted-Lowry acid/base reaction and the species formed in this reaction are the conjugate base and the conjugate acid of the initial reactants.

$$H-C\equiv N\!: \; + \; B\!: \; \longrightarrow \; {}^-\!:C\equiv N\!: \; + \; H-B^+ \qquad\qquad \text{EQ. 11.1}$$

$$H-C\equiv C-CH_3 \; + \; B\!: \; \longrightarrow \; {}^-\!:C\equiv C-CH_3 \; + \; H-B^+ \qquad\qquad \text{EQ. 11.2}$$

One of the reasons that the cyanide ion is more familiar than the methylacetylide ion is that the cyanide ion is less basic than the methylacetylide ion. This makes the cyanide ion better able to accomodate the negative charge resulting from removal of a proton. That the cyanide ion is a weaker base also makes it a weaker nucleophile than the methylacetylide ion in many reactions.

Other common sources of carbanions are **organometallic compounds** usually formed by reaction of an alkyl halide with a metal. For synthetic purposes the two most common metals are magnesium and lithium. Reaction of magnesium metal with an alkyl halide inserts the magnesium atom between the carbon atom and the halogen atom in the well–known Grignard reaction. This reaction requires one mole of magnesium per mole of the alkyl halide and is illustrated in equation 11.3. Reaction of an alkyl halide with lithium metal requires two moles of lithium metal per mole of alkyl halide. This reaction is illustrated in equation 11.4. These compounds should already be very familiar from your study last year.

$$CH_3CH_2CH_2CH_2CH_2Br + Mg(s) \longrightarrow \overset{\delta- \ \delta+}{CH_3CH_2CH_2CH_2CH_2MgBr}$$ EQ. 11.3

$$CH_3CH_2CH_2CH_2CH_2Br + 2\ Li(s) \longrightarrow \overset{\delta- \ \delta+}{CH_3CH_2CH_2CH_2CH_2Li} + LiBr$$ EQ. 11.4

These compounds react very much as if they are actually ionic compounds. This is due, of course, to the large difference in the electronegativity of the carbon atom and the metal atom that causes the carbon–metal bond to be significantly polarized. The result is that a large amount of the electron density associated with the bond is situated on the carbon atom. These compounds are not truly ionic because many of them are liquids or low–melting solids. Many are, however, so reactive that it is difficult to isolate them as the pure compound. Instead they spontaneously react in the presence of air or water. Fortunately, for their use as reagents in synthesis, this is not necessary. Many of these reagents are available as solutions that are removed from the reagent bottle by a syringe through a septum (*e.g.*, butyl lithium) or are generated *in situ* (*e.g.*, the Grignard reagents) just prior to their use.

In addition to the organolithium and organomagnesium compounds that we have just discussed, another group of organometallic compounds has been found to possess different properties that are quite useful in other synthetic reactions. These are the copper(I) organometallic compounds known as the organocuprates (sometimes also called lithium dialkylcuprates). Organocuprates are prepared by reaction of an organolithium compound with copper(I) iodide. This reaction is illustrated in equation 11.5.

$$2\ CH_3Li + CuI \longrightarrow (CH_3)_2CuLi + LiI$$ EQ. 11.5

This reagent reacts as if a carbanion were present, just as do the previously discussed lithium and magnesium compounds. However, the reactivity of these organocuprates is sufficiently different from the alkyl lithium and alkyl magnesium compounds that they have a number of special uses. Organocuprates have no reactivity toward carbonyl groups (which was, of course, a chief target for the reactions of the other organometallic reagents.) It is still a sufficiently good nucleophile, however, to react with alkyl halides to form new carbon–carbon bonds. You may recall that this selectivity is not shown by lithium and magnesium alkyls. Thus, when used with a difunctional compound, two different products are possible depending upon which organometallic reagent is used. The reactions of lithium (2–propenide) and the corresponding organocuprate with 4–bromocyclohexanone are shown as equations 11.6 and 11.7. Note that, while both reagents act as nucleophiles and attack electron–poor carbon atoms, the reactivity of the alkyl lithium is directed toward the carbonyl carbon atom while that of the organocuprate is directed toward the carbon atom to which a halogen atom is attached.

EQ. 11.6

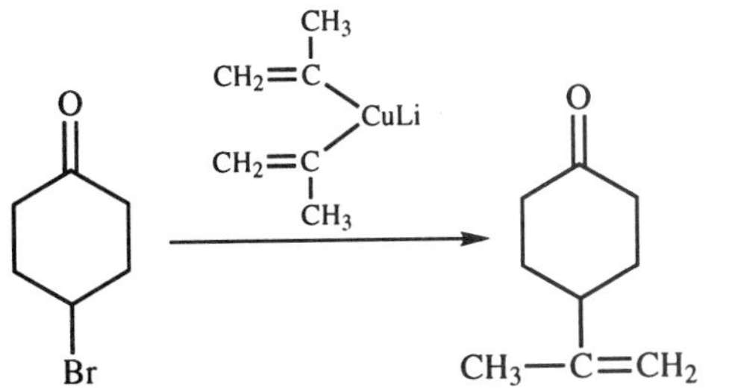

EQ. 11.7

## 11.3 THE KETO-ENOL EQUILIBRIUM: FORMATION OF CARBOANIONS BY REACTION WITH STRONG BASE

Remember that, for an acid/base reaction to go in the forward direction, the conjugate base formed in the reaction must be a weaker base than the base used to remove the proton (the reactant base). This, of course, means that the reactant acid must have a sufficiently acidic proton to react with the base chosen to remove the proton. One of the requirements for an acidic proton is that there be a weaker than normal bond between the carbon and the hydrogen atoms. This can occur when the electron pair making up the bond is polarized in the direction away from the hydrogen atom. Usually the small difference in electronegativity between carbon and hydrogen results in carbon–hydrogen bonds that are essentially non–polar. Thus hydrogen atoms bonded to carbon atoms are very rarely removed even by the strongest bases. Removal of the proton only occurs when there is some other functional group nearby to pull the electrons away or to delocalize the charge formed by the ionization of the proton. Either, or both, of these may increase the acidity of a proton directly bonded to a carbon atom enough that it is possible to remove it by reaction with a strong base.

For example, the carbonyl group is able to increase the acidity of hydrogens bonded to adjacent carbons. Since the first carbon away from the carbonyl group is said to be alpha (α) (from the first letter in the Greek alphabet) to the carbonyl group, any hydrogen atom bonded to this carbon is termed an alpha hydrogen atom. A portion of a molecule containing α and β carbon and hydrogen atoms is shown in figure 11.1. Only three of the four bonds are shown on the α and β carbon atoms because the fourth bond could be to an atom or group other than hydrogen.

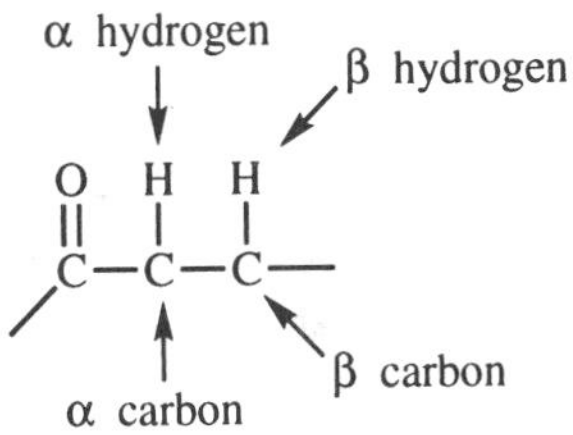

FIGURE 11.1: Alpha and beta notation.

Alpha hydrogen atoms in a carbonyl compound are located in a position that makes them particularly available for participating in a common type of equilibrium termed a keto–enol **tautomerism**. Figure 11.2 illustrates this keto–enol tautomerism for a generic carbonyl compound.

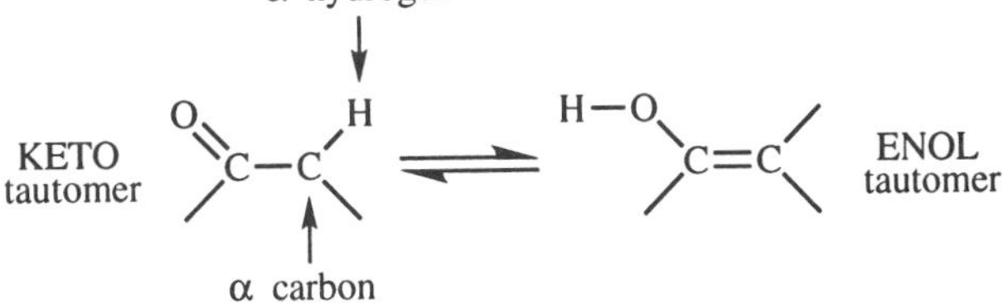

FIGURE 11.2: Tautomeric keto-enol equilibrium.

The word enol comes from the two words alkENe and alcohOL. This is an actual equilibrium reaction, not resonance, because the hydrogen atom has shifted position in the two structures. Had these been resonance forms, only the electrons would have been indicated in different positions. The two forms in this equilibrium reaction are given the special name "**tautomers**" to indicate their special relationship to one another and the type of equilibrium is termed a tautomeric equilibrium. Refer back to equation 5.24 where another example of tautomerism — the imino and amido forms of acetamide — has already been seen.

Under normal situations, the enol tautomer is present in only very small quantities and the value of the equilibrium constant for the reaction represented in figure 11.2 would be very small. For this reason the structures of ketones, aldehydes and esters are normally drawn showing only the keto tautomer. However, in some compounds, because of other functional groups in the molecule, the enol form can be present to a much larger extent in the equilibrium mixture and be very important in the reactions of the compound. Even in compounds in which the keto form is the predominate form, there is strong evidence for the existence of the enol form. We shall examine some of that evidence for such compounds in the next section. First let us examine evidence for the enol form in a compound for which the enol form is present to a much larger extent.

Many different spectroscopic methods can be used to detect the presence of the enol tautomer in a carbonyl compound. One of the most useful is H–NMR. When the H–NMR spectrum of 2,4-pentanedione in figure 11.3 is examined, it does not show the expected two singlet peaks with area ratios of 3:1 at *ca.* $\delta = 1.7$ ppm and $\delta = 3.2$ ppm respectively. Instead the H–NMR spectrum shows several other peaks. One of these in a position ($\delta = 5.5$ ppm) typical of a hydrogen atom on an unsaturated carbon atom; the other is shifted slightly from the methyl peaks and appears at *ca.* $\delta = 2.2$ ppm. These peaks are in addition to the expected two singlets ($\delta = 2.0$ ppm and $\delta = 3.6$ ppm) for the methyl groups and the methylene group situated between the two carbonyl groups of the keto form. Integration of these peaks also shows that the ratios are not simply the whole number ratios expected for simple compounds. The relative area of the peak at $\delta = 5.5$ ppm indicates that the enol form of the compound is the predominate form for 2,4 pentanedione.

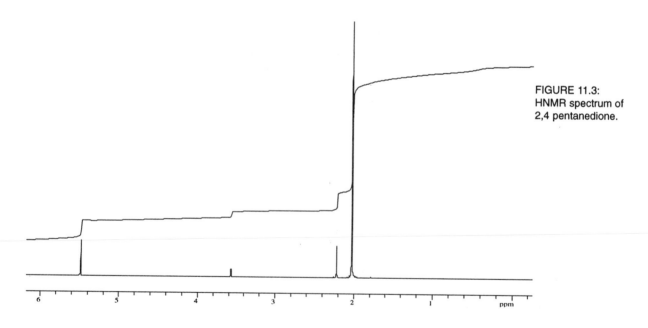

FIGURE 11.3:
HNMR spectrum of
2,4 pentanedione.

We conclude that the extra peaks come from a keto–enol tautomerism of the type that we have just postulated. However, in the case of 2,4–pentanedione, the enol form is stabilized by the presence of the second carbonyl group situated in a position to support the formation of a 6–membered, hydrogen–bonded ring. This is illustrated in figure 11.4. An even larger amount of enol form is present because each carbonyl group can participate in an identical equilibrium.

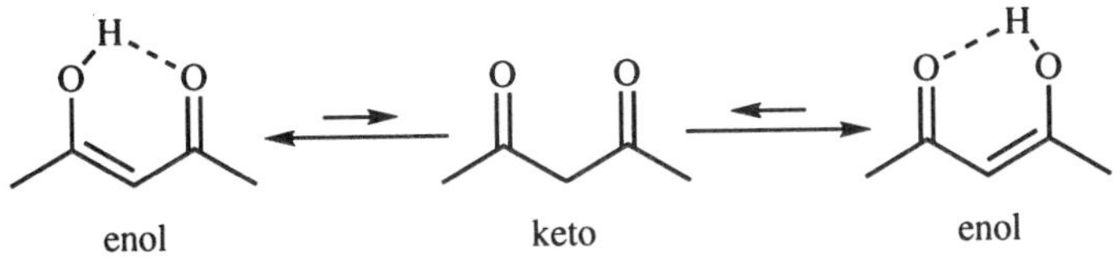

enol           keto           enol

FIGURE 11.4:
Tautomeric ring
formation in
2,4 pentanedione.

## 11.4 DEUTERIUM EXCHANGE EVIDENCE FOR KETO-ENOL EQUILIBRIUM —

If it is impossible to show easily the presence of one of the tautomeric forms of a simple ketone such as acetone, what is the evidence for a tautomeric equilibrium in these cases? One of the important pieces of evidence used in support of the enol form is the exchange of the **protium** isotope of hydrogen for **deuterium** when the simple ketone is dissolved in deuterium oxide ("heavy water"). Simple ketones undergo exchange of the alpha hydrogen atoms for deuterium when dissolved in deuterium oxide. The exchange reaction is slow under ordinary conditions but is catalyzed by the presence of either a trace of acid or base. If the protium isotope of hydrogen ($_1^1H$) is designated by H, and the deuterium isotope ($_1^2H$) by D, the first step in the chemical reaction is shown in equation 11.8. After prolonged contact with the solvent, especially if fresh $D_2O$ is continually added to keep the fraction of deuterium atoms in the $D_2O$ high, it is possible to isolate essentially pure acetone–$d_6$ ($CD_3COCD_3$).

acetone                   acetone-$d_1$

EQ. 11.8

Let us now look at the mechanism by which this occurs. In a heavy water solution containing a small amount of DCl, the carbonyl group is protonated, or in this case deuterated, by the strong acid. This step is followed by the loss of one of the alpha hydrogen atoms to regenerate the hydronium ion. The result of this series of reactions is formation of the enol containing a deuterium atom bonded to the oxygen atom. This series of reactions is shown as figure 11.5.

At this stage, a shift in the tautomeric equilibrium to generate the more stable keto form causes transfer of the deuterium atom from the oxygen to the carbon. Alternatively, transfer of a second deuterium ion to the carbon atom after nucleophilic attack on the enol double bond by a $D_3O^+$ ion, followed by transfer of a $D^+$ ion from the oxygen atom to a deuterium oxide molecule produces acetone–$d_1$. This series of reactions is shown as figure 11.6. Prolonged contact with the acidic deuterium oxide solvent eventually completes the deuteration to form acetone–$d_6$.

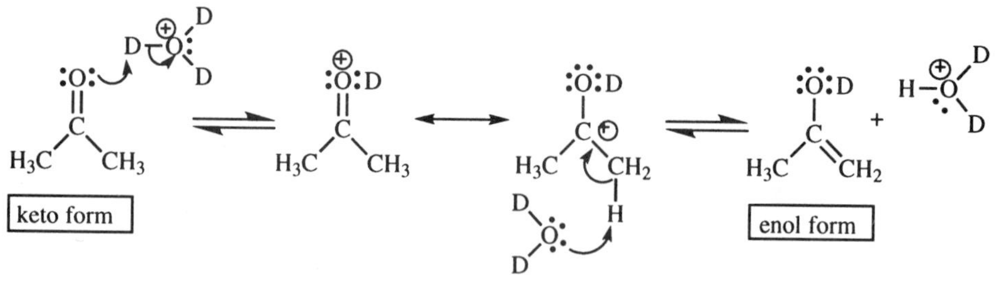

FIGURE 11.5:
Formation of the enol
in acidic solution.

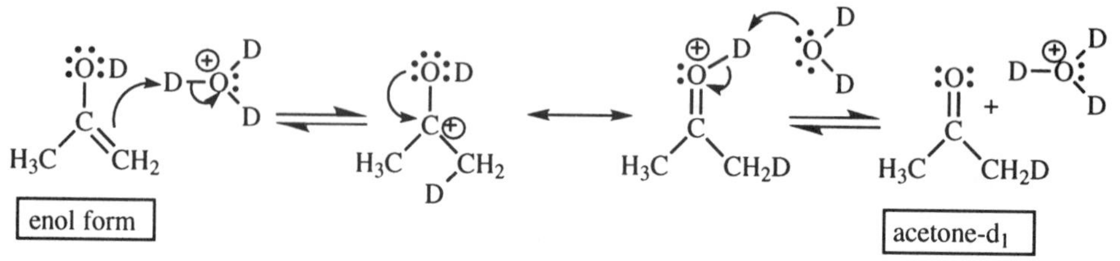

keto form

enol form

FIGURE 11.6:
Formation of
acetone-d₁ from the
enol.

enol form

acetone-d₁

The mechanism is slightly different in solution containing a small amount of
NaOD. In basic solution, the hydroxide (deuteroxide) ion first removes one of
the alpha hydrogens as a proton with the formation of HDO. This leaves
behind a carbanion. There are two resonance forms of this carbanion possible
— one in which the negative charge is localized on the carbon atom from
which the hydrogen ion was removed and another in which the negative charge
resides on the oxygen atom. The two forms differ in the formal placement of
the double bond and the negative charge. Since the oxygen atom is the more
electronegative, the second form should be the more stable. This is shown in
figure 11.7. The species formed by this series of reactions in known as an
**enolate** ion.

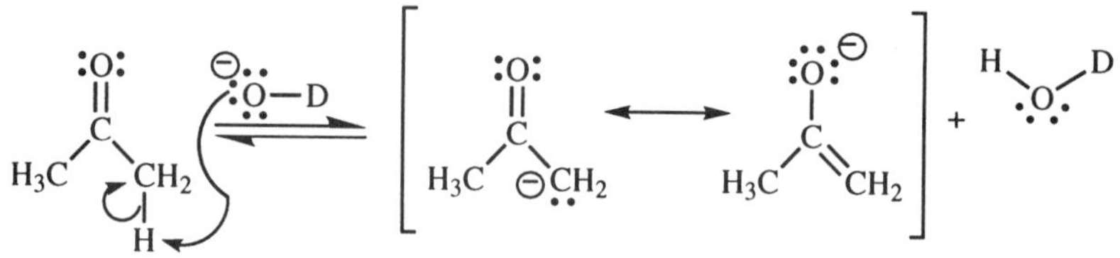

FIGURE 11.7:
Formation of the
enolate ion in basic
solution.

At this point nucleophilic attack of the double bond on a deuterium atom in
deuterium oxide, and subsequent rearrangement of the electrons in the molecule, forms the final product — acetone–d₁. This is shown in figure 11.8.
Once again, prolonged contact with the deuterium oxide solvent eventually
results in complete exchange of the protium isotope with deuterium and the
formation of acetone–d₆.

FIGURE 11.8:
Formation of
acetone-d₁ from the
enolate ion.

Before moving on, let us mention that exchange of **labile** protium isotopes with deuterium has a number of important, practical uses. One such use is, of course, the synthesis of deuterated compounds for use as solvents in HNMR. Both acetone–$d_6$ and chloroform–$d_1$ are commonly prepared this way. In a slightly less obvious use, deuterium exchange can be used to determine how many exchangeable hydrogen atoms are present in a molecule. Mass spectra of the compound are determined before and after it has been in prolonged contact with $D_2O$ containing a catalytic quantity of DCl or NaOD. The gain in weight of the molecular ion peak then shows how many exchangeable hydrogen atoms are present in the molecule. If the molecule contains a carbonyl group, for instance, this could be used to tell how many alpha hydrogens are present. This important information can be used to confirm the structure of the compound.

## 11.5 RACEMIZATION EVIDENCE FOR THE KETO-ENOL EQUILIBRIUM

Additional evidence for the keto/enol equilibrium is provided by the observation that many chiral ketones are racemized in acidic or basic solution. This can be explained if the ketone undergoes tautomeric equilibrium to the enol form. Figure 11.9 shows how the chiral ketone (R)–3–methyl–2–pentanone is racemized to the (S) isomer. In the first step the loss of a proton from C3 forms the enolate ion pictured, in both its keto and enol forms in the middle of the figure. One can see that the enol form contains two $sp^2$ hybridized carbon atoms (C2 and C3). The arrangement around C3 then goes from a tetrahedral configuration to a trigonal planar configuration and the methyl group and the C4 carbon atom become coplanar with carbon C2, C1, and the oxygen atom. Reprotonation can then occur from either side of the double bond. If the proton adds from the back side (of the drawing), the result will be the regeneration of the original R isomer while if the proton adds from the top side the S isomer will result. If addition of the proton occurs from both sides, as will be the case except in highly hindered ketones, a racemic mixture of the ketone is the result. While this reaction can certainly occur without the presence of a catalyst, in most cases reasonable **racemization** rates require the presence of a trace of strong acid or strong base as a catalyst.

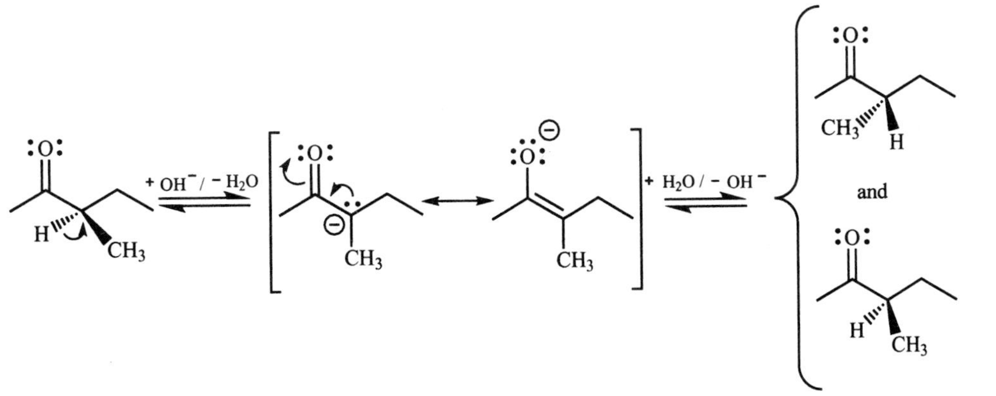

FIGURE 11.9:
Racemization of a chiral ketone in basic solution.

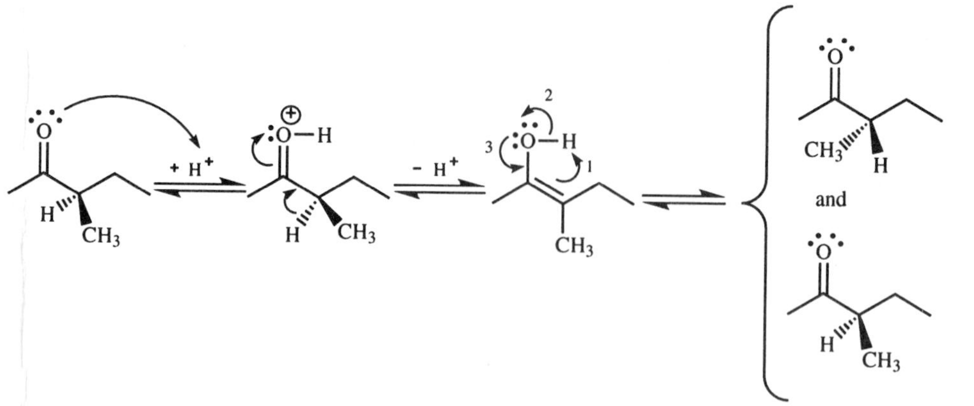

FIGURE 11.10:
Racemization of a
chiral ketone in acidic
solution.

## 11.6  HALOGENATION OF CARBONYLS

Additional evidence for the enol is provided by the halogenation of carbonyl compounds — a reaction that goes more smoothly and readily at the alpha position than halogenation at any other place in the molecule. We shall illustrate this reaction by considering the bromination of acetone. Like deuteration and racemization, the reaction is believed to occur through the enol intermediate that provides a carbon–carbon double bond for electrophilic attack by the halogen molecule. The steps in this reaction are outlined in figure 11.11. The first three lines provide an outline of the overall mechanism of the reaction. The fourth line explicitly shows the movement of the electrons during the second and third steps of the reaction. It is interesting, and important, to note that the rate of bromination is the same as the rate of deuteration. This is additional evidence that participation of a common intermediate (the enol) in both reactions. Once the enol has formed, it is quickly attacked by whatever electrophile (hydronium ion or halogen molecule) is present.

It is certainly not surprising to learn that enolate ions also undergo bromination. The series of steps in the bromination of acetone in basic solution is outlined in figure 11.12. However, in basic solution, the reaction does not typically stop here. The remaining protons are even more acidic and the tribromo product forms readily and then decomposes to the corresponding carboxylate salt. This is the basis of the iodoform test for methyl ketones in which a solid bright yellow iodoform product forms along with the carboxylate salt of the acid with one fewer carbon atom than the original methyl ketone. This reaction is illustrated in figure 11.13. Since iodine is a good oxidizing agent and is capable of oxidizing an alcohol to a ketone, secondary alcohols with the hydroxyl group next to a methyl group (e.g., isopropyl alcohol) typically give positive iodoform tests.

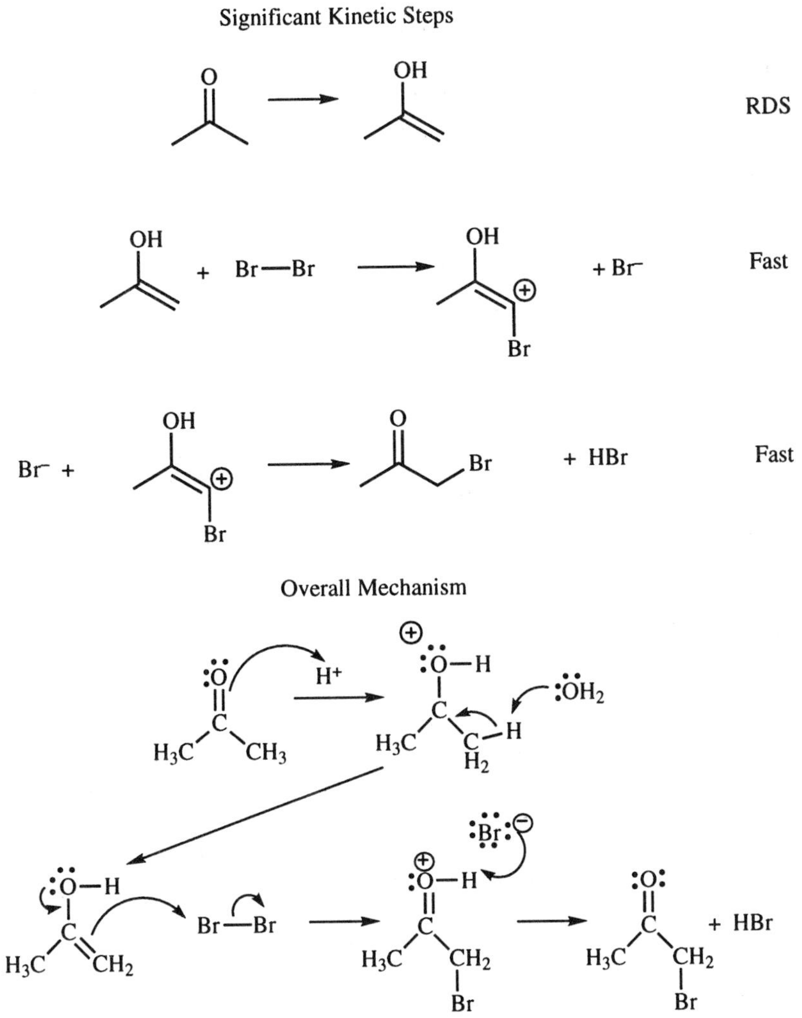

Significant Kinetic Steps

FIGURE 11.11:
Bromination of
acetone.

RDS

Fast

Fast

Overall Mechanism

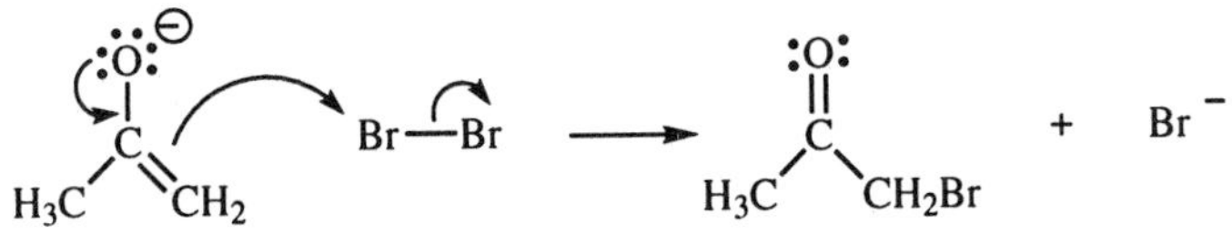

FIGURE 11.12:
Bromination of
acetone in basic
solution.

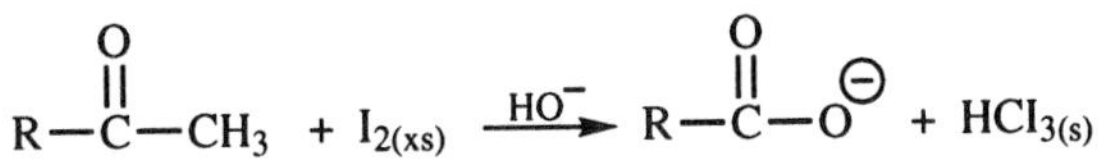

FIGURE 11.13:
Iodoform test for
methyl ketones.

## 11.7 ALKYLATION OF ENOLATES FROM KETONES

In addition to halogens, such as bromine, the alpha position of carbonyl compounds is also attacked by a variety of other electrophiles. Alkyl halides can serve as good electrophiles because of the unequal distribution of the electrons in the carbon–halogen bond that creates an electron–deficient carbon atom. (This is, of course, the same reason that makes them good subjects for nucleophilic attack.) Thus **alkylation** of carbonyl compounds at the alpha carbon atom is possible by reaction with an appropriate alkyl halide in strongly basic solution. The role of the strong base is to produce the enolate ion — an even better candidate for electrophilic attack than the enol itself. The course of a typical alkylation reaction is illustrated in figure 11.14 for the reaction of cyclopentanone with methyl iodide.

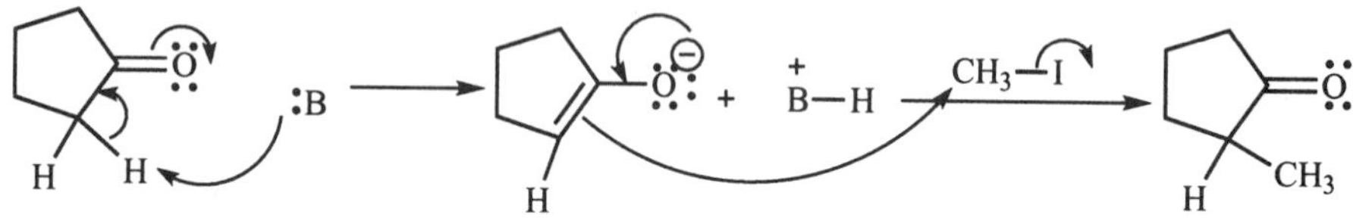

FIGURE 11.14:
Alkylation of
cyclopentanone.

It is important that a very strong base be used in this reaction because the alpha hydrogen atoms of most carbonyl compounds are only very slightly acidic ($pK_a$ ~16–25). This means that bases even stronger than the hydroxide ion must be used and that some solvent much less acidic than water be used as the solvent. One such base sometimes used is sodium amide ($NaNH_2$). In this case protonation of the amide ion by one of the alpha protons produces its conjugate acid, ammonia ($NH_3$), Ammonia is, of course, a gas and a moderately active nucleophile. Fortunately, since it is a gas, it usually escapes the solution before it can cause other problems such as producing unwanted nucleophilic substitution of the alkyl halide. Another means of preventing unwanted nucleophilic substitution reactions is to use a strong base that is not a strong nucleophile. One widely used strong base that has only weak nucleophilic properties is lithium diisopropylamide, commonly known simply by its initials, LDA. The diisopropylamide ion is an extremely strong base (its conjugate acid has a $pK_a$ ~ 36) and, also important for the formation of an enolate ion, the very small lithium ion adds stability to the enolate structure by coordinating with the negatively charged oxygen atom (figure 11.15). The lithium ion coordinates more strongly with the oxygen atom than would the other more common alkali metal ions such as sodium or potassium because it is much smaller than sodium or potassium. From the structure of LDA, shown in figure 11.16, it is apparent that the bulky isopropyl groups make it only weakly nucleophilic while still allowing it to react with the small proton — the property that makes it strongly basic.

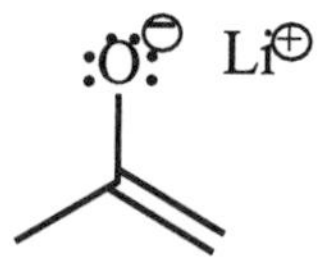

FIGURE 11.15:
Stabilization of enolate ion by lithium ion.

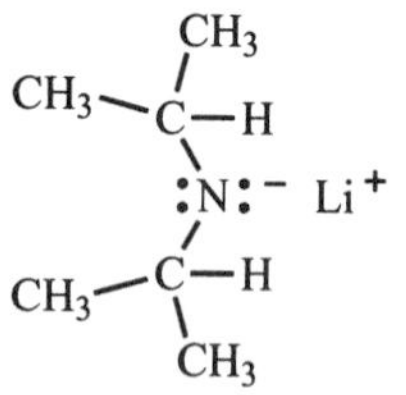

FIGURE 11.16:
Lithium
diisopropylamide.

Lithium diisopropylamide is commonly made by reacting N,N–diisopropylamine with an even stronger base, *n*–butyl lithium. This reaction is shown in equation 11.9. LDA has considerable ionic character. This makes the two pairs of non–bonding electrons on the nitrogen atom quite basic.

$$((CH_3)_2CH)_2NH + (C_4H_9)^-Li^+ \longrightarrow ((CH_3)_2CH)_2N^-Li^+ + C_4H_{10} \qquad \text{EQ. 11.9}$$

When the ketone is asymmetric, one additional question to be answered is: "Which product will be formed?" Let us consider the alkylation of the asymmetric ketone, 2–methylcyclohexanone. When this ketone is treated with methyl iodide in basic solution, two possible products could form: 2,2–dimethylcyclohexanone and 2,6–dimethylcyclohexanone. Figure 11.17 outlines the steps in the reactions that form each of these products. The enolate ion and final product in the sequence leading to 2,2-dimethylcyclohexanone and 2,6-dimethylcyclohexanone are labeled 'TE' and 'TP' (T = 'thermal') respectively and those in the sequence leading to the 2,6 isomer are labeled 'KE' and 'KP' (K = 'kinetic'). Note especially that the final product is determined by which enolate ion is formed. The product marked 'TP' is formed after removal of the hydrogen atom on carbon atom C2. One of the resonance forms of this has a carbon-carbon double bond between carbon atoms C1 and C2. To form the product marked 'KP' the hydrogen atom on carbon atom C6 must be removed. A resonance form of this species has the double bond between carbon atoms C1 and C6.

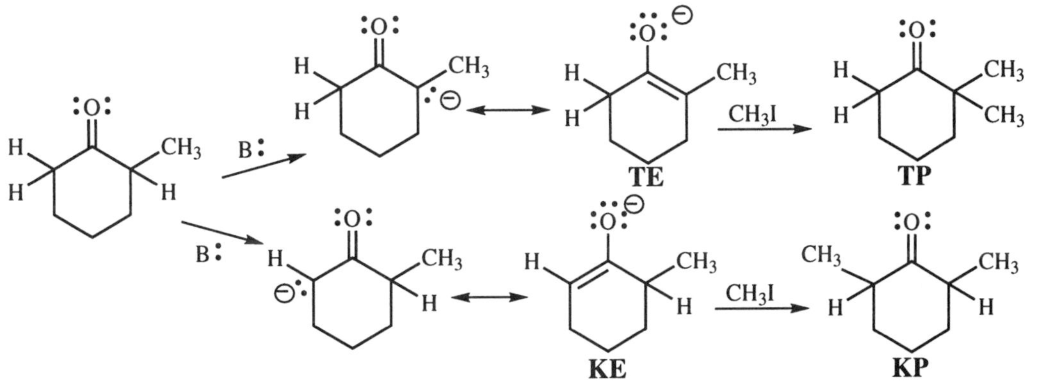

FIGURE 11.17: Alkylation of 2-methylcyclohexanone.

Which enolate ion is formed depends upon the conditions. The more thermodynamically stable enolate ion, the one marked 'TE', is the one that has the higher degree of substitution. The reasons for the greater stability of this enolate ion are the same as those that favor the production of the more substituted alkene, *i.e.*, there is more opportunity to "spread out" the charge over a larger number of atoms when alkyl groups are substituted for hydrogen atoms. Recall that alkyl groups are electron donating and thus contribute to the electron density of the $\pi$ bond. However, because this enolate ion is more highly substituted, it is also more sterically hindered — thus preventing approach of the strong base that leads to removal of the proton. This kinetically favors the production of the less highly substituted enolate ion, the one marked 'KE'. If the formation of the enolate ion is the slow step in the reaction mechanism, subsequent reaction of the enolate ion with the alkyl halide occurs before the enolate ion can rearrange into the thermodynamically more stable form. The outcome is, of course, the formation of 'KP' as the product. Figure 11.18 summarizes the relative stability of the enolate ions for this molecule.

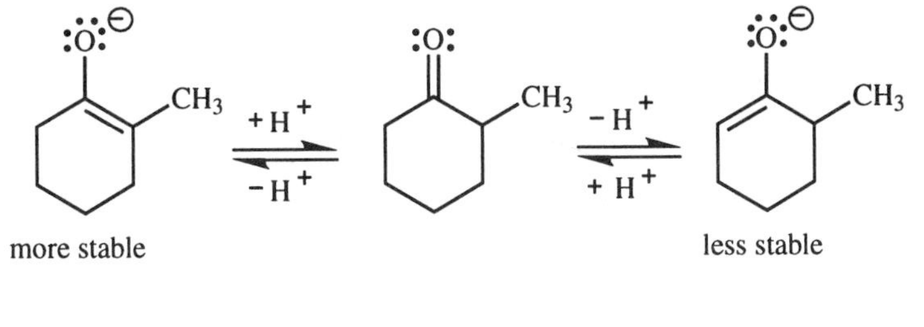

more stable                          less stable

FIGURE 11.18: Relative enolate ion stability.

Even though the less substituted enolate ion is usually the one that is kinetically favored and, therefore, the one which forms more quickly, if enough time is allowed for equilibration, the more thermodynamically favored enolate ion is the one that predominates in solution. If we understand and can control the conditions that favor each of the enolate ions, then we can control which product will predominate in the final mixture.

Since the intermediate marked 'KE' is the one formed first and the intermediate marked 'TE' is only formed after equilibration, we must avoid conditions that allow equilibration before reaction of the enolate with the alkyl halide. In general a decrease in temperature slows the equilibration of the two forms more than it slows reaction of the enolate ion with the alkyl halide. Hence, performing the reaction in an ice bath at 0 °C or a dry ice — solvent (usually acetone) bath at -78 °C will favor the production of 2,6—dimethylcyclohexanone (KP). The presence of unreacted methylcyclohexanone also favors equilibration. To reduce the concentration of unreacted methylcyclohexanone in the mixture we use larger concentrations of a stronger base so that the equilibrium conditions of the ionization reaction more strongly favor the enolate ion.

In order to favor the thermodynamically controlled product, of course, we use the opposite conditions — *i.e.*, use higher temperatures and smaller amounts of a weaker base to ensure the presence of as large a quantity of unreacted ketone as possible. This is summarized in Table 11.1.

TABLE 11.1 REACTION CONDITIONS SUMMARY

TABLE 11.1

| STRUCTURE OF PRODUCT | KINETIC | THERMODYNAMIC |
|---|---|---|
| | | |
| AMOUNT OF BASE | One molar equivalent | Catalytic amounts |
| TYPE OF BASE | Stronger base (LDA) | Weaker base $((CH_3CH_2)_3N)$ |
| TEMPERATURE | Cold (0°C to-78 °C) | Warm (>25 °C) |

That different products were formed under different conditions was attributed to a difference in the structure of the intermediate enolate ions formed during the reaction. These differences were rationalized on the basis of kinetic *vs.* thermodynamic control. However, because it is impossible to isolate the enolate intermediate, to a certain extent the explanation offered above was simply informed conjecture. Fortunately, other experiments utilizing certain silicon compounds, have permitted the isolation of two different products that provide direct confirmation of the formation of different intermediates under different reaction conditions.

These products are formed by reacting 2–methylcyclohexanone with trimethylchlorosilane $((CH_3)_3SiCl)$ under two different sets of conditions —

one set that we identified above as favoring the thermodynamic product and the other set that we identified as favoring the kinetic product. This type of evidence is available because the silicon–oxygen bond is much more stable than is the silicon–carbon bond. As a result, when the trimethylchlorosilane reacts with the enolate ion, the silicon adds to the oxygen atom rather than the carbon atom. This, then, prevents the product from undergoing tautomerism back to the ketone form and "freezes" the carbon–carbon double bond into the position it occupied in the enolate intermediate. The position can then be easily identified by standard means of determining structure.

To restate the conditions used before as they apply to this reaction, since triethylamine is a weak base (its conjugate acid has $pK_a$ ~10), deprotonation of the ketone is less complete. This allows time for equilibrium to be reached before further reaction occurs. Equilibration is also helped by running the reaction at higher temperatures. On the other hand, the use of the very strong base, LDA means that, even at low temperatures, a very high concentration of the most rapidly formed enolate ion exists. Subsequent reaction is quite rapid and the kinetic product is the one that forms. Figure 11.19 shows the chemical equation that leads to the different products. Table 11.2 summarizes the product distribution ratios under the different conditions.

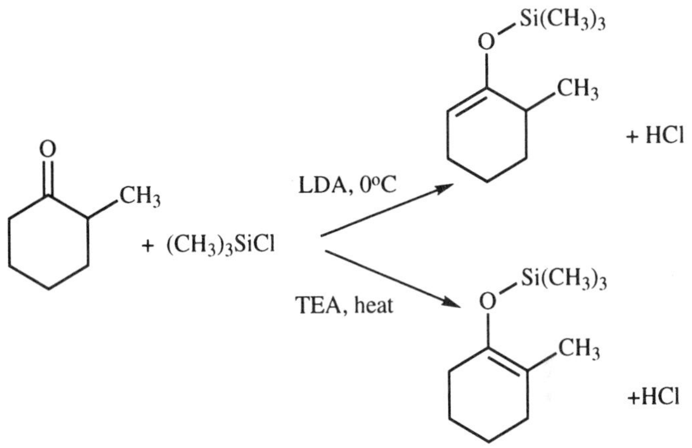

FIGURE 11.19: Reaction of methylcyclohexanone with trimethylchlorosilane.

**TABLE 11.2** PRODUCT DISTRIBUTION RATIOS

| STRUCTURE OF PRODUCT | 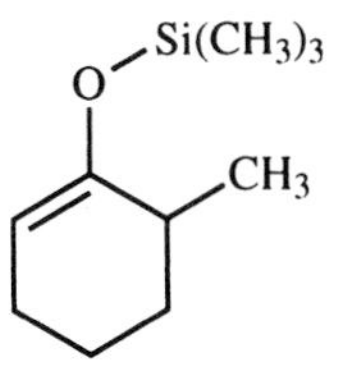 | 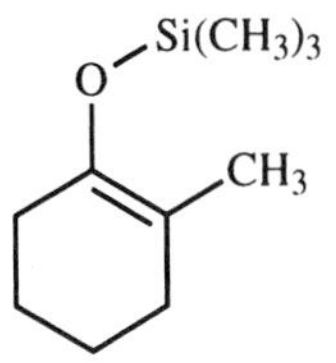 |
| --- | --- | --- |
| Triethylamine, heat | 22 | 78 |
| LDA, 0°C | 99 | 1 |

**CONCEPT CHECK**

Show how the following compounds can be prepared from the indicated starting materials

a.

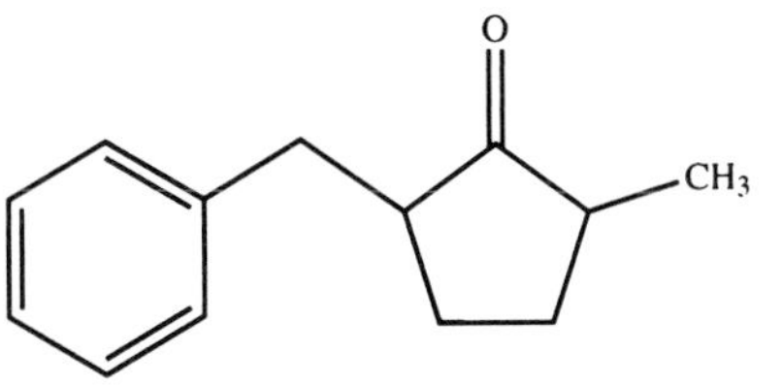

from

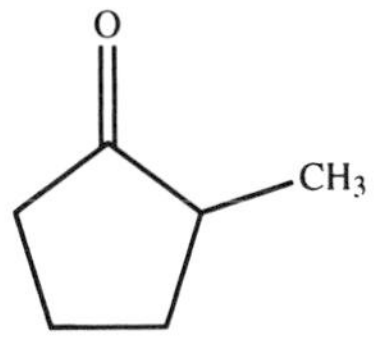

b.

from

In (a) we will need to use benzyl bromide ($C_6H_5CH_2Br$) as the other reactant. We should note that this is the kinetically controlled product since removal of the proton from the secondary carbon atom should be thermodynamically more favorable. Therefore we want to use large amounts of a very strong base to encourage rapid and complete formation of the enolate ion of the ketone and cold conditions to discourage rearrangement of the enolate ion initially formed. We should, therefore, treat the 2–methylcyclopentanone with an equimolar amount of LDA at dry ice temperatures.

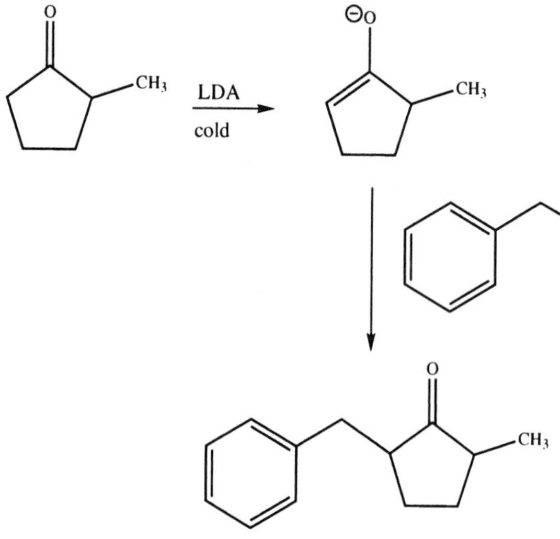

EQ. 11.10

In (b) the product is the thermodynamically controlled product formed by reaction of ethyl bromide with 2–methylcyclopentanone. Here we need the opposite conditions as in (a). We shall therefore, run the reaction at slightly elevated temparatures using only a catalytic amount of a weaker base, triethylamine.

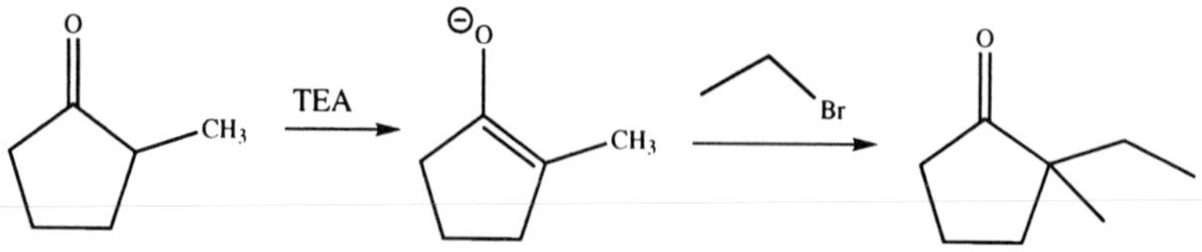

EQ. 11.11

## 11.9 ALKYLATION OF ESTERS

Even though all previous examples of enol and enolate ion formation have used ketones as the carbonyl compound, other carbonyl–containing compounds are also capable of undergoing reactions after formation of an enolate ion interme-diate. The most important of these reactions are alkylation reactions of esters. Let us illustrate using the ester, methyl cyclohexanoate.

Methyl–1–*n*–heptyl–cyclohexanecarboxylate is formed by first treating methyl cyclohexanoate with one mole equivalent of lithium diisopropylamide (LDA) in tetrahydrofuran at dry ice temperatures. After a brief reaction time to allow the deprotonation reaction to go to completion, slightly more than one mole equivalent of *n*–heptyl bromide is added to the solution and the dry ice bath is removed to allow the mixture to warm to room temperature. The reaction to form methyl–1–*n*–heptyl–cyclohexanecarboxylate is represented in equation 11.12. The student is left to work out the details of the reaction sequence as an exercise.

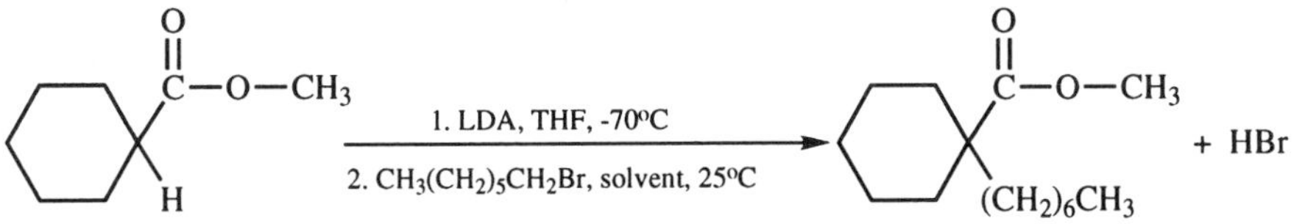

EQ. 11.12

**Lactones** can also be alkylated, provided that there is at least one proton on the alpha carbon atom. As an example, consider the addition of an *n*–butyl group to the lactone of 4–hydroxybutanoic acid. Equation 11.13 shows the course of this two step reaction. Again the student is left to work out the steps of the reaction as an exercise.

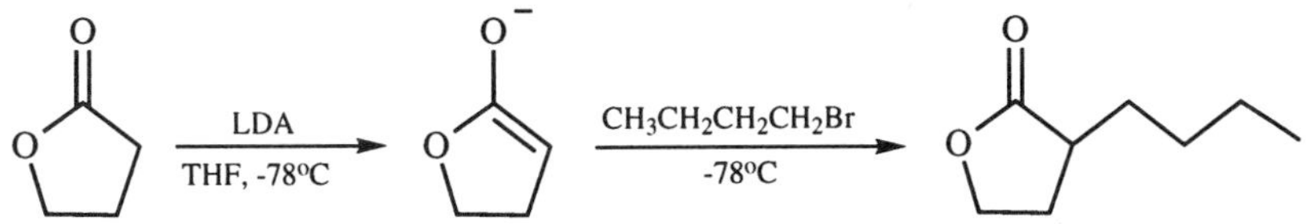

EQ. 11.13

## 11.10 β-DICARBONYLS

In section 11.3 we briefly mentioned that β–dicarbonyl compounds had the two carbonyl groups situated in a way that provided extra stability to the enol form. There are several of these compounds commonly used in synthetic applications because of the extra stability that the beta carbonyl group gives to the interme-diate structures. Table 11.3 gives the structures, IUPAC names, and common names of a number of these compounds. The last structure in the table is of a β–dicarbonyl compound involved in the fatty–acid degradation pathway.

**TABLE 11.3** COMMONLY ENCOUNTERED β–DICARBONYL COMPOUNDS

TABLE 11.3

| STRUCTURE | IUPAC NAME | COMMON NAME |
|---|---|---|
| CH₃–C(=O)–CH(H)–C(=O)–H | 3–oxobutanal | acetylacetaldehyde |
| CH₃–C(=O)–CH(H)–C(=O)–CH₃ | 2,4–pentanedione | acetylacetone |
| CH₃–C(=O)–CH(H)–C(=O)–O–CH₃ | methyl–4–oxobutanoate | methyl acetoacetate |
| CH₃–O–C(=O)–CH(H)–C(=O)–O–CH₃ | dimethyl propanedioate | dimethyl malonate |
| HO–C(=O)–CH(H)–C(=O)–OH | propanedioic acid | malonic acid |
| (CH₃)(⁻O–C=O)CH–C(=O)–S–CoA | — | methylmalonyl–CoA |

A number of these dicarbonyl compounds are biochemically important. They play key roles in a number of biochemical pathways. Both the **Krebs cycle** and the fatty acid cycle involve dicarbonyl compounds at different places. Before discussing these biochemical cycles, let us first look at a number of chemical reactions in which these compounds participate in the laboratory. As we will later see (chapter 16), biochemical reactions involving these compounds are quite similar to the more traditional chemical reactions. In many cases the only difference is the presence of an enzyme as the catalyst for the reaction.

# 11.11 ALKYLATION OF β-KETO ESTERS AND β-DIKETONES

β–dicarbonyl compounds, specifically β–diketones and β–ketoesters, undergo alkylation reactions that are similar to those of the monofunctional ketones and esters that we previously discussed. The principal difference is the ease with which one of the hydrogen atoms located on the carbon between the two carbonyl groups (the α–carbon to both carbonyls) can be removed. As we discussed in section 11.3, extra stability is conferred on the enol structure because of its ability to form a six–membered ring (figure 11.4). This additional stability means that the pK$_a$ of this hydrogen atom is considerably lower (*i.e.*, the hydrogen atom is much more acidic) than is the pK$_a$ of the a hydrogen atom on a simple carbonyl compound. Therefore weaker bases such as potassium carbonate can be used to remove the proton and provide the enolate ion intermediates that have been involved in many of the reactions. Indeed, alkylation reactions of the β–dicarbonyls were known long before similar reactions of the simple carbonyls primarily because the extremely strong bases, such as LDA, required for deprotonation of simple carbonyl compounds had not yet been discovered.

One example, the alkylation of ethyl acetoacetate, is shown in figure 11.20. Note especially the use of the weak base carbonate ion (put into the solution as potassium carbonate) to form the enolate ion that will attack the electrophilic carbon atom on the propyl bromide molecule. Note also that, even after the first proton has been removed and alkylation has occurred, the second proton is still sufficiently acidic that it, too, can be removed and a second propyl group substituted in its place.

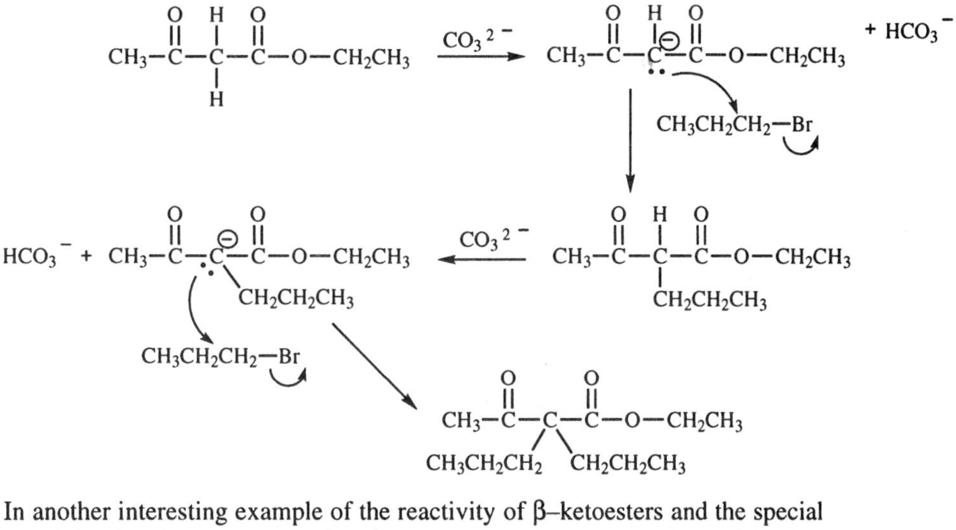

FIGURE 11.20:
Di-alkylation of ethylacetoacetate.

In another interesting example of the reactivity of β–ketoesters and the special reactivity that the two carbonyl groups convey to the compound, consider the ketoester ethyl–2–oxo–cyclohexanecarboxylate (figure 11.20). The alpha carbon atom of this molecule contains only one hydrogen atom. Because of the two carbonyl groups this hydrogen atom is sufficiently acidic that it can be removed using the base sodium ethoxide (equation 11.14). The resulting carbanion can then be methylated using methyl iodide (equation 11.15). If the resulting compound, ethyl–1–methyl–2–oxo–cyclohexanecarboxylate, is then treated with the strong base sodium ethoxide, the ethoxide ion will add to the ketone carbon atom to form an ion that then rearranges with the opening of the cyclohexane ring (equation 11.16).

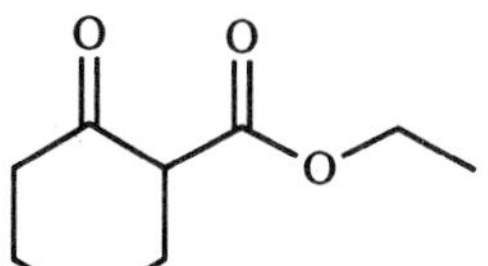

FIGURE 11.21:
Structure of ethyl-2-oxocyclohexanecar-boxylate.

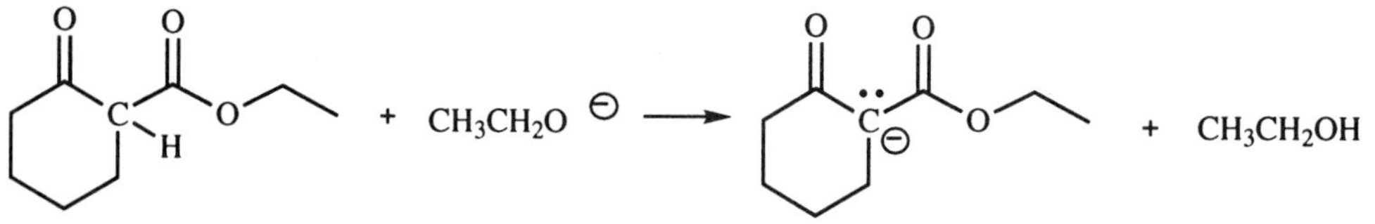

EQ. 11.14

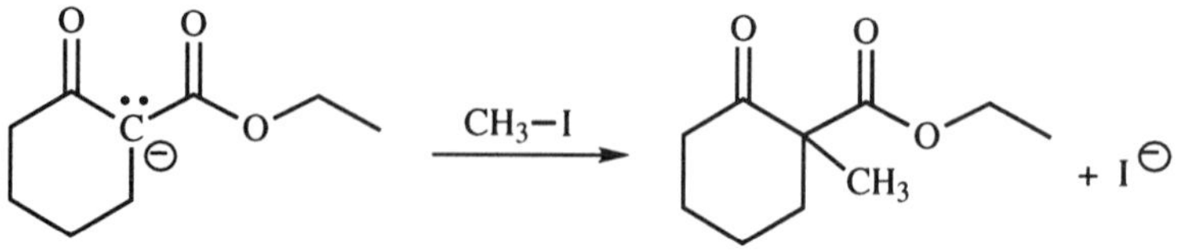

EQ. 11.15

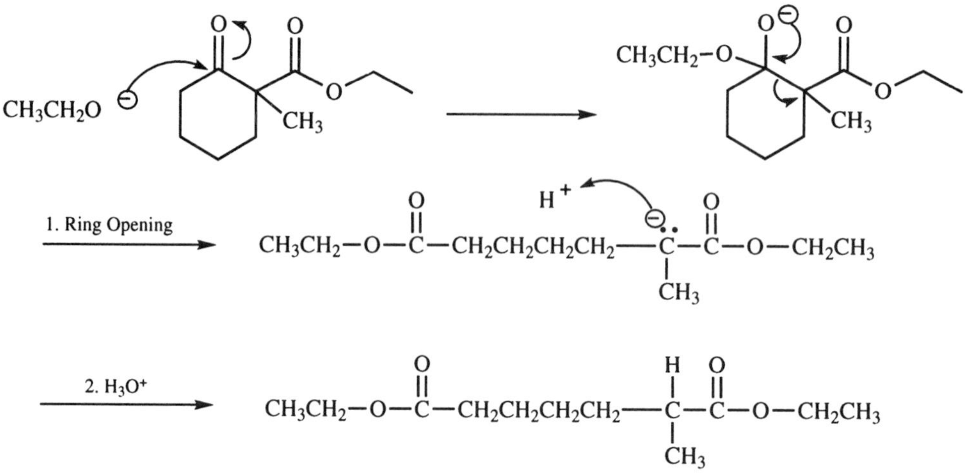

EQ. 11.16

1. Ring Opening

$$CH_3CH_2-O-\overset{O}{\overset{\|}{C}}-CH_2CH_2CH_2CH_2-\overset{\ominus}{\underset{CH_3}{\overset{|}{C}}}-\overset{O}{\overset{\|}{C}}-O-CH_2CH_3$$

2. H₃O⁺

$$CH_3CH_2-O-\overset{O}{\overset{\|}{C}}-CH_2CH_2CH_2CH_2-\overset{H}{\underset{CH_3}{\overset{|}{C}}}-\overset{O}{\overset{\|}{C}}-O-CH_2CH_3$$

## 11.12   DECARBOXYLATION OF β-KETOACIDS

The product from the reaction illustrated in figure 11.20 is an ester. Hydrolysis of this ester proceeds easily under normal ester hydrolysis conditions (section 7.5). However, unless one is extremely careful with conditions, the product that is obtained is not the acid that would result from simple hydrolysis of the ester bond. Under many conditions, the product formed is the ketone: 3–ethyl–2–hexanone. The acid has undergone a reaction known as **decarboxylation**.

It is quite difficult to prevent decarboxylation when another carbonyl group is beta to the carboxyl group. In the decarboxylation reaction, the acid simply decomposes to carbon dioxide and a ketone. The relevant portion of the reactants and products for this reaction is shown as equation 11.17.

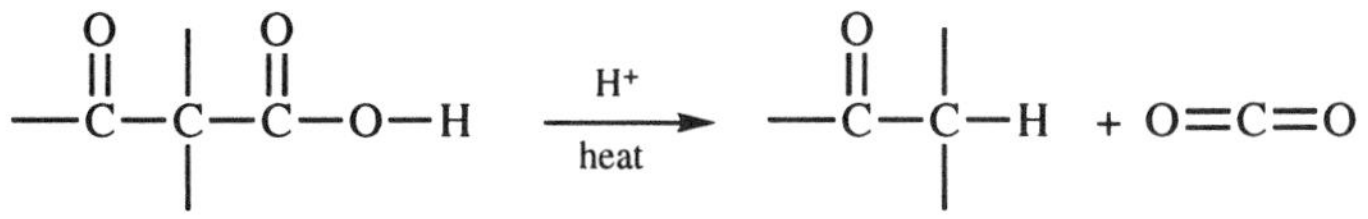

EQ. 11.17

$$\underset{\text{heat}}{\overset{H^+}{\longrightarrow}}$$

To understand how this reaction occurs, we need to recall that 5-or 6-membered ring structures normally convey additional stability to a species capable of forming them. Earlier in this chapter (section 11.3) we saw that the formation of a six–membered ring structure gave enough additional stability to the enol form of a β–diketone that it was possible to easily determine its presence. Here the formation of a six–membered, hydrogen–bonded ring stabilizes the transition state for this reaction. Note the formation of this intermediate in the reaction scheme for a generic decarboxylation reaction as shown in figure 11.22. Loss of carbon dioxide by the cyclic intermediate produces the enol form of the final ketone.

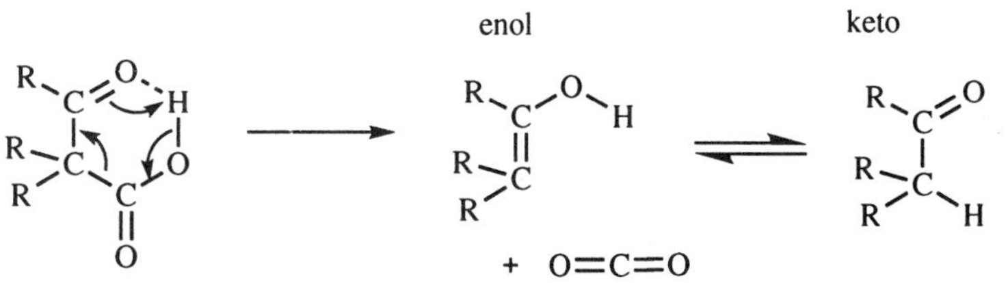

enol    keto

$+\ O{=}C{=}O$

FIGURE 11.22:
Decarboxylation
reaction scheme.

A specific example is the decarboxylation of 3-oxo-hexanoic acid shown in equation 11.18. This reaction occurs so easily that it is often difficult to isolate β–ketoacids. Isolation of the acid always requires that the solution be made at least slightly acidic to force the acid into its protonated form. Unfortunately, unless the solution is kept quite cold, acidic solutions of β–ketoacids often spontaneously decarboxylate.

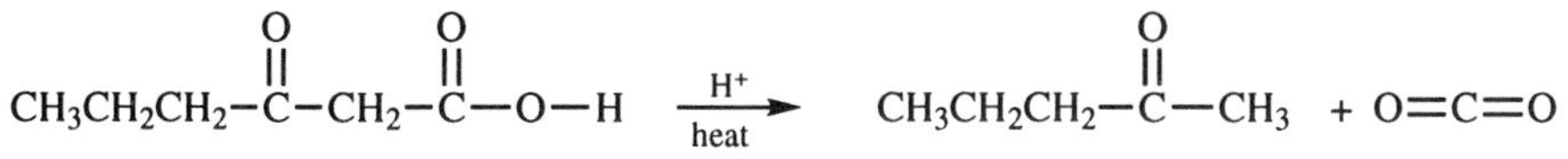

$$CH_3CH_2CH_2{-}\overset{O}{\overset{\|}{C}}{-}CH_2{-}\overset{O}{\overset{\|}{C}}{-}O{-}H \underset{\text{heat}}{\overset{H^+}{\longrightarrow}} CH_3CH_2CH_2{-}\overset{O}{\overset{\|}{C}}{-}CH_3 \ +\ O{=}C{=}O$$

EQ. 11.18

## 11.13  FORMATION OF DIANIONS

If both of the carbon atoms surrounding at least one of the carbonyl groups of a β–diketone are bonded to at least one hydrogen atom, then it is possible to form a doubly charged ion. One such example is 4–phenyl–2,4–dioxobutane whose keto–enol tautomeric equilibrium reaction is represented in figure 11.23. In this molecule, as in 2,4–pentanedione that we earlier discussed, the enol form is much more stable because of the intramolecular hydrogen–bonded cyclic structure of the enol.

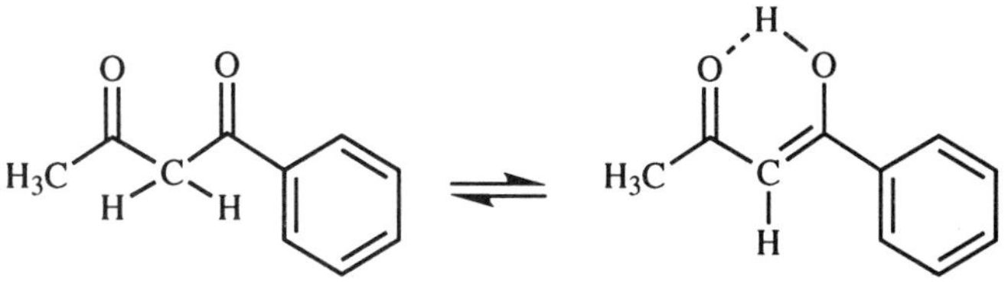

FIGURE 11.23:
Structure of 4-phenyl-
2,4-dioxobutane.

This molecule is capable of acting as a diprotic base. Although two different enolate ions could be formed — one with the charge carried by either of the oxygen atoms, the more stable enolate ion is the one formed by the carbonyl nearest the phenyl group. Formation of this enolate ion produces a conjugated system of double bonds that allows the greatest amount of charge delocalization.

Now, if a second equivalent of a very strong base (*e.g.*, LDA) is present, a second proton can be removed. This proton ultimately comes from the other carbon atom and exhibits a larger $pK_a$ (~ 20) because it is being removed from a species that already carries a negative charge.

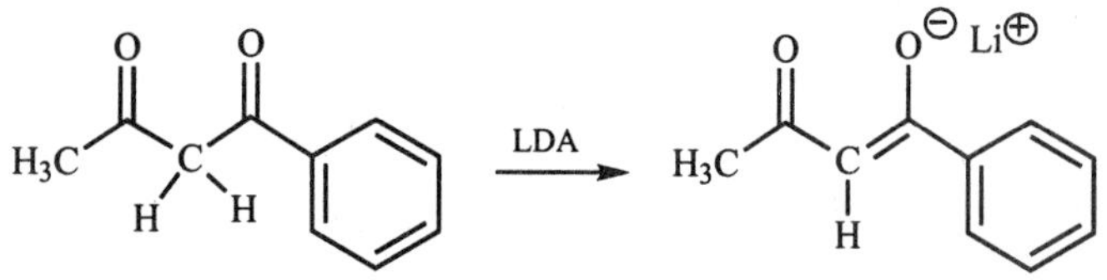

EQ. 11.19

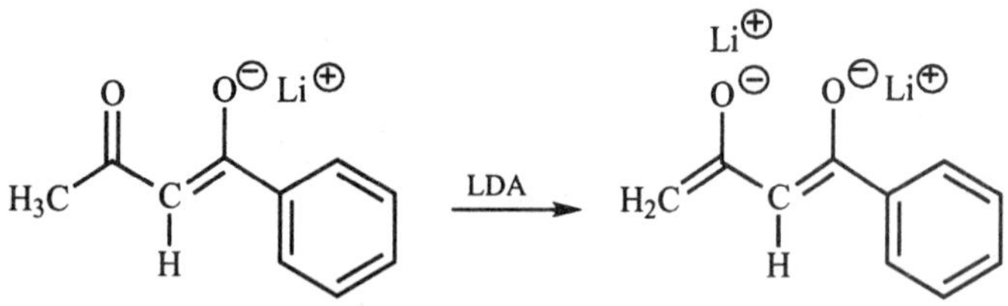

EQ. 11.20

If the dianion product in equation 11.20 is now treated with an alkyl halide, the alkyl halide can participate in an electrophilic reaction. The result is that the alkyl group is added to the former methyl group of the diketone. The course of this reaction is easiest to rationalize if it is remembered that the enolate ion is in resonance with the keto form. It is the keto form of this anion that participates in the reaction as shown in equation 11.21. Alkylation occurs at the terminal carbon rather than at the anionic oxygen site, because the carbon site is the more basic, and hence, the more nucleophilic of the two sites, because it is the less hindered of the sites at which substitution can occur, and because the nucleophilicity of the oxygen is reduced by its interaction with the lithium ion. It should, however, be noted that it is possible to produce substitution at the oxygen atom under the right set of conditions.

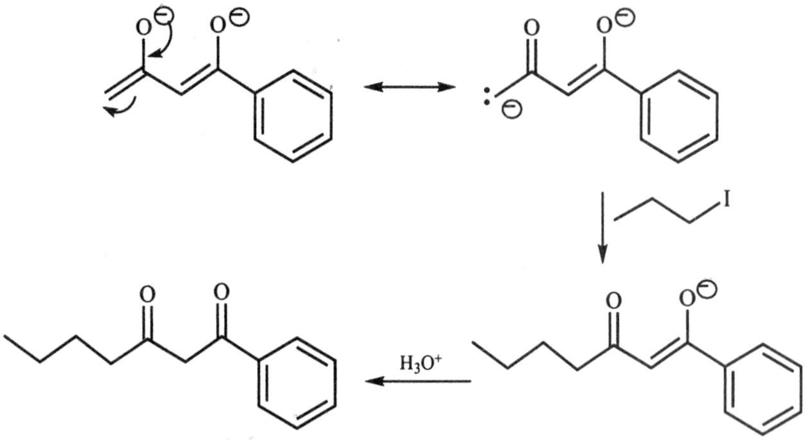

EQ. 11.21

## 11.14   CONDENSATION WITH ALDEHYDES AND KETONES: THE ALDOL CONDENSATION

In one very important class of reactions the enolate ion reacts with another carbonyl compound. These reactions are often called condensation reactions because, in most situations, two molecules "condense" to form a larger molecule. The simplest such reaction is that of acetaldehyde reacting with itself in the presence of sodium hydroxide. The name "aldol" has been given to the reaction type and this reaction is commonly known as the **aldol condensation**.

When acetaldehyde reacts with itself, the product is 3–hydroxybutanal — a compound that has both an aldehyde functional group (the "ald") and an alcohol functional group (the "ol"). Indeed, the common name for the compound formed in this particular reaction is aldol. Aldol is, however, but one of a variety of related compounds that can be made in the same way. The overall reaction for the aldol condensation forming 3–hydroxybutanal is shown in equation 11.22.

EQ. 11.22

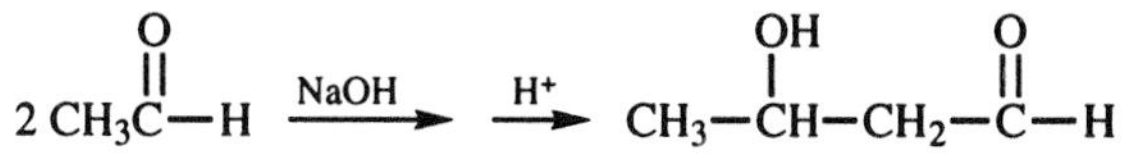

$$2\ CH_3\overset{\displaystyle O}{\overset{\|}{C}}-H \xrightarrow{\ NaOH\ } \xrightarrow{\ H^+\ } CH_3-\overset{\displaystyle OH}{\underset{|}{CH}}-CH_2-\overset{\displaystyle O}{\overset{\|}{C}}-H$$

acetaldehyde                3-hydroxybutanal

A more detailed look at the mechanism for this reaction is given in figure 11.24. Note, especially, the action of the hydroxide ion. First it is used to remove one of the alpha hydrogen atoms to form the enolate ion. After the subsequent reaction of the enolate ion with another molecule of acetaldehyde, the hydroxide ion is regenerated when the alkoxide ion removes a proton from a water molecule.

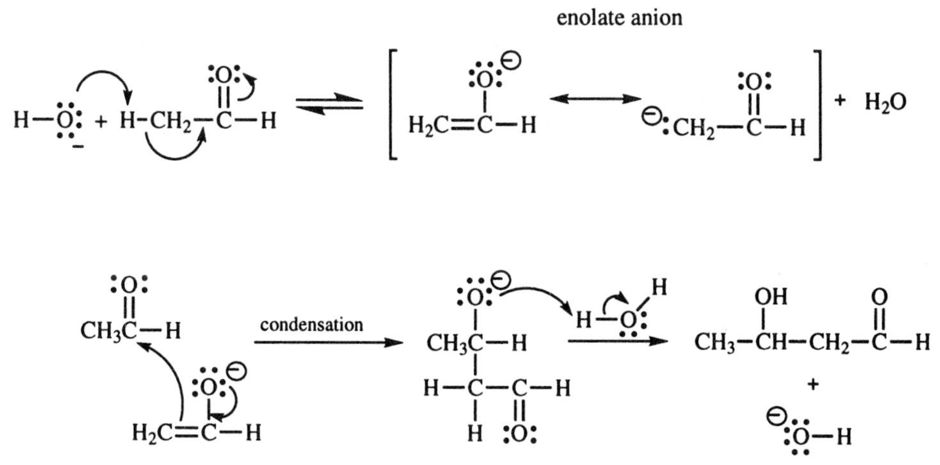

enolate anion

FIGURE 11.24:
Mechanism of the aldol condensation reaction.

If the reaction mixture contains more than one aldehyde with an alpha hydrogen atom, a mixture of products is possible. For instance, a mixture of ethanal and propanal, reacting in the presence of sodium hydroxide, will produce the mixture of four aldol products shown in table 11.4. In practice, this mixture would be extremely difficult to separate. Therefore, under most circumstances, it is impractical to use mixtures of aldehydes in this reaction.

The exception to this rule occurs when one of the compounds has no alpha hydrogen atoms and can, therefore, not form an enolate ion. For example, the reaction between benzaldehyde and propanal is shown as equation 11.23.

EQ. 11.23

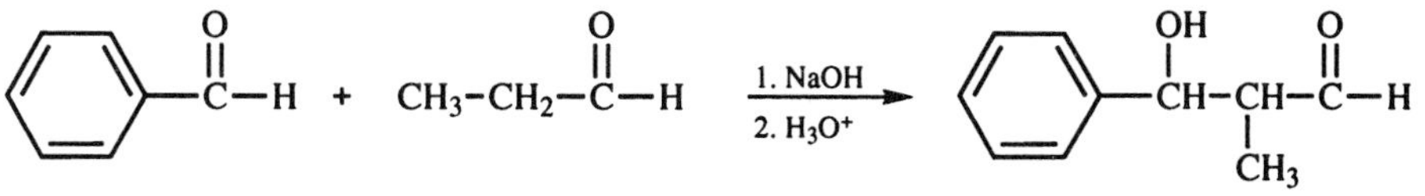

TABLE 11.4

**TABLE 11.4** ALDOL CONDENSATION PRODUCTS

| PRODUCT | NAME | ENOLATE ION DERIVED FROM | ALDEHYDE |
|---|---|---|---|
| $\underset{\displaystyle CH_3-CH-CH_2-C-H}{\overset{\displaystyle OH \qquad O}{}}$ | 3–hydroxybutanal | Ethanal | Ethanal |
| $\underset{\displaystyle CH_3-CH-CH-C-H}{\overset{\displaystyle OH \qquad O}{}}$ CH₃ | 3–hydroxy–2–methylbutanal | Propanal | Ethanal |
| $CH_3-CH_2-CH-CH_2-C-H$ (OH, O) | 3–hydroxy pentanal | Ethanal | Propanal |
| $CH_3-CH_2-CH-CH-C-H$ (OH, O) CH₃ | 3–hydroxy–2–methylpentanal | Propanal | Propanal |

However, we would expect 3–hydroxy–2–methylpentanal, formed by the condensation of propanal with itself, to be formed in relatively large amounts. Is it possible to find conditions under which we could minimize the formation of this, undesired, product, and maximize the yield of the desired product, 3–hydroxy–2–methyl–3–phenylpropanal?

To minimize formation of the self–condensation product, we need to have the enolate ion react only with benzaldehyde and not with excess propanal. This means that we wish to form the enolate ion under conditions that minimize the amount of excess propanal present. As we have seen before, we can do this by slowly adding the propanal to a solution containing at least one mole equivalent of the strong base. This ensures that there is a minimum of the unreacted propanal present. Only after the formation of the enolate ion is complete, is benzaldehyde added to the solution. When enough time has been given for reaction of the enolate ion with the benzaldehyde, acid is added to neutralize any excess base and protonate the product.

Good yields can also be obtained by condensing aldehydes and ketones — provided that the aldehyde has no alpha hydrogen atoms. For example, condensation of acetone with benzaldehyde produces the product shown in equation 11.24. This reaction is somewhat easier to accomplish without concern for multiple products because ketones are not as reactive in the aldol condensation reaction as are aldehydes. However, this particular reaction produces significant quantities of dibenzylacetone by reaction of a second benzaldehyde molecule with a proton on the terminal methyl group of the initial product.

EQ. 11.24

## CONCEPT CHECK

The following compounds were made using the aldol condensation.
Determine what carbonyl compounds served as the starting materials
for each compound.

a.

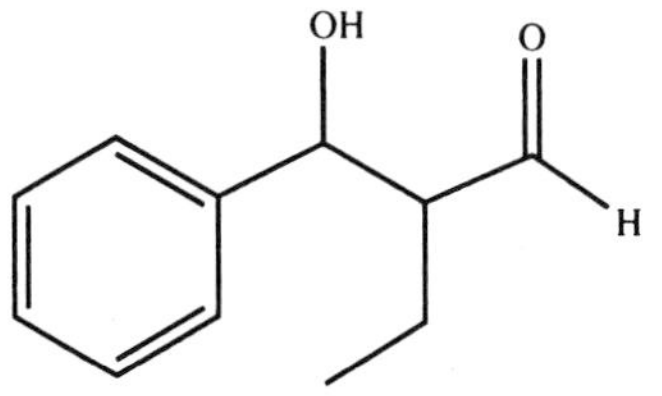

b.

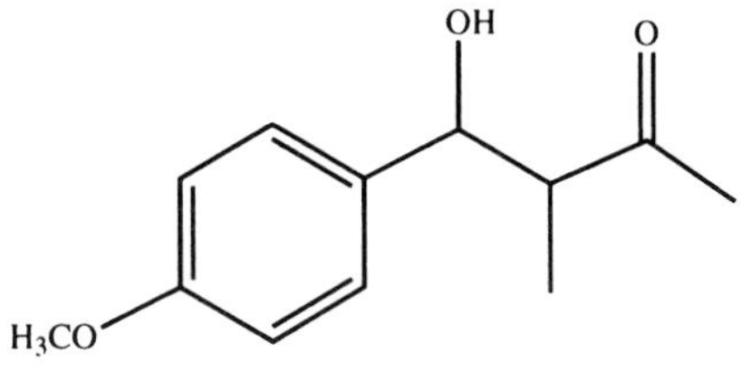

c.

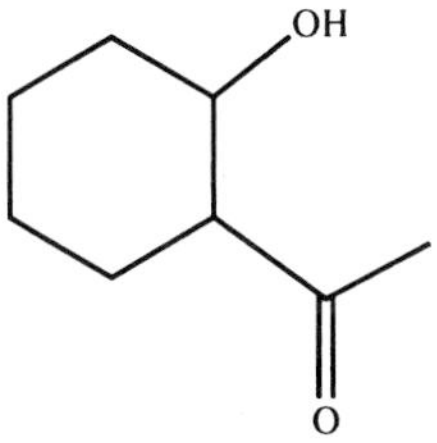

Product (a) requires the condensation of benzaldehyde and butanal.

Product (b) is made from p–methoxybenzaldehyde and 2–butanone.

Product (c) is made by an internal aldol condensation of 7–oxooctanal.

## 11.15 RETROALDOL

The aldol condensation reaction is a reversible reaction. For this reason, it is
possible to break down beta–hydroxy ketones or aldehydes into two simpler
carbonyl compounds. The reaction by which this happens is termed a "retroaldol" reaction. The simplest case of such a reaction, the reverse of the reaction
forming 3–hydroxybutanal, is shown below as equation 11.25.

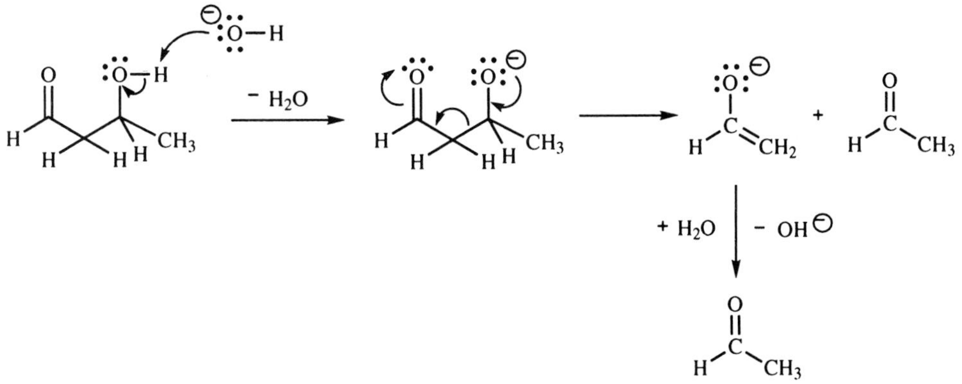

EQ. 11.25

The first step in this reaction is a simple deprotonation of the hydroxyl group to form an alkoxide ion. This ion can rearrange its electrons as shown in the equation. Such a rearrangement results in the breaking of the carbon–carbon bond and formation of acetaldehyde and the enolate ion that initially reacted to form the aldol. Finally, reprotonation and tautomerization of the enol forms the second molecule of acetaldehyde.

The aldol condensation and, therefore, the retroaldol reaction, are quite important in biochemical conversions involving the sugars. For instance, the breakdown of sucrose in the metabolic scheme involves, at one point in the metabolic pathway, the degradation of fructose–6–phosphate into two 3–carbon compounds through a retroaldol reaction.

Figure 11.25 shows the first few steps in the biochemical degradation of glucose in the metabolism of that important carbohydrate. In the first step shown, the open chain form of glucose –6–phosphate is isomerized into fructose–6–phosphate with the help of the enzyme isomerase. In the next step, a second phosphate group is added by the enzyme phosphofructokinase. In the final step shown in this figure, the enzyme aldolase catalyzes a retroaldol reaction that breaks the compound into the phosphate derivatives of dihydroxypropanal and dihydroxyacetone. The mechanistic details of this process are covered in Section 16.4.

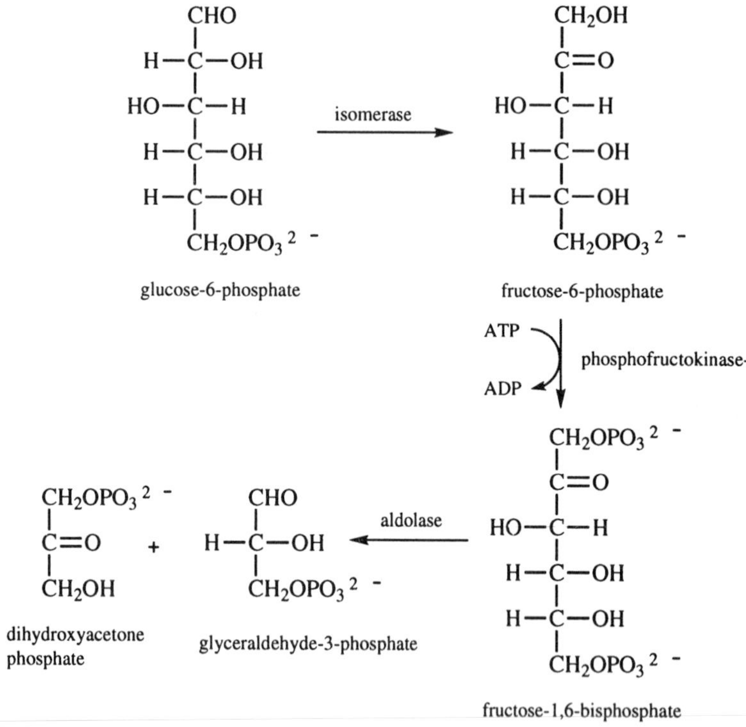

FIGURE 11.25: Degradation of glucose-6-phosphate.

## 11.16 DEHYDRATION OF ALDOLS

Another reaction that can occur under the conditions used for the aldol condensation is dehydration of the aldol product. Some β–hydroxyketones are especially susceptible to dehydration under some of the conditions commonly used for aldol condensation if the aldehyde is unreactive or sterically hindered. Dehydraion is especially common when it results in formation of highly conjugated products. One particular example is shown in figure 11.26. This reaction provides a good synthetic route for α,β unsaturated ketones.

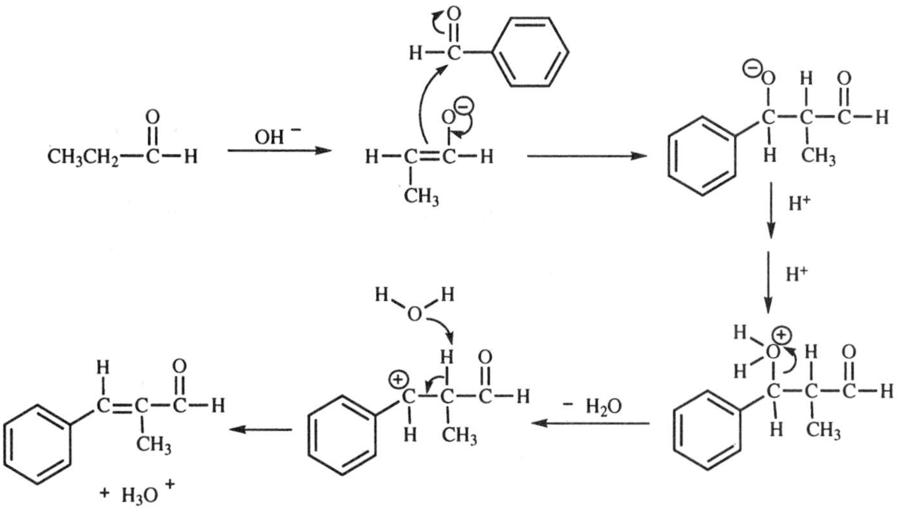


FIGURE 11.26: Dehydration of a β-hydroxyaldehyde.

## 11.17 CONDENSATION WITH ESTERS: CLAISEN CONDENSATION

Enolate ions formed from esters with available alpha hydrogen atoms are also known to undergo self–condensation to form β–keto esters. The best known of these compounds is ethyl acetoacetate — a compound that is very useful as an intermediate in the synthesis of other compounds. Equation 11.26 represents the overall reaction that occurs when ethyl acetate is first treated with sodium ethoxide and the solution is later neutralized with acetic acid. Figure 11.27 shows the steps in this reaction. Note that this reaction is similar to many other reactions we have already studied in this chapter. It is simply the reaction of an electrophile (the carbon atom of a carbonyl group) with a nucleophile (the enolate ion). Many similar esters undergo this type of reaction that is now known as the **Claisen condensation**.

ethyl acetoacetate

EQ. 11.26

$$2\ CH_3{-}\overset{:O:}{\overset{\|}{C}}{-}O{-}CH_2CH_3 \xrightarrow[2.\ CH_3COOH]{1.\ CH_3CH_2ONa} CH_3{-}\overset{:O:}{\overset{\|}{C}}{-}CH_2{-}\overset{:O:}{\overset{\|}{C}}{-}O{-}CH_2CH_3$$

$$H-XH_2-\overset{\overset{\displaystyle O}{\|}}{X}-O-XH_2XH_3 \;\;\rightleftharpoons\;\; \left[ \;\ominus XH_2-\overset{\overset{\displaystyle O}{\|}}{X}-O-XH_2XH_3 \;\longleftrightarrow\; \overset{H}{\underset{H}{>}}X=X-O-XH_2XH_3 \;\right]$$

$$\overset{\ominus OXH_2XH_3}{}$$

$$XH_3-\overset{\overset{\displaystyle O}{\|}}{X}-XH_2-\overset{\overset{\displaystyle O}{\|}}{X}-O-XH_2XH_3 \;\;\underset{-\;XH_3XH_2O^{\ominus}}{\xleftarrow{\hspace{2cm}}}\;\; XH_3-\overset{\overset{\displaystyle O^{\ominus}}{|}}{\underset{O-XH_2XH_3}{X}}-XH_2-\overset{\overset{\displaystyle O}{\|}}{X}-O-XH_2XH_3$$

$$XH_3XH_2O^{\ominus}$$

$$XH_3-\overset{\overset{\displaystyle O}{\|}}{X}-\underset{\ominus}{XH}-\overset{\overset{\displaystyle O}{\|}}{X}-O-XH_2XH_3 \;\;\rightleftharpoons\;\; XH_3-\overset{\overset{\displaystyle \ominus O}{|}}{X}=XH-\overset{\overset{\displaystyle O}{\|}}{X}-O-XH_2XH_3$$

The alpha hydrogen of the β–keto ester initially formed in the reaction is sufficiently acidic that it reacts with the base to produce the corresponding enolate
ion.  For this reason, one mole equivalent of sodium ethoxide must be used in
order to force the reaction to completion.  Once the reaction is complete, the
enolate ion and any excess sodium ethoxide is neutralized with a stronger acid.

Claisen condensations of two different esters are synthetically unimportant
because of the mixture of different products that can be formed.  However, if
one of the esters contains no alpha hydrogen atoms, then reaction with another
ester with at least one alpha hydrogen often does lead to a good yield of the
cross product.

A particularly interesting application of the Claisen condensation occurs if a
difunctional ester containing four or five methylene groups between the ester
groups is used.  In this case an intramolecular reaction occurs in which one of
the ester groups reacts with the other to form a six or seven–membered ring.
This particular reaction is referred to as a **Dieckmann condensation**.  Equation
11.27 shows the overall reaction for the Dieckmann condensation of
diethylpimelate.  The student is left to work out the details of the reaction
sequence.

$$CH_3CH_2-O-\overset{\overset{\displaystyle O}{\|}}{C}-(CH_2)_5-\overset{\overset{\displaystyle O}{\|}}{C}-O-CH_2CH_3 \quad\xrightarrow[\text{2. }H_3O^+]{\text{1. }CH_3CH_2O^-}\quad$$

EQ. 11.27

## 11.18  THE ACETOACETIC ESTER SYNTHESIS AND THE MALONIC ESTER SYNTHESIS

Both of these synthetic transformations make use of much of the chemistry presented thus far in this chapter. The acetoacetic ester synthesis is useful for the preparation of substituted ketones, while the malonic ester synthesis is used to make substituted acetic acids. In this section we will look at both of these sequences in turn while reviewing many of the concepts presented elsewhere in this chapter.

As we saw in the last section, β–keto esters can be prepared using the Claisen condensation in which two ester molecules are condensed with each other. Earlier in the chapter we saw that β–keto esters are readily alkylated at the carbon between the two carbonyls. And finally, we saw that β–keto acids are readily decarboxylated upon heating. The acetoacetic ester synthesis is really just putting all of these steps together. The steps include: (1) mono– or dialkylation of a β–keto ester (which might have been made using the Claisen condensation), (2) hydrolysis of the ester, and (3) decarboxylation of the resulting acid. Often decarboxylation accompanies hydrolysis of the ester and no separate step is required. A generalized scheme for this overall process is shown in figure 11.28.

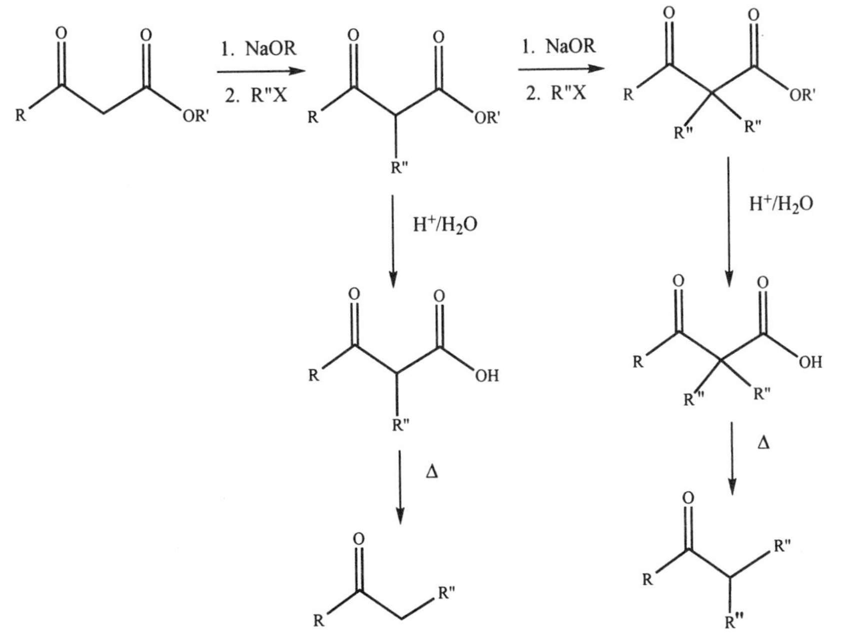

FIGURE 11.28:
Generalized process for the acetoacetic ester synthesis.

Very often the β–keto ester of choice for this process is ethyl acetoacetate, thus the name acetoacetic ester synthesis. A specific example using this starting material is shown in figure 11.29. Of course other starting materials can be used so a variety of ketones and substituted ketones can be made using these transformations.

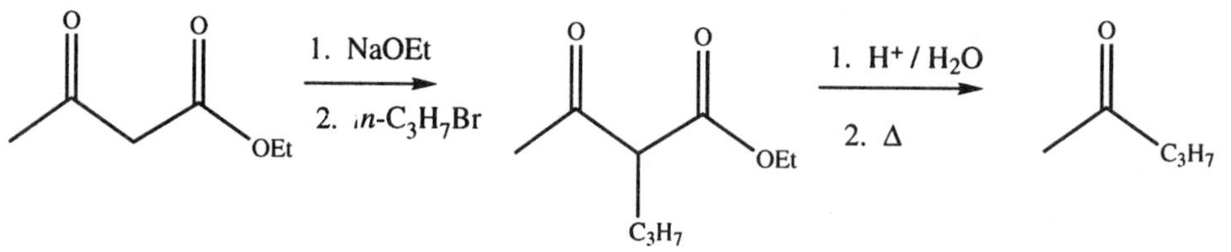

FIGURE 11.29:
Synthesis of
methyl-n-propyl
ketone via the
acetoacetic ester
sythesis.

A related reaction is the malonic ester synthesis. In this case the starting material is a diester of malonic acid (a β–diacid). The sequence of steps involved in the malonic ester synthesis is: (1) mono– or dialkylation of the malonic ester, (2) ester hydrolysis, and (3) decarboxylation by heating. A generalized example of this process is shown as figure 11.30.

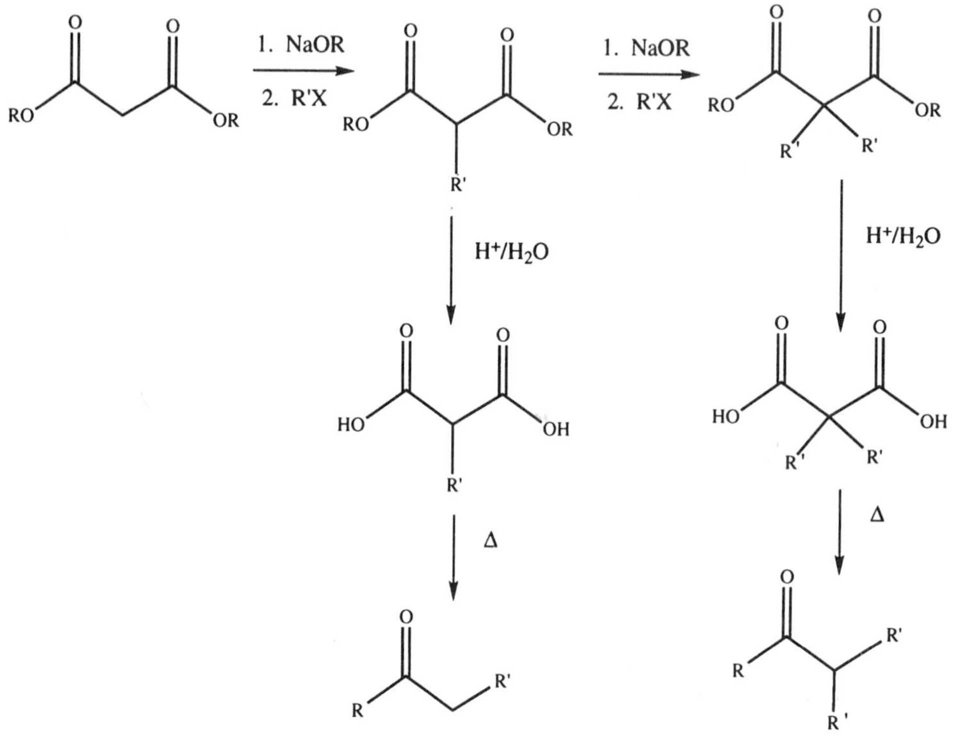

FIGURE 11.30:
The malonic ester
sythesis.

Very often the diester used in this process is dimethylmalonate or diethylmalonate. A specific example using dimethylmalonate is shown in figure 11.31.

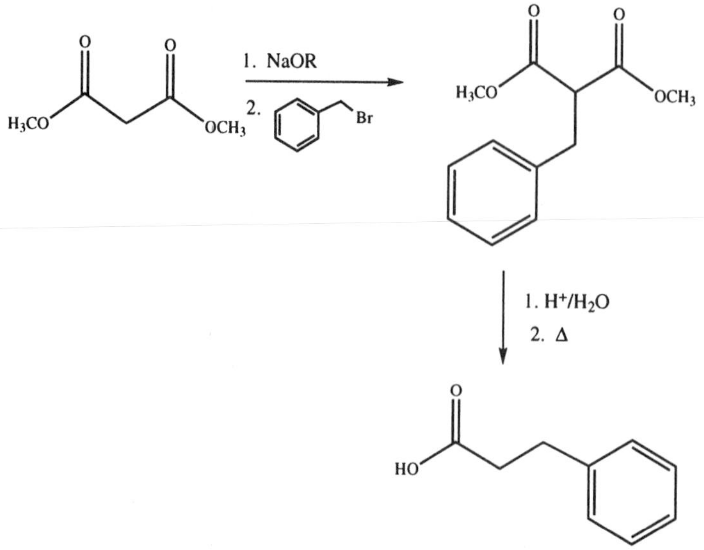

FIGURE 11.31:
Synthesis of
3-phenylpropanoic
acid via the malonic
ester sythesis.

### CONCEPT CHECK

The following compounds were made using either the acetoacetic ester synthesis of the malonic ester synthesis. Show how (no mechanisms) each compound was prepared.

a. 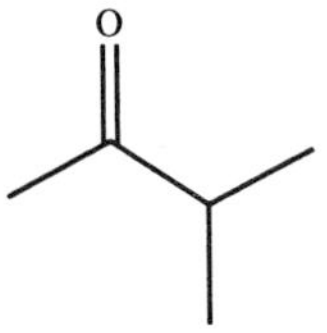

b. 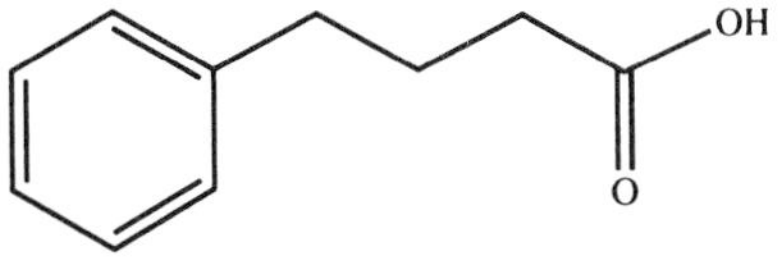

c. 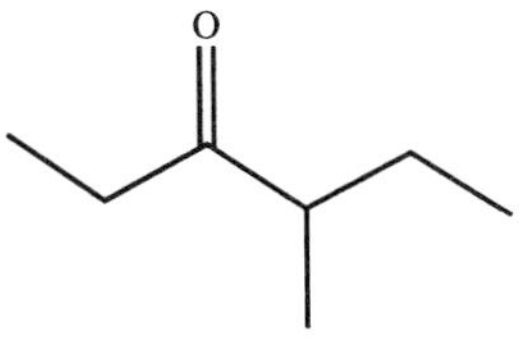

d. 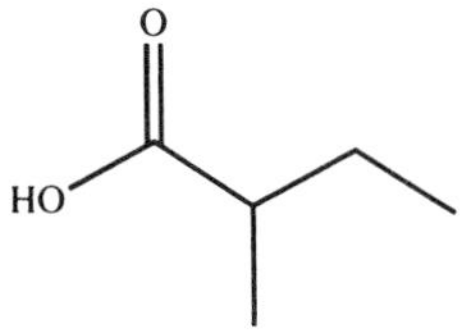

In (a) and (c) the products are substituted ketones. Therefore we shall use the acetoacetic ester synthesis to accomplish the task. Product (a) is made by adding 2 equivalents of methyl iodide to ethylacetoacetate before the usual workup. This is shown in equation 11.28. Product (c) requires the use of a different ester as the starting material in order to have an ethyl rather than a methyl group at the left side of the product as pictured. We then need to alkylate with both methyl iodide and ethyl iodide (one equivalent of each) to produce the desired product after the usual workup. This reaction is shown as equation 11.29.

Products (b) and (d) are substituterd acetic acids which are formed *via* the malonic ester synthesis. The alkyl halide used in the alkylation reaction to form the product (b) would be 3–bromo–1–phenylpropane while product (d), like (c) requires a dialkylation reaction using 1 equivalent of methyl iodide and 1 equivalent of ethyl iodide. The reactions to produce these products are shown as equations 11.30 and 11.31.

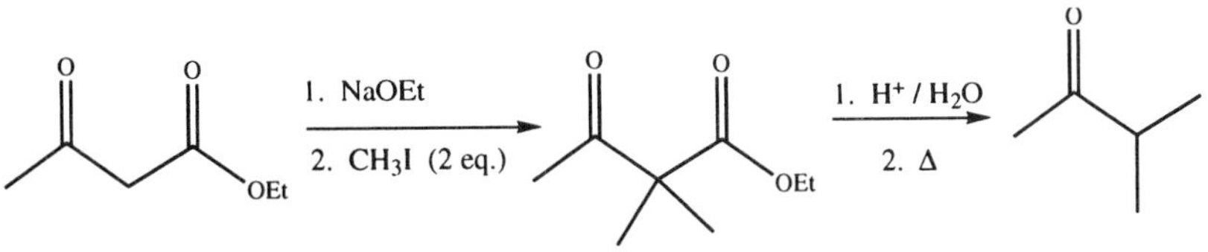
O
O
OEt
1. NaOEt
2. CH3I (2 eq.)
O
O
OEt
1. H+ / H2O
2. Δ
O

EQ. 11.28

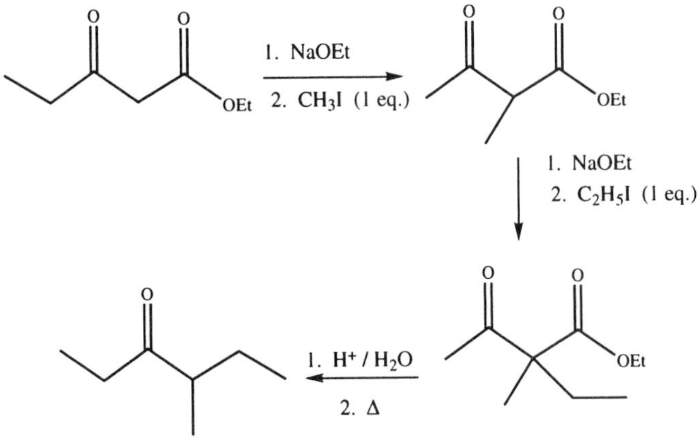
O
O
OEt
1. NaOEt
2. CH3I (1 eq.)
O
O
OEt
1. NaOEt
2. C2H5I (1 eq.)
O
O
OEt
1. H+ / H2O
2. Δ
O

EQ. 11.29

EQ. 11.30

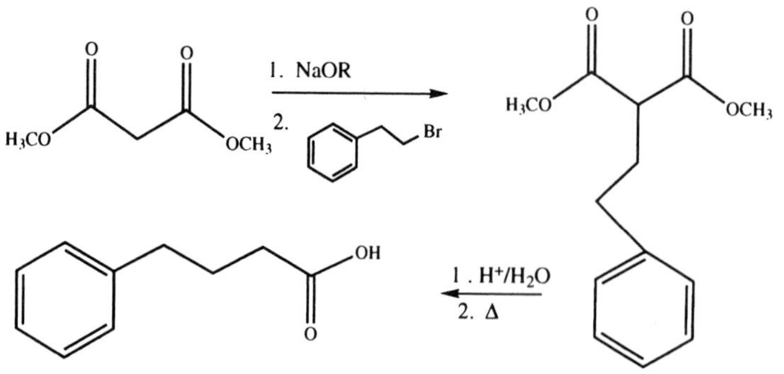

EQ. 11.31

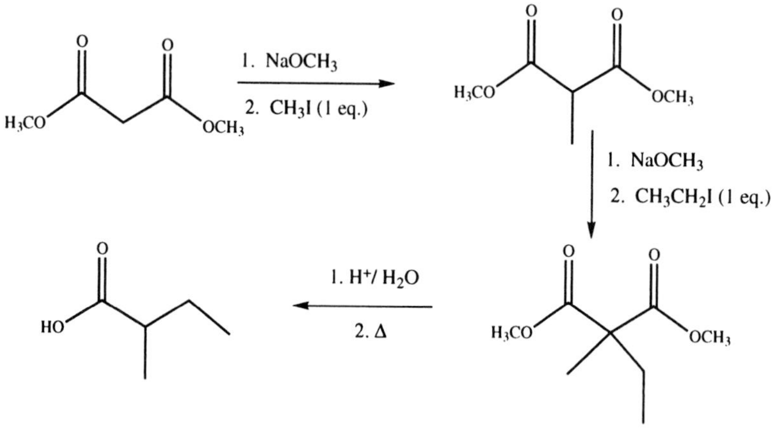

## 11.19   YLIDES—A SPECIAL TYPE OF INTERNAL SALT

When phosphines (the phosphorous analog of amines) react with alkyl halides, a compound, analogous to a quaternary ammonium salt is formed. Consider the reaction of triphenylphosphine with methyl bromide in equation 11.32.

EQ. 11.32

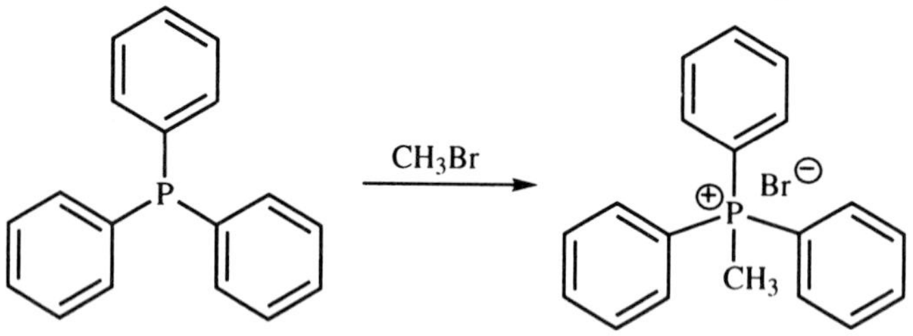

If at least one of the carbon atoms directly connected to the phosphorous atom has at least one hydrogen atom bonded to it, that proton on the carbon atom can be removed when the salt is treated with a strong base (equation 11.33.) The species resulting from the removal of the proton formally has a negatively charged carbon atom and a positively charged phosphorous atom. It is, therefore, a kind of internal salt since the same species carries both a positive and a negative charge. Compounds in which a negatively charged atom is situated

next to a positively charged atom are termed **ylides** (pronounced "ill–id").
There are other ylides other than the particular type shown here. This type of
ylide is known as a phosphonium ylide.

EQ. 11.33

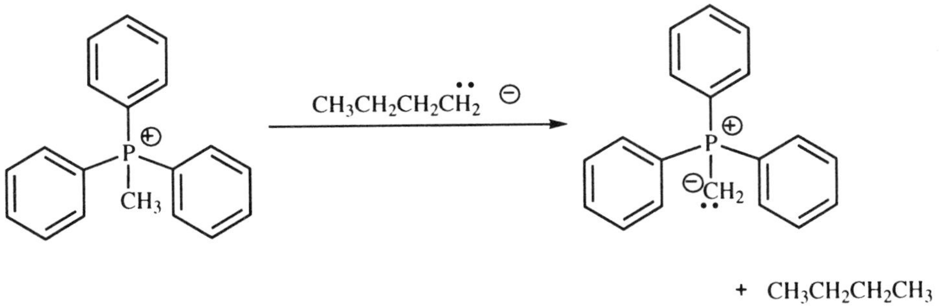

$$CH_3CH_2CH_2CH_2 \quad \ominus$$

$$+ \quad CH_3CH_2CH_2CH_3$$

Ylides are especially stable because a resonance structure exists that is formally
uncharged. In the phosphonium ylide, this structure forms when the unshared
pair of electrons in a 2p orbital on the carbon atom overlaps with an unfilled 3d
orbital on the phosphorous atom. The overlap is pictured in figure 11.32 and
the resonance form is shown in equation 11.34. Although not as good an overlap as would be produced by interaction of a p orbital with another p orbital,
there is enough energy reduction from this interaction of the two orbitals to
make the species stable.

FIGURE 11.32:
Overlap of carbon 2p
orbital with
phosphorous 3d
orbital.

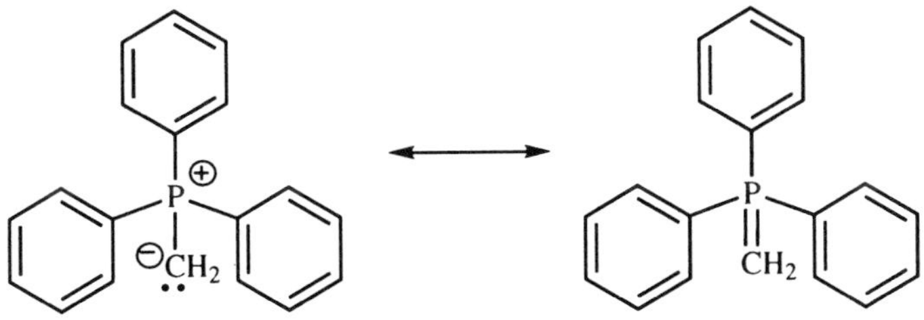

EQ. 11.34

The synthetically useful aspect of these compounds is that reaction with an
aldehyde or ketone exchanges the alkyl fragment formerly bonded to the phosphorous atom for the oxygen atom of the carbonyl group as shown in equation
11.35. This reaction is termed a **Wittig reaction**. Of course the use of an alkyl
halide different from methyl bromide will result in the placement of a different
carbon fragment in place of the former carbonyl oxygen atom. This reaction
can be synthetically quite useful.

EQ. 11.35

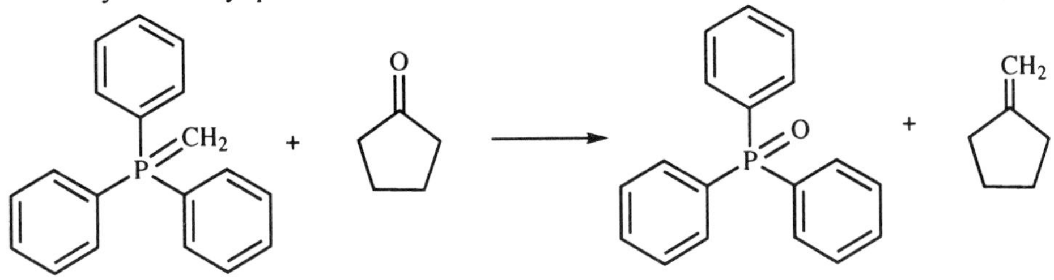

The mechanism for this reaction depends somewhat on the reaction conditions,
but both pathways begin with nucleophilic attack on the carbonyl carbon as
shown in figure 11.33. At low temperatures the 4–membered oxaphosphetane
ring is a likely intermediate. On the other hand, certain salts may promote the
ionic pathway. Either way the products are the same.

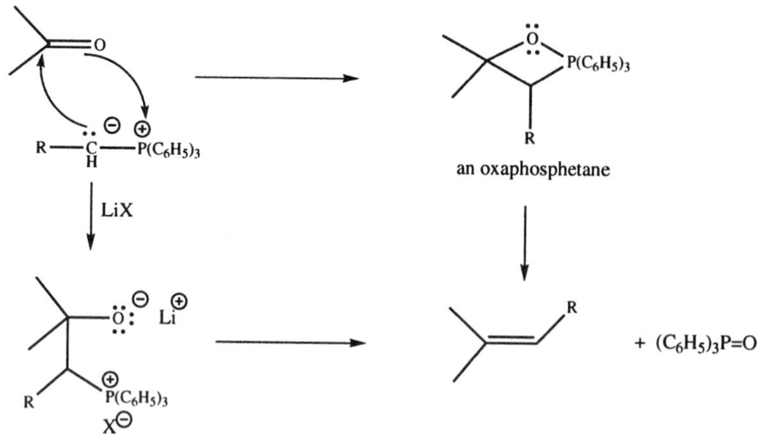

FIGURE 11.33:
Possible mechanisms
of the Wittig reaction.

an oxaphosphetane

+ (C₆H₅)₃P=O

## 11.20 THIOACETALS (1,3-DITHIANES)

There are synthetic situations in which it would be desirable to have a carbanion in which the charge is carried on the carbon atom of the carbonyl group rather than on the alpha carbon atom. Such a carbanion might be derived by removing a proton from an aldehyde *via* a reaction that might occur as pictured in equation 11.36. If such a carbanion could be formed, it would be possible to introduce a carbonyl group into a molecule *via* a nucleophilic substitution reaction. Unfortunately in most aldehydes the carbanion is formed by deprotonation at the alpha carbon atom because this proton is the more acidic.

EQ. 11.36

The solution to the problem of introducing a carbonyl functional group through a nucleophilic substitution reaction is to find a nucleophile that can undergo the desired substitution reaction and then, later, be converted to the carbonyl functional group. Such a nucleophile has been discovered by E.J. Corey and Dieter Seebach. It is the 1,3–dithiane anion (figure 11.34).

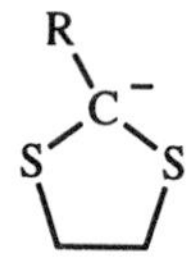

This carbanion is formed by the deprotonation of a 1,3 dithiane. These compounds are very weak acids ($pK_a$ ~ 32); therefore a very strong base is required. Typically a base such as butyl lithium is used (equation 11.37).

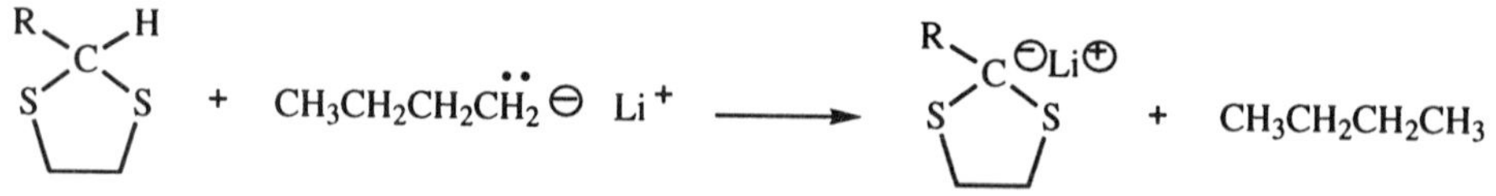

FIGURE 11.34:
The 1,3-dithiane
anion.

EQ. 11.37

These 1,3 dithiane anions are good synthetic reagents because they are good nucleophiles. Let us consider the synthesis of the symmetrical ketone, 4–heptanone from 1–bromopropane. The synthetic route to this compound starts with formation of the 1,3 dithiane that will introduce the carbonyl between the propyl groups. This is formed by reaction of 1,3 propanedithiol with formaldehyde. This reaction requires an acid catalyst.

EQ. 11.38

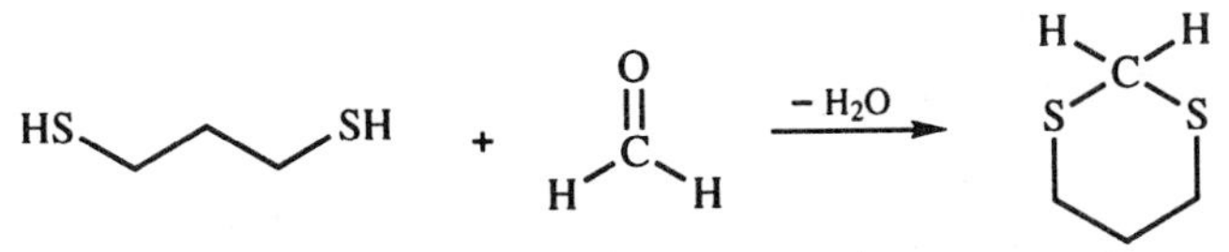

The next several steps replace the two hydrogen atoms with propyl groups. First one of the hydrogen atoms is removed by reaction with 1 mole equivalent of butyl lithium in tetrahydrofuran at −20°C. This is followed by treatment of the anion with 1 mole equivalent of propyl bromide. After this reaction is finished, a second mole equivalent of butyl lithium is added followed by a second mole equivalent of propyl bromide. The entire series of reactions is indicated in equations 11.39 — 11.42.

EQ. 11.39

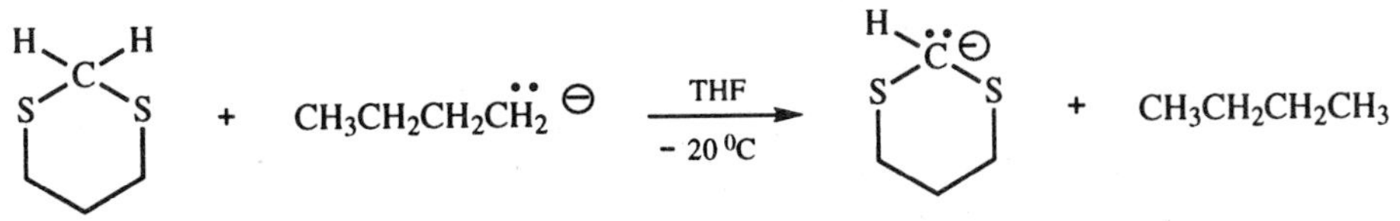

EQ. 11.40

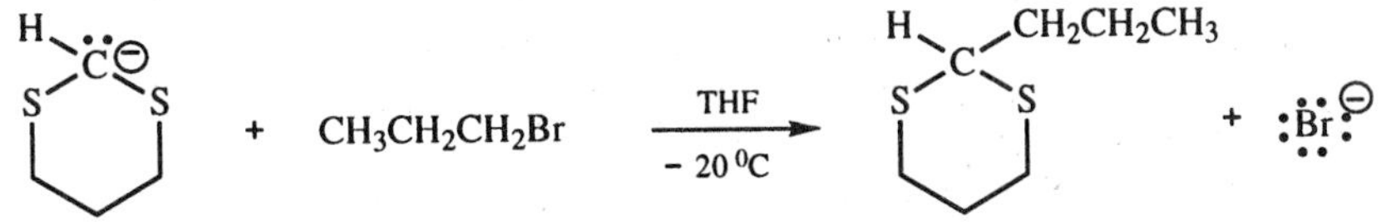

EQ. 11.41

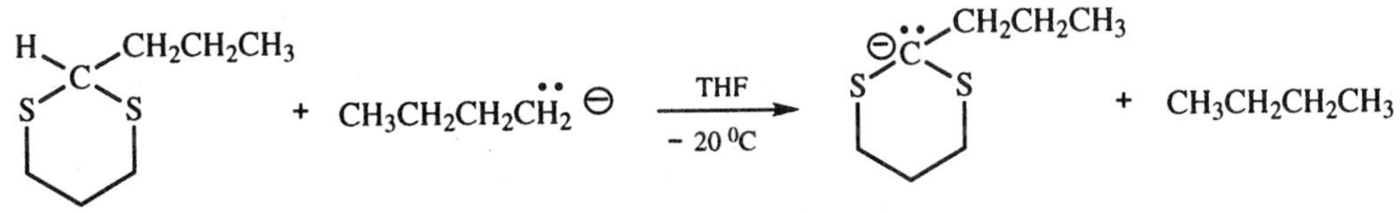

EQ. 11.42

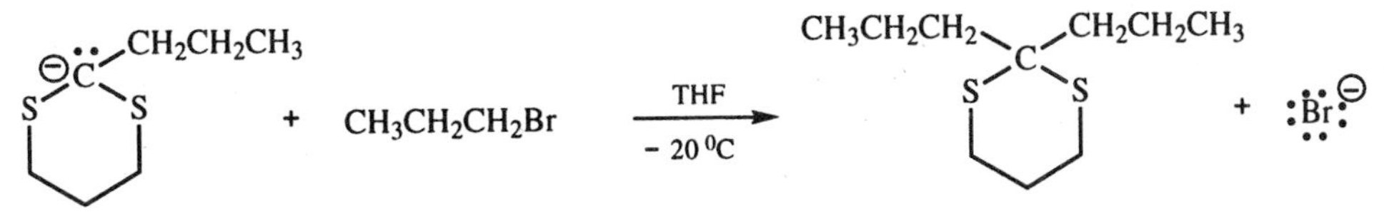

Once the 2,2–dipropyl–1,3–dithiane is produced, it is hydrolyzed in the presence of a Hg(II) catalyst to produce the ketone (equation 11.43).

EQ. 11.43

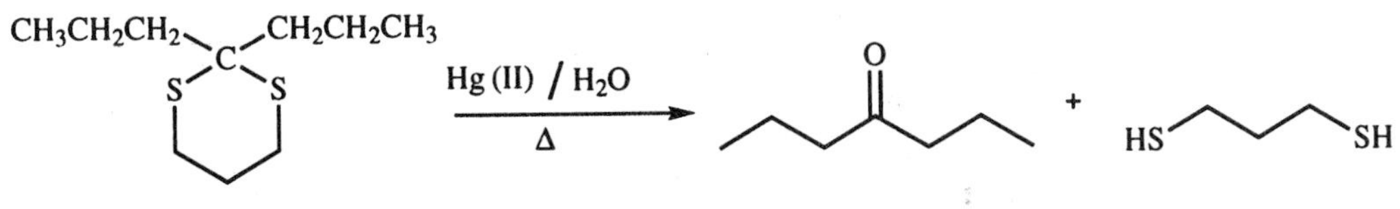

## 11.21   1,4 ADDITION TO CONJUGATED CARBONYL COMPOUNDS

Nucleophilic additions to $\alpha,\beta$ unsaturated ketones present an entirely new possibility for reaction that we have not seen before. This new reaction possibility occurs because of the **conjugation** of the carbon–carbon double bond with that of the carbonyl group. Conjugation allows delocalization of the partial positive charge on the carbonyl carbon atom over the larger unsaturated system. This delocalization is shown in figure 11.35. Note that, in addition to the partial positive charge on the carbonyl carbon atom, one of the resonance forms places a partial positive charge on the carbon atom β to the carbonyl group. This raises the possibility of formation of a carbon–carbon bond to this carbon atom as well as to the carbonyl carbon atom.

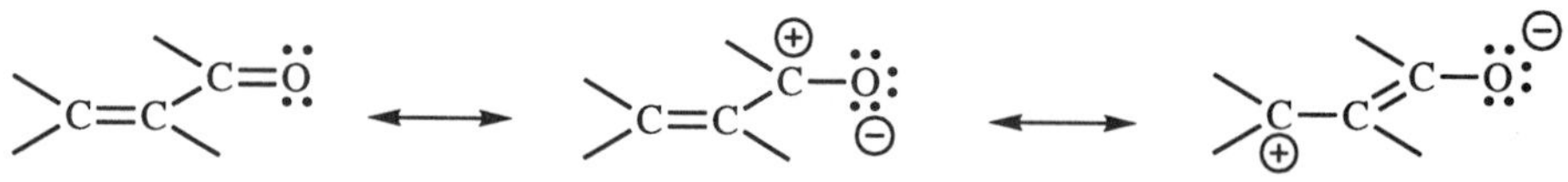

FIGURE 11.35: Delocalization of partial positive charge.

Let us compare the addition of two different electrophiles to an α,β–unsaturated ketone. First, consider the addition of the Grignard reagent, methyl magnesium iodide to 4–methyl–2–cyclohexanone. In this reaction addition occurs according to the examples learned previously — that is, addition occurs across the double bond of the carbonyl oxygen and the ultimate product is an α,β–unsaturated alcohol (equation 11.44).

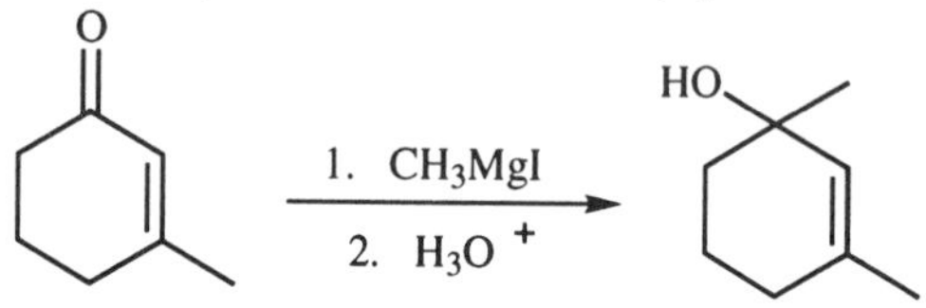

EQ. 11.44

Other nucleophiles do not give a similar product. Instead, as in the case of the organocuprates, discussed in section 11.2, addition occurs to the other (β) carbon atom carrying the partial positive charge. This is illustrated by the reaction of methyl cuprate to the same α,β unsaturated ketone (equation 11.45.)

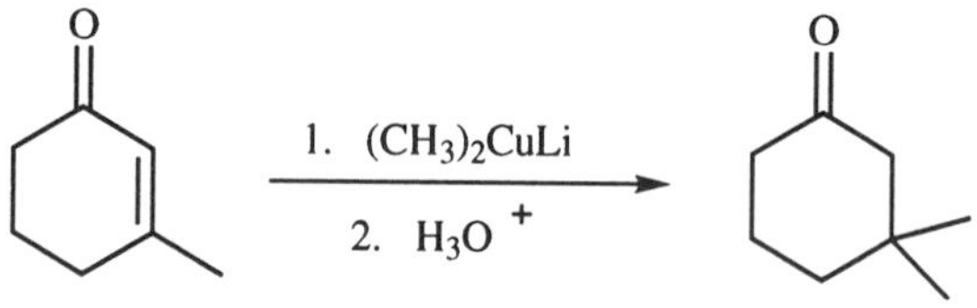

EQ. 11.45

Of further interest is the addition of enolate ions to conjugated carbonyl systems. The most interesting of these are the enolate ions derived from compounds having two carbonyl groups separated by one methylene group. Let us use methyl acetoacetate as our example. As we mentioned in section 11.17, the hydrogen atoms on the methylene group separating the two carbonyl groups are doubly activated. As a result, much weaker bases, such as carbonate ion, can be used to remove the proton to form the enolate ion. Once formed, the enolate ion can be rather easily added to an α,β unsaturated ketone. As an example, consider the reaction of methyl acetoacetate with methyl vinyl ketone (equation 11.46). Even after addition of the first ketone molecule, the remaining hydrogen atom is sufficiently acidic that it, too, can be removed and a second ketone can be added. The student is left to work out the details of the reaction sequence.

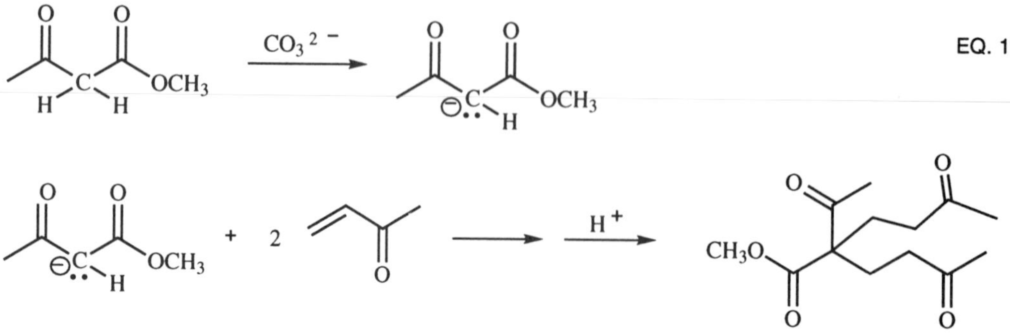

EQ. 11.46

## 11.22  THE ROBINSON ANNULATION

The Robinson annulation reaction is a special type of 1,4 addition undergone by certain α,β unsaturated ketones.  What makes it especially useful is that it can be used to form 6–membered rings.  The reaction was named for the British chemist, Sir Robert Robinson, who first showed its utility.  He won the Nobel Prize in 1947 for his work.

In the Robinson annulation reaction, an enolate serves as the nucleophile in an addition reaction with, usually, the α,β unsaturated ketone methyl vinyl ketone.  Equation 11.47 illustrates the formation of the enolate ion of 2–methyl–1,3–cyclohexanedione.  Equation 11.48 then shows the reaction of this enolate ion with methyl vinyl ketone.  Note that this is simply the 1,4 addition to a carbonyl that we saw in the last section.

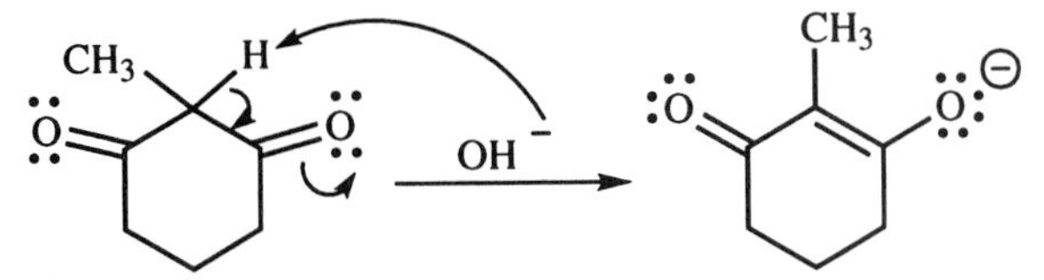

EQ. 11.47

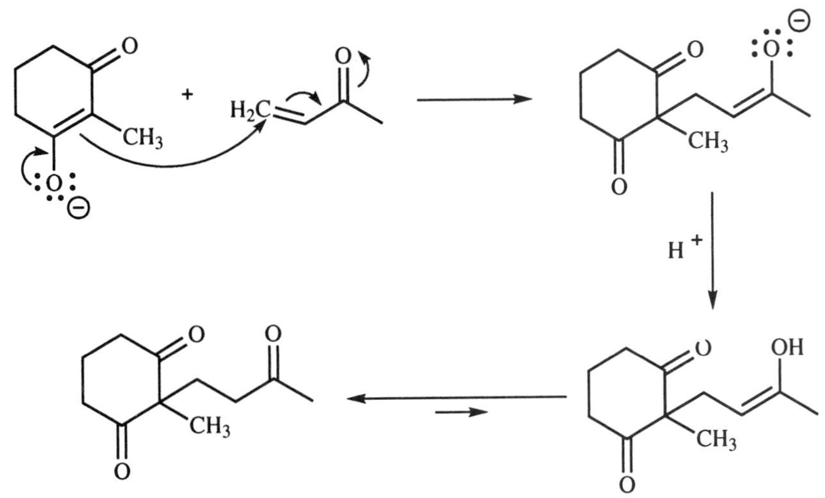

EQ. 11.48

The product of this reaction is also capable of enolate ion formation.  There are three possible enolate ions all in equilibrium with one another as shown in equation 11.49.

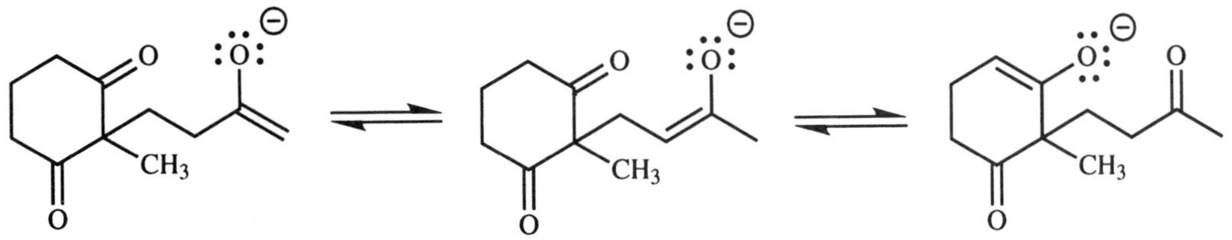

EQ. 11.49

The first of these has a resonance form on which the negative charge is carried on the terminal carbon atom:

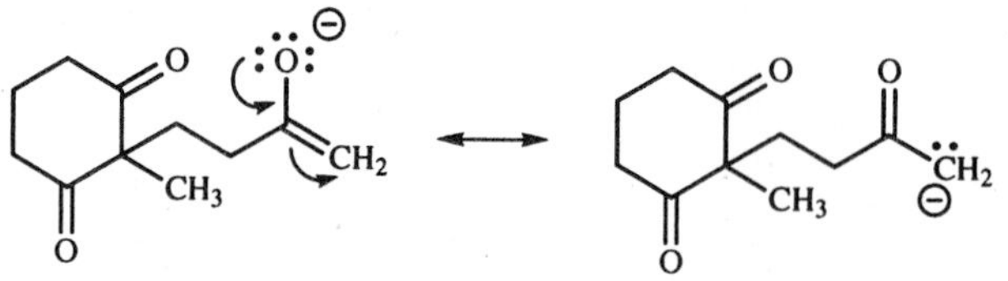

EQ. 11.50

The terminal carbon atom then participates in aldol formation with either of the two carbonyl functions on the ring to form a six–membered, fused second ring:

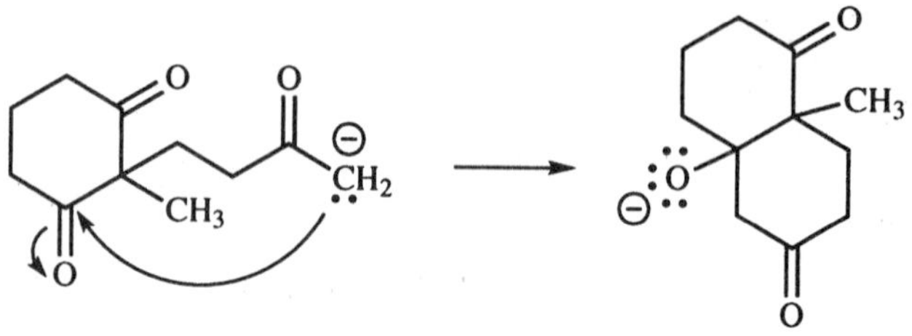

EQ. 11.51

that, after protonation easily loses water to form:

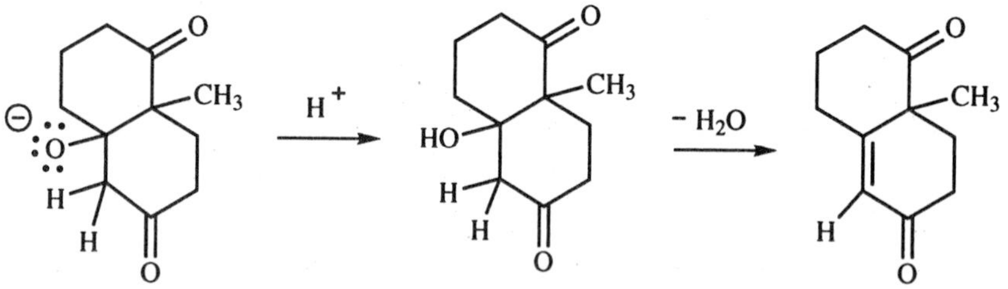

EQ. 11.52

Notice that the product is another α,β unsaturated ketone and can undergo even further reaction. Of, perhaps, more importance is that the final compound has a structure that is common in steroids. The two fused rings, including the methyl group bonded to the carbon at the juncture of the rings are two of the rings common to all steroids. Syntheses of steroid compounds, thus, often include one or more Robinson annulation reactions in the synthetic pathway. If the initial ring compound or the α,β unsaturated ketone used in the synthesis contain other substituents, further transformations can be carried out to make additional rings or other parts of the desired molecule.

In most instances the methyl vinyl ketone required for the synthesis is not added directly to the synthetic mixture. It is, rather, generated *in situ* by a chemical reaction that results in the compound. The reason for this is that methyl vinyl ketone is unstable and quickly decomposes. One convenient method of generating the compound is by reaction of the quaternary ammonium compound, 4–(N,N–diethyl–N–methyl)–2–butanone with a strong base. This compound is commonly made by reaction of 4–(N,N–diethyl)–2–butanone with methyl iodide. An example of a synthesis performed in this fashion is given in figure 11.36. In this synthesis, methyl iodide would be added to the initial keto–amine. Then, after allowing time for reaction, the ring compound and sodium ethoxide would be added. Again, after sufficient time for reaction to occur, the reaction mixture would by neutralized by addition of strong acid. The final product would then be isolated.

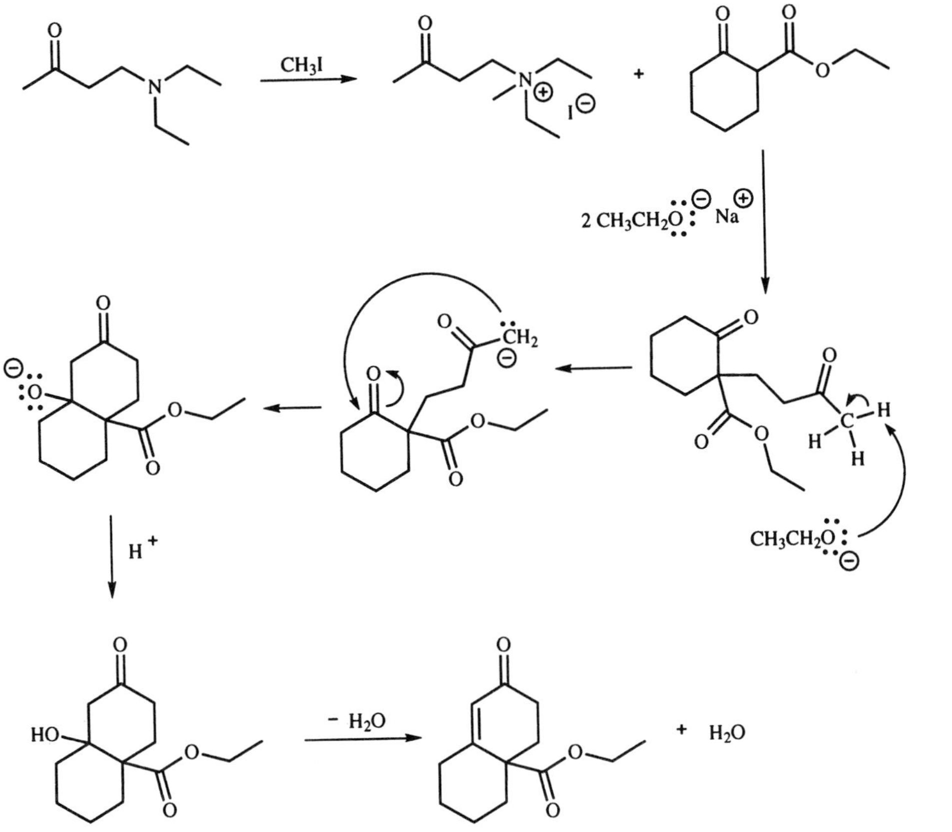

FIGURE 11.36:
Robinson annulation
with *in situ* generation
of methyl vinyl
ketone.

**Aldol condensation:** A condensation reaction between the carbonyl groups of two aldehydes to form a larger molecule containing both an aldehyde and an alcohol functional group.

**Alkylation:** Replacing an atom in a molecule with an alkyl group.

**Aprotic solvent:** A solvent containing no ionizable hydrogen atoms.

**Carbanion:** An anion on which the negative charge is formally carried on a carbon atom.

**Claisen condensation:** A condensation reaction between the carbonyl group of an ester and the carbonyl group of another acid derivative to form a larger molecule.

**Conjugation:** Two double bonds are said to be conjugated when they are separated by exactly one single bond. Conjugation can be extended as long as the multiple bonds are each separated by one single bond.

**Decarboxylation:** A reaction in which the carbon and two oxygen atoms of a carboxyl group are lost as carbon dioxide.

**Deuterium:** That isotope of the element hydrogen having one proton and one neutron.

**Dieckmann condensation:** A condensation reaction, similar to the Claisen condensation, in which two ester groups in the same molecule react to form a cyclic five– or six– member ring.

**Enol:** A tautomeric form of a carbonyl compound in which one of the hydrogen atoms on a carbon atom adjacent to the carbonyl group migrates to the oxygen atom of the carbonyl group and the carbonyl double bond migrates to become a carbon–carbon double bond.

**Enolate ion:** An ion formed by removing the proton from the oxygen atom of an enol.

*In situ:* A term meaning, literally, "in place". This term is used to mean that some reagent is formed in the same solution in which it will then be used without isolation.

**Krebs cycle:** A biochemical cycle involving the production of energy by the enzymatic oxidation of carbohydrate molecules eventually to carbon dioxide and water.

**Labile:** A term used to mean "easily removed".

**Lactone:** A cyclic ester.

**Organometallic compound:** A compound in which an organic group is bonded directly to a metal. The metal bond is highly polarized and the organic group reacts as a fairly strong nucleophile.

**Protium:** The isotope of the element hydrogen containing only one proton. This is the most common isotope of hydrogen.

**Racemization:** A term meaning the production of a recemic mixture from what was, initially, a pure chiral compound.

**Tautomerism:** A term referring to the process by which two different forms of a compound exist that differ only by the migration of a hydrogen atom from one place to another. This migration is accompanied by a rearrangement of the electrons in the bonds. This term is only applied when the two forms of the compound exist in a dynamic equilibrium.

**Tautomers:** The term applied to the two forms of a compound in tautomeric equilibrium.

**Wittig reaction:** A method of selectively forming carbon–carbon double bonds from carbonyl compounds by reaction with phosphonium ylides. The reaction was developed by the German chemist Georg Wittig.

**Ylides:** A species having a positively charged atom next to one that is negatively charged.

1. Draw and name the structures of expected products from reactions of the following carbanions with these halides.

    a. $C_6H_5CH_2Br + CH_3CH_2C\equiv CLi \rightarrow$
    b. $C_6H_5CH_2Br + NaCN \rightarrow$
    c. $CH_3CH_2CH_2CH_2Br + CH_3CH_2C\equiv CLi \rightarrow$
    d. $CH_3CH_2CH_2CH_2Br + NaCN \rightarrow$
    e. $C_2H_5Br + LiC\equiv CLi \rightarrow$

2. When sodium cyanide is used in reactions, why should the solution NEVER be allowed to become acidic?

3. Balance each of the following reactions and supply the structure of the lettered compound.

    a. 3-methyl-1-bromobutane + magnesium metal $\rightarrow$ A
        A + HCl $\rightarrow$ B + C
    b. 3-methyl-1-bromobutane + 2 lithium metal $\rightarrow$ D + E
        D + HCl $\rightarrow$ F + G
    c. $C_6H_5CH_2Br$ + magnesium metal $\rightarrow$ H
        H + $H_2SO_4$ $\rightarrow$ I + J
    d. $C_6H_5CH_2Br$ + 2 lithium metal $\rightarrow$ K + L
        L + $H_2SO_4$ $\rightarrow$ M + N
    e. acetone + methyl lithium $\rightarrow$ A
        A + $H_2O$ $\rightarrow$ B + C (in acid solution)
    f. acetone + methyl magnesium iodide $\rightarrow$ C
        C + $H_2O$ $\rightarrow$ D + E (in acid solution)
    g. 4-bromo-2-butanone + methyl magnesium $\rightarrow$ E
        E + $H_2O$ $\rightarrow$ F + G ( in acid solution)
    h. 4-bromo-2-butanone + methyl magnesium chloride $\rightarrow$ G
        G + $H_2O$ $\rightarrow$ H + I (in acid solution)
    i. Show the polarity of each of the above halogen – carbon and metal carbon bonds

4. Draw all possible keto and enol tautomers for the following carbonyl compounds.

    a. acetone
    b. cyclohexanone
    c. 2-butanone
    d. ethanal
    e. 3-methyl-2,4-pentanedione
    f. methyl 3-oxopentanoate

5. Explain the origin of each set of HNMR peaks for

    a. 1,3-diphenylpropane
    b. 2-phenylethanal

6. Write the mechanism for

    a. the formation of A from cyclopentanone with NaOD in $D_2O$
    b. the formation of B from cyclopentanone with NaOD in $D_2O$
    c. the formation of $CH_3COCH_3$ from $CD_3COCD_3$ in NaOH and $H_2O$

7. Write the mechanism for the formation of 1,1,1-tribromopropanone from acetone, HCl, and bromine.

8. Write reactions to prepare one compound from the other using any other chemicals you wish.

    a. 2-butanone from acetone
    b. 3,3-dimethyl-2,4-pentanedione from 2,4-pentanedione
    c. 2-ethylcyclohexanone from cyclohexanone
    d. ethyl propanoate from ethyl ethanoate
    e. methyl 2-phenylpropanoate from methyl 2-phenylethanoate
    f. methyl 2-phenylpropanoate from 2-phenylethanoic acid

9. Give structures of compounds A - D

    3,3-dimethyl-2-butanone + LDA $\rightarrow$ A + B
    A + 1-bromopropane $\rightarrow$ C + D

10. Write structural formulas for all the possible products when the following compounds react via the aldol condensation reaction. Give products before and after dehydration.

    a. ethanal
    b. propanal
    c. phenylethanal
    d. ethanal and propanal in equimolar amounts
    e. ethanal and benzaldehyde in equimolar amounts
    f. propanal and phenylethanal in equimolar amounts
    g. $OHC(CH_2)_5CHO$

11. Write structural formulas for all the possible products when the following compounds react via the Claisen condensation reaction.

    a. methyl ethanoate
    b. ethyl propanoate
    c. methyl phenylethanoate
    d. ethyl propanoate and ethyl ethanoate in equimolar amounts
    e. methyl phenylethanoate and methyl benzoate in equimolar amounts
    f. $C_2H_5OOC(CH_2)_4COOC_2H_5$

12. Write a mechanism by which butanal reacts with sodium hydroxide to produce 2-ethyl-3-hydroxyhexanal.

13. Write a mechanism by which 2-ethyl-3-hydroxyhexanal reacts with sodium hydroxide to produce butanal via a retroaldol condensation reaction.

14. Write the products when the salt $(C_6H_5)_3PCH_3{}^+Br^-$ is reacted with $n$-butyl lithium and each of the following:

    a. formaldehyde
    b. acetone
    c. 3-oxo-1-butene
    d. cyclohexanone-2-ene

15. Write the reactions necessary to prepare the following ylides from triphenylphosphine and any other chemicals you wish.

    a. $(C_6H_5)_3PCHCH_3$
    b. $(C_6H_5)_3PCH(CH_3)_2$
    c. $(C_6H_5)_3PCH(C_6H_5)_2$

16. Write reactions to prepare the following compounds *via* the thioacetal of benzaldehyde. Write all intermediates.

    a. $C_6H_5CH_2CH_3$
    b. $C_6H_5CH(CH_3)_2$
    c. $C_6H_5CH(CH_2C_6H_5)_2$

17. Why is 1,3-propanedithiol preferred for dithiane synthesis rather than the simpler compound methylthiol?

18. Show all reactions necessary to prepare the following compounds *via* ethyl acetoacetate and any other compound you wish.

    a. $CH_3COCH(CH_2CH_2CH_3)COOC_2H_5$
    b. $CH_3COC(CH_2CH_2CH_3)_2COOC_2H_5$
    c. 2-hexanone
    d. 3-propyl-2-hexanone
    e. 3-ethyl-2-hexanone

19. Write a mechanism for the ring opening shown in equation 11.16 (a retroaldol).

20. Predict the products of the following reactions.

a.

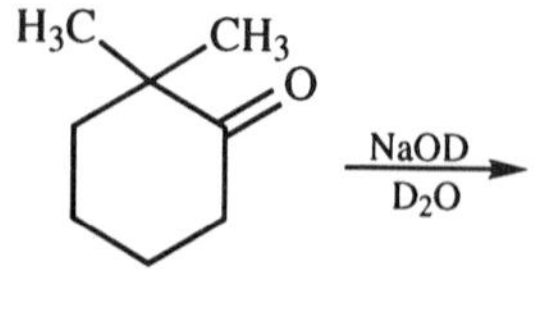

b.

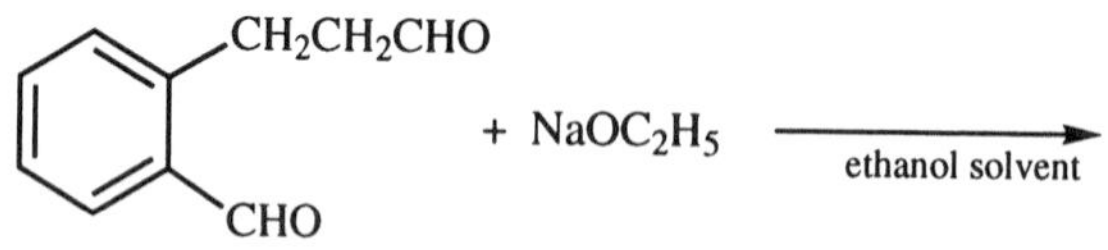

c.

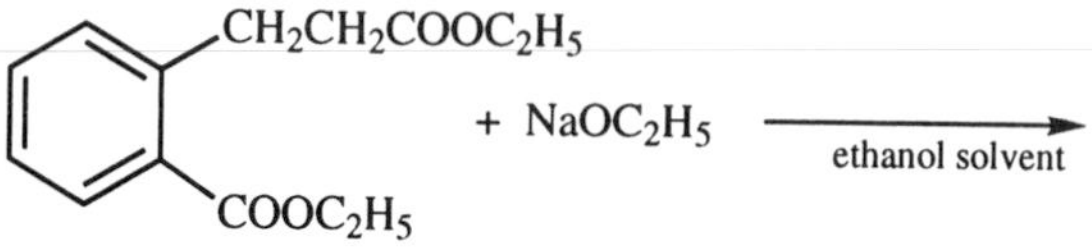

21. Use any additional chemicals necessary to prepare the indicated compound from the starting compound.

     a. propyl butanoate from 1-butanol
     b. pentanamide from 1-pentanol
     c. 3-methylbutanal from methyl 3-methylbutanoate
     d. hexanenitrile from 1-pentanol
     e. 2-chloropentane from 2-pentanone
     f. 1-bromoheptane from heptanoic acid
     g. 2-butanone from ethyne
     h. 2-hydroxy-1-cyclohexanone from 1,6-hexanediol
     i. 1,6-dibromohexane from 1,6-hexanedioic acid
     j. 1-aminobutane from 1-bromopropane

## CHAPTER ELEVEN

OVERVIEW
**Problems**

22. Dicarbonyl compounds occur in the Kreb citric acid cycle. One of the compounds has molecular weight 120 +/- 5 g/mol, and 40.68% carbon, 5.12% hydrogen, and 54.20% oxygen. The IR spectrum shows a strong absorbance at 1700 $cm^{-1}$ and a wide absorbance at 3700-2500 $cm^{-1}$. The HNMR spectrum shows a singlet at 2.2 ppm, and a singlet at 12.0 ppm. The CNMR spectrum shows two absorbances: 29 ppm and 174 ppm.
     a. Calculate the empirical formula of the compound.
     b. Calculate the molecular formula of the compound, and the IHD value.
     c. Interpret the IR spectrum.
     d. Interpret the NMR spectra, and relate this to the actual molecular structure.

23. When ethylacetate is reacted with LDA in THF solvent, followed by methyl iodide, the organic product has the following characteristics: The CNMR shows 5 unique carbon environments; the HNMR shows 4 proton environments: a quartet at 4.1 ppm, a quartet at 2.3 ppm, a triplet at 1.3 ppm, and a triplet at 1.1 ppm.
     a. Show the structure of ethyl acetate.
     b. Show the structure of the anticipated product.
     c. Relate the CNMR spectrum of the product to the structure in b.
     d. Relate the HNMR spectrum of the product to the structure in b.

24. Two molecules of acetaldehyde ($\Delta H_f$ = -192.0 kJ/mol) undergo a condensation reaction in ethanol in the presence of sodium ethoxide to form 3-hydroxy-butanal ($\Delta H_f$ = -430.5 kJ/mol) .
     a. Write the balanced reaction.
     b. What is the expected sign of the entropy change for this reaction, and why?
     c. Calculate the value of the enthalpy change for this reaction.
     d. Given that the reaction is spontaneous under standard state conditions, does entropy or enthalpy play a greater role in making the reaction proceed?

25. When the product formed in the previous reaction, 3-hydroxybutanal ($\Delta H_f$ = −430.5 kJ/mol), is exposed to NaOH, liquid water and the organic dehydration product are formed. The dehydration product has $\Delta H_f$ = -144.1 kJ/mol.

    a. Write the dehydration reaction forming 2-butenal and water.

    b. Write the dehydration reaction forming 3-butenal and water.

    c. Which reaction (a or b) is favored, and why?

    d. What is the anticipated sign of the entropy change for this reaction, and why?

    e. Calculate the enthalpy change for this reaction.

    f. Given that the reaction is spontaneous under standard state conditions, does entropy or enthalpy play a greater role in making the reaction proceed?

new bonding
interactions

# CHAPTER twelve
# 12

# MOLECULAR
# Orbitals
## SYNTHETIC IMPLICATIONS

# MOLECULAR
## Orbitals: SYNTHETIC IMPLICATIONS

## 12.1   ELECTRON DENSITY AND REACTIVITY

By this point we are very well acquainted with the way chemical reactions
occur.  In short, all chemical reactions involve interactions between molecules
and atoms.  These interactions finally bring individual atoms sufficiently close
to one another that reaction occurs.  During the reaction the electrons associated with each of the atoms involved in the reaction assume new arrangements.
All of this is driven by the energy levels of the reactants and products as the
system attempts to find a more stable energy state as products than it had as
reactants.

As chemists we want to determine if a reaction will occur between two species,
and then, if a reaction does occur, what the likely products of the reaction are.
To do this we must simply(!) determine what causes two atoms to be attracted
to one another so that they get close enough to exchange electrons.  In earlier
chapters we talked of electrophiles and nucleophiles and mentioned that certain
types of reactions could be classified as electrophilic or nucleophilic reactions.
In these we look at the reaction as if one of the reactants were seeking a nucleophilic or electrophilic center.  Chemists have built a long list of nucleophilic
and electrophilic centers within molecules from observing literally thousands of
reactions to aid the predictions of which we were speaking.

Provided that there is sufficient room for the reactant to approach, we should
be able to predict where an electrophile will attack a **substrate** molecule by
determining where the most electron–dense portions of the substrate are.
Theoreticians have helped enormously in this work in the past few years and
there are now very powerful computer programs available that will permit the
calculation of many different parameters for many different molecules and give
pictorial representation of electron densities, determine the lowest energy molecular geometries, and do many other things that aid our design of practical syntheses.  For a long time the ability to do such calculations was limited to those
who had access to very powerful  computers and the knowledge to use them.
Today such programs will run on desktop machines and are easy enough to use
that chemistry students with even a modicum of chemical knowledge can get
useful results.  In the first part of this chapter we will see the basis upon which
these programs are written.  In the latter parts of this chapter we will use some
of the results of these programs to help us understand how reactions occur and
then design syntheses to accomplish a particular objective.

## 12.2   SCHRÖDINGER'S WAVE EQUATION

Shortly after Niels Bohr developed his explanation of the structure of the atom
and shortly after deBroglie's explanations of the wave properties of electrons,
Erwin Schrödinger (1887–1961) developed a new theory of atomic structure
now known as Quantum Mechanics or Wave Mechanics.  We used the results

of quantum mechanics when we discussed the shapes and energies of the atomic orbitals (s, p, d, *etc.*) and the quantum numbers associated with those energy levels and orbitals. It is now time to discuss, in a bit more detail, the origins of these orbitals in order to help us determine the electronic structure of molecules. Fortunately, although exact solutions of these equations are very complicated, it is not difficult to get very useful information using only simple mathematics. The next few paragraphs contain mathematical expressions that are quite difficult to solve (indeed, for most real molecules, they cannot be solved without the use of approximation methods.) However, in this course, it will never be necessary to actually solve the equations to obtain useful results; we need not fear the mathematics. What will be used to obtain the information will involve only simple mathematical operations.

Schrödinger's basic equation, the one upon which all of Quantum Mechanics is based, is:

$$H\Psi = E\Psi$$

EQ. 12.1

In this equation $\Psi$ is termed the **wave function**, H is a **mathematical operator** termed a Hamiltonian operator, and E is the energy associated with the wave function. A wave function is nothing more than a mathematical function, *i.e.*, a series of mathematical terms such as $4x^2 + 3x + 5$, (the actual wave function is somewhat more complicated.) "Operator" is the mathematical term for something that carries out a transformation (operation) on a function that changes it into another function. Some everyday examples of mathematical operations include multiplication, division, differentiation, and integration. A Hamiltonian operator is simply a particular type of mathematical operator with certain mathematical properties. It is constructed using the postulates of wave mechanics and well–known principles of physics. The energy term is a simple constant.

While the details of the solutions to Schrödinger's equation are well beyond the scope of this book, the job of the theorist is to find wave functions and Hamiltonian operators that allow this equation to provide answers that correspond with experimental measurements and which lead to useful predictions of the behavior of chemical systems. For simple chemical systems containing one electron and one nucleus, the mathematical forms taken by the Hamiltonian operator are well understood by relatively simple extensions of classical physics. As additional particles are added, the Hamiltonian operator becomes more complicated and approximations must be made to allow actual solution of the equations. However the thing that is of most interest to chemists are the wave functions. Here, too, the wave functions can be developed in exact form only for very simple chemical systems and approximations must be made in real systems. Nevertheless, current methods used in quantum mechanics have become quite sophisticated and, with the vast amount of computer power now available, approximations that yield quite good solutions can be made if necessary. It is these wave functions that we can then use to provide the pictures of the electron domains (orbitals) that ultimately help us to understand chemical reactions.

Physically we understand the wave function, $\Psi$, to be a function that describes the wave–like behavior of an electron as it interacts with its surroundings. The pictures that we have previously shown (and are normally shown) of orbitals are actually plots of the integral of the square of the function:

$$\int \psi^2 d\tau$$

EQ. 12.2

where the $d\tau$ indicates that the integration takes place over all space. The result of this integration is a probability function that describes the **electron density** at various points in space. The pictures of these functions normally shown are surfaces within which it is 90% (or some other value) probable that the electron will be found when the electron is in the orbital.

Key to this entire discussion, then, is the wave function, $\Psi$. As a mathematical function, the wave function may have both positive and negative regions and, except for the wave function of the s orbitals, all wave functions do have these regions. Plots of the orbitals normally shown do not emphasize these characteristics of the wave function because they are plots of the probability distribution that is the square of the wave function. To properly understand bonding, though, it is necessary to understand this signed feature of the wave function. Figure 12.1, therefore shows a plot of the wave function, $\Psi$, itself for a $3p_z$ and a $3d_{xz}$ orbital. In this diagram the unshaded box represents the area of space in which the wave function has a positive value and the shaded box that area in which the wave function is negative. Note that, for the $3p_z$ orbital, the sign of the wave function changes from positive to negative as the orbital crosses the xy plane and that the $3d_{xz}$ orbital is positive in two and negative in two of the quadrants of the xz plane.

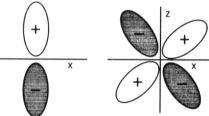

3p$_z$ orbital    3d$_{xz}$ orbital

FIGURE 12.1: Wave functions of $3p_z$ and $3d_{xz}$ orbitals.

## 12.3   BONDING THEORIES

When we first considered the forces that held atoms together in molecules, we used Lewis Dot Theory to place electron pairs between atoms and called the result a "**bond**". We did not attempt to discuss why the electrons were placed between the atoms or why such a configuration was more stable than placing the electrons elsewhere other than to, perhaps, point out that it "made sense" that the negatively charged electrons would occupy a region of space between two positively charged nuclei and that, in this position, they could form the "glue" that held the atoms together in the molecule. Despite its lack or mathematical sophistication, Lewis Dot Theory worked remarkably well for explaining many experimentally observed phenomena. Lewis Dot Theory, however, did not do well explaining multiple bonds between atoms, nor, as we saw three chapters ago, was it able to explain the geometries of certain metal species.

In our next discussion of bonding, after we had discussed atomic orbitals, we described a bond as being formed by the overlap of two atomic orbitals. In Valence Bond Theory, these atomic orbitals might be either hybridized or unhybridized orbitals. We named the bonds formed $\sigma$ and $\pi$ bonds depending upon whether the overlap of the orbitals was end–to–end ($\sigma$) or sideways ($\pi$). Valence Bond Theory, while it did somewhat better in explaining multiple bonding and explained certain geometries observed in metal complexes, still was unable to explain a number of observations: (1) it did not easily explain molecular spectra, and (2) it did not explain why the dioxygen molecule ($O_2$) is paramagnetic.

These problems of Valence Bond Theory are eliminated by a newer bonding theory called Molecular Orbital Theory. In Molecular Orbital Theory when two (or more) atoms come together to form a molecule, a set of **molecular orbitals** forms in which the electrons are reside. When all of the electrons are in the lowest possible energy state (termed the ground state), the total energy of the species is lower than the energy of the isolated atoms (in which the electrons were in atomic orbitals) and the species is stable. Molecular Orbital Theory allows the calculation of this stabilization energy once the wave functions for the molecular orbitals are known. The key to the process, then, is the formation of the molecular orbitals. The most common method of forming the molecular orbital wave functions is by the Linear Combination of Atomic Orbitals (into) Molecular Orbitals. This technique goes by the acronym **LCAO–MO** and means that the mathematical wave functions for the molecular orbitals are formed by adding or subtracting the mathematical wave functions of the atomic orbitals from which they are made. This can be mathematically expressed as:

$$\Psi_1 = \varphi_1 + \varphi_2 \qquad\qquad \Psi_2 = \varphi_1 - \varphi_2 \qquad\qquad \text{EQ. 12.3}$$

where the $\Psi$'s are molecular wave functions and the $\varphi$'s are atomic wave functions. The subscripts on the molecular wave function describe the two possible wave functions that can arise from different linear combinations of the atomic orbitals. The subscripts on the atomic orbitals refer to the different atoms involved in the bond (*i.e.*, $\varphi_1$ refers to an atomic orbital on atom 1 involved in the bond formation while $\varphi_2$ refers to an atomic orbital on atom 2).

We will eventually want to use these methods to describe the bonding in fairly complicated molecules with extended $\pi$ systems. Before we do that, however, let us look at bonding in some simple diatomic species from the standpoint of molecular orbital theory.

First let us consider the simplest possible diatomic molecule, $H_2$. We know that the electrons in the isolated hydrogen atoms exist in 1s orbitals. Figure 12.2 shows the electrons and the energy of the 1s orbitals in the separated hydrogen atoms on the outside of the diagram and the energy levels and placement of the electrons in the ground state hydrogen molecule in the center of the diagram. Note that there were two orbitals (one on each of the hydrogen atoms) before the formation of the molecular orbitals, and these two atomic orbitals were combined in such a way that two molecular orbitals were formed. Figure 12.2 also shows the shapes of these orbitals. The atomic 1s orbitals are, of course, spherical. When the wave functions for the two 1s orbitals are added, the resulting molecular orbital is the lower orbital pictured in the middle of the diagram. Note that the sign of the molecular wave function is positive everywhere, just as were the atomic 1s orbitals. When, however, the two atomic wave functions are subtracted, the resulting molecular wave function must change sign. This will happen at those points where the value of $\varphi_1$ and $\varphi_2$ are equal. This, of course, means that the value of $\Psi$ at this point will be zero. These points at which this occurs all lie in a plane perpendicular to the bond axis. This plane is termed a **nodal plane**.

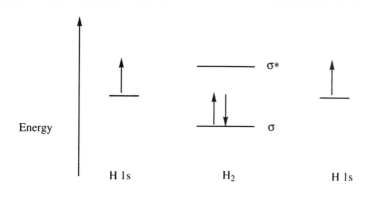

Note that the shape of the molecular orbital formed by adding the two wave
functions is such that the electrons spend most of their time between the atoms
— the situation that we have hitherto said to be a bonding situation.
Calculations, using the additive wave function, have, indeed, shown that the
energy of this orbital is lower than the isolated hydrogen 1s atomic orbital.
Electrons placed into this orbital are, therefore, stabilized, relative to electrons
in the hydrogen atom and energy is released when the bond forms. This orbital
is called a bonding orbital for that reason. Since, similar to the atomic s
orbital, this molecular orbital is circularly symmetric to the bond axis, it is
termed a σ molecular orbital. On the other hand, calculations of the energy of
the other orbital — the one formed by subtracting the two atomic 1s wave
functions, is higher than the energy of the atomic 1s orbital. This means that
electrons placed into this molecular orbital are less stable than they would be in
the isolated atoms. Such an orbital is termed an anti–bonding orbital. This
anti–bonding characteristic of the orbital is usually designated by appending an
'*' to the symbol for the orbital. In this case, the molecular orbital is circularly
symmetric to the bond axis making it, also, a σ orbital. Its complete designation is, then, a σ* orbital.

Once we have the molecular orbital energy diagram and have put the electrons
into it, we can write an electron configuration for the molecule. For the hydrogen molecule this is: $(\sigma_{1s})^2$. We can also calculate the bond order for the molecule using equation 12.4. The calculation for the hydrogen molecule is shown
as equation 12.5. For the hydrogen molecule, the bond order is 1 which we
interpret to mean that a single bond exists between the two hydrogen atoms.

$$\text{B.O.} = 0.5[(\#e^- \text{ in bonding orbitals}) - (\#e^- \text{ in anti–bonding orbitals})]$$

EQ. 12.4

$$\text{B.O.} = 0.5[(2) - (0)] = 0.5(2) = 1.0$$

EQ. 12.5

If we move to lithium, a second period element, the molecular orbital situation becomes slightly more complicated. Figure 12.3 shows the situation.

This molecule is predicted to be stable because, although the four electrons in the $\sigma_{1s}$ and $\sigma^*_{1s}$ orbitals have a net of zero stabilization energy relative to the atomic orbitals, those in the $\sigma_{2s}$ orbital are stable relative to the atomic 2s orbitals. We could similarly show that this molecule would be predicted to be stable by calculating the bond order as in equation 12.6.

$$B.O. = 0.5[(4)–(2)] = 1.0 \qquad \text{EQ. 12.6}$$

The bond order calculated for this molecule is also 1.0. Thus the dilithium crystals of Star Trek fame would exist if it were not that the metallic bonding is even more stable for lithium. The electrons in closed shells will always completely fill both the bonding and antibonding molecular orbitals formed from the atomic orbitals of that shell. The net stabilization energy of these electrons will always be zero. Likewise the net bond order for electrons initially in closed shells will be zero. Whether or not a molecule will be stable will always be determined by the electrons in the outer, or valence, shell. In the future, therefore, in order to simplify the drawings, we shall omit all closed shell electrons from our drawings.

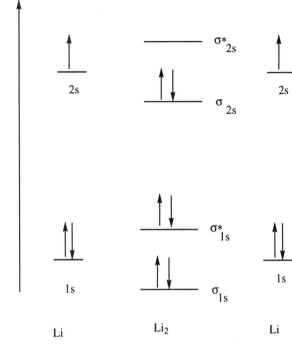

FIGURE 12.3:
Molecular orbitals of $Li_2$.

The next molecule we wish to discuss is dioxygen, $O_2$. Figure 12.4 shows the molecular orbital diagram for this molecule. It is noteworthy because here the 2p orbitals are involved in the formation of molecular orbitals. Because only one of the p orbitals has its lobes lying along the bond axis, only it is oriented in such a way that it can form a $\sigma$ bond. The overlap of the other two p orbitals must be the same type of side–to–side orverlap that we saw in Valence Bond Theory. The molecular orbitals formed by this overlap have a nodal plane along the bond axis, *i.e.*, they are not circularly symmetric about the bond axis. Such orbitals are called $\pi$ molecular orbitals. In addition to the energy level diagram, figure 12.4 also shows the shapes of the $\sigma_{2p}$ and $\pi_{2p}$ orbitals. It also clearly shows that the eight electrons from the 2p levels of the oxygen atoms will completely fill the bonding $\sigma_{2p}$ and $\pi_{2p}$ levels with two additional electrons that will be placed, spin parallel, into the anti–bonding $\pi^*_{2p}$ levels. This predicts that dioxygen will be paramagnetic and confirms experimental observations. Since there are 6 electrons in bonding orbitals and only 2 in anti–bonding orbitals (of the electrons initially in the 2p atomic levels), there is net bonding energy and the molecule is predicted to be stable. Applying equation 12.4 to this molecule yields a bond order of 2.0 — just exactly the double bond that we had always predicted for the oxygen molecule.

Finally, let us look at a molecule formed by two non–identical atoms: HF. Figure 12.5 shows the atomic and molecular orbitals for this molecule. Note especially that the atomic orbital energy of the 1s orbital on the H atom is not identical to that of the F atom 2s or 2p electrons. In general, the highest energy electrons of the more electronegative atom are somewhat lower in energy than the highest energy electron(s) of the less electronegative atom. Notice also that the bonding overlap in this case is between a 1s orbital on the H atom and one of the 2p atomic orbitals on the F atom. This overlap will form a $\sigma_{sp}$ and $\sigma^*_{sp}$ molecular orbital pair. Since the remaining two atomic 2p orbitals on the F atom are not involved in the bonding, their energy levels remain unaffected. These orbitals are also present in the molecule, are termed **non–bonding orbitals**, and are signified by 'n'. (Non–bonding electrons are not considered in the calculation of the bond order. Therefore the bond order for HF is, as expected, 1.0.

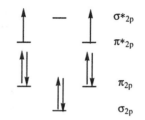

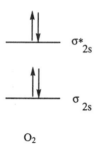

$\sigma^*_{2p}$

$\pi^*_{2p}$

$\pi_{2p}$

$\sigma_{2p}$

2p

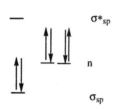

$\sigma^*_{2s}$

$\sigma_{2s}$

$O_2$

2s

O

FIGURE 12.4:
Molecular orbitals of $O_2$.

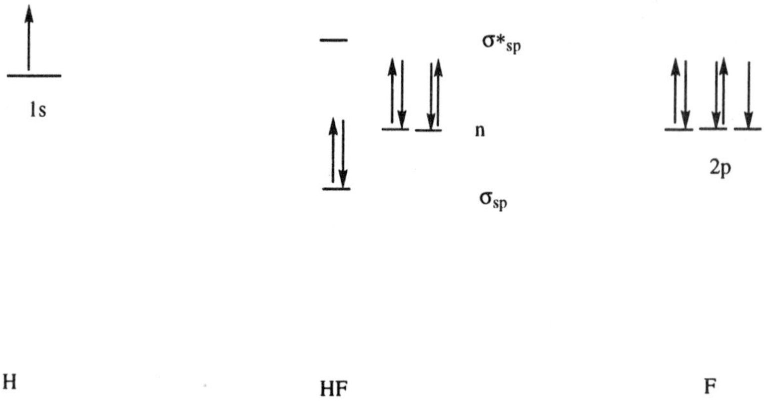

FIGURE 12.5:
Molecular orbitals of $F_2$.

1s

$\sigma^*_{sp}$

n

$\sigma_{sp}$

2p

H                                    HF                                    F

## 12.4 Π MOLECULAR ORBITALS FROM ATOMIC ORBITALS

If bond formation is to occur by orbital overlap, it is crucial that the orbitals be properly oriented to one another for the bond to form. Consider the formation of a $\pi$ bond between two $2p_z$ orbitals located on adjacent carbon atoms. Figure 12.6 shows this overlap for two different relative orientations of the $2p_z$ orbitals involved in the bond. In one case both orbitals have their positive lobes oriented upward, in the other the positive lobe of one and the negative lobe of the other point upward. When the orbitals overlap, the resulting molecular orbital wave function is simply the sum of the values of the two overlapping wave functions at each point in space. In one case the two wave functions add and the resultant molecular orbital is larger everywhere. This is the situation that we expected when we set out to determine why atoms bond. They bond because the electrons are concentrated in the area between the atoms. In the other case, the positive values of the one orbital are canceled by the negative values of the other orbital and the resulting molecular orbital is smaller everywhere and, most important, has a value of zero in one plane perpendicular to the bond axis (note that the molecular orbital will also have a value of zero in the xy plane — a result of the characteristics of the atomic $p_z$ orbitals from which it is formed.)

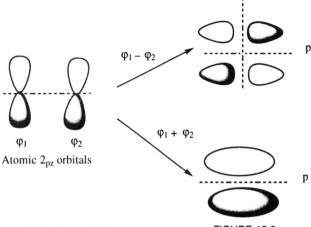

FIGURE 12.6: Overlap of $2p_z$ orbitals forming $\pi$ orbitals .

Once again we wish to emphasize that both of the molecular orbitals have nodal planes which are shown in figure 12.6 as a dashed line. One of these includes the bond axis. This is a result of the properties of the atomic p orbitals from which the molecular orbital is derived. It is this property (a nodal plane that includes the bond axis) of the molecular orbitals that causes them to be named as $\pi$ orbitals. It is even more important to note that the orbital obtained by subtracting the two atomic orbital wave functions has an additional nodal plane. This nodal plane is perpendicular to the bond axis. The effect of this additional nodal plane is to place a region of zero electron density exactly between two atoms. This is contrary to what we expect for bonding to occur. As a result, we expect this to be an unfavorable situation. Molecular orbital calculations of the energy of these orbitals bear out this expectation as well as the expectation that the other orbital describes a situation favorable to the formation of a bond.

## 12.5   THE HÜCKEL APPROXIMATION FOR MOLECULAR ORBITALS

From this point forward we shall be interested only in the $\pi$ molecular orbitals of the molecule. As we have previously mentioned, when molecules instead of atoms are involved, the only thing that changes about Schrödinger's equation is the exact form of the wave function. As we also have mentioned earlier, the most common approach to the formation of molecular orbitals from atomic orbitals is the LCAO–MO approach. For a number of reasons, too numerous and mathematically complicated to discuss at this point, it is possible to separate the molecular orbital energies into several different types that, we assume, do not influence one another. This assumption is one that has been used for a long time and seems to be justified because experimental measurements agree with the results obtained using this method. We shall, therefore, take the approach that we can separate the $\pi$ orbital system of a molecule from its $\sigma$ orbital system and consider only the $\pi$ orbitals. Energies that are calculated will be $\pi$ electron energies. Once the energies of the $\pi$ orbitals are obtained, it is possible to determine many other parameters of the molecule.

## 12.6   ENERGY LEVELS

The energies of the $\pi$ molecular orbitals are obtained by solving equation 12.1 for each of the molecular orbitals in the $\pi$ system of the molecule. Exact solutions of the equations are not possible. As a result various approximations have been made. One commonly used approximate method is the Hückel approximation that reduces the mathematics necessary for obtaining the energies to solving a determinate to find the roots of its associated polynomial. Because of these simplifications, the solutions require little more than a good working knowledge of high school algebra. For those who wish to try their hand at these solutions, Appendix C provides the details of the Hückel solution for the molecules ethene and 1,3–butadiene. Some students may wish to explore the solutions of the Hückel equations for other systems. At this point, we shall simply use the results of these calculations to discuss the shapes and energy levels of the $\pi$ orbitals in each of these systems.

The solutions to equation 12.1 for the two $\pi$ molecular orbitals of ethylene yield the two molecular orbitals and their associated energies shown in table 12.1. Solutions of the Hückel determinants provide the energies of the $\pi$ molecular orbitals in terms of a unit given the symbol $\beta$. $\beta$ is the amount of energy that is released when one electron, initially in an atomic p orbital, is placed into the $\pi$ molecular orbital. A positive value of $\beta$ indicates that the orbital is a stable, bonding orbital. The Hückel method does not provide the value of $\beta$, merely the energies of all of the $\pi$ orbital energies relative to one another in units of $\beta$. The zero energy level is taken to be the energy of the electron in the atomic p orbital. What should be of most interest to us are the energies of each orbital and the coefficients of the atomic orbital terms making up the molecular orbital. The energies of the two orbitals show that placing an electron in one of the orbitals is more stable than the unbonded situation (the zero point of energy corresponds to the unbonded situation) and placing an electron in the other orbital is less stable than the unbonded situation. Because each of the molecular orbitals has a nodal plane that includes the bond axis, they are both $\pi$ orbitals. Notice that the atomic orbitals are equally weighted (the coefficients of the atomic orbital terms describe "how much" each atomic orbital contributes to the particular molecular orbital) in both molecular orbitals. Pictures of the ethylene molecular orbitals are identical to those already shown in figure 12.6. These pictures can be considered representative of any simple, unconjugated double bond situation.

**TABLE 12.1** MOLECULAR ORBITALS AND ENERGIES FOR ETHYLENE TABLE 12.1

| MOLECULAR ORBITAL | ENERGY |
|---|---|
| $\Psi_1 = 0.71\ \varphi_1 + 0.71\varphi_2$ | $\beta$ |
| $\Psi_2 = 0.71\ \varphi_1 - 0.71\varphi_2$ | $-\beta$ |

When we move to a larger molecule, butadiene, the four carbon atoms involved in the $\pi$ electron system produce four molecular orbitals. Solution of the Hückel equations for butadiene produces the results in table 12.2. Representations of the four molecular orbitals are presented in figure 12.7. These pictures are representative of situations in which two double bonds are **conjugated** with one another.

**TABLE 12.2** MOLECULAR ORBITALS AND ENERGIES FOR BUTADIENE TABLE 12.2

| MOLECULAR ORBITAL | ENERGY |
|---|---|
| $\Psi_1 = 0.37\varphi_1 + 0.60\varphi_2 + 0.60\ \varphi_3 + 0.37\ \varphi_4$ | $1.618\ \beta$ |
| $\Psi_2 = 0.60\ \varphi_1 + 0.37\ \varphi_2 - 0.37\ \varphi_3 - 0.60\ \varphi_4$ | $0.618\ \beta$ |
| $\Psi_3 = 0.60\ \varphi_1 - 0.37\ \varphi_2 - 0.37\ \varphi_3 + 0.60\ \varphi_4$ | $-0.618\ \beta$ |
| $\Psi_4 = 0.37\ \varphi_1 - 0.60\ \varphi_2 + 0.60\ \varphi_3 - 0.37\ \varphi_4$ | $-1.618\ \beta$ |

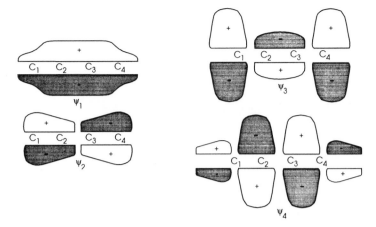

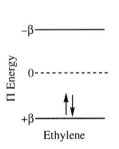

It is important to understand that the symmetry properties of butadiene are such that carbon atoms 1 and 4 must be equivalent to one another. Similarly, atoms 2 and 3 also must be equivalent to one another. This has many ramifications — most of which are beyond the scope of this book. One that is important for us here is that the symmetry of the molecule requires that, if there is a nodal plane between atoms 1 and 2, there must also be a nodal plane between atoms 3 and 4. Notice that the wave functions $\Psi_1$ and $\Psi_2$ do not have a nodal plane between those atoms, while wave functions $\Psi_3$ and $\Psi_4$ do have a nodal plane in that position. A second consequence of the symmetry of the molecule is that the absolute value of the coefficients of atomic orbitals $\varphi_1$ and $\varphi_4$ must be equal and those of $\varphi_2$ and $\varphi_3$ must also be equal. An understanding of these symmetry relationships will help with the formulation of the molecular orbitals for larger conjugated $\pi$ orbital systems.

Once we know the energies of the orbitals, it is an easy matter to construct an energy level diagram for the $\pi$ system of the molecule. These energy level diagrams look much like the atomic energy level diagrams that we drew in book 1, chapter 9. Since each molecular orbital can hold two electrons if the spins of the electrons are paired, placing the proper number of electrons into the orbitals with the lowest available energy produces the energy level diagrams for the **ground state** shown in figure 12.8. In a similar way to what we used to indicate the electron distribution in atoms, we would write the electron distribution for the ground state of ethylene as $\pi_1{}^2$ and that of butadiene as $\pi_1{}^2\pi_2{}^2$.

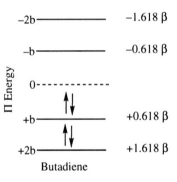

We can use these energy level diagrams to determine the $\pi$ bond energy of a molecule. Seeing that the ethylene molecule has two electrons in an orbital whose $\pi$ energy is 1 $\beta$, the total $\pi$ energy in the molecule must be 2 $\beta$. For the butadiene molecule, placing four electrons into the molecular energy diagram as shown in figure 12.8 results in a total $\pi$ energy of 4.472 $\beta$. Note that conjugation produced a bonus stabilization. A system of two conjugated $\pi$ bonds is more than twice as stable as a single $\pi$ bond.

Finally, notice that the energy of the particular orbital increases as the number of nodal planes perpendicular to the bond axis increases. This property is a general one: the more nodal planes perpendicular to the bond axis in an orbital, the higher its energy.

## 12.7 ELECTRONIC ABSORPTION SPECTRA

In the last section we calculated the energy of the molecule when the molecule had its electrons in the lowest possible energy $\pi$ molecular orbitals. This is an energy state termed the ground state. It is also possible to place the electrons in these orbitals differently. For instance, in the butadiene molecule it would have been possible to take one of the electrons out of the $\pi_2$ orbital and place it in the $\pi_3$ orbital instead to produce an electron distribution that we could characterize as $\pi_1{}^2\pi_2{}^1\pi_3{}^{*1}$. If this were done, the $\pi$ stabilization energy for the molecule would be 3.236 $\beta$. This is certainly not the lowest energy state of the molecule. It is termed an **excited state**. The molecule must gain energy when it goes from the ground to the excited state. The most common way for this to be done is to absorb a photon. Since an electron is being moved from a $\pi$ molecular orbital to a $\pi^*$ molecular orbital, this electronic transition is referred to as a $\pi \rightarrow \pi^*$ (pi to pi star) **transition**.

Absorption of a photon also provides a means by which we can determine the energy difference between the ground and excited states. If we look at the energy level diagram for the ethylene molecule in figure 12.8, we notice that the excited state is produced by moving an electron from an orbital in which its energy is $+\beta$ and placing it in an orbital in which its energy is $-\beta$. This requires a photon carrying 2 $\beta$ energy. With this information, and knowing that the ethylene absorbs at approximately 175 nm in the ultraviolet region of the spectrum, we can calculate the approximate value of $\beta$. (This is only approximate because of all of the assumptions that had to be used to calculate the energy levels.) Equations 12.7 and 12.8 present these calculations, first for the energy of the photon and then converting those values to a per mole basis. Notice that this is an enormous amount of energy — quite enough to break a typical bond.

$$E = \frac{hc}{\lambda} = \frac{(6.626 \times 10^{-34}\ \text{J s})(2.998 \times 10^8\ \text{m/s})}{175 \times 10^{-9}\ \text{m}} = 1.44 \times 10^{-18}\ \text{J} \qquad \text{EQ. 12.7}$$

$$E = (1.44 \times 10^{-18}\ \text{J})(6.022 \times 10^{23}\ \text{mol}^{-1}) = 6.84 \times 10^5\ \text{J/mol} = 684\ \text{kJ/mol} \qquad \text{EQ. 12.8}$$

Since the transition energy actually represents 2 $\beta$, the value of $\beta$ in the ethylene molecule is about 342 kJ mol$^{-1}$. Similar calculations can be done with other molecules. In all cases, the amount of energy involved in an electronic transition is very large and comparable to the bond energy in a typical bond. It should be emphasized, however, that the value of $\beta$ varies from one molecule to another and calculations of this type need to be done for each molecule. Even so, for reasons that are, again, beyond the scope of this book, the values are only approximate.

The wavelengths at which simple $\pi$ systems absorb are greatly affected by the length of the conjugated $\pi$ system of the molecule but are not greatly affected by saturated alkyl groups. For instance, the absorption of ethylene, which occurs at 174 nm, shifts to 187 nm when an ethyl group is attached to make

1–butene. In 1,4 pentadiene, a non–conjugated diene, the absorption maximum shifts only as far as 178 nm — essentially unshifted from ethylene. On the other hand, in 1,3 butadiene — a conjugated diene, the absorption shifts all the way to 217 nm. It takes a great deal less energy to promote an electron from the highest energy filled orbital to the lowest energy unfilled orbital in 1,3 butadiene than it does in ethylene. Indeed, increasing the $\pi$–electron system by one additional $\pi$ bond adds approximately 40 nm to the wavelength of absorption for the first few conjugated bonds. The long–chain, unsaturated compound, $\beta$–carotene, with 11 conjugated bonds, absorbs at 494 nm. These data are summarized in table 12.3.

TABLE 12.3

**TABLE 12.3** ABSORPTION MAXIMA FOR SELECTED UNSATURATED HYDROCARBONS

| STRUCTURE | NUMBER CONJUGATED BONDS | WAVELENGTH (nm) | TRANSITION ENERGY (kJ mol$^{-1}$) |
|---|---|---|---|
| | 1 | 175 | 684 |
| | 1 | 178 | 673 |
| | 2 | 217 | 552 |
| | 3 | 265 | 452 |
| $\beta$–Carotene | 11 | 494 | 243 |

## 12.8 PHOTOCHEMICAL REACTIONS

One of the things that a molecule can do after it absorbs light is to undergo a chemical reaction. This is not surprising considering the large amount of energy imparted to the molecule by absorption of the photon. In some cases this energy merely supplies the energy of activation necessary to get the molecule over the energy barrier and on the way to products. In many other molecules, however, the products formed after irradiation are quite different from those formed when this activation energy is supplied by heating the reaction mixture. In the situations in which there is a difference in the products that are formed, the products resulting from reaction under conditions of irradiation are termed the **photochemical products** and those resulting from the "standard way" of carrying out the reaction are termed the **thermal products**. The real difference is whether or not the reaction occurs when the reactant is in the excited state (photochemical reaction) or whether reaction occurs when the reactant is in the ground state (thermal reaction).

The simplest chemical reaction resulting from the absorption of light by an alkene is isomerization. In the ground state there is a very large energy barrier to rotation around the carbon–carbon bond axis. It is easy to see the reason for this energy barrier — rotating around the carbon–carbon bond axis requires that the $\pi$ bond break and then reform. Breaking the $\pi$ bond is such a large barrier to rotation that alkenes simply do not thermally isomerize even though one of the isomers may be more thermodynamically stable than the other isomer. However, in the excited state, when one electron is in the $\pi$ (bonding) orbital and one electron is in the $\pi^*$ (anti–bonding) orbital, there is effectively no $\pi$ bond in the molecule. The barrier to rotation is, therefore, non–existent. It is not possible to observe this with ethylene. In larger molecules, for instance, 2–butene, it would be possible. If our mental picture of the process is correct, we would expect that it that *cis*–2–butene would isomerize to *trans*–2–butene when it is irradiated by UV radiation at wavbelengths that it can absorb. This does, indeed, happen. A photochemical *cis-trans* isomerization reaction in the rhodopsin molecule is important in generating the nerve impulses involved in sight.

## 12.9    FRONTIER ORBITALS AND PHOTOCHEMICAL DIMERIZATION REACTIONS -

Of great importance in the application of molecular orbital theory to synthesis are the characteristics of the highest occupied molecular orbital and the lowest unoccupied molecular orbital. In the ground state of butadiene, the highest energy electron occupies the $\pi$ orbital whose wave function is given by $\Psi_2$. Since it is the highest energy orbital occupied, it is termed **HOMO** (**H**ighest **O**ccupied **M**olecular **O**rbital). The orbital whose wave function is given by $\Psi_3$ is the lowest energy unoccupied orbital. It is, therefore termed the **LUMO** (**L**owest **U**noccupied **M**olecular **O**rbital). Both designations just made are valid, as described, for *the ground state* of butadiene. When an electron is promoted from $\Psi_2$ to $\Psi_3$ to form the lowest energy excited state, the definitions of HOMO and LUMO must change. Under these circumstances, $\Psi_3$ becomes the HOMO and $\Psi_4$ becomes the LUMO. As we shall see, these orbitals are key to understanding what will happen to the molecule during a chemical reaction.

Theoretical work done in the 1960's led to the understanding of a major class of reactions that occur through a cyclic transition state. These reactions are **concerted reactions**; that is, they are single step reactions in which all bonding changes occur simultaneously. These reactions are best understood by the application of molecular orbital theory. For real molecules, with a large number of molecular orbitals, the use of all molecular orbitals would make the description of the changes unnecessarily complicated. Fortunately only a few molecular orbitals (HOMO and/or LUMO of each reactant) are normally involved in the reactions.

Chemical reactions involve the breaking and forming of bonds. When bonds form, the electrons participating in the bond occupy molecular orbitals that form to accommodate the electrons. In concerted reactions involving molecules with $\pi$ electrons — reactions that proceed by the formation of an intermediate transition state that involves all of the $\pi$ electrons — these temporary molecular orbitals are formed by donating electrons from the HOMO of one reactant into the LUMO of the other reactant. It is, therefore, the characteristics of these orbitals that determine the course of the reaction. These orbitals are termed **frontier orbitals** because they are on the "frontier" of the reaction.

Consider first the dimerization reaction of a simple alkene to form a cycloalkane. Such a reaction is termed a [2+2] reaction since each reactant contributes 2 π electrons to the transition state. The overall proposed reaction is shown as equation 12.9.

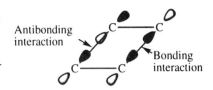

EQ. 12.9

In this reaction we might envision a cyclic transition state in which the electrons from the HOMO of one molecule interact with the LUMO of the other molecule. This would require that the two molecules approach each other with the π clouds pointing toward each other as shown in figure 12.9. In this figure the lower ethylene molecule's HOMO is pictured and the upper molecule's LUMO is pictured. Notice that the interactions include one bonding type interaction (the orbitals are in phase on the right–most two carbon atoms) and one anti–bonding interaction (the orbitals are out–of–phase on the left–most two carbon atoms). Since we are attempting to form two sigma bonds to complete the ring, it would be possible to form the bond between the two carbon atoms at the right of the figure, but not at the left of the figure. Therefore the molecule will not cyclize.

FIGURE 12.9:
Thermal interaction of the HOMO and LUMO in ethylene.

When the reaction mixture is irradiated and one of the molecules is raised into the excited state, however, the situation is quite different. The HOMO of the excited state molecule now has the electron distribution pictured at the bottom of figure 12.10. The LUMO of the other molecule is pictured at the top of figure 12.10. Now both of the interactions are in–phase (bonding) interactions and the molecule can cyclize.

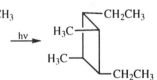

FIGURE 12.10:
Photochemical interaction of the HOMO and LUMO in ethylene.

Not only is the cyclization possible in the excited state, but the addition reaction is also stereoselective. Let us illustrate this with E–2–pentene. Figure 12.11 shows two E–2–pentene molecules approaching one another to undergo the cyclization. What is important to note is that the ethyl and methyl groups must end up on opposite sides of the ring once it has formed. There are several ways that the two molecules could approach each other. The approach shown is the one that would lead to the most stable product since it keeps the ethyl groups as far away from one another as possible. It is suggested that the student examine the stereochemistry of the products formed from E–2–pentene molecules approaching one another from different orientations as well as the stereochemistries of products formed from the photochemical cyclization of Z–2–pentene.

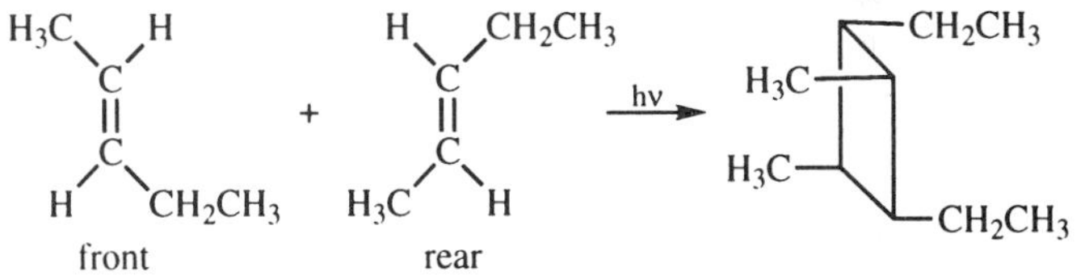

FIGURE 12.11:
Photochemical cyclization of E-2-pentene.

Reactions of this type are thought to be the reason for photochemical damage to DNA. A [2+2] cyclization involving 2 adjacent thymine residues of the DNA strand would lead to the formation of a cyclobutane ring. Since this strand of DNA would not be capable of replicating properly, at the very least the result would be the loss of genetic information stored at that point on the chain. This is pictured in equation 12.10. Fortunately, the reverse of this reaction is also a photochemical reaction and some of the UV damage to the DNA can be repaired by photochemical ring opening. If the photochemical ring opening reaction does not happen, the cell also has enzymes that can cut out and repair the damaged section of DNA. If there is not too much damage caused by UV irradiation for these repair mechanisms to handle, the function of the DNA molecule can be restored rather quickly and the organism will not suffer lasting damage. However, if the DNA is extensively damaged, or if the damaging reactions occur at the wrong time, serious consequences can result leading, even, to the death of the organism.

EQ. 12.10

## 12.10   ELECTROCYCLIC REACTIONS

**Electrocyclic reactions** are an interesting class of reactions that lend additional insight into the role of frontier orbitals in chemical reactions. In these reactions a cyclic unsaturated compound is in equilibrium with an open chain polyene of the same empirical formula. Let us look at the simplest two such reactions: the equilibrium between cyclobutene and butadiene and that between 1,3 cyclohexadiene and 1,3,5 hexatriene. Both of the cyclic compounds undergo the thermal chain–opening reactions at elevated temperatures shown in reactions 12.11 and 12.12.

EQ. 12.11

EQ. 12.12

Let us now look at these reactions from the standpoint of the molecular orbitals involved. First we shall consider the reaction in equation 12.11.

In equation 12.11, we must break a sigma bond between the carbon atoms that will later become atoms number 1 and 4 in the open chain compound. The electrons in the sigma bond must then be incorporated into a second $\pi$ bond. At the same time, the existing $\pi$ bond must migrate one carbon atom in either direction. While, on the surface, this looks to be a difficult task, in practice it is quite simple from the standpoint of the orbitals involved. Consider figure 12.12. In this figure, the sigma bond between the carbon atoms indicated with the asterisk has just been broken leaving the atomic $\pi$ orbitals. In order to form a bonding overlap with the LUMO of the double bond located on the other two atoms, each of the two carbon atoms must rotate 90° counterclockwise (as viewed down the bond axis). Because both bonds are rotating in the same direction, this motion is termed *conrotatory*. Notice that once the second $\pi$ bond forms, the orbital has the shape of the $\Psi_2$ wave function for butadiene.

This is the HOMO of the newly formed butadiene.

To understand the opening of the cyclohexadiene ring, we will need to look at the shapes of the molecular orbitals for 1,3,5 hexatriene. Figure 12.13 shows the signs of the atomic wave functions on each carbon atom in each of those molecular orbitals. In this figure $\Psi_1$ is the lowest energy (most stable) orbital and $\Psi_6$ is the highest energy orbital. Since the molecule has 6 $\pi$ electrons, the frontier orbitals for the ground state of the molecule are the $\Psi_3$ and $\Psi_4$ orbitals.

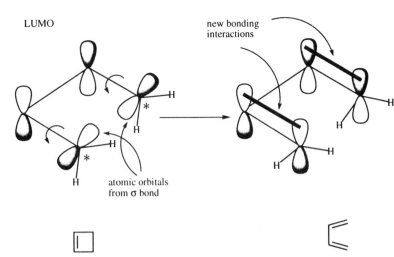

LUMO

new bonding interactions

atomic orbitals from σ bond

FIGURE 12.12: Molecular orbitals involved in the opening of the cyclobutene ring.

FIGURE 12.13: Molecular orbitals of hexatriene.

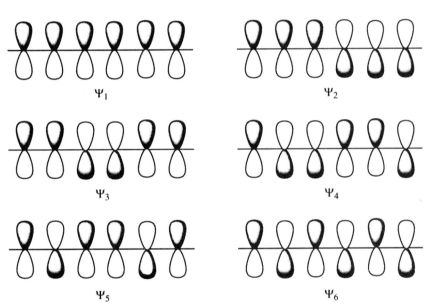

$\Psi_1$

$\Psi_2$

$\Psi_3$

$\Psi_4$

$\Psi_5$

$\Psi_6$

Let us now consider the situation with the cyclohexadiene molecule. Let us number the carbons atoms in this molecule differently than would be done by IUPAC in order to discuss the reaction. Specifically, instead of using the lowest numbers possible to designate the position of the double bonds, let us consider that the double bonds formally exist between carbons 2 and 3 and between carbons 4 and 5. Thus numbered the molecule would be named 2,4 cyclohexadiene instead of 1,3 cyclohexadiene. While incorrect according to IUPAC naming conventions, it lets us discuss the ring opening reaction without having to renumber carbon atoms in the middle of the reaction.

If we now look at the molecular orbital situation just after the sigma bond between carbon atoms number 1 and 6 break, we find the picture in figure 12.14. Carbon atoms 1 and 6 now must rotate in opposite directions (termed *disrotatory*) for the proper overlap of the atomic p orbitals with the LUMO of the diene portion of the molecule (this LUMO would have the same characteristics as $\Psi_3$ for the butadiene molecule pictured in figure 12.7). After overlap occurs, the orbital that forms is the HOMO ($\Psi_3$) of the 1,3,5–hexatriene molecule.

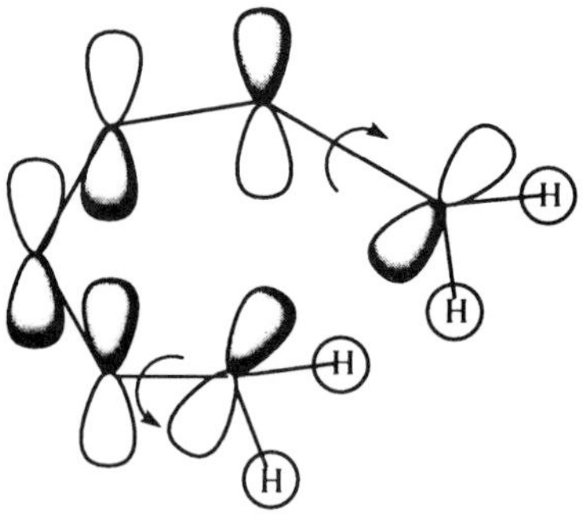

By this time you may be wondering how it is that we know that this explanation of what happens during these ring–opening reactions is, indeed, correct. The answer is, of course, that experiments were designed to test these hypotheses. Those experiments were done with substituted cyclobutenes and cyclohexadienes. The opening of the *cis*–3,4–dimethylcyclobutene has been shown to yield >99% (2Z,4E)–2,4–hexadiene while the *trans* isomer yields (2E,4E)–2,4–hexadiene. Figure 12.15 shows that this is consistent with conrotation of the carbon after σ bond breakage. In a similar fashion, *cis*–5,6–dimethyl–1,3–cyclohexadiene yields principally (2E,4Z,6E)–2,4,6–octatriene. The student should confirm this.

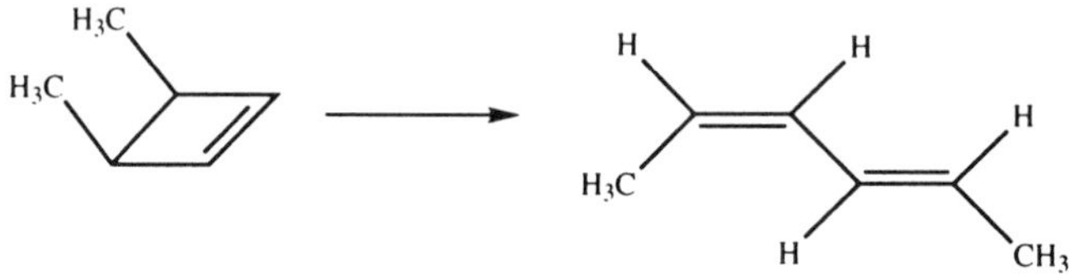

FIGURE 12.14: Molecular orbital orientation upon σ bond breaking.

FIGURE 12.15: Formation of *(2Z,4E)*-2,4-hexadiene from *cis*-3,4-dimethylcyclobutene.

Both of the reactions discussed above are equilibrium reactions. However, the equilibrium constants for the formation of the open–chain compound are quite different. The equilibrium constant for the reaction in equation 12.11 is very large. The open chain compound is favored over the cyclic compound in this case. This is, of course, because of the large amount of ring strain caused by the four–membered ring — ring strain that is increased because of the double bond incorporated into the ring. On the other hand, the equilibrium constant for equation 12.12 strongly favors the cyclic compound because, in this case, the ring is a six–membered ring.

One additional piece of evidence that helps to support the hypothesis advanced above is provided by examining the stereochemical configurations of the products formed in the photochemical cyclization reactions. Figure 12.16 summarizes both the thermal and photochemical electrocyclic reactions for the 2,3–dimethylcyclobutene isomers. Notice that the stereochemistry of the products is different. Does this make sense? Use figure 12.7 to determine whether rotation of the ends of the open chain compound must be conrotatory or disrotatory for the formation of a sigma bond between the end carbon atoms. Don't forget that, after the molecule is excited, the highest energy electron is now in $\Psi_3$. It is this orbital, now the HOMO, which determines the direction of rotation for σ bond formation. Likewise, figure 12.17 summarizes the reactions for 2,4,6 octatriene. The student is left the exercise of rationalizing the molecular orbital picture and the products for the photochemical reaction.

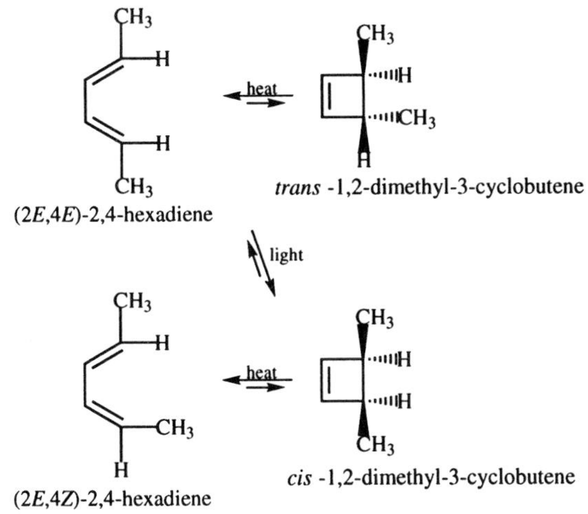

**(2E,4E)-2,4-hexadiene** ⇌ *heat* ⇌ **trans-1,2-dimethyl-3-cyclobutene**

**(2E,4Z,6E)-octatriene** ⇌ Δ ⇌ **cis-1,2-dimethyl-3,5-cyclohexadiene**

light

**(2E,4Z)-2,4-hexadiene** ⇌ *heat* ⇌ **cis-1,2-dimethyl-3-cyclobutene**

**(2E,4Z,6Z)-octatriene** ⇌ Δ ⇌ **trans-1,2-dimethyl-3,5-cyclohexadiene**

FIGURE 12.16:
Electrocyclic reaction summary for 2,4-hexadiene.

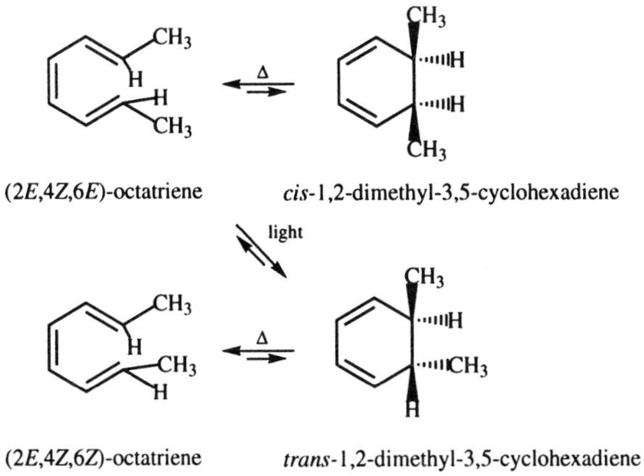

FIGURE 12.17:
Electrocyclic reaction summary for 2,4,6-octatriene.

## 12.11  CYCLOADDITION, DIELS-ALDER REACTION

A second type of cycloaddition reaction, much used in synthesis because the product contains a six–membered ring, is the **Diels–Alder reaction**. The course of this reaction also can be most easily understood from a molecular orbital picture of the reactants. In a Diels–Alder reaction a conjugated diene and a dienophile containing a double bond or triple bond react to produce a cyclic product by addition of the dienophile across the 1 and 4 positions of the diene. Although, for reasons we shall discuss later, ethylene will not work well in the reaction, for the purposes of illustration let us consider the simplest possible example of a Diels–Alder reaction, the condensation of ethylene with butadiene shown in equation 12.13. The condensation of ethylene with butadiene would form cyclohexene. Notice that this reaction shares many similarities with the dimerization reactions of alkenes that we discussed earlier in this chapter. Using the symbolism that we used at that time, the reaction would be referred to as a [4+2] cycloaddition.

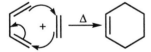

EQ. 12.5

Let us now apply our knowledge of molecular orbital theory to the reaction. We expect that the electrons in the HOMO of the diene are higher in energy than those in the HOMO level of the dienophile. This is, of course, confirmed for the present system by noting that butadiene absorbs at longer wavelength than does ethylene. We then expect that the cyclic intermediate will be formed by donation of the HOMO electrons of the diene into the LUMO of the dienophile. This is illustrated in figure 12.18 as we picture the ethylene molecule approaching from the underside of the diene. Overlap of the lobes of the ethylene LUMO with those of the lobes on the end carbon atoms of the butadiene's HOMO produces the two σ bonds that must be formed in the reaction.

The remaining two lobes of what was the diene then become the LUMO of the new double bond in the adduct.

A simple alkene such as ethylene is not sufficiently active as a dienophile to react easily in the Diels–Alder reaction. What is required is the presence of some activating group or groups. The best activating groups have electron withdrawing groups such as carbonyl groups, cyano groups, or nitro groups conjugated with the double or triple bond. Dienophiles such as maleic anhydride, fumaric and maleic acids and the dimethyl esters of maleic and fumaric acid are commonly used in Diels–Alder reactions. All of these compounds make good dienophiles because of the number and strongly electron withdrawing nature of the groups conjugated with the double bond.

FIGURE 12.18: Molecular orbitals involved in the Diels-Alder reaction.

As one would expect, the Diels–Alder reaction has a very specific stereochemistry. It is obviously a requirement of the concerted reaction that the diene be in the configuration in which the double bonds point in the same direction as shown in figure 12.19a rather than the configuration shown in figure 12.19b. This configuration is termed the *S–cis* configuration. Any other substituents on the diene molecule will remain in the same relative configuration in the adduct as they had in the diene. Thus, when (2*E*,4*E*)–2,4–hexadiene (figure 12.19c) or (2*Z*,4*Z*)–2,4–hexadiene serve as the diene, the methyl groups end up on the same side of the ring whereas reaction of (2*Z*,4*E*)–2,4–hexadiene results in the methyl groups being placed on opposite sides of the ring. It should be noted, though, that the steric hindrance would make (2*Z*,4*Z*)–2,4–hexadiene a rather poor diene for this reaction.

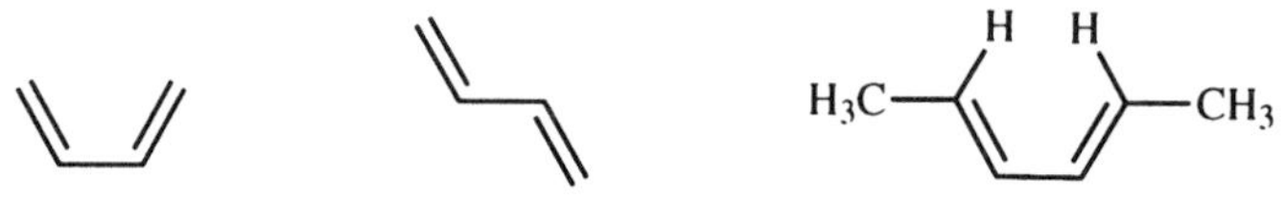

a. Reactive Configuration

b. Non-Reactive Configuration

c. (2*E*, 4*E*)-2, 4-hexadiene

FIGURE 12.19: Reactive and non-reactive diene configurations.

The same steric considerations that apply to the diene also apply to the dienophile. For instance, use of maleic acid (or anhydride) as the dienophile always results in a cyclic product that has the carbonyl groups on the same side of the ring while use of fumaric acid results in a product having the carbonyl groups on opposite sides of the ring. See equation 12.14 for the maleic acid reaction with butadiene and equation 12.15 for the fumaric acid reaction. The student should confirm the stereochemistry of these reactions by sketching out the course of the reaction using the molecular orbitals of the molecules to determine how the σ bonds will form.

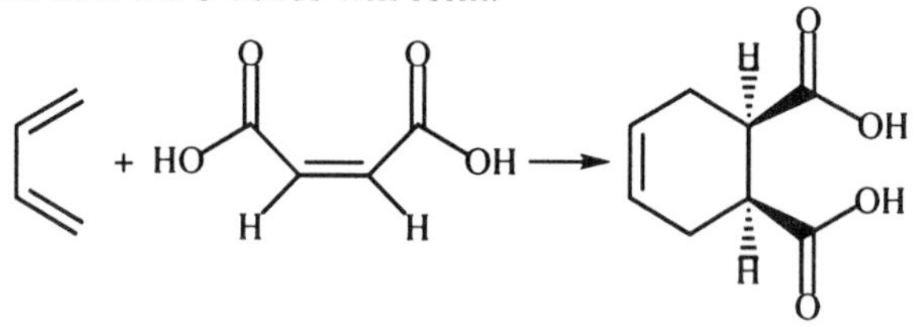

EQ. 12.14

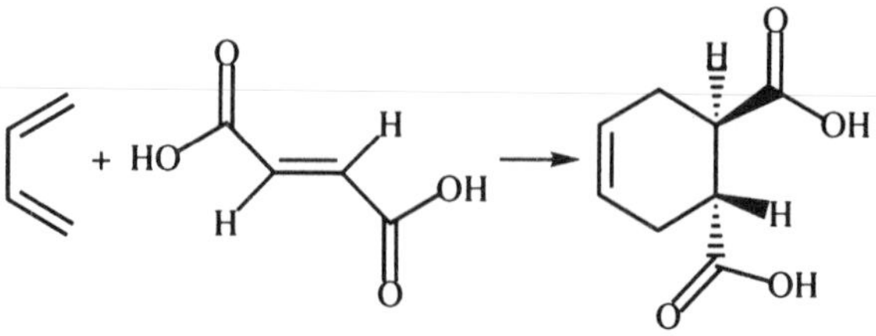

EQ. 12.15

If a dienophile is used that contains a triple bond, the cyclic product will contain two, non–conjugated double bonds in a six–membered ring. For instance, consider the reaction represented in equation 12.16. Note that the methyl groups both ended up on the same side of the ring.

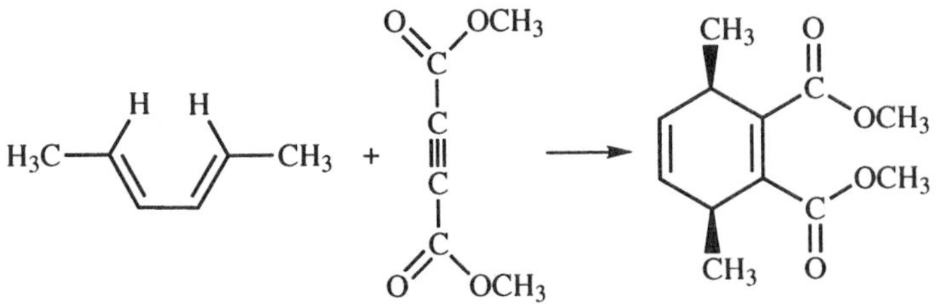

EQ. 12.16

If the diene is a cyclic diene, addition of a dienophile creates a bridged compound. The classic example of this is the dimerization of cyclopentadiene (equation 12.17) in which the same molecule acts as both diene and dienophile. Note that addition of one of the double bonds of one molecule of cyclopentadiene across the 1 and 4 positions of the other molecule creates a molecule with fused rings and a bridging group. This reaction happens slowly at room temperature. Dicyclopentadiene, the common name of the bridged compound, is sufficiently stable that the monomer must be stored in an ice–surrounded flask and used quickly after it is produced by thermal cracking of the dimer. After only a short time at room temperature, significant quantities of the monomer will have undergone the Diels Alder addition to form the dimer.

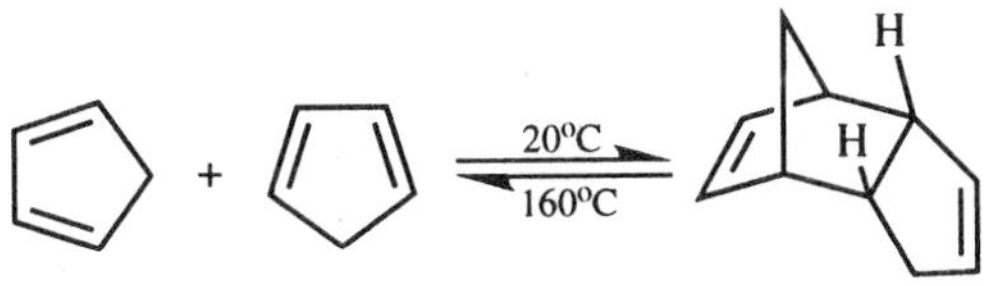

EQ. 12.17

There is yet a second configuration that would be possible for the adduct. The compound shown as the product of equation 12.17 has the added ring tucked downward away from the bridge. There should also be a configuration for the product in which the ring is folded upward and the hydrogen atoms go downward. Figure 12.20 shows both of these configurations. Notice that the configuration of the left has the ring tucked downward into the space formed by the cavity of the six–membered ring. This configuration is termed *endo* — literally "within". In the other configuration, the five–membered ring is pointed away from the cavity. Such a configuration is termed *exo* — literally "outside".

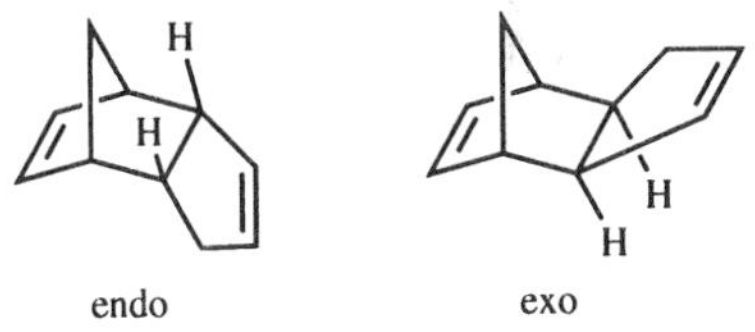

Which configuration, endo or exo, would we expect to be more stable and why? At first glance one might expect that the exo configuration would be the more stable simply on the basis of steric considerations. This does not turn out to be true. In almost all situations, the product formed is predominantly the endo product. The reason is that only an approach of the dienophile from the underside of the diene will allow the other molecular orbitals of the dienophile to interact with the orbitals of the diene. This interaction, while not leading to bond formation, stabilizes the intermediate sufficiently that the products have almost exclusively endo configuration.

endo

exo

FIGURE 12.20:
Two configurations
for
dicyclopentadiene.

**KEY WORDS**
**& Concepts**

**Anti–bonding orbital:** A molecular orbital with an energy higher than the energy of the atomic orbitals from which it was formed.

**Bonding orbital:** A molecular orbital with an energy lower than the energy of the atomic orbitals from which it was formed.

**Concerted reaction:** A reaction which occurs in one step.

**Conjugated:** A system of two or more double bonds each separated by exactly one single bond.

**Diels–Alder reaction:** A reaction in which a compound with a double bond reacts with a compound with a conjugated double bond system to form a ring. The simple compound adds to the diene across the ends of the diene system.

**Electrocyclic reactions:** A type of reaction in which a ring compound and an open–chain compound of the same empirical formula are in equilibrium with one another.

**Electron density:** A term used to describe how "concentrated" the electron(s) is (are) at a particular point in the molecule.

**Excited state:** That electronic state of a molecule or atom in which at least one electron is in an energy state above the lowest possible one. (Technically this should be termed the "electronic excited state" since there are vibrational and rotation excited states as well.)

**Frontier orbitals:** The molecular orbitals that take part in a reaction. These are the HOMO and LUMO orbitals.

**Ground state:** That electronic state of a molecule or atom in which all electrons are in their lowest possible energy states. (Technically this should be termed the "electronic ground state" since there are vibrational and rotation ground states as well.)

**HOMO:** The highest energy occupied molecular orbital in a molecule.

**LCAO–MO:** A means of formulating waves functions for molecular orbitals by making linear combinations of atomic orbital wave functions.

**LUMO:** The lowest energy unoccupied molecular orbital in a molecule.

**Mathematical operator:** A mathematical term for something that carries out a mathematical operation on a function. Common examples include multiplication and division operators.

**Molecular orbital:** An orbital that belongs only to a molecule. When the molecule ceases to exist, so does the orbital.

**Nodal plane:** A plane in the molecule on which the value of the wave function is zero. This means that the electron density in this plane must be exactly zero.

**$\pi \rightarrow \pi^*$ transition:** An electronic transition that involves the movement of an electron from a $\pi$ bonding orbital to a $\pi$ anti–bonding orbital. This movement of the electron is caused by absorption of a photon.

**$\pi$ bond:** A molecular bond formed formed by the side–to–side overlap of two or more atomic p orbitals. Because of the symmetry properties of the atomic p orbitals, all $\pi$ bonds have a nodal plane that includes the bond axis.

**Photochemical products:** Products formed when a reaction is run under illumination by a source of, usually, UV photons. Formation of these products involves participation by the excited state of one of the reactants.

**Substrate:** The reactant that is considered to be the one "acted upon" in a chemical reaction.

**Thermal products:** Products formed when a reaction is run under "normal", *i.e.*, non–irradiated, conditions. These products are formed from the electronic graound state of the molecule.

**Wave function:** A mathematical function that describes the wave–like properties of the electron as it interacts with its environment.

1. Use a computer program such as CAChe or Spartan to generate the molecular orbital diagrams and electron density maps for the following molecules:

    a. Pyridine
    b. Toluene
    c. Aniline
    d. 1,3-Cyclohexadiene
    e. Ethylene
    f. 2,3-Dimethyl-2-butene
    g. Maleic anhydride
    h. Phthalic anhydride

2. Use the results of the electron density mapping from problem 1 to predict which positions are most susceptible to electrophilic attack and which to nucleophilic attack for each of the molecules in problem 1.

3. Use a computer program to generate the molecular orbital energy level diagrams for:

    a. Ethylene $(C_2H_4)$
    b. Cyclobutadiene $(C_4H_4)$
    c. Benzene $(C_6H_6)$
    d. Cyclooctatetraene $(C_8H_8)$
    e. Cyclodecapentaene $(C_{10}H_{10})$
    f. Cyclodecahexaene $(C_{12}H_{12})$

4. Which of the molecules in problem 3 are aromatic? What do the molecular energy diagrams of these molecules have in common?

5. Use a computer program to generate the molecular orbital energy level diagrams for the following species:

    a. cyclopentadienyl anion $(C_5H_5^-)$
    b. cycloheptatrienyl cation $(C_7H_7^+)$

6. Predict whether or not the species in problem 5 are aromatic and then defend your prediction.

7. For each of the molecules in table 12.3, use a computer program to determine the difference in energy between the ground state and the lowest excited singlet state. How well does the computer program's prediction march the actual values in the chemical literature?

8. Rank the following alkenes in order of increasing wavelength in their UV spectra.

a.

b.

c.

d.

9. Three compounds all have the same molecular formula $C_5H_8$. Reaction of each of the compounds with two equivalents of hydrogen gas produces n-pentane. UV spectra reveal the following $\lambda_{max}$ values: compound A:223 nm, compound B: 224 nm, compound C: 178 nm. Suggest structural formulas for A, B and C. What ambiguity exists for A and B?

10. A compound has formula $C_6H_8$ and a $\lambda_{max}$ value of 256 nm. It decolorized two equivalents of bromine in carbon tetrachloride. The $\lambda_{max}$ value of cyclohexane is 182 nm. Suggest a structure of the unknown compound. Draw the structure of a close isomer of the sompound D and suggest a $\lambda_{max}$ value for that compound.

11. Draw the structures of the products formed when 1,3-butadiene reacts with each of the following.

a.

$H_3CO$—C with O double bond, C—$OCH_3$ with O double bond, C=C with H, H

dimethylmaleate [(Z)–dimethylbutenedioate]

b.

$H_3CO$—C with O double bond, C=C with H, C—$OCH_3$ with O double bond, H

dimethylfumarate [(E)–dimethylbutenedioate]

12. The following alkenes dimerize when the reaction mixture is irradiated with UV light. Draw the structures of the products.

   a. (*E*)-2-pentene
   b. (*Z*)-3-hexene
   c. (*Z*)-1-phenyl-1-propene
   d. Explain why these dimerizations occur upon UV radiation but not when heated.

13. What diene would give the following product?

14. What product will result from each of the following Diels-Alder reactions?

a.

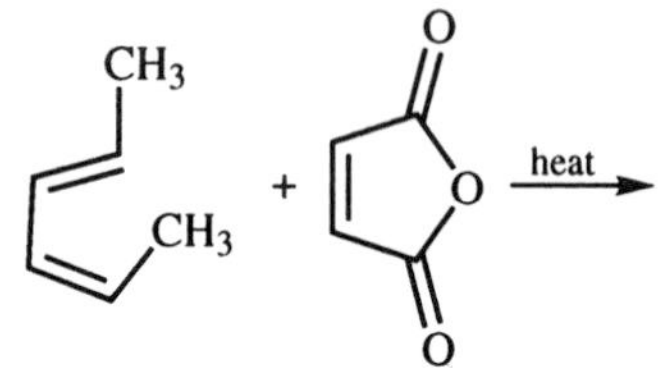

b.

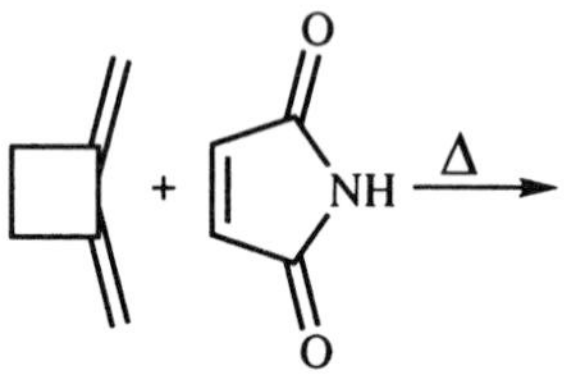

c.

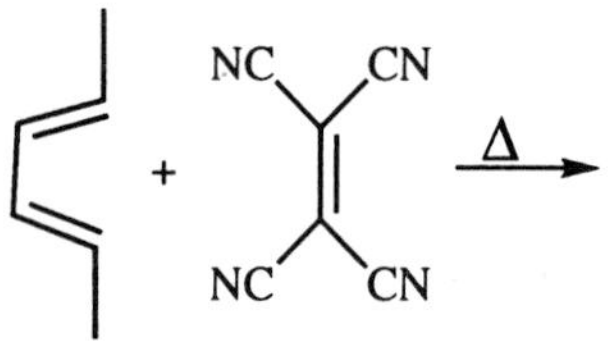

d.

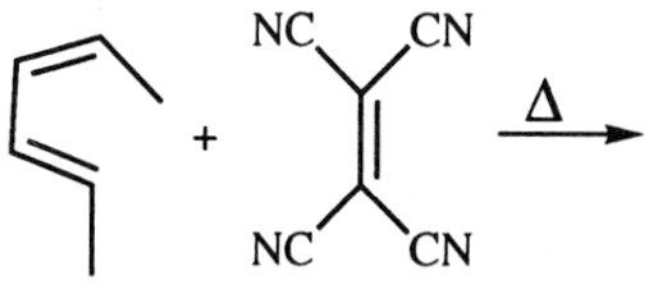

15. Which diene and dienophile would give each of the following compounds?

a.

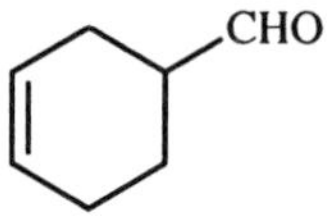

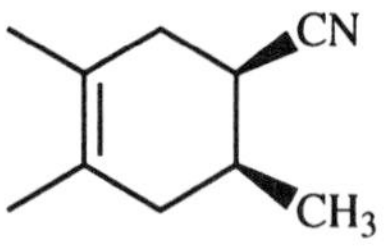

e.

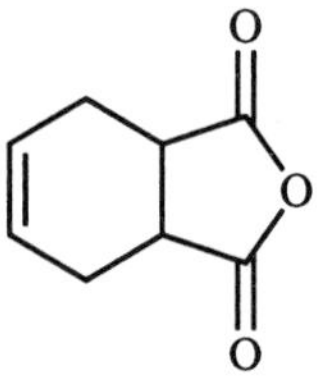

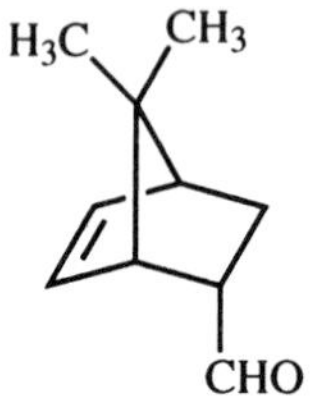

b.

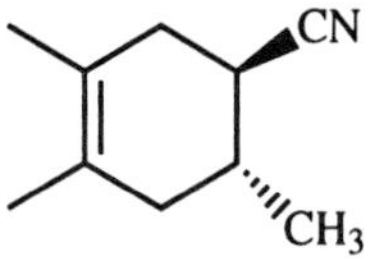

f.

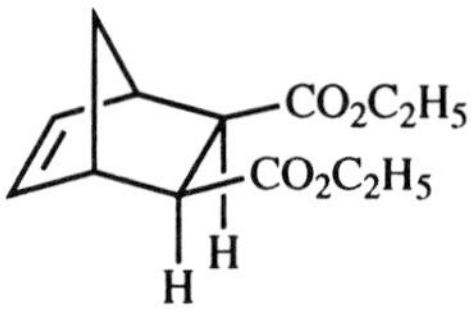

c.

g.

d.

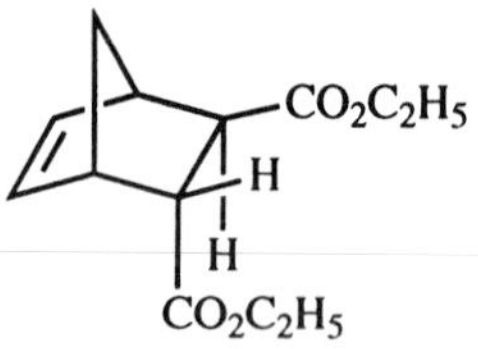

16. (Z)-1,2-diphenylethene can be produced from (E)-1,2-diphenylethene by finding a wavelength of UV light at which the E, but not the Z compound absorbs. Give a mechanism involving MOs which explains this isomerization.

17.    a. What would be the stereochemistry of the ring compound formed from (E)-dimethylcyclobutene in a thermal reaction if the ring formed (1) under con and (2) under dis conditions? (Hint: draw the LUMO)
     b. What would be the stereochemistry of the ring compound formed from (E)-dimethylcyclobutene in a photochemical reaction if the ring formed (1) under con and (2) under dis conditions?
     c. Explain why experimental results show that the reaction occurs only under thermal conditions?

18.    a. What would be the stereochemistry of the ring compound formed from (E)–5,6-dimethyl-1,3-cyclohexadiene in a thermal reaction if the ring formed (1) under con and (2) under dis conditions? (Hint: draw the LUMO)
     b. What would be stereochemistry of the ring compound formed from (E)–5,6-dimethyl-1,3-cyclohexadiene in a photochemical reaction if the ring formed (1) under con and (2) under dis conditions?
     c. Explain why experimental results show reaction occurs only under photochemical conditions.

19. Use MO theory to explain each of the following conversions.
     a. (E)–3,4-diethylcyclobutene gives (E,E)–3,5-octadiene when heated
     b. (E,E)–5,6-diethyl-1,3-cyclohexadiene gives (E,Z,E)–3,5,7-decatriene when photolyzed
     c. (E,Z,E)–3,5,7-decatriene gives (E)–5,6-diethyl-1,3-cyclohexadiene when heated

20. Use MO theory to explain each of the following conversions.
     a. (E,Z,Z)–3,5,7–decatriene gives (E)–5,6-diethyl–1,3–cyclohexadiene when heated
     b. (E)–5,6-diethyl–1,3–cyclohexadiene gives (E,Z,E)–3,5,7–decatriene when photolyzed
     c. (E,Z,E)–3,5,7–decatriene gives (E)–5,6-diethyl–1,3–cyclohexadiene when heated

21. a. Write the structures of the following endo and exo products

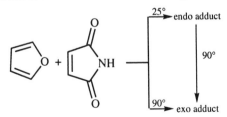

    b. Give an explanation for the experimental results summarized above:
(1) the endo adduct is formed when the reaction is run at 25°C
(2) the exo adduct is formed when the reaction is run at 90°C
(3) the endo adduct is converted to the exo adduct when heated to 90°C

22. We saw in the text that the electron configuration of $O_2$ in molecular orbital terms was: $(\sigma_{1s})^2 (\sigma^*_{1s})^2 (\sigma_{2s})^2 (\sigma^*_{2s})^2 (\sigma_{2p})^2 (\pi_{2p})^2 (\pi_{2p})^2 (\pi^*_{2p})^1 (\pi^*_{2p})^1$. The bond in $O_2$ is 0.121 nm, and has a dissociation energy of 493.7 kJ/mol.

   a. Draw the Lewis dot diagram of molecular $O_2$. How many unpaired electrons are shown in this diagram?

   b. Liquid molecular oxygen is visibly held in place between the poles of a strong magnet. Explain this in terms of the electron configuration.

   c. The species $O_2^+$ (an oxygen molecule that has lost an electron) has a bond length of 0.112 nm. How would you expect its dissociation energy to compare with that of the neutral oxygen molecule? Explain this in terms of the electron configuration of the cation.

   d. The species $O_2^-$ (an oxygen molecule that has gained an electron) has a bond length of 0.126 nm. How would you expect its dissociation energy to compare with that of the neutral oxygen molecule? Explain this in terms of the electron configuration of the anion.

23. The electron configuration of NO is $(\sigma_{1s})^2 (\sigma^*_{1s})^2 (\sigma_{2s})^2 (\sigma^*_{2s})^2 (\sigma_{2p})^2 (\pi_{2p})^2 (\pi_{2p})^2 (\pi^*_{2p})^1$. The bond length of NO is 0.115 nm. The dissociation energy of NO is 678 kJ/mol.

   a. Write the Lewis dot diagram for NO. How many unpaired electrons are shown in this diagram?

   b. Calculate the bond order for NO based upon the electron configuration.

   c. The bond length in $NO^+$ is 0.1062 nm. Calculate the bond order for $NO^+$ and explain the change in bond length based upon this difference.

24. Assume that the electron configuration of $CN^-$ is $(\sigma_{1s})^2 (\sigma^*_{1s})^2 (\sigma_{2s})^2 (\sigma^*_{2s})^2 (\sigma_{2p})^2 (\pi_{2p})^2 (\pi_{2p})^2$.

   a. Write the Lewis dot diagram for $CN^-$. How many unpaired electrons are shown in this diagram?

   b. Calculate the bond order of $CN^-$.

   c. Calculate the bond order for neutral CN.

   d. The CN bond length is 0.117 nm, while the $CN^-$ bond length is 0.114 nm. Explain.

25. Consider the compound BN.

   a. Propose a reasonable electron configuration (in molecular orbitals terms) for BN.

   b. Determine the bond order for BN.

   c. Is BN diamagnetic or paramagnetic according to the electron configuration in part a?

   d. Assume that the energy level order for $(s_{2p})$ and $(p_{2p})$ that you proposed in part a are in fact reversed. Using this altered electron configuration, is BN expected to be diamagnetic or paramagnetic? Explain.

13

# NITROGEN
# Chemistry
## HERE AN "N", THERE AN "N"

# NITROGEN CHEMISTRY
## Here an "n", there an "n"

## 13.1   THE NITROGEN CYCLE

As you probably already know, nitrogen makes up approximately 78% by volume of normal air and thus is the most prevalent component of the atmosphere. You may not already know that nitrogen forms stable compounds that range in oxidation state from –3 to +5, including all possible oxidation states in between.  Table 13.1 lists these formal oxidation states and gives the name of a compound or ion representative of each.

**TABLE 13.1**  THE VARIOUS OXIDATION STATES OF NITROGEN

TABLE 13.1

| Oxidation state | Formula | Name |
|:---:|:---:|:---:|
| – 3 | $NH_3$ | ammonia |
| – 2 | $N_2H_4$ | hydrazine |
| – 1 | $NH_2OH$ | hydroxylamine |
| 0 | $N_2$ | nitrogen |
| + 1 | $N_2O$ | nitrous oxide |
| + 2 | $NO$ | nitric oxide |
| + 3 | $NO_2^-$ | nitrite ion |
| + 4 | $NO_2$ | nitrogen dioxide |
| + 5 | $HNO_3$ | nitric acid |

The abundance of nitrogen and the wide range of its chemical compounds makes nitrogen worthy of special attention.  The use and reuse of nitrogen is crucial to plant and animal life, and is summarized in the **nitrogen cycle**, figure 13.1.

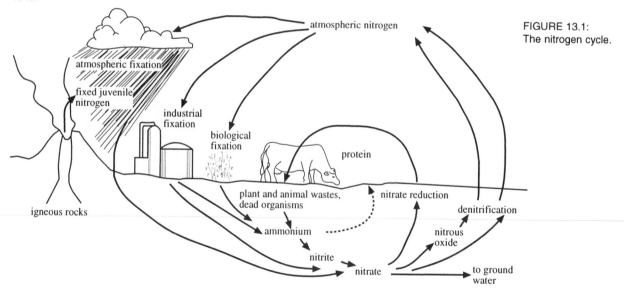

FIGURE 13.1:
The nitrogen cycle.

Diatomic nitrogen gas, the most prevalent nitrogen containing substance, is very stable. The triple bond in the nitrogen molecule requires a very large amount of energy ($942 \text{ kJ mol}^{-1}$) to break. Incorporating gas phase diatomic nitrogen as part of a larger, more complicated species is also unfavored by entropy. Thus changing molecular nitrogen into anything else is thermodynamically difficult. Nonetheless, as indicated by the nitrogen cycle, such changes are crucial to life processes. While carbon is the element that categorizes chemicals as being "organic", it can be argued that nitrogen is the signature element of living protoplasm. Proteins, after all, contain about 17% nitrogen by mass. Thus growing plants and animals both need a constant supply of nitrogen.

**Nitrogen fixation** describes several processes, both natural and artificial, that convert elemental nitrogen into nitrogen compounds. One of these processes occurs during thunderstorms. The lightning that is associated with these storms contains enough energy to initiate a reaction between molecules of nitrogen and oxygen to form nitric oxide.

$$N_2 + O_2 \longrightarrow 2\,NO \qquad \text{EQ. 13.1}$$

Since nitric oxide contains an unpaired electron, it is reactive. Nitric oxide reacts rapidly with oxygen to form nitrogen dioxide.

$$2\,NO + O_2 \longrightarrow 2\,NO_2 \qquad \text{EQ. 13.2}$$

The oxides of nitrogen act as acid anhydrides; thus they react with atmospheric water forming $HNO_3$ and $HNO_2$. Rain falling to earth typically contains both the parent acids and their conjugate bases, nitrate and nitrite ions formed after interaction with bases also present in the atmosphere. These compounds are absorbed into the soil. Subsequent enzymatic transformations then convert them into the amino acids and proteins needed by living organisms. When an organism metabolizes a compound containing nitrogen, other nitrogen compounds, such as urea, are excreted and they, in turn, can be converted back into amino acids or ammonia for further use.

Another natural means of fixing nitrogen occurs in bacteria that live in the nodules on legume (*e.g.*, soybeans) roots. These bacteria produce an enzyme known as nitrogenase. This enzyme can directly convert nitrogen from the air into ammonia. The plants absorb nitrogen through their roots and the biochemical conversion process begins again.

### SIDEBAR: COMMERCIAL FERTILIZERS

Gaseous ammonia is sometimes directly applied by farmers to their fields. More commonly, liquid or solid fertilizers are used, and nutrients other than nitrogen are included. Fertilizers are often described by a series of three numbers. A lawn fertilizer, for example, may be described as 22-6-8, while a plant food is described as 15-30-15. These numbers correspond to weight percents of certain elements or compounds. The first of the three numbers refers to nitrogen. The lawn fertilizer contains 22% nitrogen by weight. Often this is present in the form of urea [H$_2$NCONH$_2$] or ammonium phosphate [(NH$_4$)$_3$PO$_4$]. The second number in the series refers to phosphorus, specifically, the compound P$_2$O$_5$. This compound is not actually used in fertilizers, but is used as a standard. Since P$_2$O$_5$ is 43.64% phosphorus, a plant food having a "30" as its second number has a phosphorus content equivalent to 30% P$_2$O$_5$. The actual phosphorus content is thus 43.62% of 30%, or 13% by weight. Phosphorus is usually present in a fertilizer as ammonium phosphate. The third and last number in the fertilizer designation refers to potassium, specifically potassium in the form of K$_2$O. Again, this compound is a standard, and is not actually used in fertilizers. The plant food that has a "15" designation for potassium is claiming to have the equivalent of 15 grams of K$_2$O per 100 grams of fertilizer. Since K$_2$O is 83% potassium, 15% K$_2$O is equivalent to 12.4% elemental potassium. Potassium is often present in fertilizers in the form of KCl.

Neither of these natural processes can produce sufficient nitrogen per acre of crops to sustain the level of agricultural production needed to feed the world's population. Both of these natural processes of nitrogen fixation, however, have served as models for the industrial production of nitrogen-containing fertilizer. Sir William Crookes (1832–1919) (of Crookes Tube fame, *Volume 1*, Chapter 9), first recognized that the lightning-initiated formation of nitrogen oxides could be simulated industrially. Air, containing nitrogen and oxygen, was blown through an electric arc, forming the oxides of nitrogen. These were then swept into a spray of water, and nitric acid was formed. A simplified version of the process includes equations 13.1 and 13.2, followed by

$$3\,NO_2 + H_2O \longrightarrow 2\,HNO_3 + NO \qquad \text{EQ. 13.3}$$

The nitric oxide formed as a byproduct was re-introduced into the reaction stream. The nitric acid, when reacted with sodium hydroxide, formed sodium nitrate, which was used as a fertilizer.

Crookes method of nitrogen fixation was soon supplanted by an industrial process developed by Fritz Haber (1868-1934) who was awarded the Nobel Prize in 1918 for this discovery. In the **Haber process**, nitrogen and hydrogen gas are pumped over an iron-based catalytic surface and the ammonia that is formed is collected as a liquid (see figure 13.2)

The reaction occurring in the Haber process

$$N_{2(g)} + 3\,H_{2(g)} \rightleftharpoons 2\,NH_{3(g)} \qquad \text{EQ. 13.4}$$

has a molar enthalpy change of -92 kJ mol$^{-1}$. According to LeChatelier's Principle, increasing the temperature would shift the equilibrium of equation 13.4 to the left, back towards reactants. Yet a high temperature is needed to initiate the reaction and overcome the bond strength of the nitrogen-nitrogen triple bond. The industrial synthesis of ammonia occurs under conditions that help offset this problem. By running at very high pressure (200-300 atm) and by removing ammonia as it is formed, the equilibrium reaction is shifted toward the products. However, the key industrial feature is the use of an

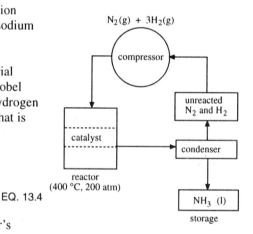

FIGURE 13.2:
The Haber process for generating ammonia.

appropriate catalyst. The iron catalyst helps to lower the activation energy, and allows the reaction to occur at a lower, more thermodynamically favorable, temperature (500-600°C). In a sense, this industrial process also follows the path previously established by nature. The nitrogenase enzyme used by nitrogen-fixing bacteria has iron atoms at the active site.

### SIDEBAR: THE NITROGENASE COMPLEX

The enzyme systems that allow certain bacteria to fix nitrogen are known as nitrogenase complexes. All such complexes contain metal atoms. Some contain only iron, some contain iron and vanadium, but most commonly they contain iron and molybdenum. The most common complex contains two separate proteins. Component I contains both iron and molybdenum. Component II has iron as the only bound metal. Both Components I and II have $M_4S_4$ subunits. In one subunit, iron is present at all four metal sites; in the other subunit, three iron atoms and one molybdenum atom are present. D.C. Rees and J. Kim have shown recently [*Science* (**260**) 792-794 (1992)] that the molecule of nitrogen is bound to the nitrogenase complex in a fashion pictured at right:

Another portion of the enzyme structure binds and hydrolyzes ATP, as a source of energy for the nitrogen fixation. The enzyme also helps shuttle electrons to the active site, since they are needed in the reduction reaction:

$$N_2 + 6\,H^+ \xrightarrow{\quad ATP \quad ADP \quad} 2\,NH_3$$

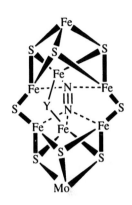

FIGURE 13.3:
Binding of $N_2$ to part of the MoFe-protein.

EQ. 13.4

Much research is being focused on the nature of enzymatic fixation of nitrogen. The industrial process involves high temperature and pressure, both of which are quite expensive. The United States annually produces nearly 20 million tons of ammonia, worth over two billion dollars. The enzymatic reduction of nitrogen, since it occurs at <u>ambient</u> temperature and pressure, offers the potential of a great economic saving.

Now that we have some background in how nitrogen is converted into useful forms for living organisms, let us turn our attention to some of the chemistry of inorganic nitrogen compounds.

## 13.2   INORGANIC OXY COMPOUNDS OF NITROGEN

Recall that nitrogen is unique among the elements in the number of stable oxidation states. The nitrogen oxides are a series of compounds in which nitrogen can be in the +1 to +5 oxidation states: $N_2O$ (+1); $NO$ (+2); $N_2O_3$ (+3); $NO_2$ (+4); and $N_2O_5$ (+5).

We begin a survey of some nitrogen oxides by looking at dinitrogen oxide, also known as nitrous oxide. Nitrous oxide is produced by the thermal decomposition of ammonium nitrate.

$$NH_4NO_{3(s)} \longrightarrow N_2O_{(g)} + 2 H_2O_{(g)} + heat$$

EQ. 13.6

You may have received dinitrogen oxide, commonly know as laughing gas, when you were at the dentist's office. Because it has a relatively high solubility in cream when the partial pressure of dinitrogen oxide is high, it is also used as the propellant in cans of whipped cream. As the cream leaves the can and reaches the atmosphere outside the can, the lower partial pressure of dinitrogen oxide in the surrounding air causes bubbles to form in the cream as the gas comes out of solution.

Nitric oxide, NO, as you have already seen, is formed when nitrogen and oxygen react in a lightening storm. It is also made industrially by the reaction of ammonia with oxygen in the presence of a catalyst. This is the first step in the **Ostwald process**, which is the industrial process by which nitric acid is prepared.

$$4 NH_{3(g)} + 5 O_{2(g)} \xrightarrow[\Delta]{catalyst} 4 NO_{(g)} + 6 H_2O_{(g)}$$

EQ. 13.7

This compound is also produced in biological systems from arginine, one of the amino acids, in an enzyme-catalyzed reaction. Among its other biological functions, it was discovered in the 1980's that nitric oxide behaves as a vasodilator and is important in the maintenance of blood vessel tone.

Recall that nitric oxide contains an unpaired electron and reacts with more oxygen to form nitrogen dioxide, $NO_2$. It is noteworthy that this is the second step in the Ostwald process.

$$2 NO_{(g)} + O_{2(g)} \longrightarrow 2 NO_{2(g)}$$

EQ. 13.8

Nitrogen dioxide, also a molecule with an unpaired electron, is the brown gas that is responsible for the yellowing often observed when bottles of aqueous nitric acid are exposed to light for long periods of time. It is also responsible for the brown haze so common in urban areas where air pollution is a problem. In this case, the internal combustion engine provides the high temperature required for the formation of nitrogen oxides from the nitrogen and oxygen in the air swept into the cylinder as the oxidant for the gasoline. Catalytic converters in automobiles are used in part to offset this pollution source.

Finally we turn our attention to nitric acid, $HNO_3$. In polluted atmospheres with an abundance of nitrogen oxides, nitric acid is an environmental nuisance, and contributes to acid rain. Nitric acid, however, is also an important industrial commodity. As we have been discussing, nitric acid is produced commercially from ammonia *via* the Ostwald process. The final step in this process is the reaction of nitrogen dioxide with water to form nitric acid.

$$3 NO_{2(g)} + H_2O_{(l)} \longrightarrow 2 HNO_{3(l)} + NO_{(g)}$$

EQ. 13.9

The nitric oxide formed during this process is, as mentioned earlier, recycled.

Nitric acid, with nitrogen in the +5 oxidation state, functions as an efficient oxidizing acid, especially when it is hot and concentrated. It is used extensively in the commercial preparation of fertilizers and explosives, and, as seen earlier, is often used as a nitrating agent in organic synthesis.

## 13.3  INORGANIC HYDROGEN COMPOUNDS OF NITROGEN

Almost everyone remembers the pungent smell of an ammonia-based cleaner. Ammonia is prepared on an industrial scale by the Haber process in which nitrogen and hydrogen are reacted in the presence of a catalyst.

$$1/2 \, N_{2(g)} + 3/2 \, H_{2(g)} \longrightarrow NH_{3(g)}$$
EQ. 13.10

Ammonia behaves as a base in aqueous solution. The ability of a base to catalyze the hydrolysis of certain greases adds greatly to ammonia's cleaning power.

$$NH_{3(aq)} + H_2O_{(l)} \rightleftharpoons NH_4^+{}_{(aq)} + OH^-{}_{(aq)}$$
EQ. 13.11

On the other hand, ammonia can also behave as an acid toward even stronger bases. The amide ion, the conjugate base of the ammonia molecule, is a stronger base than the hydroxide ion so these reactions must occur in non–aqueous solutions. The hydrogen atoms on ammonia are also capable of being removed in oxidation–reduction reactions by active metals. For example, sodium amide ($NaNH_2$) is prepared by reacting liquid ammonia with sodium metal.

Hydrazine is another common nitrogen hydride. Hydrazine is prepared from ammonia by oxidation with sodium hypochlorite. It and, more commonly, its close relative unsymmetrical dimethyl hydrazine, are used as rocket fuel in liquid fueled rockets.

$$2 \, NH_{3(aq)} + NaOCl_{(aq)} \longrightarrow N_2H_{4(aq)} + NaCl_{(aq)} + H_2O_{(l)}$$
EQ. 13.12

Hydrazine also behaves as a base in aqueous solution.

$$N_2H_{4(aq)} + H_2O_{(l)} \rightleftharpoons N_2H_5^+{}_{(aq)} + OH^-{}_{(aq)}$$
EQ. 13.13

The third compound of nitrogen and hydrogen is hydrogen azide ($HN_3$). Its water solutions are called hydrazoic acid. The compound itself is usually derived from the azide salts by treating them with strong mineral acids. For example, hydrogen azide can be prepared by reacting sodium azide with sulfuric acid. The compound behaves as a weak acid in aqueous solution. Hydrazoic acid and its salts, especially the heavy metal salts, are very unstable, shock–sensitive compounds that can explosively decompose producing nitrogen gas as one of the products. Soluble azides are potent biocides and small amounts are often added to solutions to prevent the growth of bacteria. Sometimes, when these solutions are disposed by repeatedly pouring them down the same drain, the lead drain pipes build up a coating of lead azide that can explode when the drain is repaired. Care must, therefore, always be taken in disposing of solutions containing azides. Sodium azide is also the compound being used as the inflating agent in automobile airbags. When electrically detonated, the solid sodium azide quickly decomposes releasing nitrogen gas that inflates the airbag. The reaction occurs quickly enough that the airbag is able to prevent the occupant of the car from impacting the steering wheel or dash during a collision.

## 13.4 ORGANIC NITROGEN COMPOUNDS

When we refer to organic nitrogen compounds, we normally focus on the amines. As you may recall, many organic compounds are classified based upon the number of alkyl groups attached to a particular substituent. The amines are no exception. In this case, the classification is based upon the number of alkyl groups attached to nitrogen. When one alkyl group is bonded to nitrogen we refer to the compound as a **primary amine**; two alkyl groups put the amine in the classification of a **secondary amine**; and when three alkyl groups are bonded to nitrogen the amine is classified as a **tertiary amine**. (See figure 13.4 for examples of each type of amine) Of course, as you may also remember, when four groups surround nitrogen, the nitrogen bears a positive charge, and the compound is referred to as a quaternary ammonium salt. Several different nomenclature schemes are used when naming amines. The common system of nomenclature allows us to name amines simply as alkyl amines. On the other hand, two IUPAC systems are accepted; the first names amines as alkane amines, while the second names amines as amino alkanes. Examples of each type of nomenclature are provided in figure 13.4.

| Primary Amines | Secondary Amines | Tertiary Amines |
|---|---|---|

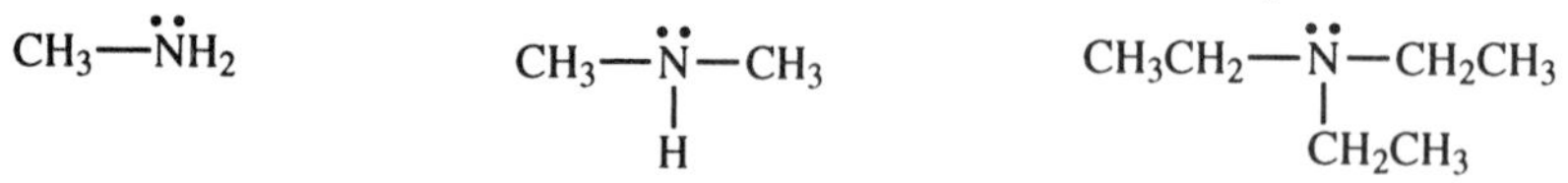

| | | |
|---|---|---|
| methylamine<br>methanamine<br>aminomethane | dimethylamine<br>N-methylmethanamine<br>methylaminomethane | triethylamine<br>N,N-diethylethanamine<br>diethylaminoethane |

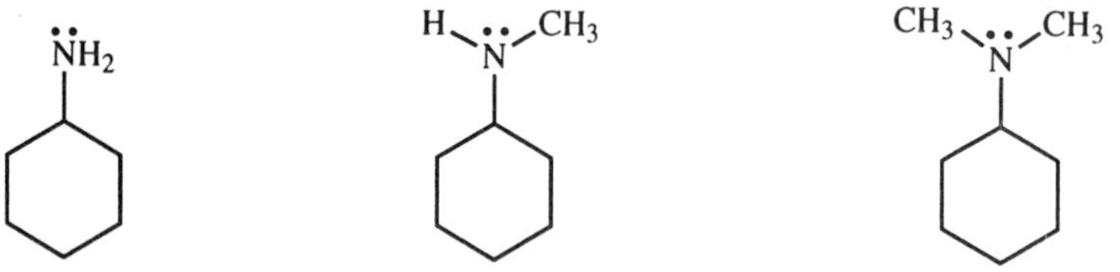

| | | |
|---|---|---|
| cyclohexylamine<br>cyclohexanamine<br>aminocyclohexane | N-methyl-cyclohexylamine<br>N-methylcyclohexanamine<br>methylaminocyclohexane | N,N-dimethyl-cyclohexylamine<br>N,N-dimethylcyclohexanamine<br>dimethylaminocyclohexane |

FIGURE 13.4: Aliphatic amine classification and nomenclature.

The aromatic amines are derivatives of aniline, the simplest of the aromatic amines, and are typically named as such.

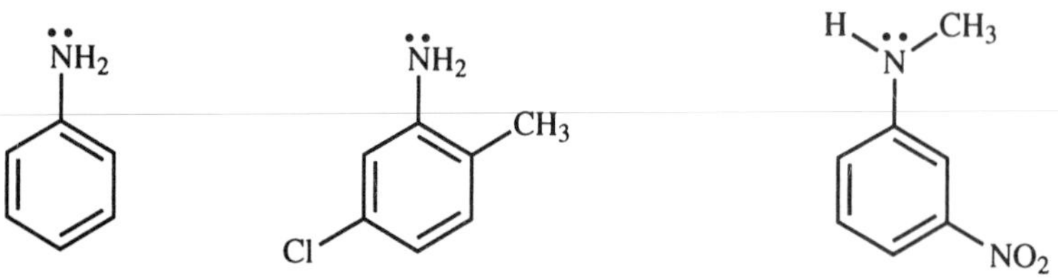

| | | |
|---|---|---|
| aniline | 5-chloro-2-methylaniline | N-methyl-3-nitroaniline |

FIGURE 13.5: Some examples of aromatic amines.

Some of the most important amine based compounds, at least from a biological perspective, are the **alpha amino acids**. These compounds, of course, are the building blocks for structural proteins and enzymes. As you probably know, there are twenty naturally occurring amino acids and nineteen of the twenty are based on the general skeleton shown in figure 13.6.

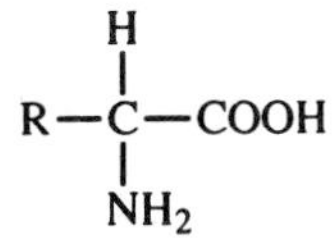

The tremendous diversity found in protein systems is the result of variation of the "R" group on the various amino acids. Four different types of "R" groups are found in the twenty amino acids. Some of the "R" groups are neutral and hydrophobic ("water hating" — non-polar); some are neutral and hydrophilic ("water loving" — polar); some are acidic; and, finally, some are basic.

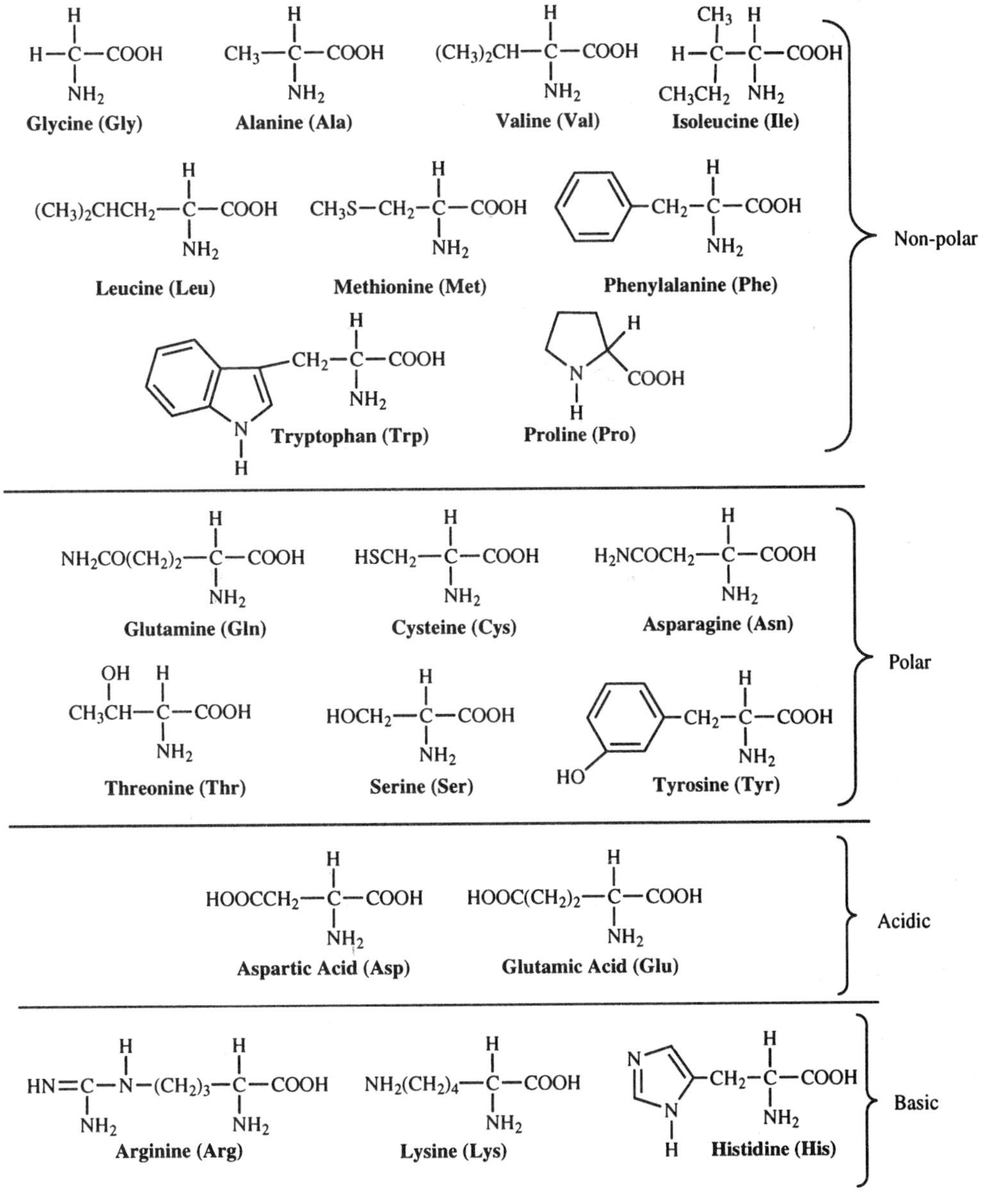

FIGURE 13.6:
The general amino acid skeleton.

FIGURE 13.7:
The twenty naturally occurring amino acids.

We should also emphasize that the twenty naturally occurring amino acids all have the *S*-configuration and they could all be drawn based upon the following sterochemically correct skeleton.

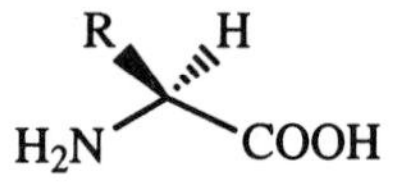

Finally, let us look briefly at various nitrogen containing heterocycles (figure 13.9). This figure contains a number of representative compounds in this category including pyridine and pyrolle, two aromatic heterocycles that we have seen before, nicotine, a component of cigarette smoke, quinine, found in tonic water, and some compounds that are frequently abused including LSD and cocaine.

Now that we have been introduced to various types of nitrogen containing compounds, let us look briefly at the acid/base properties of these compounds with special attention to the amines and amino acids.

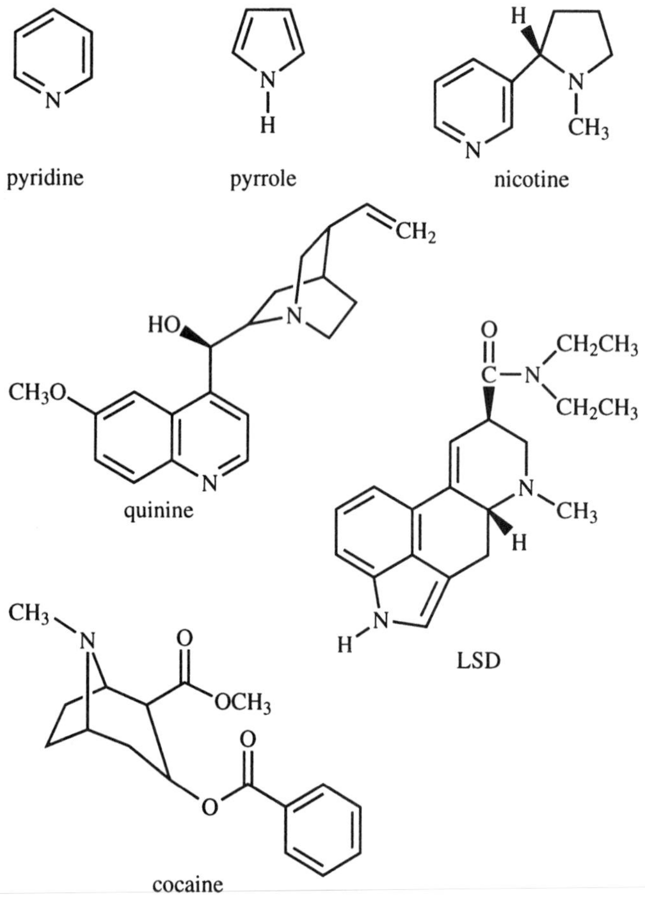

FIGURE 13.8:
Stereochemistry of the amino acids.

pyridine          pyrrole          nicotine

quinine          LSD          cocaine

FIGURE 13.9:
Various nitrogen containing heterocycles.

# 13.5 ACID/BASE CHEMISTRY OF AMINES AND AMINO ACIDS

Ammonia can undergo a Brønsted-Lowry reaction with water to produce the ammonium ion and hydroxide ions (equation 13.11). Amines are also able to undergo a similar reaction. For example, methyl amine and water react to form the methyl ammonium ion and hydroxide ion as shown in equation 13.14.

$$CH_3NH_{2(aq)} + H_2O_{(l)} \rightleftharpoons CH_3NH_3^+{}_{(aq)} + OH^-{}_{(aq)}$$

EQ. 13.14

The $pK_b$ for methyl amine is approximately 3.3 while the $pK_b$ of ammonia is around 4.7. Thus we see that methyl amine is a stronger base than ammonia. If we look at other $pK_b$ values, we note that the $pK_b$ values of other alkyl amines are similar to that of methyl amine, but that aryl amines have considerably higher $pK_b$ values. For example, aniline has a $pK_b$ value of 9.4, and is, therefore, a much weaker base. The following general trend for basicity is observed:

aryl amines < ammonia < alkyl amines

In order to explain this trend, it is helpful to look at the structure of the corresponding conjugate acids.

By looking at these structures we can see that the methyl ammonium ion is stabilized, relative to the ammonium ion, by the electron donating methyl group. On the other hand the anilinium ion is destabilized by the electron withdrawing aryl group. One way to rationalize base strength, then, is to look at the stability of the conjugate acids. A substance that has a more stable conjugate acid will be a stronger base than another that has a less stable conjugate acid.

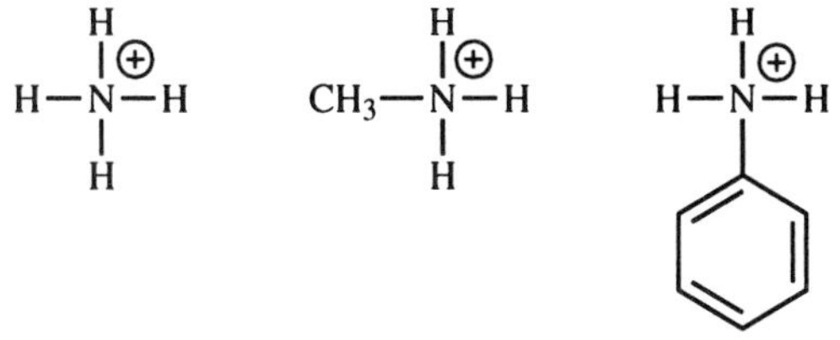

FIGURE 13.10:
The conjugate acids of ammonia, methylamine, and aniline.

Let us now turn our attention to the alpha amino acids. Rather than discuss the properties of each amino acid, let us choose two amino acids, alanine and aspartic acid, as representative of all amino acids.

Until now we have been drawing the amino acids as if both the amino and the carboxyl groups were uncharged. However, this is not the case and, in fact, the amino acids rarely exist in this state. The species that exists in solution is dependent upon the pH of that solution. The three species of alanine that are typically found in solution are shown below.

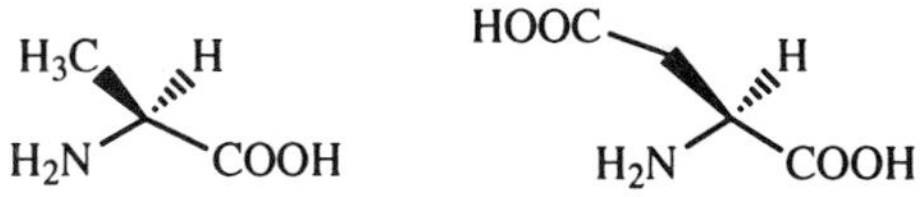

FIGURE 13.11:
Alanine and aspartic acid.

Species A is the fully protonated species and is the predominant species at low pH values (*e.g.*, pH=2). Species B is the **zwitterion** and is the predominant species at neutral pH values (*e.g.*, pH=7); this would also be the species found at physiological pH in biological systems. The zwitterion is an internally charged, but overall electrically neutral, species. The pH at which an amino acid exists exclusively as the zwitterion is known as the **isoelectric point**. As expected for an amphiprotic species, the isoelectric point of an amino acid occurs at the midpoint of two $pK_a$ values for the amino acid. Since the

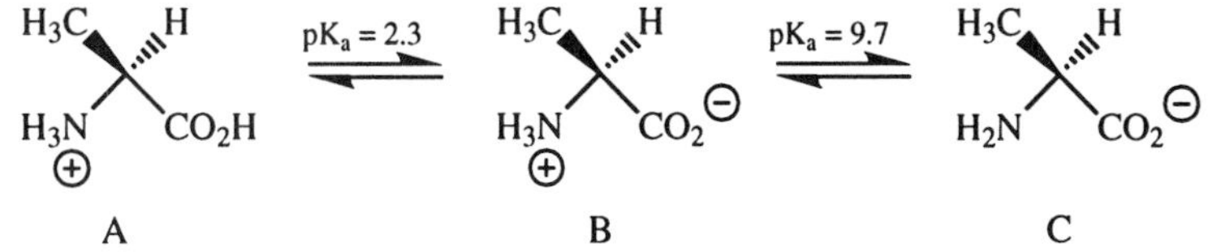

FIGURE 13.12:
Ionization of alanine.

different amino acids all have slightly different $pK_a$ values, each amino acid has its own isoelectric point. Finally, species C is the fully deprotonated species that would predominate at high pH values (*e.g.*, pH=13). At this point we should note that it is possible to use the Henderson-Hasselbach equation (section 4.5) to approximate the equilibrium concentrations of these species at pH values between adjacent $pK_a$ values.

Let us turn our attention to another of the amino acids, aspartic acid. Aspartic acid is an amino diacid and can, therefore, exist in four forms in aqueous solution depending upon the pH of the solution. The four different forms of this amino acid, are shown below.

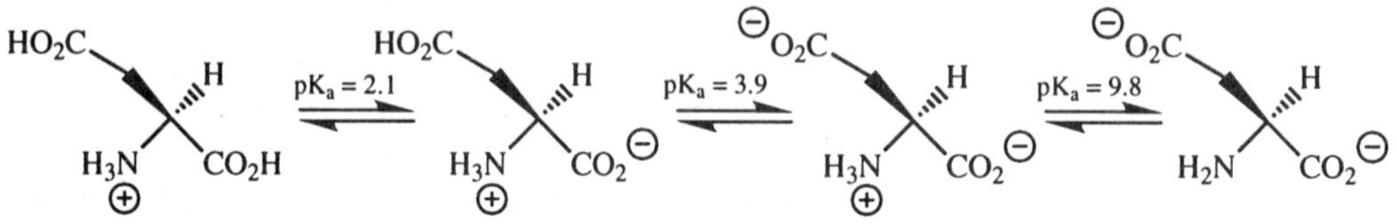

Again, note the forms that exist at the extremes of pH. Forms A and D will predominate at low and high pH respectively. The

FIGURE 13.13: Ionization of aspartic acid.

other two forms exist at intermediate pH values. Because of the second carboxylic acid group in the amino acid, the isoelectric point of this amino acid is quite low. As a result aspartic acid exists as a negatively charged species at physiological pH values. It is this second carboxylic acid group, and the negative charge that it places on the species near physiological pH values, that results in the charges on peptides and proteins in solutions that are nearly neutral. Just as the acidic amino acid, aspartic acid, has a low isoelectric point, a basic amino acid, for instance lysine, with its two amino groups, has a high isoelectric point. Proteins and peptides containing lysine, then, tend to be positively charged near pH 7. While we will discuss the structure of peptides and proteins in chapter 15, it will be important to understand that these macromolecules can be charged because of the amino acid side-chains, and that this charge can be altered by changing the pH of the environment in which they are found.

## 13.6   PREPARATION OF AMINES

Now that we have looked at the structures of some organic nitrogen compounds, let us look at the processes that can be used to prepare amines. In the next section, we will look at some of the reactions of amines.

As you recall, alkyl halides make good substrates for $S_N2$ type reactions, especially when they are reacted with good nucleophiles. Therefore, we might predict that one good way of preparing amines would be to react an alkyl halide with ammonia or another amine. For example, we might expect to be able to prepare hexylamine by reacting hexyl bromide with ammonia.

$$CH_3(CH_2)_4CH_2Br + NH_3 \longrightarrow CH_3(CH_2)_4CH_2NH_2 + HBr$$

FIGURE 13.14: Preparation of hexylamine from hexyl bromide.

In practice, however, this method is not synthetically useful because the hexylamine is actually a better nucleophile than ammonia. Thus it is difficult to stop after only one hexyl group is added to the ammonia. This makes continued alkylation a major problem. Another problem is that the alkyl halide must be a primary alkyl halide or elimination leads to a significant amount of the undesired alkene product. For example, if cyclohexyl bromide is treated with ammonia, the major product of the reaction is cyclohexene.

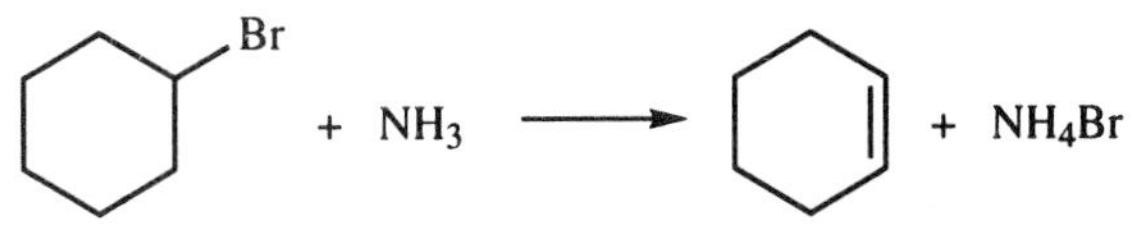

The reaction of excess ammonia with alpha halo acids is, however, a useful way of preparing alpha amino acids (figure 13.16). The electron withdrawing properties of the carboxyl group sufficiently activate the alpha carbon toward nucleophilic attack that the reaction occurs smoothly. This reaction does not produce an optically active amino acid. Rather, a racemic mixture is formed.

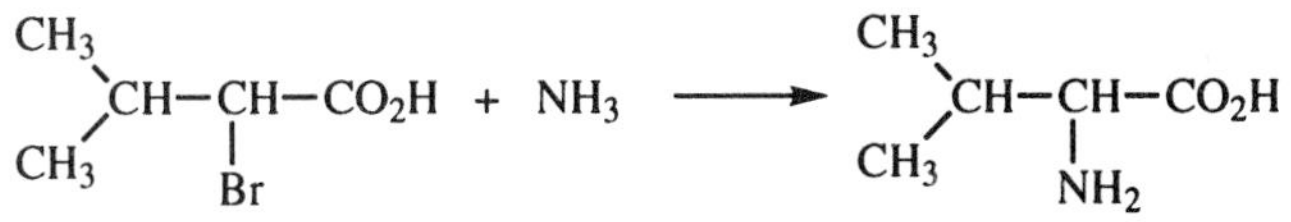

FIGURE 13.15:
The reaction between cyclohexyl bromide and ammonia.

The alpha bromo acid shown in figure 13.17 can be prepared by reaction of the corresponding acid with bromine in the presence of catalytic amounts of phosphorus trichloride. $PCl_3$ catalyzes the enolization reaction that allows halogenation to occur. This reaction is called the **Hell–Volhard–Zelinsky** reaction.

FIGURE 13.16:
Amino acid sythesis by the reaction of ammonia and an alpha bromo acid.

Another method for preparing amines involves reactions similar to those involved in enolate chemistry. These reactions make use of pthalimide as the starting material. As you no doubt recall, protons that are alpha to two carbonyl groups are especially acidic, and this position is readily deprotonated when a dicarbonyl compound is treated with base. This results in formation of an enolate ion that is readily alkylated when it is treated with an alkyl halide. This process of presented below.

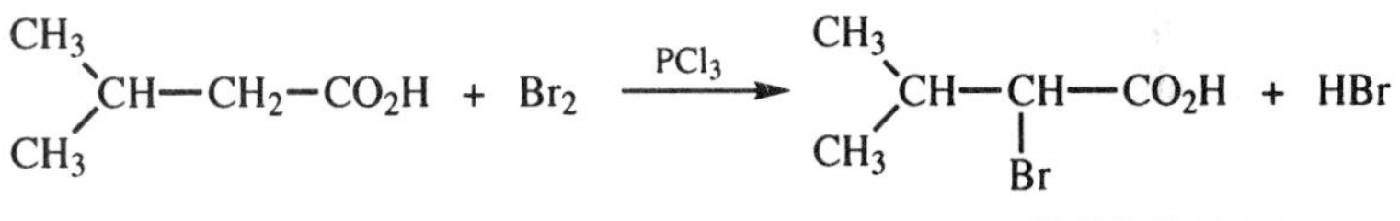

FIGURE 13.17:
Synthesis of an alpha bromo acid.

The compound, phthalimide is analogous to the β-dicarbonyl compound shown at right except that the alpha carbon is replaced by a nitrogen atom. The proton on this nitrogen atom is also acidic and can be removed by treatment with base. Treating the anion with an alkyl halide results in alkylation of the nitrogen atom and formation of an N-alkyl phthalimide. Subsequent treatment with acid, water, and heat (or base, water, and heat) leads to hydrolysis and the generation of phthalic acid or the phthalate ion and the amine. This last transformation is a hydrolysis reaction. The entire process is known as the **Gabriel synthesis** of amines.

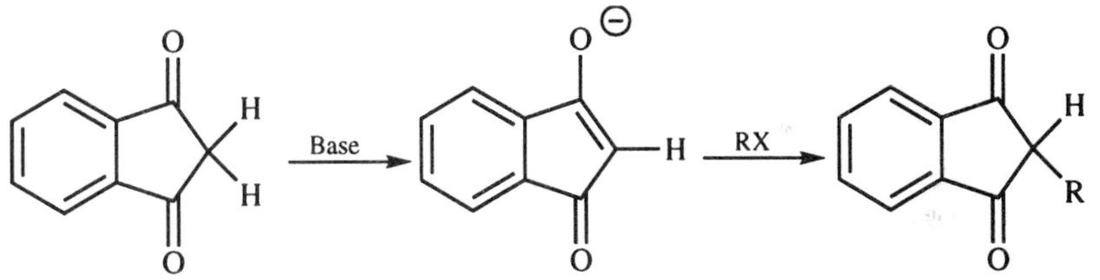

As an alternative to the hydrolysis process, the N-alkyl phthalimide can be treated with hydrazine. This results in a process known as acyl transfer. When the two acyl groups become bound to the hydrazine the amine is released. An example of this process, using N-benzyl phthalimide is shown in equation 13.15.

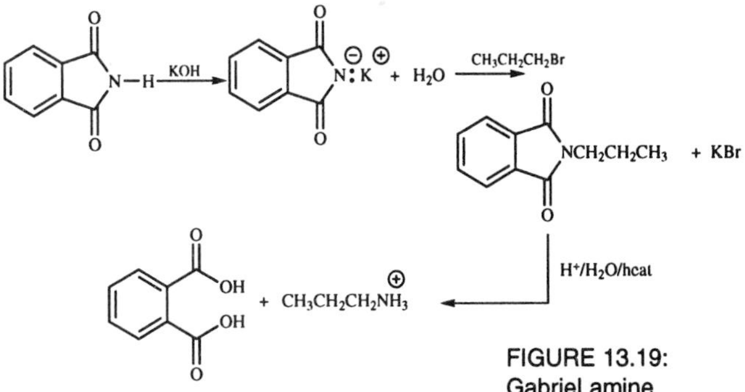

FIGURE 13.18:
Alkylation of a β-dicarbonyl compound.

FIGURE 13.19:
Gabriel amine sythesis.

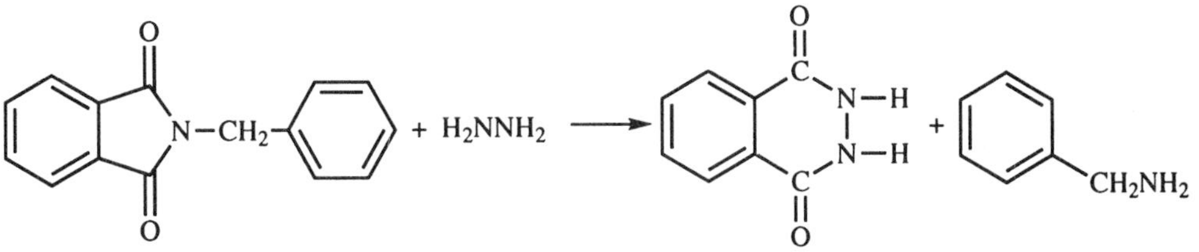

EQ. 13.15

This process can also be combined with the malonic ester synthesis to prepare amino acids. The first step in this process is the alkylation of phthalimide with α-bromo-diethyl malonate.

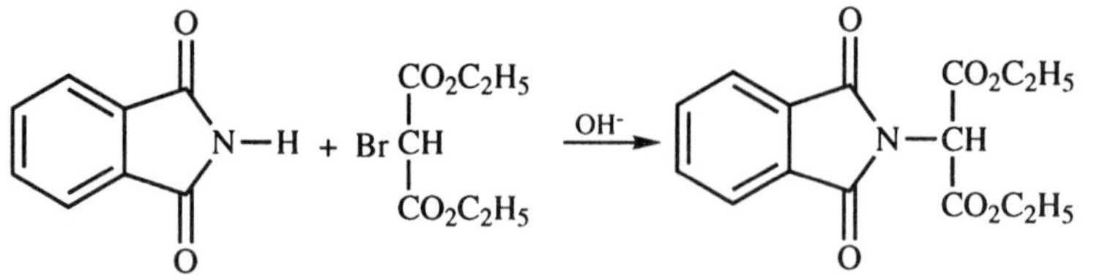

EQ. 13.16

The resulting N-phthalamidomalonic ester is then treated with additional base and an alkyl halide, isopropyl bromide in our example, in order to alkylate the α-carbon.

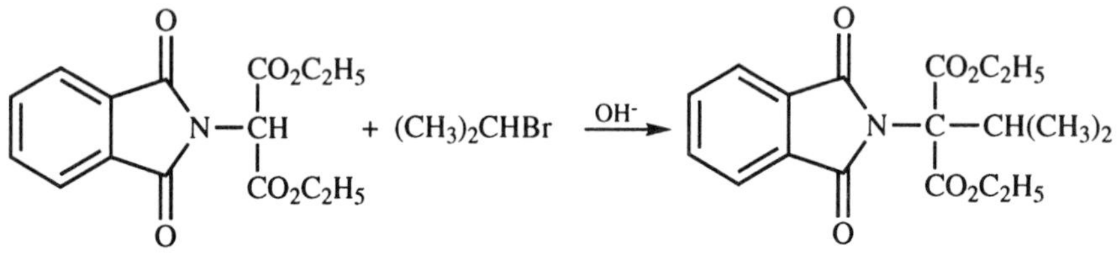

EQ. 13.17

Finally, the alkylated product is subjected to hydrolysis conditions that generate the amino acid after spontaneous decarboxylation of the intermediate β-diacid.

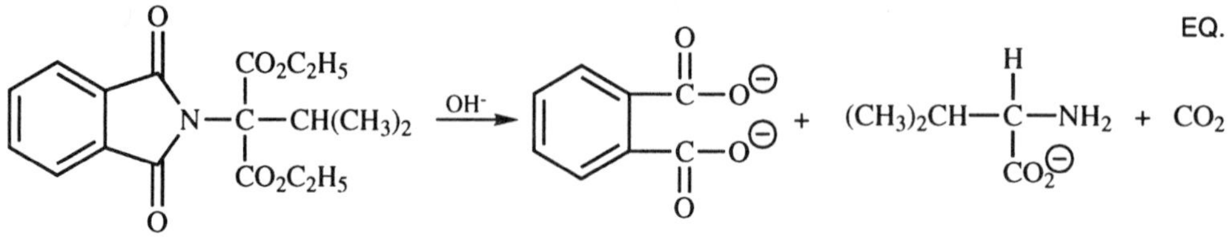

EQ. 13.18

Up to this point, we have looked at a few methods of amine preparation that involve nucleophilic substitution reactions involving amines and alkyl halides. In addition to these substitution reactions, there are a number of reductive methods that are useful for preparing amines. Nitro compounds are readily reduced to the corresponding amines. For example, nitrobenzene can be converted into aniline by treating the nitrobenzene with hydrogen and a nickel catalyst. The reduction of nitro compounds is also accomplished by treating a nitro compound with a reducing metal (e.g., Fe or Sn) in the presence of HCl. Of course to generate the free amine and not the ammonium salt, those reactions conducted in the presence of HCl must be neutralized with NaOH after the reduction of the nitro compound is finished. This method is particularly useful for the synthesis of aromatic amines.

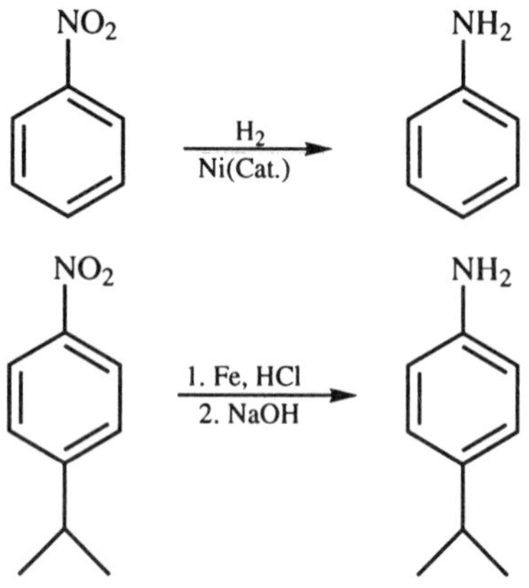

Nitriles and azides are also readily reduced to the corresponding amine with either hydrogen and a catalyst or with lithium aluminum hydride. These compounds would likely be prepared by reacting an alkyl halide with either sodium cyanide or sodium azide; so once again, we look to the alkyl halides as potential starting material for amine preparation.

FIGURE 13.20:
Amine synthesis by reduction of nitro compounds.

It is interesting to compare the two methods and note that the nitrile method results in the synthesis of an amine that is one carbon longer than the original alkyl halide while the azide method affords an amine of equal length. Thus the methods are complementary depending upon the desired product.

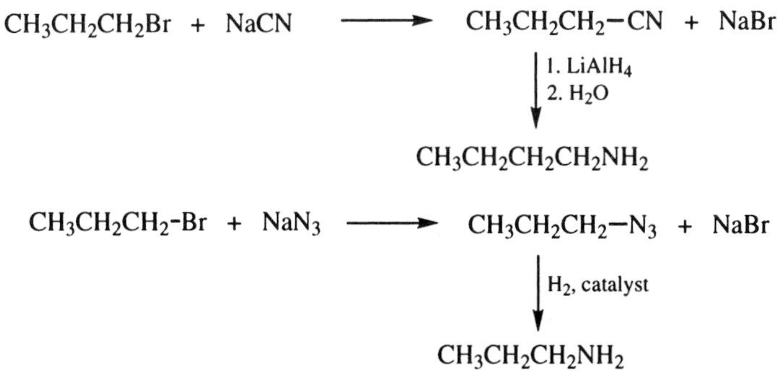

$$CH_3CH_2CH_2Br + NaCN \longrightarrow CH_3CH_2CH_2-CN + NaBr$$

1. LiAlH$_4$
2. H$_2$O

$$CH_3CH_2CH_2CH_2NH_2$$

$$CH_3CH_2CH_2\text{-}Br + NaN_3 \longrightarrow CH_3CH_2CH_2-N_3 + NaBr$$

H$_2$, catalyst

$$CH_3CH_2CH_2NH_2$$

**FIGURE 13.21:** Amine sythesis by reduction of nitriles and azides.

You might have noticed that the methods we have discussed thus far usually result in the preparation of primary amines. Amides are also readily reduced to amines with lithium aluminum hydride. One of the benefits of this process is that primary, secondary, and tertiary amines can all be prepared. The common starting material for the synthesis of amines *via* amide reduction is, of course, a carboxylic acid. The carboxylic acid is typically converted into the amide by treatment with thionyl chloride, to form the acid chloride, followed by treatment with an amine. For example, benzyl amine is prepared from benzoic acid by the sequence shown below.

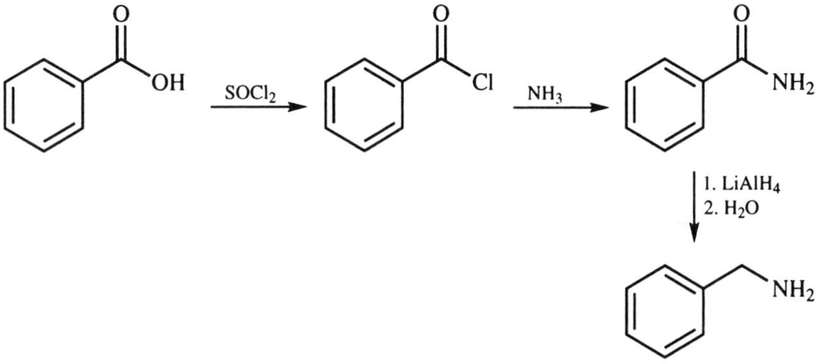

**FIGURE 13.22:** Preparation of benzyl amine from benzoic acid.

Secondary amines are prepared by reducing N-substituted amides and tertiary amines are prepared by reducing N,N-disubstituted amides.

The conversion of amides into amines by reduction with lithium aluminum hydride provides an amine that contains the same number of carbon atoms as the amide. Two other methods, the **Hofmann rearrangement** and the **Curtius rearrangement**, lead to amines that contain one fewer carbon atom than the amide.

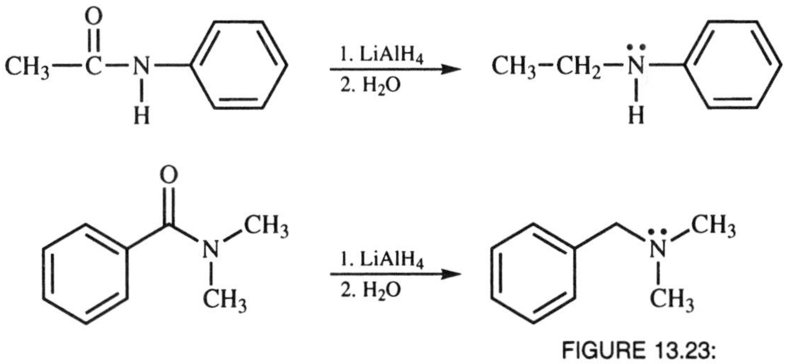

**FIGURE 13.23:** Amide reduction to prepare secondary and tertiary amines.

The Hofmann rearrangement, discovered by August Wilhelm von Hofmann (1818–1892), involves treating a primary amide with bromine in basic solution to produce the amine. The reaction is summarized below.

$$R-\overset{\overset{\text{O}}{\|}}{C}-NH_2 + Br_2 + 4\,NaOH \longrightarrow$$
$$R-NH_2 + 2\,NaBr + Na_2CO_3 + 2\,H_2O$$

EQ. 13.19

The mechanism of this reaction has been well studied and is quite instructive. We will examine the proposed mechanism and then take a look at some evidence in support of this mechanism. The first step is deprotonation of the amide.

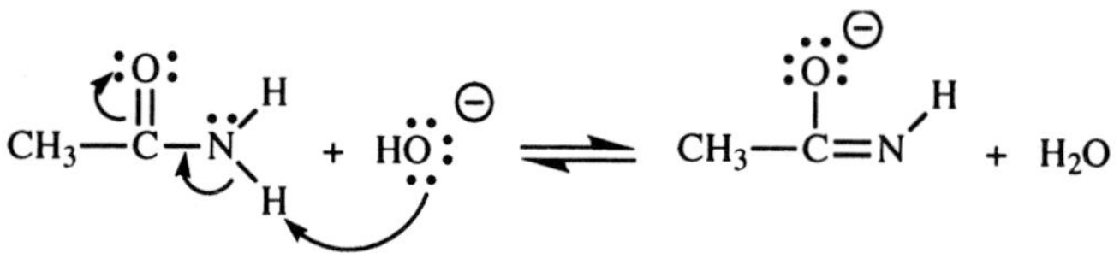

EQ. 13.20

$$CH_3-C(-N(H)(H)) + HO^- \rightleftharpoons CH_3-C=N-H + H_2O$$

The second step is formation of the bromoamide.

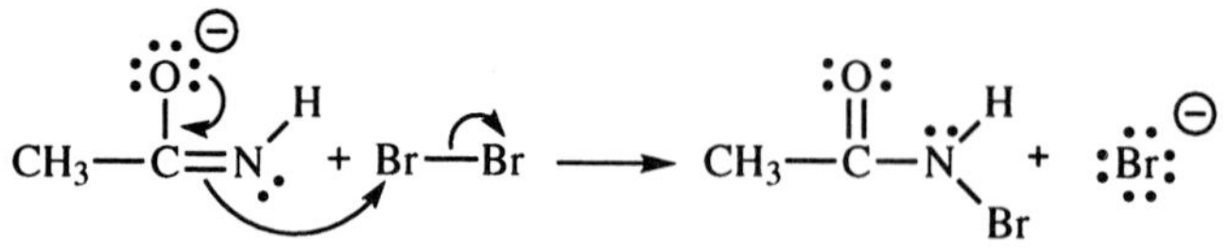

EQ. 13.21

$$CH_3-C=N + Br-Br \longrightarrow CH_3-C-N(H)(Br) + :Br:^-$$

an N-bromoamide

Once formed, the bromoamide is deprotonated.

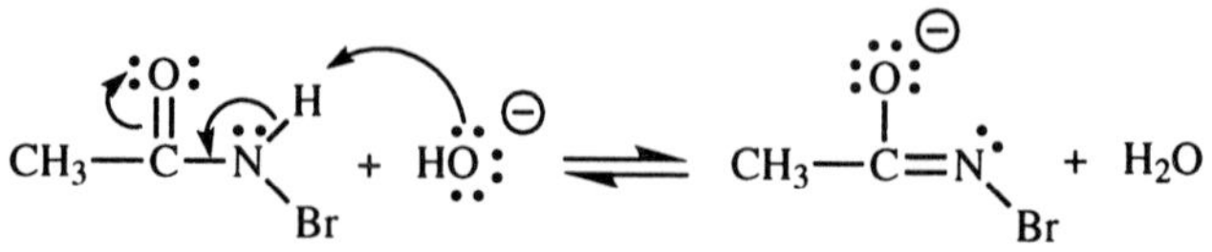

EQ. 13.22

$$CH_3-C-N(H)(Br) + HO^- \rightleftharpoons CH_3-C=N-Br + H_2O$$

In the next step an isocyanate is formed when the R group migrates to the nitrogen atom.

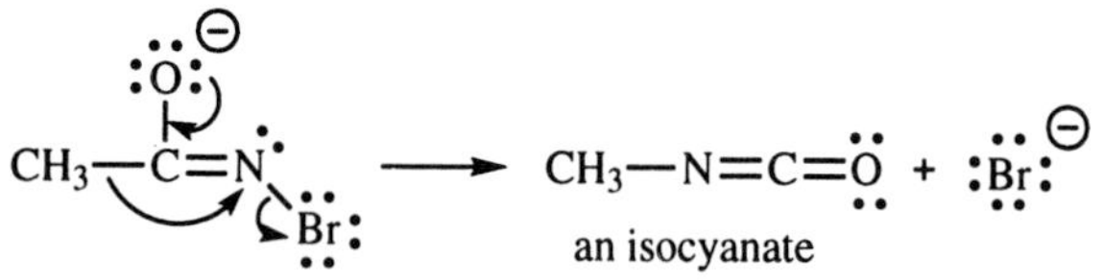

EQ. 13.23

$$CH_3-C=N-Br \longrightarrow CH_3-N=C=O + :Br:^-$$

an isocyanate

The isocyanate then reacts with base and water to form a carbamic acid.

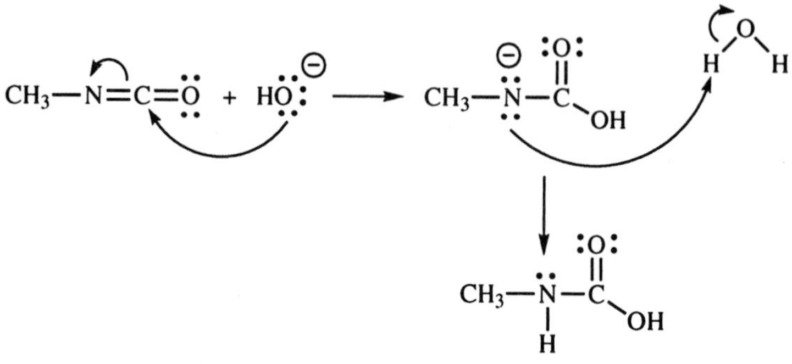

EQ. 13.24

$$CH_3-N=C=O + HO^- \longrightarrow CH_3-N-C(=O)-OH$$

$$\downarrow$$

$$CH_3-N(H)-C(=O)-OH$$

a carbamic acid

In the last step, the unstable carbamic acid is decarboxylated to form the amine.

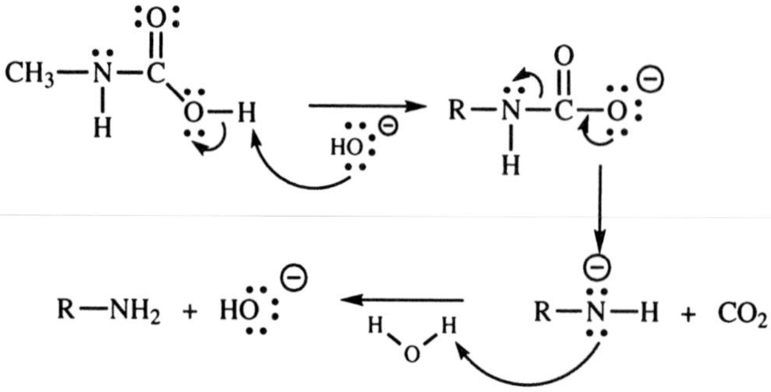

EQ. 13.25

$$CH_3-N(H)-C(=O)-O-H \xrightarrow{HO^-} R-N(H)-C(=O)-O^-$$

$$\downarrow$$

$$R-NH_2 + HO^- \longleftarrow R-N-H + CO_2$$

Let us look briefly at the evidence that has been collected in support of this mechanism. First, the reaction works only with primary amides. This makes sense since the nitrogen atom must be deprotonated twice during the process. In addition, it has been determined that secondary amines are converted to bromoamides by treatment with bromine and base. This is further evidence for the first step in the process. A second piece of evidence is that carbamates are formed if alcohols or alkoxides are used instead of water or hydroxide ion. This supports the presence of isocyanate intermediates because it is well known that isocyanates are converted to carbamates upon treatment with alcohols.

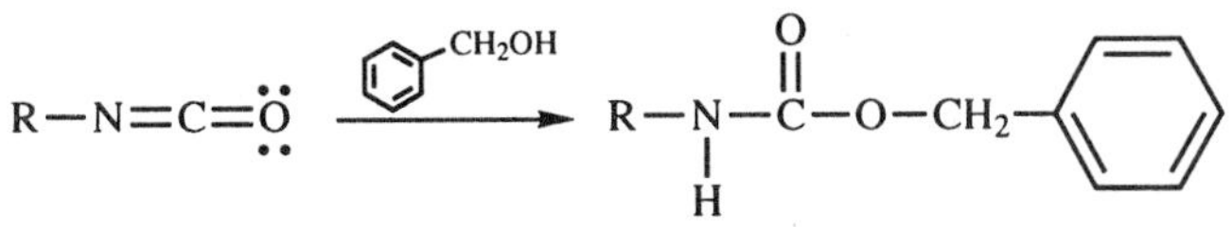

EQ. 13.26

a carbamate

Finally, if the amide is optically active, the product has the same stereochemistry as the starting amide. This supports the concerted nature of the migration step in which the carbon-nitrogen bond is formed and the carbon-carbon bond is broken.

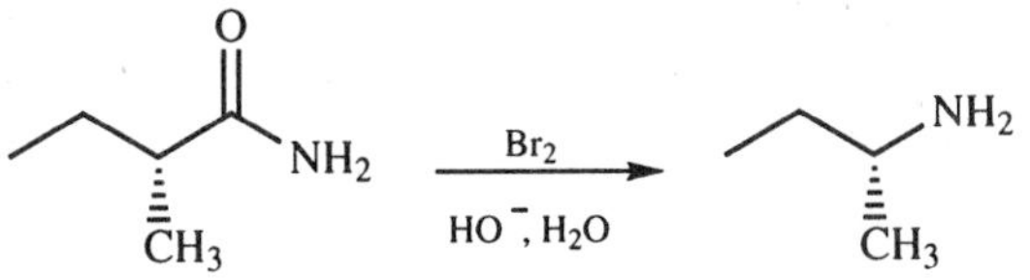

EQ. 13.27

The Curtius rearrangement, named for Theodore Curtius (1857–1928), is very similar except that it involves the rearrangement of an acyl azide, initially formed by reacting an acid chloride with sodium azide, into an isocyanate as shown at right.

Once formed, the isocyanate follows the same path to the amine as in the Hoffman rearrangement.

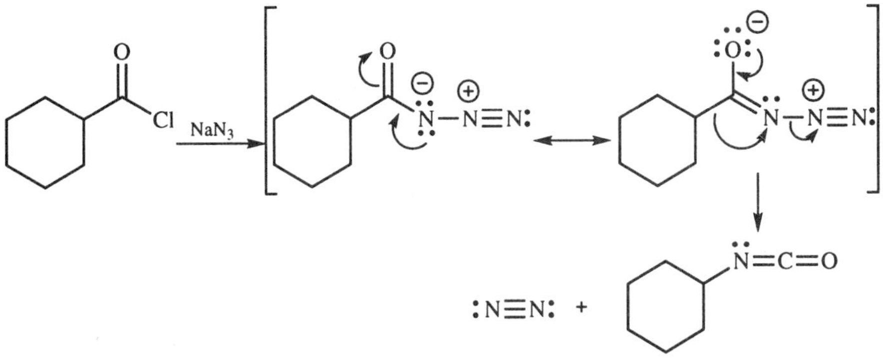

FIGURE 13.24:
The Curtius rearrangement.

## 13.7 REACTIONS OF AMINES WITH CARBONYL COMPOUNDS: IMINES AND ENAMINES

The first type of reaction we will examine is the reaction of ammonia or amines with aldehydes and ketones. The immediate product of this reaction is an **imine** (sometimes also called a Schiff base). If formation of the imine is immediately followed by reduction of the imine, the overall process is referred to as reductive amination. It is presented in a general fashion in equation 13.28.

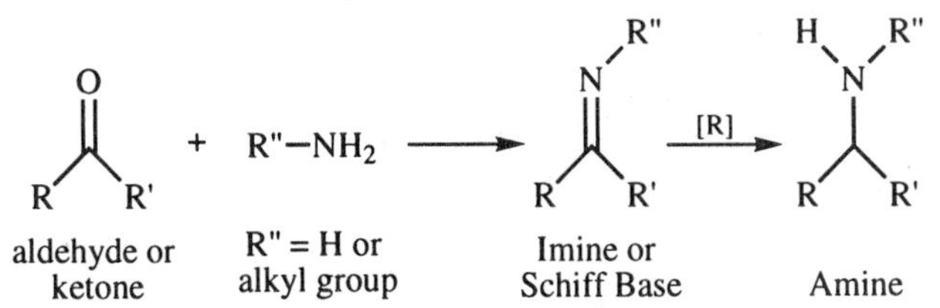

EQ. 13.28

aldehyde or ketone        R" = H or alkyl group        Imine or Schiff Base        Amine

The proposed mechanism for imine formation begins with nucleophilic attack by the primary amine on the carbonyl carbon. This is followed by protonation of the alkoxide and deprotonation of the nitrogen atom. The alcohol is protonated, turning it into a good leaving group, and a carbon nitrogen bond is formed. Finally, the nitrogen is deprotonated a second time. Imines are hydrolyzed back into carbonyl compounds by the reverse series of steps.

The carbon nitrogen double bond can be reduced a number of ways. Industrially the reaction would probably be carried out with hydrogen and a catalyst. In the laboratory the use of sodium cyanoborohydride ($NaBH_3CN$) is more common. An example of this reaction is the preparation of the N-benzyl methyl serinoate from serine methyl ester hydrochloride.

EQ. 13.29

Reactions of amine based compounds with carbonyl compounds are used extensively in classical organic qualitative analysis. The products of these reactions (called derivitives in this context) are typically solid and, therefore, are readily characterized by melting point. The melting points of many of these derivatives have been tabulated, so a simple check of the melting point allows the chemist to determine the identity of the unknown carbonyl compound. The most common derivatives are those obtained by reacting the carbonyl compound with hydroxyl amine, hydrazine, phenylhydrazine, and semicarbazide. The derivatives obtained are referred to as **oximes**, **hydrazones**, **phenylhydrazones**, and **semicarbazones** respectively. The structures of the derivatives of acetaldehyde are shown in Table 13.2.

**TABLE 13.2**: STRUCTURES OF NITROGEN DERIVATIVES                    TABLE 13.2

| REAGENT | DERIVATIVE | STRUCTURE OF REAGENT | STRUCTURE OF DERIVATIVE |
|---------|-----------|----------------------|-------------------------|
| Hydroxylamine | Oxime | $H_2N\!-\!OH$ | $CH_3\!-\!CH\!=\!N\!-\!OH$ |
| Hydrazine | Hydrazone | $H_2N\!-\!NH_2$ | $CH_3\!-\!CH\!=\!N\!-\!NH_2$ |
| Phenylhydrazine | Phenylhydrazone | 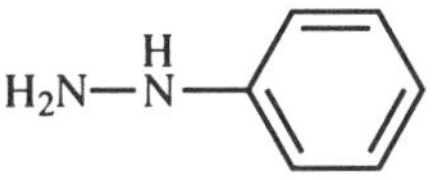 $H_2N\!-\!\overset{H}{N}\!-\!\bigcirc$ | 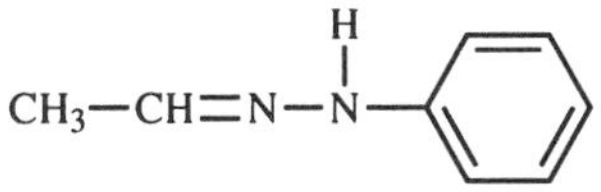 $CH_3\!-\!CH\!=\!N\!-\!\overset{H}{N}\!-\!\bigcirc$ |
| Semicarbazine | Semicarbazone | 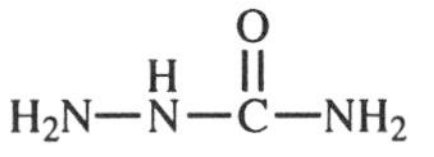 $H_2N\!-\!\overset{H}{N}\!-\!\overset{O}{\overset{\|}{C}}\!-\!NH_2$ |  $CH_3\!-\!CH\!=\!N\!-\!\overset{H}{N}\!-\!\overset{O}{\overset{\|}{C}}\!-\!NH_2$ |

In another important reaction involving amines and carbonyl compounds, the **Strecker synthesis** of amino acids utilizes a series of reactions that begin with the addition of ammonia to an aldehyde to form an imine.

Once the imine is formed, it is reacted with hydrogen cyanide to form an α–amino nitrile. Since nitriles can be readily hydrolyzed to carboxylic acids, this compound is easily converted into the desired amino acid. However, since none of the steps in the reaction sequence are stereospecific, the resulting amino acid is a racemic mixture of both stereoisomers.

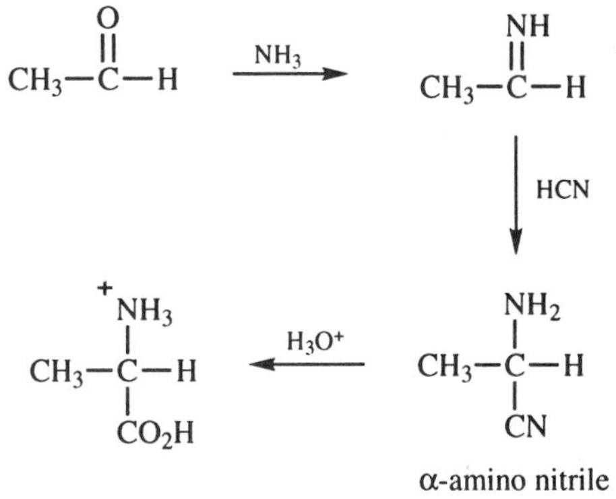

Secondary amines react with aldehydes and ketone to form **enamines** (pronounced 'een – ameenes') rather than imines. This can be understood if we look at the mechanism of this transformation.

Since the nitrogen atom does not contain a proton to be removed by the base, a proton on the α carbon atom is removed.

Enamines are the nitrogen analogs of enols and, like enols, are capable of undergoing alkylation reactions. They require somewhat more reactive alkylating agents than do enols. This is sometimes an advantage because they can be alkyated only one time. Once alkylation has occurred, the resulting iminium ion is readily converted back into a carbonyl compound by hydrolysis.

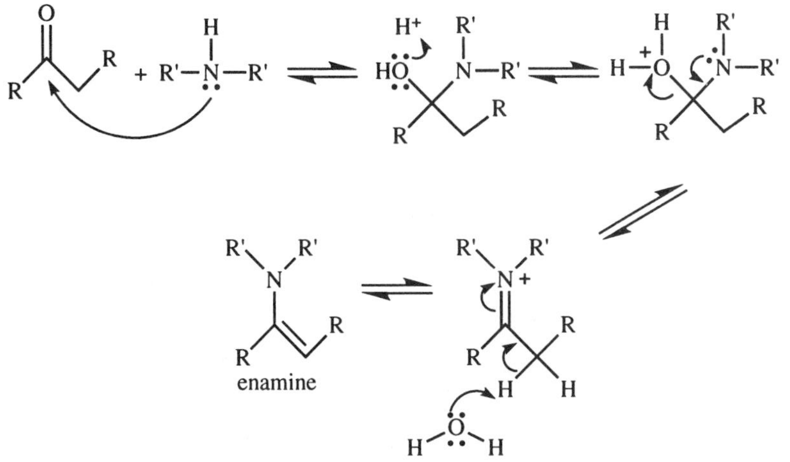

FIGURE 13.26:
The Strecker sythesis of racemic alanine.

FIGURE 13.27:
The proposed mechanism of enamine formation.

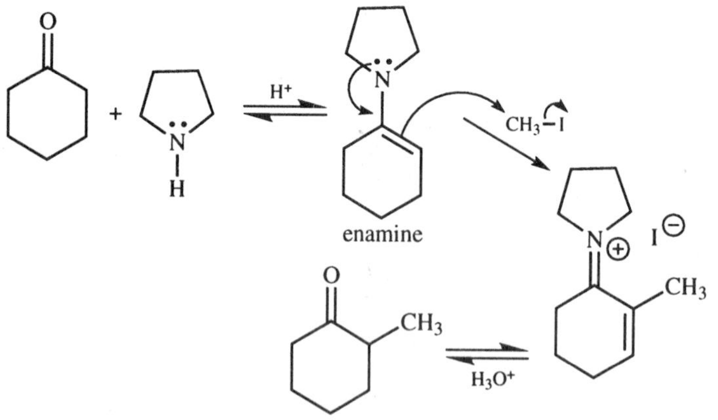

FIGURE 13.28:
Synthesis, alkylation, and hydrolysis of enamines.

Enamines can also be acylated by reacting them with acid chlorides. After hydrolysis of the intermediate iminium ion, the final product of this process is a β-dicarbonyl compound.

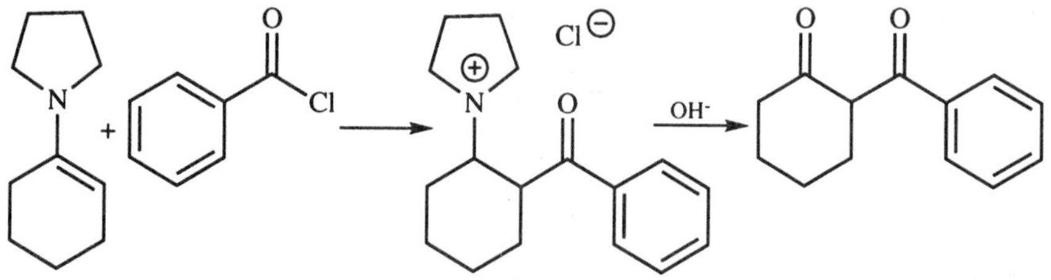

FIGURE 13.29:
Synthesis of β-dicarbonyl compounds by acylation of enamines.

## 13.8  NITROSATION OF AMINES

Nitrous acid, ($HNO_2$, but frequently written HONO),  is formed by reacting sodium nitrite with a strong acid such as HCl.  Once formed, the compound immediately begins to slowly decompose.

$$NaNO_2 + HCl \longrightarrow HONO + NaCl$$

EQ. 13.30

In the presence of additional strong acid, the nitrous acid molecule is protonated and a molecule of water is lost to form the nitrosonium ion.  This reaction is shown below.

Thus, reaction of amines with nitrous acid really involves reaction with the nitrosonium ion.

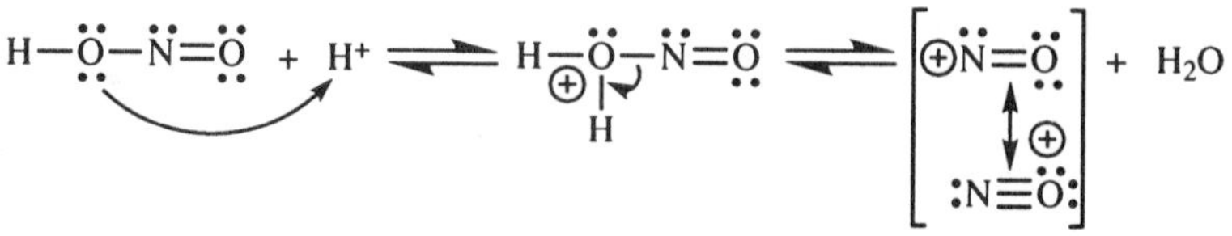

FIGURE 13.30:
Generation of the nitrosonium ion.

Primary amines react with the nitrosonium ion as shown in figure 13.31.  The nucleophilic nitrogen of the amine attacks the electrophilic nitrogen of the nitrosonium ion resulting in the formation, after deprotonation, of an N-nitrosamine.  After several more proton transfer reactions, an unstable alkanediazonium salt is formed.  It is the formation of this diazonium salt that gives the process its name: **diazotization**.  The alkanediazonium salt is unstable and rapidly loses a molecule of nitrogen.  This results in the formation of a carbocation.  The products of diazotization reactions are, therefore, those that we would expect to see from the involvement of a carbocation intermediate.  They include alkenes, formed during elimination of a proton from the carbonium ion,

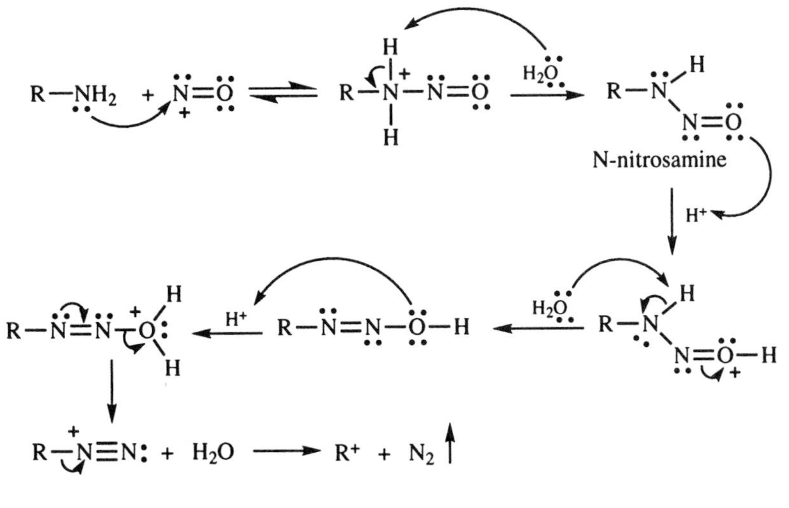

FIGURE 13.31: Reaction of primary aliphatic amines with the nitrosonium ion.

as well as alcohols and halides, formed by nucleophilic attack on the carbocation by water molecules or halide ions in the reaction mixture.

Since there is only one proton that can be removed, the reaction of the nitrosonium ion with a secondary amine stops at the stage of the nitrosamine. This reaction is shown in figure 13.32.

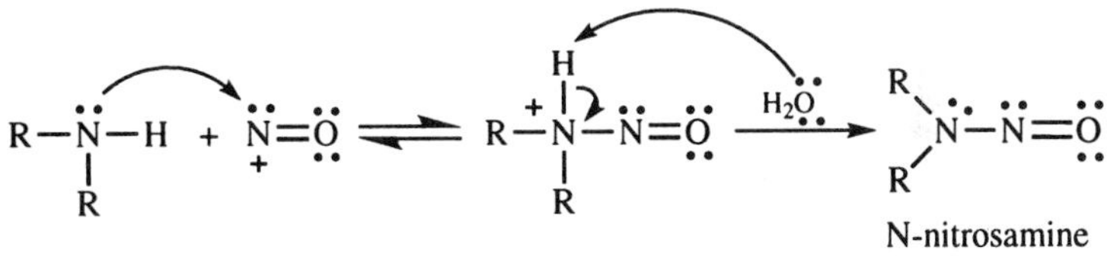

FIGURE 13.32: Reaction of secondary aliphatic amines with the nitrosonium ion.

Aromatic primary amines react with the nitrosonium ion to form the relatively stable arenediazonium ions. These compounds react with a variety of reagents to form many different functional groups. These reactions are summarized in figure 13.33.

The conversion of the arenediazonium ion to aryl chlorides, bromides, or nitriles is readily accomplished using the cuprous salts of these anions. This conversion is known as the Sandmeyer reaction. For example 4-methyl aniline is converted to 4-bromotoluene through diazotization followed by treatment with cuprous bromide.

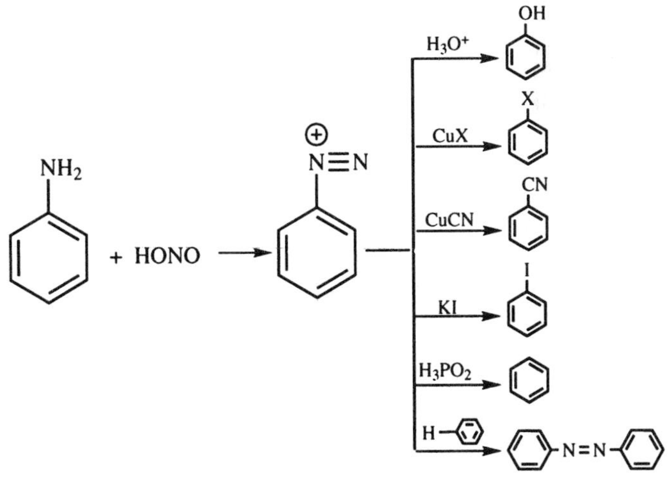

FIGURE 13.33: Various reactions of arenediazonium compounds.

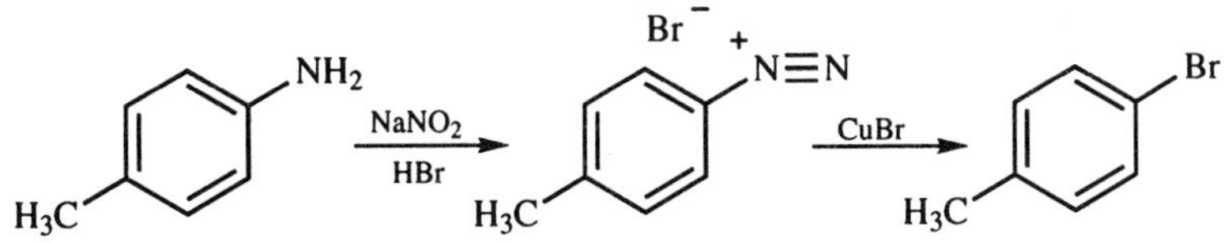

EQ. 13.31

Perhaps more interesting is the process of diazo coupling. This process results in the formation of highly colored compounds, many of which are used as dyes. Arenediazonium salts are electrophilic enough to attack strongly activated aromatic rings; *i.e.*, rings that are substituted with substituents that are strongly electron releasing, (*e.g.*, –OH). The mechanism of this process for the synthesis of Alizarin (a yellow dye) is shown below.

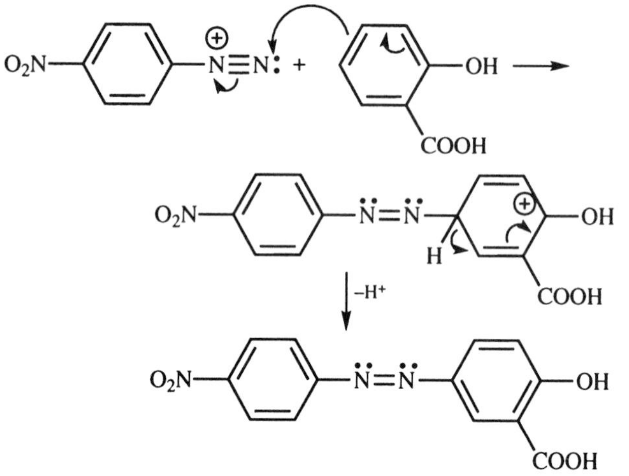


FIGURE 13.34: The sythesis of alizarin.

## 13.9 THE HOFMANN AND COPE ELIMINATIONS: ALKENES FROM AMINES

The **Hofmann elimination**, discovered in 1851 by August Wilhelm von Hofmann, is really an elimination reaction undergone by quaternary ammonium hydroxides. However, the process generally begins with an amine and takes place in three steps. The process has been used for the synthesis of alkenes as well as for the structural analysis of complex nitrogen–containing natural products.

Step one is the exhaustive methylation of the amine using methyl iodide.

$$R{-}CH_2{-}CH_2{-}NH_2 + CH_3I\ (\text{excess}) \longrightarrow R{-}CH_2{-}CH_2{-}\overset{+}{N}(CH_3)_2{-}CH_3\ \ I^- \qquad \text{EQ. 13.32}$$

The second step involves the conversion of the quaternary ammonium iodide into a quaternary ammonium hydroxide with silver oxide. The driving force for this conversion is the precipitation of the insoluble silver iodide.

$$2\ R{-}CH_2{-}CH_2{-}\overset{+}{N}(CH_3)_2{-}CH_3\ \ I^- + Ag_2O \xrightarrow{H_2O} 2\ R{-}CH_2{-}CH_2{-}\overset{+}{N}(CH_3)_2{-}CH_3\ \ HO^- + 2\ AgI\downarrow \qquad \text{EQ. 13.33}$$

Finally, in the third step, the quaternary ammonium hydroxide is heated to cause elimination.

$$R{-}CH_2{-}CH_2{-}\overset{+}{N}(CH_3)_2{-}CH_3\ \ HO^- \xrightarrow{\Delta} RCH{=}CH_2 + HO{-}N(CH_3){-}CH_3 + CH_4 \qquad \text{EQ. 13.34}$$

Let us look at the stereochemical and regiochemical nature of this process. Consider the Hofmann elimination of 2-butanamine. The possible products of such a process are 1-butene, *trans*-2-butene, and *cis*-2-butene, but only 1-butene and *trans*-2-butene are formed. Furthermore, 1-butene is the major product of this reaction. This is in direct opposition to Saytzeff's rule that states that the more substituted alkene generally predominates in elimination reactions. Thus, the formation of the less substituted alkene in an elimination reaction is often called the Hofmann orientation. Let us look more closely to see if we can rationalize these observations.

Consider the different Newman projections for the conformations that would lead to each of the possible products. As is typically the case, the proton being abstracted and the leaving group must be anti to each other.

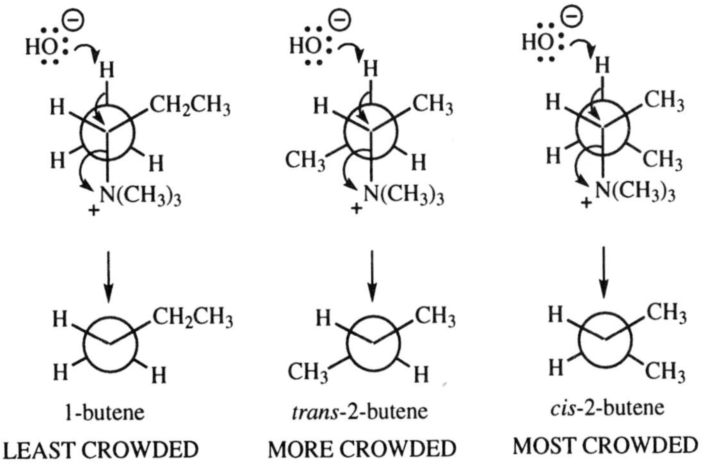

The Newman projections clearly show that the least crowded arrangement is the one that gives 1-butene as the major product. (This can be difficult to see from the 2–dimensional drawings. Students would well–advised to make models and use them to make certain they fully understand this point.) Thus, the activation energy for this process is the lowest and 1-butene would be expected to predominate. The conformation that leads to *trans*-2-butene is somwhat more crowded and the activation energy to form this product will be higher. The higher activation energy means that a smaller amount of this product is formed. The conformation leading to *cis*-2-butene is the most crowded. The activation energy for the process leading to this product is high enough that *cis*-2-butene is not observed among the products.

**FIGURE 13.35:** Newman projections for the Hofmann elimination of 2-butanamine.

The **Cope elimination** (equation 13.35) is the thermal elimination of a tertiary amine oxide. These compounds are readily prepared by oxidation of the tertiary amine using oxidizing agents such as hydrogen peroxide or *m*-CPBA (meta–chloroperoxybenzoic acid).

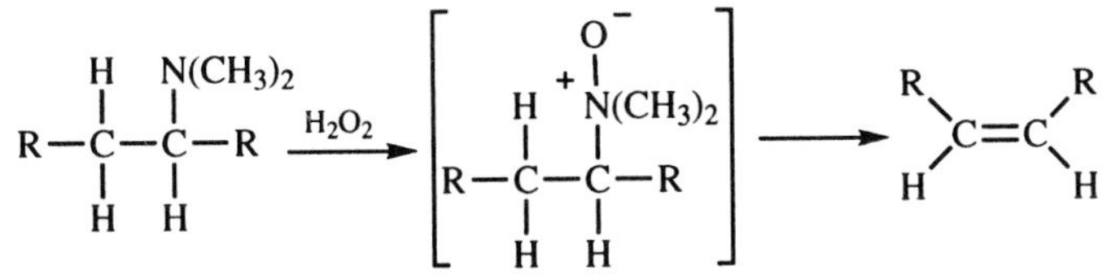

EQ. 13.35

Because the Cope elimination proceeds through a cyclic transition state, the observed stereochemistry must be *syn*. The regiochemistry is typically the same as the Hofmann elimination, that is, the less substituted alkene is formed as the major product. Since the two reactions proceed through different transition states, they are complimentary when the substrate is optically active.

# 13.10 ANOTHER LOOK AT AROMATIC SUBSTITUTION: PYRIDINE AND PYRROLE

Pyridine is an electron deficient aromatic system because of the presence of the electronegative nitrogen atom in the ring. This is most easily seen if we look at possible resonance structures for pyridine. These are shown in figure 13.36 along with a representation of the electron densities at each of the atoms in the molecule generated using molecular orbital considerations.

Looking at these resonance structures and the electron densities, we might even predict that pyridine would be resistant to electrophilic aromatic substitution due to the partial positive character of the ring. In fact, pyridine will undergo electrophilic substitution but the reaction does require forcing conditions.

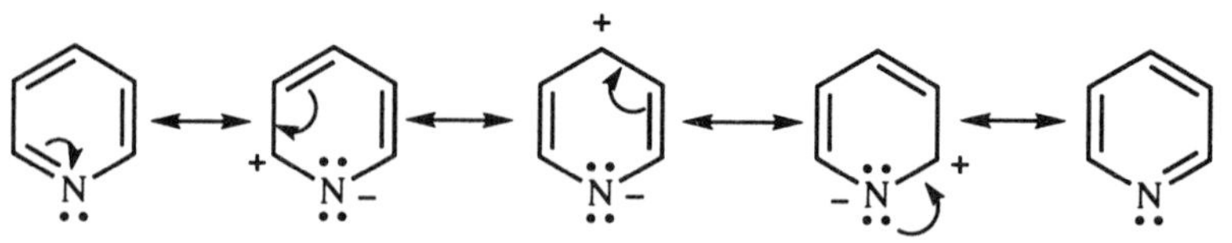

FIGURE 13.36: Resonance forms and electron densities of pyridine.

The resonance diagrams and the electrons densities allow us to predict that, if electrophilic aromatic substitution were to occur, the electrophile would attack at the three position because this position is the least electron deficient position in the ring. Note this is the only position on the ring that does not have positive character. Again, this prediction would be correct. Pyridine undergoes electrophilic aromatic substitution at the three position when treated with strong electrophiles. Shown below are the intermediates expected for nitration of pyridine at the two and three positions.

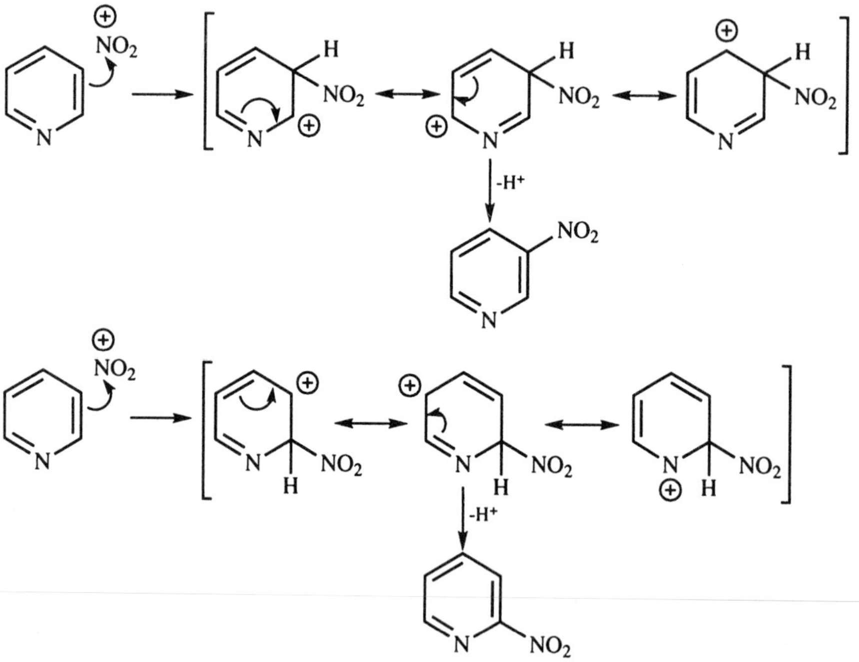

FIGURE 13.37: Nitration of pyridine.

Attack at the two position is disfavored not only because the electron density is not as high on that carbon atom but also because the resonance forms for the intermediate would place a positive charge on the more electronegative nitrogen atom.

In contrast to pyridine, pyrrole is an electron rich aromatic system and undergoes electrophilic aromatic substitution more readily than benzene. Shown below are the intermediates expected for the nitration of pyrrole.

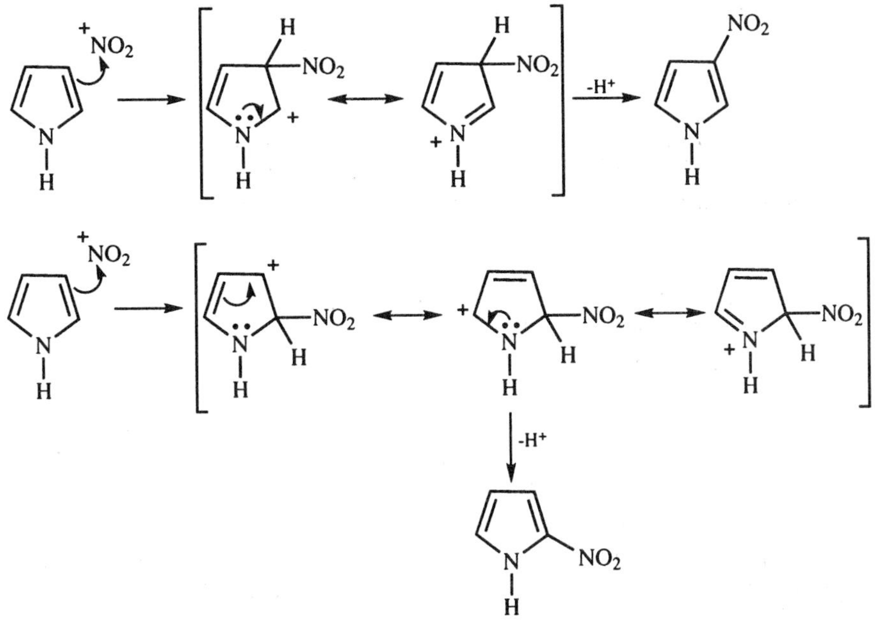

FIGURE 13.38:
Nitration of pyrrole.

In the case of pyrrole, substitution takes place primarily at the two position because the intermediate is more resonance stabilized.

As we would expect, the situation for nucleophilic aromatic substitution is exactly reversed. Pyrrole is very resistant to nucleophilic aromatic substitution while pyridine is a good substrate for this process. Figure 13.39 shows the intermediates expected when the chloropyridines are treated with methoxide.

Since attack at positions 2 and 4 allows a resonance form in which the negative charge is placed on the electronegative nitrogen atom, attack at these positions is preferred.

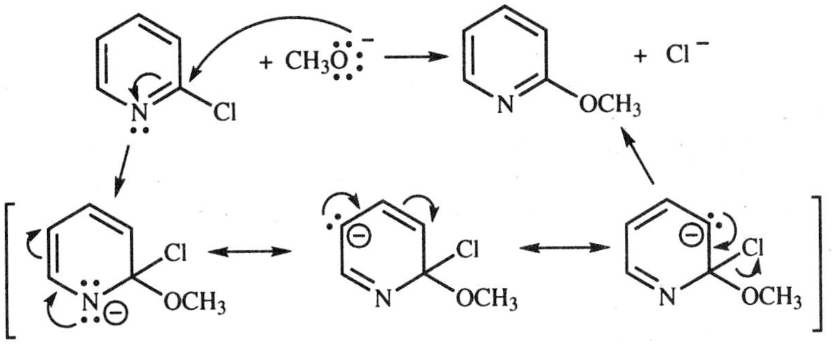

FIGURE 13.39:
An example of nucleophilic substitution on pyridine.

**Alpha amino acid:** a compound in which an amino group and a carboxyl group are bonded to the same carbon. The twenty naturally occurring amino acids are the building blocks of proteins in biological systems.

**Cope elimination:** the thermal elimination of tertiary amine oxides.

**Curtius rearrangement:** the synthesis of an amine from an acid or acid chloride through an acyl azide intermediate.

**Diazotization:** the reaction of primary aliphatic and aromatic primary amines with HONO to form diazonium salts.

**Gabriel synthesis:** the synthesis of amines and amino acids by nitrogen alkylation of phthalimide.

**Haber process:** the industrial process by which hydrogen and nitrogen are catalytically converted to ammonia.

**Hell-Volhard-Zelinsky reaction:** the synthesis of an $\alpha$-halo acid from a carboxylic acid upon treatment with $Br_2$ and catalytic $PCl_3$.

**Hofmann elimination:** the thermal decomposition of quaternary ammonium hydroxides to form alkenes. This reaction preferentially produces the less substituted alkene as the major product.

**Hofmann rearrangement:** the synthesis of an amine from an amide by treatment with bromine and aqueous base.

**Hydrazones:** a derivative of aldehydes/ketones prepared by reacting the carbonyl compound with hydrazine.

**Imine:** a compound that contains a carbon – nitrogen double bond. Imines are prepared by reacting aldehydes or ketones with primary amines.

**Isoelectric point:** the pH at which an amino acid exists exclusively as a zwitterion.

**Nitrogen cycle:** the complex process of forming, consuming, and reprocessing compounds of nitrogen.

**Nitrogen fixation:** the process of converting $N_2$ into other forms of nitrogen.

**Ostwald process:** the industrial process by which nitric acid is prepared from ammonia.

**Oximes:** a derivative of aldehydes/ketones prepared by reacting the carbonyl compound with hydroxylamine.

**Phenylhydrazones:** a derivative of aldehydes/ketones prepared by reacting the carbonyl compound with phenylhydrazine.

**Primary amine:** an amine in which the nitrogen is bonded to only one alkyl group.

**Secondary amine:** an amine in which the nitrogen is bonded to two alkyl groups.

**Semicarbazones:** a derivative of aldehydes/ketones prepared by reacting the carbonyl compound with semicarbazine.

**Tertiary amine:** an amine in which the nitrogen is bonded to three alkyl groups.

**Zwitterion:** a species that contains an equal number of positive and negative charges and is, therefore, electrically neutral. The amino acids are often found as zwitterions.

# HOMEWORK Problems

1. Complete the following reactions.

a.

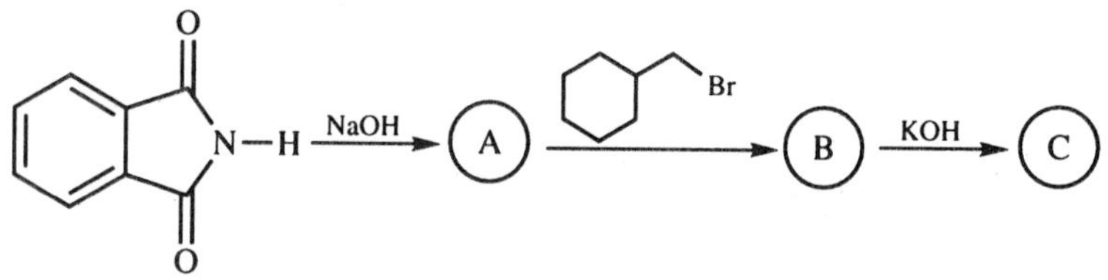


b.

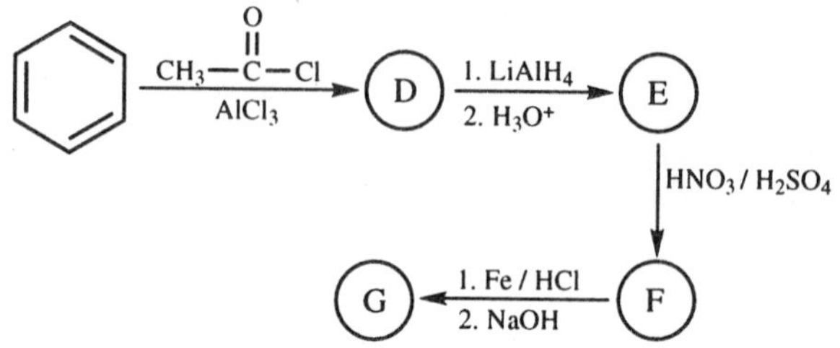


c.

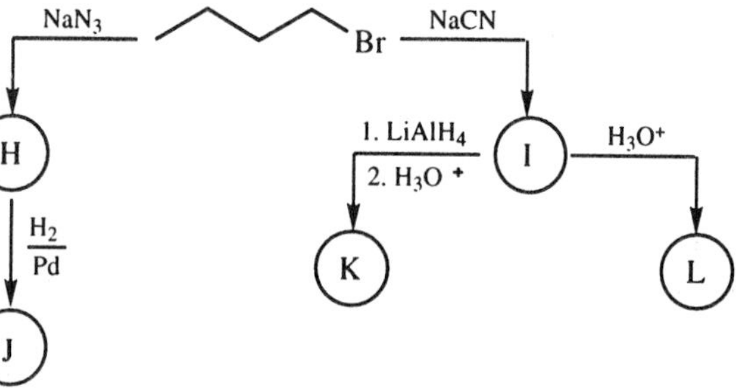


d.

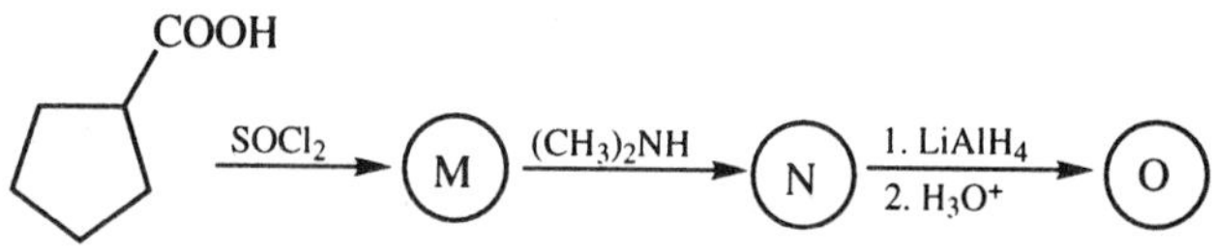


e.

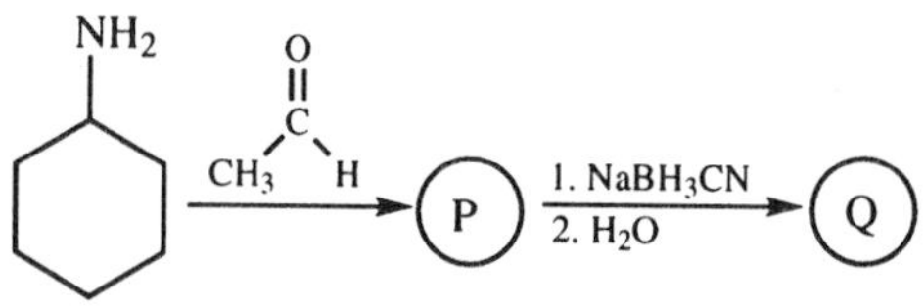


f.

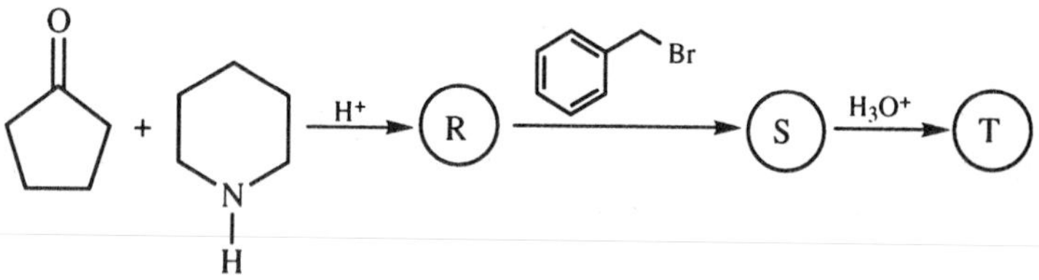


g.

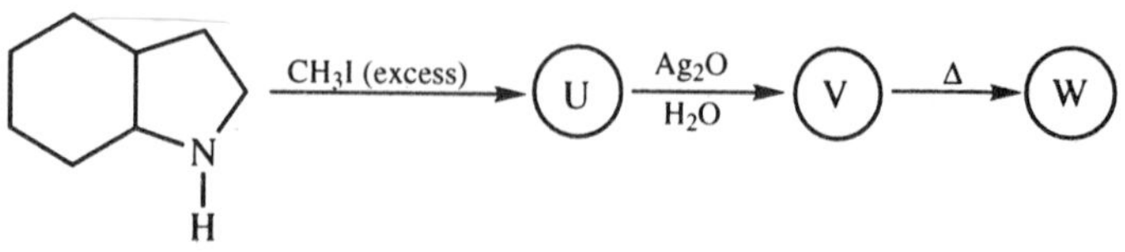

h.

2. Show how the following compounds can be prepared from the indicated starting materials.
    a. hexylamine from 1-pentanol
    b. butylamine from 1-pentanol
    c. N-methylhexanamide from hexanoic acid
    d. hexanenitrile from pentanoic acid
    e. hexylamine from hexanoic acid

3. Provide a mechanism for the following transformations.

a.

b.

c.

d.

e.

4. Draw complete mechanisms (including resonance forms) for the following transformations.

a.

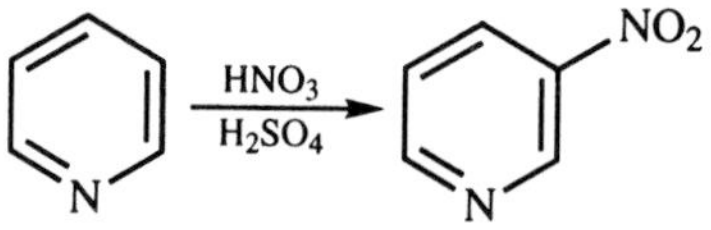

b.

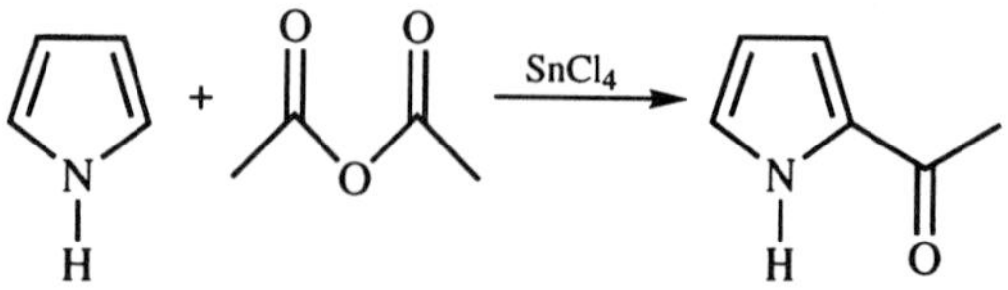

c.

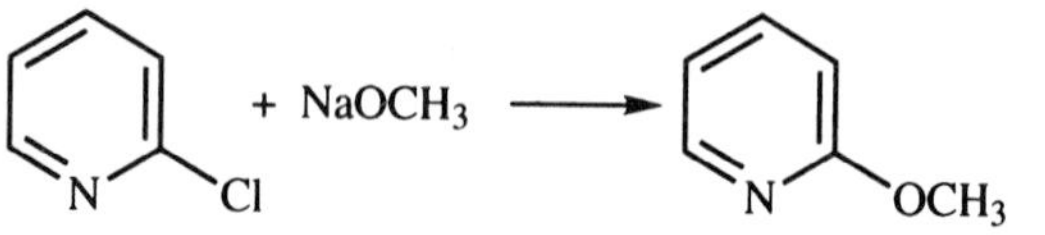

5. Provide the structure of the derivative that results from the following reactions.

a.

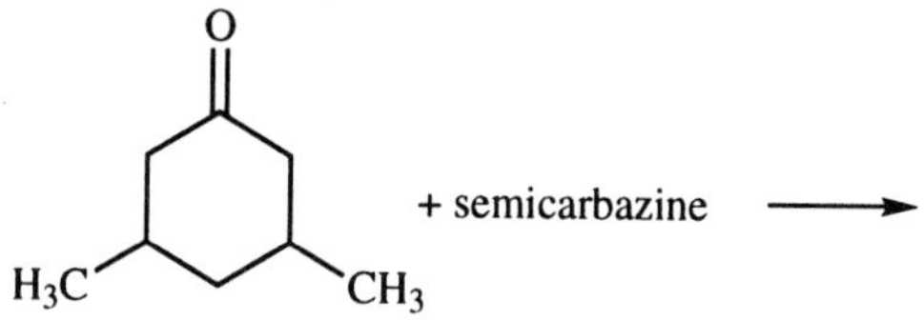

+ semicarbazine ⟶

b. 3-heptanone + $H_2N—NH_2$ ⟶

c. cyclopentanone + phenylhydrazine ⟶

d.

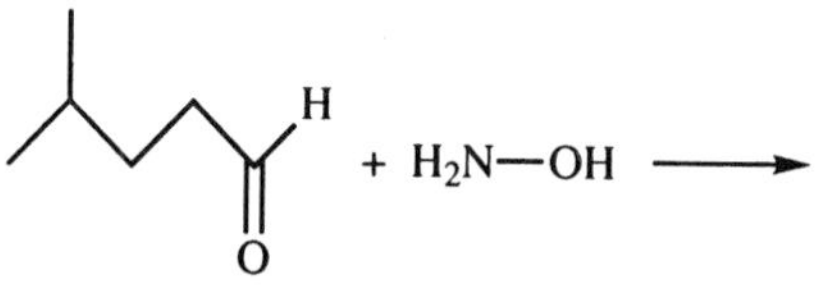

+ $H_2N—OH$ ⟶

6. Show how the following compounds can be prepared from the indicated starting material. Assume that ortho and para isomers are separable.

a. o-bromotoluene from toluene
b. p-flourotoluene from toluene
c. p-nitrobenzonitrile from benzene
d. 3,5-dibromotoluene from toluene
e.

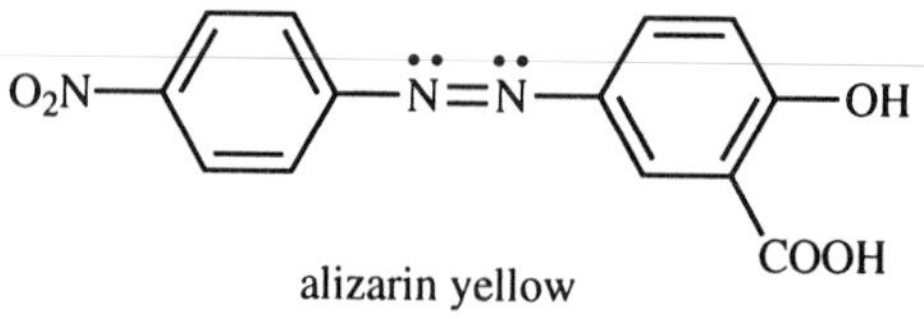

from aniline and 2-hydroxybenzoic acid

alizarin yellow

7. One mole of liquid 1,1-dimethylhydrazine $((CH_3)_2N_2H_2$, $\Delta G_f = 206.3$ kJ/mol) is a rocket fuel that reacts with two moles of liquid dinitrogentetroxide $(\Delta G_f = 97.54$ kJ/mol) to form gaseous molecular nitrogen, water vapor, and carbon dioxide gas.
   a. Write the net balanced reaction.
   b. Compute the standard state free energy change for the reaction.
   c. Calculate the pressure inside a 10.0 liter vessel due to the products formed from one mole of 1,1–dimethylhydrazine reacting completely with two moles of dinitrogen tetroxide at 500.0K, assuming ideal gas behavior

8. Nitrous oxide ($N_2O$, laughing gas) can be made by careful thermal decomposition of ammonium nitrate ($NH_4NO_3$, FW = 80.05). When 250.0 g of ammonium nitrate are decomposed, 30.0 liters of dry nitrous oxide are obtained at a temperature of 25.0°C and a pressure of $1.65 \times 10^5$ pascals.
   a. Write the balanced decomposition reaction.
   b. Calculate the moles of nitrous oxide formed.
   c. Calculate the % yield for this process.

9. Dinitrogen pentoxide ($N_2O_5$) dissolved in a liquid bromine solvent at a certain temperature decomposes to nitrogen dioxide and oxygen molecule with the following kinetics.

| t/sec | $[N_2O_5]$ |
|-------|-----------|
| 0.0 | 0.100 |
| 100.0 | 0.0819 |
| 200.0 | 0.0670 |
| 300.0 | 0.0549 |
| 500.0 | 0.0368 |
| 1000.0 | 0.0135 |

   a. Write the balanced decomposition reaction.
   b. Show that the decomposition is first order.
   c. Calculate the rate constant for the reaction.
   d. What would the concentration of $N_2O_5$ be at 30.0 minutes?

10. An unknown amine was determined to be 65.68% carbon, 15.16% hydrogen, and 19.15 % nitrogen. The IR spectrum of the amine showed two broad, distinct absorbances between 3700 and 3100 $cm^{-1}$. The HNMR showed a triplet, corresponding to 3 protons, at 0.9 ppm; a doublet; corresponding to 3 protons, at 1.1 ppm; a quintet, corresponding 2 protons, at 1.3 ppm; a broad singlet, corresponding to 2 protons, at 1.7 ppm; and a sextet, corresponding to 1 proton, at 2.8 ppm. (Note: as was the case when protons were bonded to oxygen, protons bonded to nitrogen do not participate in the normal spin coupling.)
   a. Determine the empirical formula of the amine.
   b. Interpret the IR spectrum of the amine.
   c. Propose a specific structure of the amine.
   d. Relate the peaks in the HNMR spectrum to the structure in (c).

11. An amino acid was synthesized by first reacting propanoic acid (MW=74.08, density = .993 g/ml) and bromine (MW = 159.808, density = 3.102 g/ml) in the presence of $PCl_3$. The resulting bromoacid is subsequently reacted with gaseous ammonia to form the relevant amino acid. In a specific synthesis, 25.0 ml of propanoic acid reacted with 20.0 ml of liquid bromine; the organic product at this step was then reacted with 8.00 liters of ammonia at a pressure of 735.0 torr and a temperature of $25.0^{\circ}C$. At the end of the synthesis, 18.5 g of the amino acid was obtained.

    a. Write the balanced reaction forming the bromoacid.

    b. Write the balanced reaction forming the amino acid, and name that acid.

    c. What is the theoretical yield of the bromoacid?

    d. What is the theoretical yield of the amino acid?

    e. What is the percent yield of the amino acid?

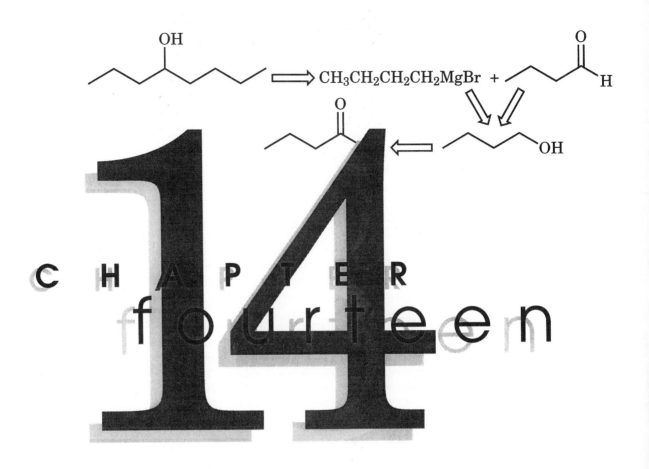

# CHAPTER fourteen

## 14

# MULTI-STEP Synthesis
AGAIN AND AGAIN UNTIL WE GET THERE

# MULTI-STEP SYNTHESIS ── 
## Again and again until we get there

## 14.1   INTRODUCTION

Thus far our emphasis on chemical reactions has been what happens to a molecule during one particular reaction.  Most of the time we have dealt with conversion of reactant to ultimate product in only one major reaction.  This has been done to concentrate attention upon learning the "tools of the trade" — those techniques in the chemist's tool box that can be used to carry out desired transformations.  A one–step reaction sequence is a very uncommon situation in the "real world" of chemistry — and even more uncommon in the world of the chemical industry.  In most situations getting from reactants to a desired product is the result of a series of very carefully planned reactions — each of which carries out only one transformation.

In this chapter we begin a discussion of syntheses that require multiple steps to go from the initial reactant to the final product.  Included in this discussion will be choosing a synthetic pathway so that only one particular chiral or geometric isomer is formed and also some other, strictly non–chemical, factors.  While studying the material in the chapter we shall have the opportunity to review many of the reactions that we studied in earlier chapters.  Along the way, reference will be made to a few modifications of already familiar reactions and a few new reactions will be introduced.

We will begin by discussing the commercial synthesis of several materials to give you a basic understanding of the process.  Since these are rather complicated, long processes involving many steps, you should not let them discourage you.  Rather you should look at the examples as just that — examples of some (hopefully interesting) syntheses that are commecially important.  The examples should also serve as a review of many of the individual reactions that we have studied.

### THE SYNTHESIS OF IBUPROFEN

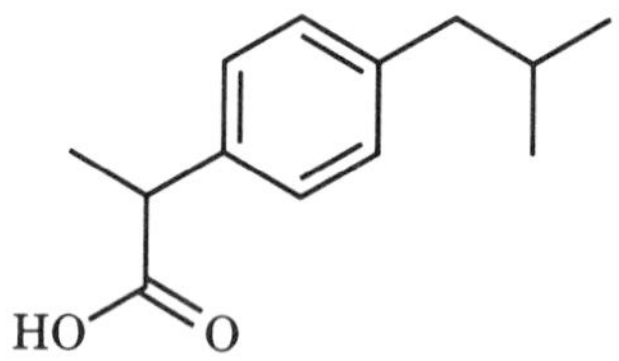

The synthesis of the commercially important pain–reliever ibuprofen provides a convenient and interesting example of the application of the many reactions studied thus far.  Ibuprofen, whose structure is shown in figure 14.1, is a sufficiently important compound that it has been synthesized by several different routes.  Let us examine two of those routes.  We shall look at the reactions involved and then discuss the reasons for choosing one route over another.

FIGURE 14.1:
Structure of
ibuprofen.

The first synthetic route to ibuprofen is illustrated in figure 14.2.  The synthesis begins with a commonly available starting material, isobutylbenzene.  The first reaction is the Friedel Crafts acylation of isobutylbenzene using acetic anhydride.  The reaction requires phosphoric acid as a catalyst.  In the second step, the newly introduced carbonyl functional group is reduced to an alcohol with sodium borohydride. Since the acetyl group that we added in step one contained one fewer carbon atoms than we need in the final product, and since we intend to acquire that additional carbon atom by using an $S_N2$ reaction to place a nitrile group where the alcohol function currently is located, we will next convert the alcohol function to a bromine atom using phosphorous tribromide. This produces the good leaving group that allows step 4, replacement of the bromine atom with a nitrile group, to proceed smoothly. Our final step is the hydrolysis of the nitrile to produce the desired carboxyl group. The total synthetic pathway consists of 6 steps (really only 5 since the step 6 is part of the product work–up from reaction 5). Although this reaction sequence could be used, for commercial success it would be better if the synthesis could be done in fewer steps.

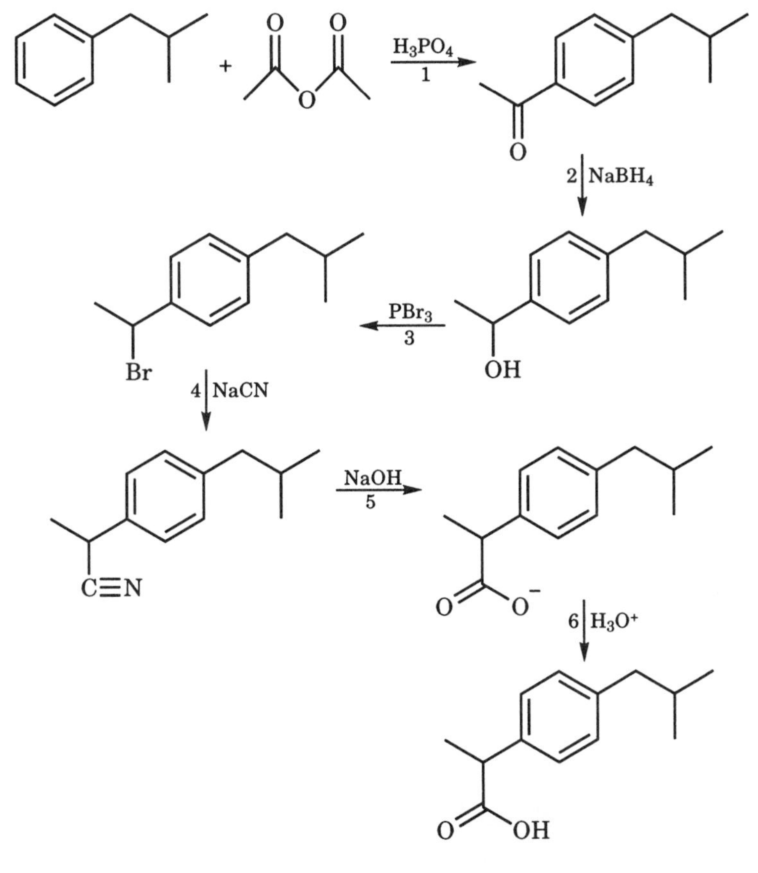

FIGURE 14.2:
One synthetic route to ibuprofen.

A shorter synthesis was developed by the Hoechst corporation and is outlined in figure 14.3. Its first step is the acylation of isobutylbenzene with acetyl chloride. The use of acetyl chloride instead of acetic anhydride means that it will not be necessary to dispose of the acetic acid formed as a by–product of the acylation step in the other synthesis. The second step is also reduction to the alcohol. However, instead of using sodium borohydride, the reduction is done by catalytic hydrogenation over a metal catalyst. This is done because hydrogen is much less expensive than is sodium borohydride. In the third step, the carboxylic acid functional group is directly formed by catalytic insertion of carbon monoxide into the carbon–oxygen bond of the alcohol. This is done over a palladium metal catalyst. This final reaction is one that we have not studied. It is not commonly used in laboratory settings for a variety of reasons. It *is* used industrially because the reactant (CO) is inexpensive and readily obtained and, as in this case, the reaction often directly produces the desired product with no additional steps that only reduce yield and are otherwise expensive to carry out.

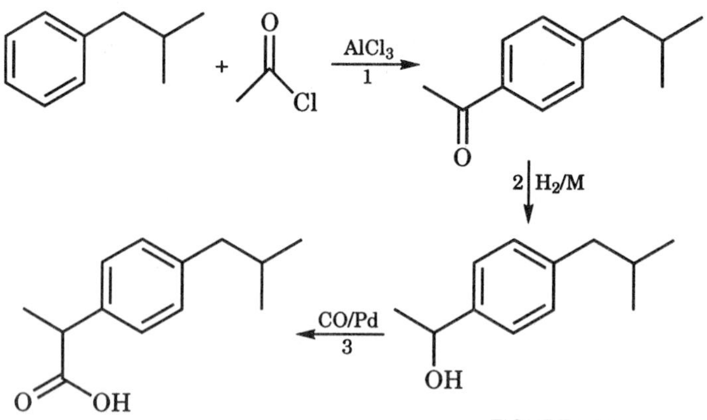

The Hoechst synthetic route to ibuprofen is much shorter, more direct, uses reagents that are less expensive, and produces less waste than does the other synthetic pathway. It does, however, use reactions that are much more difficult to accomplish in the laboratory because of the necessity of having equipment that can accommodate the high pressures required in the catalytic hydrogenation and the carbon monoxide insertion steps. These are economical on the industrial scale when the large quantities involved make the construction of large high–pressure reaction vessels and the buildings to house them more cost–effective than on the laboratory scale where the other reactions have a cost advantage.

FIGURE 14.3:
Hoechst ibuprofen sythesis.

## THE SYNTHESIS OF VASOTEC®

Our second example is another, somewhat longer, synthesis of the drug Vasotec® — one of a series of drugs used as an inhibitor of angiotensin converting enzyme (ACE). These drugs are used in treatment of hypertension. The structure of Vasotec® is given in figure 14.4.

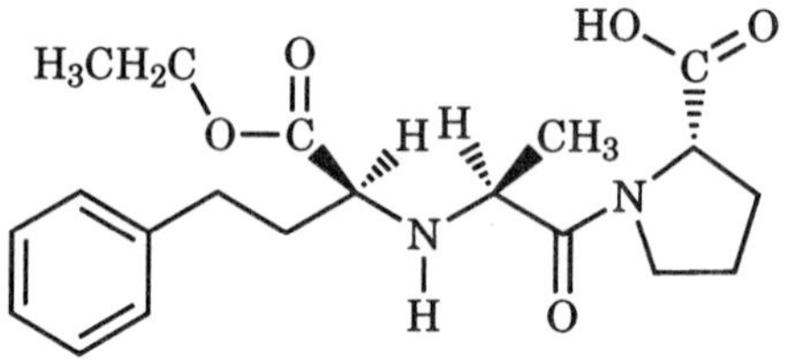

We shall look at the synthesis of this molecule starting with benzene, the amino acids alanine and proline, and several other organic compounds containing only two carbon atoms.

FIGURE 14.4:
Structure of Vasotec®.

Our route to the synthesis will be to build two relatively large portions of the molecule and then put them together at the end of the reaction sequence. The two large portions of the molecule that we shall build are approximately the left and right halves of the molecule in figure 14.4. If the bond between the nitrogen atom in the middle of the figure and the carbon atom to its left is broken, the Vasotec® molecule can be assembled from the two precursor molecules shown in figure 14.5.

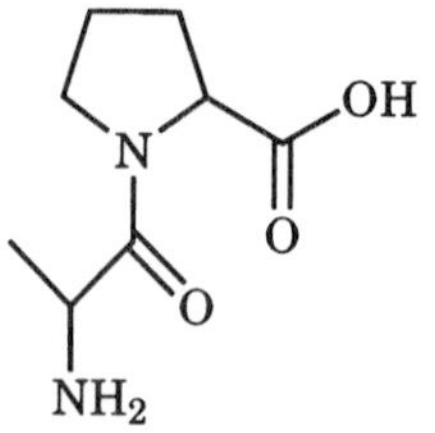

Alanylproline is made by techniques that will be discussed in more detail in a the next chapter. This molecule is a **dipeptide** synthesized from the amino acids proline and alanine. In order to make sure that the amide bond is formed between the carboxyl group on the alanine and the amine group on the proline instead of *vice versa*, we must first cover the carboxyl group on the proline and the amino group on the alanine. The protection of these functional groups is done in steps 1 and 2. Once the functional groups

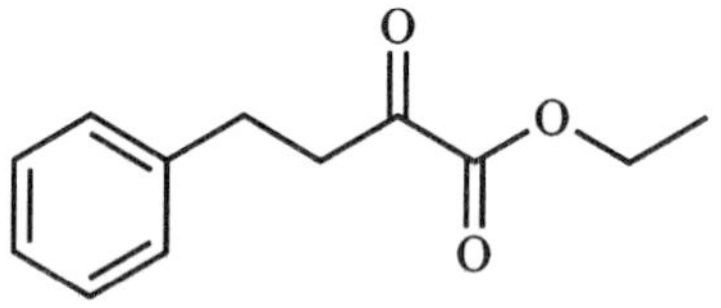

Alanylproline

Ethyl-4-phenyl-2-oxobutanoate

FIGURE 14.5:
Precursor molecules to Vasotec®.

are protected, the two molecules can be reacted under appropriate conditions to form the dipeptide (step 3). After this step, the protecting groups are removed (step 4) and the alanylproline is ready for the next part of the synthesis. The entire route is shown in figure 14.6. A complete discussion of protecting groups for pepetide synthesis is given in section 15.11.

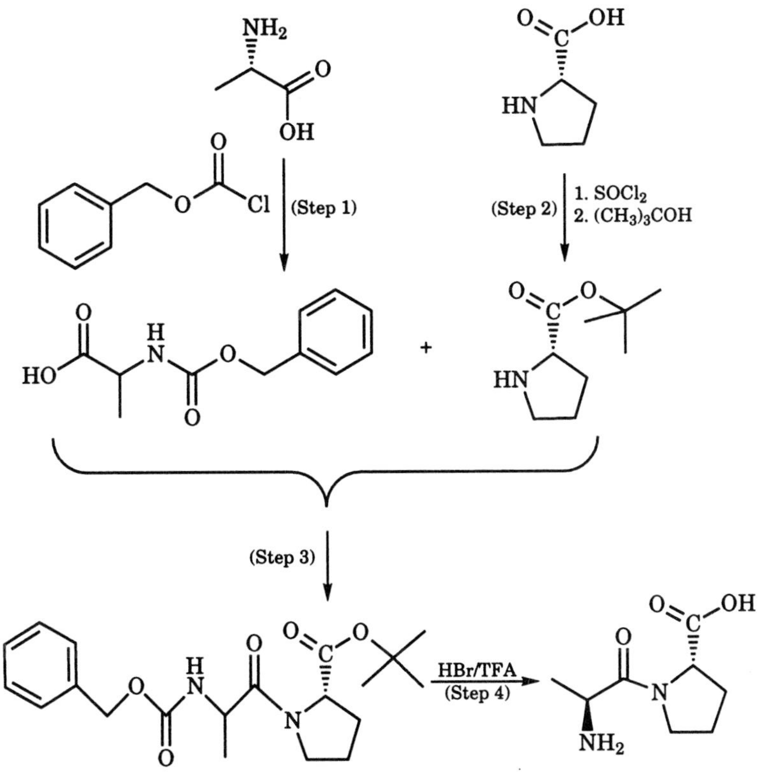

FIGURE 14.6: Synthesis of alanylproline.

Synthesis of the ethyl–4–phenyl–2–oxobutanoate from benzene and 2–carbon compounds involves a larger number of steps and is illustrated in figure 14.7. Steps 5 through 7 represent adding a 2–carbon side chain to a benzene molecule by reacting a Grignard reagent with an epoxide. This reaction is a simple extension of the Grignard reactions that we have already learned and is included in table 14.2. Steps 8 and 9 now prepare another Grignard reagent that will be reacted with an ester in step 11 to form the desired α–ketoester. We have seen these reactions, and the esterification reaction represented in step 10, previously.

The final step in the synthesis is the reductive coupling of the alanylproline with the ethyl–4–phenyl–2–oxobutanoate. This is accomplished by allowing the amine from the dipeptide to react with the ketone of the ketoester. This reaction forms an imine that is reduced *in situ* with sodium cyanoborohydride. It is important to keep a 1:1 mole ratio for alanylproline to ketoester at this point to prevent further reaction with the second carbonyl. W used sodium cyanoborohydride in chapter 13. The entire reaction is shown in figure 14.8.

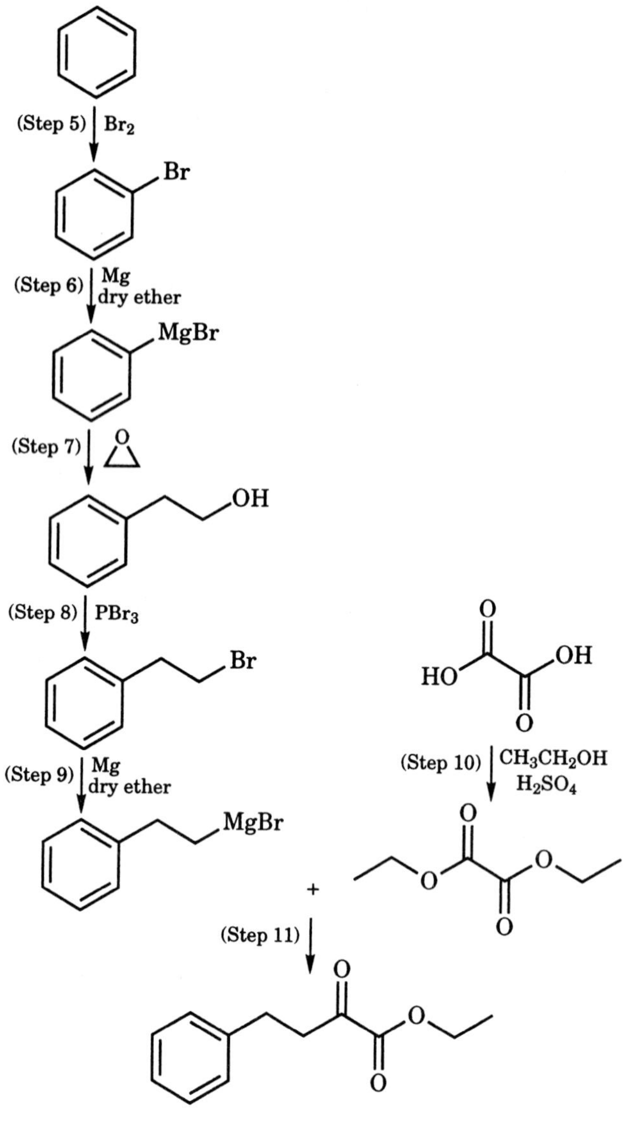

FIGURE 14.7:
Synthesis of
4-phenyl-2-oxo-
butanoic acid.

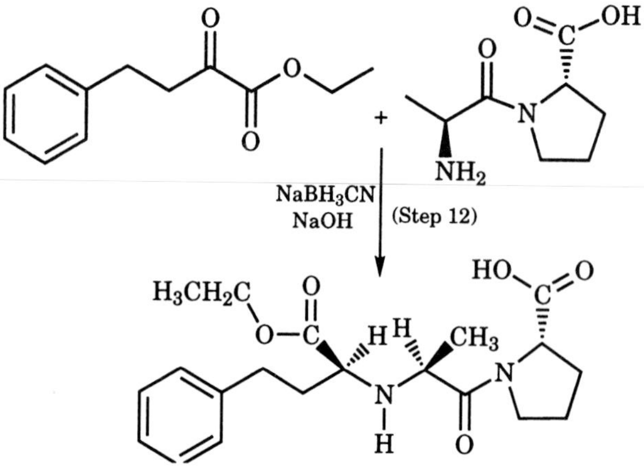

FIGURE 14.8:
Final step in the
Vasotec® synthesis.

## 14.2 REVIEW OF REACTIONS

Notice that the transformations that have just been shown as examples employ many of the same reactions you have already studied. They simply employ one reaction after another to add additional features to the starting materials until the "final" product is formed. The individual reactions that we have been studying should be considered as the chemist's "tool bag". Let us first tabulate those reactions that are in this tool bag as a review.

We shall consider the tool bag of the organic chemist as being divided into reactions that carry out three different types of conversions: (1) reactions that convert one functional group into another, (2) reactions that form a carbon–carbon bond, and (3) reactions that form rings. Table 14.1 summarizes those reactions used to interconvert functional groups. Tables 14.2 and 14.3 summarize carbon–carbon bond formation reactions and ring formation reactions respectively. Most of these reactions will be familiar ones to you. They have been discussed, and further information is available about them (*e.g.*, the mechanism) in the indicated book, chapter, and section (*i.e.*, I.5.4 means Book 1, Chapter 5, Section 4). You should take a bit of time to be certain that you are familiar with each of these reactions. In those cases where there is no book, chapter, section reference, the specific reaction is new but should be sufficiently similar to other reactions that it should be easily added to your own "tool bag".

**TABLE 14.1** FUNCTIONAL GROUP TRANSFORMATION REACTIONS

TABLE 14.1

I. Formation of an Alkane

    A. From an Alkene (I.5.2)

$$\underset{R}{\overset{R}{>}}C=C\underset{R}{\overset{R}{<}} \xrightarrow[\text{cat}]{H_2} \underset{R}{\overset{R}{>}}HC-CH\underset{R}{\overset{R}{<}}$$

    B. From an Alkyne (I.5.2)

$$RC\equiv CR \xrightarrow[\text{cat}]{H_2} RCH_2CH_2R$$

II. Formation of an Alkene

    A. From an Alcohol (I.5.3)

$$\underset{R}{\overset{R}{>}}CH\text{-}C\underset{R}{\overset{R}{<}}\text{-}OH \xrightarrow{\text{Dehydrating agent}} \underset{R}{\overset{R}{>}}C=C\underset{R}{\overset{R}{<}} + H_2O$$

    B. From an Alkyl Halide (I.5.3)

$$\underset{R}{\overset{R}{>}}CH\text{-}C\underset{R}{\overset{R}{<}}\text{-}X \xrightarrow{\text{base}} \underset{R}{\overset{R}{>}}C=C\underset{R}{\overset{R}{<}}$$

C. From an alkyne (*cis* alkene)(II.3.6)

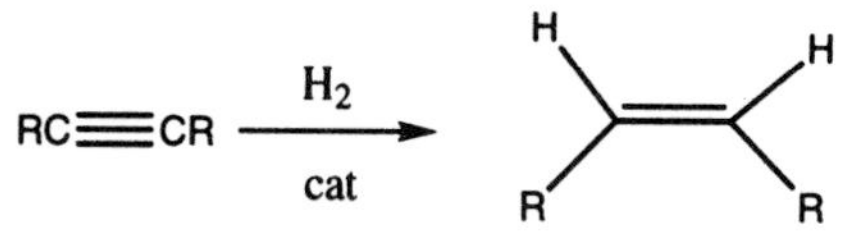

$$RC\equiv CR \xrightarrow[\text{cat}]{H_2}$$

D. From an alkyne (*trans* alkene)

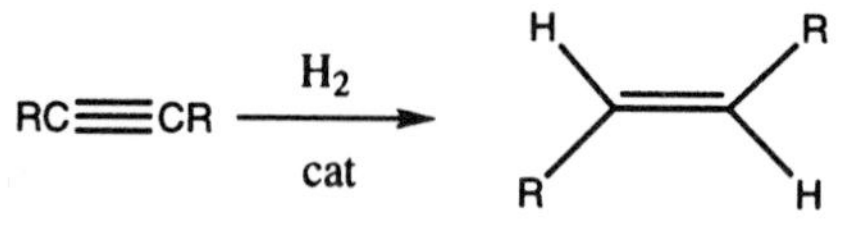

$$RC\equiv CR \xrightarrow[\text{cat}]{H_2}$$

## III. Formation of an Alkyne

A. Formation of an Alkyne from a Vicinal Dihalide  (I.5.3)

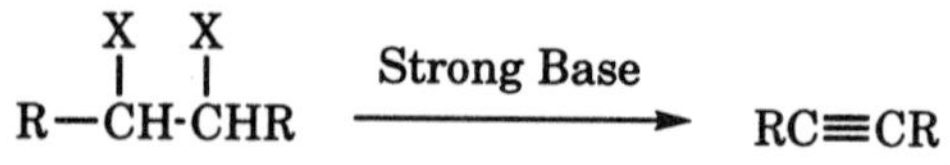

$$\underset{\displaystyle R-CH\cdot CHR}{\overset{\displaystyle X\ \ X}{\phantom{x}}} \xrightarrow{\text{Strong Base}} RC\equiv CR$$

## IV. Formation of a Halide

A. From an Alcohol  (I.8.2)

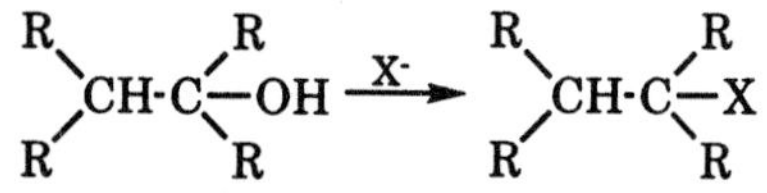

$$\text{CH-C-OH} \xrightarrow{X^-} \text{CH-C-X}$$

B. From an Alkene  (I.5.2)

$$\text{C=C} + HX \longrightarrow \text{CH-C-X}$$

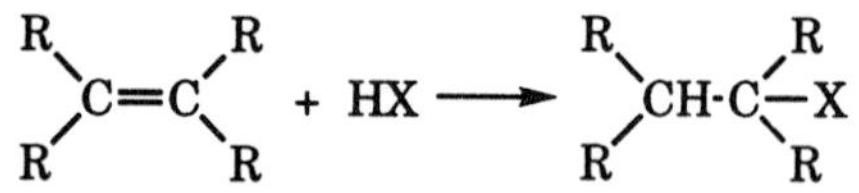

C. From an Aromatic Compound  (I.14.3)

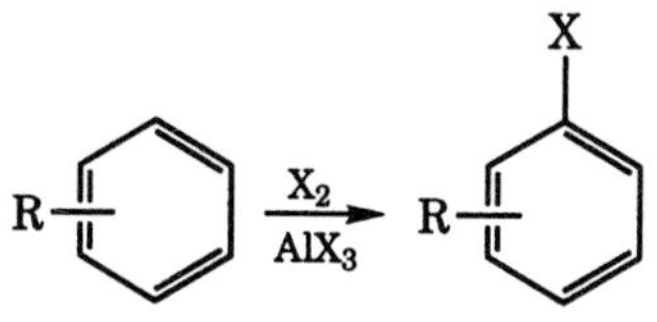

$$R\text{—}\bigcirc \xrightarrow[\text{AlX}_3]{X_2} R\text{—}\bigcirc\text{X}$$

D. Formation of an α–Haloketone  (II.11.6)

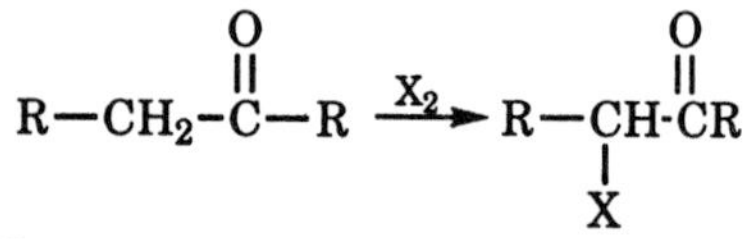

$$R-CH_2-\overset{\displaystyle O}{\overset{\displaystyle \|}{C}}-R \xrightarrow{X_2} R-\underset{\displaystyle X}{CH}\cdot\overset{\displaystyle O}{\overset{\displaystyle \|}{C}}R$$

E. From an Alkene  (I.5.2)

$$\text{C=C} \xrightarrow{X_2} X\text{-C-C-X}$$

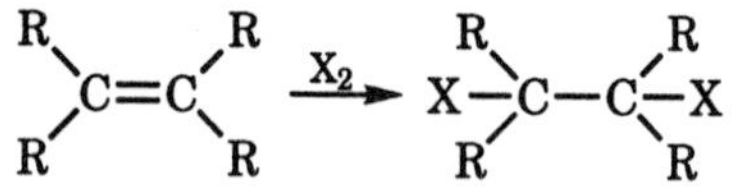

F. From an Alkyne  (I.5.2)

$$RC\equiv CR \xrightarrow{X_2} \underset{\displaystyle}{\overset{\displaystyle X\ \ X}{RC=C-R}}$$

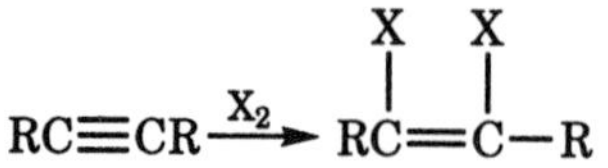

## G. From an Amine (II.3.8)

$$\text{Ph-NH}_2 \xrightarrow{\text{NaNO}_2/\text{HX}} \text{Ph-X}$$

## H. From another halide (cyanide ion is a pseudohalide)(II.8.2))

$$RX \xrightarrow{C\equiv N^-} RC\equiv N$$

## V. Formation of an Alcohol

### A. From an Alkene (II.3.6)

$$\underset{R}{\overset{R}{>}}C=C\underset{R}{\overset{R}{<}} + H_2O \xrightarrow{\text{cat}} \underset{R}{\overset{R}{>}}CH\text{-}\underset{R}{\overset{R}{C}}\text{-}OH$$

### B. From a Halide (II.8.2)

$$\underset{R}{\overset{R}{>}}CH\text{-}\underset{R}{\overset{R}{C}}\text{-}X \xrightarrow{OH^-} \underset{R}{\overset{R}{>}}HC\text{-}\underset{R}{\overset{R}{C}}\text{-}OH$$

### C. From an Aldehyde (I.6.4)

$$R-\overset{O}{\overset{\|}{C}}H \xrightarrow{\text{[Red]}} RCH_2OH$$

[Red] = LiAlH$_4$, NaBH$_4$, etc.

### D. From a Ketone (I.6.4)

$$R-\overset{O}{\overset{\|}{C}}R \xrightarrow{\text{[Red]}} R-\overset{OH}{\overset{|}{C}}HR$$

[Red] = LiAlH$_4$, NaBH$_4$, etc.

### E. From an Alkene (I.6.4)

$$\underset{R}{\overset{R}{>}}C=C\underset{R}{\overset{R}{<}} \xrightarrow{\text{[O]}} HO-\underset{R}{\overset{R}{C}}-\underset{R}{\overset{R}{C}}-OH$$

### F. From an Acid (I.6.4)

$$R-\overset{O}{\overset{\|}{C}}-OH \xrightarrow{\text{[Red]}} RCH_2OH$$

[Red] = LiAlH$_4$

### G. From an Epoxide (II.8.19)

## VI. Formation of an ether

### A. From an Alkyl Halide (II.8.18)

$$R-X \ + \ R-O^- \longrightarrow R-O-R$$

### B. From an alkene (epoxide) (I.6.4)

$$RCH=CHR \xrightarrow{Peracid} RCH\overset{O}{\triangle}CHR$$

## VII. Formation of a Ketone or Aldehyde

### A. From an Alcohol (Ketone) (I.6.4)

$$R-\underset{\underset{H}{|}}{\overset{\overset{OH}{|}}{C}}-R \xrightarrow{[O]} R-\overset{\overset{O}{\|}}{C}R$$

### B. From an Alkene (Aldehyde) (I.6.4)

$$RHC=CHR \xrightarrow{O_3} \ 2 \ R-\overset{\overset{O}{\|}}{C}-H$$

### C. From an Alkene (Ketone) (I.6.4)

$$R_2C=CR_2 \xrightarrow{O_3} \ 2 \ R-\overset{\overset{O}{\|}}{C}-R$$

### D. From an Alkyne (Ketone) (II.3.6)

$$RC\equiv CR \xrightarrow[cat]{H_2O} R-\overset{\overset{O}{\|}}{C}-R$$

### E. From a Primary Alcohol (Aldehyde) (I.6.4)

$$RCH_2OH \xrightarrow{[O]} R-\overset{\overset{O}{\|}}{C}H$$

[O] = $CrO_3$ / pyridine (non–aqueous)

## VIII. Formation of an acetal

### A. From Aldehyde and Alcohol (II.5.8)

$$R-CH_2-\overset{\overset{O}{\|}}{C}H + RCH_2OH \longrightarrow R-CH_2-\underset{\underset{OCH_2R}{|}}{\overset{\overset{OH}{|}}{C}}H$$

### B. From Ketone and Alcohol (II.5.8)

$$R-CH_2-\overset{\overset{O}{\|}}{C}R + RCH_2OH \longrightarrow R-CH_2-\underset{\underset{OCH_2R}{|}}{\overset{\overset{OH}{|}}{C}}R$$

## IX. Formation of a Carboxyl Group

### A. From an Amide (I.5.6)

$$R-\overset{\overset{\displaystyle O}{\|}}{\underset{\underset{\displaystyle H}{|}}{C}}-N-CH_2\text{-}R \;+\; H_2O \longrightarrow R-\overset{\overset{\displaystyle O}{\|}}{C}-OH \;+\; RCH_2NH_2$$

### B. From an Ester (I.5.6)

$$R-\overset{\overset{\displaystyle O}{\|}}{C}-O\text{-}CH_2\text{-}R \;+\; H_2O \longrightarrow R-\overset{\overset{\displaystyle O}{\|}}{C}-OH \;+\; RCH_2OH$$

### C. From an Alcohol (I.6.4)

$$RCH_2OH \xrightarrow{\;[O]\;} R-\overset{\overset{\displaystyle O}{\|}}{C}-OH$$

$[O] = KMnO_4,\ K_2Cr_2O_7,$ etc.

### D. From an Aldehyde (I.6.4)

$$R-\overset{\overset{\displaystyle O}{\|}}{C}H \xrightarrow{\;[O]\;} R-\overset{\overset{\displaystyle O}{\|}}{C}-OH$$

$[O] = KMnO_4,\ K_2Cr_2O_7,$ etc.

### E. From a Nitrile (II.7.5)

$$RC\equiv N \xrightarrow[\text{2. } H_3O^+]{\text{1. } OH^-} R-\overset{\overset{\displaystyle O}{\|}}{C}-OH$$

## X. Formation of an Ester

### A. From a Carboxylic Acid (I.5.3)

$$R-\overset{\overset{\displaystyle O}{\|}}{C}OH \;+\; RCH_2OH \xrightarrow{\;H^+\;} R-\overset{\overset{\displaystyle O}{\|}}{C}OCH_2R \;+\; H_2O$$

### B. From a Carboxylic Acid Anhydride (II.5.5)

$$R-\overset{\overset{\displaystyle O}{\|}}{C}-O-\overset{\overset{\displaystyle O}{\|}}{C}R \;+\; RCH_2OH \longrightarrow R-\overset{\overset{\displaystyle O}{\|}}{C}OCH_2R \;+\; R-\overset{\overset{\displaystyle O}{\|}}{C}OH$$

### C. From a Carboxylic Acid Chloride (II.5.5)

$$R-\overset{\overset{\displaystyle O}{\|}}{C}Cl \;+\; RCH_2OH \longrightarrow R-\overset{\overset{\displaystyle O}{\|}}{C}OCH_2R \;+\; HCl$$

### D. From a Cyclic Ketone (I.6.4)

### E. From a γ–Hydroxy Acid (II.5.7)

# XI. Formation of an Amine

## A. Formation of an Amine from a Nitrile   (I.6.4)

$$RC{\equiv}N \xrightarrow{\text{[Red]}} R-\underset{\underset{H}{|}}{\overset{\overset{H}{|}}{C}}-\underset{\underset{H}{|}}{\overset{\overset{H}{|}}{N}}$$

[Red] = $H_2$, $NaBH_4$, $LiAlH_4$, etc.

## B. Gabriel synthesis (II.13.6)

$$\xrightarrow[R-Br]{KOH} \xrightarrow[\Delta]{H} R-NH_2$$

# XII. Formation of an Amide

## A. From a Carboxylic Acid  (I.5.4)

$$R-\overset{\overset{O}{\|}}{C}OH + RCH_2NH_2 \longrightarrow R-\overset{\overset{O}{\|}}{C}-\underset{\underset{H}{|}}{N}CH_2R + H_2O$$

## B. From an Amine and an Acid Chloride

$$R-\overset{\overset{O}{\|}}{C}Cl + RCH_2NH_2 \longrightarrow R-\overset{\overset{O}{\|}}{C}-\underset{\underset{H}{|}}{N}CH_2R + HCl$$

# XIII. Aromatic Transformations

## A. Formation of an Aromatic Sulfonic Acid  (I.14.7)

$$R\!-\!\!\bigcirc \xrightarrow{H_2SO_4/SO_3} R\!-\!\!\bigcirc\!-SO$$

## B. Formation of a Nitro–aromatic  (I.14.5)

$$R\!-\!\!\bigcirc \xrightarrow[H_2SO_4]{HNO_3} R\!-\!\!\bigcirc\!-NO_2$$

## C. Formation of a Halo–aromatic  (I.14.3)

$$R\!-\!\!\bigcirc \xrightarrow[AlX_3]{X_2} R\!-\!\!\bigcirc\!-X$$

**TABLE 14.2** Carbon–Carbon Bond Formation Reactions

I. Aromatic Transformations

    A. Aromatic Ketone from an Acid Anhydride (Acylation)  (I.14.8)

$$R'-\bigcirc + \underset{\substack{\|\ \ \|\\ O\ \ O}}{RCOCR} \xrightarrow{H_3PO_4} R'-\bigcirc\overset{\overset{\displaystyle O}{\|}}{CR}$$

    B. Aromatic Ketone from an Acid Chloride (Acylation)  (I.14.8)

$$R'-\bigcirc + \underset{\substack{\|\\ O}}{RCOCl} \xrightarrow{AlCl_3} R'-\bigcirc\overset{\overset{\displaystyle O}{\|}}{CR}$$

    C. Alkyl–Aromatic from an Alkyl Halide (Friedel Crafts Alkylation) (I.14.4)

$$RX + R-\bigcirc \xrightarrow{AlCl_3} R-\bigcirc^{R}$$

II. Reactions of Organometallic Compounds

    A. Grignard Reagent and Aldehydes  (I.5.4)

$$RMgX + RCH_2-\underset{\substack{\|\\ O}}{CH} \longrightarrow RCH_2-\underset{\substack{|\\ H}}{\overset{\overset{\displaystyle OH}{|}}{CR}}$$

    B. Grignard Reagent and Ketone  (I.5.4)

$$RMgX + RCH_2-\underset{\substack{\|\\ O}}{CR} \longrightarrow RCH_2-\underset{\substack{|\\ R}}{\overset{\overset{\displaystyle OH}{|}}{CR}}$$

    C. Grignard Reagent and Esters

$$R-\underset{\substack{\|\\ O}}{COR} + 2\ RMgX \longrightarrow R-\underset{\substack{|\\ R}}{\overset{\overset{\displaystyle OH}{|}}{CR}}$$

    D. Grignard Reagent and Epoxide (II.8.19)

$$RMgX + \underset{R\quad R}{\overset{O}{\triangle}} \longrightarrow \underset{\substack{|\\ R}}{\overset{\overset{\displaystyle R}{|}}{R}}-\underset{\substack{|\\ R}}{C}-OH$$

E.  Grignard Reagent and Carbon Dioxide

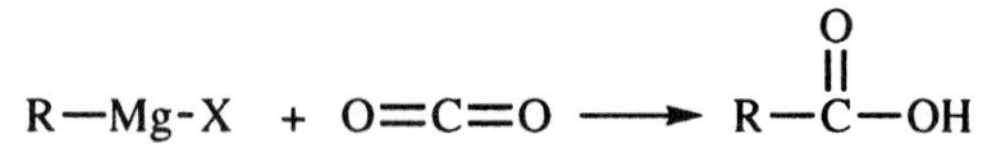

$$R\text{—}Mg\text{-}X \ + \ O\text{=}C\text{=}O \ \longrightarrow \ R\overset{\overset{O}{\parallel}}{\underset{}{C}}\text{—}OH$$

F.  Organocuprate and Unsaturated Ketone  (II.11.21)

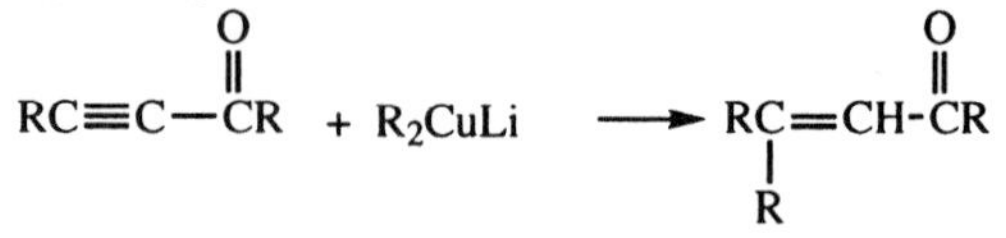

$$RC\text{≡}C\overset{\overset{O}{\parallel}}{\underset{}{C}}R \ + \ R_2CuLi \ \longrightarrow \ RC\text{=}CH\text{-}\overset{\overset{O}{\parallel}}{\underset{}{C}}R$$
$$\underset{R}{|}$$

III. Alkylations

A.  Alkylation of 1,3–Dicarbonyl Compounds  (II.11.11)

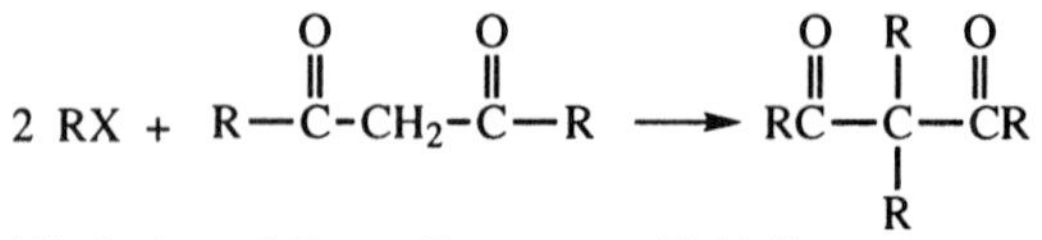

$$2 \ RX \ + \ R\overset{\overset{O}{\parallel}}{\underset{}{C}}\text{-}CH_2\text{-}\overset{\overset{O}{\parallel}}{\underset{}{C}}\text{—}R \ \longrightarrow \ RC\overset{\overset{O}{\parallel}}{\underset{\underset{R}{|}}{\overset{\overset{R}{|}}{C}}}CR$$

B.  Alkylation of Esters/Lactones  (II.11.7)

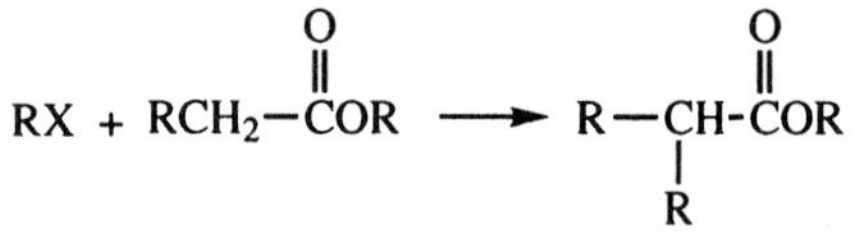

$$RX \ + \ RCH_2\overset{\overset{O}{\parallel}}{\underset{}{C}}OR \ \longrightarrow \ R\text{—}\underset{\underset{R}{|}}{CH}\text{-}\overset{\overset{O}{\parallel}}{\underset{}{C}}OR$$

C.  Alkylation of Ketones  (II.11.7)

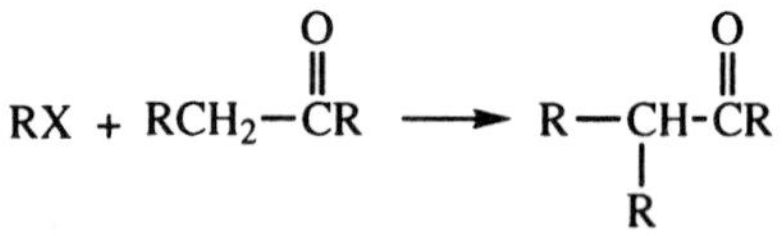

$$RX \ + \ RCH_2\overset{\overset{O}{\parallel}}{\underset{}{C}}R \ \longrightarrow \ R\text{—}\underset{\underset{R}{|}}{CH}\text{-}\overset{\overset{O}{\parallel}}{\underset{}{C}}R$$

D.  Formation of Ketones *via* Thianes  (II.11.19)

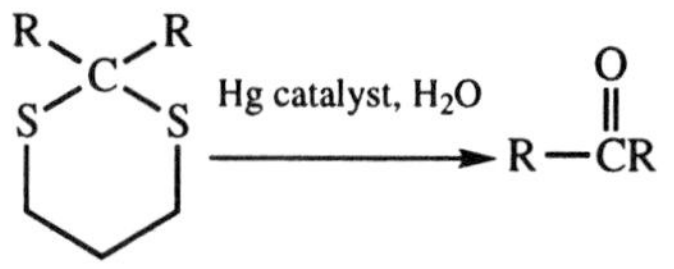

E.  Acetoacetic Ester Synthesis (II.11.18)

1. NaOEt / RBr

2. H$^+$ / H$_2$O

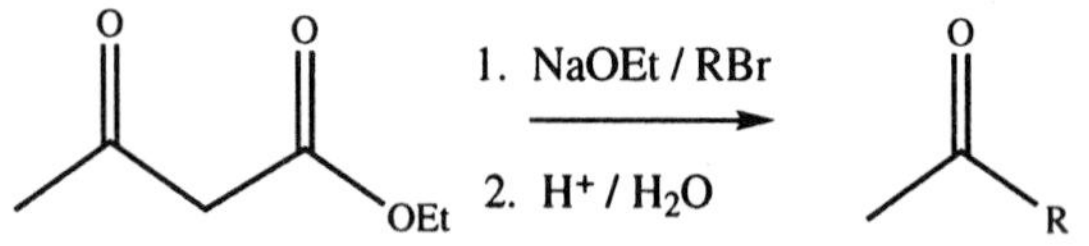

F.  Malonic Ester synthesis (II.11.18)

NaOR

RX

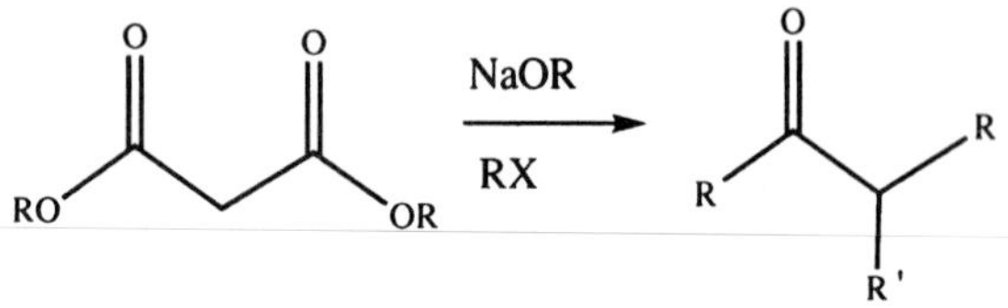

## IV. Condensations

**A. β–Hydroxyaldehydes (Aldol condensation) (II.11.14)**

$$2 \ RCH_2-\overset{\displaystyle O}{\overset{\|}{CH}} \longrightarrow RCH_2-\underset{\displaystyle R}{\overset{\displaystyle OH}{\underset{|}{\overset{|}{CH}}}}-\overset{\displaystyle H}{\underset{|}{C}}-\overset{\displaystyle O}{\overset{\|}{CH}}$$

**B. β–Ketoesters (Claissen Condensation) (II.11.17)**

$$2 \ RCH_2-\overset{\displaystyle O}{\overset{\|}{COR}} \longrightarrow R\overset{\displaystyle O}{\overset{\|}{C}}-\underset{\displaystyle R}{\overset{|}{CH}}-\overset{\displaystyle O}{\overset{\|}{COR}}$$

**C. Formation of an Alkene from a Ketone (Wittig Reaction) (II.11.19)**

## V. Miscellaneous

**A. Nitrile Formation from Alkyl Halide (II.8.2)**

$$RX \xrightarrow{C\equiv N^-} RC\equiv N$$

**TABLE 14.3** RING FORMATION REACTIONS

TABLE 14.3

### I. Cyclobutane Formation by Photochemical Cyclization

**A. 2 + 2 Reaction (II.12.9)**

**B. Electrocyclic Reaction (II.12.9)**

## II. Cyclopentane or Cyclohexane Formation

### A. Dieckmann Condensation (II.11.17)

## III. Cyclohexane Formation (only)

### A. Diels Alder Reaction (II.12.10)

### B. Robinson Annulation (II.11.22)

## 14.3  RETROSYNTHESIS

By this point you may be concerned about your ability ever to function as a chemist if doing so means having to devise long, complicated synthetic pathways such as those described above (or other, even more complicated ones, now reported commonly in the chemical literature). This despair is similar to that felt by students in many other disciplines as they, too, are concerned about the growing complexity of the tasks they face in their classes. For instance, beginning computer science students often wonder how they will ever be able to write the literally thousands of lines of computer code required by the new computer programs that are common today. Fortunately devising these long, complicated synthetic routes is not as difficult as beginning students expect when they first encounter them. The key to making any large task manageable is to break it down into individual parts. This is true in computer programming and just as true in chemistry.

There are two ways to do this: (1) begin with some commonly available, inexpensive material that looks promising as a starting material and work *toward* the final product, or (2) begin with the end product, ask how it might be made in only one chemical step, and then repeat the question over and over again until the required reagents are simple, easily available compounds. The latter approach is the one usually used by chemists in long synthesis projects because it more naturally focuses attention of the compound to be synthesized. This approach is sufficiently often used that the technique has been given its own name: **retrosynthesis**. As the technique of retrosynthesis is practiced, the chemist's attention is always on moving the synthetic pathway closer and closer to the starting materials specified in the problem or, in a real world situation, toward easily available, inexpensive materials that can be used to begin the synthesis. Let us look at a few simple examples of retrosynthesis to illustrate its use. We shall start with a relatively simple compound and move on to more and more complicated materials.

**EXAMPLE 1: SYNTHESIS OF 4-OCTANOL**

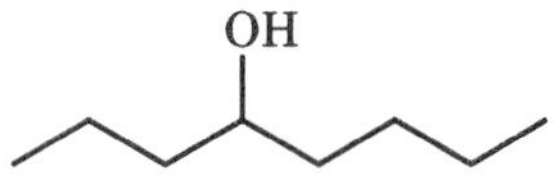

Let us first consider the synthesis of 4-octanol (figure 14.9) using as the *only* source of carbon butanoic acid, a commonly available, but vile-smelling, 4-carbon acid. One of the first things that we notice about the target compound is that it is 8 carbon atoms long and is to be made from a compound containing four carbon atoms. If we could join two of the four-carbon chains we would have the 8-carbon chain that we need. We, therefore, begin by looking for means of producing carbon-carbon bonds that will do the job for us. While there are likely to be several approaches that might work (other approaches would be the subject of the later evaluation of several different methods to determine the one most likely to work), we first think of using a Grignard reaction involving a 4-carbon halide and a 4-carbon aldehyde to form this carbon–carbon bond (table 14.2 II.A). We then write the "first step" of the retrosynthesis as shown in figure 14.10. The open arrow ( $\Longrightarrow$ ) is the convention used by chemists in writing retrosynthetic pathways to indicate that the compound on the left of the arrow is to be formed from the compound(s) on the right of the arrow. It faces the opposite direction that the synthesis will eventually go.

The next step in the retrosynthesis is to find ways of making butanal and bromobutane from butyric acid. We note that each can be made from 1–butanol (figure 14.11). (table 14.1, IV.A and VII.A)

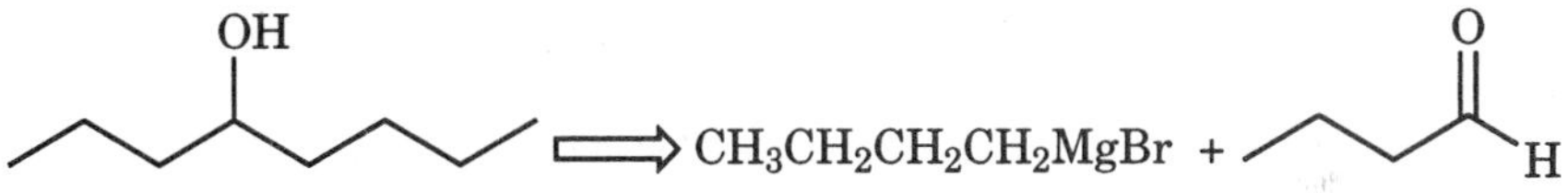

Finally, we remember that we can make an alcohol from an acid and we have solved the problem (figure 14.12) (table 14.1 V.F).

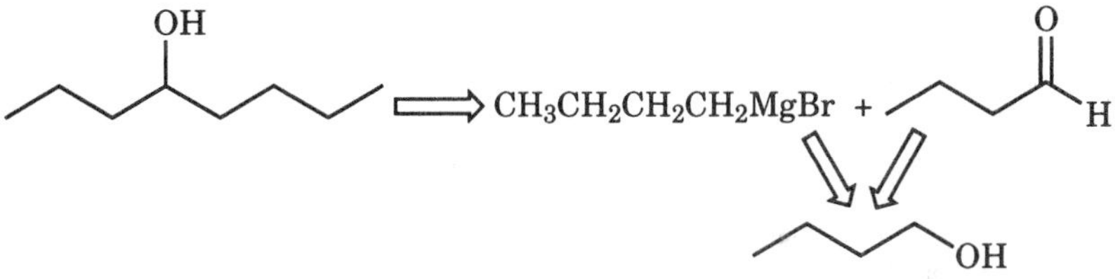

FIGURE 14.9:
4-octanol.

FIGURE 14.10:
Retrosynthesis of 4-octanol: 1.

FIGURE 14.11:
Retrosynthesis of 4-octanol: 2.

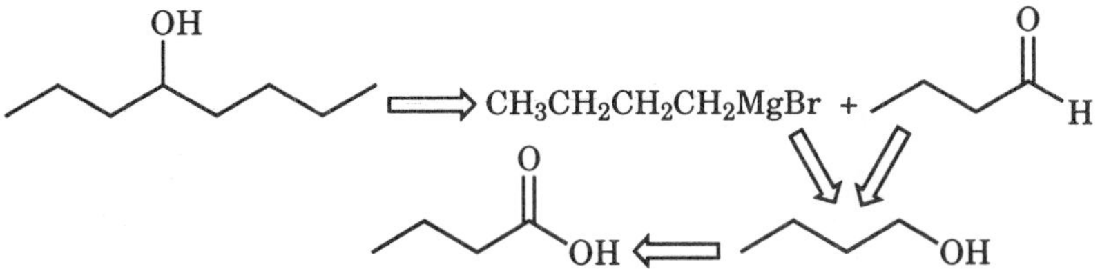

FIGURE 14.12:
Retrosynthesis of 4-octanol: 3.

The synthetic pathway, in the forward direction, would then consist of the following steps:

1:

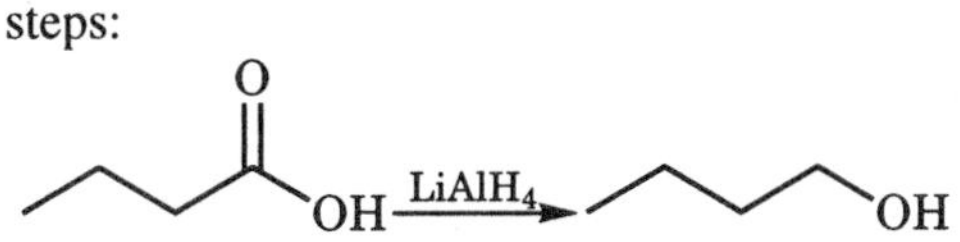

EQ. 14.1

2:

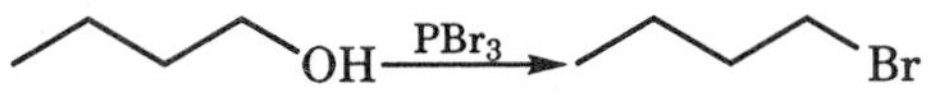

EQ. 14.2

3: 

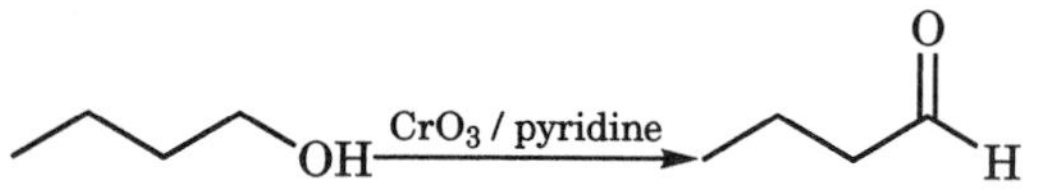

EQ. 14.3

4:

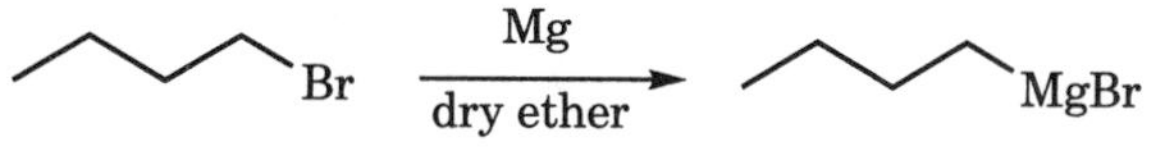

EQ. 14.4

5:

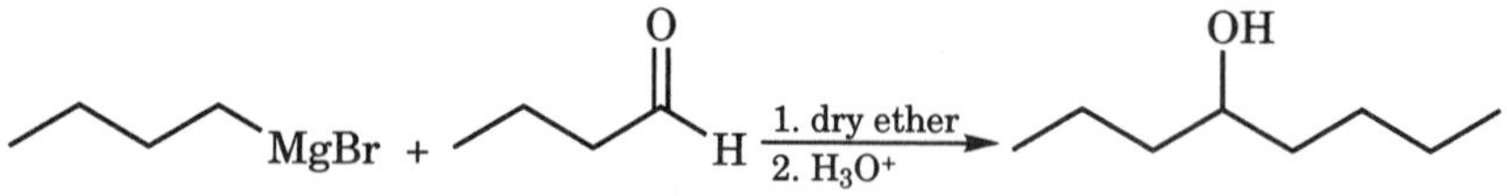

EQ. 14.5

## EXAMPLE 2: SYNTHESIS OF *CIS*-3-HEXENE

In this example we wish to synthesize *cis*-3-hexene using calcium carbide as our only source of carbon atoms. You will recall from last year that the hydrolysis of calcium carbide produces acetylene. Since the *cis* isomer was specified, we must use some type of **stereospecific synthesis** to place the alkene linkage at the proper place in the molecule. For the first step in our retrosynthesis, therefore, we might decide upon the catalytic hydrogenation of 3-hexyne because catalytic hydrogenation adds the hydrogen atoms to the same side of the unsaturated linkage (table 14.1 II.C).

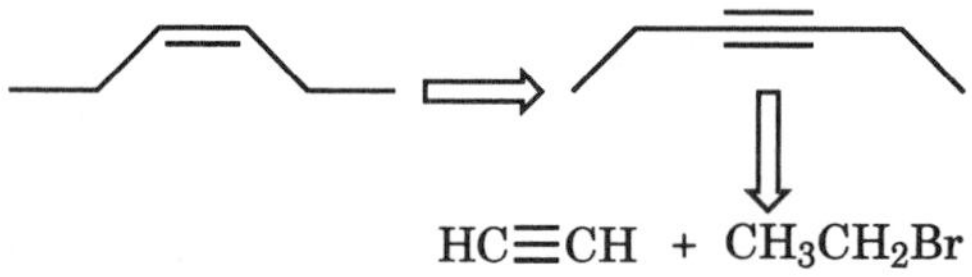

The 3-hexyne can be made by electrophilic addition of ethyl bromide to acetylene under strongly basic conditions:

The ethyl bromide can be made from ethylene and hydrogen bromide (table 14.1 IV.B):

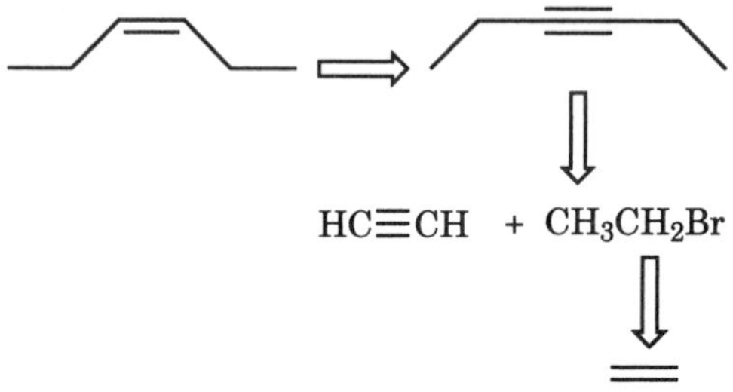

The ethylene can be made from acetylene (table 14.1 II.C):

Once again, the synthesis in the forward direction consists of:

1:   $CaC_2 + H_2O \longrightarrow HC\equiv CH$          EQ. 14.6

2:   $HC\equiv CH \xrightarrow{H_2/Pt} H_2C=CH_2$          EQ. 14.7

3:   $== \xrightarrow{HBr} CH_3CH_2Br$          EQ. 14.8

EQ. 14.9

4:   $HC\equiv CH + 2\ CH_3CH_2Br \xrightarrow{LDA}$

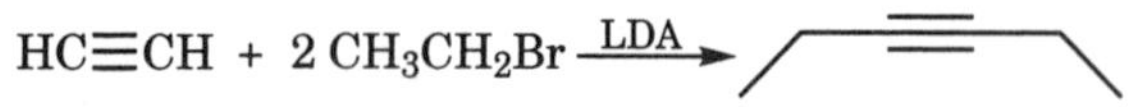

EQ. 14.10

5:

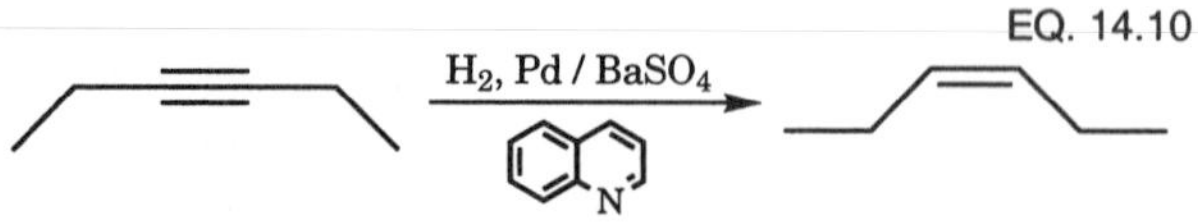

Once again the only difficult step in the entire synthesis occurs during the hydrogenation of acetylene to produce ethylene. It is necessary to control the conditions so that only one mole of hydrogen adds per mole of acetylene.

FIGURE 14.13:
Retrosynthesis of *cis*-3-hexene: 1.

$HC\equiv CH + CH_3CH_2Br$

FIGURE 14.14:
Retrosynthesis of *cis*-3-hexene: 2.

$HC\equiv CH + CH_3CH_2Br$

FIGURE 14.15:
Retrosynthesis of *cis*-3-hexene: 3.

$HC\equiv CH + CH_3CH_2Br$

$HC\equiv CH \Longleftarrow ==$

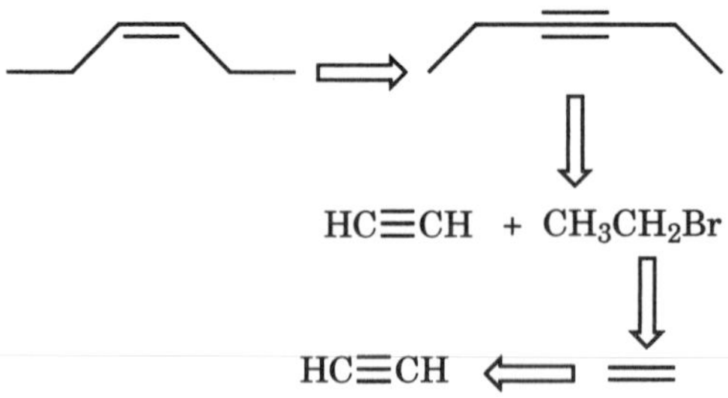

FIGURE 14.16:
Retrosynthesis of *cis*-3-hexene: 4.

Conditions that allow this reaction to stop at the *cis*-alkene involve doing the reaction in the presence of palladium on barium sulfate that has been "poisoned" with quinoline. This is known as Lindlar's catalyst. If the *trans*-alkene were the desired product, the reaction would have been accomplished using sodium metal in liquid ammonia. This now provides ways of converting alkynes into both *cis* and *trans*-alkenes.

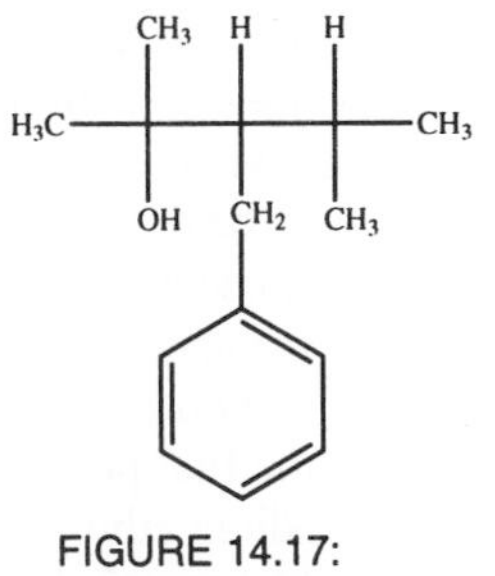

**EXAMPLE 3: SYNTHESIS OF 2,4–DIMETHYL–3–BENZYL–2–PENTANOL**

The first step in this synthesis is to determine the structure of the desired product. This is shown in figure 14.17.

One approach to the synthesis of this compound would be to recall that substituted alcohols can be made by using a ketone and a Grignard reagent (table 14.2 II.B). Our first step in the retrosynthesis would then be shown in figure 14.18.

This ketone can be made in several steps *via* a acetoacetic ester synthesis as shown in figure 14.19 (table 14.2 III.E).

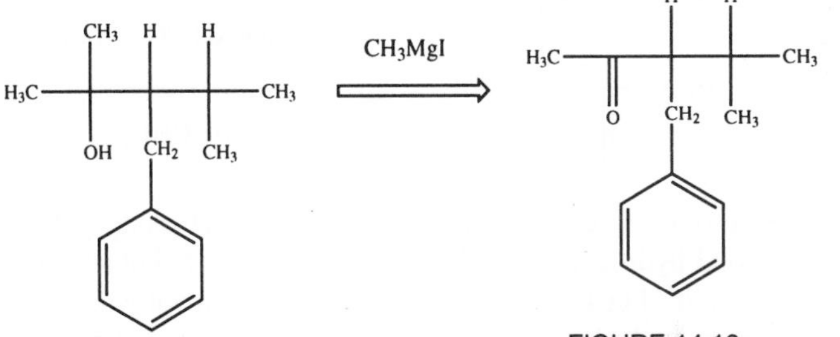

FIGURE 14.17: Structure of 2,4-dimethyl-3-benzyl-2-pentanol.

FIGURE 14.18: Retrosynthesis of 2,4-dimethyl-3-benzyl-2-pentanol: 1.

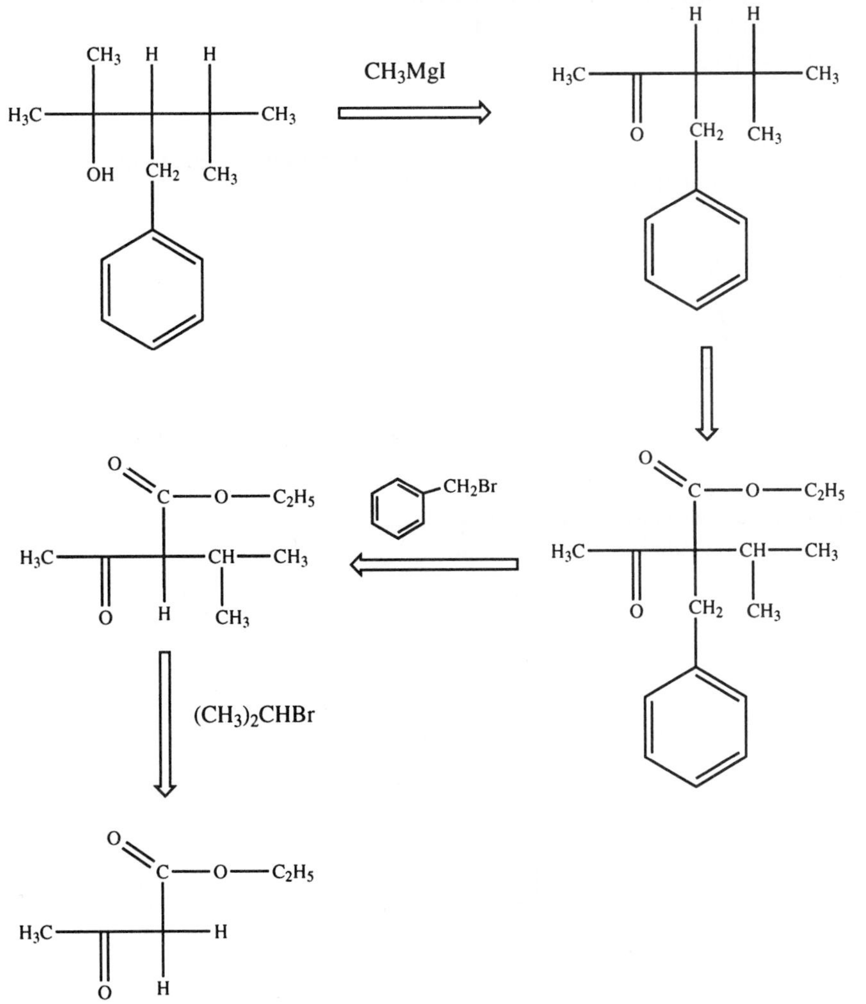

FIGURE 14.19: Retrosynthesis of 2,4-dimethyl-3-benzyl-2-pentanol: 2.

The synthesis is a relatively straight–forward application of the acetoacetic ester synthesis. It simply requires two alkylations of the acetoacetic ester molecule with one equivalent each of two different halides. Recall that these reactions must be done in the presence of a strong base to produce the enolate ion of the acetoacetic ester molecule. To prevent formation of the dianion and subsequent formation of the dialkylated product, it is especially necessary that the first step be done using a base that is not any stronger than necessary to remove the first proton and by limiting the amount of isopropyl bromide that is used. Sodium ethoxide is a strong enough base to remove the first proton, so it will be the base of choice for the first alkylation. This first step is shown in equation 14.11.

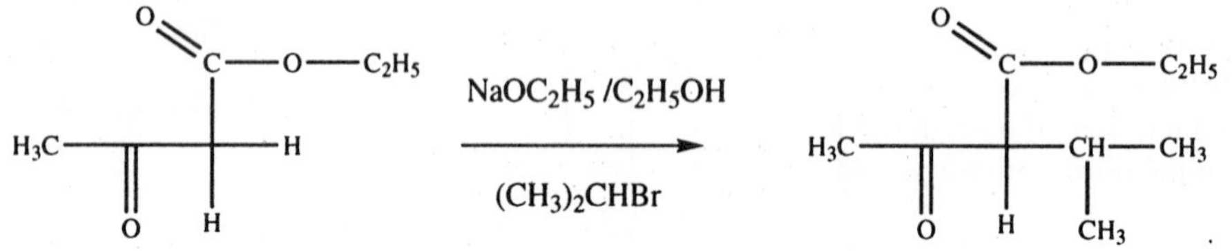

EQ. 14.11

The electron–releasing character of the isopropyl group reduces the acidity of the second hydrogen atom slightly so that a somewhat stronger base is required to remove it. For the second step of the reaction we will choose potassium t–butoxide as the base. It is somewhat less important here that only one equivalent of benzyl bromide be used because there is only one additional alkylation site, however since an excess of that reagent will not be needed to force the reaction to completion, there is no reason to employ such an excess. This reaction is shown as equation 14.12.

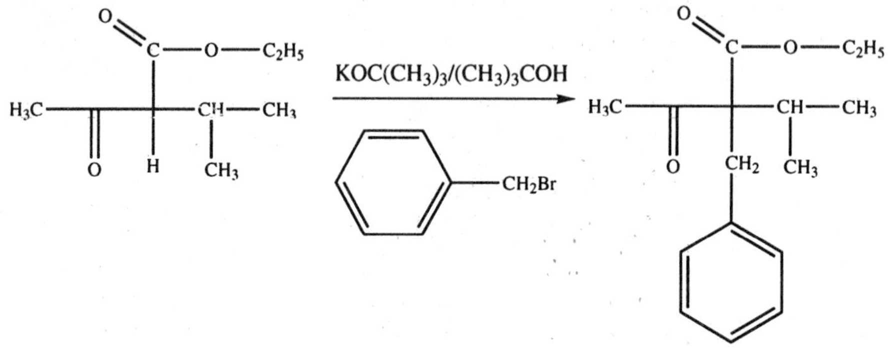

EQ. 14.12

After the second alkylation step is complete, the ester is hydrolyzed and decarboxylated by treating it with warm acid. This forms the ketone as shown in equation 14.13.

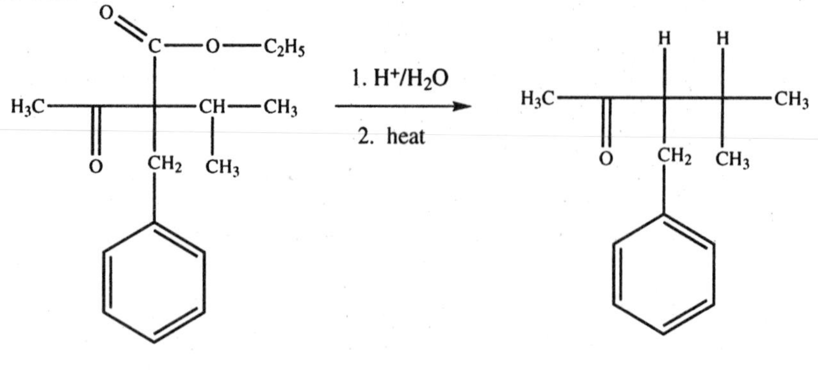

EQ. 14.13

Finally, the ketone is treated with methylmagnesium iodide in dry ether to form the tertiary alcohol in equation 14.14.

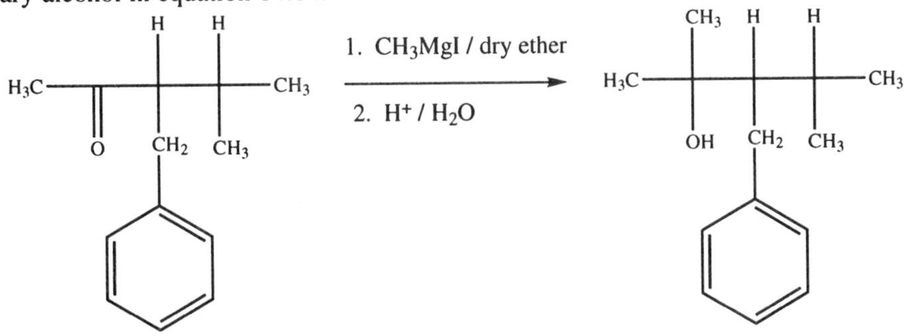

EQ. 14.14

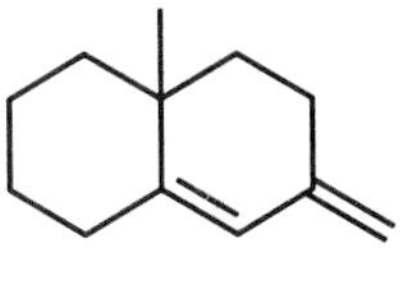

FIGURE 14.20: Structure of compound A.

**EXAMPLE 4: SYNTHESIS OF 10–METHYL–3–METHYLENE–DECALIN–4–ENE\***

\*The IUPAC rules do not unambiguously cover naming of this compound. Therefore several names are possible. The authors have chosen to use this name. Naming of such compounds is beyond the scope of this book.

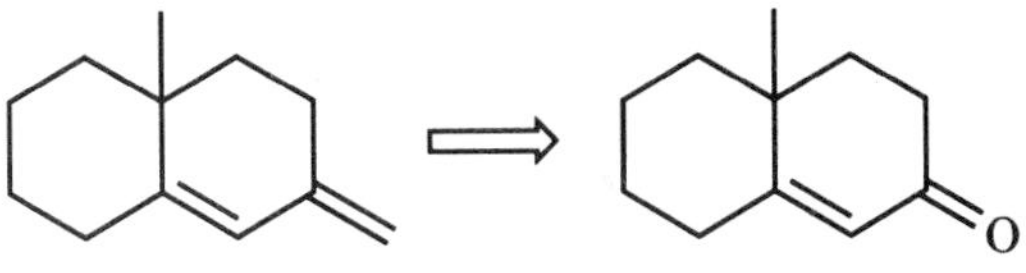

As our next example let us consider the synthesis of 10–methyl–3–methylene–decalin–4–ene (hereafter, for obvious reasons, referred to as compound A) (figure 14.20) from 1,3–pentadiene and other compounds having 4 or fewer carbon atoms.

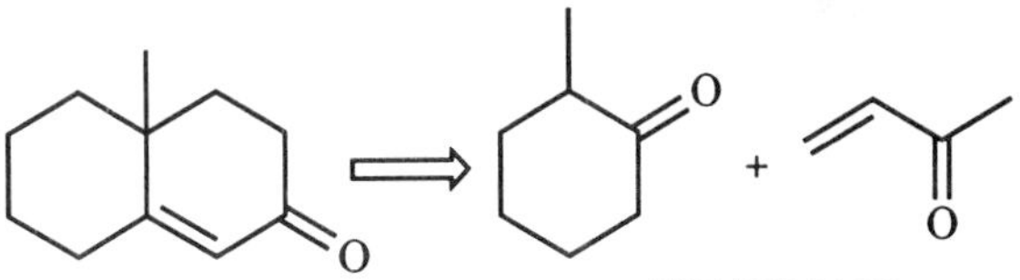

We notice that this compound has a fused ring structure that can be formed by a Robinson annulation reaction (table 14.3 III.B). As an added benefit, the Robinson annulation reaction even puts the in–ring double bond at the correct place. Let us, therefore, consider forming the double fused ring structure using that reaction. That reaction would, however, give us a cyclic α,β–unsaturated ketone as the product. The carbonyl group of the ketone would be where the exocyclic double bond should be in our desired product. We can, however, exchange a carbonyl group for a methylene group using the Wittig reaction (table 14.2 IV.C). Hence, our first step in the retrosynthesis is shown in figure 14.21:

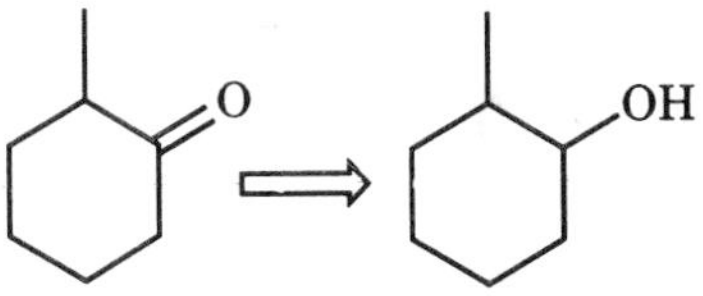

The reactants necessary for the Robinson annulation reaction that produces the cyclic ketone would be (figure 14.22):

The cyclic ketone could be made from the corresponding alcohol (table 14.1 VII.A)(figure 14.23):

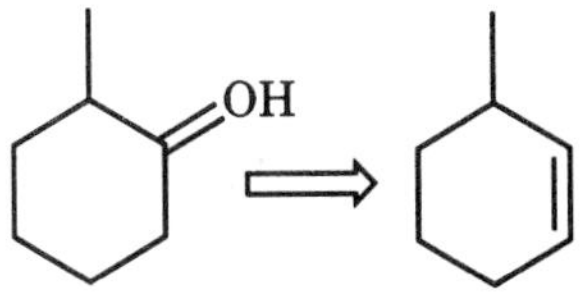

The alcohol could, in turn, be made by hydrating the alkene (table 14.1 V.A)(figure 14.24):

As a six–membered ring compound, the 2–methyl–cyclohexene can be made by a Diels Alder reaction (table 14.3 III.A)(figure 14.25):

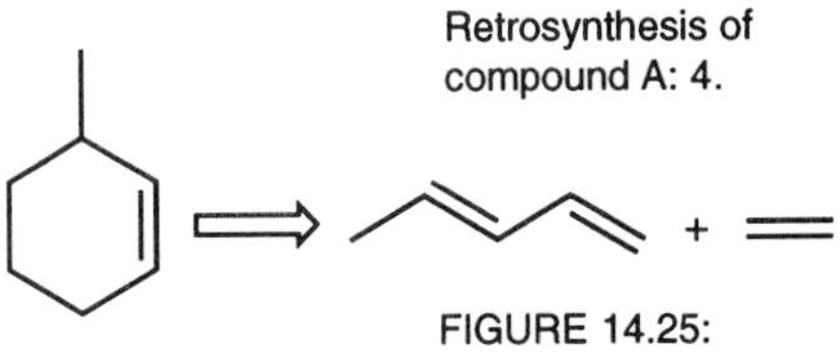

The Diels Alder reaction required in the first step of the synthesis reaction (figure 14.25) would require forcing conditions because ethylene is not a good dieneophile, but it can be accomplished. The student is left to write the forward synthesis reactions.

FIGURE 14.21: Retrosynthesis of compound A: 1.

FIGURE 14.22: Retrosynthesis of compound A: 2.

FIGURE 14.23: Retrosynthesis of compound A: 3.

FIGURE 14.24: Retrosynthesis of compound A: 4.

FIGURE 14.25: Retrosynthesis of compound A: 5.

## EXAMPLE 5: SYNTHESIS OF ENANTIOMERICALLY PURE (R)–(2–BENZYLOXYETHYL)OXIRANE

For our final example, let us consider the synthesis of a simple chiral compound in a manner that produces only the desired configuration. We shall discuss the synthesis of a molecule that has important uses in the preparation of a variety of biologically important compounds. The compound chosen for this illustration is (R)–(2–benzyloxyethyl)oxirane (figure 14.26). The one chiral center, at the juncture between the alphatic chain and the epoxide ring, will be the major problem. We will need to start with an enantiomerically pure material and be careful to use synthetic methods that retain the enantiomeric purity of that carbon or else it will be necessary to do a difficult separation of the optical isomers at the conclusion of the synthesis. Doing the separation will be costly both in time and in overall yield. As we choose our retrosynthetic pathway, therefore, we want to look for naturally occurring starting materials for the

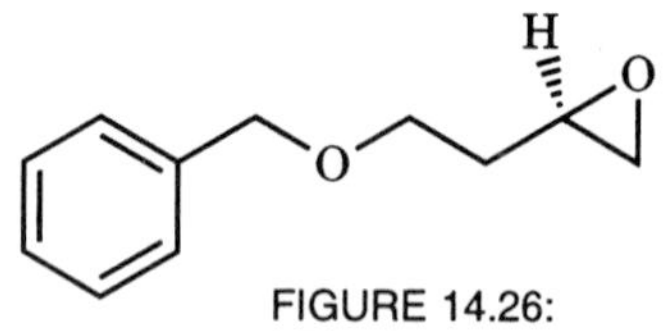

FIGURE 14.26: (R)_-(2-benzyl-oxyethyl)oxirane.

synthesis because they are often readily available in an enantiomerically pure state. In addition, when working near the chiral center, we want to use synthetic methods that are known to retain the chirality of that center.

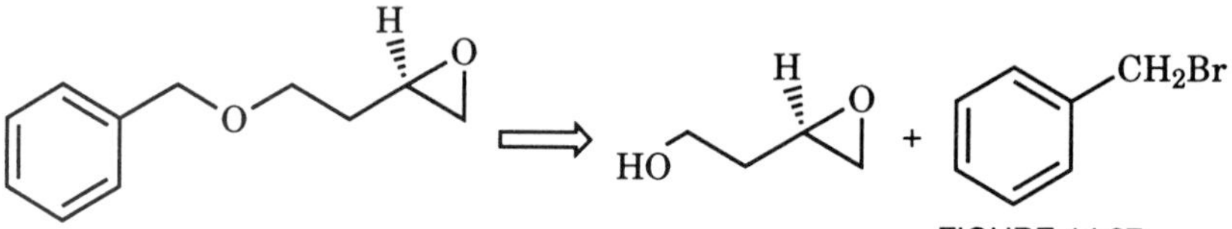

FIGURE 14.27: Retrosynthesis of (R)-(2-benzyl-oxyethyl)oxirane: 1.

We note that the compound is the benzyl ether of an alcohol that might serve as the immediate precursor. Thus, the first step in the retrosynthesis that is shown in figure 14.27:

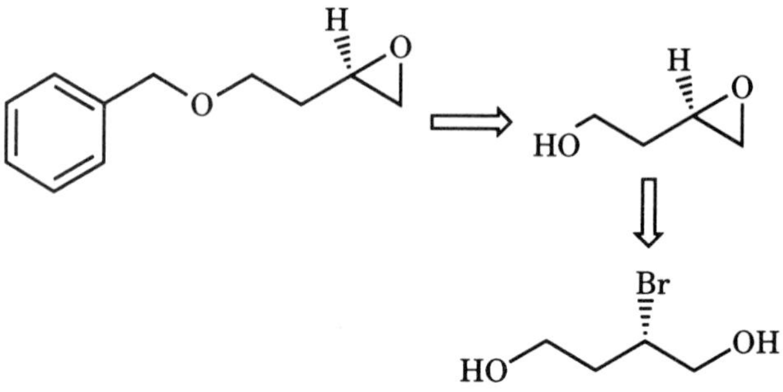

Since the formation of the ether does not involve a chiral center, we do not need to be concerned about the chemical reaction used here. One of the reagents needed for this reaction, benzyl bromide, is a common, inexpensive reagent. The other reagent, the epoxide can be produced from the α–halo diol:

FIGURE 14.28: Retrosynthesis of (R)-(2-benzyl-oxyethyl)oxirane: 2.

The chemical reaction to produce the epoxide from the alcohol does operate on a chiral center. Thus we must be careful to choose a reaction that produces the proper chirality and synthesize a diol with the proper configuration around the chiral carbon. The diol can be produced from a diacid:

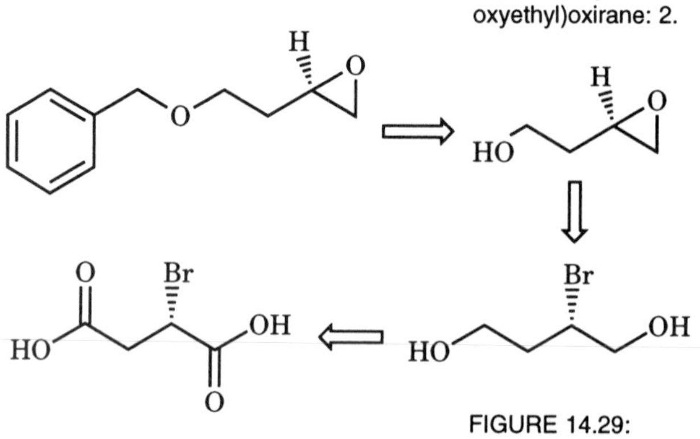

Neither of the carbon atoms involved in this step are chiral; thus we need not be concerned that the reaction used to produce the diol from the diacid be stereospecific. This diacid is very closely related to the naturally occurring amino acid, aspartic acid. They differ only in that aspartic acid has an amine group in place of the bromine atom. We need only find a reaction that we can use to convert an amine to a halide. Fortunately such a reaction does exist. Because the reaction does operate on a substituent at a chiral center, the reaction must be carefully chosen to produce the proper isomer from naturally occurring aspartic

FIGURE 14.29: Retrosynthesis of (R)-(2-benzyl-oxyethyl)oxirane: 3.

acid. Fortunately one does exist that will do the job for us. Thus we can use aspartic acid as the starting material.

All of the reactions used in this synthetic scheme have already been discussed in detail except the one used to convert the amino group of the aspartic acid to a halogen atom. As we saw in chapter 13, treatment of a primary amine with nitrite ion yields an intermediate known as a diazonium salt. This reaction, operating on aspartic acid is shown in equation 14.15.

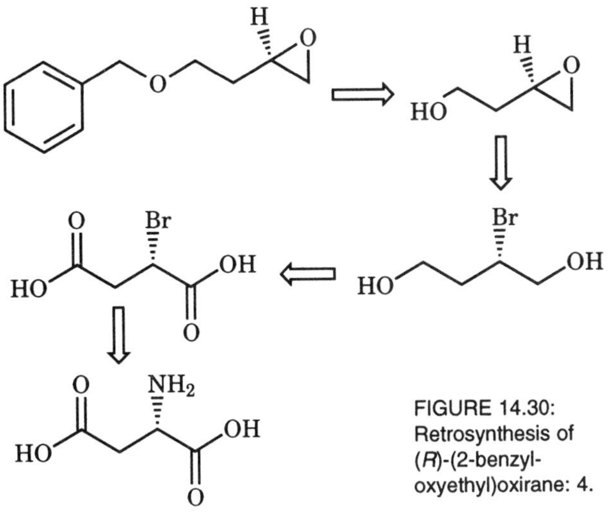

FIGURE 14.30:
Retrosynthesis of (R)-(2-benzyl-oxyethyl)oxirane: 4.

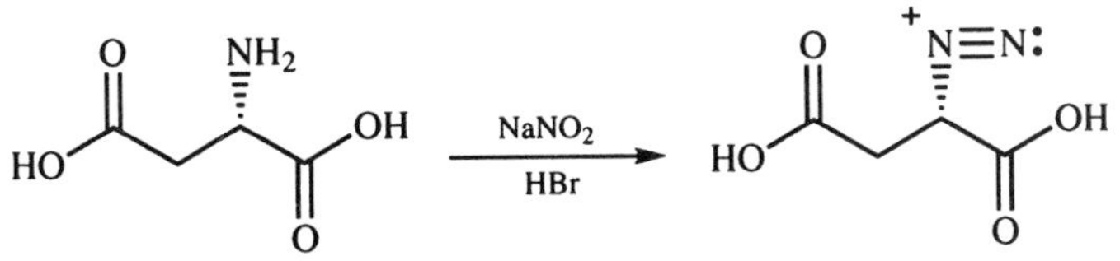

EQ. 14.15

The diazo group (—N$_2^+$) is a good leaving group and can be easily displaced by many nucleophiles. The proximity of the diazo group and the α–carbonyl group allow this reaction to proceed with retention of configuration. Therefore, when the bromide ion present in the mixture reacts with the diazonium salt, the bromine atom ends up in the same place relative to the other groups as the amine group was before the reaction. This step is shown in equation 14.16. While equations 14.15 and 14.16 have been shown as separate reactions for clarity in this discussion, the diazonium salt is never isolated and both reactions occur in the same reaction vessel. For all practical purposes, then, the two reactions are simply one.

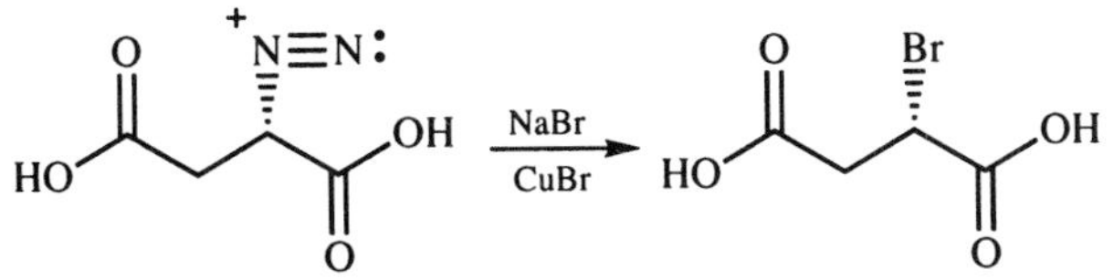

EQ. 14.16

Conversion of the two acid groups to hydroxyl groups could have been done by reaction with lithium aluminum hydride. In practice the authors of the paper from which this synthesis was taken [J.A. Frick, J.B. Klassen, A. Bathe, J.M. Abramson, and H. Rapoport, *Synthesis*, 621–623 (1992)] chose to do the reduction with borane (also known as boron hydride), a reducing agent closely related to sodium borohydride. This step is shown in equation 14.17.

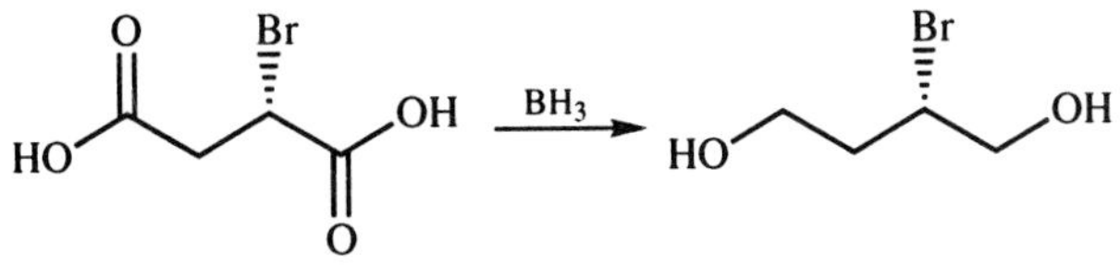

EQ. 14.17

The next step requires formation of the epoxide ring at the right side of the molecule pictured in equation 14.17. The reaction that does this is a simple nucleophilic displacement reaction. First a very strong base (sodium hydride was used in this case) is used to remove the proton from the hydroxyl group. The negatively charged oxygen atom is then a sufficiently good nucleophile to displace the bromine atom. After acidification, the epoxide is formed as shown in equation 14.18. Although there are two possible cyclic ethers that could be formed (one with each of the hydroxyl groups), a 4–membered ring including oxygen is much more sterically strained than is a 3–membered ring including an oxygen atom. Therefore the 3–membered ring forms.

A careful look at the configuration of naturally occurring aspartic acid reveals that it is the (S) configuration. The desired epoxide has the (R) configuration. Since this nucleophilic substitution reaction proceeds with inversion of the configuration at the carbon atom, this reaction will produce the correct stereochemical isomer.

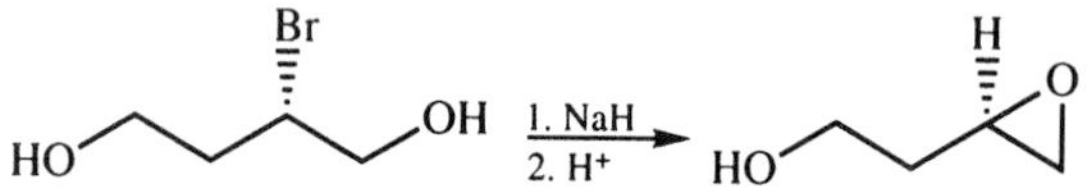

EQ. 14.18

Finally, the benzyl ether can be prepared from the remaining hydroxyl group by treating the alcohol with benzyl bromide in basic solution. This reaction is shown in equation 14.19.

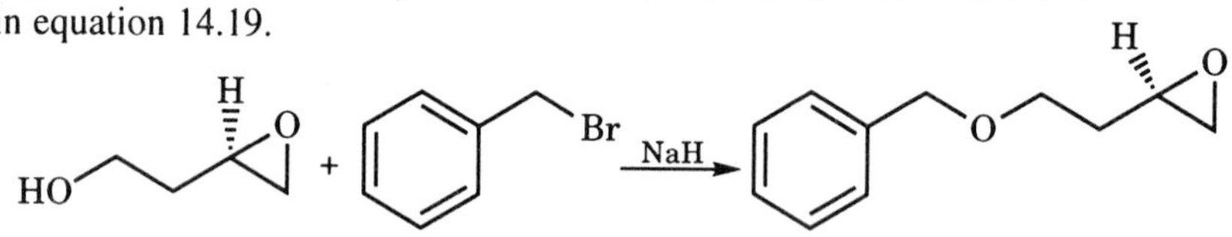

EQ. 14.19

In practice, the last two steps are carried out sequentially in the same reaction vessel without isolation of the intermediate epoxide–alcohol. This, of course, means that no acid is added (indicated as step 2 in equation 14.18) since equation 14.19 must also be done under strongly basic conditions. One simply allows sufficient time for epoxide formation *via* equation 14.18 before adding the benzyl bromide.

### CONCEPT CHECK

Devise two different syntheses to produce 2–phenyl–2–butanol from benzene, acetylene, and any additional, non–carbon–containing chemicals necessary.

Since we are to make an alcohol, we might wish to consider making it by adding a Grignard reagent to a ketone or aldehyde. This should work well because the required ketone or aldehyde is relatively easy to make from the specified starting materials. Figure 14.31 shows the route to this product using phenylmagnesium bromide and 2–butanone.

A shorter, more direct route to the same compound is shown in figure 14.32. In this route, ethyl magnesium iodide is added to acetophenone.

The second of these two syntheses is probably the better because (1) it is slightly more simple (fewer steps), and (2) the Grignard reagent is an alkyl magnesium halide which is somewhat easier to make than an aryl magnesium halide.

## 14.4 ADDITIONAL CONSIDERATIONS IN MULTI-STEP SYNTHESIS ————

There are several important considerations in the design and execution of a multi–step synthesis that deserve additional discussion. Although they have been mentioned briefly in the examples, they do need further amplification.

### USE OF PROTECTING GROUPS

It is often the case that a compound in a synthetic pathway is multi–functional. Therefore it is quite possible that a particular reagent may react with more than one of the functional groups. The result would be that the wrong product was formed or, at best, there would be a reduced yield of the desired product. In order to "steer" the reaction in the desired direction, **protecting groups** are added to inactivate those functional groups at which reaction should not occur. In order to be a "good" protecting group, the protecting group must be able to be placed on the molecule, it must survive the conditions of the reaction to be carried out on the target functional group, and it must be able to be removed at the appropriate time to restore the protected functional group — all in simple, high yield reactions. Table 14.4 lists a number of commonly used protecting groups. Examples of the use of protecting groups in the synthesis of polypeptides will be given in the next chapter.

**TABLE 14.4** COMMONLY USED PROTECTING GROUPS TABLE 14.4

| FUNCTIONAL GROUP TO BE PROTECTED | PROTECTING GROUP | COMMENTS |
|---|---|---|
| Carboxylic Acid | t–Butyl ester<br>methyl ester | removed with acid<br>removed with base |
| Aldehyde or Ketone | Acetal | Often ethylene glycol is used to form a cyclic acetal |
| Amine | N–t–butyloxycarbonyl (BOC)<br>N–carbobenzyloxy (CBZ) | removed with acid<br>removed by reduction |
| Alcohol | t-butyldimethylsilyl | t-Butyldimethylsilyl ether formed by reaction with t-butyldimethyl silyl chloride |

### STEREOSPECIFIC SYNTHESES

In many instances, especially in materials that are to be biologically–active such as drugs, the production of the proper enantiomer is extremely important. If a stereospecific synthetic route is not taken, then half of the final product will be biologically incapable of producing the desired effect. This means that half of the materials and work that go into the production are wasted. Worse yet, there are times when both of the enantiomers are biologically–active with one of them accomplishing what is desired and the other causing undesired, often toxic, side effects. Sometimes the toxicity of the "inactive" enantiomer is undiscovered until it is too late. Such was the case with thalidomide. Thalidomide, introduced in Great Britain in the 1970s as a palliative for the "morning sickness" suffered by many women during the early stages of pregnancy, was produced by a non–stereospecific method. No attempt was made to resolve the synthetic mixture into enantiomers because it was thought the other isomer was simply inactive and would be excreted without effect. The drug given to the women thus contained both enantiomers. One of the enantiomers was quite effective in reducing the effects of morning sickness. It was hailed as a wonder drug until the other isomer, unfortunately, was found to cause birth defects. The result was that a significant number of babies were born without limbs. As a consequence, the drug was recalled from the market.

To prevent such problems when one of the enantiomers is toxic or produces other undesired side effects, either a method must be found to selectively produce the desired enantiomer or a costly separation of the enantiomers must be done after the synthesis is complete. The second route to enantiomerically pure materials, of course, costs additional money because of the additional time and materials to do the separation. It also results in an overall yield for the reaction that is, at best, 50%. Therefore, whenever possible, stereospecific synthetic methods are being utilized and a great deal of research is currently being conducted into discovering stereospecific reactions and to improve the stereospecificity of known reactions.

## CHOICE OF ALTERNATE SYNTHETIC ROUTES

In many cases more than one synthetic route is available to carry out a particular synthesis. It is then up to the chemist to determine which of several routes seems the most practical means of carrying out the synthesis. In most cases there is no clear choice and the actual method will, likely, be one that is arrived at only after a process of trial and error. For instance, the method explained in example 4 of the retrosynthesis section was actually one of three similar syntheses that the authors report. Each varied from the other only slightly. The major difference was that the other two syntheses converted the amino group of the aspartic acid to a chloride rather than to a bromide. The actual choice of the method was based on the overall yield obtained. In the method presented, the overall yield was ~65% while the overall yield for the other two methods was much lower. Only by repeated trials, during which the reactants, conditions, and actual reactions are changed can the *best* method of synthesis be determined.

## CONSIDERATIONS APPLICABLE TO INDUSTRIAL SITUATIONS

Special considerations apply to large–scale industrial syntheses that are not a concern in the laboratory. Many of these, such as temperature control in large reactors, are more properly the subject of a chemical engineering course and will not be considered here. Others, such as the choice of one synthetic pathway over another do, however, deserve mention.

The ultimate choice of synthetic pathway in the laboratory is almost always a compromise between the yield of the specific reactions used, the ease with which the transformation can be accomplished, the availability and cost of the starting materials, and the identity of the waste materials produced by the synthesis. All of these are also a concern for the industrial production of any material. However, industrial production of a material also requires that the material be made at a cost that provides the company a profit after sale. The main consideration of the laboratory chemist is usually to maximize the yield of the product. While this is also a consideration of the industrial chemist and, all other things being equal, improves the "bottom line", the absolute yield of a particular synthetic route will sometimes be sacrificed in favor of another pathway that (1) uses less expensive starting materials, (2) produces "waste products" that are valuable enough to be reused or sold, or (3) produces waste products that are more "environmentally friendly" and, therefore, cost less to dispose.

**Alpha amino acid:** a compound in which the amino group and a carboxyl group are bonded to the same carbon. The twenty naturally occurring amino acids are the building blocks of proteins in biological systems.

**Cope elimination:** the thermal elimination of tertiary amine oxides.

**Curtius rearrangement:** the synthesis of an amine from an acid or acid chloride through an acyl azide intermediate.

**Diazotization:** the reaction of primary aliphatic and aromatic primary amines with HONO to form diazonium salts.

**Dipeptide:** a compound formed between two amino acids joined by an amide bond between the amino group of one and the carboxyl group of the other.

**Gabriel synthesis:** the synthesis of amines and amino acids by nitrogen alkylation of phthalimide.

**Haber process:** the industrial process by which hydrogen and nitrogen are catalytically converted to ammonia.

**Hofmann elimination:** the thermal elimination of quaternary ammonium hydroxides to form alkenes. This reaction preferentially produces the less substituted alkene as the major product.

**Hofmann rearrangement:** the synthesis of an amine from an amide by treatment with bromine and aqueous base.

**Hydrazones:** a derivative of aldehydes/ketones prepared by reacting the carbonyl compound with hydrazine.

**Imine:** a compound that contains a carbon — nitrogen double bond. Imines are prepared by reacting aldehydes or ketones with primary amines.

**Isoelectric point:** the pH at which an amino acid exists exclusively as a zwitterion.

**Nitrogen cycle:** the complex process of forming, consuming, and reprocessing compounds of nitrogen.

**Nitrogen fixation:** the process of converting $N_2$ into other forms of nitrogen.

**Ostwald process:** the industrial process by which nitric acid is prepared from ammonia.

**Oximes:** a derivative of aldehydes/ketones prepared by reacting the carbonyl compound with hydroxylamine.

**Phenylhydrazones:** a derivative of aldehydes/ketones prepared by reacting the carbonyl compound with phenylhydrazine.

**Primary amine:** an amine in which the nitrogen is bonded to only one alkyl group.

**Protecting group:** protecting groups are used when a reaction done on a multi–functional compound would affect more than one of the functional groups. In this case, the functional group at which reaction is not desired is "protected" by conversion to another functional group that will not be affected by the reaction to be used to carry out the transformation. After the reaction, the protecting group is removed to restore the original functionality.

**Retrosynthesis:** the process often used to design long, complicated synthetic pathways. It is done by imagining the synthesis backward, moving from products back to reactants.

**Stereospecific synthesis:** a synthesis designed to produce only one of the possible stereoisomers of the product. Such a synthesis avoids the necessity of doing costly, time consuming separation of the isomers if a isomerically pure material is required.

**Secondary amine:** an amine in which the nitrogen is bonded to two alkyl groups.

**Semicarbazones:** a derivative of aldehydes/ketones prepared by reacting the carbonyl compound with semicarbazine.

**Tertiary amine:** an amine in which the nitrogen is bonded to three alkyl groups.

**Zwitterion:** a species that contains an equal number of positive and negative charges and is, therefore, electrically neutral. The amino acids are often found as zwitterions.

HOMEWORK
# Problems

1. Prepare each of the following from 1-bromopentane. You may use any other compounds neessary.

  a.  1-pentene
  b.  hexanenitrile
  c.  hexanoic acid
  d.  n-propyl hexanoate
  e.  2-pentanol
  f.  2-pentanone
  g.  2-methyl-2-pentanol
  h.  N,N-diethylhexaneamide
  i.  N,N-diethylhexaneamine

2. Prepare each of the following from benzene using any other compounds neessary.

  a.  benzoic acid
  b.  phenylacetic acid
  c.  *m*-nitrobromobenzene
  d.  phenyl methyl ketone
  e.  methyl benzoate
  f.  2,4,6-tribromoaniline
  g.  N-ethylbenzamide

3. Prepare each of the foloowing from the indicated starting material. (Protecting groups may be needed.)

  a.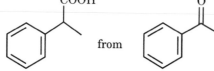

  b.

4. The following compound is a sex attractant for bollworm moths and could be useful for controlling pests. Show a synthesis using acetylene as the source of the alkene carbons and any other compounds you wish.

5. Show the preparation of the following compound. Hint: see the ibuprofen synthesis.

6. Devise a synthesis for the herbicide 2,4-D (2,4-dichlorophenoxy) acetic acid using phenol and any other compound you wish.

# CHAPTER fifteen

# 15

# Polymers
## ON AND ON AND...

# POLYMERS:
## On and on and...

**P**olymers are long chains formed by the reaction of simpler molecules called monomers. Monomer literally means "one part", and polymer means "many parts". Polymers are everywhere in nature, from the genetic code within single cells to the structural matter of the tallest trees. Polymers make life possible in a very real sense. Only in polymers do we find chemicals with both a sufficient complexity and regularity of pattern such that self–replication is possible. Many medical advances have taken advantage of the variety of properties possible with polymers. For example, polymers have made kidney dialysis and artificial hearts possible. Mankind has used naturally–occurring forms of biopolymers as food, clothing, shelter, tools, *etc.*, far back into pre–history.

The earliest scientific investigations of polymers began with those that occur in nature. Scientists sought to understand and to modify the properties of natural polymers to increase their usefulness. Natural rubber, obtained by drying the sap of certain tropical trees, has fascinating properties of resilience and is very resistant to abrasion. The water–resistant properties of rubber were used by Charles Macintosh, who layered it with cotton fabric, to make the type of rainwear that still bears his name. But rubber itself was too sticky and became too soft at high temperatures and too stiff at low temperatures for other uses. In 1839 Charles Goodyear accidentally discovered that adding about 5% sulfur into the natural rubber and heating produced a rubber with much improved properties. Today this process is known as "**vulcanization**". In the past 50 years industrial chemists have improved upon natural rubber. Today various synthetic rubbers are used when durability and resilience are important, such as in basketballs and car tires.

Serendipity was also involved when, at about the same time that Goodyear made his discovery, Christian Schonbein spilled a mixture of sulfuric and nitric acid in his lab. He wiped up the spill with a cotton apron, which he quickly rinsed and hung out before the fire to dry. Much to his surprise, the apron burst into flames. His accidental production of nitrated cellulose found an immediate military application by making an explosive known as gun cotton. Further studies improved the properties of nitrocellulose by addition of camphor to produce celluloid, a synthetic plastic. Celluloid had numerous commercial applications including the manufacture of billiard balls, which previously had to be made out of ivory. The flammability of products containing nitrocellulose, however, led to many tragic accidents. As a result other polymers have replaced nitrocellulose–containing polymers in all but a few applications (*e.g.*, ping pong balls).

Cellulose and natural rubber are the products of two important types of polymer synthesis: **condensation** and **addition**. In this chapter we will examine how the molecular structure of these and related polymers can be selected by careful synthesis, and how a polymer's structure results in the material's physical characteristics and properties.

## 15.2  THE STRUCTURE OF THE POLYMER BACKBONE

### LINEAR POLYMERS

The simplest kind of polymeric structure is **linear**. The overall length of the polymer chain is determined by the number "n" of repeating units "M" in the chain. The value of n is also known as the degree of polymerization.

$$\text{Polymeric Formula} = X{-}(M)_n{-}X$$

EQ. 15.1

The ends (X) and the repeating unit (M) may be closely related to the actual monomeric reactant.  In general, however, the monomer is chemically changed in becoming the repeating unit.  The chain forming a linear polymer has a continuous backbone, with no side branches.  The polymer, however, is not naturally oriented in a straight line, but more likely appears something like figure 15.1a.

Certain manufacturing processes and certain chemical forces can act to increase the influence of intermolecular forces such that sections of linear polymers line up alongside one another as shown in figure 15.1b.  Such structures are termed "crystalline", and the structure imparts added strength and chemical resistance to the polymer.

The formation of crystalline areas within a linear molecule results in a decrease in entropy.  It occurs only as an exothermic process.  Thus it is not surprising that the areas of **crystallinity** pictured above are lost when heated, (according to LeChatelier's principle).  Polymers can be made more crystalline by various chemical means:  using "stiff" monomer units to facilitate alignment;  incorporating polar sites in the chain to enhance intermolecular attraction;  and keeping any side groups on the chain small enough to not interfere with packing.  When polymeric material in the form of threads is desired, a physical stretching of the polymer, called "drawing", can help orient the chains to increase their crystalline character.  The spinnerets of spiders help strengthen their silk–like fiber.  Such polymers are called "fibrous".

a) coiled

b) crystalline area

FIGURE 15.1:
Structures of linear polymers.

### BRANCHED POLYMERS

It is possible to make polymers with short, irregularly spaced side chains attached to a linear backbone.  Such polymers are called **"branched"** (as opposed to linear), and are represented in figure 15.2.

The side chains seen in figure 15.2 prevent the polymer chain from folding back upon itself, or allowing other branched chains to pack closely alongside.  The result is that branched polymers are lower in crystallinity.  Branched polymers in general have lower density, lower strength, and greater flexibility than their linear counterparts.

FIGURE 15.2:
Branched polymer.

Both linear and branched polymers are generally termed **"thermoplastic"**.  That is, their shape is readily distorted with heat, but the material is not destroyed:  it can be remolded into a new shape.  This property makes linear and branched polymers a good choice for disposable containers, which can then be recycled.

## CROSSLINKED POLYMERS

Some polymers, on the other hand, are rigidly "**crosslinked**" as shown in figure 15.3.

FIGURE 15.3:
Crosslinked polymer.

Crosslinked polymers can be viewed as branched polymers with branches that connect to different main polymer chains at both ends of the short branch. Crosslinked polymers are quite strong, but rigidly inflexible. They are referred to as "**thermosetting**". Once the crosslinked polymer is manufactured (usually with heat) and put into the shape of the desired object, it can not be reheated without breaking the bonds in the polymer chain and destroying the polymer.

Some polymers that are normally linear or branched can have some of their properties fine tuned by gradually increasing the amount of crosslinking that occurs. This is most easily done when specific crosslinking agents can be used. The vulcanization of rubber by addition of sulfur is a specific example we shall consider later.

## 15.3   WAYS OF COMBINING THE MONOMERS

Historically, the synthetic routes to make polymers are divided into two categories: condensation (or step) reactions, and addition (or chain) reactions. Nearly all polymerization reactions fall into these two categories, although these categories can be further subdivided. For example, addition polymerization reactions are sometimes further classified as radical, cationic, anionic, or coordination, depending upon the specifics of the mechanism.

### CONDENSATION POLYMERIZATION

Like condensation reactions that occur in the synthesis of low molecular weight species, condensation polymerization combines two molecules into one by reacting two functional groups, usually eliminating a small molecule like water as a by–product. What makes a polymerization reaction possible is that the reactants are polyfunctional. That is, two monomeric molecules, each being polyfunctional, combine to form a larger, polyfunctional molecule. This larger molecule can continue to react at the remaining functional sites. A typical example of such a reaction would be the synthesis of a linear polyester from a diacid and a diol.

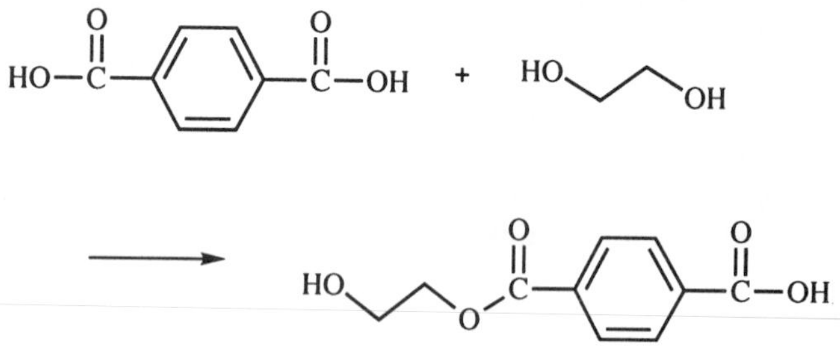

EQ. 15.2

Note that the alcohol functional group on the left end of the product molecule can react with another diacid, and the acid functional group on the right end of the product molecule can react with another diol. In this way, a linear polymer can be synthesized. If more than two functional groups are used, a three–dimensional, crosslinked product can be formed.  An example of this was the first commercial resin, Bakelite®, discovered by Leo Hendrik Baekeland (1863–1944) in 1907.  It is formed by the reaction between phenol and formaldehyde.  The ortho and para hydrogen atoms on phenol are all susceptible to reaction.  Water is split out and a cross–linking network can be formed. The first step in such a reaction is illustrated in equation 15.3.

EQ. 15.3

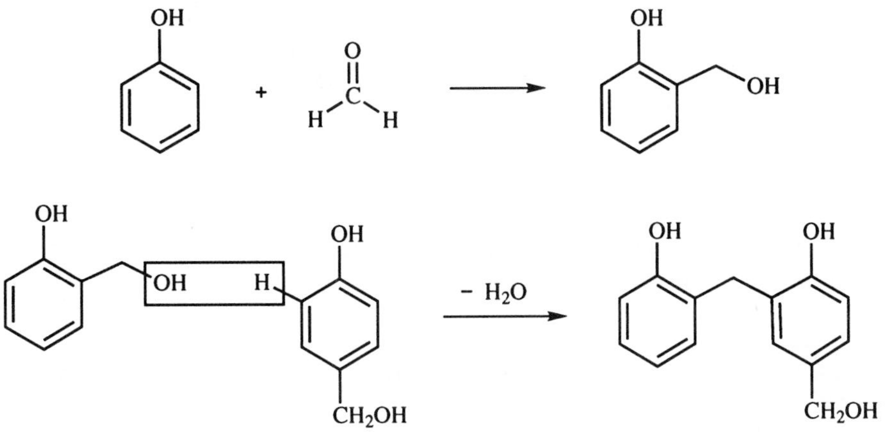

## ADDITION POLYMERIZATION

Addition to a double bond characterizes this type of polymerization. Frequently a free radical is formed in an initiation step.  The free radical adds to a multiple bond.  The pi bond opens and a sigma bond is formed between the initial radical and one of the atoms engaged in the multiple bond.  The other atom initially involved in the multiple bond ends up with an unpaired electron, and the radical attack on yet another multiply bonded monomer continues until the chain is terminated.  A typical example of this would be the benzoyl peroxide initiated polymerization of polyethylene.

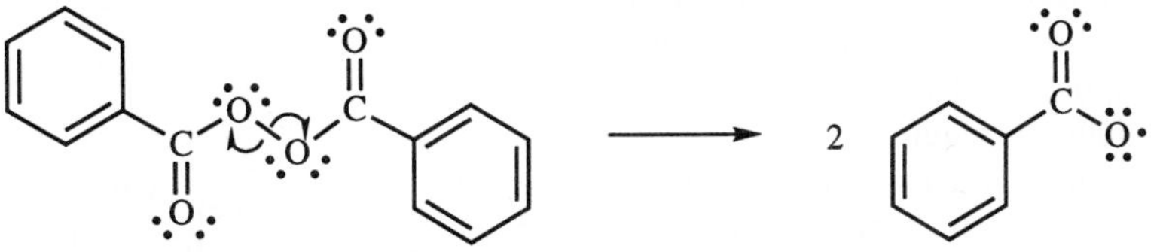

EQ. 15.4a

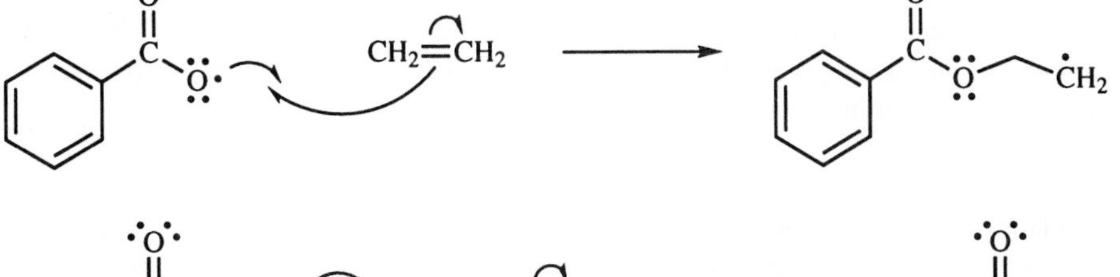

EQ. 15.4b

EQ. 15.4c

EQ. 15.4d

Addition polymerization reactions of some species will result in chains with an unsaturated backbone structure. For example, when 1,3–butadiene is used as the monomer in a radical–initiated addition reaction, the following occurs:

$$R\cdot + CH_2=C\underset{H}{\overset{H}{|}}C=CH_2 \longrightarrow RCH_2-C=C-\overset{\cdot}{C}H_2$$

EQ. 15.5

## MECHANISTIC CONSEQUENCES

The nature of the reaction mixture during synthesis, and the final polymer product, both vary dramatically when the monomers used and the specific reaction conditions are changed. However, some characteristics are clearly linked to whether the polymer formed is an addition or a condensation polymer. Many of these differences are due to the kinetics of the mechanisms by which the polymers are made.

Condensation polymers are made by a series of reactions in which covalent bonds are first broken, then made. Such reactions have sufficiently high activation energy that it can play an important part in the kinetics of polymerization. These condensation reactions, however, occur between any two monomer species in contact (*i.e.*, no specific initiation is required). Consequently, monomers disappear early in condensation polymerization reactions. Oligomers (literally, "a few parts", such as dimers, trimers, tetramers, etc.) are rapidly formed, and longer chains are formed by condensation of oligomers. Typically, by the time the average chain length of all species reaches ten units, only 1% of the initial amount of monomer remains in a condensation reaction mixture. The relatively gradual buildup of longer and longer oligomeric units means that long reaction times are essential in order to make high molecular weight condensation polymers.

Addition polymers, on the other hand, require a specific initiation. After the initiation, however, the subsequent propagation steps occur with essentially zero activation energy. Consequently, long polymer chains are formed almost at once. Since monomers cannot react directly with each other in an addition reaction, but only react with a growing radical chain, monomer concentration decreases slowly throughout the course of the reaction. At any point during the reaction the mixture is mostly high molecular weight polymer chains and

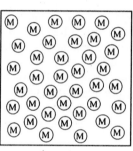

a)

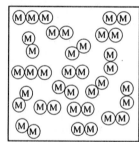

b)

FIGURE 15.4:
Reaction composition for condensation polymerization a) initially, and b) during reaction.

monomers. Growing chains comprise a very small (*ca.* $10^{-8}$) fraction of the reaction mixture. Longer reaction times may give a higher yield of polymer, but do not significantly affect the average molecular weight of the polymer. The length of addition polymers is instead related to the frequency of initiation and termination.

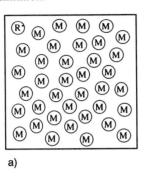

a)

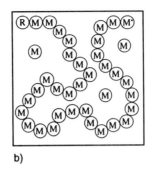

b)

FIGURE 15.5:
Reaction composition for addition polymerization
a) initially, and
b) during reaction.

## 15.4 ADDITION POLYMERS: MECHANISMS AND FINAL STRUCTURES

### ANIONIC POLYMERIZATION

An alternate to the mechanism producing polyethylene, shown earlier in equation 15.5, is known as anionic polymerization. Here the chain is initiated by the reaction of an anion with the monomer. The growing chain is itself an anion, and terminates when the anion reacts with a cation (*e.g.*, $H^+$ in an aqueous solution) to form a stable species. The example below shows polyethylene formed via an anionic mechanism, using methyl magnesium bromide (a Grignard reagent) as the initiator.

$$CH_3Br \xrightarrow[\text{ether}]{Mg} {}^-\!:CH_3 + Mg^{2+} + Br^- \qquad \text{EQ. 15.6a}$$

$$^-\!:CH_3 + CH_2{=}CH_2 \longrightarrow {}^-\!:CH_2CH_2CH_3 \qquad \text{EQ. 15.6b}$$

$$^-\!:CH_2CH_2CH_3 + CH_2{=}CH_2 \longrightarrow {}^-\!:CH_2CH_2CH_2CH_2CH_3 \qquad \text{EQ. 15.6c}$$

$$^-\!:CH_2CH_2(CH_2CH_2)_nCH_3 + H^+ \longrightarrow CH_3CH_2(CH_2CH_2)_nCH_3 \qquad \text{EQ. 15.6d}$$

Step a shows initiation, steps b and c show propagation, and step d shows termination.

### CATIONIC POLYMERIZATION

The method of cationic polymerization of ethylene begins with a cation attack on the monomer. The growing chain is itself a cation and the chain terminates when hydroxide (from the aqueous solution) combines with the carbocation to yield the neutral polymer.

$$H^+ + CH_2{=}CH_2 \longrightarrow {}^+CH_2CH_3 \qquad \text{EQ. 15.7a}$$

$$^+CH_2CH_3 + CH_2{=}CH_2 \longrightarrow {}^+CH_2CH_2CH_2CH_3 \qquad \text{EQ. 15.7b}$$

$$^+CH_2CH_2CH_2CH_3 \quad + \quad CH_2{=}CH_2 \quad \longrightarrow \quad {}^+(CH_2CH_2)_2CH_2CH_3 \qquad \text{EQ. 15.7c}$$

$$CH_3CH_2(CH_2CH_2)_n\overset{+}{C}H_2 \quad + \quad OH^- \quad \longrightarrow \quad CH_3CH_2(CH_2CH_2)_nCH_2OH \qquad \text{EQ. 15.7d}$$

Again, step a shows initiation, steps b and c show propagation, and step d shows termination.

## FREE RADICAL POLYMERIZATION

The mechanism in equation 15.4 shows a chain polymerization initiated by a radical formed from benzoyl peroxide. The growing chain in this mechanism was itself another radical, and termination occurred when another radical species combined with the chain radical. If the mechanism were to occur in exactly this fashion, with no other steps, linear polyethylene (high–density polyethylene, HDPE) would result. In practice, however, free radical polymerization normally yields branched polyethylene, known as low–density polyethylene, or LDPE.

One explanation as to how branches occur in LDPE includes a reaction where a hydrogen atom is abstracted by a growing radical chain from a polymer chain that has already been terminated.

$$R_1CH_2\overset{\bullet}{C}H_2 \quad + \quad R_2CH_2CH_2R_3 \quad \longrightarrow \quad R_1CH_2CH_3 \quad + \quad R_2\overset{\bullet}{C}HCH_2R_3 \qquad \text{EQ. 15.8}$$

$R_1CH_2\overset{\bullet}{C}H_2$ represents a growing chain, $R_2CH_2CH_2R_3$ represents a previously completed polymer chain, $R_1CH_2CH_3$ represents the termination of the previously growing chain, and $R_2\overset{\bullet}{C}HCH_2R_3$ represents the reactivated polymer with a radical site on an otherwise already complete backbone. The reaction of $R_2\overset{\bullet}{C}HCH_2R_3$ with additional monomeric ethylene units forms the branch in the chain.

So far, only the formation of polyethylene has been addressed in these examples. Modification of the monomer unit from $CH_2{=}CH_2$ to $CH_2{=}CHX$ provides a vast number of alternate polymers, some of which are listed in table 15.1. Further substitution for other of the hydrogen atoms provides additional variations, some of which are listed in table 15.2

Note the wide variety of properties and applications seen in these polymers, most of which are formed from simple modifications of ethylene. The substitution of a chlorine atom for one of the hydrogens, for example, results in PVC instead of polyethylene. PVC is quite different from polyethylene in many respects. PVC is less flammable and has greater strength than polyethylene. The strength is thought to be due to the greater electronegativity of chlorine, which indirectly acts to bring the polymer chains closer together, increasing crystallinity and thus accounting for PVC's strength.

**TABLE 15.1** ADDITION POLYMERS FORMED FROM $CH_2=CHX$

TABLE 15.1

| STRUCTURE | NAME | COMMON NAME & APPLICATION |
|---|---|---|
| $X = H$; | polyethylene | LDPE; films, coatings HDPE; containers, toys |
| $X = CH_3$; | polypropylene | Herculon®; beakers, milk cartons, outdoor carpeting |
| $X = C_6H_5$; | polystyrene | clear plastic cups, foam plastic cups, cast transparent parts |
| $X = CN$; | polyacrylonitrile | Orlon®, Acrylon®; carpets, knitware, wool substitute |
| $X = Cl$; | polyvinyl chloride | PVC; pipes, siding, floor tile, garden hoses |
| $X = O_2CCH_3$; | polyvinyl acetate | PVA; adhesives, chewing gum resin |

**TABLE 15.2** OTHER ADDITION POLYMERS

TABLE 15.2

| STRUCTURE | NAME | COMMON NAME & APPLICATION |
|---|---|---|
| $-[CH_2-CCl_2]_n-$ | poly(vinylidene chloride) | Saran® |
| $-[CF_2-CF_2]_n-$ | poly(tetrafluoroethylene) | Teflon®; non–stick coatings |
| 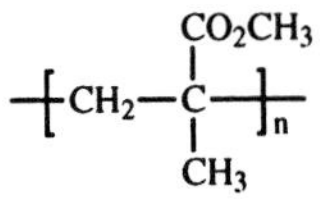 $-[CH_2-C(CO_2CH_3)(CH_3)]_n-$ | poly(methyl methacrylate) | Plexiglass®, Lucite®; acrylic resins, latex paints |
| 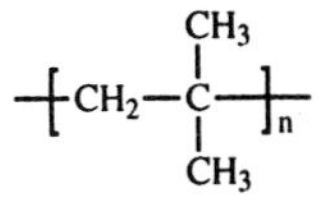 $-[CH_2-C(CH_3)(CH_3)]_n-$ | polyisobutylene | butyl rubber, inner tubes |
| $-[CH_2-CH=CH-CH_2]_n-$ | polybutadiene | synthetic rubber |
| 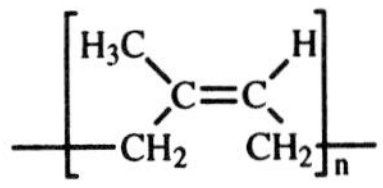 poly(cis–1,3–isoprene) structure | poly (cis–1,3–isoprene) | natural rubber |
| 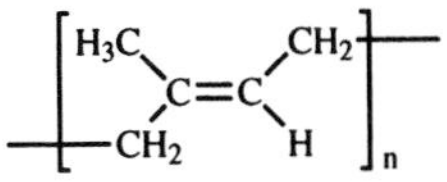 poly(trans–1,3–isoprene) structure | poly(trans–1,3–isoprene) | gutta percha, golf balls |
| 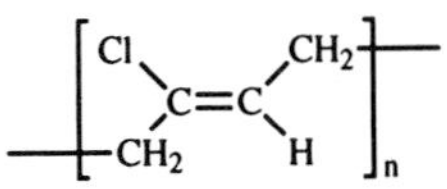 polychloroprene structure | polychloroprene | neoprene, synthetic rubber |

## POLYMER TACTICITY

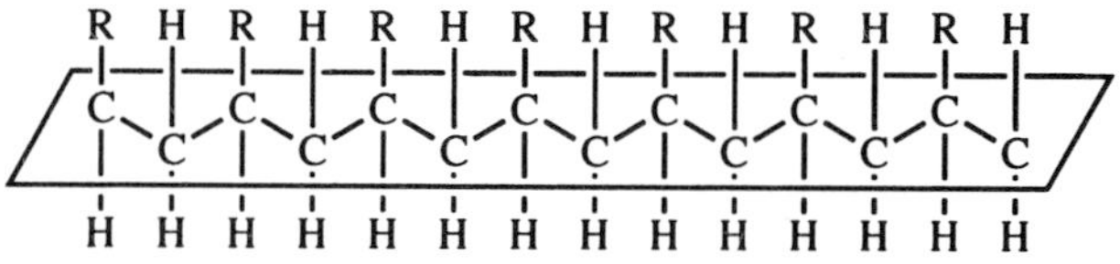

Some carbons in the chain formed from $CH_2=CHR$ are optically active. Figure 15.6 shows all R substituents on the same side of the chain when straightened out and put into the all *trans* zig–zag conformation. This conformation is known as **isotactic**.

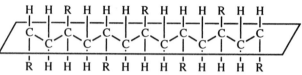

FIGURE 15.6:
Isostatic polymer.

Figure 15.7 shows a regular alternate arrangement of the R groups, with every other R appearing in the equivalent position. This conformation is known as **syndiotactic**. Both isotactic and syndiotactic arrangements are known as stereoregular. Normal free radical formation of substituted polyethylenes, however, results in non–stereoregular arrangements. These polymers are described as being **atactic**, and an example is illustrated in figure 15.8, showing no regularity of the structure around the optically active carbon.

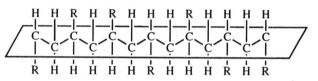

FIGURE 15.7:
Syndiotatic polymer.

The stereoregularity of the polymer can greatly influence its physical properties. Atactic polypropylene, for example, lacks the strength of isotactic polypropylene. It is thought that the closer packing possible with the isotactic structure may make it more crystalline, and thus stronger.

FIGURE 15.8:
Atatic polymer.

## GEOMETRIC ISOMERS OF UNSATURATED ADDITION POLYMERS

When monomers with more than one double bond enter into a chain polymerization, a double bond may remain in the carbon backbone. Equation 15.9 shows the addition reaction that forms polybutadiene:

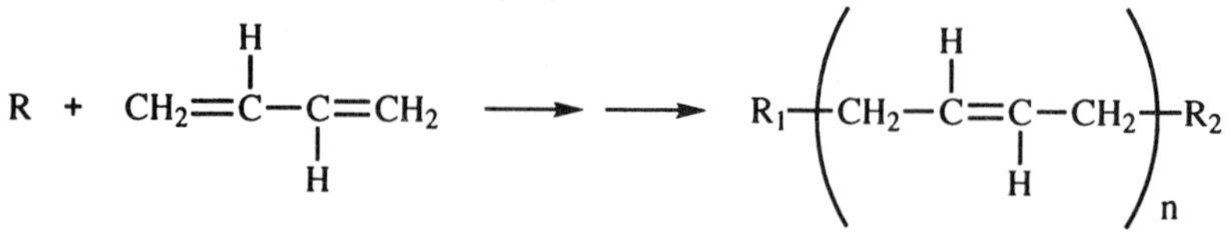

$$R + CH_2{=}C{-}C{=}CH_2 \longrightarrow \longrightarrow R_1\left(CH_2{-}C{=}C{-}CH_2\right)_n R_2$$

EQ. 15.9

The monomeric unit in polybutadiene can have either a *cis* or *trans* orientation, as seen in Fig. 15.9

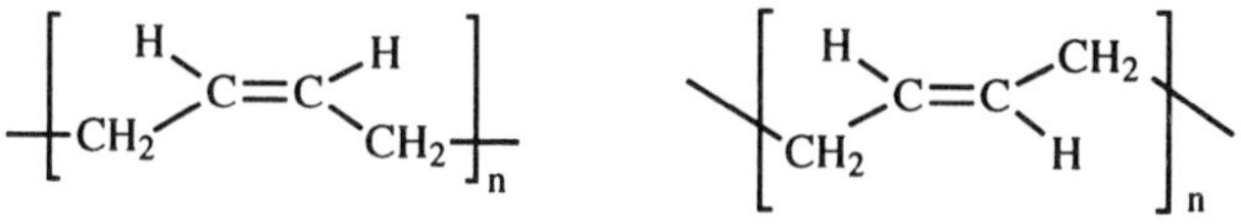

The best known case of a polymer with an unsaturated carbon backbone is natural rubber. By the mid–nineteenth century, chemists knew that both natural rubber and gutta percha were made of units of isoprene, $C_5H_8$, shown in figure 15.10. Although attempts at polymerization of isoprene formed high molecular weight species, the polymeric products had no similarity to either natural rubber or gutta percha. Figures 15.10 and 15.11 show the stereoregularity of these natural polymers.

FIGURE 15.9:
*cis*- and *trans*-polybutadiene.

When chemists tried to polymerize isoprene, each chain had a mixture of *cis* and *trans* orientations, and the synthetic polymers had properties inferior to the natural all *cis* or all *trans* products. Finally in 1955, Ziegler and Natta produced polymerization catalysts that solved two problems: how to make polymers with a saturated carbon backbone having the desired tacticity, and how to make polymers with an unsaturated carbon backbone having the desired geometric regularity.

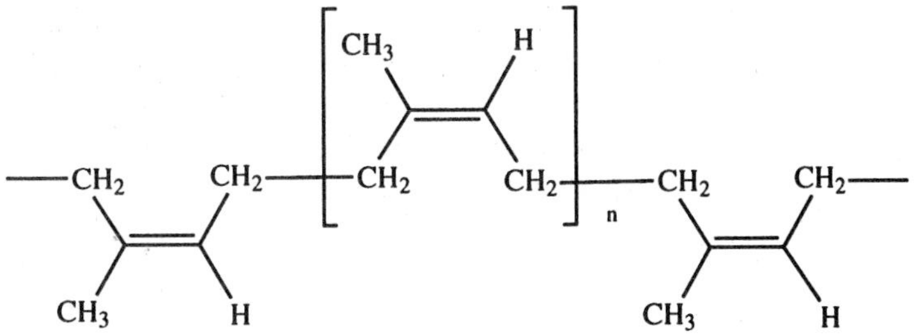

FIGURE 15.10: Structure of natural rubber, poly (*cis*-1,3-isoprene).

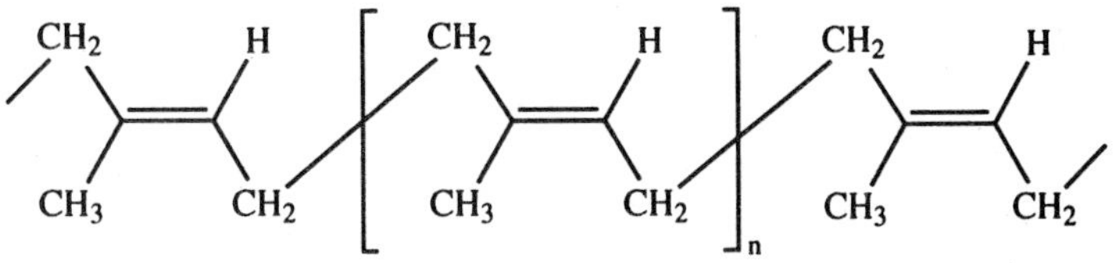

FIGURE 15.11: Structure of gutta percha, poly (*trans*-1,3-isoprene).

## Coordination Polymerization

Ziegler–Natta catalysts are made by mixing alkyls of Groups I–III with halides of transition metals of Groups IV–VIII. One of the earliest useful catalysts of this type was a complex formed between aluminum alkyls and titanium chloride. The original catalysts had the disadvantage of reacting violently with air. The choice of catalyst and reaction conditions allows great selectivity in the form of the polymer. Polyethylene can be made almost completely linear. Polypropylene can be made either isotactic or syndiotactic depending on the choice of catalyst. It is still difficult, however, to select the stereochemical arrangement of monomers in polymers made from larger olefinic monomers. Perhaps the most notable achievement was the catalyst's ability to produce almost exclusively *cis*–1,3– or *trans*–1,3–polyisoprene. The result of this discovery is that, for some time now, the majority of "natural rubber" has been synthetically produced rather than produced from the sap of rubber trees.

How the individual Ziegler–Natta catalysts work is not completely understood. In some cases, it seems that the transition metal is the active site. Figure 15.12 shows a possible mechanism for the addition of a propylene unit to a growing polypropylene chain, using an active site on a titanium atom.

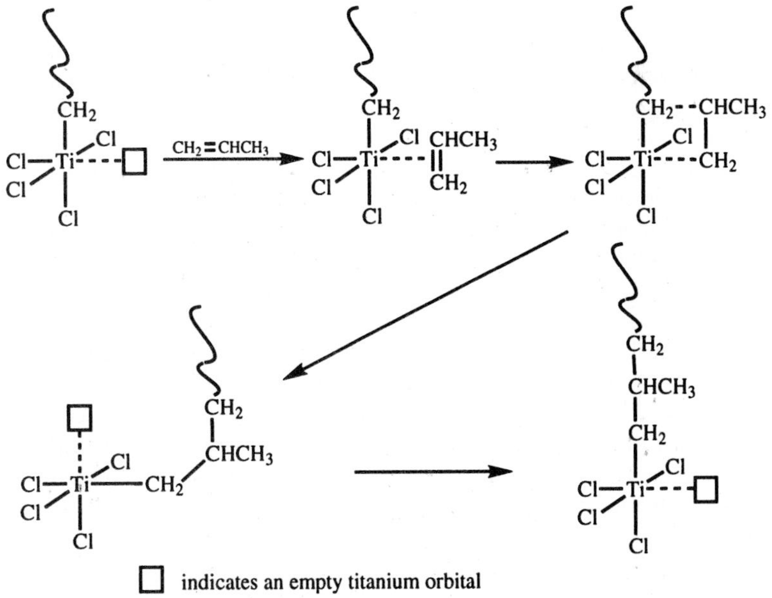
indicates an empty titanium orbital

FIGURE 15.12: Ziegler-Natta catalysis mechanism.

Another possible mechanism utilizes both the transition metal and the main group metal — titanium and aluminum in the following example:

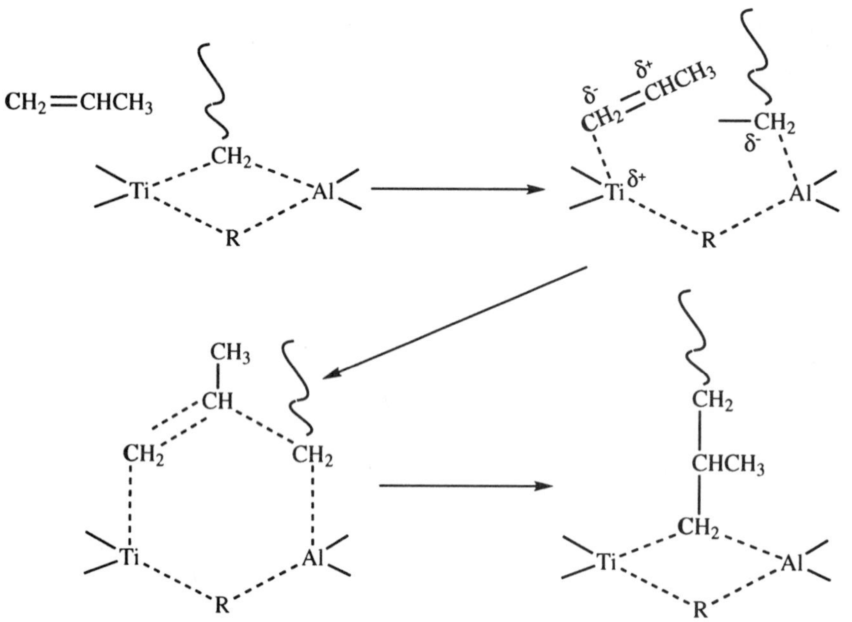

FIGURE 15.13:
Ziegler-Natta
catalysis mechanism.

## 15.5   A CLOSER LOOK AT ELASTOMERS

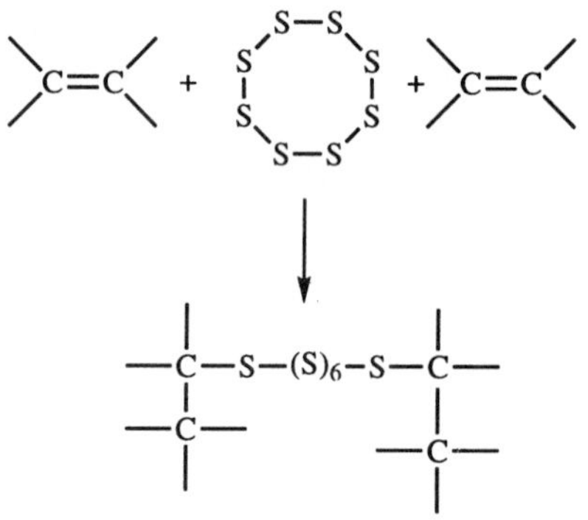

Elastic materials, known as elastomers, have physical properties that can best be understood in terms of their molecular conformation. Natural rubber, with the molecular structure shown in figure 15.10, is normally wrapped into a random coil, like that pictured in figure 15.1. Multiple polymer chains are likely intertwined at this stage. When stretched, elastomers uncoil without breaking any covalent bonds, and the lengthened coils roughly align with one another. This alignment is important for the strength of the elastomer. Gutta percha, with a *trans* orientation, cannot align well and as a result is not elastic. Natural rubber, with an all *cis* structure, is more easily aligned and therefore quite elastic. The stretched chains, although oriented, have little attraction for one another and do not maintain this crystalline order once the stress is removed. Instead, they return to the coiled state driven by the increase in entropy this randomization causes. But with untreated rubber, if the temperature is too high or the stress more than moderate, the individual polymer chains may slip past one another under stress and permanently deform or break. Vulcanization of rubber helps to solve this problem. The most common natural allotrope of sulfur consists of rings with the formula $S_8$. When heated, these rings open to form a linear diradical chain of sulfur atoms. These reactive chains of sulfur attack the double bonds in the polymer chains, causing a crosslinking to occur, as in figure 15.14. These crosslinked molecules prevent the permanent deformation of the rubber that might otherwise occur under great stress. Nature utilizes sulfur in the same fashion, creating sulfur crosslinks with cystine. Wool, a strong, natural polymeric fiber, is a prime example.

FIGURE 15.14:
Vulcanization
cross-linking in
natural rubber.

## 15.6 CONDENSATION POLYMERS: MECHANISMS AND PROPERTIES ———

There are numerous types of condensation polymers. A few of the more common types are treated in this section.

### POLYAMIDES

A large class of condensation polymers forms by reaction between a carboxylic acid functional group and an amine group. Industrial polyamides are called nylons, and are frequently made by the reaction of diacids with diamines, splitting out water to form amide linkages.

$$H_2N(CH_2)_6NH_2 \quad + \quad HO-\overset{\overset{O}{\|}}{C}-(CH_2)_4-\overset{\overset{O}{\|}}{C}-OH \quad \longrightarrow$$

EQ. 15.10

$$\left[ -N-(CH_2)_6-\underset{H}{\overset{H}{N}}-\overset{\overset{O}{\|}}{C}-(CH_2)_4-\overset{\overset{O}{\|}}{C}- \right]$$

In the equation above, adipic acid (1,6–hexanedioic acid) reacts with hexamethylene diamine (1,6–diaminohexane). Repeated reaction of the remaining end functional groups form linear thermoplastics. The nylons have numerical designation, with the carbon chain length of the diamine represented by the first number, the carbon chain length of the acid the second number. The polymer formed from the example in equation 15.10 is called nylon 66. If sebacic acid (a ten–carbon dioic acid) reacted with hexamethylene diamine, nylon 610 would be formed. Amino acids, which contain both groups within the same molecule, can also form polyamides. The polyamides formed by the naturally occurring α-amino acids are called **proteins**. Wool is one example of a protein. These biochemical examples will be examined at the end of this chapter. Industrially, Nylon can also be produced from synthetic amino acids. These amino acids are ω–amino acids, including ω–aminocaproic acid, shown reacting below to form Nylon 6, and ω–amino–undecanoic acid, which forms Nylon 11. Nylon 6, Nylon 11, Nylon 66 and Nylon 610 are the leading commercial polyamides.

$$NH_2-(CH_2)_5-\overset{\overset{O}{\|}}{C}-OH \quad \xrightarrow[\text{polymerization}]{-\ H_2O} \quad \left[ -\underset{H}{\overset{H}{N}}-(CH_2)_5-\overset{\overset{O}{\|}}{C}-\underset{H}{\overset{H}{N}}-(CH_2)_5-\overset{\overset{O}{\|}}{C}- \right]$$

EQ. 15.11

Research into polyamides began about 1928 when W.H. Carothers (1896–1937) of DuPont reacted ω–aminoacids to form waxy polymers with about 30 monomer units. Carothers realized that the concentration of water built up during the reaction, and that ultimately a thermodynamic equilibrium might be limiting the length of polymer chains formed. By distilling the water out of the reaction mixture, Carothers managed to get polyamides with molecular weights of about 25,000 Daltons (Dalton is a unit of molecular weight equivalent to an amu — the name honors John Dalton). The product formed, however, was still too weak for practical purposes. One day, Julian Hill, one of Carothers' co-workers, was playing with a ball of the high molecular weight nylon 610. He stuck a stirring rod into it and pulled a piece away. The material was sticky, and as he pulled, a thread formed. He noted that the thread had a silky appearance. When he tried to break the thread it had great strength. The great strength of these Nylon fibers is the result of hydrogen bonding between adjacent chains, as seen in figure 15.15.

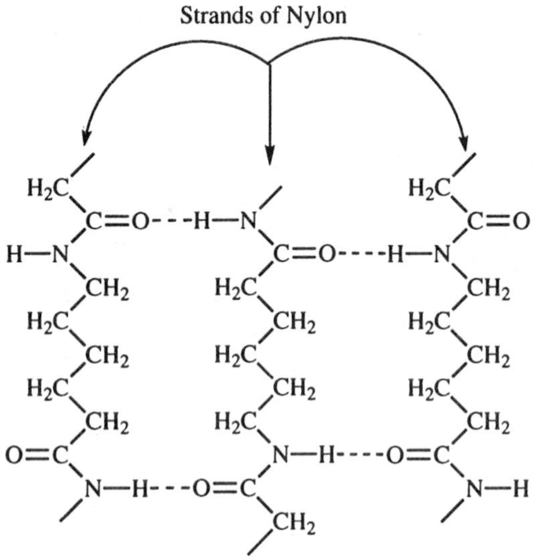

These hydrogen bonds are of intermediate strength, which is just what a fiber needs to be: not so weak as to allow slippage (making the thread too weak), but not so strong as to remove all elasticity (making the thread too brittle). As mentioned earlier, spiders and silkworms do a similar drawing process to strengthen the silk–like threads they produce.

In commercial production of nylons, the acid chloride functional group is frequently subsituted for the regular carboxylic acid in order to make the acid more reactive. Of course, HCl gas rather than water is split out when the reaction occurs:

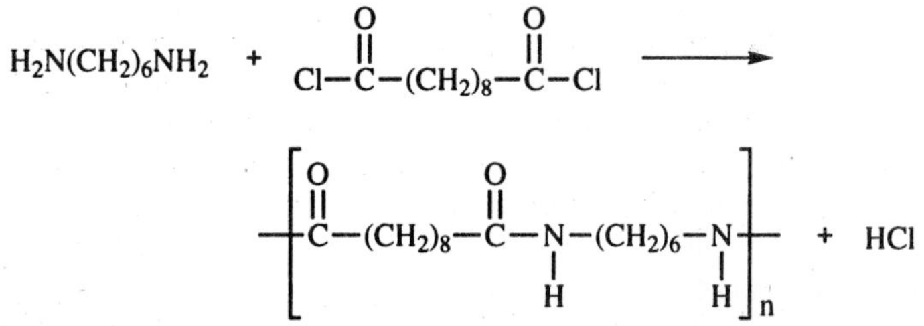

FIGURE 15.15: Hydrogen bonding in nylon chains.

$$H_2N(CH_2)_6NH_2 \; + \; Cl-\overset{O}{\overset{\|}{C}}-(CH_2)_8-\overset{O}{\overset{\|}{C}}-Cl \longrightarrow$$

$$\left[\overset{O}{\overset{\|}{C}}-(CH_2)_8-\overset{O}{\overset{\|}{C}}-\underset{H}{\overset{}{N}}-(CH_2)_6-\underset{H}{\overset{}{N}}\right]_n \; + \; HCl$$

EQ. 15.12

## POLYESTERS

The most important polyester today is poly(ethylene terephthalate). It has a variety of trade names, including Dacron®. This thermoplastic compound is readily recycled, and has the polymer recycle code number 1. It is used in two–liter plastic soda bottles, and in a variety of other containers. It can be synthesized from ethylene glycol and terephthalic acid, as shown in equation 15.13.

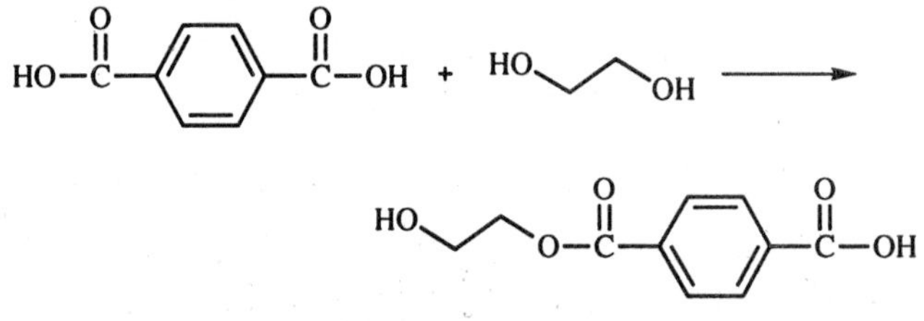

EQ. 15.13

As is the case with low molecular weight esterification, polyesters are made with acid catalysis. Again, water is formed as a byproduct of the reaction, and must be removed to increase polymer yield. The preparation of some poly-esters suffers from a number of manufacturing problems due to low solubility of the dicarboxylic acid in the solvent used. This happens to be the case for terephthalic acid, so in practice the synthesis of Dacron® involves reacting eth-ylene glycol with dimethyl terephthalate, as shown in equation 15.14.

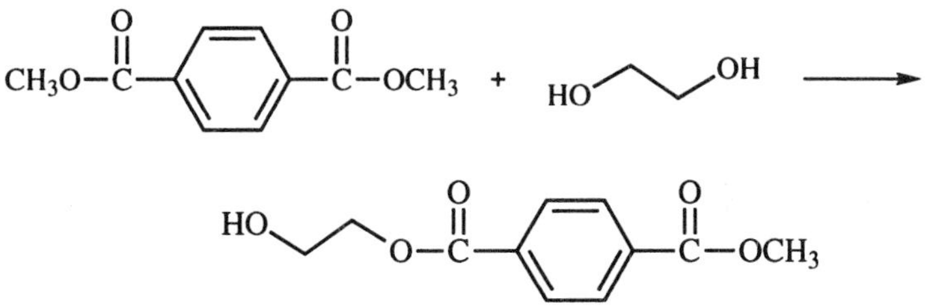

EQ. 15.14

$$CH_3O-\overset{\overset{\text{O}}{\|}}{C}-\left(\text{benzene ring}\right)-\overset{\overset{\text{O}}{\|}}{C}-OCH_3 \quad + \quad HO-\underset{\qquad}{\quad}-OH \quad \longrightarrow$$

$$HO-\underset{\qquad}{\quad}-O-\overset{\overset{\text{O}}{\|}}{C}-\left(\text{benzene ring}\right)-\overset{\overset{\text{O}}{\|}}{C}-OCH_3$$

In this case, the equilibrium is shifted toward the polymer by removing the methanol formed in the reaction. For the manufacture of polyesters, acid chlorides are not used to react with glycols, because competing side reactions yield low molecular weight products.

Alkyd resins are a class of polyesters used in the production of paints and coatings. When ethylene glycol reacts with sebacic acid, a linear alkyd is formed. It imparts flexibility when added to the natural resins often used for coatings, like shellac. Glycerol is sometimes used in place of ethylene glycol to form hard resins. Glycerol, with three reaction sites, provides three–dimensional cross–linking. Acid anhydrides are often substituted for the regular carboxylic acids in industrial production. The first step of the reaction between maleic anhydride and glycerol to form a three dimensional crosslinked polyester resin is shown below. The polymer continues to grow because the additional hydroxyl groups of the glycerol and the (newly formed) carboxyl group of the maleic anhydride continue forming ester linkages by addition.

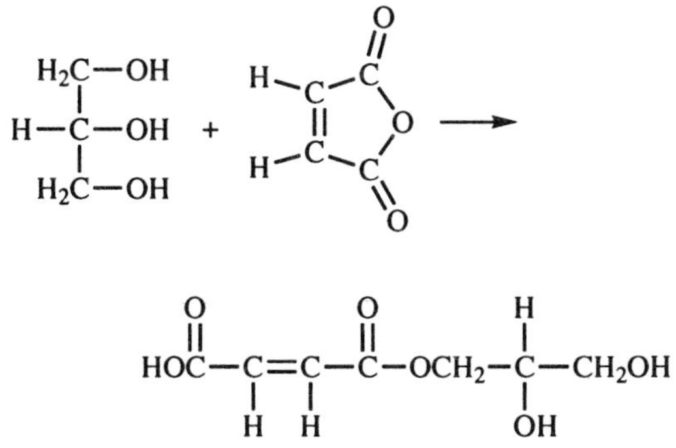

EQ. 15.15

$$\begin{array}{l} H_2C-OH \\ | \\ H-C-OH \\ | \\ H_2C-OH \end{array} \quad + \quad \begin{array}{c}\text{(maleic anhydride)}\end{array} \quad \longrightarrow$$

$$HO\overset{\overset{\text{O}}{\|}}{C}-\underset{\underset{H}{|}}{C}=\underset{\underset{H}{|}}{C}-\overset{\overset{\text{O}}{\|}}{C}-OCH_2-\underset{\underset{OH}{|}}{\overset{\overset{H}{|}}{C}}-CH_2OH$$

## Polyurethanes

We shall consider polyurethanes within the class of condensation polymers because the reaction mechanism, shown below, is clearly similar to that of other condensation polymers. The only difference is that no small molecule is split out when the monomers react. These reactions are sometimes called rearrangement polymerizations.

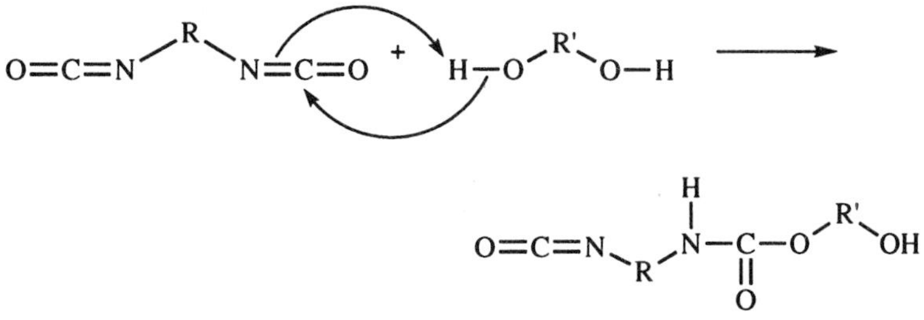

EQ. 15.16

$$O=C=N-R-N=C=O \quad + \quad H-O-R'-O-H \quad \longrightarrow$$

$$O=C=N-R-\underset{\underset{\text{O}}{\|}}{\overset{\overset{H}{|}}{N}}-C-O-R'-OH$$

The diisocyanate reacts with a diol, yielding a molecule with reactive functional groups on both ends for repeated reaction.

Spray polyurethanes are used commercially as a transparent protective coating. Foam rubber is formed by adding some water to the reaction mixture. Water reacts with the diisocyanate to release bubbles of carbon dioxide gas into the middle of the rapidly forming polymer. These bubbles are trapped in the polymer matrix and result in a lightweight foam that we call foam rubber.

### Polycarbonates

As a final example of organic condensation polymers, we consider polycarbonates. One of these, Lexan®, has numerous unique properties: it is transparent, heat resistant, and incredibly shatter resistant. It is used as bulletproof "glass". Polycarbonates can be made by reacting phosgene with a diol such as ethylene glycol. In the process HCl is split out.

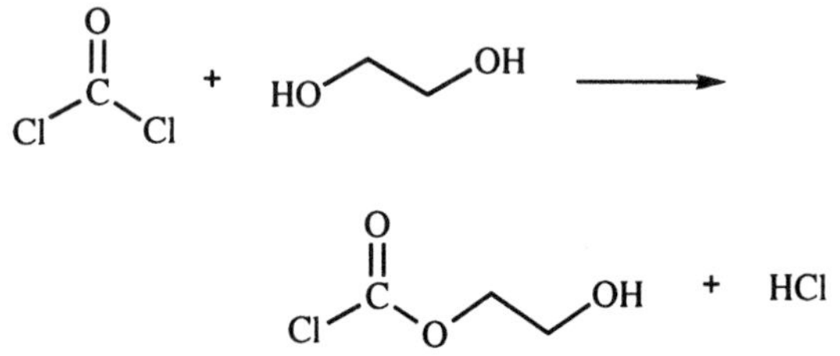

EQ. 15.17

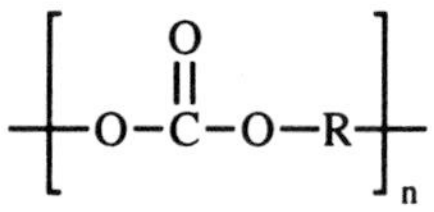

After the first stage of the reaction, shown in equation 15.17, both ends remain active for additional monomer addition. The term "carbonate" is due to the repeating structure in the polymer:

FIGURE 15.16: Polycarbonate repeating unit.

## 15.7 INORGANIC POLYMERS

Thus far in this chapter the polymers have been organic polymers, with carbon comprising most, if not all, of the backbone. The large majority of synthetic polymers are, indeed, organic. There are, however, important and interesting examples of polymers without a carbon backbone. Many inorganic polymers have synthesis mechanisms, structures and properties that compliment those we have already seen. In this section our survey includes several examples of inorganic polymers.

### POLYTHIAZYL

The simplest of inorganic polymers may be polymeric sulfur nitride, polythiazyl. It is formed from $S_2N_2$, a nearly square–planar covalent molecule, which polymerizes to form linear, zig–zag chains of the formula $(SN)_n$.

This polymer crystallizes in oriented fibers, and the solid has a metallic sheen. The metal–like character is also seen in the ability of the polythiazyl chains to carry electric current. Polythiazyl has the potential for use in advanced technologies, since it is a superconductor and it behaves as a one–dimensional metal: it carries current only along the direction of the oriented chains, but not perpendicular to the chains.

## POLYPHOSPHAZINES

Almost one hundred years ago, chemists discovered that hexachlorotriphosphazine could form an elastic polymer with the formula $(-PCl_2N-)_n$ shown in figure 15.17

The phosphorous–nitrogen backbone contains double bonds, which were previously seen to be crucial to the elastic properties of natural rubber. In fact, polyphosphazine chains can be crosslinked like natural rubber, to fine–tune the elasticity. Interest in the original polymer was initially short–lived when it was found to hydrolyze when exposed to water. Further work, however, showed that chlorine atoms could be replaced with a variety of other groups by a variety of reactions. Some of these alternate side groups make the polymer water–resistant. All provide a wide assortment of other properties.

FIGURE 15.17: Polyphosphazine synthesis.

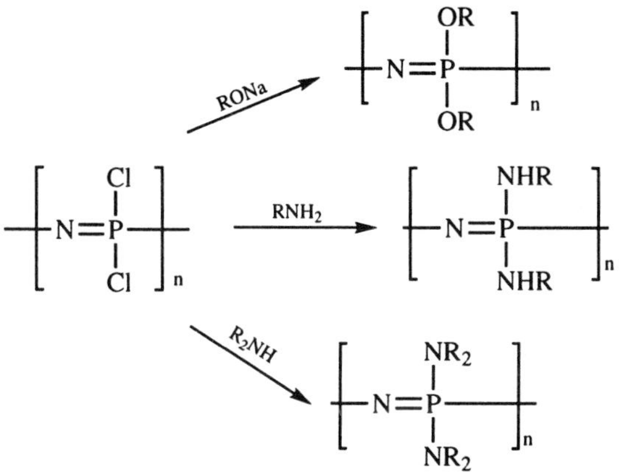

EQ. 15.18

The most widespread industrial use of polyphosphazines is as a flame retardant when incorporated with other (flammable) polymers. Various polyphosphazines are used in waterproofing and for O–rings. They are also used in the rubber hose used to deliver gasoline from the gasoline pump to the car. When the side group is $-O-CH_2-CF_3$, a strong, flexible, water–repellent and biocompatible polymer is formed. This particular polymer is considered a strong candidate for making replacement blood vessels. Other polyphosphazines chains have had biologically active molecules added as side groups. The hope is that these may be used to introduce drugs directly into the area of the body to be treated (e.g., to deliver some forms of chemotherapy directly to a tumor site), by grafting the drug onto a large, chemically inert polymer backbone. The polymer will keep the drug where it is needed since the polymer is too large to diffuse through cell membranes. The therapeutic agent can then slowly hydrolyze and become active, while the remaining unhydrolyzed dose stays where it will eventually be used. Such a delivery system could increase the drug's potency while reducing unwanted and sometimes dangerous side effects.

## SILICA, SILICATES AND SILICONES

Given carbon's success in making polymers, it is tempting to consider its close chemical relative, silicon, as an alternative. But silicon differs from carbon in many of its properties. Unlike carbon, silicon has difficulty catenating to form chains with only silicon–silicon bonds. Silanes ($Si_nH_{2n+2}$) are the silicon equivalent of alkanes. Linear silanes are known with up to 8 silicon atoms. Of these only mono silane $SiH_4$ and disilane $Si_2H_6$ are indefinitely stable at room temperature. Silanes become increasingly unstable for $n > 2$. Only poorly characterized, branched silanes exist for $n>8$. Silanes of any length also spontaneously burn when exposed to air. The combustion reactions of alkanes and of silanes are both thermodynamically spontaneous. The high activation energy (and thus slow kinetics) for the combustion of hydrocarbons like polyethylene keeps them air stable. The activation energy in the corresponding combustion for silanes is considerably lower. Attempting to produce usable polymers with chains of silicon–silicon bonds is almost certainly a wasted effort. In addition, there would be real experimental problems with formation of silicon polymers by polymerization reactions similar to those used to form polyethylene. Silicon–silicon double bonds, undiscovered until 1981, are very unstable and, therefore, compounds containing them are unsuitable for use as monomers in polymerization reactions.

Introducing oxygen along with silicon, however, provides a wide variety of inorganic polymeric structures. This combination is quite common, since oxygen and silicon are the two most abundant elements in the earth's crust, comprising 46.6% and 27.7% by mass respectively. Carbon dioxide, $CO_2$, is a gas, forming discrete molecules even in the solid state. Its silicon analog, silica, $SiO_2$, is really a family of very high melting solids, with a basic structural unit of $SiO_4$ tetrahedra, shown in figure 15.18.

The $SiO_4$ tetrahedra are fused in an extended fashion to produce structures like that of quartz, shown at right in figure 15.19.

## SILICATES

In addition to the regular three–dimensional silica network, linear chains and sheets of anions of silicon and oxygen readily form. Starting with the basic tetrahedral $SiO_4$ structure of figure 15.18, one silicon atom can share two of the oxygen atoms surrounding it with adjacent silicon atoms. This results in a chain with the repeating unit $(SiO_3)^{2-}$, pictured below. The linear chains lie parallel, held together by the cations that lie between them.

These chains form a class of minerals called pyroxenes. One such mineral is spodumene, $LiAl(SiO_3)_2$. Spodumene is an important source of lithium, which has been shown to have definite psychotherapeutic value to some individuals. Thus the purported mental health benefit of "taking the waters" at certain spas in Europe, may in fact have some scientific basis, since the water has a high concentration of lithium due to the presence of spodumene deposits near the springs.

Two linear chains may cross–link by having half of the silicon atoms share three oxygens, while the other half share two. This linked double chain has the formula $(Si_4O_{11}^{6-})_n$, and is pictured at the right.

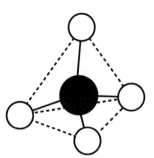

FIGURE 15.18:
$SiO_4$ tetrahedron structure.

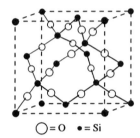

$\bigcirc = O \quad \bullet = Si$

FIGURE 15.19:
Structure of quartz, $SiO_2$.

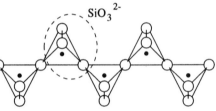

$SiO_3^{2-}$

$\bigcirc = O \quad \bullet = Si$

FIGURE 15.20:
Linear $SiO_3^{2-}$ chain.

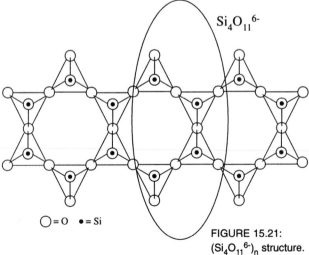

$Si_4O_{11}^{6-}$

$\bigcirc = O \quad \bullet = Si$

FIGURE 15.21:
$(Si_4O_{11}^{6-})_n$ structure.

These cross–linked chains occur in minerals known as amphiboles, some of which, known collectively as asbestos, are quite fibrous. These materials proved to be excellent thermal insulators, and have found widespread commercial use. Both pyroxenes and amphiboles have great strength along the chain of fused tetrahedra. But the forces between the chains are relatively weak. Thus asbestos can be easily cleaved along the axis parallel to the chains, leading to airborne particles. Some airborne asbestos particles have been shown to be related to lung cancer, and the safe removal of asbestos from older buildings has become a severe problem.

## SILICONES

Silicones are synthetic polymers with the general structure shown below. The name was intended to show the similarity with the ketone functional group in carbon chemistry. While there is some similarity in geometric structure, the type of reactivity and bonding in silicones and ketones are quite different. The C–O and C=O bond dissociation energies are 358 and 803 kJ/mole respectively, while the Si–O and Si=O bond dissociation energies are 464 and 640 kJ/mole. The size mismatch from forming a pi bond between elements from different periods leads to a weaker pi bond, and to the tendency of the silicones to form chains of $-Si-O-$, as opposed to discrete molecules with a Si=O functional group.

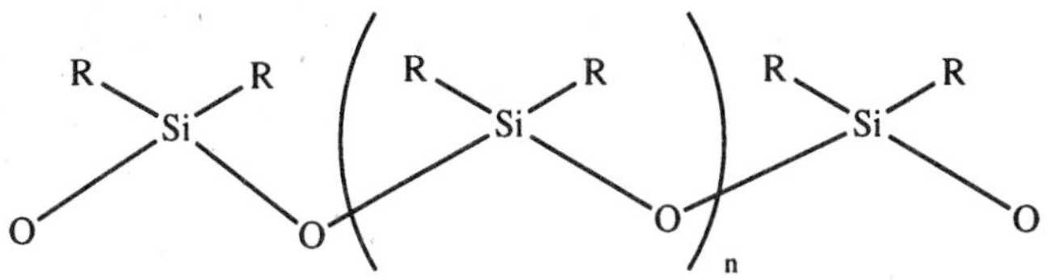

FIGURE 15.22:
General structure of a silicone polymer.

The synthesis of silicones begins with silicon and an organic halide RX.

$$Si(s) + 2\,RX \longrightarrow R_2SiX_2 \qquad\qquad \text{EQ. 15.19}$$

$$2\,H_2O + R_2SiX_2 \longrightarrow R_2Si(OH)_2 + 2\,HX \qquad\qquad \text{EQ. 15.20}$$

The silicone polymers, with the general structure shown in figure 15.22, generally have non–polar organic groups R along the chain of silicon and oxygen, and this makes silicones water–resistant. Since both the Si–O and Si–C bonds are rather strong, silicones are generally chemically unreactive and thermally stable. Silicones were thought to be totally biocompatible, and were used as components in artificial hearts and for cosmetic surgery until some trials began to show adverse effects for some patients.

By varying the specific R groups and the length of the silicon–oxygen (silicone) chain, a variety of material properties result. Lower molecular weight silicones form oils and greases. For example, when R is the methyl group and chain lengths are kept short, the silicone will be an oil. Unlike their organic counterparts, the viscosity of these silicones does not vary much with temperature, and these silicones thus work as lubricants over a wide range of temperatures. Higher molecular weight silicones form rubbery solids, which again compare favorably to organic elastomers under extreme temperature conditions. The door gaskets for refrigerators are generally silicon rubber. The famous first footprint on the moon was made with a silicon rubber boot, designed for temperature extremes.

Silicon rubbers can be made by having R groups that can crosslink chains. This can be done at the time of the production of the polymer, or the precursors can be processed in such a way that the silicone oil becomes a rubber when exposed to the moisture of the air. Silicone bathtub caulk, for example, cures in air, by hydrolyzing as shown in the reaction below, splitting off acetic acid and water.

EQ. 15.22

Silly Putty® is a mixture of silicone oil and silicone rubber; that is, it has a mixture of short chain length species and long chain length species that are partially crosslinked. The resulting mixture has an unusual combination of properties: when deformed suddenly, it shows the resistance to stress characteristic of elastomers; but when slowly stretched, it shows the low viscosity of a fluid.

## 15.8   COPOLYMERS AND ADDITIVES

Often the physical characteristics of a polymer made from any single monomeric unit are not ideal. There are two ways to modify these characteristics. One is to put in relatively small amounts of non–polymeric material (additives) that produce the desired characteristics. The other approach is to modify the polymer itself, by incorporating two (or more) distinct monomeric species into the polymer chain, resulting in a copolymer.

### ADDITIVES

Additives are used in nearly every commercial polymer, fulfilling a variety of functions. Making a polymer heat–resistant is a common problem. As an example, automobile tires made exclusively of rubber can experience a tremendous amount of heat buildup during highway driving as a result of friction.

Neither natural rubber nor synthetic elastomers maintain their desired physical properties at extremely high temperatures. Instead, they would become quite sticky and lose much of the stiffness and strength. To solve this problem, carbon black (a good thermal conductor) is added to help dissipate the heat. The black color of automobile tires is due to the presence of carbon black; natural and synthetic elastomers tend to be white or light amber in the absence of this additive. Thermal decomposition can also be a problem for poly (vinyl chloride), PVC. Lead salts proved quite effective additives for increasing the thermal stability of PVC. However, since PVC was widely marketed as a safer, lead–free replacement of older metal plumbing, the lead salts were soon replaced by antimony salts, which have proved also to have the desired effect of increasing the thermal stability of the PVC.

Outdoor furniture is often made with polypropylene webbing. When this was initially marketed, there was an extreme failure rate never previously noted in the laboratory. This was found to be due to degradation of the polymer chains caused by exposure to ultraviolet radiation from the sun. This eventually caused the webbing to break under load. To solve this problem, material that absorbs ultraviolet light much better than the surrounding polymer is added. This could be titanium (IV) oxide (commonly used to brighten paints for the same reason) or substituted benzophenones.

If a plastic can orient into crystalline regions too readily, brittleness can become a problem. Incorporation of "plasticizers" can slow down or eliminate this orientation into crystals and make the material far more flexible. The characteristic "new car smell" is due to high–molecular–weight esters added to make the polymeric material on the inside of the car more resilient. But, as your nose tells you, these additives are volatile materials. Eventually the plasticizers evaporate, leaving behind a more brittle polymeric material. Some plasticizers that have been used industrially include: dibutyl phthalate, dibutyl sebacate, dioctyl phthalate, diethyl maleate, dioctyl adipate and tributyl phosphate. Some low molecular weight poly (propylene glycol) esters are now also used as plasticizers. The figure below is a cartoon rendering of how plasticizers can increase flexibility by reducing the crystallinity that causes brittleness.

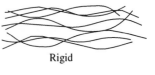
Rigid

Less Rigid

FIGURE 15.23: Representation of plasticizers reducing crystallinity.

Other additives are used to make the polymeric material flame retardant. The most widely used additive for this purpose is antimony trioxide. However, this additive is effective only for polymers such as PVC that contain chlorine. It is thought that the actual flame–retarding molecule is antimony oxychloride. Methods by which flame–retardants work can vary, but they often involve charring the polymer, removing the contact between the air and the unreacted polymer. The addition of radical scavengers in the polymer can also help fight the free radical chain process occurring in fires. The latter compounds are also useful as antioxidants, which inhibit the oxidation of the polymer under normal conditions. Finally, additives can be added as biocides to prevent the buildup of bacteria on the polymer. Arsenic compounds are often added for this purpose.

Earlier, we saw that changing a component of the monomer unit could produce very different products: *e.g.*, PVC versus polyethylene. The combination of different monomer units within the same polymer can cause similar changes in properties. A copolymer made by combining vinyl chloride ($CH_2$=CHCl) and vinylidene chloride ($CH_2$=$CCl_2$) in the same chain produces a product known as Saran®. Films of this copolymer have an ability to cling to itself. This property is also due to an increased attraction between chains caused by the combined chlorine substituents.

Other than a random order, there are three possible organizations of monomer units A and B within a copolymer chain. These are shown below as (a) alternating, (b) block, and (c) graft copolymers.

Generally, condensation polymers formed from two difunctional comonomers A and B will have a random ordering. Addition reactions involving comonomers $M_1$ and $M_2$, and the radical species $M_1\cdot$ and $M_2\cdot$ are expected to have the four following propagation steps in the polymerization mechanism:

—ABABABABABABABABABABABABABAB—

(a) Alternating Copolymer Organization

—AAAAAAAAAAAAAABBBBBBBBBBBBBB—

(b) Block Copolymer Organization

```
    B           B           B
    B           B           B
    B           B           B
    B           B           B
    |           |           |
—AAAAAAAAAAAAAAAAAAAAAAAAAAAAA—
```

(c) Graft Copolymer Organization

FIGURE 15.24: Polymer chain organizations.

$$-M_1\cdot + M_1 \longrightarrow -M_1M_1\cdot \qquad R = k_{11}[-M_1\cdot][M_1] \qquad \text{EQ. 15.23a}$$

$$-M_1\cdot + M_2 \longrightarrow -M_1M_2\cdot \qquad R = k_{12}[-M_1\cdot][M_2] \qquad \text{EQ. 15.23b}$$

$$-M_2\cdot + M_1 \longrightarrow -M_2M_1\cdot \qquad R = k_{21}[-M_2\cdot][M_1] \qquad \text{EQ. 15.23c}$$

$$-M_2\cdot + M_2 \longrightarrow -M_2M_2\cdot \qquad R = k_{22}[-M_2\cdot][M_2] \qquad \text{EQ. 15.23d}$$

A random copolymerization occurs when the growing radical chain ($-M_1\cdot$ or $-M_2\cdot$) adds to either comonomer $M_1$ or $M_2$ with equal probability. That is, $k_{11} \approx k_{12}$, and $k_{21} \approx k_{22}$. For example, if $M_1$ = styrene and $M_2$ = 2–vinylthiophene, a random copolymer is formed. On the other hand, if $k_{11} \ll k_{12}$, and if $k_{21} \gg k_{22}$, then a regularly alternating copolymer forms. An example of this type occurs if $M_1$ = styrene and $M_2$ = diethyl fumarate. With these monomers, the copolymer strongly favors an alternating structure as shown in figure 15.25.

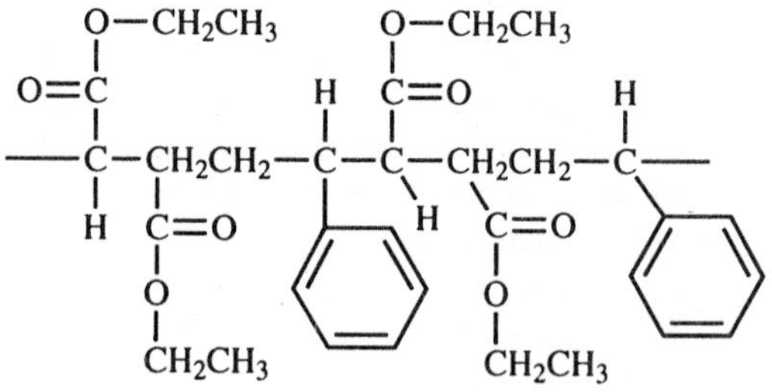

Most radically initiated addition polymerizations occur somewhere in between these first two extremes: that is, there is often a small preference for alternating addition.

If $k_{11} \gg k_{12}$ and $k_{21} \ll k_{22}$, then a block copolymer forms; this is the least common situation for simple chain growth in a mixture containing both $M_1$ and $M_2$. As we shall soon see, addition block copolymers are also produced by other means. For condensation polymers, making block copolymers is relatively easy. A block polyester copolymer can be prepared by making polyester A

FIGURE 15.25: Alternating copolymer of styrene and diethylfumarate.

and polyester B separately. Then the two polymers are mixed and allowed to react further so that a structure like 15.24(b) results. For another example, consider B to represent a polyester chain of intermediate length, with reactive hydroxyl groups on both ends. These polyester chains can then be connected through urethane linkages using a diisocyanate. The reaction of 2,4–toluene diisocyanate with the polyester chain B is shown in the following equation. Highly elastic, lightweight materials like Spandex® or Lycra®, can be made in this fashion, with long, flexible chains connected with short, stiff chains.

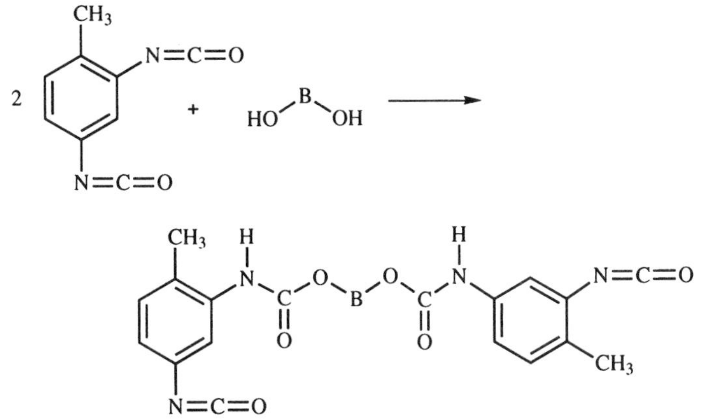

EQ. 15.24

$$2 \quad \text{(2,4-toluene diisocyanate: } CH_3, N{=}C{=}O, N{=}C{=}O) \quad + \quad HO-B-OH \quad \longrightarrow$$

$$CH_3,\ H{-}N{-}C({=}O){-}O{-}B{-}O{-}C({=}O){-}N{-}H,\ N{=}C{=}O,\ CH_3,\ N{=}C{=}O$$

Addition block copolymers can be made by breaking one of the A—X bonds in the block homopolymer $X–A(A)_nA–X$ to form a primary radical species $(X–A(A)_nA\cdot)$ in the presence of monomer B. Careful repetition of this process can produce several alternating runs of $(A)_x(B)_y(A)_z$, *etc.* For example, propylene and ethylene can be either randomly copolymerized or block copolymerized, and different properties result. The random copolymer yields a better transparent film, but the block copolymer is more impact resistant, and is used for injection molding.

Finally, graft copolymers with the structure shown in figure 15.24(c) are usually formed by radical methods with unsaturated addition homopolymers. The most common examples of this are various synthetic elastomers made by grafting styrene, acrylonitrile or acrylonitrile–styrene copolymers onto a backbone of polybutadiene. A radical can be formed at various sites along the polybutadiene chain by a radical chain transfer, by ultraviolet radiation, ionizing radiation, redox initiation, or other methods. The co–monomer then begins polymerizing at these sites. Examples of these and other copolymers are shown in table 15.3.

TABLE 15.3

**TABLE 15.3** SOME COMMERCIAL COPOLYMERS

| MONOMER 1 | MONOMER 2 | MONOMER 3 | PRODUCT /USE |
|---|---|---|---|
| acrylonitrile | butadiene | | nitrile rubber |
| acrylonitrile | vinyl chloride | | Dynel® (clothing) |
| vinyl chloride | vinylidene chloride | | Saran® (film for wrapping food) |
| styrene | butadiene | | SB rubber (tires) |
| isobutylene | isoprene | | butyl rubber (inner tubes) |
| vinyl chloride | vinyl acetate | | vinylite (shower curtains) |
| acrylonitrile | butadiene | styrene | ABS rubber (luggage, crash helmets) |

## 15.9 BIOPOLYMERS: AN OVERVIEW

As you are probably by now well aware, the structural and functional systems in most organisms, including humans, are composed of polymers of simple organic molecules such as carbohydrates and amino acids. These larger molecules and macromolecules are referred to as biopolymers and will be the focus of the remainder of this chapter. We will examine each class of polymer in some detail and attempt to connect back to much of the chemistry that you have learned up to this point. In the next chapter we will also look at how some of these macromolecules behave in living systems.

## 15.10 CARBOHYDRATE BASED BIOPOLYMERS

**THE SUGAR MONOMER**

While we briefly discussed the chemistry associated with carbohydrates previously (section 5.10), we should probably examine carbohydrates and some of the chemistry associated with this class of compound in a more systematic fashion.

Sugars are classified both by the number of carbons that comprise the longest carbon chain and by the functional group of the most highly oxidized carbon atom. For example, a three–carbon sugar might exist as an **aldotriose** (D–glyceraldehyde) or a ketotriose (dihydroxyacetone). The term triose obviously comes from the fact that the sugar molecule contains three carbons and the prefixes, aldo– and keto– are derived from the functional group on the carbon atom in the highest oxidation state.

You are probably wondering where the "D" in D–glyceraldehyde originates. This designation dates back to early carbohydrate chemistry and refers to the stereochemistry about the chiral carbon in glyceraldehyde. Note that in D–glyceraldehyde, the hydroxyl group is on the right. L-glyceraldehyde would have the hydroxyl group on the left. Since, by convention, sugar molecules are drawn so that the carbon with the highest oxidation state is near the top of the molecule, we can generalize and say that all sugars that have this same configuration about the chiral center closest to the bottom are also of the D–configuration. For example, if we look at the structures of two common aldotetroses we note that the hydroxyl group on the chiral center drawn closest to the bottom are both to the right. Since this is the same configuration as in D–glyceraldehyde, these aldotetroses are also dextrorotatory. In fact most of the sugars commonly found in nature have the D–configuration.

Other important sugars include the aldopentose, D–ribose; the aldohexoses, D–glucose and D–galactose; the ketohexose, D–fructose; and, of course, the deoxysugar, D–2–deoxyribose.

It is important to realize that the name of each sugar provides us with the relationship of each of the hydroxyl groups relative to the others. For example, the name ribose implies that all three of the hydroxyl groups are on the same side of the carbon chain. So, in the structure of L–ribose, all of the hydroxyl groups are on the left side of the carbon chain while, as we have seen, in D–ribose the hydroxyl groups are on the right side of the carbon chain.

In chapter 5 you were also introduced to the ring forms of sugars and we looked in some detail at the six–membered ring form of glucose. This structure is called the *pyranose* form of glucose. The name is derived from the oxygen containing heterocycle, pyran, and can be drawn as a flat ring. An even better drawing, since it more closely represents the actual structure, would be of the molecule in the chair conformation.

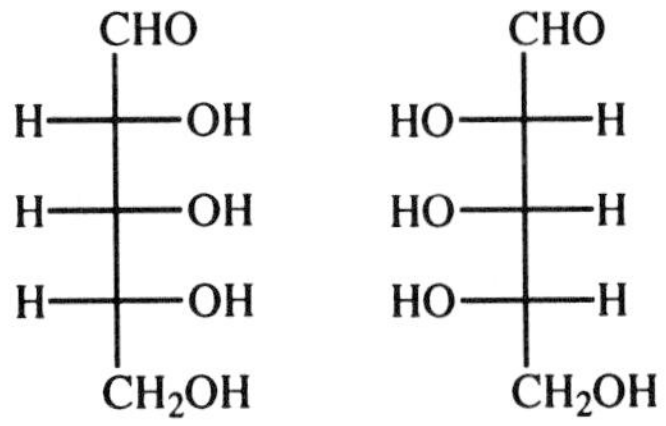

It may not be readily apparent how the transformation from the Fisher projection to the Haworth projection is accomplished. Therefore, we will look at this transformation in some detail. The first step is to turn the Fisher projection on its side; then the appropriate bonds are bent. After rotation about the $C_4$–$C_5$ bond, the hemiacetal linkage is formed. This process is illustrated in figure 15.31.

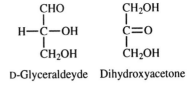

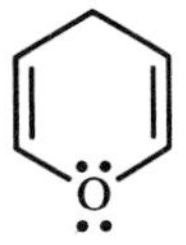

D-Glyceraldeyde    Dihydroxyacetone

FIGURE 15.26:
The structure of two three carbon sugars.

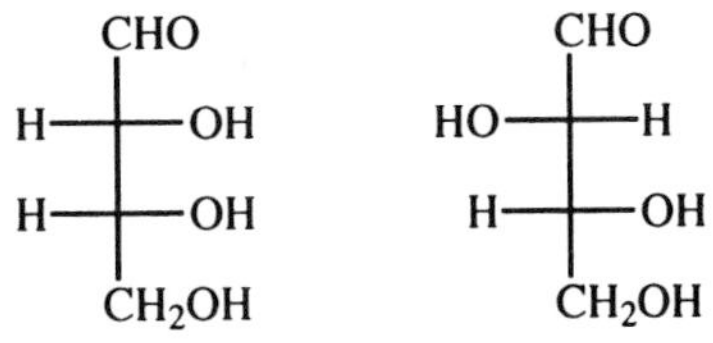

D-Erythrose          D-Threose

FIGURE 15.27:
Two aldotetroses.

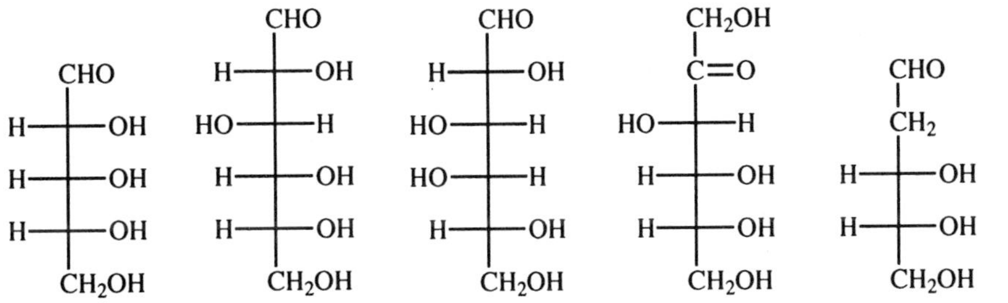

D-Ribose    D-Glucose    D-Galactose    D-Fructose    D-2-Deoxyribose

FIGURE 15.28:
Other common sugars of biological importance.

FIGURE 15.29:
D– and L–ribose.

Pyran

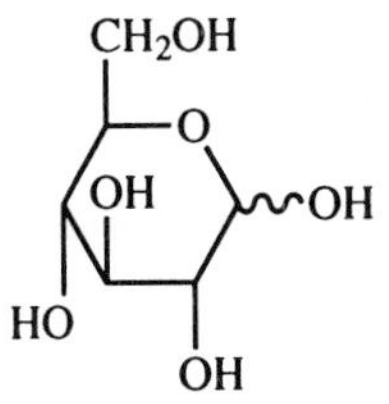

D-Glucose

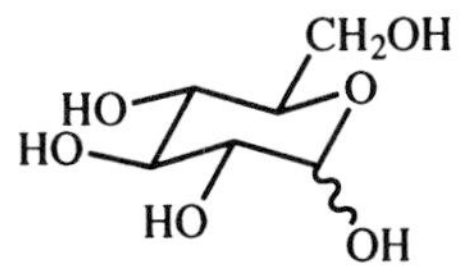

FIGURE 15.30:
The pyranose forms of D–glucose.

It is interesting to note that
D–Fructose typically forms a five,
rather than a six, membered hemiacetal. This is known as the *furanose*
form; this name is derived from the
five–membered aromatic
oxygen–containing heterocycle,
furan.

Before we go on to look at the structure of some di– and
*polysaccharides*, we should introduce
a few more terms. If we look back at
figure 15.30 we notice that the
hydroxyl group that is part of the
hemiacetal is drawn in a manner that
does not indicate stereochemistry.
The carbon to which this hydroxyl group is attached is called the *anomeric*
*carbon*. You may recall that the carbonyl group is planar and can, therefore, be
attacked from either the top face or the bottom face when the hemiacetal forms.
If the attack is from the top face then the hydroxyl group will end up pointing
down. This is known as the alpha form. If the attack is from the bottom face
then the hydroxyl group will end up pointing up. This is known as the beta
form. The two stereoisomers are related as *anomers*.

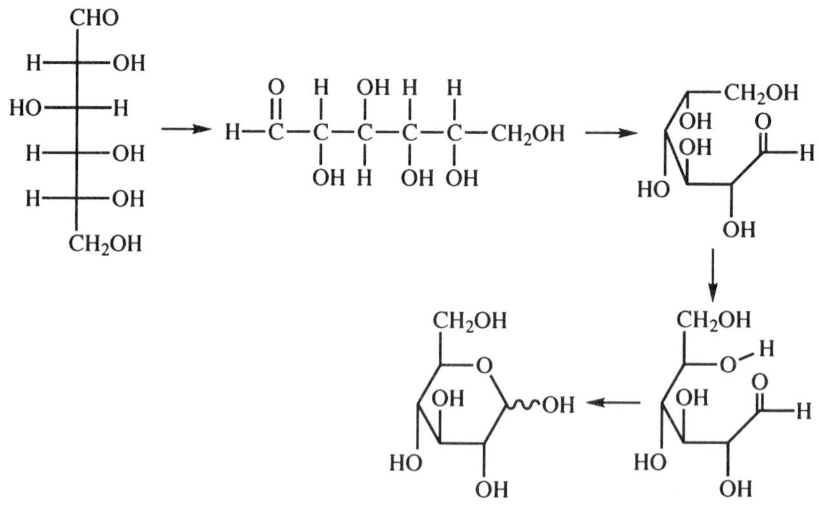

FIGURE 15.31:
The conversion of
Fischer projections to
Hayworth projections.

## THE GLYCOSIDIC BOND AND DISACCHARIDES

When a sugar molecule, or any hemiacetal for that matter, reacts with a
molecule of alcohol the hemiacetal becomes an acetal. Consider the
reaction of β–D–glucopyranose with ethanol, in the presence of trace
amounts of acid catalyst.

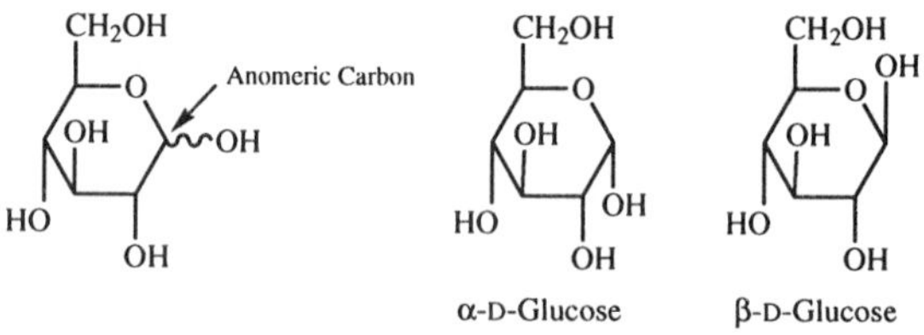

α-D-Glucose    β-D-Glucose

FIGURE 15.32:
The anomers of
D–glucose.

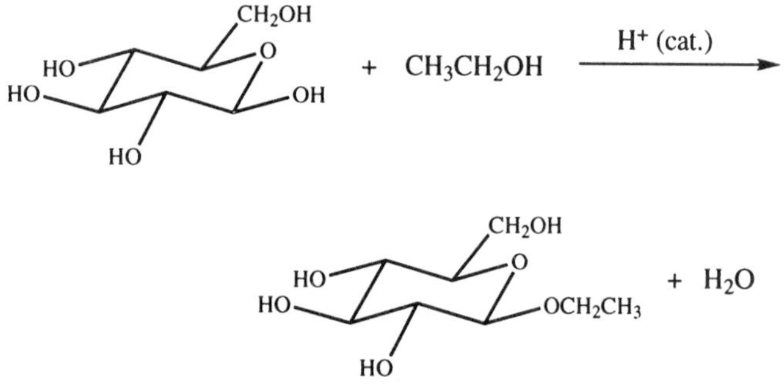

EQ. 15.25

The acetals of carbohydrates are called *glycosides* and the bonds that connect
the sugar with another moiety are called *glycosidic bonds*. Although, in this
case the sugar is attached to an ethyl group, it is also possible, and common,
that two sugars can bond to each other in this manner. When two simple sugar
units are joined by a glycosidic bond, the new unit is called a *disaccharide*.
Some of the more common disaccharides include: maltose, a disaccharide
obtained when starch is hydrolyzed; lactose, a disaccharide commonly found in
milk and often referred to as "milk sugar"; and sucrose or common table sugar.

If we look carefully at these three disaccharides we will see differences. Maltose is composed of two glucose units joined by an α–glycosidic bond between carbon 1 of the first glucose unit and carbon 4 (often labeled 4′) of the second glucose unit.

Lactose is composed of a galactose unit and a glucose unit. These two monomers are joined by a β–glycosidic bond between carbon 1 of galactose and carbon 4 (often labeled 4′) of glucose.

Finally, sucrose consists of a glucose unit joined to a fructose unit by an α–glycosidic bond between carbon 1 of glucose and carbon 2 of fructose.

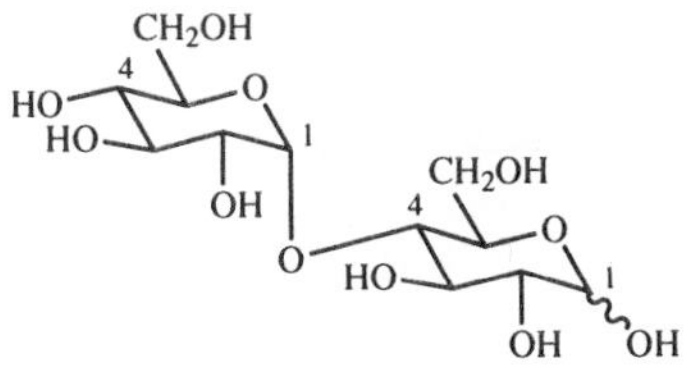

FIGURE 15.33:
The structure of
maltose.

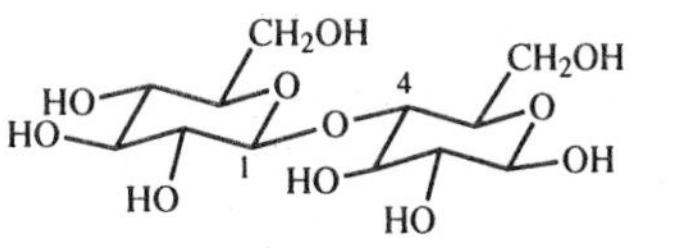

FIGURE 15.34:
The structure of
lactose.

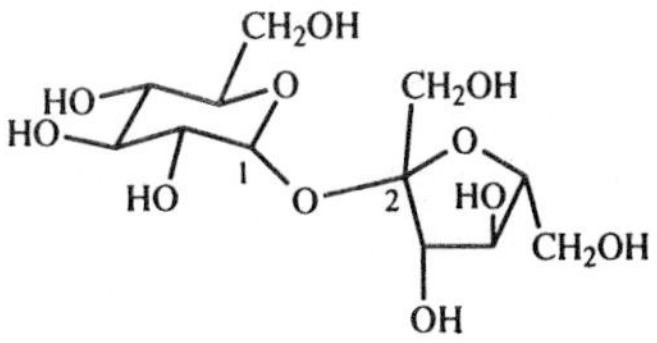

FIGURE 15.35:
The structure of
sucrose.

## POLYSACCHARIDES

The polysaccharides are composed of many carbohydrate units linked by glycosidic bonds. In this section we will examine the structure of a few biologically relevant polysaccharides that are composed of polyglucose. We begin with cellulose, a structural material in plants. Cellulose consists of many glucose units joined by β–1,4–glycosidic bonds.

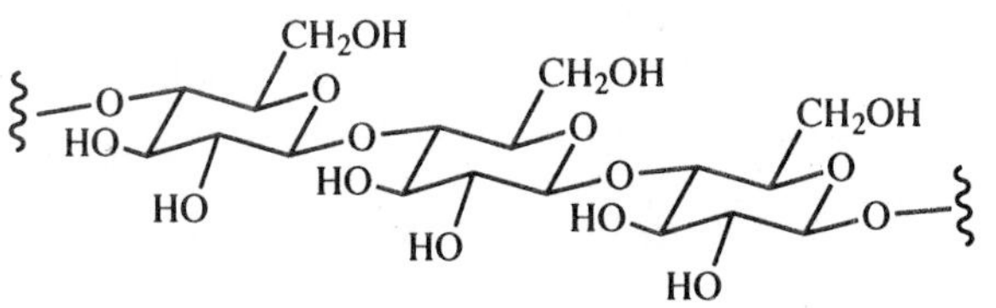

FIGURE 15.36:
A small portion of
cellulose.

Part of the structural nature of cellulose comes from its ability to form extensive networks of hydrogen bonds. It turns out that this hydrogen bond network gives rise to a well–defined **secondary structure**, a left–handed α helix.

Amylose, one of the components of starch (the other is amylopectin) differs from cellulose only in that the glycosidic bonds are α rather than β.

As you know, humans are not capable of digesting cellulose, but they can digest amylose. That is because, while we have the enzyme α–galactosidase,

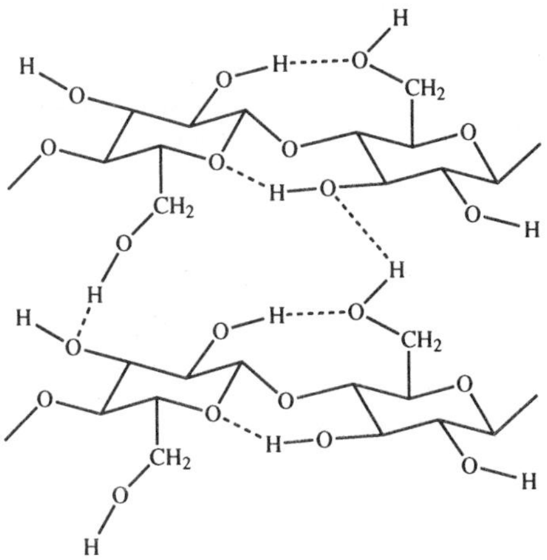

FIGURE 15.37:
Hydrogen bonding in
cellulose.

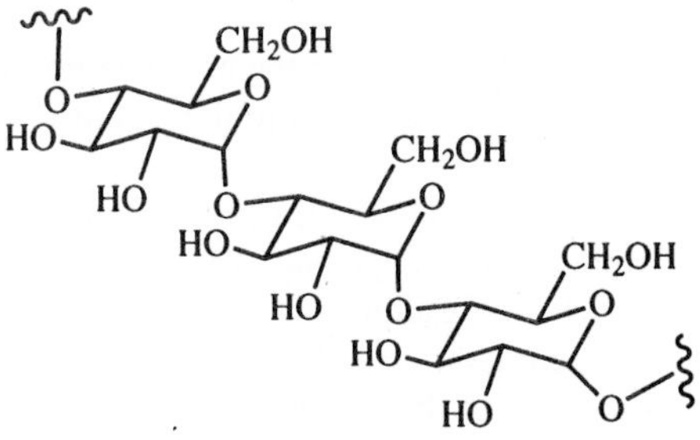

FIGURE 15.38:
A small portion of
amylose.

we lack the enzyme β–galactosidase. This means that we are capable of hydrolyzing the α–1,4–glycosidic bonds of amylose, but not the β–1,4–glycosidic bonds of cellulose. Certain bacteria that are capable of hydrolyzing the β–1,4–glycosidic bond are found in the digestive tracts of cows and termites, thus these animals are capable of utilizing the glucose contained in cellulose as a food source because of their symbiotic relationships with the bacteria.

Amylopectin is similar to amylose in that it contains units linked by α–1,4–glycosidic bonds. However, in addition to the α–1,4 bonds, additional α–1,6–glycosidic bonds cause branch formation throughout its structure.

The last polysaccharide we wish to mention is glycogen. Glycogen is a polysaccharide with a structure almost identical to amylopectin. There are two differences. First, glycogen has a higher molecular weight, and second, glycogen is much more highly branched. Thus glycogen is more complex than amylopectin. Glycogen functions as a glucose storage polymer in humans, as well as in many other animals. Glucose that is not immediately required for energy is converted into glycogen for future use.

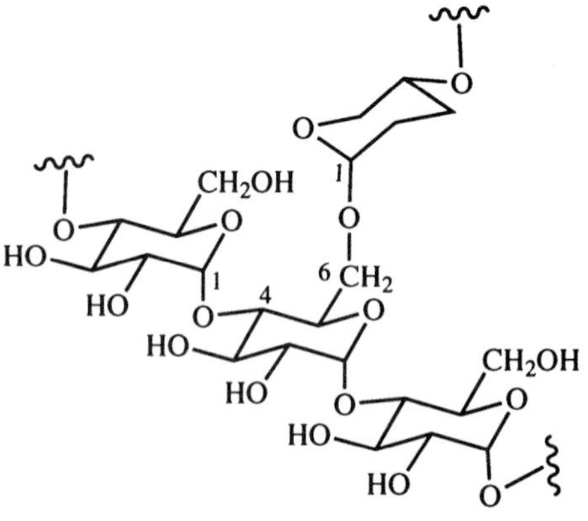

FIGURE 15.39:
A small portion of amylopectin.

## 15.11  AMINO ACID BASED BIOPOLYMERS

We now turn our attention to peptides and proteins. As we have seen in chapter 13, amino acids are the building blocks of these biopolymers. Since we have already spent time discussing these building blocks, we can get right into the study of the structure of peptides and proteins.

### THE STRUCTURE OF PEPTIDES AND PROTEINS

Peptides and proteins are chains of amino acids linked by the amide, or peptide, bond. The amide bond consists of a carbonyl carbon atom bonded to a nitrogen atom and is depicted in the figure at right.

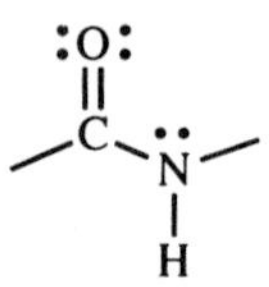

To understand the stereochemistry of the amide bond, it is important to understand that the α–carbon atom of each of the amino acid residues, the carbonyl group, and the nitrogen and hydrogen atoms of the amino group are co–planar. This can only be the case if resonance occurs involving the two structures shown in figure 15.41. One of the resonance forms is the carbonyl form — the one usually drawn of the amide bond. In this form the nitrogen, oxygen, and carbon atoms attached to the carbonyl carbon atom must be co–planar. The other resonance structure involves a carbon–nitrogen double bond. This requires that the hydrogen atom and both carbon atoms bonded to the nitrogen atom must be coplanar. The result of the resonance is that the conformation in which all six atoms are coplanar is the most stable, and therefore the preferred, configuration. Biochemists have developed a shorthand method of referring to the result of this resonance stabilization by talking of a "planar amide bond" even though, technically, a bond cannot be planar.

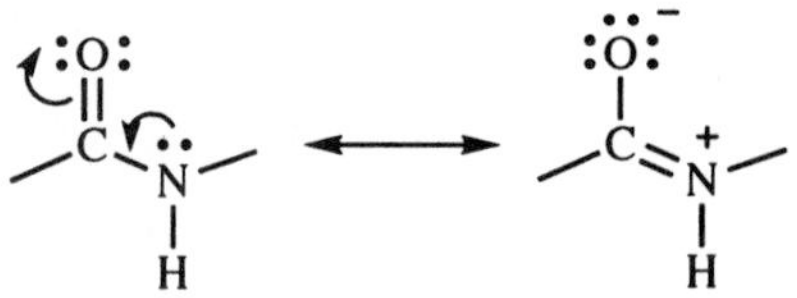

By convention, peptide chains are drawn such that the free amino end is to the left and the free carboxyl end is to the right. For example, consider the follow-

FIGURE 15.40:
The amide bond.

FIGURE 15.41:
Resonance forms of the amide bond.

ing tripeptide composed of serine, alanine, and tyrosine shown in figure 15.43. This shows serine as the amino terminus of this tripeptide, alanine as the internal amino acid, and tyrosine as the carboxy terminus of the tripeptide.

Protein chains may contain from as few as 50 amino acid residues to as many as well over 1000 amino acid residues. Thus we might expect a great variability in protein structure based not only upon the length of a particular protein, but also upon the types of amino acid residues present in a protein. Let us now take a look at some typical proteins and begin to explore the structure of proteins.

We begin by looking at hemoglobin, a rather complex oxygen transport protein that consists of four subunits and is referred to as a tetramer. The way in which these four subunits interact with each other, and thus the way they are arranged in space with respect to one another, is termed the quaternary structure of the protein. In order for a protein to have quaternary structure, it must consist of more than one amino acid chain or subunit. Quaternary structure is the highest, most complex level of protein structure.

The tertiary structure of a protein is the three dimensional arrangement of the amino acid chain. We can look at the tertiary structure of myoglobin, an oxygen storage protein that is similar in structure to a single subunit of hemoglobin.

If you look carefully at the myoglobin molecule you will note areas of well–defined structure within the overall structure of the molecule. These areas of localized three dimensional structure are referred to as the secondary structure of the protein.

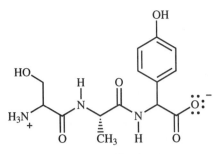

FIGURE 15.42:
A tripeptide.

FIGURE 15.43:
The structure of hemoglobin.

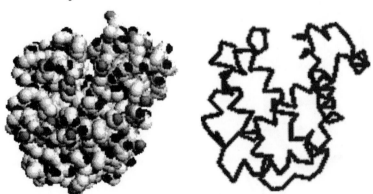

FIGURE 15.44:
The structure of myoglobin.

FIGURE 15.45:
The α-helix.

Two of the most common types of secondary structure are the α–helix and the β–pleated sheet. The right–handed α–helix arises from hydrogen bonding between the carbonyl oxygen of one amino acid residue with a hydrogen attached to the amide nitrogen four amino acid residues away. Note that a right–handed helix has a counterclockwise rotation as it rises and that the amino acid side chains point out, away from the helix. The distance covered by one complete turn, known as the rise of the helix, is approximately 5.4 angstroms.

The β–pleated sheet is the other most common secondary structure found in proteins. This secondary structure arises when the chain of amino acids runs side–by–side for some length. In this situation, hydrogen bonds can form between the carbonyl oxygen on one side and the amide hydrogen atom on the other side. If the chains are oriented such that the sides are both running from the amino terminus to the carboxy terminus, the structure is known as the parallel β–pleated sheet. If, on the other hand, one side runs from the amino terminus to the carboxy terminus and the other side runs from the carboxy terminus to the amino terminus, the structure is known as the antiparallel β–pleated sheet. The antiparallel structure is slightly more stable because the carbonyl oxygen atoms and the amide hydrogen atoms are better aligned for hydrogen bond formation.

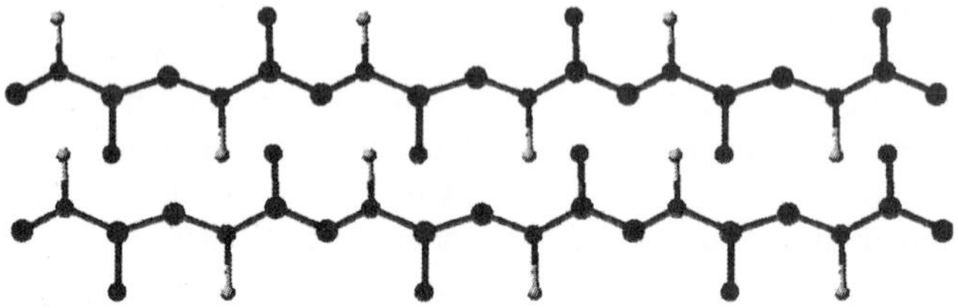

**FIGURE 15.46:**
One of the β-pleated sheet structures.

The figure below summarizes the important interactions that determine secondary and tertiary structure. These include: hydrogen bonding, hydrophobic interactions, electrostatic interactions, and disulfide bonds.

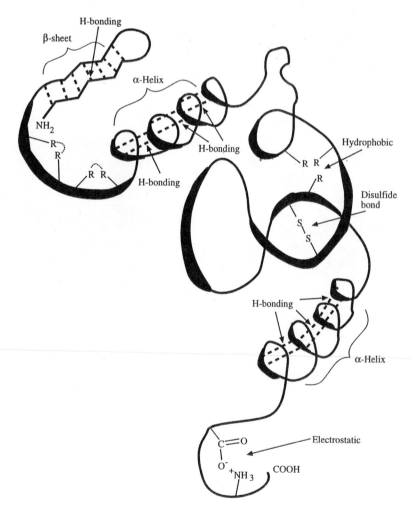

**FIGURE 15.47:**
Some important interactions for determining secondary and tertiary structure in proteins.

Finally, the primary structure is simply the sequence of amino acids that comprise an individual protein chain. It is important to note that the primary sequence of a protein is the determining factor for the shape of a given protein, and ultimately for the function of that protein.

## PEPTIDE SYNTHESIS

Let us now look at peptide synthesis and, in particular, the things that we need to be considered carefully when synthesizing peptides. Consider the combination of the two amino acids, alanine and valine, shown below. Amide bonds form between amino groups and carboxyl groups. Thus, if alanine and valine were simply mixed under the appropriate conditions, we would expect to obtain a rather messy mixture of four different dipeptides: Ala–Val, Val–Ala, Ala–Ala, and Val–Val. As you can imagine, the situation would become even more complex if we tried to make a tri– or tetrapeptide by simply mixing the amino acid monomers. That being the case, we need to rely on protecting group chemistry to have any hope of synthesizing peptides of known composition in any useful quantities.

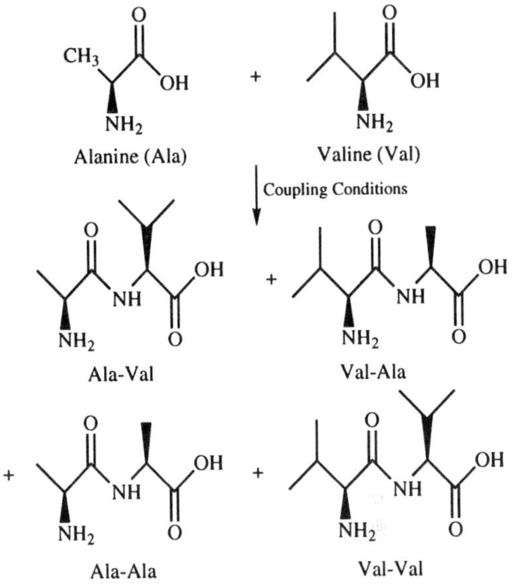

We will look first at the protecting groups for the amino group and then consider the protecting groups for the carboxylic acids. As we discussed in chapter 14, a good protecting group must be both added and removed easily and in good yield. Two nitrogen protecting groups that meet these criteria are the benzyloxycarbonyl (CBZ) group and the t–butyloxycarbonyl (BOC) group. The CBZ–group is added to an amino acid by reacting the amino acid with benzyl chloroformate (CBZ–Cl).

FIGURE 15.48: Possible dipeptide formation between two amino acids.

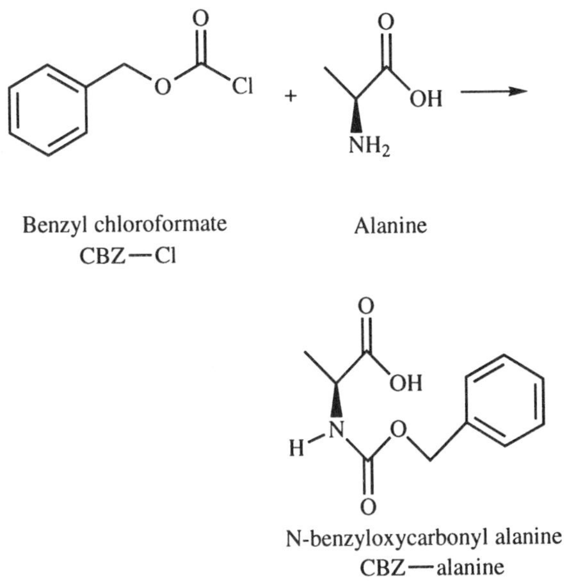

EQ. 15.26

Benzyl chloroformate
CBZ—Cl

Alanine

N-benzyloxycarbonyl alanine
CBZ—alanine

The BOC–group is added by reacting the amino acid with either *t*–butyloxycar-bonyl chloride (BOC–Cl) or di–*t*–butyldicarbonate (BOC–anhydride).

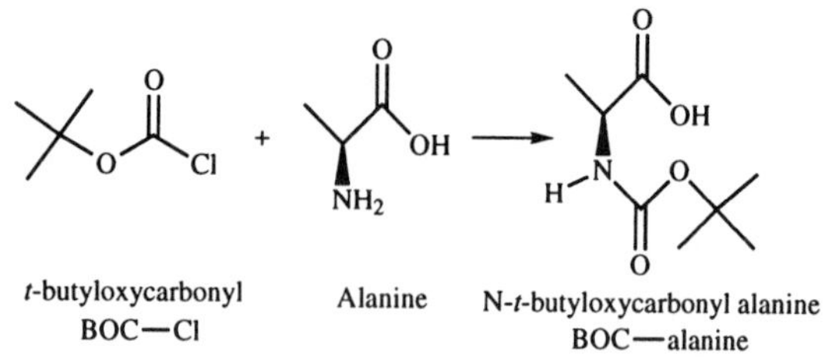

*t*-butyloxycarbonyl
BOC—Cl

Alanine

N-*t*-butyloxycarbonyl alanine
BOC—alanine

EQ. 15.27

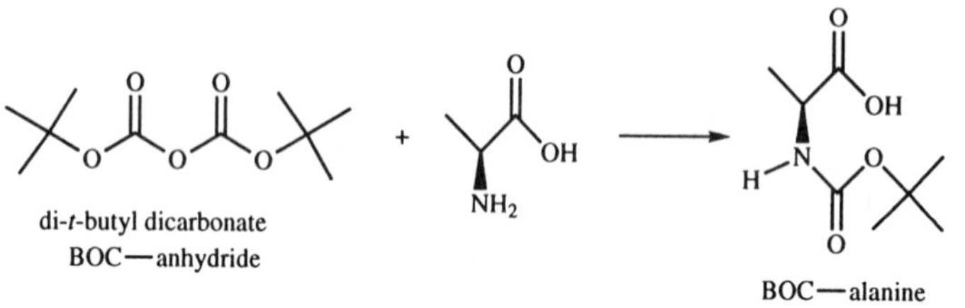

di-*t*-butyl dicarbonate
BOC—anhydride

BOC—alanine

EQ. 15.28

The CBZ–group is easily removed by hydrogenolysis in the presence of a pal-ladium catalyst.

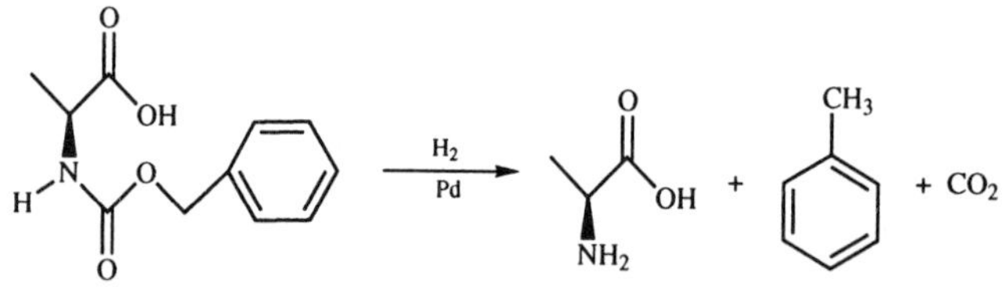

$+ CO_2$

EQ. 15.29

The BOC–group is removed by treatment with dilute HCl or trifluoroacetic acid (TFA).

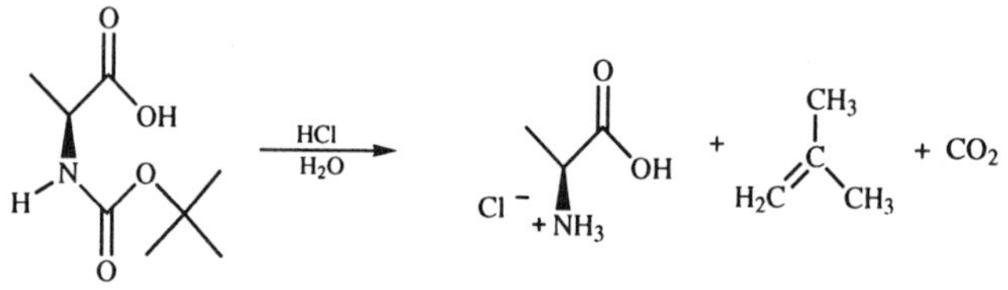

$+ CO_2$

EQ. 15.30

$+ CO_2$

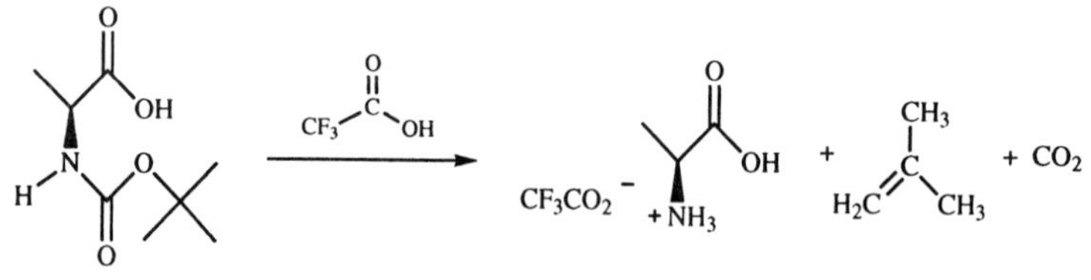

EQ. 15.31

Let us now look at some of the methods used to protect the carboxyl group. The most common type of carboxyl protecting group is the ester. Various esters including methyl, ethyl, and *t*–butyl are routinely used. Treatment of the amino acid with methanol and anhydrous HCl produces the methyl ester hydrochloride of the amino acid. The free amine must be generated by treat-ment with base prior to peptide coupling. Alternatively, the methyl ester can be prepared directly by treating the amino acid with methyl iodide and sodium carbonate. These reactions are shown in equations 15.32 and 15.33.

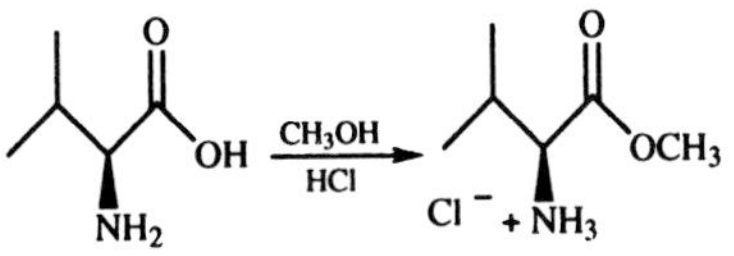

EQ. 15.32

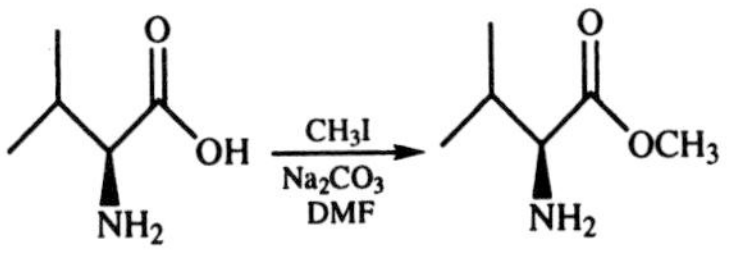

EQ. 15.33

On the other hand, ethyl ester tosylate salts are commonly prepared by refluxing the amino acid with ethanol and catalytic amount of $p$–toluenesulfonic acid, in benzene as shown in equation 15.34.

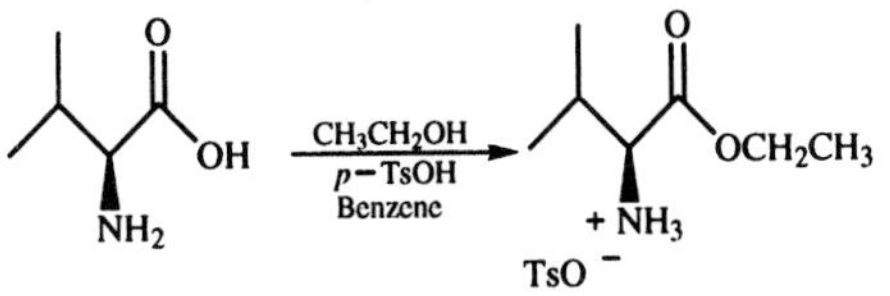

EQ. 15.34

The formation of $t$–butyl esters is somewhat more complicated. The least vigorous method requires prior protection of the amino group. After the nitrogen has been protected, reaction with $t$–butanol and dicyclohexylcarbodiimide (DCC) results in the formation of the $t$–butyl ester. The mechanism of this reaction will be discussed shortly. The amine group must, of course, be deprotected prior to the peptide coupling reaction.

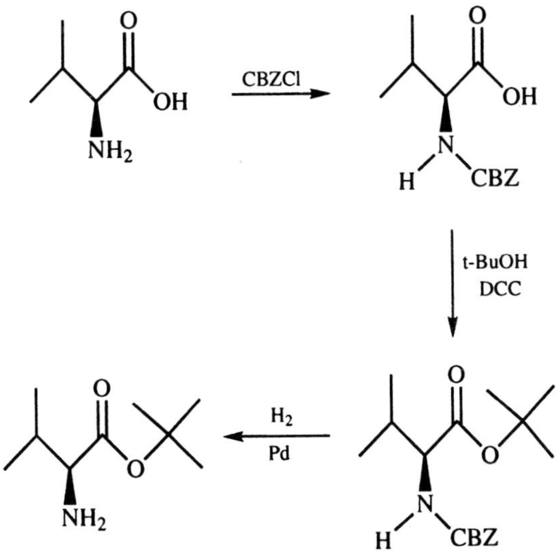

FIGURE 15.49:
The preparation of $t$-butyl amino esters.

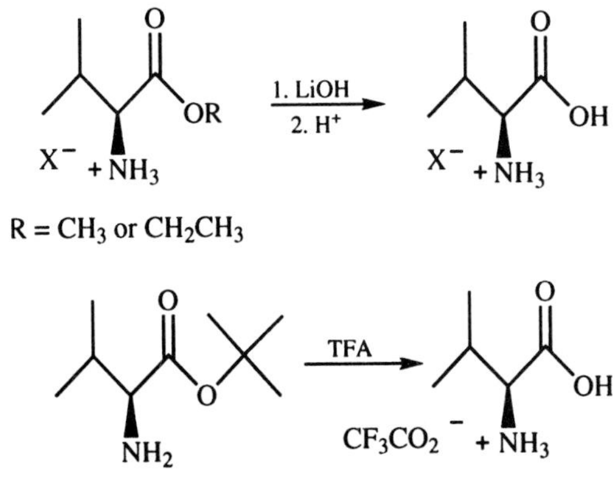

FIGURE 15.50:
Various carboxyl group deprotections.

R = CH$_3$ or CH$_2$CH$_3$

All of these carboxyl protecting groups are easily removed. Both the methyl ester and the ethyl ester are hydrolyzed by aqueous base. The $t$–butyl ester is usually cleaved with TFA (trifluoroacetic acid).

Now that we have a good understanding of amino acid protecting groups we can look at an efficient method for preparing a dipeptide. If we want to form the dipeptide Ala–Val, we will need to react an amino protected Ala with a carboxyl protected Val.  Consider reacting N–CBZ–alanine with the *t*–butyl ester of valine as shown in equation 15.35.

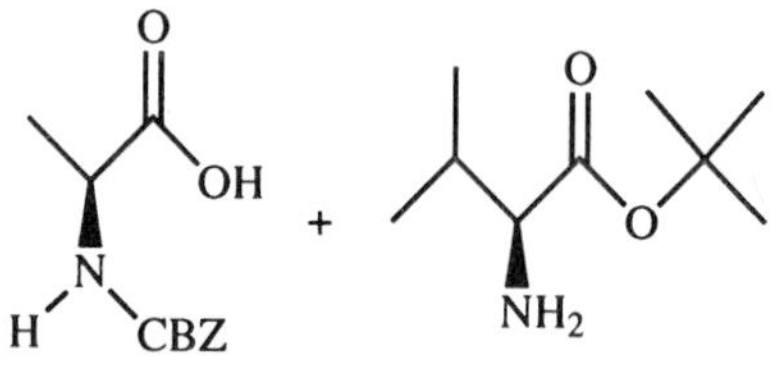

EQ. 15.35

Doing the reaction as shown would lead to very little product formation.  Even though we have taken care of the problem of undesired coupling reactions, we need to activate the carboxyl group in order to get reasonable yields of peptides.  As you know, the hydroxyl group, by itself, is a poor leaving group.  Therefore, we must turn the hydroxyl group into a good leaving group so the amide bond will form readily under mild conditions.  Two methods are commonly used; the mixed anhydride method and the DCC method.

In the mixed anhydride method, the N–protected amino acid is treated with a reagent that reacts readily with the carboxyl group to form a mixed anhydride.  Ethyl chloroformate is commonly used for this purpose.  Then the nucleophile, in this case a carboxyl protected amino acid, is added and coupling occurs.  Subsequent deprotection of the amino and carboxyl groups completes the synthesis of the dipeptide.

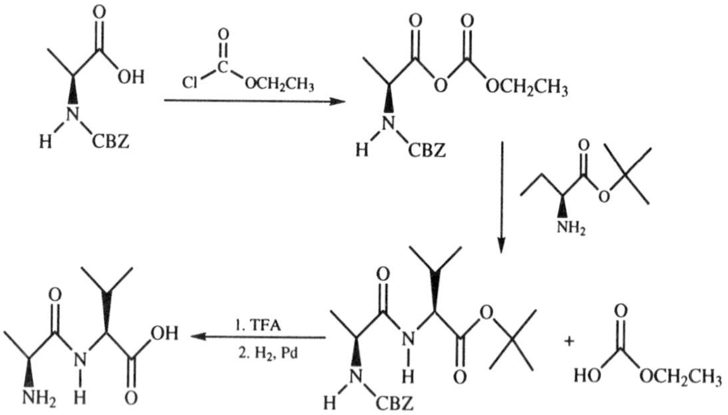

In the other method of peptide bond formation, DCC activates the carboxyl group by turning the hydroxyl group into an excellent leaving group, the stable diamide cyclohexylurea.  Again, deprotection is necessary to complete the synthesis.

FIGURE 15.51:
The mixed anhydride method for peptide synthesis.

These methods are solution phase methods for peptide synthesis and they work fairly well for preparing short peptides.  However, if a longer peptide (up to about 50 residues) is desired, or if two peptides are coupled, a much better method is the Merrifield solid–phase peptide synthesis.  This method is now automated and instruments can be purchased that do most of the work.  This method was developed by R. Bruce Merrifield (1921– ) who was awarded the Nobel Prize in Chemistry for his work in this area in 1984.  The method uses a resin as the solid support.  The carboxyl end of the desired peptide is attached to the resin and amino acid residues are added one–at–a–time using standard peptide chemistry.

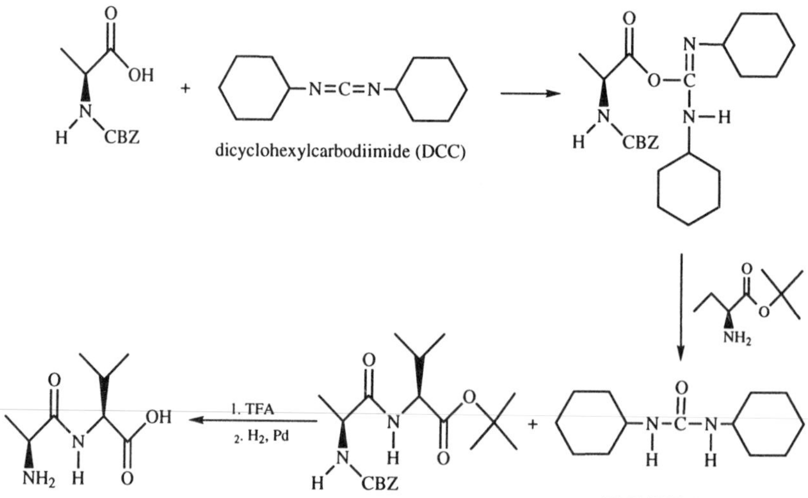

dicyclohexylcarbodiimide (DCC)

FIGURE 15.52:
The DCC method of peptide synthesis.

The use of the solid support allows all impurities to be rinsed out of the reaction mixture between each peptide coupling reaction. This results in higher yield couplings and a more pure end product. After addition of the last amino acid in the sequence, the completed protein is simply detached from the resin by standard methods. Figure 15.53 illustrates one cycle of this process.

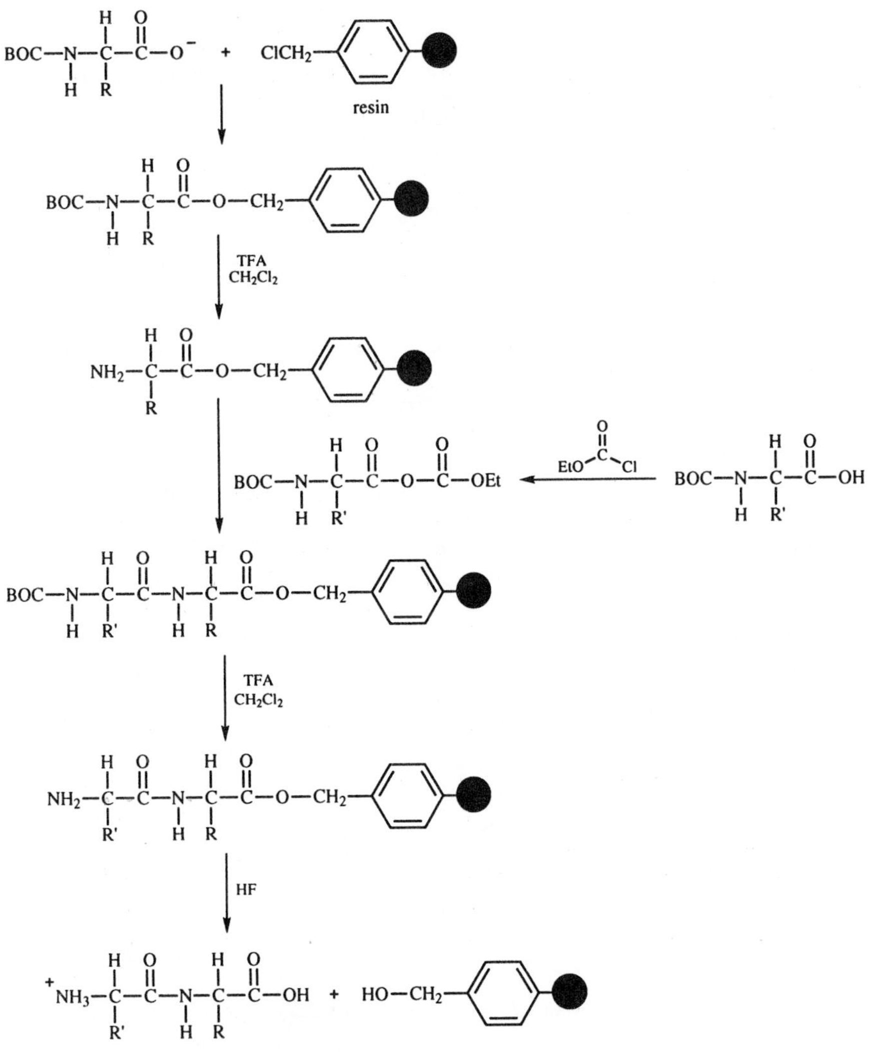

FIGURE 15.53:
Solid phase peptide synthesis.

## 15.12   NUCLEIC ACID BASED BIOPOLYMERS

### THE NUCLEIC ACID BASES: AROMATIC HETEROCYCLES REVISITED

As you will recall from our study of aromatic compounds, the nucleic acid bases are all aromatic heterocycles. This general class of compounds consists of the pyrimidine bases, and the purine bases. Each of these has the basic skeletal structure of the simplest compound in the series: pyrimidine and purine, respectively.

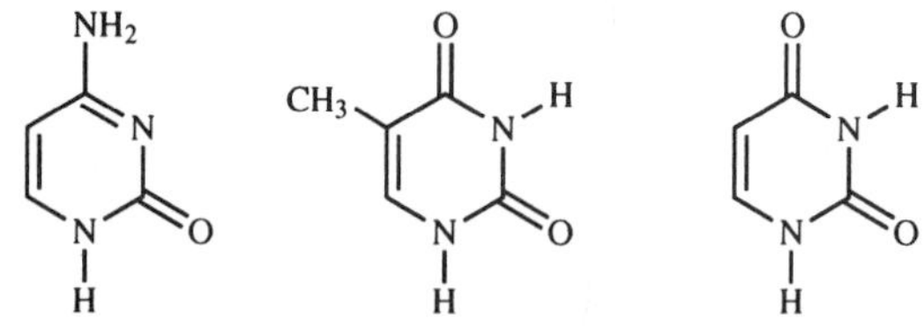

**Cytosine** **Thymine** **Uracil** **Adenine** **Guanine**

FIGURE 15.54:
The pyrimidine bases.

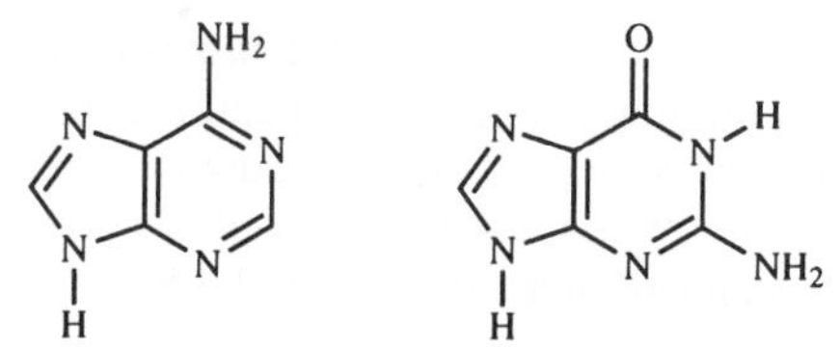

FIGURE 15.55:
The purine bases.

It is important to note that these bases all have a variety of *tautomers*. Recall that tautomeric structures are interconverted by intramolecular proton transfer reactions and exist in equilibrium with each other. We last saw tautomeric forms in chapter 11 when we noted that simple carbonyl compounds containing protons $\alpha$ to the carbonyl group can tautomerize between the keto form and the enol form. This same type of tautomerism is also possible with the nucleic acid bases. The tautomers of thymine and guanine are shown below.

It is interesting to note that Watson and Crick failed to recognize the possibility of tautomerism for a long time. This likely delayed their progress on the determination of the structure of DNA.

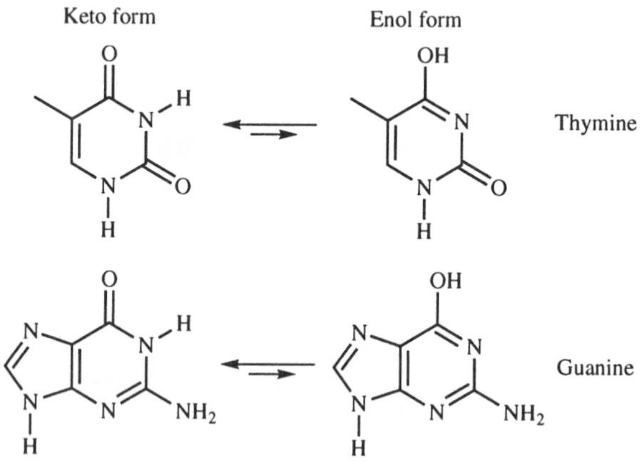

**NUCLEOSIDES AND NUCLEOTIDES: WHICH IS WHICH?**

As we saw in section 15.10, alcohols react with sugars at the anomeric position to form glycosidic bonds. In that case the product is an acetal. Amines can also react with sugars at the anomeric position to form a class of compounds known as *hemiaminals*. When the amine is a nucleic acid base and the sugar is either ribose or 2–deoxyribose, the term *nucleoside* is applied to designate the product of the reaction. A nucleoside is formed by the reaction of N–9 of a purine or N–1 of a pyrimidine with the anomeric carbon on ribose of 2–deoxyribose. The common nucleosides are shown in figure 15.57.

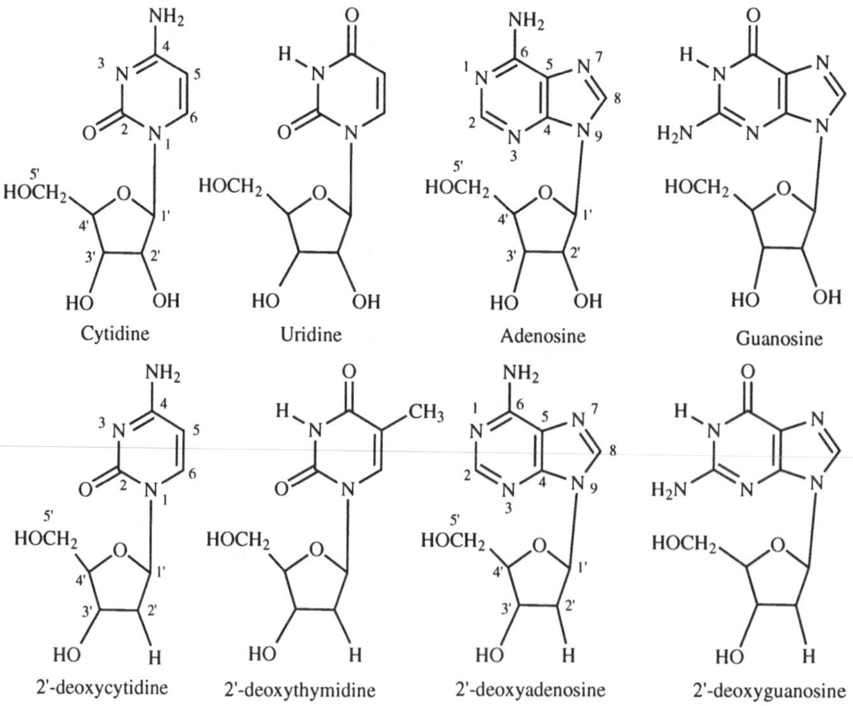

Keto form   Enol form

Thymine

Guanine

FIGURE 15.56:
Tautomerism of
nucleic acid bases.

**Cytidine** **Uridine** **Adenosine** **Guanosine**

FIGURE 15.57:
Nucleosides and
deoxynucleosides.

**2'-deoxycytidine** **2'-deoxythymidine** **2'-deoxyadenosine** **2'-deoxyguanosine**

If one of the hydroxyl groups on the sugar of the nucleoside is esterified with phosphoric acid, the structure is known as a **nucleotide**. The **ribonucleotides** are esterified at the 5′–OH group of ribose while the **deoxyribonucleotides** are esterified at the 5′–OH group of deoxyribose. A nucleotide can also be called nucleoside–5′–monophosphate. The structures of the nucleotides are shown below.

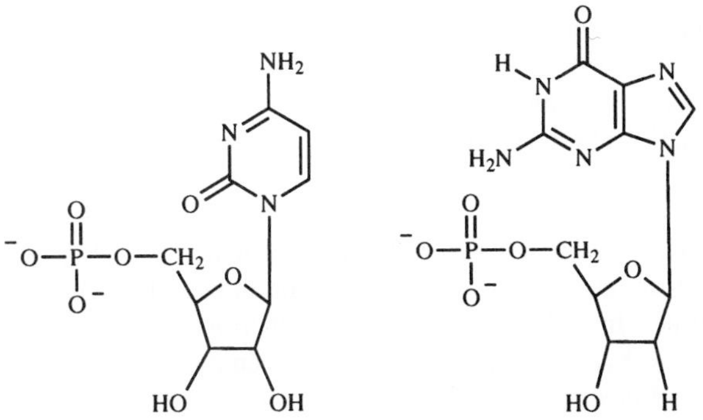

### POLYNUCLEOTIDE SYNTHESIS: AN ELEMENTARY INTRODUCTION

Let us take a brief look at the fundamentals of polynucleotide synthesis by considering a polynucleotide consisting of N residues. If we want to add another nucleotide to the chain you might think that some sort of dehydration reaction would occur to simply add the nucleoside–5′–monophosphate monomer to the already existing polynucleotide chain. This reaction is, however, not thermodynamically favorable. In order to add an additional nucleotide in a thermodynamically favorable way, the polynucleotide chain must react with a nucleoside–5′–triphosphate as shown in the reaction scheme in figure 15.59. The overall $\Delta G°'$ for the process is approximately –6 kJ/mol.

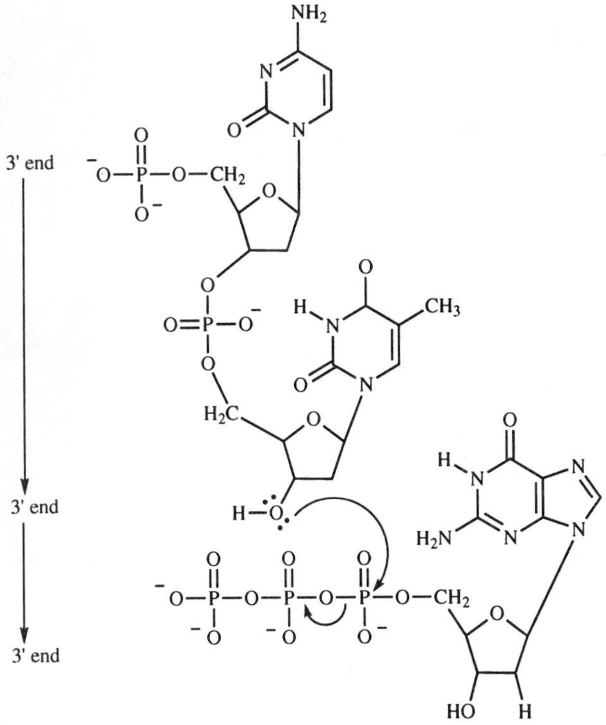

### NUCLEIC ACIDS: THIS TIME A DOUBLE HELIX

Just as peptides and proteins have primary and secondary structure, and directionality, so do the nucleic acids. The primary structure of a nucleic acid is, of course, the sequence of nucleotides that comprise any individual nucleic acid. As we saw, by convention the directionality of proteins is considered to be from the free amino to the free carboxyl. That of nucleic acids is considered to be in the 3′ to 5′ direction. This means that the chain of nucleotides contains an unreacted hydroxyl group at the 3′–end of the chain and a free phosphate group at the 5′–end of the chain. Please note, however, that by convention the sequence of nucleotides, in a given chain, is usually written with the 5′–end at the left and the 3′–end at the right. For example, consider the following sequence of nucleotides containing the bases adenine (A), cytosine (C), thymine (T), guanine (G), adenine (A), and cytosine (C) in that order

ACTGAC

This particular short segment of nucleic acid chain has a free phosphate group on a 2′–deoxyadenosine–5′–monophosphate unit at the 5′–end and a free 3′–hydroxyl group on the 2′–deoxycytidine–5′–monophosphate unit at the other end.

In 1953 James Watson and Francis Crick published their landmark paper in which they proposed the double helix as the secondary structure of DNA. The

Cytidine-5'-monophosphate

2'-deoxyguanosine-5'-monophosphate

FIGURE 15.58:
Ribonucleotides and deoxyribonucleotides.

FIGURE 15.59:
Polynucleotide synthesis.

double helix consists of two strands of DNA held together by hydrogen bonds between pairs of nucleic acid bases and by van der Waals interactions between the bases on either side of any particular base.

The hydrogen bonds that hold the nucleic acid strands together form between a pyrimidine and a purine. In DNA the hydrogen bonds form between thymine and adenine or between cytosine and guanine. In RNA thymine is replaced by uracil.

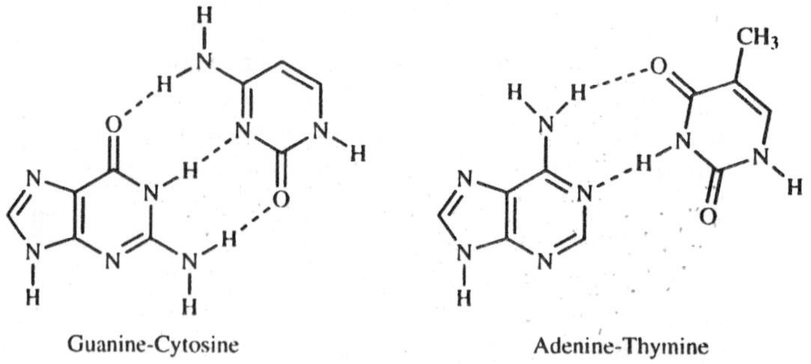

You should note that the base pairs occupy the inside of the double helix while the sugar–phosphate backbone is along the outside of the double helix. This helps to hold the structure together in the aqueous environment of the cell. The following rendering of DNA highlights some of these aspects of structure.

This concludes our look at the synthesis of polymers that are essential for life. In the next chapter we shall discuss some of the key chemical reactions in which these polymers participate.

Guanine-Cytosine

Adenine-Thymine

FIGURE 15.60:
Base pairing in DNA.

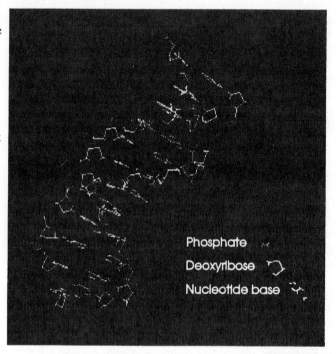

Phosphate

Deoxyribose

Nucleotide base

FIGURE 15.61:
Section of DNA.

**Addition polymer**: a polymer for which the chain grows by addition to a multiple bond in the monomer.

**Aldohexose**: a six–carbon carbohydrate that contains an aldehyde as the most oxidized functional group.

**Aldopentose**: a five–carbon carbohydrate that contains an aldehyde as the most oxidized functional group.

**Aldotetrose**: a four–carbon carbohydrate that contains an aldehyde as the most oxidized functional group.

**Aldotriose**: a three–carbon carbohydrate that contains an aldehyde as the most oxidized functional group.

**α–helix**: a common secondary structure found in proteins. The α helix forms as a result of hydrogen bonding between a carbonyl oxygen and an amide hydrogen atom four amino acid residues away.

**Anomeric carbon**: the hemiacetal carbon of carbohydrates.

**Anomers**: stereoisomers of carbohydrates that differ from one another only in configuration at the anomeric carbon.

**Atactic**: a configuration of a polymer in which the substituents are randomly arranged both above and below the plane formed by the backbone.

**β–pleated sheet**: a common secondary structure found in proteins. The β pleated sheet results from hydrogen bonding between sections of the amino acid running alongside one another.

**Branched polymer**: a polymer chain with side chains randomly attached to the main backbone.

**Condensation polymer**: a polymer in which the chain grows by reaction of functional groups on the monomers, usually by splitting out a small molecule (*e.g.*, water) as the monomers link to form the growing chain.

**Copolymer**: an addition polymer synthesized from two or more distinct monomers.

**Crosslinked polymer**: a polymer in which branches from the backbone fuse to other backbones, resulting in a rigid, three dimensional structure (see thermosetting.)

**Crystallinity**: the degree to which sections of polymer chains align in a parallel fashion. Crystallinity increases the strength of the bulk polymer.

**Deoxyribonucleotides**: the nucleotides present in DNA. The are based on the 2–deoxyribose sugar unit.

**Disaccharide**: a molecule consisting of two sugar units joined by a glycosidic bond.

**Furanose**: a five–membered ring present in some carbohydrates. The name furanose is derived from the aromatic heterocycle furan.

**Glycosides**: the acetals of carbohydrates.

**Glycosidic bonds**: the bond that attaches a sugar unit to another moiety. Typically this is the bond that attaches two sugar molecules to one another.

**Hemiaminal:** a functional group in which a carbon atom is singly bonded to both nitrogen and oxygen.

**Isotactic**: a configuration of a polymer in which the substituents are alternately arranged above and below the plane formed by the backbone.

**Ketohexose**: a six–carbon carbohydrate that contains a ketone as the most oxidized functional group.

**Linear polymer**: a polymer chain with no branches.

**Nucleoside**: an aminal formed by the reaction of N–9 of a purine or N–1 of a pyrimidine with the anomeric carbon on ribose or 2–deoxyribose.

**Nucleotide**: a nucleoside in which one of the hydroxyl groups (usually the hydroxyl group on C–5) is esterified with phosphoric acid.

**Peptide**: a relatively short sequence of amino acids linked by amide bonds.

**Polysaccharide**: a group of many sugar units joined by glycosidic bonds.

**Primary structure**: the primary structure of any biopolymer is the sequence of monomer units that comprise the particular polymer. For example, in proteins, the primary structure is the chain of amino acid residues.

**Proteins**: a long chain of amino acids linked by amide bonds. Proteins are both structural (*e.g.*, keratin) and functional (*e.g.*, acetylcholinesterase) in biological systems.

**Pyranose**: a six–membered ring present in some carbohydrates. The name pyranose is derived from the heterocycle pyran.

**Quaternary structure**: the three dimensional structure formed by two or more protein subunits.

**Ribonucleotides**: the nucleotides found in RNA.

**Secondary structure**: the local spatial arrangement of a biopolymer. Examples include the α helix and the β pleated sheet.

**Syndiotactic**: a configuration of a polymer in which the substituents are all above (or below) the plane formed by the backbone.

**Tautomers**: compounds, related by intramolecular proton transfer reactions, in equilibrium with each other.

**Tertiary structure**: the overall three dimensional structure of a protein containing only one subunit.

**Thermoplastic**: a polymer that softens and flows when heated.

**Thermosetting**: a polymer that degrades and decomposes when heated.

**Vulcanization**: a process of connecting two polymer chains (*e.g.*, natural rubber) by adding a crosslinking agent (*e.g.*, sulfur).

1. Draw the structure of the repeating unit in an addition polymer that is made from the monomer propylene.

2. What is the degree of polymerization (n) for polypropylene that has a molecular weight of 100,000?

3. Draw the structure of the repeating unit in an addition polymer that is made from the monomer vinyl acetate.

4. What is the degree of polymerization (n) for polyvinylacetate that has a molecular weight of 100,000?

5. Draw the structure of the repeating unit in a condensation polymer that is made from 1,4–butanedioic acid and ethylene glycol.

6. What is the degree of polymerization (n) for the polymer drawn in #5, assuming that it has a molecular weight of 100,000?

7. What is the expected %C, %H, %O in the polymer shown in #5?

8. Draw the structure of the repeating unit in a condensation polymer that is made from 1,5–pentanediamine and oxalic acid.

9. What is the degree of polymerization (n) for the polymer drawn in #8, assuming that it has a molecular weight of 100,000?

10. What is the %C, %H, %O and %N in the polymer shown in #8?

11. Draw the structure(s) and give the name(s) of the monomer(s) used to make the polymer shown below:

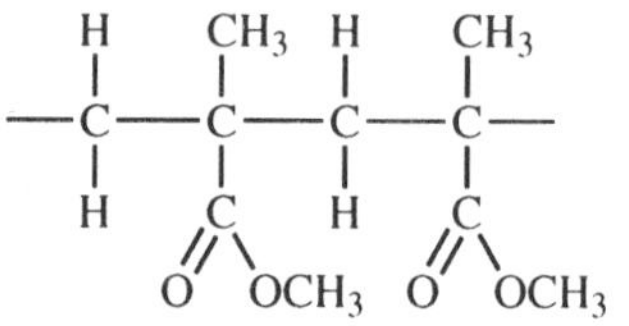

12. Draw the structure(s) and give the name(s) of the monomer(s) used to make the polymer shown below:

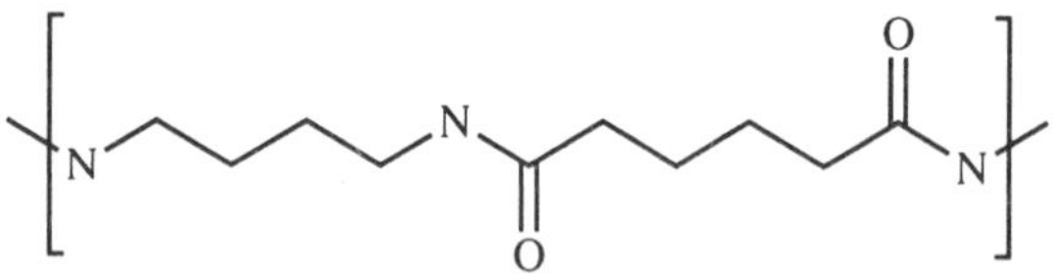

13. Kevlar®, a bulletproof plastic, has the structure shown below. Draw the structure(s) and give the name(s) of the monomer(s) used to make Kevlar®.

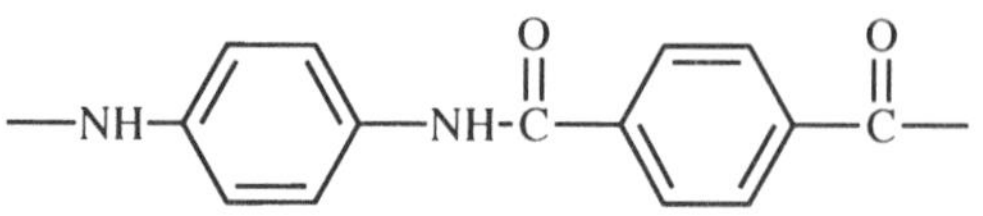

14. The strength of the material is critical for a particular application. Would you use a linear or a branched polymer? Explain your reasoning with reference to the structure of the polymer chain.

15. Low density and flexibility are critical for a particular application,. Would you use a linear or a branched polymer? Explain your reasoning with reference to the structure of the polymer chain.

16. In making a condensation polymer from formaldehde and phenol, the phenol is most reactive at positions ortho and para to the hydroxyl group. Explain this in terms of a likely reaction mechanism.

17. A spontaneous polymerization reaction converts numerous small monomer molecules into fewer, significantly larger molecules. What can be said about the probable entropy and enthalpy changes for spontaneous polymerization reactions?

18. Polyesters are formed *via* a condensation reaction wherein a small molecule such as water is produced along with the polymer. In commercial operations, the reaction mixture is often heated and water distilled off from the mixture. Explain the reason for this.

19. In most polymer applications, chains with a higher degree of polymerization "n" have more desirable characteristics than shorter length chains. Polymeric addition reactions are often started by adding a certain amount of initiator "I". If "M" represents the concentration of monomer in the polymerization reaction, the expression below is frequently found to be true. First, explain in a few words what the influence of increasing the amount of initiator is on the average polymeric chain length that results. Then explain the kinetic reason for this effect.

$$n = k[M][I]^{-1/2}$$

20. Cotton fabrics have a much different "feel" to the body than do nylon fabrics. Part of this has to do with the different water–absorbing characteristics of each. Does cotton or nylon absorb water more readily? Explain why based upon a comparison of the structure of the repeating unit.

21. Melamine® is an aromatic cyclic trimer of urea formed in a condensation reaction. Show the overall reaction that forms melamine from urea.

22. Formica® is synthesized from melamine and formaldehyde, reacting in a 1:3 ratio. Show the structure of a representative portion of the product, and comment on the product's properties based upon this molecular structure.

23. Draw structures of the following
   a. the Fischer projection of L-glucose
   b. an epimer of D-glucose
   c. the furanose form of D-fructose
   d. the pyranose form of D-galactose
   e. two D-galactose units linked by an α-1,4-glycosidic bond
   f. a D-glucose molecule linked to a molecule of D-galactose by a β–1,4–glycosidic bond

24. In your own word define:
    a. primary protein structure
    b. secondary protein structure
    c. tertiary protein structure
    d. quaternary protein structure

25. Draw out all the reactions for the synthesis of the tripeptide
Ser–Gly–Asp starting from serine, glycine, and aspartic acid.

26. The guanine-cytosine base pair is more stable than the adenine thymine
base pair. What reason(s) can you offer for this extra stability?

# CHAPTER sixteen

# 16

# BIOCHEMICAL
# Catalysis
PUTTING IT ALL TOGETHER

BIOCHEMICAL CATALYSIS

# Putting it all together

## 16.1  INTRODUCTION TO ENZYME CATALYZED REACTIONS

In this chapter we want to begin to understand how enzymes exert their catalytic effects.  That is, how does an enzyme help to convert a certain starting material, called the **substrate**, into the desired product(s).  First, we will review reaction energetics.  Then we will apply our knowledge of energetics to biochemical systems.  Let us begin by examining the conversion of a substrate, S, into a product, P.

$$S \rightarrow P$$

EQ. 16.1

If we assume that the conversion of S to P is thermodynamically favored, that is that the reaction has a negative $\Delta G$, a simple progress of reaction diagram for this reaction would probably look something like the one in figure 16.1:

As we saw in section 6.5, while thermodynamically favored, the conversion of substrate to products may not proceed at a measurable rate because the **activation energy barrier** is just too high for the reaction to occur.  This is often the case for biochemical reactions.  How do biochemical systems solve this problem?  The answer is, of course, that biochemical systems contain enzymes.

An enzyme is a protein molecule whose purpose is to serve as a biochemical **catalyst**.  Remember that catalysts are substances that participate in reactions by altering the reaction pathway (mechanism) between starting material and product(s).  As a catalyst, the enzyme responsible for the conversion of substrate into product works by lowering the activation energy barrier separating the two compounds.  A simple progress of reaction diagram that compares the enzyme–catalyzed reaction with the uncatalyzed reaction looks something like figure 16.2:

It is easily seen from this diagram that the activation energy for the enzyme–catalyzed reaction is substantially less than that of the uncatalyzed reaction.  Please note that the addition of the enzyme does not alter the overall free energy change for the reaction.

Let us now examine, in a bit more detail, how the addition of the enzyme alters the path taken from substrate to product.  The general reaction scheme for an enzyme–catalyzed process is:

$$E + S \longrightarrow E{\cdot}S \longrightarrow E + P$$

EQ. 16.2

In the enzyme–catalyzed reaction, the enzyme and the substrate first interact to form the **enzyme—substrate complex** (E·S).  In the second step of the general scheme, the product and free enzyme are formed.  A bit of activation energy is

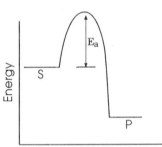

FIGURE 16.1:
Energy vs. extent of reaction for an exothermic reaction.

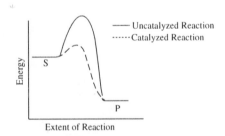

FIGURE 16.2:
Simple energy profiles for an enzyme catalyzed reaction compared to the corresponding uncatalyzed reaction.

464    CHAPTER SIXTEEN    BIOCHEMICAL CATALYSIS

required for the enzyme — substrate interaction step. However, typically the largest amount of the total activation energy for the process is needed to form the product(s) and free enzyme from the enzyme–substrate complex. The progress of reaction diagram for the process is shown below.

Note that the overall activation energy is lower for the enzyme–catalyzed reaction than for the reaction in the absence of enzyme.

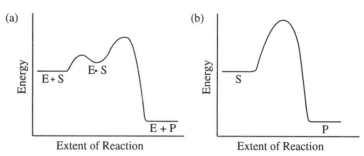

FIGURE 16.3:
a) energy profile for an enzyme catalyzed reaction, and b) energy profile for the same reaction in the absence of enzyme.

## 16.2 HOW DO ENZYMES EXERT THEIR CATALYTIC POWER?

Now that we understand that enzymes work by altering the pathway from substrates to products in such a way to reduce the activation energy for the reaction, we turn our attention to trying to understand what methods enzymes use to accomplish this task. There are several factors that are employed by enzymes that result in decreased activation energy. We shall list some of these in this section and then look at several specific mechanisms to see how these work in concert with one another. Some of the factors enzymes use to exert their remarkable catalytic effects include:

•**Orientation and Proximity Effects**: The enzyme orients the substrate close to, and in the correct orientation for reaction with, other substrates or residues within the enzyme.

•**Acid-Base Catalysis**: Certain amino acid residues (*e.g.*, histidine, aspartic acid, glutamic acid, and lysine) are capable of proton donation (acid catalysis) or proton abstraction (base catalysis).

•**Covalent Catalysis**: Certain other amino acid residues (*e.g.,* serine, cysteine, and lysine) actually form covalent bonds with substrates during the conversion of substrate into product.

•Stablilization of Intermediates: This occurs because certain **amino acid residues** are capable of stabilizing intermediates through intermolecular interactions, chiefly hydrogen bonding. By stabilizing the intermediate, the transition state leading to that intermediate is also stabilized (energy decreased) and the overall activation energy for the process is reduced.

In the next sections we will look at the mechanisms of three different enzymes in an effort to understand how these factors work together. It is important to understand that these are representative of many varied enzymatic mechanisms and that not every enzyme will utilize all of these factors.

# 16.3   THE CATALYTIC MECHANISM OF TRIOSE PHOSPHATE ISOMERASE

Triose phosphate **isomerase** is the enzyme responsible for catalyzing the inter–conversion of dihydroxyacetone phosphate and glyceraldehyde–3–phosphate as part of the glycolytic pathway.  The structures of these two compounds are given in figure 16.4. Triose phosphate isomerase is remarkably efficient and works by interconverting the ketone and the aldehyde functional groups *via* an **enediol** intermediate.

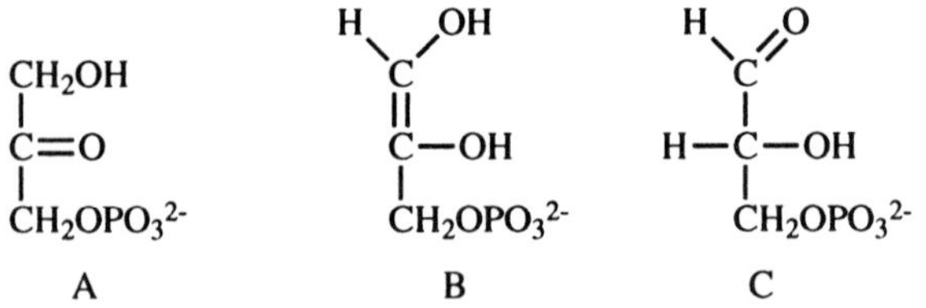

The important amino acids at the **active site** of triose phosphate isomerase are glutamate and histidine, both of which function as acid-base catalysts.  We will examine this conversion going in the direction from dihydroxyacetone phosphate into glyceraldehyde–3–phosphate.  In the first step of this process glutamate abstracts a proton from the dihydroxyacetone phosphate resulting in the formation of an **enediolate**.  Here, the glutamate acts as a base catalyst.

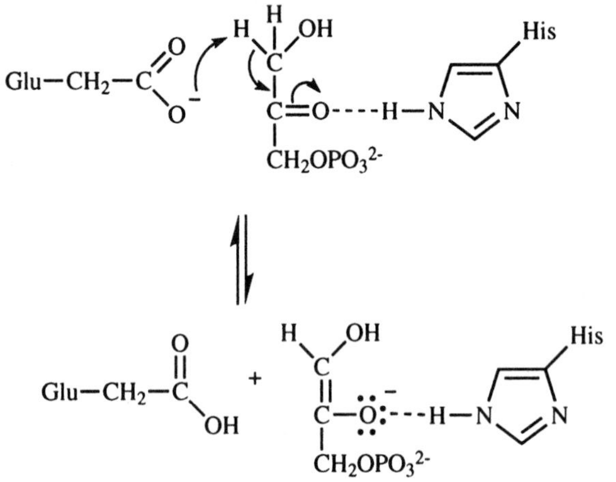

A

$$CH_2OH$$
$$C=O$$
$$CH_2OPO_3^{2-}$$

B

C

FIGURE 16.4:
The structure of a) dihydroxyacetone phosphate, b) the enediol intermediate, and c) glyceralde-hyde-3-phosphate.

EQ. 16.3

In the next step, the enediolate abstracts a proton from histidine.  Here, histidine acts as an acid catalyst.

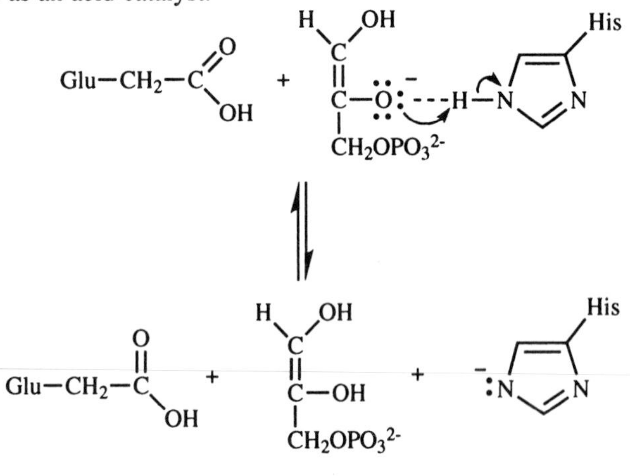

EQ. 16.4

Next, the histidine, acting as a base catalyst, abstracts the hydroxylic proton from the **enediol** intermediate.  This results in the formation of a second enediolate.

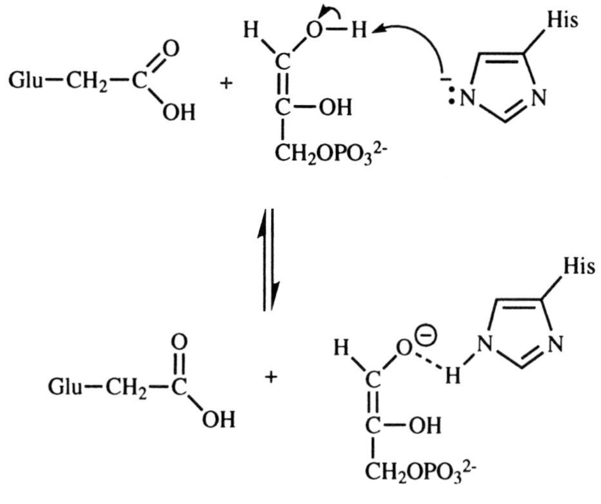

EQ. 16.5

Glu—CH$_2$—C(=O)OH  +  [enediol: H—C—O—H ... :N imidazole His, C—OH, CH$_2$OPO$_3^{2-}$]

$$\rightleftharpoons$$

Glu—CH$_2$—C(=O)OH  +  [H—C—O$^{\ominus}$ ... H—N imidazole His, C—OH, CH$_2$OPO$_3^{2-}$]

Finally, the glutamate donates a proton to the enediolate and the product, glyceraldehyde–3–phosphate is formed. In this step the glutamate acts as an acid catalyst.

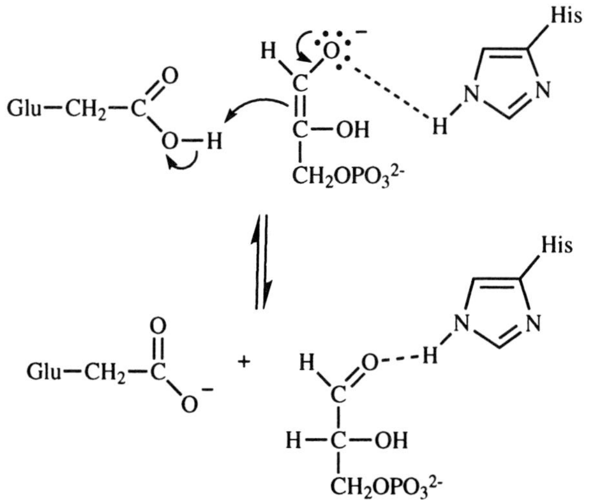

EQ. 16.6

Glu—CH$_2$—C(=O)O—H  +  [H—C=(:O:$^-$) ... His imidazole H—N, C—OH, CH$_2$OPO$_3^{2-}$]

$$\rightleftharpoons$$

Glu—CH$_2$—C(=O)O$^-$  +  [H—C—O---H—N imidazole His, H—C—OH, CH$_2$OPO$_3^{2-}$]

You should note that each step of the reaction is reversible so the enzyme is also able to catalyze the conversion of glyceraldehyde-3-phosphate into dihydroxyacetone phosphate. Thus the direction of the equilibrium is, in part, dependent upon the relative concentrations of the two compounds.

## 16.4  THE CATALYTIC MECHANISM OF ALDOLASE

We next turn our attention to the mechanism of aldolase. Aldolase is the enzyme responsible for catalyzing the conversion of fructose–1,6–bisphosphate into dihydroxyacetone phosphate and glyceraldehyde–3–phosphate. This reaction is also part of the **glycolytic pathway**.

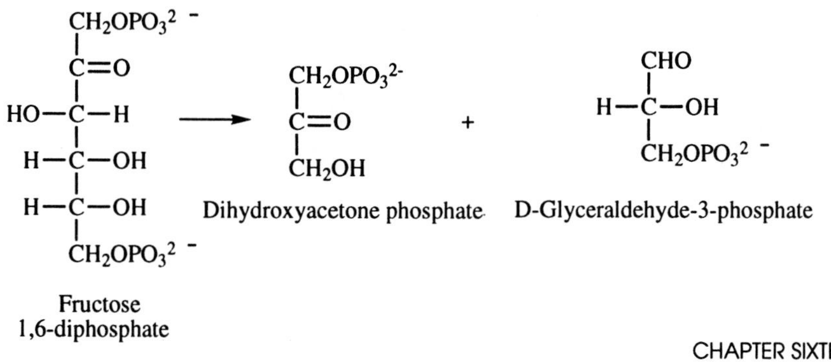

CH$_2$OPO$_3^{2-}$
|
C=O
|
HO—C—H                    CH$_2$OPO$_3^{2-}$              CHO
|                 $\longrightarrow$   |            +      |
H—C—OH                    C=O                          H—C—OH
|                         |                            CH$_2$OPO$_3^{2-}$
H—C—OH                    CH$_2$OH
|
CH$_2$OPO$_3^{2-}$      Dihydroxyacetone phosphate    D-Glyceraldehyde-3-phosphate

Fructose
1,6-diphosphate

**FIGURE 16.5:**
The aldolase catalyzed conversion of fructose-1,6-bisphosphate into dihydroxyacetone phosphate and glyceraldehyde-3-phosphate.

In section 11.14 you learned that the aldol condensation results in the formation of a β–hydroxy carbonyl compound. You were also introduced to the retro–aldol reaction in section 11.15. The retro–aldol condensation is the transformation of a β-hydroxy carbonyl compound back into the corresponding carbonyl compounds from which the aldol was derived. In the present case these would be dihydroxyacetone phosphate and glyceraldehyde–3–phosphate. If we look carefully at the structure of fructose–1,6–bisphosphate, we see the β–hydroxy ketone functionality typical of an aldol condensation product.

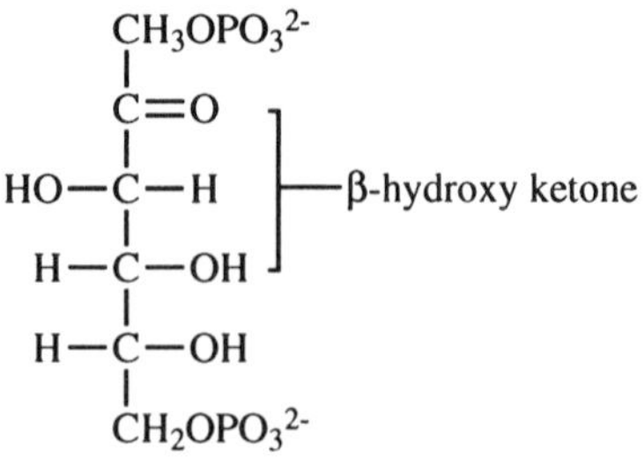

The important amino acid residues in the active site of aldolase are lysine, which acts as a covalent catalyst, and tyrosine which acts at different times in the mechanism as both a base and an acid catalyst.

The first step in this mechanism involves nucleophilic attack on the ketone portion of fructose–1,6–bisphosphate by the terminal amino group from lysine. As we have seen, primary amines react with ketones to form **imines**, otherwise known as **Schiff bases**, after loss of a water molecule. This results in the covalent attachment of the substrate to the enzyme via the **iminium linkage**.

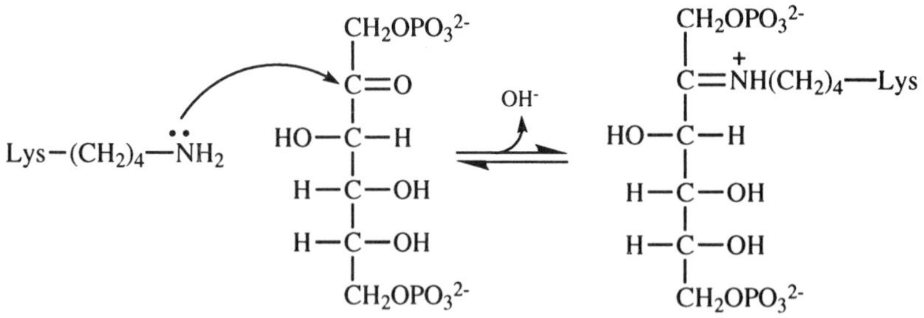

FIGURE 16.6: Fructose-1,6-bisphosphate with emphasis on the β-hydroxy ketone.

EQ. 16.7

In the next step, the phenolate anion of tyrosine removes the hydroxylic proton from the alcohol in the β position. In so doing, the C3–C4 bond is broken. This results in formation of glyceraldehyde–3–phosphate. It is this step in the mechanism that is the actual retro–aldol reaction. The enamine of dihydroxyacetone phosphate remains bound to the enzyme.

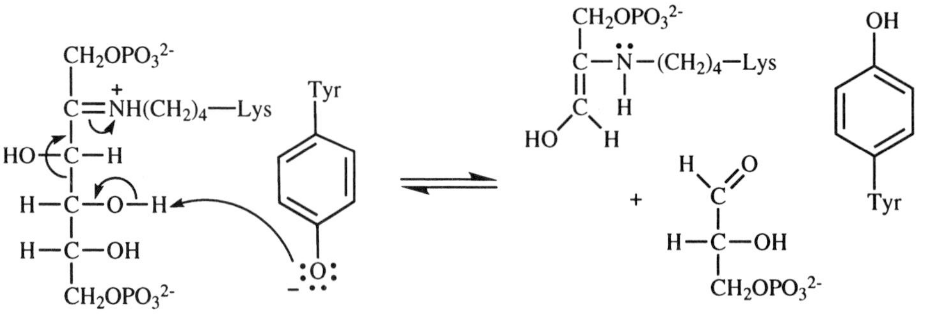

EQ. 16.8

The iminium ion is then reformed by protonation of the enamine intermediate.

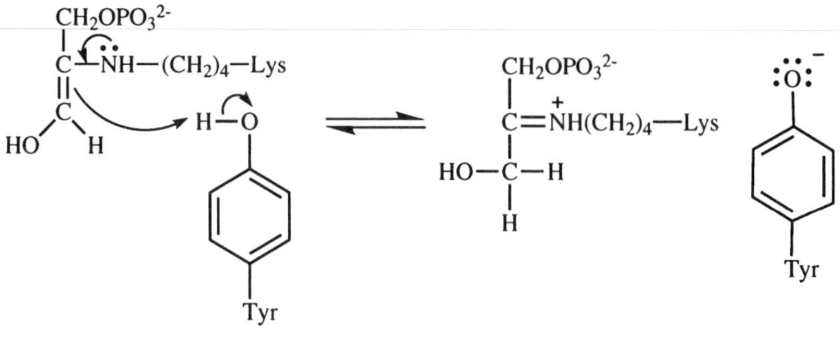

EQ. 16.9

$$\begin{array}{ccc}
\mathrm{CH_2OPO_3^{2-}} & & \mathrm{CH_2OPO_3^{2-}} \\
| & & | \\
\overset{+}{\mathrm{C}}\!=\!\mathrm{NH(CH_2)_4\!-\!Lys} & \xrightleftharpoons{\;H_2O\;} & \mathrm{C}\!=\!\mathrm{O} \quad + \quad \mathrm{H_2N\!-\!(CH_2)_4\!-\!Lys} \\
| & & | \\
\mathrm{CH_2OH} & & \mathrm{CH_2OH}
\end{array}$$

EQ. 16.10

In the final step, the iminium ion is hydrolyzed to generate dihydroxyacetone phosphate and the free enzyme.

As we saw in the preceding section, the dihydroxyacetone phosphate is converted into glyceraldehyde–3–phosphate and continues through the glycolytic pathway.

## 16.5 THE CATALYTIC MECHANISM OF CHYMOTRYPSIN

Chymotrypsin is probably the most well understood of all enzymes because its mechanism has been examined in great detail. In this section, we will see how many different modes of catalysis operate in concert with one another in this enzyme. We will see examples of acid base catalysis, covalent catalysis, stabilization of intermediates (transition states), and specificity.

Chymotrypsin belongs to a enzyme family know as the **serine proteases**. These enzymes, as the name protease suggests, are responsible from **peptide bond** cleavage. While many enzymes fall into this classification, each enzyme cleaves only a particular type of peptide bond. That is, each different serine protease has specificity for cleaving peptide bonds after certain amino acid residues. In the case of chymotrypsin, peptide bond cleavage typically takes place after a large **hydrophobic** amino acid such as phenylalanine, tyrosine or tryptophan. This specificity comes from the presence of **specificity pockets** present within the active site of the individual serine proteases. These specificity pockets "recognize" certain amino acids and allow the peptide to bind to the enzyme in such a way that peptide bond hydrolysis is facilitated. This is done by holding the substrate so that the peptide bond is close to the residues within the active site that are responsible for cleavage.

Another feature that the serine proteases have in common is the presence of the so-called **catalytic triad**. This is a series of three amino acid residues; aspartic acid, histidine, and serine that act together to facilitate the formation of both tetrahedral intermediates that form during peptide bond hydrolysis. We will see the specificity of this interaction as we examine the mechanism of chymotrypsin–catalyzed peptide hydrolysis in detail.

The first step in the mechanism of peptide bond hydrolysis is nucleophilic attack of the bound substrate by the serine oxygen. This process is facilitated by abstraction of the hydroxylic proton of the serine residue by the histidine residue. The resulting **imidazolium ion** is stabilized by interaction with the negatively charged aspartate reside. Nucleophilic attack on the amide carbonyl leads to covalent bond formation between the serine oxygen and the carbonyl carbon atom of the amide,. This the first tetrahedral intermediate in the mechanism.

EQ. 16.11

His
CH₂

Ser
CH₂

Asp—CH₂—C

Tetrahedral
Intermediate

It is important to note that this tetrahedral intermediate is stabilized by hydrogen bonding interactions with amide N-H groups as illustrated in figure 16.7. This stabilization is a major factor in the catalytic power of chymotrypsin. The region of the enzyme that contains these N-H groups is known as the **oxanion hole**.

Eventually, the tetrahedral intermediate collapses simultaneously with hydrolysis of the peptide bond. This results in cleavage of the amide carbon-nitrogen bond and the release of the amino portion of the peptide from the enzyme. The carboxy portion of the peptide remains covalently attached to the chymotrypsin as an acyl enzyme.

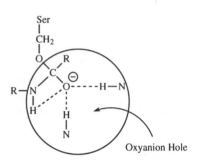

FIGURE 16.7:
Stabilization of the tetrahedral intermediate by chymotrypsin.

EQ. 16.12

His
CH₂

Ser
CH₂

Asp—CH₂—C

His
CH₂

Ser
CH₂

Asp—CH₂—C   +   R—NH₂

Acyl Enzyme

In the next step, base catalyzed nucleophilic attack of water on the acyl enzyme results in formation of the second tetrahedral intermediate. We again see stabilization of the intermediate. This time it is the imidazolium ion that is stabilized by the nearby aspartate residue. This second tetrahedral intermediate is also stabilized by hydrogen bonding in the oxanion hole.

EQ. 16.13

Collapse of the second tetrahedral intermediate results from hydrolytic cleavage of the covalent bond between the serine hydroxyl on the enzyme and the car-boxy carbon of the substrate. The enzyme is now available for the hydrolysis of other peptide bonds.

These three examples have shown the varied means that enzymes use to exert their tremendous catalytic power In addition, the authors of this book hope that you have been able to see the relationship of the chemistry learned in the earlier chapters to the chemistry of biological processes. You should now have the basic understanding of chemistry needed for further study in whatever area(s) you choose to pursue. We wish you well in these endeavors.

**Acid–Base Catalysis:** catalysis by certain amino acid residues that work by transferring a proton to or from the substrate.

**Activation energy barrier:** an increased energy state of the reactants which must be traversed as reactants become products. This occurs because an intermediate must be formed and/or bonds must be broken before bonds are formed to make the product.

**Active site:** the specific site on an enzyme at which the catalytic event occurs.

**Amino acid residues:** the portion of the amino acids remaining after formation of the peptide bond by removal of water between two amino acids.

**Catalyst:** a substance that affects the rate of a chemical reaction by altering the mechanism by which the reaction occurs.

**Catalytic triad:** a series of three amino acid residues; aspartic acid, histidine, and serine that act together to facilitate the formation of tetrahedral intermediates formed during peptide bond hydrolysis.

**Covalent catalysis:** catalysis occurring *via* covalent bond formation between the enzyme and the substrate.

**Enediol:** a molecule having two hydroxy groups and an unsaturated carbon–carbon linkage. See enediolate.

**Enediolate:** an negatively charged ion formed by removing a proton from one of the hydroxyl groups of an enediol. See enediol.

**Enzyme:** a special type of protein that serves as a catalyst for a biochemical reaction.

**Enzyme—substrate complex:** a complex formed between an enzyme and a substrate as the first step in an enzyme–catalyzed reaction.

**Gycolytic pathway:** a biochemical pathway involved in conversion of glucose into energy.

**Hydrophobic:** a term that refers to the "water hating" portion of a molecule. As typically used in biochemistry this is the portion of a protein containing only uncharged amino acid residues.

**Imidazolium ion:** a positive ion formed by protonating the five–membered heterocyclic ring compound imidazole.

**Imine:** a compound with a carbon–nitrogen double bond. Also called a Schiff base.

**Iminium linkage:** linkage of two moieties through an imine bond.

**Isomerase:** an enzyme whose function is to produce an isomer of the substrate.

**Orientation effects:** catalytic effects that occur because the enzyme holds the substrate in a particular position during the catalytic event.

**Oxyanion hole:** the region of the enzyme, chymotrypsin, containing a number of N—H linkages on the amino acid resides that function to stabilize intermediates through hydrogen bonding.

**Peptide bond:** the bond between two amino acids in a protein. This is otherwise known as an amide bond in organic chemistry.

**Proximity effects:** catalytic effects that occur because the enzyme holds the substrate in close proximity to another species during the catalytic event.

**Schiff bases:** a compound having a carbon–nitrogen double bond. Also called an imine.

**Serine proteases:** a class of enzymes responsible for peptide bond cleavage. Each individual serine protease cleaves peptide bonds following a certain amino acid(s).

**Specificity pockets:** portions on an enzyme at or near the active site whose function is to "recognize" certain portions of the substrate.

**Substrate:** the reactant in an enzyme–catalyzed reaction.

1. In your own words, briefly explain how enzymes work.

2. Draw a step–by–step mechanism for the conversion of glyceraldehyde–3–phosphate into dihydroxyacetone phosphate.

3. Papain is a cysteine protease found in papaya that contains a cysteine rather than a serine residue in the active site. Propose a mechanism for papain catalyzed amide bond hydrolysis.

4. Propose a general mechanism for the enzyme–catalyzed isomerization of glucose–6–phosphate to fructose–6–phosphate.

glucose-6-phosphate          fructose-6-phosphate

| SUBSTANCE | PHASE | $\Delta \overline{H}_f^{\,\circ}$ (kJ/mol) | $\Delta \overline{G}_f^{\,\circ}$ (kJ/mol) | $\overline{S}^{\,\circ}$ J/K mol |
|---|---|---|---|---|
| **Aluminum** | | | | |
| Al | s | 0 | 0 | 28.3 |
| Al | g | 326 | 286 | 164.4 |
| $Al_2O_3$ | s | -1676 | -1582 | 50.92 |
| $AlF_3$ | s | -1504 | -1425 | 66.44 |
| $AlCl_3$ | s | -704.2 | -628.9 | 110.7 |
| $AlCl_3.6H_2O$ | s | -2692 | n.a. | n.a. |
| $Al_2S_3$ | s | -724 | -492.4 | n.a. |
| $Al_2(SO_4)_3$ | s | -3440.8 | -3100.1 | 239 |
| **Antimony** | | | | |
| Sb | s | 0 | 0 | 45.69 |
| Sb | g | 262 | 222 | 180.2 |
| $Sb_4O_6$ | s | -1441 | -1268 | 221 |
| $SbCl_3$ | g | -314 | -301 | 337.7 |
| $SbCl_5$ | g | -394.3 | -334.3 | 401.8 |
| $Sb_2S_3$ | s | -175 | -174 | 182 |
| $SbCl_3$ | s | -382.2 | -323.7 | 184 |
| SbOCl | s | -374 | n.a. | n.a. |
| **Arsenic** | | | | |
| As | s | 0 | 0 | 35v |
| As | g | 303 | 261 | 174.1 |
| $As_4$ | g | 144 | 92.5 | 314 |
| $As_4O_6$ | s | -1313.9 | -1152.5 | 214 |
| $As_2O_5$ | s | -924.87 | -782.4 | 105 |
| $AsCl_3$ | g | -258.6 | -245.9 | 327.1 |
| $As_2S_3$ | s | -169 | -169 | 164 |
| $AsH_3$ | g | 66.44 | 68.91 | 222.7 |
| $H_3AsO_4$ | s | -906.3 | n.a. | n.a. |
| **Barium** | | | | |
| Ba | s | 0 | 0 | 66.9 |
| Ba | g | 175.6 | 144.8 | 170.3 |
| BaO | s | -558.1 | -528.4 | 70.3 |
| $BaCl_2$ | s | -860.06 | -810.9 | 126 |
| $BaSO_4$ | s | -1465 | -1353 | 132 |

| Substance | Phase | $\Delta \bar{H}_f^\circ$ (kJ/mol) | $\Delta \bar{G}_f^\circ$ (kJ/mol) | $\bar{S}^\circ$ J/K mol |
|---|---|---|---|---|
| **Beryllium** | | | | |
| Be | s | 0 | 0 | 9.54 |
| Be | g | 320.6 | 282.8 | 136.17 |
| BeO | s | -610.9 | -581.6 | 14.1 |
| **Bismuth** | | | | |
| Bi | s | 0 | 0 | 56.74 |
| Bi | g | 207 | 168 | 186.90 |
| $Bi_2O_3$ | s | -573.88 | -493.7 | 151 |
| $BiCl_3$ | s | -379 | -315 | 177 |
| $Bi_2S_3$ | s | -143 | -141 | 200 |
| **Boron** | | | | |
| B | s | 0 | 0 | 5.86 |
| B | g | 562.7 | 518.8 | 153.3 |
| $B_2O_3$ | s | -1272.8 | -1193.7 | 53.97 |
| $B_2H_6$ | g | 36 | 86.6 | 232.0 |
| $B(OH)_3$ | s | -1094.3 | -969.01 | 88.83 |
| $BF_3$ | g | -1137.3 | -1120.3 | 254.0 |
| $BCl_3$ | g | -403.8 | -388.7 | 290.0 |
| $B_3N_3H_6$ | l | -541.0 | -392.8 | 200 |
| $HBO_2$ | s | -794.25 | -723.4 | 40 |
| **Bromine** | | | | |
| $Br_2$ | l | 0 | 0 | 152.23 |
| $Br_2$ | g | 30.91 | 3.142 | 245.35 |
| Br | g | 111.88 | 82.429 | 174.91 |
| $BrF_3$ | g | -255.6 | -229.5 | 292.4 |
| HBr | g | -36.4 | -53.43 | 198.59 |
| **Cadmium** | | | | |
| Cd | s | 0 | 0 | 51.76 |
| Cd | g | 112.0 | 77.45 | 167.64 |
| CdO | s | -258 | -228 | 54.8 |
| $CdCl_2$ | s | -391.5 | -344.0 | 115.3 |
| $CdSO_4$ | s | -933.28 | -822.78 | 123.04 |
| CdS | s | -162 | -156 | 64.9 |

| SUBSTANCE | PHASE | $\Delta \bar{H}_f^{\,\circ}$ (kJ/mol) | $\Delta \bar{G}_f^{\,\circ}$ (kJ/mol) | $\bar{S}^{\,\circ}$ J/K mol |
|---|---|---|---|---|
| **Calcium** | | | | |
| Ca | s | 0 | 0 | 41.6 |
| Ca | g | 192.6 | 158.9 | 154.78 |
| $CaH_2$ | s | -189. | -150. | 42. |
| CaO | s | -635.5 | -604.2 | 40 |
| $Ca(OH)_2$ | s | -986.59 | -896.76 | 76.1 |
| $CaSO_4$ | s | -1432.7 | -1302.3 | 107 |
| $CaSO_4.2H_2O$ | s | -2021.1 | -1795.7 | 194.0 |
| $CaCO_3$ | s (calcite) | -1206.9 | -1128.8 | 92.9 |
| $CaSO_3.2H_2O$ | s | -1762 | -1565 | 184 |
| **Carbon (Inorganic)** | | | | |
| C | s (graphite) | 0 | 0 | 5.740 |
| C | s (diamond) | 1.897 | 2.900 | 2.38 |
| C | g | 716.681 | 671.289 | 157.987 |
| CO | g | -110.52 | -137.15 | 197.56 |
| $CO_2$ | g | -393.51 | -394.36 | 213.6 |
| $CCl_4$ | l | -135.4 | -65.27 | 216.4 |
| $CCl_4$ | g | -102.9 | -60.63 | 309.7 |
| $CS_2$ | l | 89.70 | 65.27 | 151.3 |
| $C_2N_2$ | g | 308.9 | 297.4 | 241.8 |
| HCN | l | 108.9 | 124.9 | 112.8 |
| HCN | g | 135 | 124.7 | 201.7 |
| **Chlorine** | | | | |
| $Cl_2$ | g | 0 | 0 | 222.96 |
| Cl | g | 121.68 | 105.70 | 165.09 |
| ClF | g | -54.48 | -55.94 | 217.8 |
| $ClF_3$ | g | -163 | -123 | 281.5 |
| $Cl_2O$ | g | 80.3 | 97.9 | 266.1 |
| $Cl_2O_7$ | l | 238 | n.a. | n.a. |
| $Cl_2O_7$ | g | 272 | n.a. | n.a. |
| HCl | g | -92.307 | -95.299 | 186.80 |
| $HClO_4$ | l | -40.6 | n.a. | n.a. |
| **Chromium** | | | | |
| Cr | s | 0 | 0 | 23.8 |
| Cr | g | 397 | 352 | 174.4 |
| $Cr_2O_3$ | s | -1140 | -1058 | 81.2 |
| $CrO_3$ | s | -589.5 | n.a. | n.a. |
| $(NH_4)_2Cr_2O_7$ | s | -1807 | n.a. | n.a. |

| SUBSTANCE | PHASE | $\Delta \bar{H}_f^\circ$ (kJ/mol) | $\Delta \bar{G}_f^\circ$ (kJ/mol) | $\bar{S}^\circ$ J/K mol |
|---|---|---|---|---|
| **Cobalt** | | | | |
| Co | s | 0 | 0 | 30.0 |
| CoO | s | -237.9 | -214.2 | 52.97 |
| $Co_3O_4$ | s | -891.2 | -774.0 | 103 |
| $Co(NO_3)_2$ | s | -420.5 | n.a. | n.a. |
| **Copper** | | | | |
| Cu | s | 0 | 0 | 33.15 |
| Cu | g | 338.3 | 298.5 | 166.3 |
| CuO | s | -157 | -130 | 42.63 |
| $Cu_2O$ | s | -169 | -146 | 93.14 |
| CuS | s | -53.1 | -53.6 | 66.5 |
| $Cu_2S$ | s | -79.5 | -86.2 | 121 |
| $CuSO_4$ | s | -771.36 | -661.9 | 109 |
| $Cu(NO_3)_2$ | s | -303 | n.a. | n.a. |
| **Fluorine** | | | | |
| $F_2$ | g | 0 | 0 | 202.7 |
| F | g | 78.99 | 61.92 | 158.64 |
| $F_2O$ | g | -22 | -4.6 | 247.3 |
| HF | g | -271 | -273 | 173.67 |
| **Hydrogen** | | | | |
| $H_2$ | g | 0 | 0 | 130.57 |
| H | g | 217.97 | 203.26 | 114.60 |
| $H_2O$ | l | -285.83 | -237.18 | 69.91 |
| $H_2O$ | g | -241.82 | -228.59 | 188.71 |
| $H_2O_2$ | l | -187.8 | -120.4 | 110 |
| $H_2O_2$ | g | -136.3 | -105.6 | 233 |
| HF | g | -271 | -273 | 173.67 |
| HCl | g | -92.307 | -95.299 | 186.80 |
| HBr | g | -36.4 | -53.43 | 198.59 |
| HI | g | 26.5 | 1.7 | 206.48 |
| $H_2S$ | g | -20.6 | -33.6 | 205.7 |
| $H_2Se$ | g | 30 | 16 | 218.9 |

| SUBSTANCE | PHASE | $\Delta \overline{H}_f^\circ$ (kJ/mol) | $\Delta \overline{G}_f^\circ$ (kJ/mol) | $\overline{S}^\circ$ J/K mol |
|---|---|---|---|---|
| Iodine | | | | |
| $I_2$ | s | 0 | 0 | 116.14 |
| $I_2$ | g | 62.438 | 19.36 | 260.6 |
| I | g | 106.84 | 70.283 | 180.68 |
| IF | g | 95.65 | -118.5 | 236.1 |
| ICl | g | 17.8 | -5.44 | 247.44 |
| IBr | g | 40.8 | 3.7 | 258.66 |
| $IF_7$ | g | -943.9 | -818.4 | 346 |
| HI | g | 26.5 | 1.7 | 206.48 |
| Iron | | | | |
| Fe | s | 0 | 0 | 27.3 |
| Fe | g | 416 | 371 | 180.38 |
| $Fe_2O_3$ | s | -824.2 | -742.2 | 87.40 |
| $Fe_3O_4$ | s | -1118 | -1015 | 146 |
| $Fe(CO)_5$ | l | -774.0 | -705.4 | 338 |
| $Fe(CO)_5$ | g | -733.9 | -697.26 | 445.2 |
| $FeSeO_3$ | s | -1200 | n.a. | n.a.- |
| FeO | s | -272 | n.a. | n.a. |
| FeAsS | s | -42 | -50 | 120 |
| $Fe(OH)_2$ | s | -569.0 | -486.6 | 88 |
| $Fe(OH)_3$ | s | -823.0 | -696.6 | 107 |
| FeS | s | -100 | -100 | 60.29 |
| $Fe_3C$ | s | 25 | 20 | 105 |
| Lead | | | | |
| Pb | s | 0 | 0 | 64.81 |
| Pb | g | 195 | 162 | 175.26 |
| PbO | s (yellow) | -217.3 | -187.9 | 68.70 |
| PbO | s (red) | -219.0 | -188.9 | 66.5 |
| $Pb(OH)_2$ | s | -515.9 | n.a. | n.a. |
| PbS | s | -100 | -98.7 | 91.2 |
| $Pb(NO_3)_2$ | s | -451.9 | n.a. | n.a. |
| $PbO_2$ | s | -277 | -217.4 | 68.6 |
| $PbCl_2$ | s | -359.4 | -314.1 | 136 |
| Lithium | | | | |
| Li | s | 0 | 0 | 28.0 |
| Li | g | 155.1 | 122.1 | 138.67 |
| LiH | s | -90.42 | -69.96 | 25 |
| LiOH | s | -487.23 | -443.9 | 50.2 |
| LiF | s | -612.1 | -584.1 | 35.9 |
| $Li_2CO_3$ | s | -1215.6 | -1132.4 | 90.4 |

| Substance | Phase | $\Delta \bar{H}_f^\circ$ (kJ/mol) | $\Delta \bar{G}_f^\circ$ (kJ/mol) | $\bar{S}^\circ$ J/K mol |
|---|---|---|---|---|
| **Magnesium** | | | | |
| Mg | s | 0 | 0 | 32.5 |
| $MgBr_2$ | s | -517.6 | | |
| $MgCO_3$ | s | -1110 | -1030 | 65.7 |
| $MgCl_2$ | s | -641.8 | -592.3 | 89.5 |
| $MgF_2$ | s | -1102 | -1049 | 57.24 |
| MgO | s | -601.8 | -569.6 | 27. |
| $Mg(OH)_2$ | s | -924.7 | -833.7 | 63.14 |
| $MgSO_4$ | s | -1278 | -1174 | 91.6 |
| **Manganese** | | | | |
| Mn | s | 0 | 0 | 32.0 |
| Mn | g | 281 | 238 | 173.6 |
| MnO | s | -385.2 | -362.9 | 59.71 |
| $MnO_2$ | s | -520.03 | -465.18 | 53.05 |
| $Mn_2O_3$ | s | -959.0 | -881.2 | 110 |
| $Mn_3O_4$ | s | -1388 | -1283 | 156 |
| **Mercury** | | | | |
| Hg | l | 0 | 0 | 76.02 |
| Hg | g | 61.317 | 31.85 | 174.8 |
| HgO | s (red) | -90.83 | -58.555 | 70.29 |
| HgO | s (yellow) | -90.46 | -57.296 | 71.1 |
| $HgCl_2$ | s | -224 | -179 | 146 |
| $Hg_2Cl_2$ | s | -265.2 | -210.78 | 192 |
| HgS | s (red) | -58.16 | -50.6 | 82.4 |
| HgS | s (black) | -53.6 | -47.7 | 88.3 |
| $HgSO_4$ | s | -707.5 | n.a. | n.a. |
| **Nitrogen** | | | | |
| $N_2$ | g | 0 | 0 | 191.5 |
| N | g | 472.704 | 455.579 | 153.19 |
| NO | g | 90.25 | 86.57 | 210.65 |
| $NO_2$ | g | 33.2 | 51.3 | 239.9 |
| $N_2O$ | g | 82.05 | 104.2 | 219.7 |
| $N_2O_3$ | g | 83.72 | 139.4 | 312.2 |
| $N_2O_4$ | g | 9.16 | 97.89 | 304.29 |
| $N_2O_5$ | g | 11.3 | 115.6 | 355.7 |
| $NH_3$ | g | -46.11 | -16.5 | 192.3 |

| SUBSTANCE | PHASE | $\Delta \bar{H}_f^{\circ}$ (kJ/mol) | $\Delta \bar{G}_f^{\circ}$ (kJ/mol) | $\bar{S}^{\circ}$ J/K mol |
|---|---|---|---|---|
| $N_2H_4$ | l | 50.63 | 149.2 | 121.2 |
| $N_2H_4$ | g | 95.4 | 159.3 | 238.4 |
| $NH_4NO_3$ | s | -365.6 | -184.0 | 151.1 |
| $NH_4Cl$ | s | -314.4 | -201.5 | 94.6 |
| $NH_4Br$ | s | -270.8 | -175 | 113 |
| $NH_4I$ | s | -201.4 | -113 | 117 |
| $NH_4NO_2$ | s | -256 | n.a. | n.a. |
| $HNO_3$ | l | -174.1 | -80.79 | 155.6 |
| $HNO_3$ | g | -135.1 | -74.77 | 266.2 |

Oxygen

| | | | | |
|---|---|---|---|---|
| $O_2$ | g | 0 | 0 | 205.138 |
| $O$ | g | 249.17 | 213.75 | 160.95 |
| $O_3$ | g | 143 | 163 | 238.8 |

Phosphorous

| | | | | |
|---|---|---|---|---|
| $P$ | s | 0 | 0 | 41.1 |
| $P$ | g | 58.91 | 24.5 | 280.0 |
| $P_4$ | g | 314.6 | 278.3 | 163.08 |
| $PH_3$ | g | 5.4 | 13 | 210.1 |
| $PCl_3$ | g | -287 | -268 | 311.7 |
| $PCl_5$ | g | -375 | -305 | 364.5 |
| $P_4O_6$ | s | -1640 | n.a. | n.a. |
| $P_4O_{10}$ | s | -2984 | -2698 | 228.9 |
| $HPO_3$ | s | -948.5 | n.a. | n.a. |
| $H_3PO_2$ | s | -604.6 | n.a. | n.a. |
| $H_3PO_3$ | s | -964.4 | n.a. | n.a. |
| $H_3PO_4$ | s | -1279 | -1119 | 110.5 |
| $H_3PO_4$ | l | -1267 | n.a. | n.a. |
| $H_4P_2O_7$ | s | -2241 | n.a. | n.a. |
| $POCl_3$ | l | -597.1 | -520.9 | 222.5 |
| $POCl_3$ | g | -558.48 | -512.96 | 325.3 |

Potassium

| | | | | |
|---|---|---|---|---|
| $K$ | s | 0 | 0 | 63.6 |
| $K$ | g | 90.00 | 61.17 | 160.23 |

| Substance | Phase | $\Delta \overline{H}_f^{\circ}$ (kJ/mol) | $\Delta \overline{G}_f^{\circ}$ (kJ/mol) | $\overline{S}^{\circ}$ J/K mol |
|---|---|---|---|---|
| KF | s | -562.58 | -533.12 | 66.57 |
| KCl | s | -435.868 | -408.32 | 82.68 |

### Silicon

| Substance | Phase | $\Delta \overline{H}_f^{\circ}$ (kJ/mol) | $\Delta \overline{G}_f^{\circ}$ (kJ/mol) | $\overline{S}^{\circ}$ J/K mol |
|---|---|---|---|---|
| Si | s | 0 | 0 | 18.8 |
| Si | g | 455.6 | 411 | 167.9 |
| $SiO_2$ | s | -910.94 | -856.67 | 41.84 |
| $SiH_4$ | g | 34 | 56.9 | 204.5 |
| $H_2SiO_3$ | s | -1189 | -1092 | 130 |
| $H_4SiO_4$ | s | -1481 | -1333 | 190 |
| $SiF_4$ | g | -1614.9 | -1572.7 | 282.4 |
| $SiCl_4$ | l | -687.0 | -619.90 | 240 |
| $SiCl_4$ | g | -657.01 | -617.01 | 330.6 |
| SiC | s | -65.3 | -62.8 | 16.6 |

### Silver

| Substance | Phase | $\Delta \overline{H}_f^{\circ}$ (kJ/mol) | $\Delta \overline{G}_f^{\circ}$ (kJ/mol) | $\overline{S}^{\circ}$ J/K mol |
|---|---|---|---|---|
| Ag | s | 0 | 0 | 42.55 |
| Ag | g | 284.6 | 245.7 | 172.89 |
| $Ag_2O$ | s | -31.0 | -11.2 | 121 |
| AgCl | s | -127.1 | -109.8 | 96.2 |
| $Ag_2S$ | s | -32.6 | -40.7 | 144.0 |

### Sodium

| Substance | Phase | $\Delta \overline{H}_f^{\circ}$ (kJ/mol) | $\Delta \overline{G}_f^{\circ}$ (kJ/mol) | $\overline{S}^{\circ}$ J/K mol |
|---|---|---|---|---|
| Na | s | 0 | 0 | 51.0 |
| Na | g | 108.7 | 78.11 | 153.62 |
| $Na_2O$ | s | -415.9 | -377 | 72.8 |
| NaCl | s | -411.00 | -384.03 | 72.38 |

### Sulfur

| Substance | Phase | $\Delta \overline{H}_f^{\circ}$ (kJ/mol) | $\Delta \overline{G}_f^{\circ}$ (kJ/mol) | $\overline{S}^{\circ}$ J/K mol |
|---|---|---|---|---|
| S | s (rhombic) | 0 | 0 | 31.8 |
| S | g | 278.80 | 238.27 | 167.75 |
| $SO_2$ | g | -296.83 | -300.19 | 248.1 |
| $SO_3$ | g | -395.7 | -371.1 | 256.6 |
| $H_2S$ | g | -20.6 | -33.6 | 205.7 |
| $H_2SO_4$ | l | -813.989 | 690.101 | 156.90 |
| $H_2S_2O_7$ | s | -1274 | n.a. | n.a. |
| $SF_4$ | g | -774.9 | -731.4 | 291.9 |
| $SF_6$ | g | -1210 | -1105 | 291.7 |
| $SCl_2$ | l | -50 | n.a. | n.a. |
| $SCl_2$ | g | -20 | n.a. | n.a. |
| $S_2Cl_2$ | l | -59.4 | n.a. | n.a. |
| $S_2Cl_2$ | g | -18 | -32 | 331.4 |
| $SOCl_2$ | l | -246 | n.a. | n.a. |
| $SOCl_2$ | g | -213 | -198 | 309.7 |

| Substance | Phase | $\Delta \bar{H}_f^\circ$ (kJ/mol) | $\Delta \bar{G}_f^\circ$ (kJ/mol) | $\bar{S}^\circ$ J/K mol |
|---|---|---|---|---|
| SO$_2$Cl$_2$ | l | -394 | n.a. | n.a. |
| SO$_2$Cl$_2$ | g | -364 | -320 | 311.8 |

**Tin**

| | | | | |
|---|---|---|---|---|
| | s | 0 | 0 | 51.55 |
| Sn | g | 302 | 267 | 168.38 |
| SnO | s | -286 | -257 | 56.5 |
| SnO$_2$ | s | -580.7 | -519.7 | 52.3 |
| SnCl$_4$ | l | -511.2 | -440.2 | 259 |
| SnCl$_4$ | g | -471.5 | -432.2 | 366 |

**Titanium**

| | | | | |
|---|---|---|---|---|
| Ti | s | 0 | 0 | 30.6 |
| Ti | g | 469.9 | 425.1 | 180.19 |
| TiO$_2$ | s | -944.7 | -889.5 | 50.33 |
| TiCl$_4$ | l | -804.2 | -737.2 | 252.3 |
| TiCl$_4$ | g | -763.2 | -726.8 | 354.8 |

**Tungsten**

| | | | | |
|---|---|---|---|---|
| W | s | 0 | 0 | 32.6 |
| W | g | 849.4 | 807.1 | 173.84 |
| WO$_3$ | s | -842.87 | -764.08 | 75.90 |

**Zinc**

| | | | | |
|---|---|---|---|---|
| Zn | s | 0 | 0 | 41.6 |
| Zn | g | 130.73 | 95.178 | 160.87 |
| ZnO | s | -348.3 | -318.3 | 43.64 |
| ZnCl$_2$ | s | -415.1 | -369.43 | 111.5 |
| ZnS | s | -206.0 | -201.3 | 57.7 |
| ZnSO$_4$ | s | -982.8 | -874.5 | 120 |
| ZnCO$_3$ | s | -812.78 | -731.57 | 82.4 |

**Carbon (Organic)**

| | | | | |
|---|---|---|---|---|
| CH$_4$ | g | -74.81 | -50.75 | 186.15 |
| CH$_3$OH | l | -238.7 | -166.4 | 127 |
| CH$_3$OH | g | -200.7 | -162.0 | 239.7 |
| CHCl$_3$ | l | -134.5 | -73.72 | 202 |
| CHCl$_3$ | g | -103.1 | -70.37 | 295.6 |
| C$_2$H$_2$ | g | 226.7 | 209.2 | 200.8 |
| C$_2$H$_4$ | g | 52.26 | 68.12 | 219.5 |
| C$_2$H$_6$ | g | -84.68 | -32.9 | 229.5 |
| CH$_3$COOH | l | -484.5 | -390 | 160 |
| CH$_3$COOH | g | -432.25 | -374 | 282 |
| C$_2$H$_5$OH | l | -277.7 | -174.9 | 161 |

| Substance | Phase | $\Delta \bar{H}_f^\circ$ (kJ/mol) | $\Delta \bar{G}_f^\circ$ (kJ/mol) | $\bar{S}^\circ$ J/K mol |
|---|---|---|---|---|
| $C_2H_5OH$ | g | -235.1 | -168.6 | 282.6 |
| $C_3H_8$ | g | -103.85 | -23.49 | 269.9 |
| $C_6H_6$ | g | 82.927 | 129.66 | 269.2 |
| $C_6H_6$ | l | 49.028 | 124.50 | 172.8 |
| $CH_2Cl_2$ | l | -121.5 | -67.32 | 178 |
| $CH_2Cl_2$ | g | -92.47 | -65.9 | 270.1 |
| $CH_3Cl$ | h | -80.83 | -57.40 | 234.5 |
| $C_2H_5Cl$ | l | -136.5 | -59.41 | 190.8 |
| $C_2H_5Cl$ | g | -112.2 | -60.46 | 275.9 |

# APPENDIX Ⓑ SELECTED EQUILIBRIUM CONSTANTS

## I. SOLUBILITY PRODUCTS

| NAME | FORMULA | $K_{sp}$ |
|------|---------|----------|
| Aluminum hydroxide | $Al(OH)_3$ | $2.00 \times 10^{-32}$ |
| Barium arsenate | $Ba_3(AsO_4)_2$ | $7.76 \times 10^{-51}$ |
| Barium carbonate | $BaCO_3$ | $8.13 \times 10^{-9}$ |
| Barium chromate | $BaCrO_4$ | $2.40 \times 10^{-10}$ |
| Barium fluoride | $BaF_2$ | $1.70 \times 10^{-6}$ |
| Barium iodate | $Ba(IO_3)_2$ | $1.51 \times 10^{-9}$ |
| Barium oxalate | $BaC_2O_4$ | $2.29 \times 10^{-8}$ |
| Barium sulfate | $BaSO_4$ | $1.07 \times 10^{-10}$ |
| Beryllium hydroxide | $Be(OH)_2$ | $7.08 \times 10^{-22}$ |
| Bismuth iodide | $BiI_3$ | $8.13 \times 10^{-19}$ |
| Bismuth phosphate | $BiPO_4$ | $1.29 \times 10^{-23}$ |
| Bismuth sulfide | $Bi_2S_3$ | $1.00 \times 10^{-97}$ |
| Cadmium arsenate | $Cd_3(AsO_4)_2$ | $2.19 \times 10^{-33}$ |
| Cadmium hydroxide | $Cd(OH)_2$ | $5.89 \times 10^{-15}$ |
| Cadmium oxalate | $CdC_2O_4$ | $1.51 \times 10^{-8}$ |
| Cadium sulfide | $CdS$ | $7.76 \times 10^{-27}$ |
| Calcium arsenate | $Ca_3(AsO_4)_2$ | $6.76 \times 10^{-19}$ |
| Calcium carbonate | $CaCO_3$ | $8.71 \times 10^{-9}$ |
| Calcium fluoride | $CaF_2$ | $3.98 \times 10^{-11}$ |
| Calcium hydroxide | $Ca(OH)_2$ | $5.50 \times 10^{-6}$ |
| Calcium iodate | $Ca(IO_3)_2$ | $6.46 \times 10^{-7}$ |
| Calcium oxalate | $CaC_2O_4$ | $2.57 \times 10^{-9}$ |
| Calcium phosphate | $Ca_3(PO_4)_2$ | $2.00 \times 10^{-29}$ |
| Calcium sulfate | $CaSO_4$ | $1.91 \times 10^{-4}$ |
| Cerium (III) hydroxide | $Ce(OH)_3$ | $2.00 \times 10^{-20}$ |
| Cerium (III) iodate | $Ce(IO_3)_3$ | $3.24 \times 10^{-10}$ |
| Cerium (III) oxalate | $Ce_2(C_2O_4)_3$ | $3.02 \times 10^{-29}$ |
| Chromium (II) hydroxide | $Cr(OH)_2$ | $1.00 \times 10^{-17}$ |
| Chromium (III) hydroxide | $Cr(OH)_3$ | $6.03 \times 10^{-31}$ |
| Cobalt (II) hydroxide | $Co(OH)_2$ | $2.00 \times 10^{-16}$ |
| Cobalt (III) hydroxide | $Co(OH)_3$ | $1.00 \times 10^{-43}$ |

| NAME | FORMULA | $K_{sp}$ |
|------|---------|-----|
| Copper (II) arsenate | $Cu_3(AsO_4)_2$ | $7.59 \times 10^{-36}$ |
| Copper (I) bromide | $CuBr$ | $5.25 \times 10^{-9}$ |
| Copper (I) chloride | $CuCl$ | $1.20 \times 10^{-6}$ |
| Copper (II) iodate | $Cu(IO_3)_2$ | $7.41 \times 10^{-8}$ |
| Copper (I) iodide | $CuI$ | $5.13 \times 10^{-12}$ |
| Copper (I) sulfide | $Cu_2S$ | $2.00 \times 10^{-47}$ |
| Copper (II) sulfide | $CuS$ | $8.91 \times 10^{-36}$ |
| Copper (I) thiocyanate | $CuSCN$ | $4.79 \times 10^{-15}$ |
| Gold (III) hydroxide | $Au(OH)_3$ | $5.50 \times 10^{-46}$ |
| Iron (III) arsenate | $FeAsO_4$ | $5.75 \times 10^{-21}$ |
| Iron (II) carbonate | $FeCO_3$ | $3.47 \times 10^{-11}$ |
| Iron (II) hydroxide | $Fe(OH)_2$ | $7.94 \times 10^{-16}$ |
| Iron (III) hydroxide | $Fe(OH)_3$ | $3.98 \times 10^{-38}$ |
| Lead arsenate | $Pb_3(AsO_4)_2$ | $4.07 \times 10^{-36}$ |
| Lead bromide | $PbBr_2$ | $3.89 \times 10^{-5}$ |
| Lead carbonate | $PbCO_3$ | $3.31 \times 10^{-14}$ |
| Lead chloride | $PbCl_2$ | $1.58 \times 10^{-5}$ |
| Lead chromate | $PbCrO_4$ | $1.82 \times 10^{-14}$ |
| Lead fluoride | $PbF_2$ | $3.72 \times 10^{-8}$ |
| Lead iodate | $Pb(IO_3)_2$ | $2.57 \times 10^{-13}$ |
| Lead oxalate | $PbC_2O_4$ | $4.79 \times 10^{-10}$ |
| Lead oxide | $PbO$ | $1.20 \times 10^{-15}$ |
| Lead sulfate | $PbSO_4$ | $1.58 \times 10^{-8}$ |
| Lead sulfide | $PbS$ | $7.94 \times 10^{-28}$ |
| Magnesium arsenate | $Mg_3(AsO_4)_2$ | $2.09 \times 10^{-20}$ |
| Magnesium carbonate | $MgCO_3$ | $1.00 \times 10^{-5}$ |
| Magnesium fluoride | $MgF_2$ | $6.46 \times 10^{-9}$ |
| Magnesium hydroxide | $Mg(OH)_2$ | $1.20 \times 10^{-11}$ |
| Magnesium oxalate | $MgC_2O_4$ | $1.00 \times 10^{-8}$ |
| Manganese (II) hydroxide | $Mn(OH)_2$ | $1.91 \times 10^{-13}$ |
| Mercury (I) bromide | $Hg_2Br_2$ | $5.75 \times 10^{-23}$ |
| Mercury (I) chloride | $Hg_2Cl_2$ | $1.29 \times 10^{-18}$ |
| Mercury (I) iodide | $Hg_2I_2$ | $4.47 \times 10^{-29}$ |
| Mercury (II) oxide | $HgO$ | $3.02 \times 10^{-26}$ |
| Mercury (I) sulfate | $Hg_2SO_4$ | $7.41 \times 10^{-7}$ |
| Mercury (II) sulfide | $HgS$ | $3.98 \times 10^{-53}$ |
| Mercury (I) thiocyanate | $Hg_2(SCN)_2$ | $3.02 \times 10^{-20}$ |
| Nickel arsenate | $Ni_3(AsO_4)_2$ | $3.09 \times 10^{-26}$ |
| Nickel carbonate | $NiCO_3$ | $6.61 \times 10^{-9}$ |
| Nickel hydroxide | $Ni(OH)_2$ | $6.46 \times 10^{-18}$ |
| Nickel sulfide | $NiS$ | $3.02 \times 10^{-19}$ |

| Name | Formula | $K_{sp}$ |
|---|---|---|
| Palladium (II) hydroxide | $Pd(OH)_2$ | $1.00 \times 10^{-31}$ |
| Platinum (II) hydroxide | $Pt(OH)_2$ | $1.00 \times 10^{-35}$ |
| Radium sulfate | $RaSO_4$ | $4.27 \times 10^{-11}$ |
| Silver arsenate | $Ag_3AsO_4$ | $1.00 \times 10^{-22}$ |
| Silver bromate | $AgBrO_3$ | $5.75 \times 10^{-5}$ |
| Silver bromide | $AgBr$ | $5.25 \times 10^{-13}$ |
| Silver carbonate | $Ag_2CO_3$ | $8.13 \times 10^{-12}$ |
| Silver chloride | $AgCl$ | $1.78 \times 10^{-10}$ |
| Silver chromate | $Ag_2CrO_4$ | $2.45 \times 10^{-12}$ |
| Silver cyanide | $Ag[Ag(CN)_2]$ | $5.01 \times 10^{-12}$ |
| Silver iodate | $AgIO_3$ | $3.02 \times 10^{-8}$ |
| Silver iodide | $AgI$ | $8.32 \times 10^{-17}$ |
| Silver oxalate | $Ag_2C_2O_4$ | $3.47 \times 10^{-11}$ |
| Silver oxide | $Ag_2O$ | $2.57 \times 10^{-8}$ |
| Silver phosphate | $Ag_3PO_4$ | $1.29 \times 10^{-20}$ |
| Silver sulfate | $Ag_2SO_4$ | $1.58 \times 10^{-5}$ |
| Silver sulfide | $Ag_2S$ | $2.00 \times 10^{-49}$ |
| Silver thiocyanate | $AgSCN$ | $1.00 \times 10^{-12}$ |
| Strontium carbonate | $SrCO_3$ | $1.10 \times 10^{-10}$ |
| Strontium chromate | $SrCrO_4$ | $3.63 \times 10^{-5}$ |
| Strontium fluoride | $SrF_2$ | $2.82 \times 10^{-9}$ |
| Strontium iodate | $Sr(IO_3)_2$ | $3.31 \times 10^{-7}$ |
| Strontium oxalate | $SrC_2O_4$ | $1.58 \times 10^{-7}$ |
| Strontium sulfate | $SrSO_4$ | $3.80 \times 10^{-7}$ |
| Thallium (I) bromate | $TlBrO_3$ | $8.51 \times 10^{-5}$ |
| Thallium (I) bromide | $TlBr$ | $3.39 \times 10^{-6}$ |
| Thallium (I) chloride | $TlCl$ | $1.70 \times 10^{-4}$ |
| Thallium (I) chromate | $Tl_2CrO_4$ | $9.77 \times 10^{-13}$ |
| Thallium (I) iodate | $TlIO_3$ | $3.09 \times 10^{-6}$ |
| Thallium (I) iodide | $TlI$ | $6.46 \times 10^{-8}$ |
| Thallium (I) sulfide | $Tl_2S$ | $5.01 \times 10^{-21}$ |
| Tin (II) oxide | $SnO$ | $1.41 \times 10^{-28}$ |
| Tin (II) sulfide | $SnS$ | $1.00 \times 10^{-25}$ |
| Titanium (III) hydroxide | $Ti(OH)_3$ | $1.00 \times 10^{-40}$ |
| Titanium (IV) hydroxide) | $TiO(OH)$ | $1.00 \times 10^{-29}$ |
| Zinc arsenate | $Zn_3(AsO_4)_2$ | $1.29 \times 10^{-28}$ |
| Zinc carbonate | $ZnCO_3$ | $1.41 \times 10^{-11}$ |
| Zinc ferrocyanide | $Zn_2[Fe(CN)_6]$ | $4.07 \times 10^{-16}$ |
| Zinc hydroxide | $Zn(OH)_2$ | $1.20 \times 10^{-17}$ |
| Zinc oxalate | $ZnC_2O_4$ | $2.82 \times 10^{-8}$ |
| Zinc phosphate | $Zn_3(PO_4)_2$ | $9.12 \times 10^{-33}$ |
| Zinc sulfide | $ZnS$ | $1.00 \times 10^{-21}$ |

## II ACID DISSOCIATION CONSTANTS

| NAME | FORMULA | $K_a$ |
|------|---------|-------|
| Acetic Acid | $HC_2H_3O_2$ | $1.74 \times 10^{-5}$ |
| Arsenic | $H_3AsO_4$ | $6.46 \times 10^{-3}$ |
| $K_{a2}$ | | $1.15 \times 10^{-7}$ |
| $K_{a3}$ | | $3.16 \times 10^{-12}$ |
| Arsenious | $H_3AsO_3$ | $5.13 \times 10^{-10}$ |
| Benzoic | $HC_7H_5O_2$ | $6.31 \times 10^{-5}$ |
| Boric | $H_3BO_3$ | $5.75 \times 10^{-10}$ |
| Bromoacetic | $HC_2H_2BrO_2$ | $1.26 \times 10^{-3}$ |
| Carbonic | $H_2CO_3$ | $4.47 \times 10^{-7}$ |
| $K_{a2}$ | | $4.68 \times 10^{-11}$ |
| Chloroacetic | $HC_2H_2ClO_2$ | $1.38 \times 10^{-3}$ |
| Chromic | $H_2CrO_4$ | $9.55$ |
| $K_{a2}$ | | $3.16 \times 10^{-7}$ |
| Citric | $H_3C_6H_5O_7$ | $7.41 \times 10^{-4}$ |
| $K_{a2}$ | | $1.70 \times 10^{-5}$ |
| $K_{a3}$ | | $3.98 \times 10^{-7}$ |
| EDTA | $H_6C_{10}H_{12}N_2O_8{}^{2+}$ | $1.26 \times 10^{-1}$ |
| $K_{a2}$ | | $2.51 \times 10^{-2}$ |
| $K_{a3}$ | | $1.02 \times 10^{-2}$ |
| $K_{a4}$ | | $2.14 \times 10^{-3}$ |
| $K_{a5}$ | | $6.92 \times 10^{-7}$ |
| $K_{a6}$ | | $5.50 \times 10^{-11}$ |
| Eriochrome Black T | | $3.98 \times 10^{-7}$ |
| $K_{a2}$ | | $7.76 \times 10^{-11}$ |
| $K_{a3}$ | | $3.24 \times 10^{-12}$ |
| Formic | $HCHO_2$ | $1.78 \times 10^{-4}$ |
| Hydrazoic | $HN_3$ | $1.91 \times 10^{-5}$ |
| Hydrocyanic | $HCN$ | $6.17 \times 10^{-10}$ |
| Hydrofluoric | | $1.48 \times 10^{-11}$ |
| Hydrosulfuric | | $9.77 \times 10^{-8}$ |
| $K_{a2}$ | | $1.10 \times 10^{-15}$ |
| Hypochlorous | | $3.39 \times 10^{-7}$ |
| Iodic | $HIO_3$ | $1.62 \times 10^{-1}$ |
| Lactic | $HC_3H_5O_3$ | $1.38 \times 10^{-4}$ |
| Malonic | $H_2C_3H_2O_4$ | $1.38 \times 10^{-3}$ |
| $K_{a2}$ | | $2.00 \times 10^{-6}$ |

| NAME | FORMULA | $K_a$ |
|---|---|---|
| *n*–Butanoic | $HC_4H_7O_2$ | $1.51 \times 10^{-5}$ |
| Nitrilotriacetic | | $3.24 \times 10^{-3}$ |
| $K_{a2}$ | | $1.86 \times 10^{-10}$ |
| $K_{a3}$ | | $7.76 \times 10^{-13}$ |
| Nitrous | $HNO_2$ | $1.58 \times 10^{-3}$ |
| *o*–Phthalic | $H_2C_8H_4O_4$ | $1.12 \times 10^{-3}$ |
| $K_{a2}$ | | $3.89 \times 10^{-6}$ |
| Oxalic | $H_2C_2O_4$ | $6.46 \times 10^{-2}$ |
| $K_{a2}$ | | $6.17 \times 10^{-5}$ |
| PAN | $H_2In$ | $1.26 \times 10^{-2}$ |
| $K_{a2}$ | | $6.31 \times 10^{-13}$ |
| Phenol | $HC_6H_5O$ | $1.05 \times 10^{-10}$ |
| Phosphoric | $H_3PO_4$ | $7.41 \times 10^{-3}$ |
| $K_{a2}$ | | $6.17 \times 10^{-8}$ |
| $K_{a3}$ | | $4.79 \times 10^{-13}$ |
| Phosphorous | | $2.00 \times 10^{-7}$ |
| $K_{a2}$ | | $1.95 \times 10^{-13}$ |
| Picric | $HC_6H_2N_3O_7$ | $5.13 \times 10^{-1}$ |
| Propanoic | $HC_3H_5O_2$ | $1.35 \times 10^{-5}$ |
| Salicylic | $HC_7H_5O_3$ | $1.07 \times 10^{-3}$ |
| Succinic | $H_2C_4H_4O_4$ | $6.17 \times 10^{-5}$ |
| $K_{a2}$ | | $2.29 \times 10^{-6}$ |
| Sulfamic | $HSO_3NH_2$ | $1.02 \times 10^{-1}$ |
| Sulfurous | $H_2SO_3$ | $1.74 \times 10^{-2}$ |
| $K_{a2}$ | | $6.17 \times 10^{-8}$ |
| Tartaric | $H_2C_4H_4O_6$ | $9.12 \times 10^{-4}$ |
| $K_{a2}$ | | $4.27 \times 10^{-5}$ |

### III. BASE DISSOCIATION CONSTANTS

| NAME | FORMULA | $K_b$ |
|---|---|---|
| Ammonia | $NH_3$ | $1.82 \times 10^{-5}$ |
| Aniline | $C_6H_5NH_2$ | $3.98 \times 10^{-10}$ |
| Diethylamine | $(C_2H_5)_2NH$ | $8.51 \times 10^{-4}$ |
| Dimethylamine | $(CH_3)_2NH$ | $5.89 \times 10^{-4}$ |
| Ethanolamine | $HOCH_2CH_2NH_2$ | $3.16 \times 10^{-5}$ |
| Ethylamine | $C_2H_5NH_2$ | $4.68 \times 10^{-4}$ |
| Ethylenediamine | $NH_2CH_2CH_2NH_2$ | $1.51 \times 10^{-7}$ |
| $K_{b2}$ | $H_2NCH_2CH_2NH_2$ | $9.12 \times 10^{-5}$ |

| Name | Formula | $K_b$ |
|------|---------|-------|
| Hydrazine | $N_2H_4$ | $9.77 \times 10^{-7}$ |
| Hydroxlyamine | $HONH_2$ | $9.55 \times 10^{-9}$ |
| Methylamine | $CH_3NH_2$ | $5.25 \times 10^{-4}$ |
| Methylethylamine | $(CH_3)(C_2H_5)NH$ | $7.08 \times 10^{-4}$ |
| Piperidine | $C_5H_{10}NH$ | $1.58 \times 10^{-3}$ |
| n-Propylamine | $C_3H_7NH_2$ | $5.50 \times 10^{-4}$ |
| Pyridine | $C_5H_5N$ | $1.70 \times 10^{-9}$ |
| THAM | $C_4H_9O_3NH_2$ | $1.20 \times 10^{-6}$ |
| Triethylamine | $(C_2H_5)_3N$ | $5.89 \times 10^{-4}$ |
| Trimethylamine | $(CH_3)_3N$ | $6.31 \times 10^{-5}$ |

## IV. Dissociation Constants of Amino Acids

| Amino Acid | $K_{a1}$ | $K_{a2}$ | $K_{a3}$ |
|------------|----------|----------|----------|
| Alanine | $4.47 \times 10^{-3}$ | $1.35 \times 10^{-10}$ | |
| Asparagine | $7.24 \times 10^{-3}$ | $1.91 \times 10^{-9}$ | |
| Glutamine | $6.76 \times 10^{-3}$ | $9.77 \times 10^{-10}$ | |
| Glycine | $4.47 \times 10^{-3}$ | $1.66 \times 10^{-10}$ | |
| Isoleucine | $4.79 \times 10^{-3}$ | $1.78 \times 10^{-10}$ | |
| Leucine | $4.68 \times 10^{-3}$ | $1.78 \times 10^{-10}$ | |
| Methionine | $6.31 \times 10^{-3}$ | $8.91 \times 10^{-10}$ | |
| Phenylalanine | $6.31 \times 10^{-3}$ | $4.90 \times 10^{-10}$ | |
| Proline | $1.12 \times 10^{-2}$ | $2.29 \times 10^{-11}$ | |
| Serine | $6.46 \times 10^{-3}$ | $6.17 \times 10^{-10}$ | |
| Threonine | $8.13 \times 10^{-3}$ | $7.94 \times 10^{-10}$ | |
| Tryptophan | $4.47 \times 10^{-3}$ | $4.68 \times 10^{-10}$ | |
| Valine | $1.35 \times 10^{-3}$ | $1.91 \times 10^{-10}$ | |

### Acidic Amino Acids

| Amino Acid | $K_{a1}$ | $K_{a2}$ | $K_{a3}$ |
|------------|----------|----------|----------|
| Aspartic Acid | $1.02 \times 10^{-2}$ | $1.02 \times 10^{-4}$ | $1.00 \times 10^{-10}$ |
| Glutamic Acid | $5.89 \times 10^{-3}$ | $3.80 \times 10^{-5}$ | $1.12 \times 10^{-10}$ |

### Basic Amino Acids

| Amino Acid | $K_{a1}$ | $K_{a2}$ | $K_{a3}$ |
|------------|----------|----------|----------|
| Arginine | $1.51 \times 10^{-2}$ | $1.02 \times 10^{-9}$ | $3.31 \times 10^{-13}$ |
| Cysteine | $1.95 \times 10^{-2}$ | $4.37 \times 10^{-9}$ | $1.70 \times 10^{-11}$ |
| Histidine | $2.00 \times 10^{-2}$ | $9.55 \times 10^{-7}$ | $8.32 \times 10^{-10}$ |
| Lysine | $9.12 \times 10^{-3}$ | $8.32 \times 10^{-10}$ | $2.04 \times 10^{-11}$ |
| Tyrosine | $6.76 \times 10^{-3}$ | $6.46 \times 10^{-10}$ | $3.39 \times 10^{-11}$ |

# APPENDIX Ⓒ HÜCKEL MOLECULAR ORBITAL THEORY

## SECTION C.A THE SECULAR DETERMINATE

There are many steps in the derivation of the method we will be using that have been omitted from this book because of the complexity of the mathematics involved and the lack of time and space. Those who desire to see the derivation of the Hückel methods are encouraged to read about these derivations in the references listed at the end of this appendix.

The energies of the $\pi$ electron system for a molecule are obtained by solving what is known as the secular equation for the molecule. The secular equation is a polynomial obtained by expansion of the secular determinate. The secular determinate, in turn, is written from a basis set of the atomic orbitals belonging to the atoms of the molecule. In all cases, we consider only the p electron system which means that the basis set consists of unhybridized atomic $2p_z$ orbitals or linear combinations of these $2p_z$ orbitals (by convention the $p_z$ orbitals are those involved in p bonding.) We will also consider that all of the $2p_z$ orbitals are parallel to each other, *i.e.*, that one xy plane serves all carbon atoms. Provided that the $\pi$ bonds are conjugated, this assumption will be valid. If the bonds are not conjugated, making this assumption affects only the ease with which the solution is obtained, it does not affect the results.

Consider first the molecule ethene: $C_2H_4$. The carbon skeleton of the molecule is: $C_1$——$C_2$

We first proceed to set up the secular determinate for the ethylene molecule. To do this we write the basis set of atomic orbitals across the top and down the left edge of a 2 X 2 determinate:

$$
\begin{array}{c c c}
 & \varphi_1 & \varphi_2 \\
\varphi_1 & \Big| & \Big| \\
\varphi_2 & \Big| & \Big|
\end{array}
$$

The elements of the determinate describe the interactions of the atomic orbitals. There are two different types of interactions that give rise to three different values to be entered (only two will be seen in the case of ethylene). All of the interactions will be of the type $\varphi_i\varphi_j$. An "x" is entered into the secular determinate for the case of $i = j$. Where $i \neq j$ there are two possibilities. If atoms i and j are adjacent to one another (and, therefore, their $p_z$ orbitals directly interact), a value of "1" is entered into the secular determinate. If atoms i and j are not adjacent, a value of "0" is entered. Following these guidelines, the secular determinate for ethylene becomes:

$$
\begin{array}{c c c}
 & \varphi_1 & \varphi_2 \\
\varphi_1 & X & 1 \\
\varphi_2 & 1 & X
\end{array}
$$

The energies for the $\pi$ electron orbitals are now obtained by setting the secular determinate equal to 0 and solving the polynomial equation obtained:

$$\begin{array}{c} \varphi_1 \\ \\ \varphi_2 \end{array} \begin{vmatrix} X & 1 \\ \\ 1 & X \end{vmatrix} = 0$$

EQ. C.1

The secular equation is easily determined by remembering the value of a determinate is obtained by cross–multiplication, thus:

$$\begin{vmatrix} A & B \\ \\ C & D \end{vmatrix} = AD - CB$$

The secular determinate in equation C.1 is then:

$$x^2 - 1 = 0$$

EQ. C.2

from which we obtain the solutions to the equation: $. x = \pm \sqrt{1} \; ; \; x = \pm 1$

Each root of the secular equation represents the energy of a different molecular orbital. In the Hückel method x stands for an energy of $\alpha - \beta$, where $\alpha$ is the energy of an atomic p orbital and $\beta$ is termed the bond energy (or the resonance energy). If we take the value of $\alpha$ to be 0, then the energy levels of the p electron orbitals in the ethylene molecule, relative to the energy levels of the electrons in the atomic $\pi$ orbitals, are $\pm\beta$ . $\beta$ is the energy released when an electron is placed into the orbital. This means that $\beta$ is a negative number that makes the more stable orbital the one at an energy of $+ \beta$ (negative values of x). The energy level diagram for the p electron system of the ethylene molecule in its ground state is shown in figure C.1. Since one of the orbitals ($+ \beta$ energy) is more stable than the atomic $\pi$ orbital, it is termed a bonding orbital because an electron is more stable in it than in the atomic orbital. The other orbital is termed a non–bonding orbital since placing an electron in is a less stable state than keeping the electron in the atomic orbital. Placing two electrons into the bonding orbital is the most stable situation. This is termed the ground state for ethylene and provides a total $\pi$ electron bond energy of 2 $\beta$.

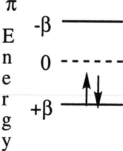

FIGURE C.1: $\pi$ electron energy levels for ground state ethylene.

Moving to a slightly more complicated molecule, let us now look at 1,3 butadiene, a conjugated diene. In order to apply Hückel techniques to this molecule we must consider it to be in the totally planar configuration — the configuration that would be necessary for the conjugation of the double bonds. The configuration of the butadiene molecule is:

The secular determinate can now be written for this molecule: $C_1 \text{------} C_2 \text{------} C_3 \text{------} C_4$

$$
\begin{array}{c|cccc}
 & \varphi_1 & \varphi_2 & \varphi_3 & \varphi_4 \\
\hline
\varphi 1 & X & 1 & 0 & 0 \\
\varphi 2 & 1 & X & 1 & 0 \\
\varphi 3 & 0 & 1 & X & 1 \\
\varphi 4 & 0 & 0 & 1 & X
\end{array}
$$

The simplest means of converting this secular determinate into the corresponding secular equation is to expand it by minors along the first row elements. Section C.E, at the end of this appendix, presents the method of expansion by minors for those not familiar with this method of expansion. The first step of the process provides two smaller 3 X 3 determinates:

$$
(-1)^{(1+1)} X \begin{vmatrix} X & 1 & 0 \\ 1 & X & 1 \\ 0 & 1 & X \end{vmatrix} \qquad + (-1)^{(1+2)} 1 \begin{vmatrix} 1 & 1 & 0 \\ 0 & X & 1 \\ 0 & 1 & X \end{vmatrix}
$$

This can be rewritten as:

$$
X \begin{vmatrix} X & 1 & 0 \\ 1 & X & 1 \\ 0 & 1 & X \end{vmatrix} \qquad -1 \begin{vmatrix} 1 & 1 & 0 \\ 0 & X & 1 \\ 0 & 1 & X \end{vmatrix}
$$

Since elements $A_{13}$ and $A_{14}$ of the original determinate each have the value of zero, it is not necessary to write those terms. The two resulting 3 X 3 determinates can each be expanded by minors to produce 2 X 2 determinates:

$$
X \left\{ (-1)^{(1+1)} X \begin{vmatrix} X & 1 \\ 1 & X \end{vmatrix} \quad + (-1)^{(1+2)} 1 \begin{vmatrix} 1 & 1 \\ 0 & X \end{vmatrix} \right\}
$$

$$
+ (-1) \left\{ (-1)^{(1+1)} 1 \begin{vmatrix} X & 1 \\ 1 & X \end{vmatrix} \quad + (-1)^{(1+2)} 1 \begin{vmatrix} 0 & 1 \\ 0 & X \end{vmatrix} \right\}
$$

This can be simplified to:

$$
X \left\{ X \begin{vmatrix} X & 1 \\ 1 & X \end{vmatrix} \quad + (-1) \begin{vmatrix} 1 & 1 \\ 0 & X \end{vmatrix} \right\}
$$

$$
+ (-1) \left\{ (1) \begin{vmatrix} X & 1 \\ 1 & X \end{vmatrix} \quad + (-1) \begin{vmatrix} 0 & 1 \\ 0 & X \end{vmatrix} \right\}
$$

Which can be even further simplified:

$$X^2 \begin{vmatrix} X & 1 \\ 1 & X \end{vmatrix} \qquad -X \begin{vmatrix} 1 & 1 \\ 0 & X \end{vmatrix}$$

$$- \begin{vmatrix} X & 1 \\ 1 & X \end{vmatrix} \qquad + \begin{vmatrix} 0 & 1 \\ 0 & X \end{vmatrix}$$

Finally the secular equation can be written:

$$X^2 \, ( \, X^2 - 1 \, ) \, - X \, ( \, X - \, 0 \, ) \, - (X^2 - 1) \, + \, (0 - 0) \, = \, 0 \qquad \text{EQ. C.3}$$

which, after algebraic simplification, eventually becomes:

$$X^4 - 3 \, X^2 + 1 = 0 \qquad \text{EQ. C.4}$$

The easiest means of solving this is to note that only even powers of 2 appear in the equation. This allows us to set $Y = X^2$ and rewrite the equation as:

$$Y^2 - 3 \, Y + 1 = 0 \qquad \text{EQ. C.5}$$

This allows us to determine that:

$$Y = \frac{3 \pm \sqrt{3^2 - 4}}{2} = \frac{3 \pm \sqrt{5}}{2} = 2.618 \, ; \, 0.382 \qquad \text{EQ. C.6}$$

which provides four roots:

$$X = \pm \sqrt{2.618} = \pm 1.618 \text{ and}$$
$$X = \pm \sqrt{0.382} = \pm 0.618$$

Figure C.2 shows the $\pi$ energy level diagram for butadiene.

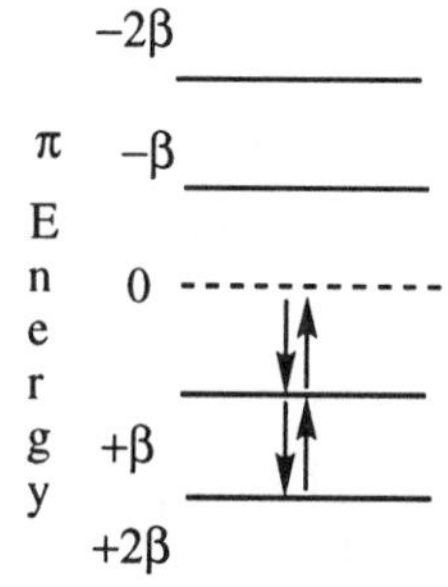

FIGURE C.2: $\pi$ Electron Energies Level Diagram for 1,3 Butadiene

With two electrons in each of the 1.618 $\beta$ and 0.618 $\beta$ energy levels, the total $\pi$ electron stabilization energy of butadiene is 4.472 $\beta$. Note that this $\pi$ electron energy is more than twice as great as that which would be expected if we were to consider butadiene as simply two ethylene molecules. The extra 0.472 $\beta$ energy is the result of delocalization of the electron cloud that allows each electron to interact with all four carbon atoms. This will become even more clear when we look at the molecular orbitals themselves.

Unfortunately as molecules become larger, the size of the secular determinate increases proportionately and the tedium of solving it grows exponentially. Fortunately there are a number of techniques that can be used to reduce the tedium of the solution. These techniques are beyond the scope of the present course and will not be covered here. They have been utilized by many of the computer programs that are easily accessible to beginning students. This means that it is possible for the results of these advanced methods to be used even before the means of solving the equations themselves are mastered.

### SECTION C.B MOLECULAR ORBITALS

Once the roots to the secular equations have been found, the next step is to determine the particular weighting factors for each atomic p orbital in each of the molecular orbitals. This is easily done because the weighting factors for each of the orbitals in the base set are the minors of the first row elements of the secular determinate. The final molecular orbital is the sum of the base orbitals multiplied by their respective minors. Let us look at the ethylene molecule first.

The molecular orbitals will take the form:

$$Y = (1) \, |x| \, \varphi_1 \; + \; (-1) \, |x| \, \varphi_2 \qquad\qquad \text{EQ. C.7}$$

The $1|x|$ in front of the $\varphi 1$ is the minor of the first element in the top row of the secular determinate; $(-1)|x|$ is the minor of the second element in the top row of the secular determinate.

Since there are two different energy levels, we can obtain two different molecular orbitals:

| x | Molecular Orbital |
|---|---|
| $-1$ | $\Psi_1 = -\, \varphi_1 - \varphi_2 = \varphi_1 + \varphi_2$ |
| $1$ | $\Psi_2 = \varphi_1 - \varphi_2$ |

These orbitals are pictured in Figure C.3.

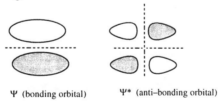

$\Psi$ (bonding orbital)     $\Psi^*$ (anti–bonding orbital)

FIGURE C.3: Molecular Orbitals of Ethylene

Although not necessary in order to see the relative contributions of each atomic orbital to the molecular orbital, it is customary to normalize each of the molecular orbitals. This ensures that the probability of finding the electron in the orbital does come out to a value of one when the integration shown in Equation 12.2 is carried out. The process involves multiplying the entire molecular orbital by a normalizing factor. This factor is defined in equation C.8. In this equation the $c_i$s are the coefficients of the atomic orbitals.

$$N = \left( \sum_{i=1}^{n} c_i^2 \right)^{-1/2}$$

EQ. C.8

Hence, for the ethylene molecule the final, normalized molecular orbitals are:

$$Y_1 = \frac{1}{\sqrt{2}} (\varphi_1 + \varphi_2)$$

$$Y_2 = \frac{1}{\sqrt{2}} (\varphi_1 - \varphi_2).$$

The situation is the same for the butadiene molecule. The only real difference is the slight additional complexity of the minors of the secular determinate because of the higher order of that determinate. Table C.1 lists the atomic orbitals and the minors of the secular determinate that are their coefficients.

**TABLE C.1**  COEFFICIENTS FOR BUTADIENE MOLECULAR ORBITALS

TABLE C.1

| ATOMIC ORBITAL | DETERMINATE MINOR | | | VALUE |
|---|---|---|---|---|
| $\varphi_1$ | X | 1 | 1 | $X^3 - 2X$ |
| | 1 | X | 1 | |
| | 0 | 1 | X | |
| $\varphi_2$ | 1 | 1 | 0 | $-(X^2 - 1)$ |
| | 0 | X | 1 | |
| | 0 | 1 | X | |
| $\varphi_3$ | 1 | X | 0 | X |
| | 0 | 1 | 1 | |
| | 0 | 0 | X | |
| $\varphi_4$ | 1 | X | 1 | $-1$ |
| | 0 | 1 | X | |
| | 0 | 0 | 1 | |

The molecular orbitals for butadiene can be written:

$$\Psi_n = (X^3 - 2X) \varphi_1 - (X^2 - 1) \varphi_2 + X \varphi_3 - \varphi_4$$

EQ. C.9

This provides the set of coefficients shown in Table C.2:

**TABLE C.2**  BUTADIENE MOLECULAR ORBITALS (UN–NORMALIZED)

TABLE C.2

| X | UN–NORMALIZED MOLECULAR ORBITAL | | | | | | | |
|---|---|---|---|---|---|---|---|---|
| $-1.618$ | $\Psi_1 =$ | $-1.00$ | $\varphi_1$ | $-1.62$ | $\varphi_2$ | $-1.62$ | $\varphi_3$ | $-1.00$ $\varphi_4$ |
| $-0.618$ | $\Psi_2 =$ | $1.00$ | $\varphi_1$ | $+0.62$ | $\varphi_2$ | $-0.62$ | $\varphi_3$ | $-1.00$ $\varphi_4$ |
| $0.618$ | $\Psi3 =$ | $-1.00$ | $\varphi_1$ | $+0.62$ | $\varphi_2$ | $+0.62$ | $\varphi_3$ | $-1.00$ $\varphi_4$ |
| $1.618$ | $\Psi4 =$ | $1.00$ | $\varphi_1$ | $-1.62$ | $\varphi_2$ | $+1.62$ | $\varphi_3$ | $-1.00$ $\varphi_4$ |

These yield the following molecular orbitals after normalization (and multiplication by − 1 if necessary to make the first coefficient positive):

**TABLE C.3** NORMALIZED BUTADIENE MOLECULAR ORBITALS

$$\Psi_1 = \quad 0.37 \;\; \varphi_1 \quad + 0.60 \;\; \varphi_2 \quad + 0.60 \;\; \varphi_3 \quad \; 0.37 \;\; \varphi_4$$
$$\Psi_2 = \quad 0.60 \;\; \varphi_1 \quad + 0.37 \;\; \varphi_2 \quad - 0.37 \;\; \varphi_3 \quad - 0.60 \;\; \varphi_4$$
$$\Psi3 = \quad 0.60 \;\; \varphi_1 \quad - 0.37 \;\; \varphi_2 \quad - 0.37 \;\; \varphi_3 \quad + 0.60 \;\; \varphi_4$$
$$\Psi4 = \quad 0.37 \;\; \varphi_1 \quad - 0.60 \;\; \varphi_2 \quad + 0.60 \;\; \varphi_3 \quad - 0.37 \;\; \varphi_4$$

These orbitals are pictured in figure C.4. Notice that the number of nodes (places across which the sign of the orbital changes sign) in the orbital increases as the energy increases. This is in accord with other ideas that higher energy orbitals will have more nodes than lower energy orbitals and is in agreement with what has been the case with atomic orbitals as well.

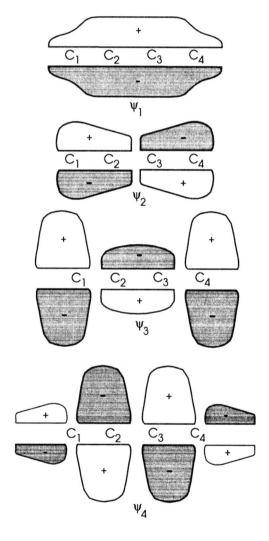

FIGURE C.4  Molecular Orbitals of Butadiene

## Section C.C  Electron Densities

While the pictures of the molecular orbitals in figures C.3 and C.4 do give a good picture of the electron distribution within the molecule, they do not give us a value to place on the electron density. Fortunately, if we know the wave functions for the molecular orbitals and the electron distribution within the molecule, it is a very simple process to calculate the electron density at each of the carbon atoms. The equation that permits the calculation of the electron density at a particular carbon atom is:

$$q_i = \sum_{j=1}^{n} n_i c_{ij}^2 \qquad \qquad \text{EQ. C.10}$$

where:         $q_i$ is the electron density at carbon atom i,

               $n_i$ is the number of electrons in MO j

and      $c_{ij}$ is the coefficient of carbon atom i in MO j.

The summation is carried out over all occupied MOs in the molecule.

Let us illustrate by applying the equation to the calculation of the electron density at the carbon atoms in the ground state of butadiene. The electron configuration of the ground state of butadiene is: $\Psi_1^2 \Psi_2^2$ . Substituting the numbers into equation C.10 gives us:

$$q_1 = 2(.37)^2 + 2(.60)^2 = 0.994 \qquad \qquad \text{EQ. C.11}$$

Had we carried more significant figures in the calculation of the coefficients, this value would have been exactly 1.0. Repeating this calculation for all of the carbon atoms shows, as expected, that the electron density is equal on all four carbon atoms in the ground state. This changes dramatically in some of the excited states, however. Consider the situation of the excited state whose electron configuration is: $\Psi_1^2 \Psi_2 \Psi_4$ .

If we now substitute these values into equation C.10 we obtain:

$$q_1 = 2(.37)^2 + 1(.60)^2 + 1(.37)^2 = 0.771 \qquad \qquad \text{EQ. C.12}$$

$$q_2 = 2(.60)^2 + 1(.37)^2 + 1(-0.60)^2 = 1.22 \qquad \qquad \text{EQ. C.13}$$

Notice that the electron densities are no longer equal on the two carbon atoms. This means that, while the attack of an electrophile on a butadiene molecule in the ground state may not show a preference for any specific carbon atom, if the molecule is raised to this excited state, there should be a definite preference for attack at the second carbon atom (provided that the lifetime of the excited state was long enough for reaction to occur).

## Section C.D  Π Bond Order Calculations

Calculation of the $\pi$ bond order is another very simple matter once the coefficients of the atomic orbitals are known. The equation is very similar to that used to determine the electron density:

$$B_{i,j} = \sum_{k=1}^{n} n_k c_{ik} c_{jk}$$

EQ. C.11

where:     $B_{i,j}$ is the $\pi$ bond order between atoms i and j
           $n_i$ is the number of electrons in MO k
           $c_i$ is the coefficient of carbon atom i in MO k,
           $c_j$ is the coefficient of carbon atom j in MO k

Let us illustrate by calculating the p bond order between carbon atoms 1 and 2 and also carbon atoms 2 and 3 in the ground state and in the first excited state for butadiene. These calculations are shown in table C.3.

TABLE C.3  Π BOND ORDER CALCULATIONS

| ATOMS | STATE | Π BOND ORDER | CALCULATION |
|-------|-------|--------------|-------------|
| 1,2 | Ground | 0.89 | $B_{1,2}=(2)(.37)(.60) + (2)(.60)(.37)$ |
| 2,3 | Ground | 0.45 | $B_{2,3}=(2)(.60)(.60) + (2)(.37)(-.37)$ |
| 1,2 | Excited | 0.22 | $B_{1,2}=(2)(.37)(.60) + (1)(.60)(-.37) + (1)(.60)(-.37)$ |
| 2,3 | Excited | 0.72 | $B_{2,3}=(2)(.60)(.60) + (1)(.37)(-.37) + (1)(.37)(-.37)$ |

Notice that these calculations indicate that there is a partial $\pi$ bond between atoms 2 and 3 in the ground state. This reinforces our idea of "conjugation". Notice, also, that the $\pi$ bond order changes drastically in the excited state.

SECTION C.E  EXPANSION OF A DETERMINATE BY MINORS

The solution of a 2 X 2 determinate is accomplished by multiplying the top–left and bottom–right elements and subtracting from that product the product of the bottom–left and top–right elements. This is sometimes referred to as the difference of the diagonals. This is illustrated by the following:

$$\begin{vmatrix} A & B \\ C & D \end{vmatrix} = AD - CB$$

In principle this can be extended to 3 X 3 and larger determinates, however the number of terms and the meaning of "diagonal" becomes more confusing in these cases. In the opinion of this author, solution of determinates larger than 2 X 2 is best accomplished by expansion of the determinate by minors.

In order to expand a determinate by minors, we first pick any row or any column. The value of the determinate is the sum of the elements of that row or column multiplied by the minors of those elements. This is best illustrated by an example. Consider the 4 X 4 determinate:

$$\begin{vmatrix} A & B & C & D \\ E & F & G & H \\ I & J & K & L \\ M & N & O & P \end{vmatrix}$$

The value of this determinate is:

A(Minor of A) + B(Minor of B) + C(Minor of C) + D(Minor of D). It could also be B(Minor of B) + F(Minor of F) + J(Minor of J) + N(Minor of N). The minor of any element of an $n^{th}$ order determinate is $(-1)^{(i+j)}$ times the determinate of $(n-1)^{th}$ order formed by striking the elements in the row and column of the element. For instance, the minors of A and B in the above example are:

$$\begin{vmatrix} F & G & H \\ J & K & L \\ N & O & P \end{vmatrix} \quad\text{and}\quad -\begin{vmatrix} E & G & H \\ I & K & L \\ M & O & P \end{vmatrix}$$

Minor of element A            Minor of element B.

Note that the minor of element A is positive since A is in the first row and first column which makes $(-1)^{(1+1)} = +1$ but the minor of element B is negative since $(-1)^{(1+2)} = -1$.

Complete expansion of the 4 X 4 determinate will then produce:

$$A\begin{vmatrix} F & G & H \\ J & K & L \\ N & O & P \end{vmatrix} - B\begin{vmatrix} E & G & H \\ I & K & L \\ M & O & P \end{vmatrix} + C\begin{vmatrix} E & F & H \\ J & I & L \\ M & N & P \end{vmatrix} - D\begin{vmatrix} E & F & G \\ I & J & K \\ M & N & O \end{vmatrix}$$

Each of the 3 X 3 determinates can then be further expanded until only 2 X 2 determinates are left.

While this appears as if it will require a great deal of work, the solution of secular determinates is not as bad as it would appear because many of the elements of these determinates are zero. Judicious choice of a row or column to use for the expansion (one that contains the largest number of terms whose value is zero) can decrease the amount of tedious work necessary for the solution.

**REFERENCES**

1. Streitweiser, Jr., Andrew, *Molecular Orbital Theory for Organic Chemists*, John Wiley & Sons, Inc., New York, 1961, Chaps. 1–2.

2. P. Zeegers, *J. Chem. Ed.* **74**, 299 (1997)

**Appendix**  **Answers to Selected Problems**

1.4  267.5 kJ

1.7a. Nitrous acid  1/2 $H_2(g)$ + 1/2 $N_2(g)$ + $O_2(g)$ → HONO

1.7d. 3 $C_{(graphite)}$ + 1/2 $O_2(g)$ + 3 $H_2(g)$ → CH3CH2CHO(l)

1.13  −53.76 kJ

1.16  − 379.07 kJ/mol

1.19  −276 kJ/mol

1.21  738.8 L

1.27c  102.6 mL

1.27d  −23.5 kJ

2.7  + 70.4 kJ/mol

2.9a  +

2.9d  +

2.12a  −174.4 kJ/mol; −111.8 J/K mol

2.13 a  −141.0 kJ/mol

2.16a  4.73 kJ/mol

2.16b  0.148

2.19c  −65.5 kJ/mol  −33.0 J/K mol

2.22c  −40.0 kJ/mol

2.22d  31.26 L

3.3  1.4;  −0.834 kJ/mol

3.5c  Both centers are *R*

3.5f  Both centers are *R*

3.9a  3,4–dimethylhexane

3.9 d  (*S*)–3,4–dimethyl–3–hexanol

3.11

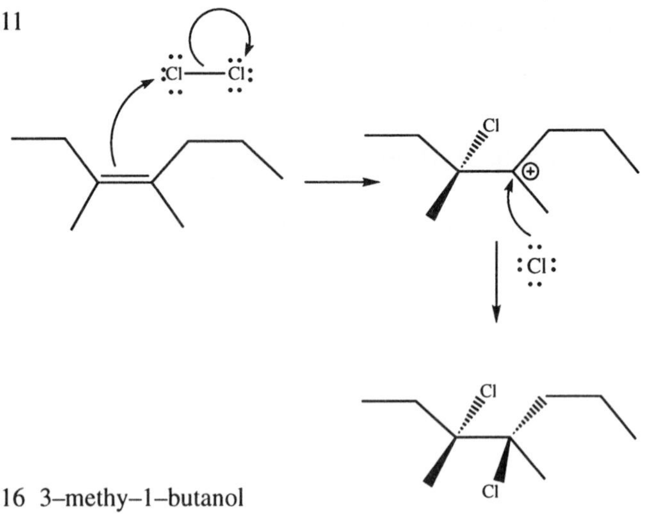

3.16  3–methy–1–butanol

3.17a  2–Butene

3.17b  −1.41 kJ/mol

4.2  7.2 x $10^{-4}$

4.5b  12.00

4.5d  4.00

4.6a  2.68

4.6c  3.17

4.7a  11.33

4.7b  11.51

4.8b  2.27

4.8e  5.51

4.9b  8.38

4.9f  calculated answer is 6.84, but some assumptions are incorrect and real answer would be slightly higher than 7 since it is a base

4.12  6.3 x $10^{-10}$

4.14  9.80

4.16  120 mL of lactic acid solution and 130 mL of NaOH solution

4.18  11.34

**4.20** $[H_3Arg^{2+}] = 1.2 \times 10^{-7}$; $[H_2Arg^+] = 0.181$; $[HArg] = 0.019$; $[Arg^-] = 6.3 \times 10^{-7}$; $[H^+] = 1.0 \times 10^{-8}$; $[OH^-] = 1.0 \times 10^{-6}$

**4.25b**

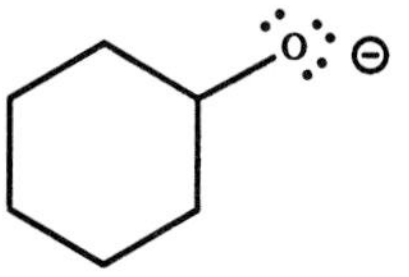

No other resonance forms

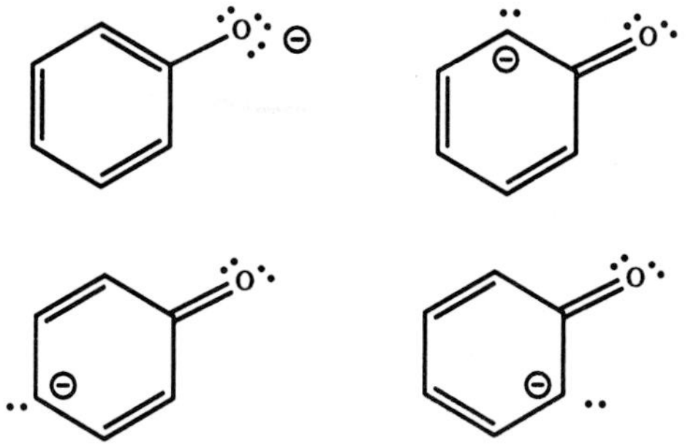

**4.27b** $-5$ kJ/mol; $K = 7.5$

**5.1b**

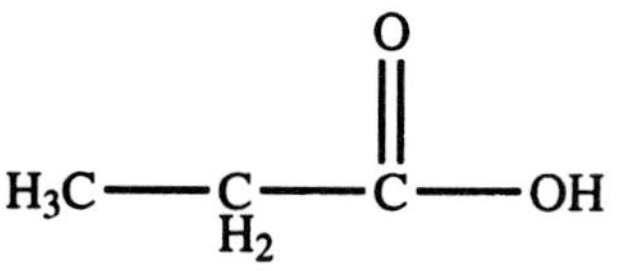

**5.1f**

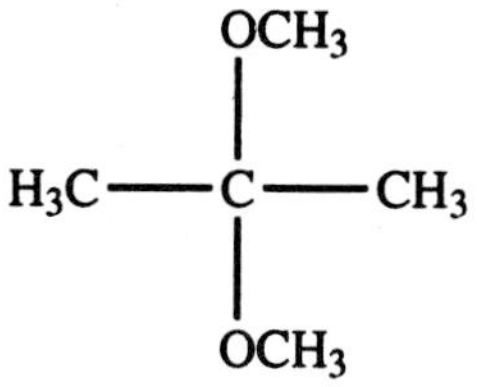

**5.3b** $CH_3ONO_2$

**5.4b**

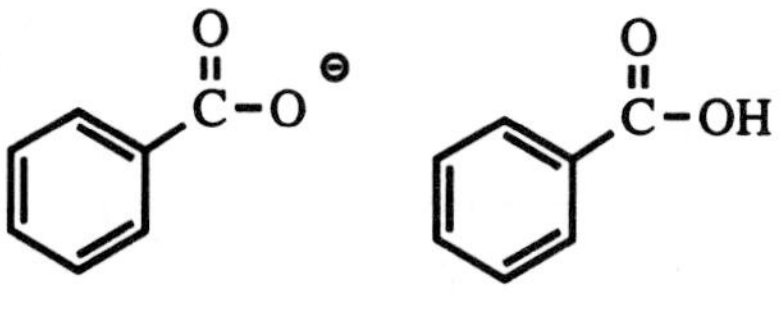

**5.7a**

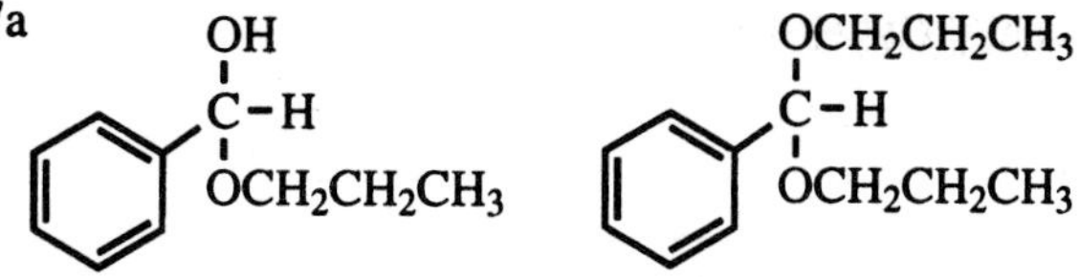

**5.11** 51%

**5.15b** $-112.0$ kJ/mol

**5.15c** $4.3 \times 10^{19}$

6.3  $6.0 \times 10^{-5}$  mol l$^{-1}$ sec$^{-1}$

6.7a  2.97 **M**

6.10  174 kJ/mol   $2.15 \times 10^{25}$

6.14  92%

6.18  Rate = $5.6 \times 10^5$ [H$_2$SeO$_3$][I$^-$]$^3$[H$^+$]$^2$

6.26b  $8.77 \times 10^{10}$  186 kJ/mol

6.26c  $2.21 \times 10^{-22}$

7.14a  H$_2$O$_2$ + 2 I$^-$ 2 H$^+$ Æ I$_2$ + 2 H$_2$O

7.14b  Rate = k[H$_2$O$_2$][I$^-$]

7.16a  OCl$^-$ + I$^-$ → Cl$^-$ + OI$^-$

7.16b  Rate = $\dfrac{k_2 k_1 [OCl^-][I^-][H_2O]}{k_{-1}[OH^-]}$

7.23b  321.5 kJ/mol

7.23c  Rate = k[*cis*–stilbene]

8.1b  1–(bromomethyl)–cyclohexane > 1–bromocyclohexane > 1–bromo–1–methylcyclohexane

8.2a  2–bromo–2–methylbutane > 2–bromo–3–methylbutane > 1–bromo–3–methylbutane

8.5a  NaOCH$_3$

8.5e  NaCN

8.6a  2–butene

8.6d  1-methylcyclohexene.

8.7

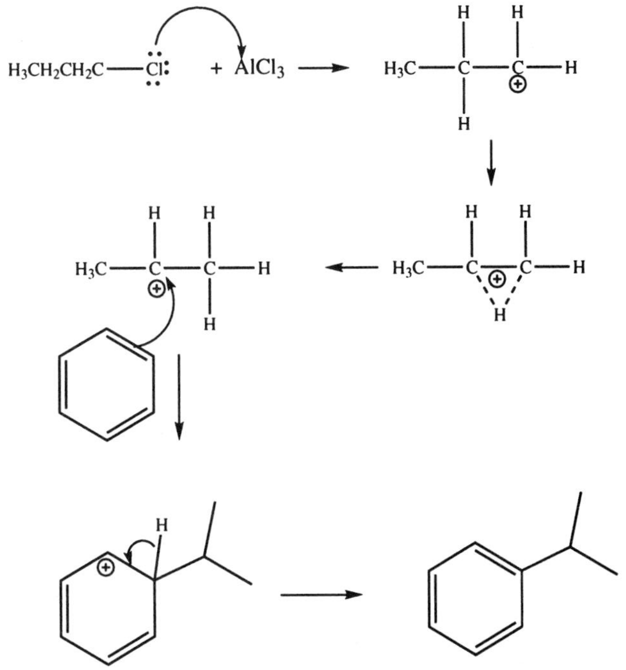

$H_3CH_2CH_2C\!-\!\ddot{C}l:$ + $\overset{+}{A}lCl_3$ $\longrightarrow$ $H_3C\!-\!\overset{\displaystyle H}{\underset{\displaystyle H}{C}}\!-\!\overset{\displaystyle H}{\overset{+}{C}}\!-\!H$

$\downarrow$

$H_3C\!-\!\overset{\displaystyle H}{\underset{\displaystyle H}{\overset{+}{C}}}\!-\!\overset{\displaystyle H}{\underset{\displaystyle H}{C}}\!-\!H$ $\longleftarrow$ $H_3C\!-\!\overset{\displaystyle H}{\overset{+}{C}}\!-\!\overset{\displaystyle H}{\underset{\displaystyle H}{C}}\!-\!H$

$\downarrow$

$\longrightarrow$

8.11c  0.24 mol L$^{-1}$ sec$^{-1}$

8.15a  $H_3C\!-\!C\!\equiv\!\overset{\ominus}{C}$ $\longrightarrow$ $H_3C\!-\!C\!\equiv\!C\!-\!CH_2CH_2CH_2OH$

$\downarrow$

$H_3C\!-\!C\!\equiv\!C\!-\!CH_2CH_2CH_2CN$ $\longleftarrow$ $H_3C\!-\!C\!\equiv\!C\!-\!CH_2CH_2CH_2Cl$

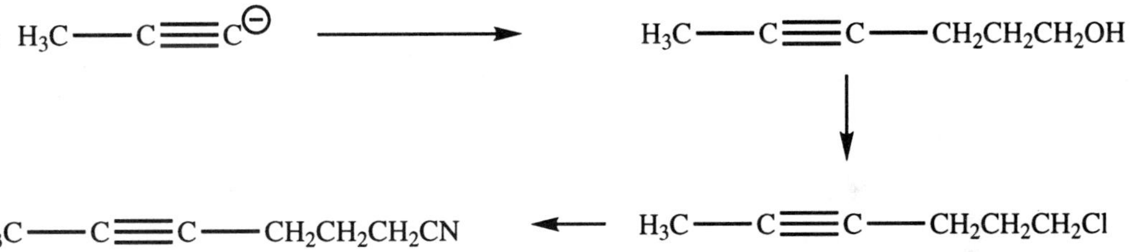

8.20  6.3 x 10$^{-3}$ L mol$^{-1}$ sec$^{-1}$

9.2

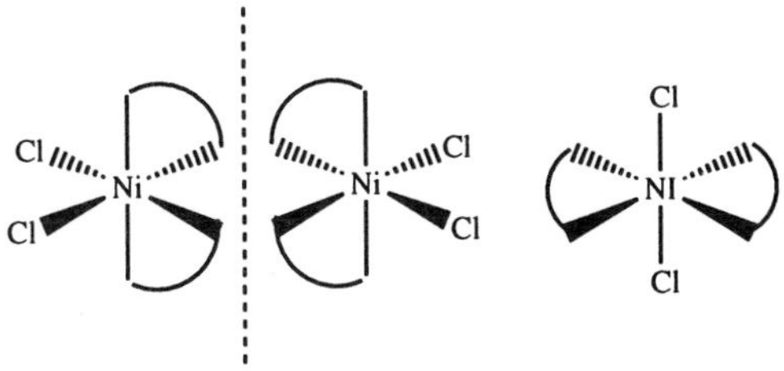

9.7a  potassium [hexafluoroferrate (III)]

9.7e  [Hexafluorosiliconate(IV)] ion

9.7g  [Hexamminechromium(III)][hexachlorocobaltate(III)]

**9.8e**

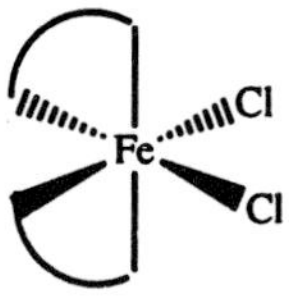

F with Fe center, two Cl ligands (axial/equatorial), bidentate ligand

**9.8h**

$$\left[ \begin{array}{c} \text{(Si center structure with F, OH, H}_2\text{O bridges)} \end{array} \right]^{2-}$$

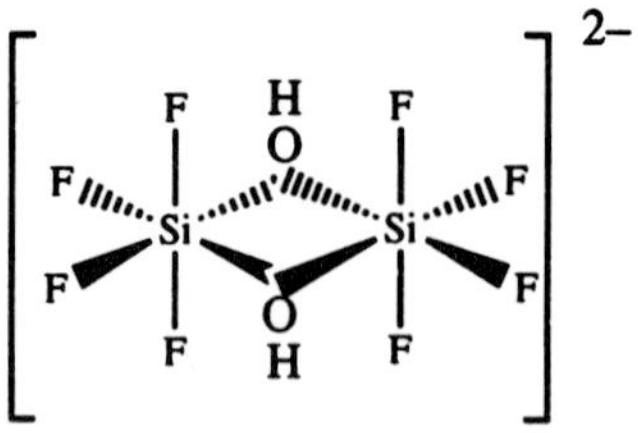

**9.9** The aluminum ion is the more stable

**9.12a** 5

**9.15**

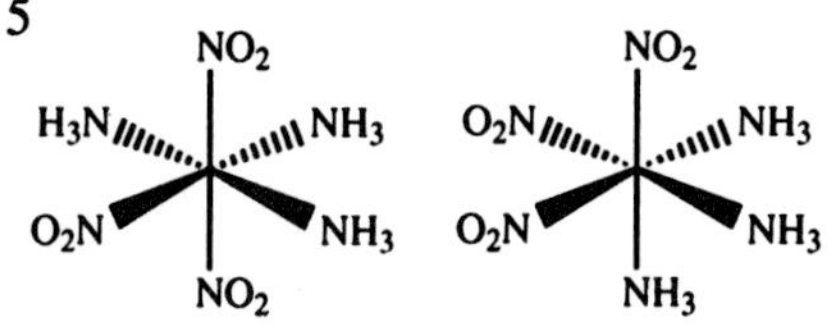

*mer*          *fac*

**9.18b** [Diamminwtetrachloroplatinum(IV)]

**9.20d** $2\,Cr^{3+} + 10\,OH^- + 3\,H_2O_2 \rightarrow 8\,H_2O + 2\,CrO_4^{\,2-}$

**9.23a** $[Co(NH_3)_5Cl]^{2+} + OH^- \rightarrow [Co(NH_3)_5OH]^{2+} + Cl^-$

**9.23b** Rate $= \dfrac{kK\big[[Co(NH_3)_55Cl]^{2+}\big][OH^-]}{[H_2O]}$

but, since $[H_2O]$ is constant,

Rate $= k\big[[Co(NH_3)_55Cl]^{2+}\big][OH^-]$

**10.1c**

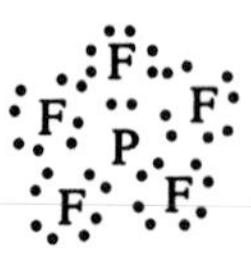

trigonal byprimidal, same, $dsp^3$ all $\sigma$

**10.1f**

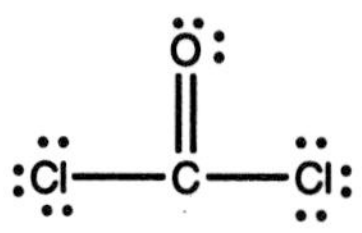

$:\!Cl\!-\!C\!-\!Cl\!:$ with O double bonded to C

trigonal planar, same, $sp^2$, 3 $\sigma$, 1 $\pi$

10.1g

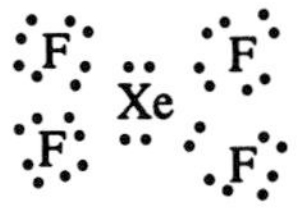

octahedral, square planar, $d^2sp^3$, all $\sigma$

10.2c  $d^2sp^3$; 0; $24D_q$-P; $4D_q$

10.2e  $dsp^2$; 0

10.2g  $dsp^3$; 4

10.6c  3

10.6d  0

10.8a

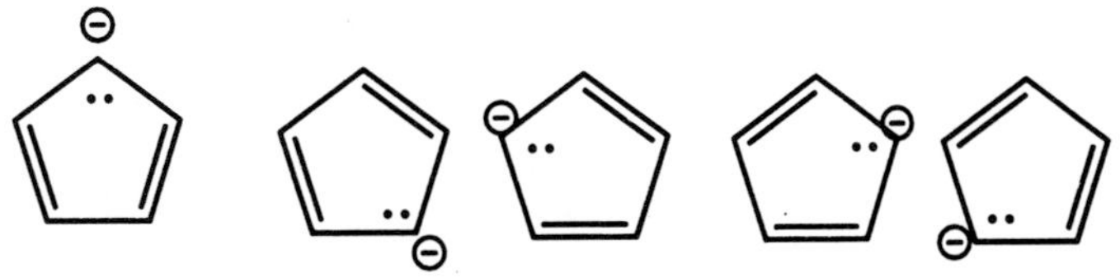

12.22a  0

12.22c  stronger bond, higher dissociation energy

11.1d  $CH_3CH_2CH_2CH_2CN$  pentanenitrile or n-butyl cyanide

11.3 d  K = LiBr    L = $C_6H_5CH_2Li$    M = toluene    N = $LiHSO_4$ or other way around

11.3 h  G = magnesium chloride salt of 4-bromo-2-methyl-2-butanol;  H = 4-bromo-2-methyl-2-butanol;  I = MgBr(OH) or other way around

11.9

A= $(CH_3)_3CCCH_2Li$ (with O double bond)    B= $CH_3-\overset{CH_3}{\underset{|}{C}H}-NH-\overset{CH_3}{\underset{|}{C}H}-CH_3$

C= $(CH_3)_3CCCH_2CH_2CH_2CH_3$ (with O double bond)    D= LiBr

11.10  $CH_3CH(OH)CH_2CHO$; $CH_3CH=CHCHO$
$CH_3CH_2CH(OH)CH(CH_3)CHO$; $CH_3CH_2CH=C(CH_3)CHO$
$CH_3CH(OH)CH(CH_3)CHO$; $CH_3CH=C(CH_3)CHO$
$CH_3CH_2CH(OH)CH_2CHO$; $CH_3CH_2CH=CHCHO$

11.10g

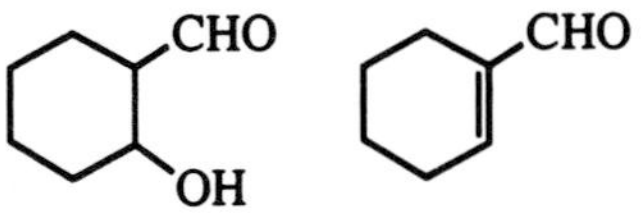

**11.11c** $C_6H_5CH(COOCH_3)COCH_2C_6H_5$

**11.14a** ethene

**11.15b** $(C_6H_5)_3P + CH_3CHBrCH_3 \longrightarrow (C_6H_5)_3\overset{+}{P}\text{-}CH(CH_3)\overset{-}{C}H_3 \ Br$

$(C_6H_5)_3\overset{+}{P}\text{-}CH(CH_3)CH_3 \ \overset{-}{Br} + CH_3CH_2CH_2CH_2Li \longrightarrow (C_6H_5)_3P=CH(CH_3)CH_3$

**11.16c** PhCHO + HSCH₂CH₂CH₂SH $\xrightarrow{\text{acid}}$ (dithiane) $\xrightarrow{CH_3CH_2CH_2CH_2Li}$

(dithiane anion) $\xrightarrow{BrCH_2Ph}$ (dithiane) $\xrightarrow[\text{methanol, heat}]{HgCl_2, H_2O}$ PhCHCH₂Ph

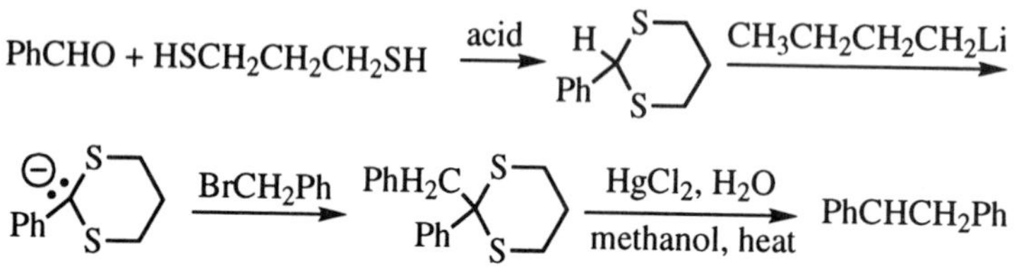

**11.18c**

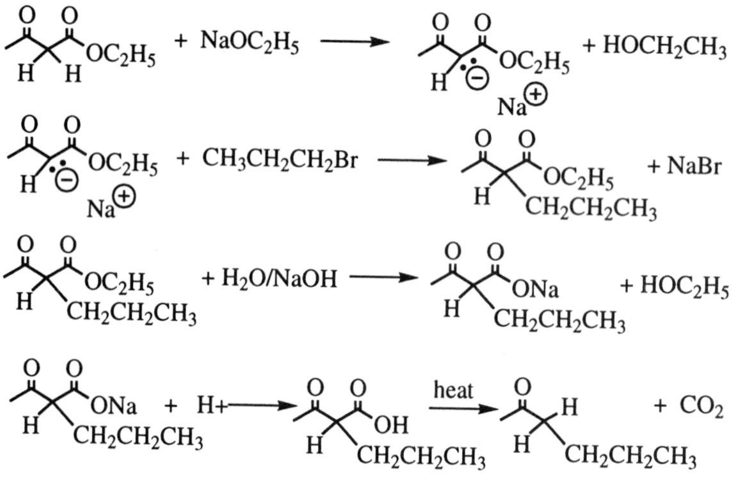

(malonate ester) + NaOC₂H₅ $\longrightarrow$ (anion) + HOCH₂CH₃

(anion) + CH₃CH₂CH₂Br $\longrightarrow$ (alkylated ester) + NaBr

(alkylated ester) + H₂O/NaOH $\longrightarrow$ (sodium carboxylate) + HOC₂H₅

(sodium carboxylate) + H+ $\longrightarrow$ (acid) $\xrightarrow{\text{heat}}$ (aldehyde) + $CO_2$

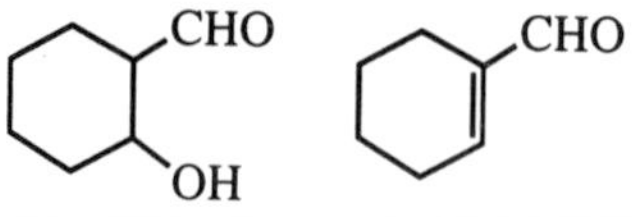

**11.20b** (cyclohexane with CHO and OH) (cyclohexene with CHO)

**11.21e** (ketone) + LiAlH₄ $\xrightarrow{H+/H_2O}$ (alcohol) $\xrightarrow{SOCl_2}$ (chloride)

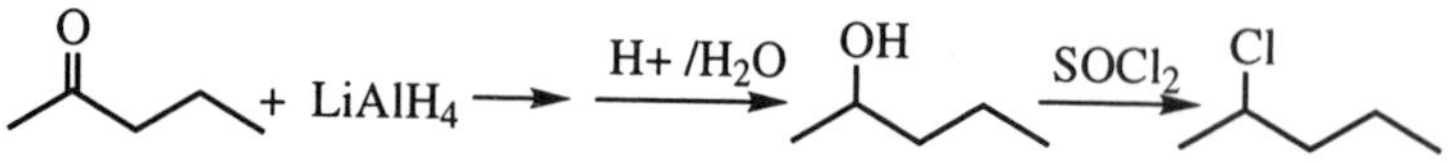

**11.21j** 1-bromopropane $\xrightarrow{\text{NaCN}}$ butanenitrile $\xrightarrow{\text{LiAlH4}}$ 1-aminobutane

**12.11b**

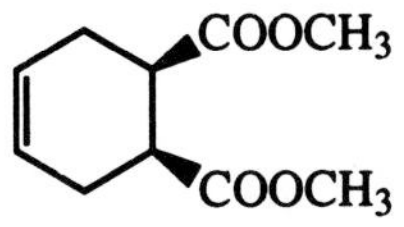

**12.12c**

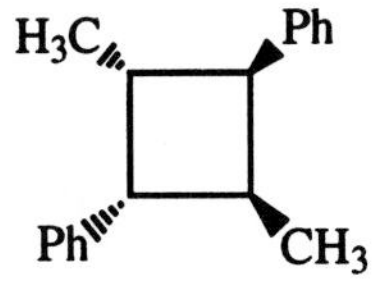

**12.14b**

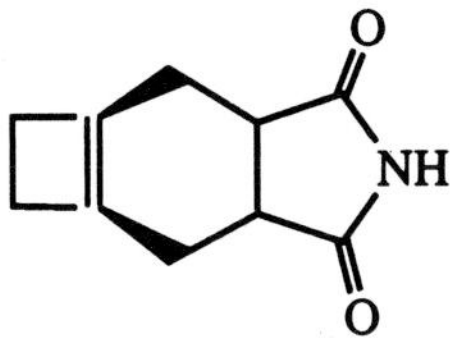

**12.15b**

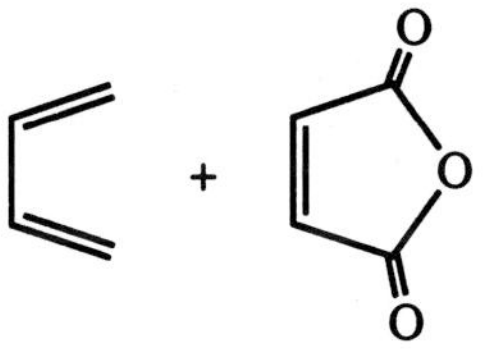

**12.15g**

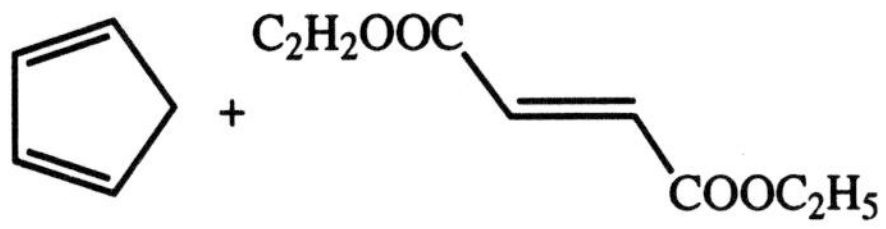

**13.1d**

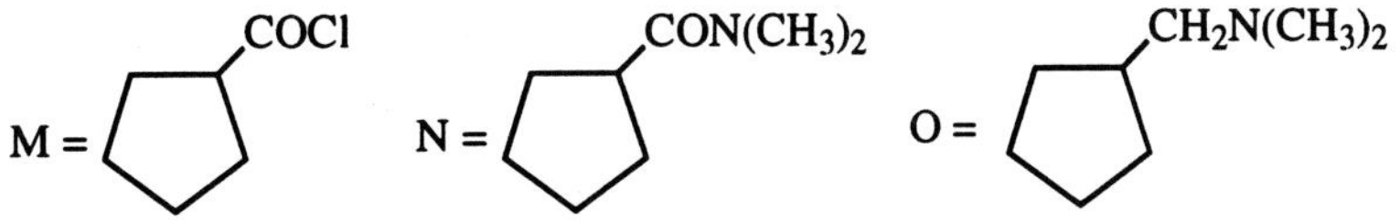

M =     N =     O =

**13.2d**

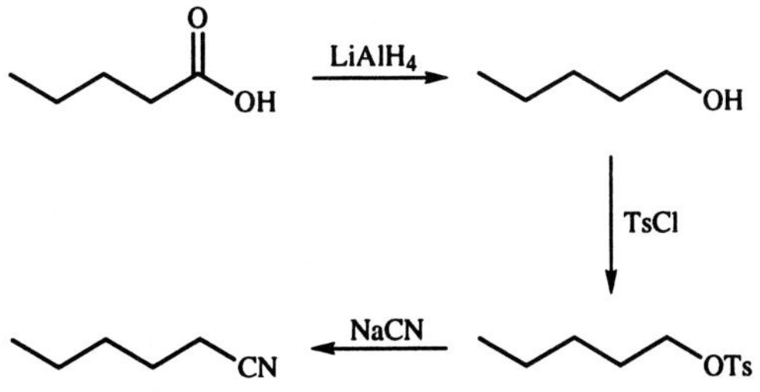

**13.3e**

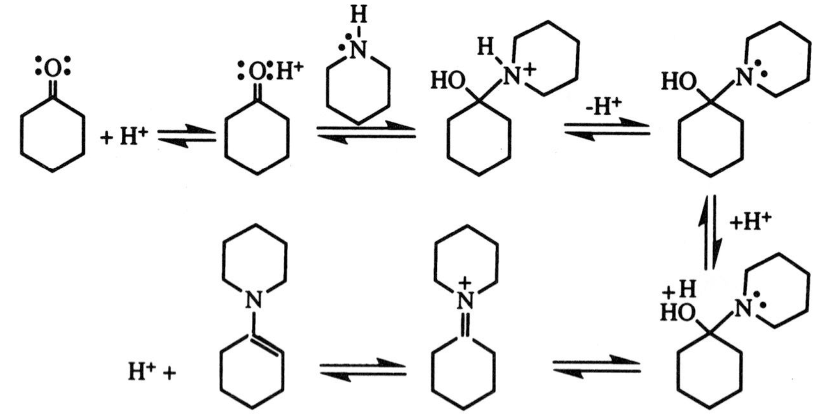

**13.4c**

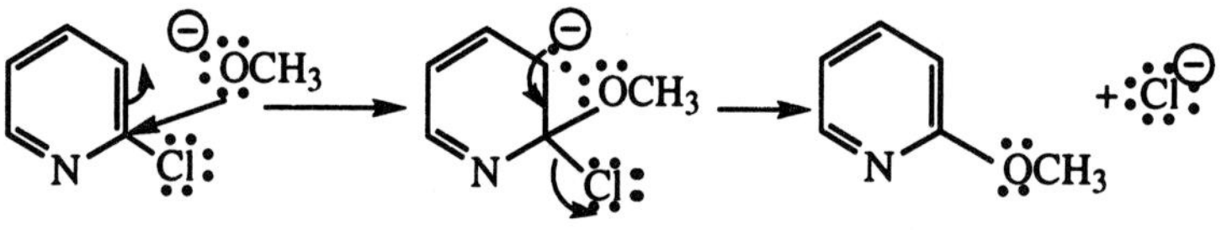

**13.5c**

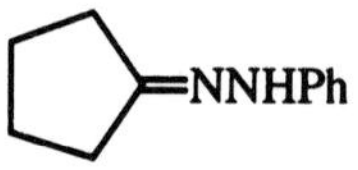

=NNHPh

**13.6d**

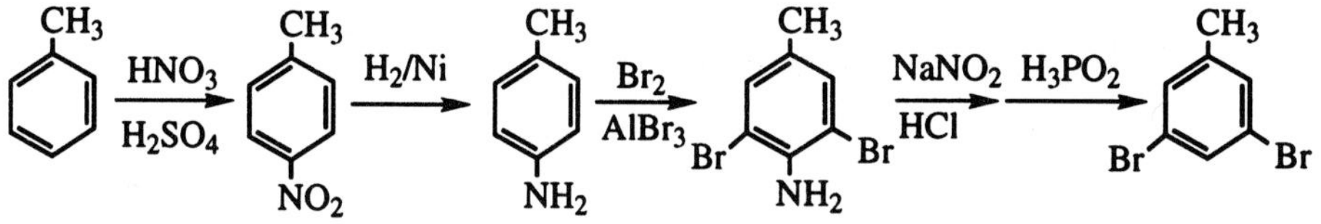

**13.7a** $CH_3NHNHCH_3 + 2 N_2O_4 \rightarrow 2 N_2 + 2 CO_2 + 4 H_2O$

**13.7c** 8.2 Atm

**14.1d**

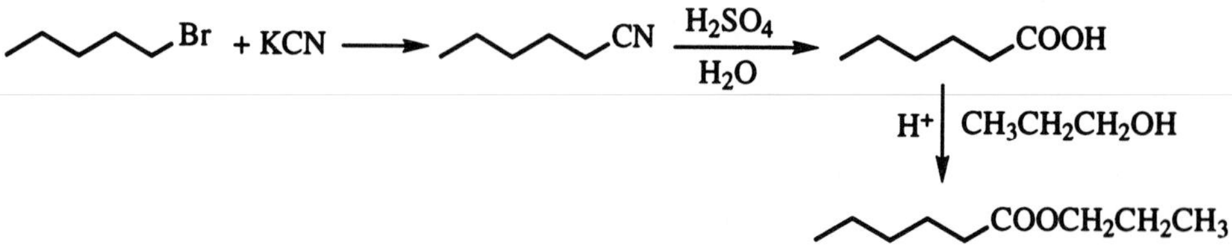

**14.1i**

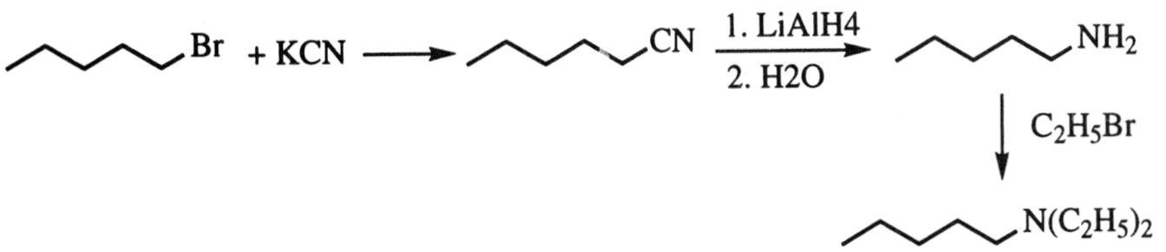

**14.2d**

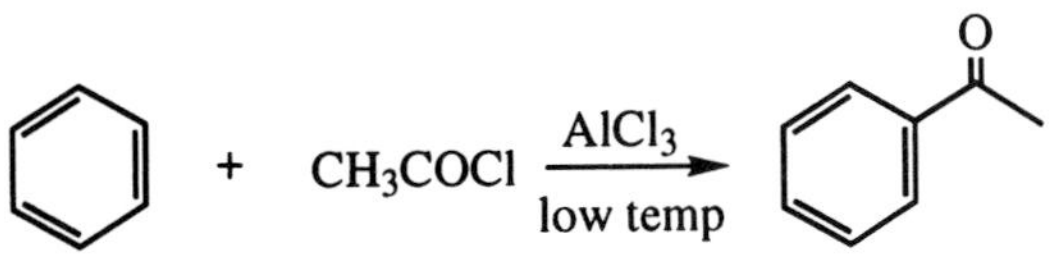

**14.2g**

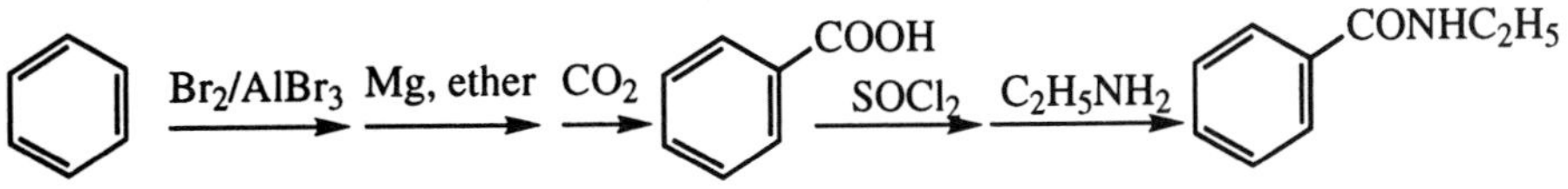

**14.6**

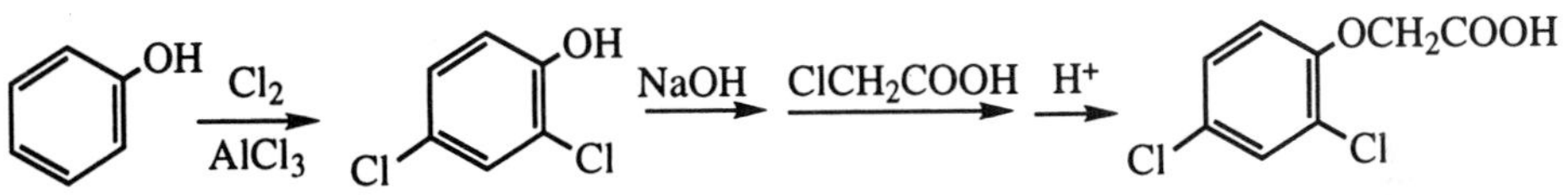

**15.2**  ~ 2400

**15.11**  $CH_3CH=CCOOCH_3$

**15.20**  Cotton, much more opportunity for hydrogen bonding with OH groups

## APPENDIX **E** THE SI SYSTEM

The SI (or metric) System has the following base units:

| PHYSICAL QUANTITY | NAME OF UNIT | SYMBOL |
|---|---|---|
| Length | meter | m |
| Mass | kilogram | kg |
| Time | second | s |
| Temperature | Kelvin | K |
| Electric Current | ampere | amp |
| Luminous Intensity | candela | cd |
| Amount of substance | mole | mol |

The SI (or metric) System includes the following derived units:

| PHYSICAL QUANTITY | NAME OF UNIT | SYMBOL |
|---|---|---|
| Energy | joule | $J \ (kg \ m^2 \ s^{-2})$ |
| Frequency | hertz | $Hz \ (cycles \ s^{-1})$ |
| Force | newton | $N \ (kg \ m \ s^{-2})$ |
| Pressure | pascal | $Pa \ (kg \ m^{-1} \ s^{-2})$ |
| Power | watt | $W \ (J \ s^{-1})$ |
| Electric Charge | coulomb | $C \ (amp \ s)$ |
| Electric Potential | volt | $V \ (J \ C^{-1})$ |

The following prefixes and symbols are used in modifying SI units:

| PREFIX/SYMBOL | MULTIPLIER | PREFIX/SYMBOL | MULTIPLIER |
|---|---|---|---|
| exa(E) | $10^{18}$ | atto(a) | $10^{-18}$ |
| peta(P) | $10^{15}$ | femto(f) | $10^{-15}$ |
| tera(T) | $10^{12}$ | pico(p) | $10^{-12}$ |
| giga(G) | $10^{9}$ | nano(n) | $10^{-9}$ |
| mega(M) | $10^{6}$ | micro(m) | $10^{-6}$ |
| kilo(k) | $10^{3}$ | milli(m) | $10^{-3}$ |
| hecto(h) | $10^{2}$ | centi(c) | $10^{-2}$ |
| deka(d) | $10^{1}$ | deci(d) | $10^{-1}$ |

APPENDIX  OTHER COMMON UNITS

The following are units in common use, and their relation to corresponding SI units:

| PHYSICAL QUANTITY | COMMON UNIT | VALUE IN SI UNITS |
|---|---|---|
| Length | angstrom | $10^{-10}$ m |
| | inch | 0.0254 m |
| Mass | pound (avdp) | 0.45359 kg |
| | amu | $1.6606 \times 10^{-24}$ g |
| Temperature | °C | (K - 273.15) |
| Energy | calorie | 4.184 J |
| | electron volt | $1.6022 \times 10^{-19}$ J |
| Pressure | atm | 101,325 Pa |
| | bar | $10^5$ Pa |
| | torr | 133.32 Pa |
| Volume | liter | $10^{-3}$ m$^3$ |
| Concentration | Molarity | mol L$^{-1}$; mol dm$^{-3}$ |
| Density | | g cm$^{-3}$ |

APPENDIX  **SELECTED FUNDAMENTAL CONSTANTS**

| | |
|---|---|
| Avogadro's Number | $6.022045 \times 10^{23}$ mol$^{-1}$ |
| Electron Charge | $1.602189 \times 10^{-19}$ C |
| Electron Mass | $9.109534 \times 10^{-28}$ g |
| Proton Mass | $1.672648 \times 10^{-24}$ g |
| Neutron Mass | $1.674954 \times 10^{-24}$ g |
| Speed of Light | $2.997925 \times 10^{8}$ m s$^{-1}$ |
| Planck's Constant | $6.626176 \times 10^{-34}$ Js |
| | |
| Gas Constant | $8.31441$ J mol$^{-1}$ K$^{-1}$ |
| | $0.0820578$ L atm mol$^{-1}$ K$^{-1}$ |
| | $1.98717$ cal mol$^{-1}$ K$^{-1}$ |

# APPENDIX  THE CHEMICAL ELEMENTS

| ATOMIC NUMBER | NAME | MASS | ELECTRON CONFIGURATION |
|---|---|---|---|
| 1 | hydrogen | 1.00794 | $1s^1$ |
| 2 | helium | 4.002602 | $1s^2 = [He]$ |
| 3 | lithium | 6.941 | $[He]\ 2s^1$ |
| 4 | beryllium | 9.012182 | $[He]\ 2s^2$ |
| 5 | born | 10.811 | $[He]\ 2s^2\ 2p^1$ |
| 6 | carbon | 12.011 | $[He]\ 2s^2\ 2p^2$ |
| 7 | nitrogen | 14.00674 | $[He]\ 2s^2\ 2p^3$ |
| 8 | oxygen | 15.9994 | $[He]\ 2s^2\ 2p^4$ |
| 9 | fluorine | 18.998403 | $[He]\ 2s^2\ 2p^5$ |
| 10 | neon | 20.1792 | $[He]\ 2s^2\ 2p^6 = [Ne]$ |
| 11 | sodium | 22.989768 | $[Ne]\ 3s^1$ |
| 12 | magnesium | 24.3050 | $[Ne]\ 3s^2$ |
| 13 | aluminum | 26.981539 | $[Ne]\ 3s^23p^1$ |
| 14 | silicon | 28.0855 | $[Ne]\ 3s^23p^2$ |
| 15 | phosphorus | 30.973762 | $[Ne]\ 3s^23p^3$ |
| 16 | sulfur | 32.066 | $[Ne]\ 3s^23p^4$ |
| 17 | chlorine | 35.4527 | $[Ne]\ 3s^23p^5$ |
| 18 | argon | 39.948 | $[Ne]\ 3s^23p^6 = [Ar]$ |
| 19 | potassium | 39.0983 | $[Ar]\ 4s^1$ |
| 20 | calcium | 40.078 | $[Ar]\ 4s^2$ |
| 21 | scandium | 44.955910 | $[Ar]\ 4s^2\ 3d^1$ |
| 22 | titanium | 47.88 | $[Ar]\ 4s^2\ 3d^2$ |
| 23 | vanadium | 50.9415 | $[Ar]\ 4s^2\ 3d^3$ |
| 24 | chromium | 51.9961 | $[Ar]\ 4s^1\ 3d^5$ * |
| 25 | manganese | 54.93805 | $[Ar]\ 4s^2\ 3d^5$ |
| 26 | iron | 55.847 | $[Ar]\ 4s^2\ 3d^6$ |
| 27 | cobalt | 58.93320 | $[Ar]\ 4s^2\ 3d^7$ |
| 28 | nickel | 58.6934 | $[Ar]\ 4s^2\ 3d^8$ |
| 29 | copper | 63.546 | $[Ar]\ 4s^1\ 3d^{10}$ * |
| 30 | zinc | 65.39 | $[Ar]\ 4s^2\ 3d^{10}$ |
| 31 | gallium | 69.723 | $[Ar]\ 4s^2\ 3d^{10}4p^1$ |
| 32 | germanium | 72.61 | $[Ar]\ 4s^2\ 3d^{10}4p^2$ |
| 33 | arsenic | 74.92159 | $[Ar]\ 4s^2\ 3d^{10}4p^3$ |
| 34 | selenium | 78.96 | $[Ar]\ 4s^2\ 3d^{10}4p^4$ |
| 35 | bromine | 79.904 | $[Ar]\ 4s^2\ 3d^{10}4p^5$ |
| 36 | krypton | 83.80 | $[Ar]\ 4s^2\ 3d^{10}4p^6 = [Kr]$ |
| 37 | rubidium | 85.4678 | $[Kr]\ 5s^1$ |
| 38 | strontium | 87.62 | $[Kr]\ 5s^2$ |
| 39 | yttrium | 88.90585 | $[Kr]\ 5s^2\ 4d^1$ |

| Atomic Number | Name | Mass | Electron Configuration |
|---|---|---|---|
| 40 | zirconium | 91.224 | [Kr] $5s^2$ $4d^2$ |
| 41 | niobium | 92.90638 | [Kr] $5s^1$ $4d^4$ * |
| 42 | molybdenum | 95.94 | [Kr] $5s^1$ $4d^5$ * |
| 43 | technetium | (98)** | [Kr] $5s^2$ $4d^5$ |
| 44 | ruthenium | 101.07 | [Kr] $5s^1$ $4d^7$ * |
| 45 | rhodium | 102.90550 | [Kr] $5s^1$ $4d^8$ * |
| 46 | palladium | 106.42 | [Kr] $5s^0$ $4d^{10}$ * |
| 47 | silver | 107.8682 | [Kr] $5s^1$ $4d^{10}$ * |
| 48 | cadmium | 112.411 | [Kr] $5s^2$ $4d^{10}$ |
| 49 | indium | 114.82 | [Kr] $5s^2$ $4d^{10}$ $4p^1$ |
| 50 | tin | 118.710 | [Kr] $5s^2$ $4d^{10}$ $4p^2$ |
| 51 | antimony | 121.757 | [Kr] $5s^2$ $4d^{10}$ $4p^3$ |
| 52 | tellurium | 127.60 | [Kr] $5s^2$ $4d^{10}$ $4p^4$ |
| 53 | iodine | 126.90447 | [Kr] $5s^2$ $4d^{10}$ $4p^5$ |
| 54 | xenon | 131.29 | [Kr] $5s^2$ $4d^{10}$ $4p^6$ = [Xe] |
| 55 | cesium | 132.9054 | [Xe] $6s^1$ |
| 56 | barium | 137.327 | [Xe] $6s^2$ |
| 57 | lanthanum | 138.9055 | [Xe] $6s^2$ $5d^1$ |
| 58 | cerium | 140.115 | [Xe] $6s^2$ $5d^1$ $4f^1$ |
| 59 | praseodymium | 140.90765 | [Xe] $6s^2$ $5d^0$ $4f^3$ * |
| 60 | neodymium | 144.24 | [Xe] $6s^2$ $5d^0$ $4f^4$ * |
| 61 | promethium | (145)** | [Xe] $6s^2$ $5d^0$ $4f^5$ * |
| 62 | samarium | 150.36 | [Xe] $6s^2$ $5d^0$ $4f^6$ * |
| 63 | europium | 151.965 | [Xe] $6s^2$ $5d^0$ $4f^7$ * |
| 64 | gadolinium | 157.25 | [Xe] $6s^2$ $5d^1$ $4f^7$ |
| 65 | terbium | 158.92534 | [Xe] $6s^2$ $5d^0$ $4f^9$ * |
| 66 | dysprosium | 162.50 | [Xe] $6s^2$ $5d^0$ $4f^{10}$ * |
| 67 | holmium | 164.93032 | [Xe] $6s^2$ $5d^0$ $4f^{11}$ * |
| 68 | erbium | 167.26 | [Xe] $6s^2$ $5d^0$ $4f^{12}$ * |
| 69 | thulium | 168.93421 | [Xe] $6s^2$ $5d^0$ $4f^{13}$ |
| 70 | ytterbium | 173.04 | [Xe] $6s^2$ $5d^0$ $4f^{14}$ * |
| 71 | lutetium | 174.967 | [Xe] $6s^2$ $5d^1$ $4f^{14}$ |
| 72 | hafnium | 178.49 | [Xe] $6s^2$ $5d^2$ $4f^{14}$ |
| 73 | tantalum | 180.9479 | [Xe] $6s^2$ $5d^3$ $4f^{14}$ |
| 74 | tungsten | 183.85 | [Xe] $6s^2$ $5d^4$ $4f^{14}$ |
| 75 | rhenium | 186.207 | [Xe] $6s^2$ $5d^5$ $4f^{14}$ |
| 76 | osmium | 190.2 | [Xe] $6s^2$ $5d^6$ $4f^{14}$ |
| 77 | iridium | 192.22 | [Xe] $6s^2$ $5d^7$ $4f^{14}$ |
| 78 | platinum | 195.08 | [Xe] $6s^1$ $5d^9$ $4f^{14}$ * |
| 79 | gold | 196.96654 | [Xe] $6s^1$ $5d^{10}$ $4f^{14}$ |
| 80 | mercury | 200.59 | [Xe] $6s^2$ $5d^{10}$ $4f^{14}$ |
| 81 | thallium | 204.3833 | [Xe] $6s^2$ $5d^{10}$ $4f^{14}$ $6p^1$ |
| 82 | lead | 207.2 | [Xe] $6s^2$ $5d^{10}$ $4f^{14}$ $6p^2$ |
| 83 | bismuth | 208.98037 | [Xe] $6s^2$ $5d^{10}$ $4f^{14}$ $6p^3$ |
| 84 | polonium | (209)** | [Xe] $6s^2$ $5d^{10}$ $4f^{14}$ $6p^4$ |

| Atomic Number | Name | Mass | Electron Configuration |
|---|---|---|---|
| 85 | astatine | (210)** | [Xe] $6s^2$ $5d^{10}$ $4f^{14}$ $6p^5$ |
| 86 | radon | (222)** | [Xe] $6s^2$ $5d^{10}$ $4f^{14}$ $6p^6$ = [Rn] |
| 87 | francium | 223.0197*** | [Rn] $7s^1$ |
| 88 | radium | 226.0254 | [Rn] $7s^2$ |
| 89 | actinium | 227.0278*** | [Rn] $7s^2$ $6d^1$ |
| 90 | thorium | 232.0381 | [Rn] $7s^2$ $6d^2$ $5f^0$ * |
| 91 | protactinium | 231.03588 | [Rn] $7s^2$ $6d^1$ $5f^2$ |
| 92 | uranium | 238.0289 | [Rn] $7s^2$ $6d^1$ $5f^3$ |
| 93 | neptunium | 237.0482*** | [Rn] $7s^2$ $6d^1$ $5f^4$ |
| 94 | plutonium | (240)** | [Rn] $7s^2$ $6d^0$ $5f^6$ * |
| 95 | americium | 243.0614*** | [Rn] $7s^2$ $6d^0$ $5f^7$ * |
| 96 | curium | (247)** | [Rn] $7s^2$ $6d^1$ $5f^7$ |
| 97 | berkelium | (248)** | [Rn] $7s^2$ $6d^0$ $5f^9$ * |
| 98 | californium | (250)** | [Rn] $7s^2$ $6d^0$ $5f^{10}$ * |
| 99 | einsteinium | 252.083*** | [Rn] $7s^2$ $6d^0$ $5f^{11}$ * |
| 100 | fermium | 257.0951*** | [Rn] $7s^2$ $6d^0$ $5f^{12}$ |
| 101 | mendelevium | (257)** | [Rn] $7s^2$ $6d^0$ $5f^{13}$ |
| 102 | nobelium | 259.1009*** | [Rn] $7s^2$ $6d^0$ $5f^{14}$ |
| 103 | lawrencium | 292.11 | [Rn] $7s^2$ $6d^0$ $5f^{14}$ $7p^1$ * |
| 104 | rutherfordium | 261.11*** | [Rn] $7s^2$ $6d^2$ $5f^{14}$ * |
| 105 | dubnium | 262.114*** | [Rn] $7s^2$ $6d^3$ $5f^{14}$ * |
| 106 | seaborgium | 263.118*** | [Rn] $7s^2$ $6d^4$ $5f^{14}$ * |
| 107 | bohrium | 262.12*** | [Rn] $7s^2$ $6d^5$ $5f^{14}$ * |
| 108 | hassium | | [Rn] $7s^2$ $6d^6$ $5f^{14}$ * |
| 109 | meitnerium | | [Rn] $7s^2$ $6d^7$ $5f^{14}$ * |

n.b.  elements 104-109 named according to IUPAC's 1997 proposal

*    exceptional electron configuration

**    average mass of the stable isotopes of radioactive element

***    exact mass of the most stable isotope of radioactive element

# Periodic Table of the Elements

| IA | | | | | | | | | | | | | | | | | VIIIA |
|---|---|---|---|---|---|---|---|---|---|---|---|---|---|---|---|---|---|
| 1 H 1.00794 | IIA | | | | | | | | | | | IIIA | IVA | VA | VIA | VIIA | 2 He 4.00260 |
| 3 Li 6.941 | 4 Be 9.01218 | | | | | | | | | | | 5 B 10.81 | 6 C 12.011 | 7 N 14.0067 | 8 O 15.9994 | 9 F 18.998403 | 10 Ne 20.1797 |
| 11 Na 22.98977 | 12 Mg 24.305 | IIIB | IVB | VB | VIB | VIIB | VIIIB | | | IB | IIB | 13 Al 26.98154 | 14 Si 28.0855 | 15 P 30.97376 | 16 S 32.066 | 17 Cl 35.453 | 18 Ar 39.948 |
| 19 K 39.0983 | 20 Ca 40.078 | 21 Sc 44.9559 | 22 Ti 47.88 | 23 V 50.9415 | 24 Cr 51.996 | 25 Mn 54.9380 | 26 Fe 55.847 | 27 Co 58.9332 | 28 Ni 58.69 | 29 Cu 63.546 | 30 Zn 65.39 | 31 Ga 69.72 | 32 Ge 72.61 | 33 As 74.9216 | 34 Se 78.96 | 35 Br 79.904 | 36 Kr 83.80 |
| 37 Rb 85.4678 | 38 Sr 87.62 | 39 Y 88.9059 | 40 Zr 91.224 | 41 Nb 92.9064 | 42 Mo 95.94 | 43 Tc (98) | 44 Ru 101.07 | 45 Rh 102.9055 | 46 Pd 106.42 | 47 Ag 107.8682 | 48 Cd 112.41 | 49 In 114.82 | 50 Sn 118.710 | 51 Sb 121.757 | 52 Te 127.60 | 53 I 126.9045 | 54 Xe 131.29 |
| 55 Cs 132.9054 | 56 Ba 137.33 | 57 La 138.9055 | 72 Hf 178.49 | 73 Ta 180.9479 | 74 W 183.85 | 75 Re 186.207 | 76 Os 190.2 | 77 Ir 192.22 | 78 Pt 195.08 | 79 Au 196.9665 | 80 Hg 200.59 | 81 Tl 204.383 | 82 Pb 207.2 | 83 Bi 208.9804 | 84 Po (209) | 85 At (210) | 86 Rn (222) |
| 87 Fr (223) | 88 Ra 226.0254 | 89 Ac 227.0278 | 104 Rf (261) | 105 Db (262) | [106] Sg (263) | [107] Bh (262) | [108] Hs (265) | [109] Mt (266) | | | | | | | | | |

| 58 Ce 140.12 | 59 Pr 140.9077 | 60 Nd 144.24 | 61 Pm (145) | 62 Sm 150.36 | 63 Eu 151.96 | 64 Gd 157.25 | 65 Tb 158.9254 | 66 Dy 162.50 | 67 Ho 164.9304 | 68 Er 167.26 | 69 Tm 168.9342 | 70 Yb 173.04 | 71 Lu 174.967 |
|---|---|---|---|---|---|---|---|---|---|---|---|---|---|
| 90 Th 232.0381 | 91 Pa 231.0359 | 92 U 238.0289 | 93 Np 237.048 | 94 Pu (244) | 95 Am (243) | 96 Cm (247) | 97 Bk (247) | 98 Cf (251) | 99 Es (252) | 100 Fm (257) | 101 Md (258) | 102 No (259) | 103 Lr (260) |